Mechanics and Control of Large Flexible Structures

Edited by
John L. Junkins
Texas A&M University
College Station, Texas

Volume 129
PROGRESS IN ASTRONAUTICS AND AERONAUTICS

A. Richard Seebass, Editor-in-Chief
University of Colorado at Boulder
Boulder, Colorado

Published by the American Institute of Aeronautics and Astronautics, Inc.
370 L'Enfant Promenade, SW, Washington, DC 20024-2518.

American Institute of Aeronautics and Astronautics, Inc.
Washington, DC

Library of Congress Cataloging in Publication Data

Mechanics and control of large flexible structures/edited by John L. Junkins.
p.cm.—(Progress in astronautics and aeronautics;v.129)
Includes bibliographical references.
1. Large space structures (Astronautics) 2. Astrodynamics.
I. Junkins, John L. III. Series.
TL507.P75 vol. 129 90-594
[TL940]
629.1 s--dc20
[629.4'11]
ISBN 0-930403-73-8

Progress in Astronautics and Aeronautics

Table of Contents

Part 2. Control, Stability Analysis, and Optimization

Preface

The merger of mechanics and controls of structures into a unified field has been the product of research by hundreds of individuals over the past decade. Although much work has been done, the documentation remains diffusely distributed in the research literature and a few books. The present monograph provides a forum less constraining than the usual journal format: we have been permitted generous latitude in volume and style. The result is broad coverage of the field with tutorial discussions of the literature, analytical methods, and numerical/experimental results treated in a single source. We trust that this book will serve to accelerate future research and applications by making a large fraction of the current methodology broadly accessible. As is evident from the results presented herein, many new insights and methods have evolved as a consequence of research on the interface of the historically disjoint disciplines of mechanics and controls; these results are of enormous conceptual and practical importance. An especially significant feature of the problems studied is the family of difficulties raised by the three sources of nondeterminism: a) high dimensionality, b) nonlinearity, and c) uncertainty. Uncertainties arise not only from the usual parametric errors, measurement errors, and random external disturbances, but perhaps more importantly because the highly correlated near-deterministic components of the physics of complicated structural systems are never perfectly modeled in practice. While these model-related difficulties could be viewed as a curse, we should instead recognize them as a source of blessings.

The imperfect mathematical model lies at the heart of the mechanics and control of structures. The pursuit of answers to the multipart question, that is, How do we make these complicated structural systems work in spite of model uncertainty, nonlinearity, and high dimensionality?, has been the primary stimulus underlying the intense effort by structural control designers and structural mechanics analysts to understand, apply, and help evolve each other's methodology. Upon reviewing the elegant developments on system modeling, identification, optimization, and robust control design, we believe the reader will agree that the fruits born of the quest to solve these problems have indeed proven a multidimensioned blessing. The most significant educational spinoff has been the evolution of a substantial cadre of researchers who have excellent expertise both in structural mechanics and control systems.

This book is divided into three major parts: Part I consists of Chapters 1–9 and considers structural modeling, multibody dynamics, and system identification issues. Part II, including Chapters 10–17, deals with theory and computational methods for feedback control design, order reduction, stability of time varying systems, and nonlinear optimization. Finally, Part III, consisting of Chapters 18–25, addresses a) analytical and experimental methods for control/structure interaction, b) feedback control law design for flexible systems undergoing large motions, and c) tether connected systems. Each part contains one or more review chapters that broadly address major subsets of the research literature, as well as several more sharply focused chapters on recent methodology and experimental results. While we do not claim we have achieved global coverage of the field, we do believe that this volume will provide a broad base from which to launch many applications and future research initiatives.

The field continues to evolve rapidly, but appears to be nearing its most exciting epoch. It is apparent on many fronts that emphasis has recently shifted sharply toward mission studies and realization of operational hardware/software. Driven by the impending realization of Space Station Freedom and other large orbiting structures, there is a new-found immediacy in our research that has generated a strong acceleration toward more sophisticated ground-based implementations and plans for near-term on-orbit experiments. Concurrent advances in massively parallel and inexpensive, compact computers, in conjunction with development of new sensors and actuators, all point to a golden age of experimental research. Several examples discussed in this book show that proof-of-concept experiments carried out in inexpensive ground-based laboratories generate not only a basis for evaluating in the maturity of the technology, but more importantly, generate useful insights for further analytical/theoretical developments.

I anticipate that basic analytical research in mechanics and control of structures should and will continue at an aggressive pace, and will be periodically refueled and reoriented as we wrestle to resolve the challenges posed by particular implementations and mission objectives. As is evident throughout this book, there remain many difficult and perhaps unresolvable conceptual/theoretical issues. The field has recently shown signs of maturity, however, as evidenced by the fact that many historically analytical researchers have recently become directly and successfully active in experimental implementations. This has led to significant experimental demonstrations and verifications of recently developed methodology, as is evidenced in Part III of this book. The expected acceleration of this trend should serve to contribute immensely to both the practical and theoretical aspects of mechanics and control. The most important consequence should be to shorten the elapsed time from analytical conception to operational hardware.

This book is the culmination of extensive parallel research and a year of collaborative effort by three dozen excellent researchers. I was willing to undertake the task of editing this book only because I was confident that I could convince a sufficient number of outstanding colleagues to help me make a broad and penetrating contribution to the literature on mechanics and control of structures. It may be of interest to discuss briefly the evolution of this book.

In response to a December 1988 invitation from A. Richard Seebass (on behalf of the AIAA Editorial Committee), I agreed to serve as editor for this volume. Based on a biased perspective developed during two decades of my own research and three years as associate editor of the *Journal of Guidance, Control, and Dynamics,* I selected a list of topics and initially invited over 30 individuals to contribute material based on their recent work. These individuals were selected because of their expertise and because my experience indicated that they could communicate their research contributions in a tutorial fashion; all but three of those invited accepted my invitation. Over the course of the 10-month period beginning in January 1989, 30 chapter drafts were written, subjected to editorial and peer reviews, and went through at least one round of revisions. A number of the originally invited chapters did not survive the review process and/or could not be completed on the aggressive nine-month schedule we adopted to ensure that this book would be completed while the work was still current. Obviously, the schedule required that the authors write largely about the more mature fruits of their recent work. However, the topics on which the authors were invited to write, for the most part, represent a focus of their current research efforts. The reader will be pleased to find that these researchers have done an excellent job of developing a tutorial synthesis of original material drawn from their work over the past few years and a significant collection of their most recent, heretofore unpublished results. As a consequence, we converged to a large volume with outstanding depth, breadth, and currency. We trust the reader will agree that the technical quality is exceptionally high.

There is insufficient space here to mention all individuals who contributed to the publication of this book. Completing this project would have been impossible without the excellent cooperation of the chapter authors; appropriate acknowledgment of many other contributions are made in the individual chapters. On behalf of the authors, I am pleased to express our appreciation for the outstanding work of Becky Masters. Her editorial work and extensive communications were largely responsible for keeping this large project very nearly on schedule; she undertook this effort in addition to her normal duties as my staff assistant. The support of the AIAA editorial staff, and especially by John Newbauer is appreciated. I am pleased to thank Martin Summerfield for his support as Editor-in-Chief of the Series. I am appreciative of the original invitation from Editor Richard Seebass that led to my initiation of this effort; I am delighted that we have converged to a successful completion of this project! Finally, the significant support of Texas A&M University is most appreciated.

John L. Junkins
George J. Eppright Professor
Aerospace Engineering Department
Texas A&M University
November 17, 1989

Part 1. Structural Modeling, Identification, and Dynamic Analysis

Chapter 1

Recent Literature on Structural Modeling, Identification, and Analysis

Roy R. Craig Jr.*
The University of Texas at Austin, Austin, Texas

Introduction

THE authors of the excellent 1984 survey article on the dynamics and control of large space structures[1] conclude a brief history of the topic with the statement, "The solution, maturation, and flight testing of these problems lie in the future." Indeed, in addition to regular conferences of technical societies, over the past decade there have been numerous workshops and conferences concerned with the dynamics and control of large space structures (e.g., Refs. 2–16). Special monographs on the dynamics and control of large space structures have begun to appear.[17,18] The subject has matured to the point that the space station program has now progressed through several design iterations[19,20] to the present stage of completion of a preliminary loads analysis[21] and a preliminary control/structure interaction study,[22] and the era of on-orbit testing of this class of space structure has already begun.[23] Although it is not by any means complete or bias-free, this chapter surveys literature from 1980 to the present related to the topics of modeling, identification, and analysis of large space structures. Bibliographies such as Ref. 24 can provide the reader with further references on the topic.

The list of individual topics that could be covered under the rubric of "modeling of large space structures" is extensive. The list includes distributed-parameter vs discrete-parameter models (e.g., continuum vs finite-element models), multi-flexible-body dynamics, on-orbit dynamics, large-angle maneuvers, nonlinearity, global and/or local behavior, actuator/structure coupling, waves vs modes, joints, damping, model reduction,

*John J. McKetta Energy Professor, Department of Aerospace Engineering and Engineering Mechanics.

time-varying geometry, second-order model vs first-order model, uncertainty, optimization, and substructuring. Several of these topics will be treated in this chapter, whereas others, such as multibody dynamics and large-angle maneuvers, are covered in some detail elsewhere in this volume.

Since the First International Modal Analysis Conference in 1982, system identification and model verification as related to structures has grown into a mature discipline,[25] and controls-related literature on system identification has likewise abounded.[18,26,27] The numerous relevant topics include the following: ground testing vs on-orbit testing, sensors and actuators, support conditions, time-domain algorithms vs frequency-domain algorithms, model error localization and model updating methods, component vs system testing, modes vs parameters, and scale modeling. This chapter will emphasize topics related to verifying mathematical models of large space structures.

Finally, on the topic of analysis, this chapter will only briefly survey recent research related to eigensolvers and dynamic response solvers for large-order finite-element-based mathematical models. Related topics such as multibody dynamics formulations and solvers are treated elsewhere in this volume.

Mathematical Modeling of Large Space Structures

From the structures standpoint the most important question to be asked in creating a mathematical model for use in a structural dynamics and/or control/structure interaction (CSI) study is "What is the purpose of the model?" During the Apollo space program, analytical models of varying complexity from beam models to shell models were used in studying Apollo Saturn V space vehicle dynamics.[28] For example, the NASA document that specifies the structural/mechanical interfaces and requirements for Space Shuttle payloads indicates that, for the Orbiter/payload coupled loads analysis, the "typical payload mathematical model is in the form of Craig-Bampton modes or mass and stiffness matrices [with a] maximum of 300 degrees of freedom (DOF) for dedicated mission; 100 DOF is typical."[29] Figure 1 illustrates the present Space Station Freedom, together with a detailed truss representation and a beam representation.[20] With recent advances in computational capability has come the ability to solve static analysis structures problems with detailed finite-element models having more than 50,000 DOF.[30] On the other hand, continuum models of large lattice structures may satisfy specific modeling requirements.[31,32] To facilitate integrated design of large space structures such as the space station, a solids modeling capability has been integrated with finite-element modeling and other features to form an integrated design package.[20,33] Special truss modules are available in this software as "elements."

From the controls standpoint, plant (i.e., dynamic system) models are generally expressed in state-variable, or first-order, form. For example,

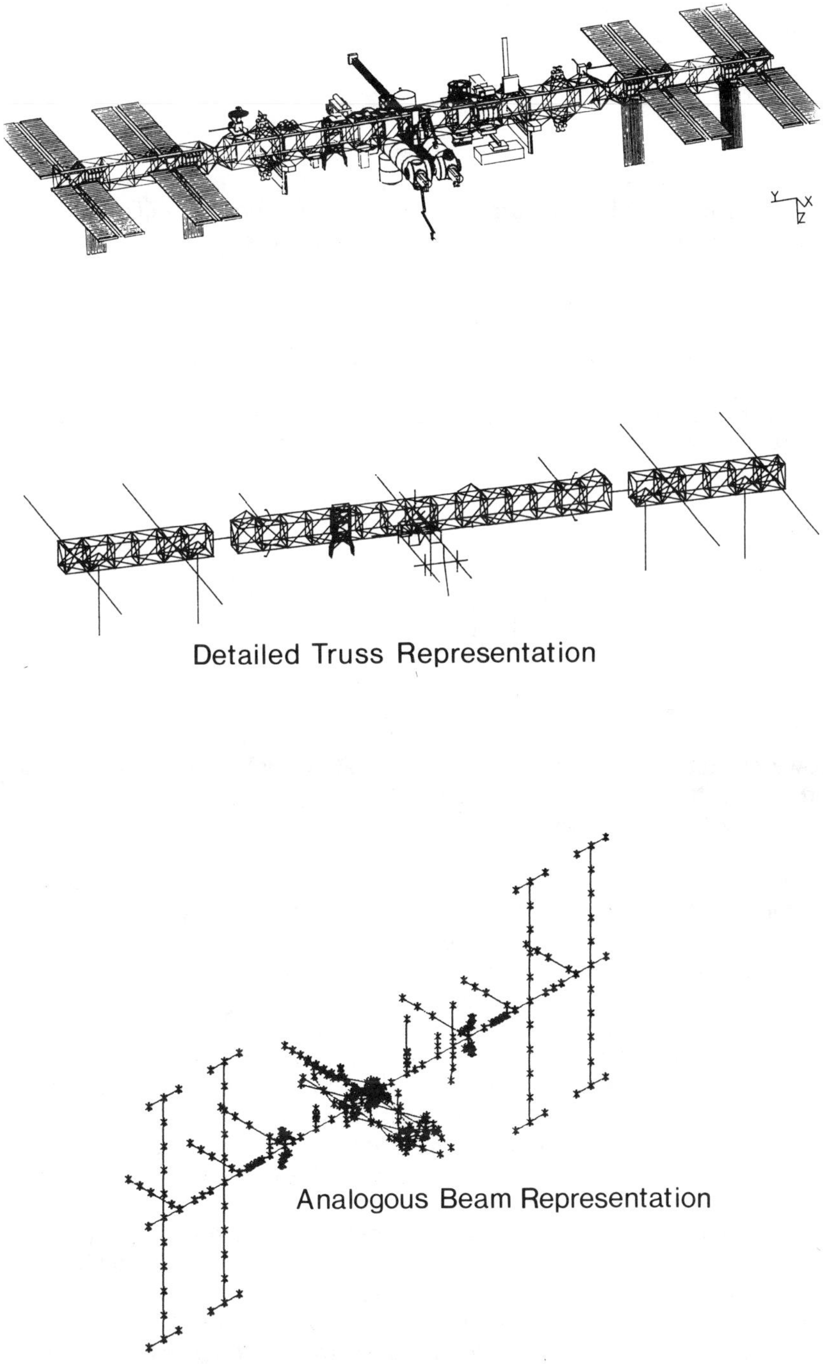

Fig. 1 Finite-element models of Space Station Freedom.

with z as the state variable vector, u the input vector, and y the measurement vector, the classical first-order form is

$$\dot{z} = Az + Bu \tag{1}$$

$$y = Cz \tag{2}$$

On the other hand, structural dynamicists are most familiar with the second-order form on the equations of motion for structures, i.e.,

$$M\ddot{x} + D\dot{x} + Kx = Fu \tag{3}$$

In controls applications the second-order model is frequently converted to a first-order model by using normal modes (eigenvectors) of the undamped system and assuming modal damping.[34] A more general form is given by the equations

$$z = \begin{Bmatrix} x \\ \dot{x} \end{Bmatrix} \tag{4}$$

$$\begin{bmatrix} D & M \\ M & 0 \end{bmatrix} \begin{Bmatrix} \dot{x} \\ \ddot{x} \end{Bmatrix} + \begin{bmatrix} K & 0 \\ 0 & -M \end{bmatrix} \begin{Bmatrix} x \\ \dot{x} \end{Bmatrix} = \begin{Bmatrix} F \\ 0 \end{Bmatrix} u \tag{5}$$

$$y = [E \quad G] \begin{Bmatrix} x \\ \dot{x} \end{Bmatrix} \tag{6}$$

However, there are some controls applications where the second-order form might be preferable to the first-order form, and some recent literature has adopted that approach.[35–38]

Continuum Models

It is quite common for a spacecraft with flexible appendages to be modeled as a rigid body to which beams are attached that are modeled as distributed-parameter components.[39,40] The appendage dynamics is frequently then reduced to a finite number of degrees of freedom by invoking "assumed modes,"[41,42] in most cases in the form of normal modes of free vibration.[43] More general applications of control of flexible structures (COFS) can also be formulated in terms of distributed-parameter systems and then reduced to a finite number of degrees of freedom by assumed modes expansions.[42,44,45] In some instances "exact" solutions to partial differential equation models are obtained.[46,47]

Whereas the preceding references to continuum, or distributed-parameter, models assume knowledge of structural stiffness and mass distributions, significant progress has been made in reducing large lattice-type structures to equivalent continua. In a recent paper Noor[32] reviews the development of continuum models for large repetitive lattice structures. Figure 2 shows continuum models for beamlike and platelike lattice structures.

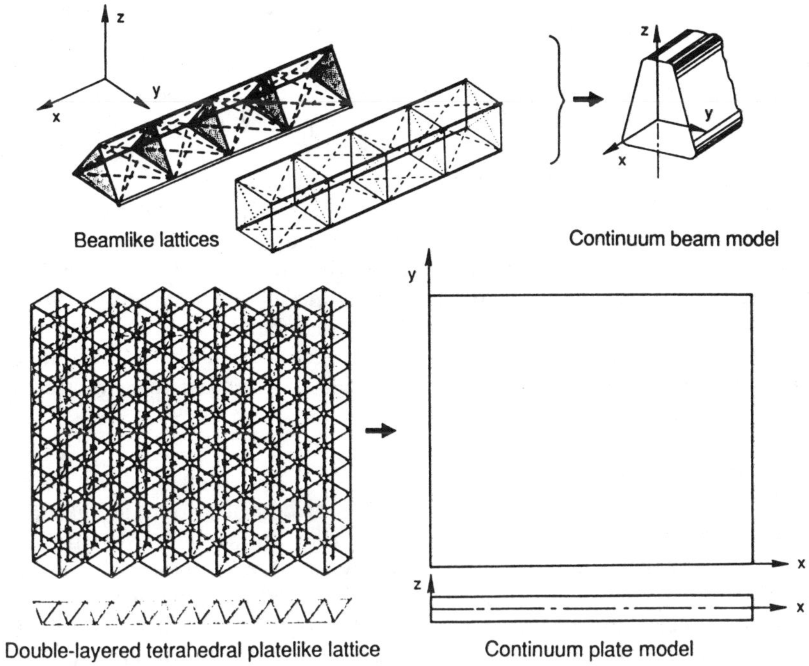

Fig. 2 Continuum models for beamlike and platelike lattice structures.

Several approaches have been proposed to construct equivalent continuum beam and plate models. In one approach a unit cell is isolated, and, by imposing characteristic deformation patterns on the cell, equivalent beam or plate continuum properties are obtained. Examples of this approach can be found in Refs. 48–51. Damping is incorporated in the continuum model in Ref. 52. Starting with stiffness matrices of individual truss members, Ref. 53 employs a finite-difference approach to derive continuum properties for beamlike space trusses. This approach has been called the "discrete field method." In Ref. 54 a continuum model is developed by carrying out the following three steps: 1) all sets of parallel members are identified, 2) unidirectional "effective continuum" properties are derived for each of these sets, and 3) orthogonal transformations are finally used to determine the contribution of each set to the "overall effective continuum" properties of the structure. This approach is followed in Refs. 55 and 56 to study the behavior of several example structures.

Several authors have chosen finite-element-based model reduction techniques for lattice-type space structures. Reference 57 develops three reduction techniques based on reduced-order modeling of beams starting with finite-element models of basic cells. The box truss components implemented in Ref. 20 are similar. In Ref. 58 sandwich plate and thin shell models are developed and implemented directly within a finite-element program.

Energy equivalence, together with polynomial expansions for displacements of the equivalent continuum, form the basis of the methods described in, for example, Refs. 32 and 59. Control/structure interaction examples based on this type of continuum model are studied in Refs. 60 and 61. Reference 62 employs continuum properties developed by this approach, together with a matrix version of Kane's equations, to develop a mathematical model for dynamics of the space station.

A distributed-element program for dynamic analysis of piecewise continuous structures is described in Ref. 63. This approach makes use of the concept of mechanical impedance and thus operates in the frequency domain. Similar approaches to continuum modeling of structures are the subject of a recent workshop.[64] An exact member theory was applied to vibration and buckling of general periodic lattices in Refs. 65 and 66 and was extended to handle transient response of lattice structures in Ref. 67.

Model Order Reduction

The reduction of the large-order models of complex systems, like large space structures, to a sufficiently low order in such a way that the important dynamic characteristics of the system are preserved is a nontrivial task. Large-order finite-element "stress" models of structures are traditionally reduced to smaller order for use in dynamics analyses by the Guyan-Irons reduction method[28] or by some application of substructuring (component synthesis).[29,68] On the other hand, models for use in control system design and implementation will probably demand even further reduction, and a combination of the previously mentioned model reduction techniques with those presently employed in the controls community[69–76] will need to be employed. The issues to be addressed include the following: "What coordinates should be used?" "What order-reduction procedure should be employed?" "How is the required model order to be determined, and how are the retained/reduced coordinates to be selected?" References 76–78 address this last question.

Substructuring; Component Synthesis

Substructuring, also referred to as component synthesis or component mode synthesis (CMS),[28,79] is frequently employed for analyzing the dynamics of aerospace structures.[29] The justification for this lies in the fact that components of an aerospace system (e.g., the Space Shuttle Orbiter plus payloads, a satellite body plus solar panels) are frequently designed by different engineering contractors; thus, there exists a need for communicating and assembling accurate component models of reasonable order. Furthermore, dynamic analysis and testing may be required on system configurations comprised of different collections of components (e.g., the space station at various stages of assembly). Order reduction is usually incorporated as an integral part of component synthesis. The use of parallel computers to carry out substructure computations is discussed in the computational techniques subsection later.

Substructure analysis may be performed in either the time domain or the frequency domain, depending on the application. Reference 68 reviews both frequency-domain and time-domain methods. Examples of frequency-domain substructuring may be found in Refs. 64, 80, and 81 and hybrid methods have also been used.[82] Although the literature on substructuring applied to damped systems is not as extensive as that dealing with undamped systems, methods for dealing with damped systems have been developed.[83-89] References 83 and 84 employ the state-space (first-order) formulation. Figure 3a shows a structure (system) divided into substructures (components), and Fig. 3b shows the labeling of interior and boundary coordinates. Figure 3c illustrates several types of "component modes." Component modes (e.g., fixed-interface normal modes, free-interface normal modes, constraint modes, attachment modes) are first developed for each component of the system. Then interface compatibility is enforced to arrive at a set of coupled-system equations of motion, i.e., the system model. Details of the procedure may be found in Refs. 28, 68, and 90, for example.

Let the undamped equations of motion of component be partitioned according to interior coordinates i and boundary coordinates b:

$$\begin{bmatrix} m_{ii} & m_{ib} \\ m_{bi} & m_{bb} \end{bmatrix} \begin{Bmatrix} \ddot{x}_i \\ \ddot{x}_b \end{Bmatrix} + \begin{bmatrix} k_{ii} & k_{ib} \\ k_{bi} & k_{bb} \end{bmatrix} \begin{Bmatrix} x_i \\ x_b \end{Bmatrix} = \begin{Bmatrix} f_i \\ f_b \end{Bmatrix} \tag{7}$$

The component displacement coordinates x are related to a reduced set of generalized coordinates p through the Ritz-type transformation

$$x = \Psi p \tag{8}$$

where Ψ contains the component "modes" (e.g., constraint modes and fixed-interface normal modes in the case of the Craig-Bampton method). As an example, constraint modes are given by the equation

$$\Psi_c = \begin{bmatrix} -k_{ii}^{-1} k_{ib} \\ I \end{bmatrix} \tag{9}$$

Although it is not uncommon for a truncated set of free-interface modes of a component to be employed in dynamic analysis, it has been shown repeatedly that serious errors may be introduced as a result of the truncation and that the effects of the truncated modes must be included in the component reduced-order model.[91-93] The residual flexibility effects are captured by supplementing the truncated set of free-interface modes with attachment modes (see Ref. 28).

Substructuring methods such as those used in the Space Shuttle program[29] can be used for modeling many large space structures. Computational enhancements are possible through the use of parallel processing. The fixed-interface method developed by Hurty and modified by Craig and Bampton and the residual flexibility method developed by MacNeal and extended by Rubin (see Refs. 28 and 68) will continue to be widely used.

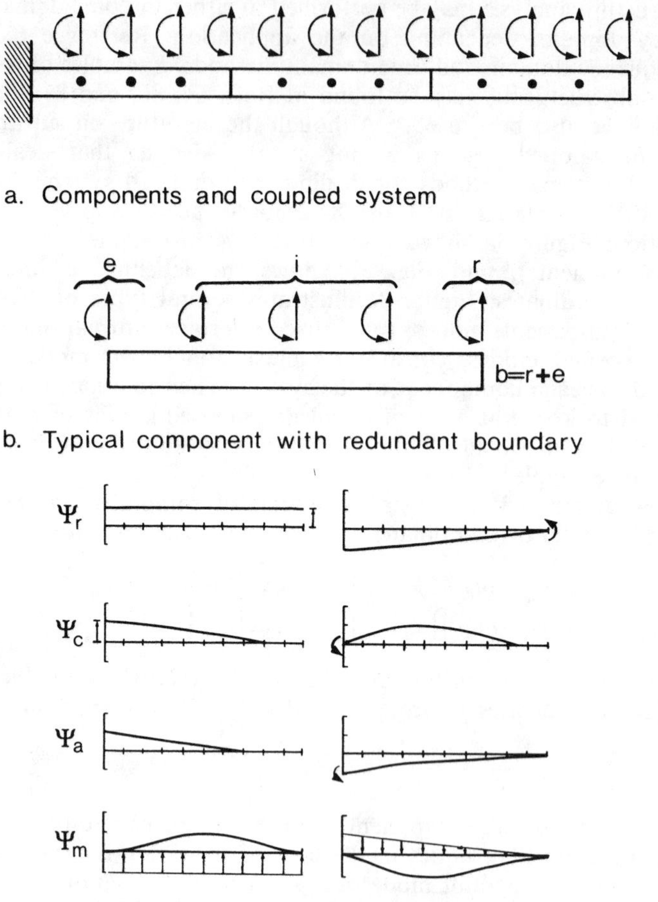

a. Components and coupled system

b. Typical component with redundant boundary

c. Examples of component modes of a beam element

Fig. 3 Illustration of component modes.

References 28, 91, and 94 provide information on the relative accuracy of various component mode synthesis methods, and Refs. 95–97 discuss some computational features of the Craig-Bampton method. References 92 and 93 stress the importance of incorporating residual flexibility effects when free-interface modes are used. Component synthesis methods employing different types of component modes (e.g., Lanczos modes) may also be useful.[90,98]

Computational Techniques

Parallel processing promises to become a key element in structural analysis.[16,30,99–103] Parallel processing via a recursive formulation may have significant impact on multibody formulations of flexible space structures.[104–107] Substructure methods lend themselves to parallelization, with versions of subspace iteration and the Lanczos method being particularly suited to parallelization.[90,108–112] Reference 113 discusses the use of dedicated parallel hardware for simulating the dynamics of flexible space structures.

Waves vs Modes

One study of vibration modes of a space station model reveals the presence of over 150 normal modes with frequencies below 5 Hz.[21] Some researchers have suggested that a wave propagation approach should be considered for describing the dynamics of structures with such high modal density.[114,115] References 116–120 discuss various aspects of dynamics and control based on waves, and in Ref. 121 a novel wave propagation approach is employed to determine the minimum time required to perform a prescribed control maneuver. Finally, Ref. 122 is representative of recent research on finite-element methods for analysis of wave propagation in structures.

Localization: Local vs Global Behavior

Mode shapes (eigenvectors) can be extremely sensitive to small perturbations in structural properties. This phenomenon has been studied with reference to the behavior of crystal lattices and turbine blades, as well as with reference to the dynamics of large space structures. Periodic lattice structures and multipanel solar arrays are two sources of potential localization problems. The reader is directed to Refs. 123–128.

Localization has been observed experimentally in structures with closely spaced modes,[129] and similar problems arising from high modal density were observed during the modal testing of a generic ¼-scale Space Station truss,[130] with global modes and local modes of individual truss members interspersed.

Nonlinearity: Joints

A thorough review of the topic of nonlinearity in relationship to modeling of large space structures is beyond the scope of this chapter; specifically nonlinear multibody dynamics problems are discussed in detail elsewhere. Reference 131 provides a recent overview of computational mechanics issues related to nonlinearities in the dynamics and control of space structures, and Ref. 132 discusses the geometrically nonlinear behavior of a truss. A force state mapping representation of the dynamics of nonlinear joints is described in Refs. 133 and 134, and Refs. 135 and 136 also discuss experimental determination of appropriate models of dynamic behavior of joints. A large number of papers have been devoted to the modeling and analysis of space structures whose behavior is strongly influenced by joints.[57,137–142]

Damping

The subject of active structural damping through the use of feedback control is treated elsewhere in this volume. Passive damping has been the subject of numerous conferences[143] and individual technical papers.[144,145]

Modeling Errors and Parameter Uncertainty

No mathematical model of any real structure, whether a distributed-parameter model or a discrete-parameter model, can ever perfectly characterize the dynamics of the structure. Vibration tests (so-called ground vibration tests, or modal tests) have been employed in the aerospace industry to "validate" mathematical models of aircraft and spacecraft, but the fact remains that model error always exists. And the problem of model validation will be far more difficult for large space structures of the type under discussion.[18,146–148] The topic of model validation and model updating will be treated at greater length later in this chapter.

Those who model large space structures that will be subject to control/structure interaction should note the theme of a recent workshop on model error concepts and compensation[149]:

> The theme of the Workshop is most easily stated in the deceptively simple phrase—*the modeling problem and the control problem are not independent*! The phrase implicitly asserts that the systems modeller cannot 'sensibly' proceed in a systematic and scientific way to produce a model for the purpose of controller design without taking direct account of the *structure of the control scheme* to be designed (e.g. feedback, feedforward, decoupling, optimal, adaptive . . .), the *changes in performance* to be achieved by the scheme (e.g. bandwidth, . . .) and assessing the magnitude and effect of *modeling errors* on model-based predictions.

The topics of model error compensation (e.g., Ref. 150), robust controller design (e.g., Refs. 70 and 151), and adaptive controller design (e.g., Ref. 152) have been treated extensively in the controls literature, and structural dynamicists have recently sought to rigorously assess the predictive accuracy of structural dynamic models (e.g., Refs. 153 and 154). However, the previously given workshop theme still poses a definite challenge to those responsible for creating, reducing, and verifying large space structure models, especially in light of the difficulties attendant to on-ground and on-orbit testing of these structures.

System Identification; Model Verification and Model Updating

It will be absolutely essential for some form of dynamic testing to be performed to validate the mathematical models used to predict and control the dynamics of large space structures on orbit. Since ground vibration tests (analog methods prior to 1970 and digital computer-based testing since 1970) have been employed to validate mathematical models of aerospace vehicles for half a century, and since the aerospace community has been in the forefront of development and application of modal testing and mathematical model verification procedures,[155] much of the technology necessary for system identification of large space structures is available. However, since large space structures cannot be fully assembled and tested on the ground, as has been the practice for spacecraft and launch vehicles up until now, many new issues must be taken into consideration. A space station structural characterization experiment has been planned,[21,147,156] and an eight-member task committee has written a monograph discussing issues related to on-orbit testing.[18] References 130, 146, 148, and 157–159 discuss issues related to testing of large space structures. During the development of several space vehicle systems, scale model vibration tests were performed in order to validate design features.[28] Thus, although full-scale tests cannot be performed on large space structures in a 1-*g* environment, scale model tests are highly desirable.[160–162]

The topic of identification of large space structures encompasses many issues: ground-based testing vs on-orbit testing (on-line vs off-line in the latter case), time domain vs frequency domain, selection and location of sensors and actuators, physical parameter vs modal parameter, model error localization, model updating, system model vs component model, scaling, and boundary conditions. This section will provide an overview of the state of the art in modal testing and model verification and updating and will provide references to the literature on scale model testing in 1 *g* and to the literature on on-orbit damage assessment.

Experimental Modal Analysis

A wealth of papers on modal testing and mathematical model verification can be found in the proceedings of the International Modal Analysis Conferences,[25] and the theory and applications of modal analysis are the subject of a textbook,[163] a monograph,[164] and a major technical report.[165] References 166–168 survey the topic and provide extensive bibliographies. Test procedures ranging from multiple-exciter harmonic (sine-dwell) testing to multiple-exciter uncorrelated random testing have been employed in recent aircraft and spacecraft qualification tests.[9,155,169] Recent advances related to sine-dwell testing are cited in Refs. 170–173. Parameter estimation algorithms are described in Refs. 73, 166, 174–181. Reference 182 provides a mathematical correlation of modal parameter identification methods, and Ref. 183 provides a statistical comparison of methods.

Model Verification: Model Error Localization and Model Updating

In Ref. 184 Hasselman observes that there are several ways to approach verification of the mathematical model of a structure, and he notes that, regardless of the approach, several basic questions must be addressed: "What is compared? What is adjusted? What is agreement? And by what criteria is a model qualified? In Ref. 185 he describes the scope of the large space structure model verification (parameter estimation) task as follows:

Parameter estimation can only be successful whenever the structure of the model itself is correct. Considerable effort will be required to ensure that it is. It is envisioned that such an effort would entail ground vibration testing of structural components to verify component models, ground testing of subassemblies to verify the modeling of component interfaces, developmental space testing of prototype structures to verify zero-*g* models, pre-operational flight testing to verify models of as-built structures, and continuous monitoring to track any changes in the structure which might occur during operation. Substructuring and modal reduction will undoubtedly play an important role in this modeling and model verification process.

A key question that arises in model verification is, "Which model?" In Refs. 186 and 187, Berman discusses the general field of technology in which a linear mass, damping, and stiffness matrix (MDK) model of a structure is identified using data obtained from dynamic tests of the structure. Since, in dynamic tests response measurements are rarely obtained at more than a hundred or so coordinates, some order reduction is necessary, and parameters of a "full-order" MDK model cannot be obtained. This order reduction usually involves some form of substructuring, or component mode synthesis. Reference 188 is based on free-interface/residual flexibility component models, and Ref. 189 compares free-interface/residual flexibility and fixed-interface test-based models. An ASME monograph[190] brings together some of the issues faced in combined experimental/analytical modeling of dynamic structural systems. Included are overviews of modal testing and measurement and of combined experimental/analytical modeling, together with several case histories. This monograph emphasizes substructure testing and model verification related to substructure models.

In performing a vibration test to obtain experimental data for use in verifying a mathematical model, important issues are the choice of support or suspension provided for the test article and the relationship of the test configuration to the mathematical model being verified. Reference 191 discusses test/analysis correlation under rigid, flexible, and hybrid interface conditions, and Ref. 192 treats component synthesis of structures with "sloppy" joints. A Space Shuttle payload is treated as a component whose mathematical model is to be interfaced with an Orbiter mathematical model. Shuttle payload tests have been conducted with fixed (actually, very stiff) interface supports,[193–197] with mass-loaded interfaces,[198] and with free interfaces.[199–202] Verification of the mathematical model in the vicinity of the payload/Orbiter interfaces is crucial, and the variations in test support conditions reflect the search for methods that will adequately verify the interface portion of the payload mathematical models. In the testing of large space structure components, the support conditions will also be very important. References 148 and 203–206 describe a multiple-boundary-condition test (MBCT) procedure, and the procedure is applied to coupled truss substructures in Ref. 207 as a means of correcting model errors in the critical interface regions.

One aspect of model verification is the determination of which portions of the model to update, i.e., which parameters to modify. References that deal specifically with model error localization include Refs. 208–218.

Some of the questions that have been addressed in the literature on model updating include the following: How is the order of the mathematical model reconciled with the order of the experimental model? How are "errors" assigned to the respective "models," i.e., errors in the mathematical model and errors in the experimental data and the reduction of the experimental data? Must the experimental data be forced to produce modes that are orthogonal with respect to some analytical mass matrix? Are updated models valid if they do not represent, for instance, physically realizable connectivities in the structure? References 186 and 219–236 reflect a growing interest in the development of model verification/model updating procedures. Reference 235 provides a comparison of four model update methods.

Scale Modeling of Large Space Structures

In order to test large space structures in a 1-*g* environment, it will be necessary to use scale models. Scaling requirements and suspension system designs are discussed in Refs. 146, 160, 162, and 237–239.

Damage Detection

Some modal test techniques and model error detection techniques may be useful in on-orbit damage detection. Several references related to this topic include Refs. 217, 240, and 241.

On-Orbit System Identification

For the most part the references in this section have addressed issues that are relevant for ground testing of scale models and components of large space structures and for on-orbit, off-line testing. Reference 18 should be consulted for an extensive discussion of on-orbit identification of large space structures, and Ref. 27 can be consulted for information on off-line and on-line system identification methods in general.

Concluding Remarks

As may be obvious from the fact that few of the references cited in this chapter specifically address modeling, identification, and analysis of large space structures per se, much work lies ahead in developing the tools necessary for characterizing the dynamics of large orbiting structures.

Acknowledgments

The preparation of this chapter was supported by the NASA Lyndon B. Johnson Space Center. Appreciation is expressed to those who generously collected and sent copies of their papers and reports, and to T. J. Su, who assisted in the preparation of this chapter.

References

[1]Nurre, G. S., Ryan, R. S., Scofield, H. N., and Sims, J. L., "Dynamics and Control of Large Space Structures," *Journal of Guidance, Control, and Dynamics*, Vol. 7, No. 5, Sept.–Oct. 1984, pp. 514–526.

[2]Meirovitch, L. (ed.), *Dynamics and Control of Large Structures*, Virginia Polytechnic Institute and State University, Blacksburg, VA, 1979, No. 1–7, 1989.

[3]Lightner, E. B. (ed.), "Structural Dynamics and Control of Large Space Structures," NASA CP-2187, Oct. 1980.

[4]Koprivar, F. III, and Boyer, W. J., (eds.), "Large Space Systems Technology," NASA CP-2168, Nov. 1980, NASA CP-2215, Nov. 1981.

[5]Brumfield, M. L. (ed.), "Structural Dynamics and Control of Large Space Structures," NASA CP-2266, Jan. 1982.

[6]Pinson, L. D., Amos, A. K., and Venkayya, V. B. (eds.), "Modeling, Analysis, and Optimization Issues for Large Space Structures," NASA CP-2258, May 1982.

[7]Taylor, L. (ed.) "NASA SCOLE Workshops," NASA Langley Research Center, Hampton, VA, 1984, No. 1–4, 1987, No. 4.

[8]Huckins, E. K. III, and Boyer, W. J. (eds.), "Large Space Antenna Systems Technology—1984," NASA Langley Research Center, Hampton, VA, Dec. 4–6, 1984.

[9]Stavrinidis, C., Mamode, A., and Bergmann, H. (eds.), "Spacecraft Structures," ESA SP-238, April 1986; "Spacecraft Structures and Mechanical Testing," ESA SP-289, Jan. 1989, European Space Agency, Noordwijk, The Netherlands.

[10]Ryan, R. S., and Scofield, H. N. (eds.), "Structural Dynamics and Control Interaction of Flexible Structures," NASA CP-2467, April 1986.

[11]"NASA/DOD CSI Technology Conference," NASA CP-2447, edited by R. L. Wright, Nov. 1986, No. 1–3, NASA CP-3041, edited by J. R. Newson, Feb. 1989, No. 3.

[12]"USAF/NASA Workshop on Model Determination for Large Space Systems," Vols. 1–3, Jet Propulsion Lab., California Institute of Technology, Pasadena, CA, JPL D-5574, March 22–24, 1988.

[13]Raney, J. P. (ed.), *AIAA SDM Issues of the International Space Station, a Collection of Technical Papers*, AIAA, Washington, DC, April 1988.

[14]Amos, A. K. (ed.), "Sixth Air Force Forum on Space Structures," Air Force Office of Scientific Research, April 1988.

[15]Taylor, L. (ed.), "Workshop on Computational Aspects in the Control of Flexible Systems," NASA TM-101578, July 1988.

[16]Man, G. K. (ed.), "NASA, NSF, DoD 3rd Annual Conference on Aerospace Computational Control," Jet Propulsion Lab., California Institute of Technology, Pasadena, CA, 1989.

[17]Atluri, S. N., and Amos, A. K., *Large Space Structures: Dynamics and Control*, Springer-Verlag, Berlin, 1988.

[18]Denman, E., Hasselman, T., Juang, J.-N., Junkins, J., Udwadia, F., Venkayya, V., and Kamat, M., "Identification of Large Space Structures on Orbit," Air Force Rocket Propulsion Lab., Edwards AFB, CA, TR-86-054, Sept. 1986.

[19]*Engineering and Configurations of Space Stations and Platforms*, Noyes, Park Ridge, NJ, 1985.

[20]Lindenmoyer, A. J., and Habermeyer, J. A., "An Automated, Integrated Approach to Space Station Structural Modeling," AIAA Paper 89-1342, April 1989, pp. 1591–1597.

[21]Burickman, K., Draper, C., Havelka, J., Kim, H. M., Shein, S. L., and Partin,

J., "Space Station Freedom Preliminary Loads Analysis," McDonnell Douglas Space Systems, Houston, TX, Final Rept., Contract NAS9-18200, Feb. 1989.

[22]Singh, S. K., and Lindenmoyer, A. J., "Preliminary Control/Structure Interaction Study of Coupled Space Station Freedom/Assembly Work Platform/Orbiter," AIAA Paper 89-0543, Jan. 1989.

[23]Taylor, L. W., Jr., and Williams, J. L., "Distributed Parameter Modeling of the Structural Dynamics of the Solar Array Flight Experiment," *AIAA Guidance, Navigation and Control Conference*, AIAA, New York, 1987, pp. 959–974.

[24]Wright, R. L., and Seward, S. K., "Technology for Large Space Systems—A Bibliography with Indexes Category 05—Structural Dynamics and Control," NASA SP-7046(17), Oct. 1987.

[25]DeMichele, D. J., and Galione, K. A. (eds.), "Proceedings of International Modal Analysis Conferences," Union College, Schenectady, NY, and Society for Experimental Mechanics, Bethel, CT, 1982–1989.

[26]Montgomery, R. C., Shenhar, J., and Williams, J. P., "On-Line Identification and Attitude Control for SCOLE," AIAA Paper 87-2459, Aug. 1987, pp. 950–958.

[27]Sinha, N. K., and Kuszta, B., *Modeling and Identification of Dynamic Systems*, Van Nostrand Reinhold, New York, 1983.

[28]Craig, R. R., Jr., *Structural Dynamics—An Introduction to Computer Methods*, Wiley, New York, 1981.

[29]"Space Shuttle Payload Design and Development—Structural/Mechanical Interfaces and Requirements," NASA Lyndon B. Johnson Space Center, Houston, TX, NSTS 20052, Vol. 8, Rev. C, June 1988.

[30]Storaasli, O. O., Hguyen, D. T., and Agarwal, T. K., "Parallel-Vector Solution of Large-Scale Structural Analysis Problems on Supercomputers," AIAA Paper 89-1259, April 1989, pp. 859–867.

[31]Noor, A. K., "Assessment of Current State of the Art in Modeling Techniques and Analysis Methods for Large Space Structures," *Modeling, Analysis, and Optimization Issues for Large Space Structures*, NASA CP-2258, May 1982, pp. 5–32.

[32]Noor, A. K., and Mikulas, M. M., "Continuum Modeling of Large Lattice Structures: Status and Projections," *Large Space Structures: Dynamics and Control*, edited by S. N. Atluri and A. K. Amos, Springer-Verlag, Heidelberg, FRG, 1988, pp. 1–34.

[33]Baker, M., Chiger, H. D., Habermeyer, J. A., Hipol, P. J., Johnson, C. L., and DeRyder, L. J., "Space Station Multidisciplinary Analysis Capability—IDEAS**2," AIAA Paper 86-0954, May 1986, pp. 421–434.

[34]Miller, D. F., Venkayya, V. B., and Tischler, V. A., "Integration of Structures and Controls—Some Computational Issues," *Proceedings of the 24th IEEE Conference on Decision and Control*, Institute of Electrical and Electronics Engineers, New York, Dec. 1985, pp. 924–931.

[35]Belvin, W. K., and Park, K. C., "On the State Estimation of Structures with Second Order Observers," AIAA Paper 89-1241, April 1989, pp. 721–727.

[36]Hamdan, A. M. A., and Nayfeh, A. H., "Measures of Modal Controllability and Observability for First- and Second-Order Linear Systems," *Journal of Guidance, Control, and Dynamics*, Vol. 12, No. 3, May–June 1989, pp. 421–428.

[37]Hashemipour, H. R., and Laub, A. J., "Kalman Filtering for Second-Order Models," *Journal of Guidance, Control, and Dynamics*, Vol. 11, No. 2, March–April 1988, pp. 181–186.

[38]Su, T.-J., and Craig, R. R., Jr., "Model Reduction and Control of Flexible Structures Using Krylov Subspaces," AIAA Paper 89-1237, April 1989, pp. 691–700.

[39]Kane, T. R., Likins, P. W., and Levinson, D. A., *Spacecraft Dynamics*, McGraw-Hill, New York, 1983.

[40]Wie, B., and Bryson, A. E., Jr., "Pole-Zero Modeling of Flexible Space Structures," *Journal of Guidance, Control, and Dynamics*, Vol. 11, No. 6, Nov.–Dec. 1988, pp. 554–561.

[41]Hablani, H. B., "Constrained and Unconstrained Modes: Some Modeling Aspects of Flexible Spacecraft," *Journal of Guidance, Control, and Dynamics*, Vol. 5, No. 2, March–April 1982, pp. 164–173.

[42]Meirovitch, L., *Elements of Vibration Analysis*, McGraw-Hill, New York, 1975.

[43]Kane, T. R., and Levinson, D. A., "Formulation of Equations of Motion for Complex Spacecraft," *Journal of Guidance and Control*, Vol. 3, No. 2, March–April 1980, pp. 99–112.

[44]Meirovitch, L., "Control of Distributed Structures," *Large Space Structures: Dynamics and Control*, edited by S. N. Atluri and A. K. Amos, Springer-Verlag, Heidelberg, FRG, 1988, pp. 195–212.

[45]Meirovitch, L., and Oz, H., "Modal-Space Control of Distributed Gyroscopic Systems," *Journal of Guidance, Control, and Dynamics*, Vol. 3, No. 2. March–April 1980, pp. 140–150.

[46]Schaechter, D. B., "Estimation of Distributed Parameter Systems: Some Closed-Form Solutions," *Journal of Guidance, Control, and Dynamics*, Vol. 9, No. 4, July–Aug. 1986, pp. 408–412.

[47]Taylor, L. W., and Naidu, D. S., "Experience in Distributed Parameter Modeling of the Spacecraft Control Laboratory (SCOLE) Structure," AIAA Paper 87-0895, April 1987, pp. 330–343.

[48]Necib, B., and Sun, C. T., "Analysis of Truss Beams Using a High Order Timoshenko Beam Finite Element," *Journal of Sound and Vibration*, Vol. 30, No. 1, April 8, 1989, pp. 149–159.

[49]Sun, C. T., and Liebbe, S. W., "A Global-Local Approach to Solving Vibration of Large Truss Structures," AIAA Paper 86-0872, May 19–21, 1986, Pt. 2, pp. 248–255.

[50]Sun, C. T., and Kim, B. J., "Continuum Modeling of Periodic Truss Structures," *Damage Mechanics and Continuum Modeling*, American Society of Civil Engineers, New York, 1985, pp. 57–71.

[51]Sun, C. T., Kim, B. J., and Bogdanoff, J. A., "On the Derivation of Equivalent Simple Models for Beam- and Plate-Like Structures in Dynamic Analysis," AIAA Paper 81-0624-CP, April 1981, pp. 523–532.

[52]Sun, C. T., and Juang, J.-N., "Modeling Global Structural Damping in Trusses Using Simple Continuum Models," *AIAA Journal*, Vol. 24, Jan. 1986, pp. 144–150.

[53]Renton, J. D., "The Beam-Like Behavior of Space Trusses," *AIAA Journal*, Vol. 22, Feb. 1984, pp. 273–280.

[54]Nayfeh, A. H., and Hefzy, M. S., "Continuum Modeling of the Mechanical and Thermal Behavior of Discrete Large Structures," *AIAA Journal*, Vol. 19, June 1981, pp. 766–773.

[55]Aswani, M., Au, N. N., and Lin, S. R., "Development of an Analytical Model for Large Space Structures," AIAA Paper 82-0648, May 1982, Pt. 1, pp. 41–48.

[56]Saw, K. C., and Tamma, K. K., "Reduced Modeling and Analysis of Large Repetitive Space Structures via Continuum/Discrete Concepts," AIAA Paper 86-1022, May 1986, pp. 720–729.

[57]Chapman, J. M., and Shaw, F. H., "Equivalent Beam Modeling Using Numerical Reduction Techniques," NASA CP-2467, April 1986, Pt. 1, pp. 567–594.

[58]Haug, E., Dowlatyari, P., and Dubois, J., "Equivalent Continuum Analysis of Large Space Trusses," *ESA Spacecraft Structures*, European Space Agency, CP-238, April 1986, pp. 57–65.

[59]Dow, J. O., Su, Z. W., Feng, C. C., and Bodley, C., "Equivalent Continuum Representation of Structures Composed of Repeated Elements," *AIAA Journal*, Vol. 23, Oct. 1985, pp. 1564–1569.

[60]Berry, D. T., Yang, T. Y., and Skelton, R. E., "Dynamics and Control of Lattice Beams Using Simplified Finite Element Models," *Journal of Guidance, Control, and Dynamics*, Vol. 8, No. 5, Sept.–Oct. 1985, pp. 612–619.

[61]Lamberson, S. E., and Yang, T. Y., "Integrated Design of Space Structures Using Lattice Plate Finite Elements," *Journal of Guidance, Control, and Dynamics*, Vol. 9, No. 4, July–Aug. 1986, pp. 478–484.

[62]Thomas, S., and Stubbs, N., "Dynamic Analysis of the Space Station Truss Structure Based on a Continuum Representation," AIAA Paper 89-1280, April 1989, pp. 1062–1068.

[63]Poelaert, D., "DISTEL, A Distributed Element Program for Dynamic Modelling and Response Analysis of Flexible Structures," *Dynamics and Control of Large Structures, Proceedingss of the Fourth VPI&SU/AIAA Symposium*, Virginia Polytechnic Institute and State University, Blacksburg, VA, 1983, pp. 319–337.

[64]Polaert, D. (ed.), *Modal Representation of Flexible Structures by Continuum Methods*, European Space Agency, Noordwijk, The Netherlands, June 1989.

[65]Anderson, M. S., and Williams, F. W., "BUNVIS-RG: Exact Frame Buckling and Vibration Program, with Repetitive Geometry and Substructuring," *Journal of Spacecraft and Rockets*, Vol. 24, No. 4, July–Aug. 1987, pp. 353–361.

[66]Anderson, M. S., and Williams, F. W., "Natural Vibration and Buckling of General Periodic Lattice Structures," *AIAA Journal*, Vol. 24, Jan. 1986, pp. 163–169.

[67]Anderson, M. S., "Transient Response of Lattice Structures Based on Exact Member Theory," AIAA Paper 89-1337, April 1989, pp. 1546–1554.

[68]Craig, R. R., Jr., "A Review of Time-Domain and Frequency-Domain Component-Mode Synthesis Methods," *International Journal of Analytical and Experimental Modal Analysis*, Vol. 2, No. 2, April 1987, pp. 59–72.

[69]Bacon, B. J., and Schmidt, D. K., "Multivariable Frequency-Weighted Order Reduction," *Journal of Guidance, Control, and Dynamics*, Vol. 12, No. 1, Jan.–Feb. 1989, pp. 97–107.

[70]Bernstein, D. S., and Hyland, D. C., "Optimal Projection for Uncertain Systems (OPUS): A Unified Theory of Reduced-Order, Robust Control Design," *Large Space Structures: Dynamics and Control*, edited by S. N. Atluri and A. K. Amos, Springer-Verlag, Heidelberg, FRG, 1988, pp. 263–302.

[71]De Villemagne, C., and Skelton, R. E., "Controller Reduction Using Canonical Interactions," *IEEE Transactions on Automatic Control*, Vol. 33, No. 8, Aug. 1988, pp. 740–750.

[72]De Villemagne, C., and Skelton, R. E., "Model Reductions Using a Projection Formulation," *International Journal of Control*, Vol. 46, No. 6, 1987, pp. 2141–2169.

[73]Inman, D. J., *Vibration—With Control, Measurement, and Stability*, Prentice-Hall, Englewood Cliffs, NJ, 1989.

[74]Panossian, H. V., "Model Order Reduction Techniques in Large Space Structure Applications," AIAA Paper 88-2467, April 1988, Pt. 3, pp. 1733–1741.

[75]Skelton, R. E., "Order Reduction for Models of Space Structures Using Modal Cost Analysis, *Journal of Guidance, Control, and Dynamics*, Vol. 5, No. 4, July–Aug. 1982, pp. 351–357.

[76]Yousuff, A., and Skelton, R. E., "Controller Reduction by Component Cost Analysis," *IEEE Transactions on Automatic Control*, Vol. AC-29, No. 6, June 1984, pp. 520–530.

[77]Blelloch, P. A., and Carney, K. S., "Modal Selection in Structural Dynamics," *Proceedings of the 7th International Modal Analysis Conference*, Union College, Schenectady, N.Y., 1989, pp. 742–749.

[78]Zhao, G., and Sinha, N. K., "Mode Selection in Aggregated Models," *Large Scale Systems, Theory and Applications*, Vol. 4, No. 2, April 1983, pp. 209–216.

[79]Meirovitch, L., *Computational Methods in Structural Dynamics*, Sijthoff & Noordhoff, Alphen aan den Rijn, The Netherlands, 1980.

[80]Geering, H. P., "New Methods in Substructuring," AIAA Paper 80-0799 CP, 1980, Pt. 2, pp. 801–808.

[81]Hagood, N. W., and Crawley, E. F., "A Frequency Domain Analysis for Damped Space Structures," AIAA Paper 89-1381, April 1989, pp. 1936–1946.

[82]Trubert, M., and Peretti, L., "A Cost-Effective Component Modes Analysis for Shuttle Payloads Using a Combination of Frequency Domain and Time Domain Approaches," AIAA Paper 85-0733, April 1985, pp. 391–403.

[83]Béliveau, J.-G., and Soucy, Y., "Damping Synthesis Using Complex Substructure Modes and a Hermitian System Representation," *AIAA Journal*, Vol. 23, No. 12, Dec. 1985, pp. 1952–1956.

[84]Craig, R. R., Jr., and Ni, Z., "Component Mode Synthesis for Model Order Reduction of Non-Classically-Damped Systems," *Journal of Guidance, Control, and Dynamics*, Vol. 12, No. 4, July–Aug. 1989, pp. 577–584.

[85]Ghlaim, K. H., and Martin, K. F., "Reduced Component Modes in a Damped System," *Proceedings of the 2nd International Modal Analysis Conference*, Union College, Schenectady, NY, 1984, pp. 683–686.

[86]Hale, A. L., "Substructure Synthesis and its Iterative Improvement for Large Nonconservative Vibratory Systems," *AIAA Journal*, Vol. 22, Feb. 1984, pp. 265–272.

[87]Hale, A. L., and Bergman, L. A., "The Dynamic Synthesis of General Non-Conservative Structures from Separately Identified Substructure Models," *Journal of Sound and Vibration*, Vol. 98, No. 3, 1985, pp. 431–446.

[88]Kubomura, K., "Component Mode Synthesis for Damped Structures," *AIAA Journal*, Vol. 25, May 1987, pp. 740–745.

[89]Lips, K. W., and Vigneron, F. R., "Damping Synthesis for a Spacecraft Using Substructure and Component Data," *Journal of Spacecraft and Rockets*, Vol. 23, No. 2, March–April 1986, pp. 158–164.

[90]Craig, R. R., Jr., and Hale, A. L., "Block-Krylov Component Synthesis Method for Structural Model Reduction," *Journal of Guidance, Control, and Dynamics*, Vol. 11, No. 6, Nov.–Dec. 1988, pp. 562–570.

[91]Benfield, W. A., Bodley, C. S., and Morosow, G., "Modal Synthesis Methods," *Symposium on Substructure Testing and Synthesis*, NASA Marshall Space Flight Center, AL, 1972.

[92]Flanigan, C. C., "A Residual Flexibility Approach to Launch Vehicle Liftoff Loads Analysis," AIAA Paper 89-1336, April 1989, pp. 1541–1545.

[93]Klahs, J. W., and Townley, G. E., "Determining Component Loads and Stresses with Improved System Modeling Techniques," *Proceedings of the 3rd International Modal Analysis Conference*, Union College, Schenectady, NY, 1985, pp. 941–947.

[94]Kammer, D. C., and Baker, M., "A Comparison of the Craig-Bampton and Residual Flexibility Methods for Component Substructure Representation," AIAA Paper 85-0817, April 1985, pp. 699–705.

[95]Engels, R. C., "A Solution to the Craig-Bampton Eigenvalue Problem for Multicomponent Structures," *International Journal of Analytical and Experimental Modal Analysis*, Vol. 1, No. 2, April 1986, pp. 19–25.

[96]Prakash, B. G., and Prabhu, M. S. S., "Reduction Techniques in Dynamic Substructures for Large Problems," *Computers & Structures*, Vol. 22, No. 4, 1986, pp. 539–552.

[97]Shu, C.-F., and Chu, F.-H., "A General Superelement Approach by the Component Mode Synthesis Method," *Proceedings of the 2nd International Modal Analysis Conference*, Union College, Schenectady, NY, 1984, pp. 676–682.

[98]Allen, J. J., "A Component Synthesis Method Using Lanczos Vectors," *Proceedings of the 6th International Modal Analysis Conference*, Union College, Schenectady, NY, 1988, pp. 507–513.

[99]Noor, A. K., and Atluri, S. N., "Advances and Trends in Computational Structural Mechanics," *AIAA Journal*, Vol. 25, July 1987, pp. 977–995.

[100]Noor, A. K., "Parallel Processing in Finite Element Structural Analysis," *Engineering with Computers*, Vol. 3, No. 4, 1988, pp. 225–241.

[101]Noor, A. K. (ed.), "Parallel Computations and Their Impact on Mechanics," *ASME Winter Annual Meeting*, AMD-Vol. 86, American Society of Mechanical Engineers, New York, 1987.

[102]Noor, A. K., and Peters, J. M., "A Partitioning Strategy For Efficient Nonlinear Finite Element Dynamics Analysis On Multiprocessor Computers," *Computers & Structures*, Vol. 31, No. 5, 1989, pp. 795–810.

[103]Storaasli, O. O., and Bergan, P., "Nonlinear Substructuring Method for Concurrent Processing Computers," *AIAA Journal*, Vol. 25, June 1987, pp. 871–876.

[104]Bae, D. S., and Haug, E. J., "A Recursive Formulation for Constrained Mechanical System Dynamics: Part I, Open Loop Systems," *Mechanics of Structures and Machines*, Vol. 15, No. 3, 1987, pp. 359–382.

[105]Bae, D. S., and Haug, E. J., "A Recursive Formulation for Constrained Mechanical System Dynamics: Part II, Closed Loop Systems," *Mechanics of Structures and Machines*, Vol. 15, No. 4, 1987–88, pp. 481–506.

[106]Bae, D. S., Kuhl, J. G., and Haug, E. J., "A Recursive Formulation for Constrained Mechanical System Dynamics: Part III, Parallel Processor Implementation," *Mechanics of Structures and Machines*, Vol. 16, No. 2, 1988, pp. 249–269.

[107]Hollerbach, J. M., "A Recursive Lagrangian Formulation of Manipulator Dynamics and a Comparative Study of Dynamics Formulation Complexity," *IEEE Transactions on Systems, Man, and Cybernetics*, Vol. 10, Nov. 1980, pp. 730–736.

[108]Akl, F., and Morel, M., "Eigensolution of Finite Element Problems in a Completely Connected Parallel Architecture,"AIAA Paper 89-1395, April 1989, pp. 2054–2063.

[109]Bennighof, J. K., and Wu, J.-Y., "A Substructure-Based Parallel Algorithm for Structure Transient Response Calculations,"AIAA Paper 89-1338, April 1989.

[110]Bostic, S. W., and Fulton, R. E., "Implementation of the Lanczos Method for Structural Vibration Analysis on a Parallel Computer," AIAA Paper 86-0930, May 1986, Pt. 2, pp. 400–410.

[111]Nour-Omid, B., and Regelbrugge, M. E., "Lanczos Method for Dynamic Analysis of Damped Structural Systems," *Proceedings of the 6th International Modal Analysis Conference*, Union College, Schenectady, NY, 1988, pp. 495–500.

[112]Storaasli, O. O., Bostic, S., Patrick, M., Mahajan, U., and Ma, S., "Three Parallel Computation Methods for Structural Vibration Analysis," AIAA Paper 88-2391, April 1988, pp. 1401–1411.

[113]Gluck, R., "Hope for Simulating Flexible Spacecraft," *Aerospace America*, Vol. 24, Nov. 1986, pp. 40–44.

[114]Von Flotow, A. H., "The Acoustic Limit of Control of Structural Dynamics," *Large Space Structures: Dynamics and Control*, edited by S. N. Atluri and A. K. Amos, Springer-Verlag, Heidelberg, FRG, 1988, pp. 213–237.

[115]Zak, M., "Dynamical Response to Pulse Excitations in Large Space Structures,"*Large Space Structures: Dynamics and Control*, edited by S. N. Atluri and A. K. Amos, Springer-Verlag, Heidelberg, FRG, 1988, pp. 133–163.

[116]Meirovitch, L., and Bennighof, J. K., "Control of Traveling Waves in Flexible Structures," *Dynamics and Control of Large Structures, Proceedings of the 5th VPI&SU/AIAA Symposium*, Virginia Polytechnic Institute and State University, Blacksburg, VA, 1985, pp. 163–182.

[117]Von Flotow, A. H., "A Travelling Wave Approach to the Dynamic Analysis of Large Space Structures,"AIAA Paper 83-0964, May 1983, Pt. 2, pp. 509–519.

[118]Von Flotow, A. H., "Traveling Wave Control for Large Spacecraft Structures," *Journal of Guidance, Control, and Dynamics*, Vol. 9, No. 4, July–Aug. 1986, pp. 462–468.

[119]Von Flotow, A. H., and Schaefer, B., "Wave-Absorbing Controllers for a Flexible Beam," *Journal of Guidance, Control, and Dynamics*, Vol. 9, No. 6, Nov.–Dec. 1986, pp. 673–680.

[120]Williams, J. H., Jr., "Wave Propagation in Large Space Structures," 6th Air Force Forum on Space Structures, Atlanta, GA, April 7–8, 1988.

[121]Bennighof, J. K., and Boucher, R. L., "An Investigation of the Time Required for Control of Structures," *Dynamics and Control of Large Structures, Proceedings of the 6th VPI&SU/AIAA Symposium*, Virginia Polytechnic Institute and State University, Blacksburg, VA. 1987, pp. 391–406.

[122]Iura, M., "Space-Time Finite Element Analysis of Traveling Wave of Rod and Beam Structures," *Computers & Structures* (to be published).

[123]Bendiksen, O. O., "Mode Localization Phenomena in Large Space Structures," *AIAA Journal*, Vol. 25, Sept. 1987, pp. 1241–1248.

[124]Cornwell, P. J., and Bendiksen, O. O., "Localization of Vibrations in Large Space Reflectors," *AIAA Journal*, Vol. 27, Feb. 1989, pp. 219–226.

[125]Kissel, G. J., "Localization in Disordered Periodic Structures," AIAA Paper 87-0819, April 1987, pp. 1046–1055.

[126]Pierre, C., "Localized Free and Forced Vibrations of Nearly Periodic Disordered Structures," AIAA Paper 87-0774, April 1987, pp. 186–197.

[127]Pierre, C., "Localized Vibrations of Disordered Multispan Beams: Theory and Experiment," *AIAA Journal*, Vol. 25, Sept. 1987, pp. 1249–1257.

[128]Pierre, C., and Cha, P. D., "Strong Mode Localization in Nearly Periodic Disordered Structures," *AIAA Journal*, Vol. 27, Feb. 1989, pp. 227–241.

[129]Craig, R. R., Jr., "Modal Vector Estimation for Closely-Spaced-Frequency Modes," *Proceedings of the 3rd International Modal Analysis Conference*, Union College, Schenectady, NY, 1985, pp. 672–678.

[130]Pappa, R. S., and Juang., J.-N., "Studies of Modal-Identification Performance Using Hybrid Data," *International Journal of Analytical and Experimental Modal Analysis*, Vol. 2, No. 2, April 1987, pp. 99–108.

[131]Atluri, S. N., and Iura, M., "Nonlinearities in the Dynamics and Control of Space Structures: Some Issues for Computational Mechanics," *Large Space Structures: Dynamics and Control*, edited by S. N. Atluri and A. K. Amos, Springer-Verlag, Heidelberg, FRG, 1988, pp. 35–70.

[132]Abrate, S., and Sun, C. T., "Dynamic Analysis of Geometrically Nonlinear Truss Structures," *Computers & Structures*, Vol. 17, No. 4, 1983, pp. 491–497.

[133]Crawley, E. F., and Aubert, A. C., "Identification of Nonlinear Structural Elements by Force-State Mapping," *AIAA Journal*, Vol. 24, Jan. 1986, pp. 155–162.

[134]Crawley, E. F., and O'Donnell, K. J., "Identification of Nonlinear System Parameters in Joints Using the Force-State Mapping Technique," AIAA Paper 86-1013, May 1986, pp. 659–667.

[135]Prucz, J., "Experimental Characterization of Passively Damped Joints for Space Structures," *Journal of Spacecraft and Rockets*, Vol. 23, No. 6, Nov.–Dec. 1986, pp. 568–575.

[136]Srinivasan, A. V., "Dynamic Friction," *Large Space Structures: Dynamics and Control*, edited by S. N. Atluri and A. K. Amos, Springer-Verlag, Heidelberg, FRG, 1988, pp. 179–194.

[137]Chapman, J. M., Shaw, F. H., and Russell, W. C., "Nonlinear Transient Analysis of Joint Dominated Structures," AIAA Paper 87-0892, April 1987, pp. 568–577.

[138]Ferri, A. A., "Modeling and Analysis of Nonlinear Sleeve Joints of Large Space Structures," *Journal of Spacecraft and Rockets*, Vol. 25, No. 5, Sept.–Oct. 1988, pp. 354–360.

[139]Foelsche, G. A., Griffin, J. H., and Bielak, J., "Transient Response of Joint-Dominated Space Structures: A New Linearization Technique," *AIAA Journal*, Vol. 26, Oct. 1988, pp. 1278–1285.

[140]Hunt, D., "Dynamic Analysis of Structures with Friction Forces at Sliding Junctures," *Journal of Spacecraft and Rockets*, Vol. 21, No. 2, March–April 1984, pp. 175–179.

[141]Jezequel, L., "Modal Synthesis of Large Structures with Nonlinear Joints from Vibration Tests," *Proceedings of the Second International Conference on Recent Advances in Structural Dynamics*, Univ. of Southampton, UK, 1984, pp. 281–295.

[142]Mercadal, M., and Vander Velde, W. E., "Analysis of Limit Cycles in Control Systems for Joint Dominated Structures," *Journal of Guidance, Control, and Dynamics*, Vol. 12, No. 3, May–June 1989, pp. 388–395.

[143]Rogers, L. (ed.), "DAMPING 89," AFWAL/FIBAA Wright-Patterson AFB, OH, Feb. 8–10, 1989.

[144]Ashley, H., "On Passive Damping Mechanisms in Large Space Structures," *Journal of Spacecraft and Rockets*, Vol. 21, No. 5, Sept.–Oct. 1984, pp. 448–455.

[145]Chen, G.-S., and Wada, B. K., "Passive Damping for Space Truss Structures," AIAA Paper 88-2469, April 1988, pp. 1742–1749.

[146]Ashley, H., "Some Considerations on Earthbound Dynamic Testing of Large Space Structures," AIAA Paper 86-0908, May 1986, Pt. 2, pp. 362–375.

[147]Johnson, J. W., "Space Station Freedon Structural Characterization Experiment," NASA Space Station Freedom Office, April 19, 1989.

[148]Wada, B. K., "Extension of Ground-Based Testing for Large Space Structures," *Journal of Spacecraft and Rockets*, Vol. 23, No. 2, March–April 1986, pp. 184–188.

[149]Skelton, R. E., and Owens, D. H. (eds.), *Model Error Concepts and Compensation, Proceedings of the IFAC Workshop*, Pergamon, Oxford, UK, 1986.

[150]Skelton, R. E., "Model Error Concepts in Control Design," *International Journal of Control*, Vol. 49, No. 5, 1989, pp. 1725–1753.

[151]Hyland, D. C., and Bernstein, D. S., "MEOP Control Design Synthesis: Optimal Qualification of the Major Design Tradeoffs," NASA CP-2467, April 22–24, 1986, Pt. II, pp. 1033–1070.

[152]Kosut, R. L., "Adaptive Control of Large Space Structures: Uncertainty Estimation and Robust Control Calibration," *Large Space Structures: Dynamics and Control*, edited by S. N. Atluri and A. K. Amos, Springer-Verlag, Heidelberg, FRG, 1988, pp. 303–321.

[153]Hasselman, T. K., "Effects of Model Deficiencies on Parameter Estimation," *USAF/NASA Workshop on Model Determination for LSS*, Jet Propulsion Lab., California Institute of Technology, Pasadena, CA, 1988.

[154]Hasselman, T. K., and Chrostowski, J. D., "Evaluation of Predictive Accuracy in Structural Dynamic Models," *Proceedings of the 7th International Modal Analysis Conference*, Union College, Schenectady, NY, 1989, pp. 360–366.

[155]Chen, J.-C., "Evaluation of Spacecraft Modal Test Methods," *Journal of Spacecraft and Rockets*, Vol. 24, No. 1, Jan.–Feb. 1987, pp. 52–62.

[156]"Space Station Structural Characterization Experiment, Phase B, Selected Studies," NASA Langley Research Center, Final Review, Contract NAS9-18200, McDonnell Douglas Space Systems Company: Space Station Division, Houston, TX, Aug. 1989.

[157]Belvin, W. K., and Edighoffer, H. H., "Dynamic Analysis and Experiment Methods for a Generic Space Station Model," *Journal of Spacecraft and Rockets*, Vol. 24, No. 3, May–June 1987, pp. 270–276.

[158]Hanks, B. R., and Pinson, L. D., "Large Space Structure Testing Challenges," *Astronautics and Aeronautics*, Vol. 21, No. 10, Oct. 1983, pp. 34–40.

[159]Moore, C. J., Townsend, J. S., and Ivey, E. W., "Space Station Structures and Dynamics Test Program," NASA TP-2710, March 1987.

[160]Gronet, M. J., Crawley, E. F., and Allen, B. R., "Design, Analysis, and Testing of a Hybrid Scale Structural Dynamic Model of a Space Station," AIAA Paper 89-1340, April 1989, pp. 1567–1575.

[161]McGowan, P. E., "Considerations in the Design and Development of a Space Station Scale Model," NASA CP-2467 Pt. I, April 1986, pp. 215–245.

[162]McGowan, P. E., "Dynamic Scale Model Technology Program Technical Status & Relation to Space Station Dynamics," NASA Langley Research Center, Jan 10, 1989.

[163]Ewins, D. J., *Modal Testing: Theory and Practice*, Wiley, New York, 1984.

[164]Chu, D. F. H. (ed.), *Modal Testing and Model Refinement*, AMD-Vol. 59, American Society of Mechanical Engineers, New York, 1983.

[165]Allemang, R. J., Brown, D. L., and Rost, R. W., "Experimental Modal Analysis and Dynamic Component Synthesis," Flight Dynamics Lab., Air Force Wright Aero. Lab, Wright-Patterson AFB, OH, AFWAL-TR-87-306911, Dec. 1987.

[166]Allemang, R. J., "Experimental Modal Analysis, *Modal Testing and Model Refinement*, AMD-Vol. 59, American Society of Mechanical Engineers, New York, 1983, pp. 1–30.

[167]Allemang, R. J., "Experimental Modal Analysis Bibliography," *Proceedings of the 1st International Modal Analysis Conference*, Union College, Schenectady, NY, 1982, pp. 714–726.

[168]Allemang, R. J., and Brown, D., "Multiple-Input Experimental Modal Analysis—A Survey," *International Journal of Analytical and Experimental Modal Analysis*, Vol. 1, No. 1, Jan. 1986, pp. 37–44.

[169]Chen, J.-C., and Wada, B. K., "Sine Dwell or Broadband Methods for Modal Testing," *AIAA Journal*, Vol. 26, June 1988, pp. 733–737.

[170]Hunt, D. L., "A Modern Approach for Modal Testing Using Multiple Input Sine Excitation," AIAA Paper 87-0964, April 1987, pp. 1016–1031.

[171]Juang, J.-N., and Wright, J. R., "A Multi-Point Force Appropriation Method Based Upon a Singular Value Decomposition Approach," *Proceedings of the 7th International Modal Analysis Conference*, Union College, Schenectady, NY, 1989, pp. 317–323.

[172]Kabe, A. M., "Efficient Direct Measurement Mode Survey Test Procedure," *AIAA Journal*, Vol. 26, Aug. 1988, pp. 1003–1010.

[173]Williams, R., and Vold, H., "Multiphase-Step-Sine Method for Experimental Modal Analysis," *International Journal of Analytical and Experimental Modal Analysis*, Vol. 1, No. 2, April 1986, pp. 26–34.

[174]Craig, R. R., Jr., Chung, Y. T., and Blair, M. A., "State-Space Formulation of Multi-Shaker Modal Analysis," *Proceedings of the 5th International Modal Analysis Conference*, Union College, Schenectady, NY, 1987, pp. 1069–1077.

[175]Creamer, N. G., and Junkins, J. L., "Identification Method for Lightly Damped Structures," *Journal of Guidance, Control, and Dynamics*, Vol. 11, No. 6, Nov.–Dec. 1988, pp. 571–576.

[176]Deblauwe, F., Brown, D. L., and Allemang, R. J., "The Polyreference Time Domain Method," *Proceedings of the 5th International Modal Analysis Conference*, Union College, Schenectady, NY, 1987, pp. 832–845.

[177]Ibrahim, S. R., "A Review of Time Domain Modal Test Methods and Applications," *ESA Spacecraft Structures*, European Space Agency, CP-238, April 1986, pp. 205–211.

[178]Juang, J.-N., and Pappa, R. S., "An Eigensystem Realization Algorithm for Modal Parameter Identification and Model Reduction," *Journal of Guidance, Control, and Dynamics*, Vol. 8, No. 5, Sept.–Oct. 1985, pp. 620–627.

[179]Favunmi, J. A., "Spectral Basis Theory for the Identification of Structural Dynamic Systems," *AIAA Journal*, Vol. 26, June 1988, pp. 726–732.

[180]Leuridan, J., and Vold, H., "A Time Domain Linear Model Estimation Technique for Multiple Input Modal Analysis Modal Testing and Model Refinement," *Modal Testing and Model Refinement*, AMD-Vol. 59, American Society of Mechanical Engineers, New York, Nov. 1983, pp. 51–62.

[181]Richardson, M. H., and Formenti, D. L., "Parameter Estimation from Frequency Response Measurements Using Rational Fraction Polynomials," *Proceedings of the 1st International Modal Analysis Conference*, Union College, Schenectady, NY, 1982, pp. 167–181.

[182]Juang, J.-N., "Mathematical Correlation of Modal-Parameter-Identification Methods Via System Realization Theory," *International Journal of Analytical and Experimental Modal Analysis*, Vol. 2, No. 1, Jan. 1987, pp. 1–18.

[183]Turunen, R., "Statistical Performance of Modal Parameter Estimation Methods," *Proceedings of the 5th International Modal Analysis Conference*, Union College, Schenectady, NY, 1987, pp. 26–36.

[184]Hasselman, T. K., "A Perspective on Dynamic Model Verification," *Modal Testing and Model Refinement*, AMD-Vol. 59, American Society of Mechanical Engineers, Nov. 1983, pp. 101–117.

[185]Hasselman, T. K., and Chrostowski, J. D., "Parameter Estimation for Large Space Structures," *Optimization Issues in the Design and Control of Large Space Structures*, American Society of Civil Engineers, NY, May, edited by M. Kamat, 1985, pp. 1–15.

[186]Berman, A., "Nonunique Structural System Identification," *Proceedings of the 7th International Modal Analysis Conference*, 1989, pp. 355–357.

[187]Berman, A., "System Identification of Structural Dynamic Models—Theoretical and Practical Bounds," AIAA Paper 84-0929, May 1984, Pt. 2, pp. 123–129.

[188]Carnes, T. G., Martinez, D. R., Tucker, M. D., and Lauffer, J. P., "Modal Analysis of a Shell-Payload Structure Using Test Data," *Proceedings of the 4th International Modal Analysis Conference*, Union College, Schenectady, NY, 1986, pp. 646–655.

[189]Baker, M., "Component Mode Synthesis Methods for Test-Based Rigidly Connected Flexible Components," *Journal of Spacecraft and Rockets*, Vol. 23, No. 3, May–June 1986, pp. 316–322.

[190]Martinez, D. R., and Miller, A. K. (eds.), *Combined Experimental/Analytical Modeling of Dynamic Structural Systems*, AMD-Vol. 67, American Society of Mechanical Engineers, New York, 1985.

[191]Breitbach, E., "Modal Synthesis—Modal Correction—Modal Coupling," *CISM Courses and Lectures*, edited by H. G. Nafke, International Center for Mechanical Sciences, No. 272, Springer-Verlag, Vienna, New York, 1982, pp. 321–348.

[192]Blackwood, G. H., and von Flotow, A. H., "Experimental Component Mode Synthesis of Structures with Sloppy Joints," AIAA Paper 88-2411, April 1988, pp. 1565–1575.

[193]Black, I., O'Keefe, J. M., Williams, R., Bennett, J. K., and Marler, D., "Modal Survey of the Intelsat VI Cradle and Carrier Structure for a Shuttle-Launched Spacecraft," *Proceedings of the 5th International Modal Analysis Conference*, Union College, Schenectady, NY, 1987, pp. 478–487.

[194]Chen, J., Rose, T., Trubert, M., Wada, B., and Shaker, F., "Modal Test/Analysis Correlation for the Centaur G Prime Launch Vehicle," *Journal of Spacecraft and Rockets*, Vol. 24, No. 5, Sept.–Oct. 1987, pp. 423–429.

[195]Chung, Y.-T., and Simonian, S. S., "Test/Analysis Correlation for the Gamma Ray Observatory," *Proceedings of the 7th International Modal Analysis Conference*, Union College, Schenectady, NY, 1989, pp. 1239–1247.

[196]Flanigan, C. C., "Test/Analysis Correlation of the STS Centaur Using Design Sensitivity and Optimization Methods," *Proceedings of the 5th International Modal Analysis Conference*, Union College, Schenectady, NY, 1987, pp. 99–107.

[197]Stroeve, A., "Modal Survey Testing of the Combined Release and Radiation Effects Satellite—A Shuttle Payload," *Proceedings of the 5th International Modal Analysis Conference*, Union College, Schenectady, NY, 1987, pp. 1084–1090.

[198]Coleman, A. D., Driskill, T. C., Anderson, J. B., and Brown, D. L., "A Mass-Additive Technique for Modal Testing as Applied to the Space Shuttle Astro-1 Payload," *Proceedings of the 6th International Modal Analysis Conference*, Union College, Schenectady, NY, 1988, pp. 154–159.

[199]Blair, M. A., and Vadlamudi, N., "Constrained Structural Dynamic Model Verification Using Free Vehicle Suspension Testing Methods," AIAA Paper 88-2359, April 1988, pp. 1187–1193.

[200]Blair, M. A., and Vadlamudi, N., "Hubble Space Telescope—Space Shuttle Interface Dynamic Verification Test," *Proceedings of the 7th International Modal Analysis Conference*, Union College, Schenectady, NY, 1989, pp. 657–663.

[201]Brillhart, R. D., Hunt, D. L., Flanigan, C. C., Guinn, R., and Hull, R., "Transfer Orbit Stage Modal Survey—Part 1—Measurement of Free-Free Modes and Residual Flexibility," *Proceedings of the 7th International Modal Analysis Conference*, Union College, Schenectady, NY, 1989, pp. 1150–1156.

[202]Brillhart, R. D., Hunt, D. L., Flanigan, C. C., Guinn, R., and Hull, R., "Transfer Orbit Stage Modal Survey—Part 2—Model Correlation," *Proceedings of the 7th International Modal Analysis Conference*, Union College, Schenectady, NY, 1989, pp. 1157–1161.

[203]Glasner, R. J., Kuo, C. P., and Wada, B. K., "Multiple Boundary Condition Testing Error Analysis," AIAA Paper 89-1162, April 1989, pp. 12–20.

[204]Kuo, C. P., and Wada, B. K., "Multiple Boundary Condition Test (MBCT): Identification with Mode Shapes," AIAA Paper 88-2353, April 1988, pp. 1132–1142.

[205]Kuo, C. P., and Wada, B. K., "Nonlinear Sensitivity Coefficients and Corrections in System Identification," AIAA Paper 86-0967, May 1986, pp. 576–584.

[206]Wada, B. K., "Multiple Boundary Condition Tests (MBCT) for Verification of Large Space Structures," AIAA Paper 86-0905, May 1986, Pt. 2, pp. 336–341.

[207]Wada, B. K., Kuo, C. P., and Glaser, R. J., "Multiple Tests Concept for Improved Validation of Large Space Structure Mathematical Models," *56th Shock and Vibration Symposium*, 1985, pp. 1–8.

[208]Chou, C.-M., O'Callahan, J. C., and Wu, C. H., "Localization of Test/Analysis Structural Model Errors," AIAA Paper 89-1244, April 1989, pp. 748–752.

[209]Ewins, D. J., He, J., and Lieven, N., "A Review of the Error Matrix Method (EMM) for Structural Dynamic Model Comparison," *ESA Spacecraft Structures and Mechanical Testing*, European Space Agency, CP-289, Jan. 1989, pp. 55–62.

[210]Fissette, E., and Ibrahim, S., "Error Location and Updating of Analytical Dynamic Models Using a Force Balance Method," *Proceedings of the 6th International Modal Analysis Conference*, Union College, Schenectady, NY, 1988, pp. 1063–1070.

[211]Lapierre, H., and Ostiguy, G., "Structural Model Verification with LQO Theory," AIAA Paper 88-2360, April 1988, pp. 1194–1201.

[212]Lin, C. S., "Location of Modeling Errors Using Modal Test Data," AIAA Paper 89-1240, April 1989, pp. 713–720.

[213]O'Callahan, J. C., and Chou, C.-M., "Localization of Model Errors in Optimized Mass and Stiffness Matrices Using Modal-test Data," *International Journal of Analytical and Experimental Modal Analysis*, Vol. 4, No. 1, Jan. 1989, pp. 8–14.

[214]Ojalvo, I. U., and Pilon, D., "Diagnostics for Geometrically Locating Structural Math Model Errors from Modal Test Data," AIAA Paper 88-2358, April 1988, pp. 1174–1186.

[215]Ojalvo, I. U., Ting, T., Pilon, D., and Twomey, W., "Practical Suggestions for Modifying Math Models to Correlate with Actual Modal Test Results," *Proceedings of the 7th International Modal Analysis Conference*, Union College, Schenectady, NY, 1989, pp. 347–354.

[216]Shepard, G. D., "Spatial Distribution of Model Error Based on Analytical/Experimental Frequency Discrepancies," *Proceedings of the 5th International Modal Analysis Conference*, Union College, Schenectady, NY, 1987, pp. 1665–1668.

[217]Smith, S. W., and McGowan, P. E., "Locating Damaged Members in a Truss Structure Using Modal Test Data: A Demonstration Experiment," AIAA Paper 89-1291, April 1989.

[218]Ting, T., Ojalvo, I. U., and Chen, T. L. C., "A Robust Modal Correlation Procedure for Large-Scale Structures," *Proceedings of the 7th International Modal Analysis Conference*, Union College, Schenectady, NY, 1989, pp. 650–656.

[219]Baruch, M., "Correction of Stiffness Matrix Using Vibration Tests," *AIAA Journal*, Vol. 20, March 1982, pp. 441–442.

[220]Baruch, M., "Methods of Reference Basis for Identification of Linear Dynamic Structures," *AIAA Journal*, Vol. 22, April 1984, pp. 561–564.

[221]Baruch, M., and Zemel, Y., "Mass Conservation in the Identification of Space Structures," AIAA Paper 89-1239, April 1989, pp. 710–712.

[222]Caesar, B., "Updating System Matrices Using Modal Test Data," *Proceedings of the 5th International Modal Analysis Conference*, Union College, Schenectady, NY, 1987, pp. 453–459.

[223]Caesar, B., and Peter, J., "Direct Update of Dynamic Mathematical Models from Modal Test Data," *AIAA Journal*, Vol. 25, Nov. 1987, pp. 1494–1499.

[224]Caesar, B., "Analysis/Test Correlation and Update of Dynamic Mathematical Models," *ESA Spacecraft Structures*, European Space Agency, CP-238, April 1986, pp. 191–193.

[225]*CORDS2—User's Manual*, Version 1.0, Structural Dynamics Research Corp., San Diego, CA, 1988.

[226]He, J., and Ewins, D. J., "Analytical Stiffness Matrix Correction Using Measured Vibration Modes," *International Journal of Analytical and Experimental Modal Analysis*, Vol. 1, No. 3, July 1986, pp. 9–14.

[227]Hüners, H., "Applying Modal Survey Test Results for Updating Dynamics Mathematical Models of Spacecraft Structures," *ESA Spacecraft Structures*, European Space Agency, CP-238, April 1986, pp. 195–201.

[228]Ibrahim, S. R., "A Direct Two Response Approach for Updating Analytical Dynamic Models of Structures with Emphasis on Uniqueness," *ESA Spacecraft Structures and Mechanical Testing*, European Space Agency, CP-289, Jan. 1989, pp. 309–315.

[229]Kabe, A. M., "Stiffness Matrix Adjustment Using Mode Data," *AIAA Journal*, Vol. 23, Sept. 1985, pp. 1431–1436.

[230]Kammer, D. C., "An Optimum Approximation for Residual Stiffness in Linear System Identification," AIAA Paper 87-0813, April 1987, pp. 277–287.

[231]Kammer, D. C., "Test-Analysis-Model Development Using an Exact Modal Reduction," *International Journal of Analytical and Experimental Modal Analysis*, Vol. 2, No. 4, Oct. 1987, pp. 174–179.

[232]Niedbal, N., and Klusowski, E., "Structural Dynamic Models Combined with Measured Normal Mode Parameters," AIAA Paper 88-2313, April 1988, pp. 850–859.

[233]Niedbal, N., and Klusowski, E., "Updating a Finite-Element Model by Means of Normal Mode Parameters," *ESA Spacecraft Structures and Mechanical Testing*, European Space Agency, CP-289, Jan. 1989, pp. 47–53.

[234]Snyder, V. W., "Structural Modification and Modal Analysis," *International Journal of Analytical and Experimental Modal Analysis*, Vol. 1, No. 1, Jan. 1986, pp. 45–52.

[235]Stiles, P. A., and Kosmatka, J. G., "Comparison of Model Refinement Techniques," AIAA Paper 89-1279, April 1989, pp. 1052–1061.

[236]Zimmerman, D. C., and Widengren, M., "Model Correction Using a Symmetric Eigenstructure Assignment Technique," AIAA Paper 89-1382-CP, April 1989, pp. 1947–1954.

[237]Cooley, V. M., Juang, J. N., and Ghaemmaghami, P., "Design of Ground Test Suspension Systems for Verification of Flexible Space Structures," *Dynamics and Control of Large Structures, Proceedings of the 6th VPI&SU Symposium*, Virginia Polytechnic Institute and State University, Blacksburg, VA, 1987, pp. 713–725.

[238]Kienholz, D. A., Crawley, E. F., and Harvey, T. J., "Very Low Frequency Suspension Systems for Dynamic Testing," AIAA Paper 89-1194, April 1989, pp. 327–336.

[239]Shih, C.-F., "Verification of Large Beam-Type Space Structures," *Journal of Spacecraft and Rockets*, Vol. 24, No. 5, Sept.–Oct. 1987, pp. 469–473.

[240]Chen, J.-C., and Garba, J. A., "On-Orbit Damage Assessment for Large Space Structures," *AIAA Journal*, Vol. 26, Sept. 1988, pp. 1119–1126.

[241]West, W. M., "Illustration of the Use of Modal Assurance Criterion to Detect Structural Changes in an Orbiter Test Specimen," *Proceedings of the 4th International Modal Analysis Conference*, 1986, pp. 1–6.

Bibliography

Hablani, H. G., "Modal Analysis of Gyroscopic Glexible Spacecraft: A Continuum Approach," *Journal of Guidance, Control, and Dynamics*, Vol. 5, No. 5, Sept.–Oct. 1982, pp. 448–457.

Kabe, A. M., "Multi-Shaker Random Mode Testing," *Journal of Guidance, Control, and Dynamics*, Vol. 7, No. 6, Nov.–Dec. 1984, pp. 740–746.

Lee, U., "Dynamic Continuum Modeling of Beamlike Space Structures Using Finite Element Matrices," AIAA Paper 89-1383, April 1989, pp. 1955–1962.

Meirovitch, L., and Quinn, R. D., "Equations of Motion for Maneuvering Flexible Spacecraft," *Journal of Guidance, Control, and Dynamics*, Vol. 10, No. 5, Sept.–Oct. 1987, pp. 453–465.

Smith, S. W., and Beattie, C. A., "Secant-Method Adjustment for Structural Models," AIAA Paper 89-1278, April 1989, pp. 1041–1051.

Spanos, P. D., Cao, T. T., Nelson, D. A. R., Jr., and Hamilton, D. A., "Efficient Loads Analyses of Shuttle Payloads Using Dynamic Models with Linear or Nonlinear Interfaces," AIAA Paper 89-1278, April 1989, pp. 414–424.

Wright, R. L. (ed.), "NASA/DOD Control/Structure Interaction Technology," NASA CP-2447, Nov. 1986, Pts. 1 and 2.

Chapter 2

Orthogonal Projection Approach to Multibody Dynamics

H. Flashner*
University of Southern California, Los Angeles, California

Introduction

MODERN aerospace systems, such as large space structures, are extremely complex assemblies of rigid and flexible subsystems. In most cases the dynamic models of the individual subsystems are known and the dynamic model of the total system needs to be assembled. The assembly approach is especially appealing when dynamic models for control system design are formulated, since it allows changing interconnections and including the control system in the model. Because of the high order of the resulting system, especially in the case of flexible structures, and the complexity of the interconnections, the subsystem assembly process can be achieved only by numerical methods. Thus, the question of efficiency and numerical stability of the procedure is of paramount importance.

When the vibration of complex large space structures is modeled, the subsystem assembly deals with coupling sets of linear differential equations. This problem can be solved without approximation and can be interpreted as *modal synthesis*. The modal synthesis approach has been widely used to obtain modal representation of complex flexible structures.[1-25] However, since, in modeling space structures assembly interconnection of substructures with masses and stiffnesses of different orders of magnitude is required, numerical difficulties are often encountered.

When large-angle motion of interconnected multibody systems is modeled, it is difficult to obtain an analytical model of the complete system, and one must use simulation for analysis and performance evaluation. When formulated in terms of Lagrange multipliers, the simulation involves

*Associate Professor, Department of Mechanical Engineering.

integration of a set of ordinary differential equations subject to algebraic constraints.[26] Because there are often numercial difficulties with these methods, reduction techniques were suggested to convert the equations to ones that can be solved using ordinary differential equation solvers.[27] Hence, an important consideration is the choice of independent coordinates.[25–31]

The choice of independent coordinates selection can be viewed as an iterative extension of the modal synthesis formulation to nonlinear systems. In this paper we present the orthogonal projection approach to modeling linear and nonlinear multibody systems. We start with an analysis of linear systems and present the nonlinear numerical simulation as an iterative extension of the linear analysis. The approach is demonstrated on a number of examples that show the efficiency and numerical stability of the method.

Analysis of Linear Systems

Consider a dynamic system whose motion is formulated using n coordinates $q_1, q_2, \ldots, q_n$. Let T and V be the kinetic and potential energies of the system, respectively. Moreover, assume that the n coordinates q_i, $i = 1, \ldots, n$ are not independent but are related by m constraint equations that relate the coordinates and possibly their time derivatives and time. In this paper we shall consider holonomic time-invariant, i.e, scleronomic,[32] constraints:

$$\Phi_j(q) = 0, \qquad j = 1, \ldots, m \tag{1}$$

where $m < n$ and $q = [q_1, \ldots, q_n]^T$.

Using Lagrange's formulation of the equations of motion, the constraint equations can be incorporated in the equations of motion by introducing the concept of Lagrange multipliers,[33,34] as follows:

$$\frac{\mathrm{d}}{\mathrm{d}t}\left(\frac{\partial L}{\partial \dot{q}_i}\right) - \frac{\partial L}{\partial q_i} = N_i + \sum_{j=1}^{m} \lambda_j B_{ji}, \qquad i = 1, \ldots, n \tag{2}$$

where $L = T - V$ is the system's Lagrangian; λ_j, $j = 1, \ldots, m$ are Lagrange multipliers that can be interpreted as constraint forces[34]; $N_i(t)$ are generalized forces acting on the system; and $B_{ji} = \partial \Phi_j / \partial q_i$, $j = 1, \ldots, m$ are the elements of the Jacobian matrix of $\Phi = [\Phi_1, \ldots, \Phi_m]^T$.

For finite-dimensional linear dynamic systems considered in this section, the kinetic and potential energies are given by the following quadratic forms:

$$T = \tfrac{1}{2}\dot{q}(t)^T M \dot{q}(t), \qquad V = \tfrac{1}{2} q(t)^T K q(t) \tag{3}$$

where $M \in R^{n \times n}$ is a positive-definite constant mass matrix, and $K \in R^{n \times n}$

is a constant stiffness matrix that is positive-definite or positive-semidefinite. The equations of motion as formulated in Eq. (2) are then given by

$$M\ddot{q}(t) + Kq(t) = D^T \boldsymbol{F}(t) + B^T \boldsymbol{\lambda}(t) \tag{4}$$

where $D \in R^{n \times l}$ is a matrix of influence coefficients, $\boldsymbol{F}(t) \in \boldsymbol{R}^l$ is a vector of applied forces, and $\boldsymbol{\lambda}(t) \in \boldsymbol{R}^m$ is a vector of generalized constraint forces. We also assume that the coordinates $q(t)$ are related by a set of m linear constraint equations; thus,

$$Bq(t) = 0 \tag{5}$$

where $B \in R^{m \times n}$ is a constant matrix with $m < n$. Therefore, the matrix B in Eq. (4) is also constant. Equations (4) and (5) define the dynamic behavior of the system. Our task is to find a set of independent coordinates, i.e., generalized coordinates, and rewrite the equations of motion in terms of these coordinates.

Assume for the time being that the constraint equations in Eq. (5) are linearly independent, i.e., rank$(B) = m$. Then B can be decomposed using the QR decomposition as follows[35,38]:

$$B^T = Q\begin{bmatrix} R \\ 0 \end{bmatrix} \tag{6}$$

where $Q \in R^{n \times n}$ is an orthogonal matrix, and $R \in R^{m \times m}$ is an upper triangular matrix with $R_{ii} > 0$.

Define a new coordinate vector $\boldsymbol{\rho}(t)$ related to $q(t)$ by the transformation

$$q(t) = Q\boldsymbol{\rho}(t) \tag{7}$$

When Eq. (4) is multiplied by Q^T, the dynamic equation of motion [Eq. (4)] and the associated constraint equations can be rewritten in terms of the new coordinates $\boldsymbol{\rho}(t)$:

$$M^*\ddot{\boldsymbol{\rho}}(t) + K^*\boldsymbol{\rho}(t) = D^{*T}\boldsymbol{F}(t) + \begin{bmatrix} R \\ 0 \end{bmatrix}\boldsymbol{\lambda}(t) \tag{8}$$

$$[R^T \quad 0]\boldsymbol{\rho}(t) = 0 \tag{9}$$

where $M^* = Q^TMQ$, $K^* = Q^TKQ$, and $D^* = DQ$. Subdivide the vector $\boldsymbol{\rho}(t)$ and Q as follows:

$$\boldsymbol{\rho}(t) = \begin{bmatrix} \boldsymbol{\rho}_c(t) \\ \boldsymbol{\rho}_u(t) \end{bmatrix}, \qquad Q = [Q_c \quad Q_u] \tag{10}$$

where $\boldsymbol{\rho}_c(t) \in R^m$, $\boldsymbol{\rho}_u(t) \in R^{n-m}$. Rank$(B) = m$; therefore, also rank$(R) = m$. Therefore, from Eq. (9) we get

$$\boldsymbol{\rho}_c(t) = 0 \quad \forall t \geqslant 0 \tag{11}$$

Hence,

$$\ddot{\rho}_c(t) = \dot{\rho}_c(t) = \rho_c(t) = 0 \quad \forall t \geqslant 0 \tag{12}$$

Rewrite Eq. (8) as follows:

$$\begin{bmatrix} M^*_{cc} & M^*_{cu} \\ M^*_{uc} & M^*_{uu} \end{bmatrix} \begin{bmatrix} \ddot{\rho}_c(t) \\ \ddot{\rho}_u(t) \end{bmatrix} + \begin{bmatrix} K^*_{cc} & K^*_{cu} \\ K^*_{uc} & K^*_{uu} \end{bmatrix} \begin{bmatrix} \rho_c(t) \\ \rho_u(t) \end{bmatrix} = \begin{bmatrix} D^{*T}_c \\ D^{*T}_u \end{bmatrix} F + \begin{bmatrix} R \\ 0 \end{bmatrix} \lambda(t) \tag{13}$$

where

$$M^*_{cc} = Q^T_c M Q_c, \qquad M^*_{cu} = Q^T_c M Q_u, \qquad M^*_{uu} = Q^T_u M Q_u \tag{14a}$$

$$K^*_{cc} = Q^T_c K Q_c, \qquad K^*_{cu} = Q^T_c K Q_u, \qquad K^*_{uu} = Q^T_u K Q_u \tag{14b}$$

$$D^*_c = D Q_c, \qquad D^*_u = D Q_u \tag{14c}$$

Using Eq. (12), Eq. (13) can be rewritten in terms of the $n - m$ independent coordinates $\rho_u(t)$ that form a set of generalized coordinates:

$$M^*_{uu}\ddot{\rho}_u(t) + K^*_{uu}\rho_u(t) = D^{*T}_u F(t) \tag{15}$$

Note that the unknown Lagrange multipliers $\lambda(t)$ are automatically eliminated from the equations of motion written in terms of the unconstrained coordinates $\rho_u(t)$. This occurs because the constrained degrees of freedom are eliminated by an appropriately chosen orthogonal coordinate transformation [see Eq. (7)] and not by solving for dependent coordinates and constraint forces. The procedure has a geometric interpretation. The constraint equation [Eq. (5)] defines an $n - m$-dimensional plane in an n-dimensional space. The solution of the dynamic equation [Eq. (4)], i.e., the trajectory of the system, must lie in this plane in order to satisfy the constraints. The coordinate transformation [Eq. (7)] defines a new coordinate system in which $n - m$ coordinates lie in the constraint plane $P[\rho_u(t)]$, and m coordinates are orthogonal to $P[\rho_c(t)]$ (see Fig. 1). It is therefore clear that, in order for the constraints to be satisfied for all times, Eq. (12) must hold, and the set of unconstrained coordinates $\rho_u(t)$ spans the space of the trajectory of the system. The constraint forces $\lambda(t)$ are always orthogonal to the constraint surface P (Ref. 32); thus, they do not affect any of the coordinates $\rho_u(t)$ that lie on P. This results in automatic elimination of $\lambda(t)$ from the equations of motion.

In the previous derivation the assumption was that the constraint equations are linearly independent, i.e., that B is of full rank. If this is not the case, numerical difficulties might arise if B is ill-conditioned. In this case a QR decomposition with pivoting can be calculated bringing B^T to the form

$$B^T P = Q \begin{bmatrix} R & T \\ 0 & 0 \end{bmatrix} \tag{16}$$

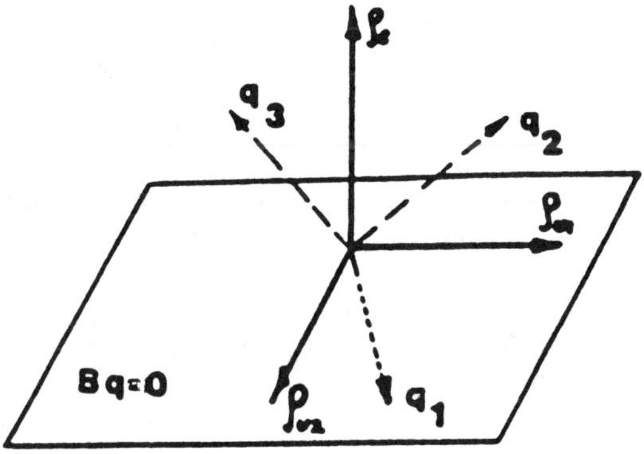

Fig. 1 Coordinate systems for linearly constrained system.

where r is the rank of B, $P \in R^{m \times m}$ is a permutation matrix, $Q \in R^{n \times n}$ is an orthogonal matrix, $R \in R^{r \times r}$ is an upper triangle matrix, and $T \in R^{r \times (m-r)}$. Using the decomposition in Eq. (16), the dynamic equation given in Eq. (15) hold, but the vector of generalized coordinates is of dimension $r < m$.

Although in practice the QR decomposition is adequate in finding the orthogonal transformation Q (Ref. 35), it might be unreliable in some cases of ill-conditioning. In these cases the orthogonal coordinate transformation can be found using the singular value decomposition (SVD).[35,37,38] Using SVD to decompose B^T results in

$$B^T = U \begin{bmatrix} \Sigma \\ 0 \end{bmatrix} V^T \tag{17}$$

where $U \in R^{n \times n}$ and $V \in R^{m \times m}$ are orthogonal matrices, and

$$\Sigma = \mathrm{diag}(\sigma_1, \dots, \sigma_r), \qquad \sigma_i > 0, \qquad i = 1, \dots, r \tag{18}$$

Defining an orthogonal coordinate transformation as in Eq. (7) but using the matrix U,

$$q(t) = U\rho(t) \tag{19}$$

the equations of motion and the constraints take the form

$$M^*\ddot{\rho}(t) + K^*\rho(t) = D^{*T}F(t) + \begin{bmatrix} \Sigma \\ 0 \end{bmatrix} V^T \lambda(t) \tag{20}$$

$$V[\Sigma^T \quad 0]\rho(t) = 0 \tag{21}$$

where $M^* = U^T MU$, $K^* = U^T KU$, and $D^* = DU$. Dividing the vector of coordinates $\boldsymbol{\rho}(t)$ into $\boldsymbol{\rho}_c(t)$ and $\boldsymbol{\rho}_u(t)$ as shown earlier, we get the unconstrained equation of motion [Eq. (15)]. However, in this case the dimension of the constrained space, i.e., the dimension of the plane P, is r that might be smaller than the number of constraints. The SVD algorithm guarantees that separation between dependent and independent coordinates is possible even in ill-conditioned cases.

As mentioned earlier, the constraint forces are eliminated from the equations of motion by the orthogonal coordinate transformation. In some applications, such as modal synthesis, these forces are of interest since they represent interaction forces between interconnected bodies. Substituting $\ddot{\boldsymbol{\rho}}_u(t)$ from Eq. (15) into Eq. (13) and using Eq. (12) yields

$$\boldsymbol{\lambda}(t) = (R^T)^{-1} b(t) \tag{22}$$

where

$$b(t) = (ED_u^{*T} - D_c^*)F(t) + (K_{cu}^* - EK_{uu}^*)\boldsymbol{\rho}_u(t), \qquad E = M_{cu}^*(M_{uu}^*)^{-1}$$

It should be noted that $(M_{uu}^*)^{-1}$ always exists since M is positive-definite. Also, since R is upper triangular, $\boldsymbol{\lambda}(t)$ can be computed by backsubstitution without actually inverting R. If SVD is used, the constraint forces are computed by

$$\boldsymbol{\lambda}(t) = V_1 \Sigma^{-1} b(t) \tag{23}$$

where $\Sigma^{-1} = \text{diag}(1/\sigma_1, ..., 1/\sigma_r)$ and V_1 contains the first r columns of V. Note that here no backsubstitution is needed. However, SVD requires considerably more computational effort than QR decomposition.[35,36,42,43] Therefore, if it is known a priori that the constraint equations are linearly independent, QR decomposition should be preferred to find the orthogonal coordinate transformation.

Application to Modal Synthesis

Consider a mechanical system undergoing a general motion in six-dimensional space, i.e., three directions of translation and three angles of rotation. We assume that the system consists of a number of substructures whose dynamic characteristics are given. Each subsystem can be either a rigid or elastic body and in this case can be viewed as possessing an infinite number of degrees of freedom. We assume that each elastic-body equations have been discretized using a numerical technique such as the finite-element method. Then the dynamics of the structure is described by the motion of a finite number of points on the structure (nodes in case of finite-element discretization). This is equivalent to assuming that the motion of each substructure can be described by a *finite* number of degrees of freedom.

Assuming small motion, linear elasticity, and neglecting damping effects, the equations of motion of a substructure m can then be written as

$$M^m \ddot{q}^m(t) + K^m q^m(t) = \boldsymbol{F}^m(t) \tag{24}$$

where $M^m \in R^{n_m \times n_m}$ is the generalized mass matrix; $K^m \in R^{n_m \times n_m}$ is the stiffness matrix; $\boldsymbol{q}^m(t) = [q^m_{1,1}, q^m_{1,2}, ..., q^m_{6,N_m}]^T$ and $\boldsymbol{F}^m(t) = [F^m_{1,1}, ..., F^m_{6,N_m}]^T$, where $q^m_{i,j}$ and $F^m_{i,j}$ are generalized displacements and generalized forces at location i and direction j, respectively; N_m is the number of discretization points; and $n_m = 6N_m$ is the number of degrees of freedom.

Without loss of generality it is assumed that the substructures $m = 1, ..., m_s$ underwent modal analysis and that an alternative dynamic representation to Eq. (24) is provided in terms of modal frequencies $\omega^m_1, ..., \omega^m_{n_m}$ and eigenvectors $\boldsymbol{\phi}^m_1, ..., \boldsymbol{\phi}^m_{n_m}$:

$$\ddot{\eta}^m(t) + \Omega^m \eta^m(t) = \Phi^{mT} \boldsymbol{F}^m(t), \qquad m = 1, ..., m_s \tag{25}$$

$$q^m(t) = \Phi^m \eta^m(t), \qquad m = 1, ..., m_s \tag{26}$$

where the following relationships hold:

$$q^m(t) \in R^{n_m}, \qquad \eta^m(t) \in R^{n_m}, \qquad \boldsymbol{F}^m(t) \in R^{l_m}$$

and the following relationships hold:

$$\Phi^{mT} K^m \Phi^m = \Omega^m, \qquad \Phi^{mT} M^m \Phi^m = I \tag{27}$$

where

$$\Omega^m = \mathrm{diag}[(\omega^m_1)^2, ..., (\omega^m_{n_m})^2], \qquad \Phi^m = \begin{bmatrix} \Phi^{(1)m}_{1,1} & \cdots & \Phi^{(n)m}_{1,1} \\ \Phi^{(1)m}_{1,2} & \cdots & \Phi^{(n)m}_{1,2} \\ \vdots & \cdots & \vdots \\ \Phi^{(1)m}_{N_m,6} & \cdots & \Phi^{(n)m}_{N_m,6} \end{bmatrix}$$

where ω^m_i is the ith natural frequency of the structure, and $\Phi^{(k)m}_{i,j}$ is the modal deflection at location (node) i and direction j of mode k.

It should be noted that the model of the substructure can be given by various representations. The component dynamic representation can be given by functions that satisfy some substructure eigenvalue problem, such as *free-free* or *constrained* modes, or by any complete set of admissible functions.[25] For the derivation here we assume modal representation; however, the approach is valid also for any other dynamic description. The substructure model need not be full order to avoid lengthy computations. In most cases the substructure model is reduced according to some criterion. In the design of control systems for large flexible structures, for example, only substructure modes that affect the combined structure's modes within the closed-loop bandwidth need to be considered.

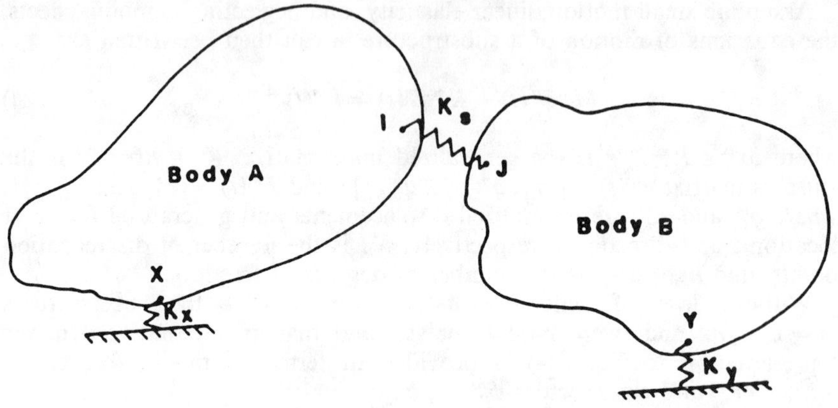

Fig. 2 Interconnections between two bodies and inertial frame.

Assume that it is required to find the dynamic model of a system consisting of interconnected bodies A and B as shown in Fig. 2. The interconnection is in discrete locations I and J, on bodies A and B, respectively. At the interface there is no relative motion between the bodies in directions $j = \{j_1, ..., j_F\}$, which will be referred to as "constrained directions." The bodies are also connected by linear springs along the directions $l = \{l_1, ..., l_H\}$. In addition, each one of the bodies may have connections to an inertial frame (*connection to ground*) at joints X and Y, respectively. If the inertial connections exist, they are rigid in directions $\alpha = \{\alpha_1, ..., \alpha_S\}$ and $\gamma = \{\gamma_1, ..., \gamma_R\}$, and can be compliant in directions $\beta = \{\beta_1, ..., \beta_U\}$ and $\delta = \{\delta_1, ..., \delta_V\}$, respectively, on bodies A and B.

The definitions given earlier allow fairly complicated interface topologies. Consider, for example, an interconnection that is gimballed about x and z axes and compliant by a torsional spring about the y axis and translational spring along the z axis; then $j = \{1,2\}$ and $l = \{3,5\}$. If A is rigidly connected to the ground, then $\alpha = \{1,2,3,4,5,6\}$. If, on the other hand, A is on elastic foundation that applies force in z direction, then $\alpha = \{1,2,4,5,6\}$ and $\beta = \{3\}$.

To further simplify derivations, we adopt the following notation. Let the set $r = \{r_1, ..., r_p\}$ denote a set of directions at node S of body G; then we denote a partial matrix $\Phi^G_{S,r}$ as

$$\Phi^G_{S,r} = \begin{bmatrix} \Phi^{G(1)}_{S,r_1} & \cdots & \Phi^{G(n)}_{S,r_1} \\ \vdots & \vdots & \vdots \\ \Phi^{G(1)}_{S,r_p} & \cdots & \Phi^{G(n)}_{S,r_p} \end{bmatrix}$$

It should be noted that the formulation presented in the following allows multiple points of interconnection between substructures, thus allowing modeling of closed topological configurations. However, to avoid cumbersome notation, we shall assume interface at a single joint.

The equations of motion for the interconnected bodies A and B in modal coordinates are given by

$$\ddot{\eta}^A(t) + \Omega^A \eta^A(t) = \Phi_{I,j}^{A\,T} F_I^A(t) + \Phi_{X,\alpha}^{A\,T} F_X^A(t)$$

$$+ \Phi_{I,l}^{A\,T} K_S[\Phi_{J,l}^B \eta^B(t) - \Phi_{I,l}^A \eta^A(t)] - \Phi_{X,\beta}^A K_X \Phi_{X,\beta}^A \eta^A(t) + \Phi^{A^T} F^A(t) \tag{28}$$

and

$$\ddot{\eta}^B(t) + \Omega^B \eta^B(t) = \Phi_{J,j}^{B\,T} F_J^B(t) + \Phi_{Y,\gamma}^{B\,T} F_Y^B(t)$$

$$+ \Phi_{J,l}^{B\,T} K_S[\Phi_{I,l}^A \eta^A(t) - \Phi_{J,l}^B \eta^B(t)] - \Phi_{Y,\gamma}^B K_X \Phi_{Y,\gamma}^B \eta^B(t) + \Phi^{B^T} F^B(t) \tag{29}$$

where $F_I^A(t), F_J^B(t)$ are forces acting at the interface on bodies A and B, respectively; $F_X^A(t), F_Y^B(t)$ are inertial connection forces acting on A and B, respectively; K_S is a stiffness matrix of the flexible interface connection; and K_X, K_Y are the stiffness matrices of the flexible connection to inertial frame.

In addition to the dynamic equations, we have algebraic relations that result from mechanics laws and kinematics. From Newton's third law we have

$$F_I^A(t) = -F_J^B(t) \tag{30}$$

and from compatibility of motion at the interface we get

$$\Phi_{I,j}^A \eta^A(t) = \Phi_{J,j}^B \eta^B(t) \tag{31}$$

Eliminate $F_J^B(t)$ from Eq. (28) using Eq. (30) and define the following variables:

$$q(t) = \begin{bmatrix} \eta^A(t) \\ \eta^B(t) \end{bmatrix}, \qquad B = \begin{bmatrix} \Phi_{I,j}^A & -\Phi_{J,j}^B \\ \Phi_{I,\alpha}^A & 0 \\ 0 & \Phi_{Y,\gamma}^B \end{bmatrix}, \qquad \lambda(t) = \begin{bmatrix} F_I^A(t) \\ F_X^A(t) \\ F_Y^B(t) \end{bmatrix}$$

$$D = [\Phi^A \quad \Phi^B], \qquad F(t) = \begin{bmatrix} F^A(t) \\ F^B(t) \end{bmatrix}$$

$$K = \begin{bmatrix} \Omega^A + \Omega_{I,l}^{A\,T} K_S \Phi_{I,l}^A + \Phi_{X,\beta}^{A\,T} K_X \Phi_{X,\beta}^A & -\Phi_{I,l}^{A\,T} K_S \Phi_{J,l}^B \\ -\Phi_{I,l}^{B\,T} K_S \Phi_{J,l}^A & \Omega^B + \Phi_{I,l}^{B\,T} K_S \Phi_{I,l}^B + \Phi_{Y,\delta}^{B\,T} K_X \Phi_{Y,\delta}^B \end{bmatrix}$$

Then the dynamic equations of motion of a system consisting of bodies A and B are given by

$$\ddot{q}(t) + Kq(t) = D^T F(t) + B^T \lambda(t) \tag{32}$$

$$Bq(t) = 0 \tag{33}$$

These equations are of the general form of linear constrained systems given in Eqs. (4) and (5) but with $M = I$. It is obvious that one can obtain the dynamic representation of any number of interconnected bodies by applying sequentially the procedure described earlier and by forming the matrices B, D, and K for all of the substructures. The resulting equations will be of the form given in Eqs. (32) and (33). Using Eqs. (15) and (22) the unconstrained equations of motion and the constraint forces are given by

$$\ddot{\boldsymbol{\rho}}_u(t) + K^*_{uu}\boldsymbol{\rho}_u(t) = D^{*T}_u \boldsymbol{F}(t) \tag{34}$$

$$\lambda(t) = R^{-1}[K^*_{cu}\boldsymbol{\rho}_u(t) - D^{*T}_c \boldsymbol{F}(t)] \tag{35}$$

The natural frequencies and mode shapes of the combined system are found by solving an eigenvalue problem of dimension n_c,

$$(-\mu I + K^*_{uu})\boldsymbol{u} = 0 \tag{36}$$

with

$$n_c = \sum_{i=1}^{m_s} n^i - \sum_{i=1}^{m_s - 1} m^i_C - \sum_{i=1}^{m_s} m^i_I \tag{37}$$

where n^i is the number of modes of body i, m^i_C is the number of constrianed directions between bodies i and $i+1$, and m^i_I is the number of inertially constrained directions for body i. Solving Eq. (36) results in modal frequencies of the combined structure ω^c_i, $i = 1, ..., n_c$ and eigenvectors $\boldsymbol{u}^i$, $i = 1, ..., n_c$. Let $\Phi_u = [\boldsymbol{u}^1, ..., \boldsymbol{u}^{n_c}]$; then the modal matrix of the combined system is given by

$$\Phi_c = D^*_u \Phi_u \tag{38}$$

It should be observed that the eigenvalue problem in Eq. (36) is a standard eigenvalue problem with only one symmetric matrix. This is in contrast to some existing modal synthesis approaches that lead to a generalized eigenvalue problem, i.e., the problem with two symmetric matrices.

When some of the subsystems are modeled as rigid bodies, the procedure presented here can be directly applied by writing rigid-body equations of small motion in modal form. A formalism for obtaining these equations, given the mass properties of the body, is given in the Appendix.

Another advantage of the proposed method is the ability to efficiently select the modes to be included in the modal synthesis. To demonstrate this feature, assume that the columns of B were ordered according to the relative importance of the modes. For example, one may determine the relative importance of mode j according to the value of g_j given by

$$g_j = \sum_{i=1}^{m} |b_{ij}| \tag{39}$$

This criterion expresses the amplification of the constraint forces at the interface for mode j, needed to make the motion compatible with the kinematic constraints. Other mode selection criteria are also possible.[24] Then we can divide the modes into groups of similar importance and consecutively include additional modes in the synthesis until convergence is achieved.

Based on this discussion we propose the following procedure. Let n be the total number of modes of the subsystems, i.e.,

$$n = \sum_{i=1}^{m_s} n_i$$

Assume that the modes were divided into N groups according to some criterion such as Eq. (39), and in each group there are N^i modes such that

$$n = \sum_{i=1}^{N} N^i$$

Select ϵ_ω, ϵ_ϕ and perform the following steps:

Step 1: Set $i = 1$, $e_\phi = 2\epsilon_\phi$, $e_\omega = 2\epsilon_\omega$, $n = N^i$.
Step 2: Perform a QR decomposition of B^T.
Step 3: If $i \neq 1$, add N^i columns to B and update QR decomposition.
Step 4: Find the modal frequencies matrix $\Omega^{(i)}$ and modal matrix $\Phi^{(i)}$.
Step 5: If $i \neq 1$, compute $e_\omega = \|\Omega^{(i)} - \Omega^{(i-1)}\|$ and $e_\phi = \|\Phi^{(i)} - \Phi^{(i-1)}\|$.
Step 6: If $e_\omega \leqslant \epsilon_\omega$ and $e_\phi \leqslant \epsilon_\phi$ or $i = N$, stop. Otherwise, set $i = i + 1$ and go to step 3.

The addition of modes to be included in modal synthesis implies addition of columns to the constraint matrix B and therefore recomputation of the QR decomposition of B. One of the advantages of the QR decomposition is that adding columns to B requires only *updating* of the decomposition but *not* recomputation.[35,37,39] The updating requires computational effort proportional to the number of *added* modes and not to the *total* number of modes. It should be noted that one economizes the computational effort by ordering the *rows* of B according to their magnitudes. This implies different contribution of the interface joints to the kinematic constraints. Again, only updating and not recomputation of the entire QR decomposition is required.

Computational Examples of the Modal Synthesis Algorithm

Example 1: Synthesis Using Analytically Computed Modes

The following examples will use the analytic free-free modal representation of an Euler-Bernoulli beam to demonstrate the accuracy of the method, its ability to model systems with different boundary conditions, and the ease in which parametric studies can be conducted. Finally, a simple but realistic model of a flexible spacecraft is developed.

Consider the motion of a Euler-Bernoulli beam whose deflection is denoted by $y(x,t)$. Assume that the load can be represented by concentrated forces $F_j(t)$ and bending moments $T_j(t)$ at locations a_j, $j = 1, ..., N$; assume also that we are interested in both deflections and slopes at the locations a_j. Let the generalized coordinates $q(t)$ and acting forces $\boldsymbol{F}(t)$ be given by

$$q(t) = [y(a_1,t), y'(a_1,t), ..., y(a_N,t), y'(a_N,t)]^T$$

$$\boldsymbol{F}(t) = [F_1, T_1, F_2, T_2, ..., F_N, T_N]^T$$

and assume that the eigenfunctions of the problem are $\phi_i(x)$, $i = 1, ...$ Then the equations of motion of the beam in terms of the n first modes can be written in the standard form of Eqs. (25) and (26) with the modal matrix given by

$$\Phi = \begin{bmatrix} \phi_1(a_1) & \cdots & \phi_n(a_1) \\ \phi_1'(a_1) & \cdots & \phi_n'(a_1) \\ \vdots & \cdots & \vdots \\ \phi_1(a_N) & \cdots & \phi_n(a_N) \\ \phi_1'(a_N) & \cdots & \phi_n'(a_N) \end{bmatrix} \tag{40}$$

For a free-free beam with constant mass per unit length m, length L, and rigidity EI, the eigenfunctions are given by[40]

$$\phi_1(x) = \frac{1}{\sqrt{mL}}, \qquad \phi_2(x) = \sqrt{\frac{12}{mL^2}}, \qquad \omega_i = 0, \qquad i = 1,2 \tag{41}$$

$$\phi_i(x) = \frac{1}{\sqrt{mL}} \{[\cos\beta_i(x) + \cosh\beta_i(x)] - A_i[\sin\beta_i(x) + \sinh\beta_i(x)]\} \tag{42}$$

$$\omega_i = \beta_i^2 \sqrt{EI/m}, \qquad i = 3, 4, ... \tag{43}$$

where

$$A_i = (\cos\beta_i L - \cosh\beta_i L)/(\sin\beta_i L - \sinh\beta_i L) \tag{44}$$

and where $\beta_i L$ is the ith solution of the equation $\cosh\beta L \cos\beta L = 1$. The solutions of this equation are $\beta_3 L = 1.506$, and the rest of the solutions can be very accurately approximated by $\beta_i L = (2i - 1)\pi/2$.

Simply Supported Beam

Using the free-free modes of a beam, the modal frequencies of a beam simply supported at both ends were computed. Table 1 shows the computed modal frequencies as a function of the number of modes employed in the synthesis and compares them to the analytic solution. As can be seen from the table, the accuracy of the method is satisfactory even for a small number of synthesis modes (less than 3% for the fourth frequency with four synthesis modes). It should be noted that in this application the approach can be interpreted as the *boundary collocation method.*[40]

Table 1 Modal frequencies for a simply supported beam

	Six synthesis modes		Ten synthesis modes		Fifteen synthesis modes		Analytic
Mode no.	Frequency, rad/s	Error, %	Frequency, rad/s	Error, %	Frequency, rad/s	Error, %	Frequency, rad/s
1	9.8968	0.273	9.8741	0.046	9.8711	0.015	9.8686
2	39.7202	0.612	39.5260	0.121	39.4894	0.028	39.4784
3	91.1296	2.593	89.1912	0.411	88.9459	0.134	88.8261
4	162.0683	2.634	158.7132	0.506	158.1244	0.134	157.9132
5			249.4651	1.104	247.5173	0.314	246.7405
6			359.3924	1.150	356.3004	0.280	355.3053
7			495.1305	2.446	486.5783	0.631	483.6103
8			646.0314	2.270	634.7301	0.486	631.6543
9					807.8251	1.043	799.4373
10					994.0043	0.731	986.9603

Cantilevered Beam with End Mass and Inertia

The configuration considered is shown in Fig. 3. If formulated analytically, the boundary conditions of the problem are eigenvalue dependent and a closed-form analytic solution is impossible. The modal frequencies were generated by attaching the free-free beam rigidly to an inertial frame on one end, thus imposing cantilevered boundary conditions, and attaching the other end to a rigid body with mass m_a and inertia I_a represented in modal form (see the Appendix). The modal frequencies of the system for various values of m_a and I_a are shown in Table 2, and the first natural frequency as a function of end inertia is given in Fig. 4. This example demonstrates that using the proposed approach parametric studies are extremely convenient and numerically stable.

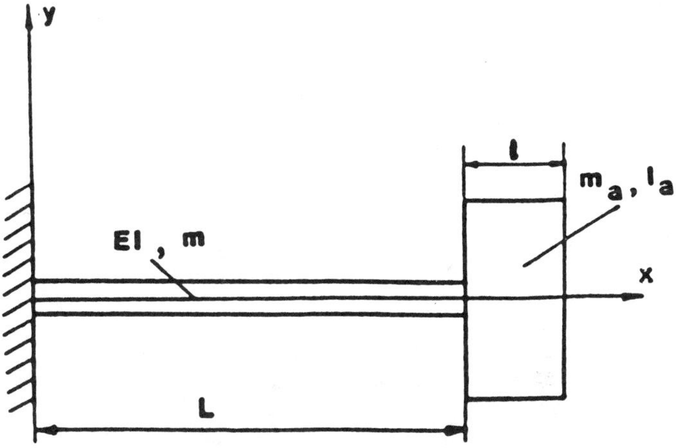

Fig. 3 Cantilevered beam with tip mass and inertia.

Table 2 Modal frequencies of a beam with attachment

Mode no.	$I_a = 0.1, m_a = 1$ Frequency, rad/s	$I_a = 10, m_a = 1$ Frequency, rad/s
1	1.31459	0.31444
2	6.86537	3.14728
3	26.92810	25.33934
4	68.35274	67.37706
5	130.95701	130.23043
6	215.32775	214.71989
7	319.85012	319.32078
8	447.12635	446.64372
9	593.44567	593.00134
10	764.10327	763.68401

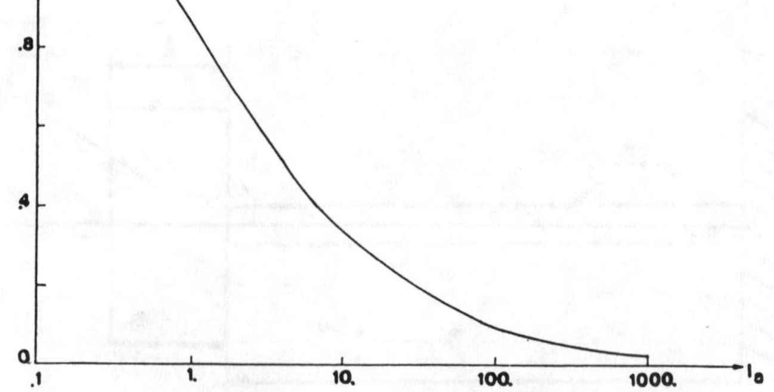

Fig. 4 Frequency of first natural frequency vs tip inertia.

Example 2: Modeling of a Spacecraft

In this example we apply the method to model a spacecraft whose components were modeled by a finite-element method and have undergone modal analysis. The system, shown in Fig. 5, consists of the power extension package (PEP) attached to the Shuttle Orbiter by the remote manipulator system (RMS). In this example we assume the RMS to be fully extended and all its joints locked. It is assumed that at any given position the β gimbal is locked and the solar array is allowed to rotate about the α axis. The modal representation of the PEP consisted of 36 free-free modes (6 rigid body and 30 flexible). The RMS and the Shuttle were modeled as one body by six rigid body and six flexible modes. Both models were obtained by using the finite-element program TRWSAP,[41] and each body was described by modal frequencies and mode shapes. Table 3 shows the results of the modal synthesis. Generation of the first 30 modes of the combined system of the Cyber 175 took about 5 s of CPU time. The cost of the run was about 30 times less than the cost of running a finite-element program for the combined system. It should be noted that the method of orthogonal decomposition did not encounter numerical difficulties despite the difference in orders of magnitude of the modal matrices of the two combined bodies. A standard modal synthesis program encountered severe singularity problems when solving for the dependent coordinates.

Versatility of the synthesis approach is demonstrated in Table 4. It shows the modal frequencies of the PEP and RMS + Shuttle system when both

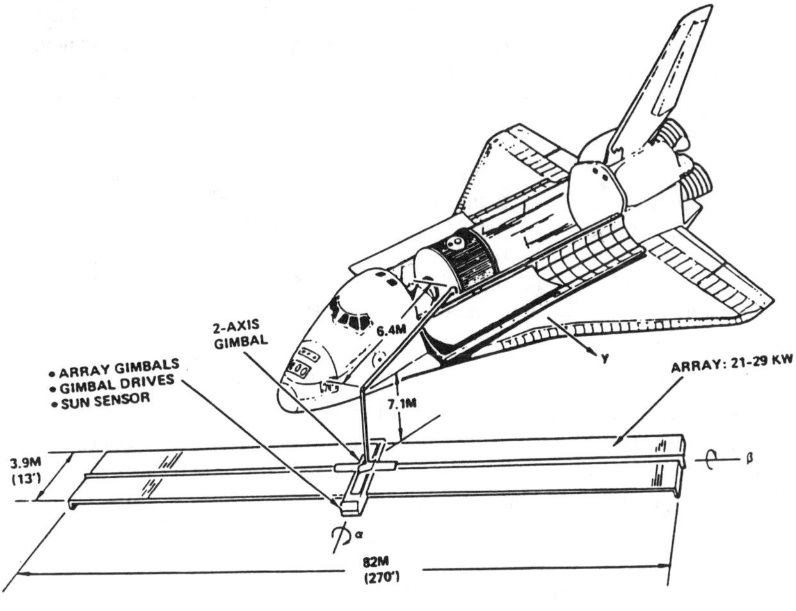

Fig. 5 Shuttle, RMS, and PEP of example 2.

Table 3 Modes of PEP and Orbiter + RMS system

Mode no.	PEP modes Frequency, rad/s	Orbiter + RMS modes Frequency, rad/s	Combined spacecraft modes Frequency, rad/s
1–6	Rigid body	Rigid body	Rigid body
7	0.4549	1.3967	PEP rotation
8	0.4995	1.5733	0.3050
9	0.6069	11.7847	0.3178
10	0.6308	18.2218	0.4359
11	0.7583	34.9489	0.4994
12	0.8815	36.7088	0.6071
13	0.9902		0.6292
14	1.2604		0.6309
15	1.2604		0.7969
16	1.2880		0.8545
17	1.2880		0.8829
18	1.3031		1.0347
19	1.7215		1.2605
20	1.7404		1.2605

Table 4 Modal frequencies of PEP and Orbiter + RMS system for different gimbal angles

Mode no.	$\alpha = 0, \beta = 0$ Frequency, rad/s	$\alpha = 90, \beta = 0$ Frequency, rad/s	$\alpha = 0, \beta = 90$ Frequency, rad/s	$\alpha = 90, \beta = 90$ Frequency, rad/s
1–6	Rigid body	Rigid body	Rigid body	Rigid body
7	0.30314	0.26637	0.29385	0.26119
8	0.31031	0.31065	0.31402	0.30820
9	0.31498	0.40973	0.32188	0.32130
10	0.43906	0.43906	0.43607	0.43288
11	0.60710	0.54575	0.60710	0.55747
12	0.60767	0.60713	0.60798	0.60713

the α- and β-gimbals are locked; i.e., the PEP is cantilevered to the RMS for various relative rotations of the PEP with respect to the RMS. The modal data for the PEP in the various configurations was generated by appropriate rotations of the PEP coordinate system and transformation of the modal data prior to synthesis.

Application to Multibody Simulation

In the case of a system of bodies undergoing large motion, the resulting constraints are nonlinear. As a result, orthogonal projection decoupling by a *constant* linear transformation (i.e., matrix) between the constrained and unconstrained coordinates is impossible. However, such a decoupling is possible *locally* for every point on the trajectory by locally linearizing the constraint equations at points along the trajectory and then using the approach presented in the first section to eliminate constraint forces and constrained coordinates from the equations of motion. Since the dynamic equations in terms of the unconstrained coordinates are valid only locally, a numerical procedure for enforcing the constraints must be included during integration. The simulation algorithm also includes an update procedure for the orthogonal projection transformation at any point along the trajectory, thus resulting in elimination of constraint forces and reduction of the number differential equations to the number of degrees of freedom along the entire trajectory.

The equations of motion of general multibody can be written using Eq. (2) as follows:

$$M(q)\ddot{q}(t) + B(q)\boldsymbol{\lambda}(t) = f(q,\dot{q}) \tag{45}$$

subject to the constraints given in Eq. (1), namely,

$$\Phi(q) = 0 \tag{46}$$

Assume that, at time $t = t^0$, $q(t^0) = q^0$ and assume that q^0 satisfies the constraints. Then, at time $t = t^0 + \Delta t$, $q = q(t)$, taking small Δt we can assume $\Delta q = q - q^0$ to be small. Noting that $B(q)$ is the Jacobian matrix of Φ and using Taylor series expansion we have

$$\Phi(q) = B(q^0)(q - q^0) = 0 \tag{47}$$

The geometric interpretation of Eq. (47) is that for short time intervals the trajectory at time $t = t^0$ moves on a plane tangent to the manifold defined by Eq. (46). This can also be seen by dividing Eq. (47) by Δt and letting $\Delta t \to 0$. Then

$$B(q^0)\dot{q}(t) = 0 \tag{48}$$

Or the velocity vector is always in the tangent plane to the constraint manifold.

Using the transformation in Eq. (7) and Eq. (12), Eq. (47) yields

$$\boldsymbol{\rho}_c(t) = \boldsymbol{\rho}(t^0) = \text{const} \tag{49}$$

$$\dot{\boldsymbol{\rho}}_c(t) = \ddot{\boldsymbol{\rho}}_c(t) = 0 \tag{50}$$

And defining local coordinates $\xi(t) = \boldsymbol{\rho}(t) - \boldsymbol{\rho}^0$, where $\boldsymbol{\rho}^0 = \boldsymbol{\rho}(t^0)$ we get

$$\xi(t) = \mathbf{0} \tag{51}$$

Assuming again that Δq is small and substituting $\boldsymbol{\rho}_c(t)$ and its derivatives from Eqs. (49) and (50), the constraint forces can be eliminated and the dynamic Eq. (45) can be approximately (for short time intervals) written in terms of the unconstrained coordinates similarly to the linear case [see Eq. (15)]:

$$Q_u^T M[\boldsymbol{\rho}_u(t)] Q_u \ddot{\boldsymbol{\rho}}_u(t) = Q_u^T f[\dot{\boldsymbol{\rho}}_u(t), \boldsymbol{\rho}_u(t)] \tag{52}$$

If there is a need to compute the constraint forces, they can be computed as they were in the linear case [see Eq. (22)]:

$$\boldsymbol{\lambda}(t) = Q_c^T \{M Q_u \ddot{\boldsymbol{\rho}}_u(t) - f[\dot{\boldsymbol{\rho}}_u(t), \boldsymbol{\rho}_u(t)]\} \tag{53}$$

Note that the dynamic Eq. (52) has only $n - m$ differential equations in the $n - m$ unknowns $\boldsymbol{\rho}_u(t)$. Also note that these equations are an approximation to the equations of motion [Eq. (45)] for short time intervals Δt. Even for short Δt the constraint equations might be violated. To enforce the constraint Eq. (46) at time t, we fix $\boldsymbol{\rho}_u(t)$ at the value obtained by integration of Eq. (52) and solve the set of nonlinear algebraic constraint Eq. (46) for the coordinates $\boldsymbol{\rho}_c(t)$; i.e., we solve for $\boldsymbol{\rho}_c(t)$ the equation

$$\Phi[\boldsymbol{\rho}_u(t), \boldsymbol{\rho}_c(t)] = 0 \tag{54}$$

Since we perform only a short time integration of Eq. (52), the numerical solution of Eq. (54) using, for example, Newton's method will have a high convergence rate if the initial guess is $\boldsymbol{\rho}(t^0)$. Note also that the Jacobian required by Newton's method can be extracted from $B(q^0)$. It should be noted that the update of the orthogonal transformation does not have to be performed at every successive point on the trajectory. If the Jacobian does not change much, the transformation needs not be updated. A simple criterion to decide if updating is needed is given in Refs. 30 and 31. Based on the discussion presented earlier, the following numerical simulation procedure is proposed.

Assume that at time $t = t_0$ the initial conditions q_0 and $\dot{q}_0$ are compatible with the constraints (47) and (48), respectively. Then perform the following steps:

Step 0: Set $t = t_0$, $q = q_0$, $\dot{q} = \dot{q}_0$.

Step 1: Decompose $B(q_0)$ using QR decomposition according to Eq. (6).

Step 2: Transform the coordinates q to $\boldsymbol{\rho}$ using Eq. (7) and get $\boldsymbol{\rho}_u(t)$ and $\boldsymbol{\rho}_c(t)$.

Step 3: Integrate Eq. (52) from t to $t = t + \Delta t$ to get $\boldsymbol{\rho}_u(t)$ at time $t = t + \Delta t$.

Step 4: Using Newton's method, solve the constraint Eq. (54) to get $\boldsymbol{\rho}_c(t)$ at time $t = t + \Delta t$.

Step 5: Using Eq. (7) transform back $\boldsymbol{\rho}$ to $q(t)$ to get the physical $q(t)$ at time $t = t + \Delta t$. Compute $\boldsymbol{\lambda}(t)$ from Eq. (53) if required.

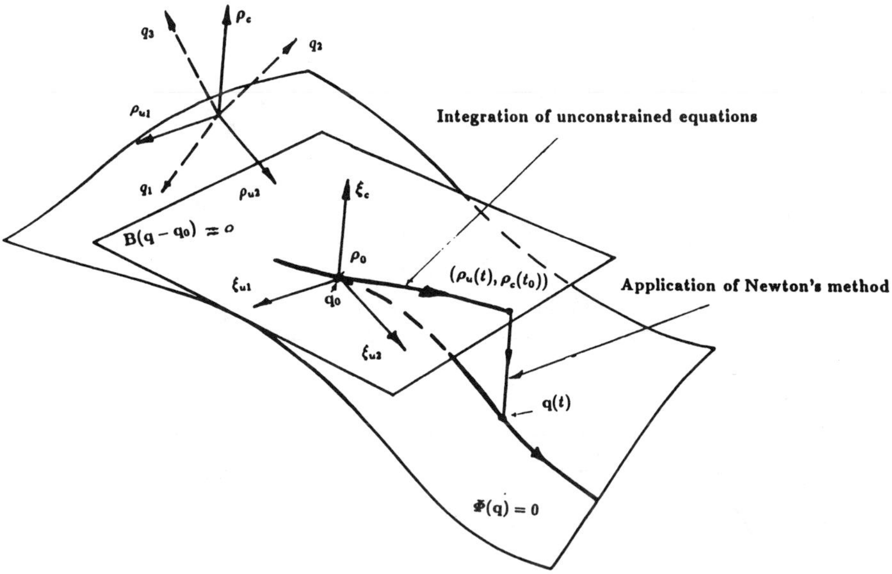

Fig. 6 Geometric interpretation of multibody simulation algorithm.

Step 6: Check the criterion for updating of QR decomposition. If the QR decomposition needs to be updated, go to step 1. If not, go to step 2. Repeat steps 1–6 until final time is reached.

A geometric interpretation of this procedure is given in Fig. 6.

Appendix: Modal Representation of Rigid-Body Motion

We present a modal form of the equations of motion of a rigid body. In this form rigid subsystems can be included in the modal synthesis process in the same way as a flexible body that is modeled by a finite-element method and that underwent modal analysis. Consider a motion of a rigid body shown in Fig. A1: $\boldsymbol{\rho}_C(t)$ is a radius vector from the origin of an inertial frame to the body's center of mass, $\boldsymbol{\rho}_P(t)$ is the radius vector to point P on the body, and $\boldsymbol{\theta}(t)$ is a vector of rotation angles for the body. Assume small displacements, small rotations, and small angular velocities. Consider the case in which the body is acted by a force $\boldsymbol{F}_P(t)$ and torque $T_P(t)$ applied at point P. Then the equations of motion are given by

$$M\ddot{\boldsymbol{\rho}}_C(t) = \boldsymbol{F}_P(t) \tag{A.1}$$

$$I_C\ddot{\boldsymbol{\theta}}(t) = T_P(t) + A(\boldsymbol{\rho})\boldsymbol{F}_P(t) \tag{A.2}$$

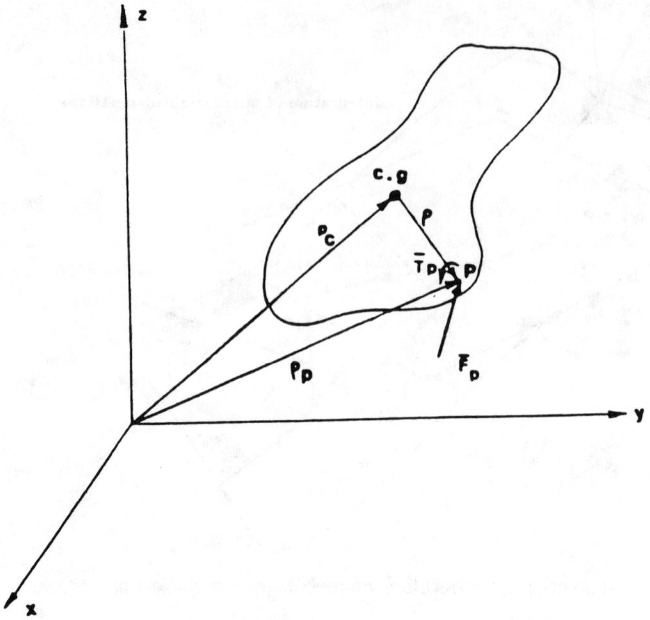

Fig. A1 Rigid body position and orientation.

where

$$A(\rho) = \begin{bmatrix} 0 & -\rho_3 & \rho_2 \\ \rho_3 & 0 & -\rho_1 \\ -\rho_2 & \rho_1 & 0 \end{bmatrix}$$

$M = \mathrm{diag}(m,m,m)$, where m is the mass of the body, I_C is the body's moment of inertia matrix about the center of gravity, and $\boldsymbol{\rho} = [\rho_1, \rho_2, \rho_3]^T$, $\boldsymbol{\rho} = \boldsymbol{\rho}_P(t) - \boldsymbol{\rho}_C(t)$.

The kinematic relations for a rigid body undergoing small motion can be written as

$$\begin{bmatrix} \boldsymbol{\rho}_P(t) \\ \boldsymbol{\theta}(t) \end{bmatrix} = \begin{bmatrix} I & A^T(\boldsymbol{\rho}) \\ 0 & I \end{bmatrix} \begin{bmatrix} \boldsymbol{\rho}_C(t) \\ \boldsymbol{\theta}_C(t) \end{bmatrix} \tag{A.3}$$

Using Eqs. (A.1–A.3) results in a modal representation of a small motion of a point P on a rigid body acted upon by forces and moments acting at this point:

$$\ddot{\eta}_R = \Phi_R^T \begin{bmatrix} \boldsymbol{F}_P(t) \\ T_P(t) \end{bmatrix} \tag{A.4}$$

$$q_R = \Phi_R \eta_R(t) \tag{A.5}$$

where

$$q_R,\eta_R(t) \in R^6, \qquad q_R = \begin{bmatrix} \rho_P(t) \\ \theta(t) \end{bmatrix}, \qquad \Phi_R = \begin{bmatrix} \sqrt{M^{-1}} & A^T(\rho)(L^T)^{-1} \\ 0 & L^{-1} \end{bmatrix}$$

where $\sqrt{M^{-1}} = \text{diag}(1/\sqrt{m},1/\sqrt{m},1/\sqrt{m})$, and L is the Choleski factor of I_C, i.e., $L^TL = I_C$. When the rotational motion is about the principal axes, then $L = \text{diag}(1/\sqrt{I_1},1/\sqrt{I_2},1/\sqrt{I_3})$. When a number of points with coordinates with respect to the center of mass ρ^i, $i = 1, ..., n$ are considered, the modal matrix Φ_R is given by

$$\Phi_R = \begin{bmatrix} \sqrt{M^{-1}} & A^T(\rho^1)(L^T)^{-1} \\ 0 & L^{-1} \\ \sqrt{M^{-1}} & A^T(\rho^2)(L^T)^{-1} \\ 0 & L^{-1} \\ \vdots & \vdots \\ \sqrt{M^{-1}} & A^T(\rho^n)(L^T)^{-1} \\ 0 & L^{-1} \end{bmatrix} \tag{A.6}$$

References

[1]Hurty, W. C., "Vibration of Structural Systems by Component Mode Synthesis," *ASCE, Journal of the Engineering Mechanics Division*, Vol. 86, Aug. 1960, pp. 51–69.

[2]Hurty, W. C., "Dynamic Analysis of Structural Systems Using Component Modes," *AIAA Journal*, Vol. 3, April 1965, pp. 678–685.

[3]Gladwell, G. M. L., "Branch Mode Analysis of Vibrating Systems," *Journal of Sound and Vibration*, Vol. 1, Jan. 1964, pp. 41–59.

[4]Benfield, W. A., and Hruda, R. F., "Vibration Analysis of Structures by Component Mode Substitution," *AIAA Journal*, Vol. 9, July 1971, pp. 1255–1261.

[5]Banjan, R. L., Feng, C. C., and Jaszlics, I. J., "Vibration Analysis of Complex Structural Systems by Modal Substitution," *Shock and Vibration Bulletin*, Vol. 39, Jan. 1969, pp. 99–105.

[6]Goldman, R. L., "Vibration Analysis by Harmonic Partitioning," *AIAA Journal*, Vol. 7, June 1969, pp. 1152–1154.

[7]Hou, S. N., "Review of Modal Synthesis Techniques and a New Approach," *Shock and Vibration Bulletin*, Vol. 40, July 1969, pp. 25–30.

[8]Craig, R. R., Jr., and Bampton, M. C., "Coupling of Substructures for Dynamic Analysis," *AIAA Journal*, Vol. 6, July 1968, pp. 1313–1319.

[9]Dowell, E. H., "Free Vibrations of Linear Structures with Arbitrary Support Conditions," *Journal of Applied Mechanics*, Vol. 38, Sept. 1971, pp. 595–600.

[10]Dowell, E. H., "Free Vibrations of an Arbitrary Structure in Terms of Component Modes," *Journal of Applied Mechanics*, Vol. 39, Sept. 1972, pp. 727–732.

[11]Dowell, E. H., "On Some General Properties of Combined Dynamical Systems," *Transactions of ASME*, Vol. 46, March 1979, pp. 206–209.

[12]Edge, B. L., Meyer, P. G., and Pierce, G. A., "An Analysis Technique for Composite Structures Subject to Dynamic Loads," *Journal of Applied Mechanics*, Vol. 38, March 1971, pp. 118–124.

[13]Klein, L., "Transverse Vibrations of Nonuniform Beams," *Journal of Sound and Vibration*, Vol. 37, Dec. 1974, pp. 491–505.

[14]Klein, L., "Vibrations of Constrained Plates by Rayleigh-Ritz Method Using Lagrange Multipliers," *Quarterly Journal of Applied Mathematics*, Vol. 30, Jan. 1977, pp. 51–70.

[15]Grief, R., and Mittendorf, S. C., "Structural Vibrations and Fourier Series," *Journal of Sound and Vibration*, Vol. 48, Sept. 1976, pp. 113–122.

[16]Mittendorf, S. C., and Grief, R., "Vibrations of Segmented Beams by Fourier Series Component Mode Method," *Journal of Sound and Vibration*, Vol. 55, Dec. 1977, pp. 431–441.

[17]Kerstens, J. G. M., "Vibrations of Rectangular Plates Supported at an Arbitrary Number of Points," *Journal of Sound and Vibration*, Vol. 65, Aug. 1979, pp. 493–504.

[18]Kerstens, J. G. M., "Vibration of Complex Structures: The Modal Constraint Method," *Journal of Sound and Vibration*, Vol. 76, June 1981, pp. 467–480.

[19]Kerstens, J. G. M., "Vibration of Modified Discrete Systems: The Modal Constraint Method," *Journal of Sound and Vibration*, Vol. 83, July 1982, pp. 81–92.

[20]Hale, A. L., and Meirovitch, L., "A General Substructure Synthesis Method for the Dynamic Simulation of Complex Structures," *Journal of Sound and Vibration*, Vol. 69, March 1980, pp. 309–326.

[21]Flashner, H., "A Procedure to Combine Modal Data of Two Flexible Bodies," TRW Space Technology Group, Redondo Beach, CA, Rept. IRAD/80.5-018, 1980.

[22]Flashner, H., "An Algorithm to Compute a Linear Representation of *n* Interconnected Flexible and Rigid Bodies," TRW Space Technology Group, Redondo Beach, CA, Rept. IRAD/81.5-016, 1981.

[23]Flashner, H., "Dynamic Representation of Linearly Constrained Dynamic Systems and its Application to Modal Synthesis," TRW Space Technology Group, Redondo Beach, CA, Rept. IRAD/82.5-018, 1982.

[24]Flashner, H., "An Orthogonal Decomposition Approach to Modal Synthesis," *International Journal for Numerical Methods in Engineering*, Vol. 23, Feb. 1986, pp. 471–493.

[25]Meirovitch, L., *Computational Methods in Structural Dynamics*, Sijthoff & Noordhoff, Alphen aan den Rijn, The Netherlands, 1980.

[26]Nikravesh, P. E., "Some Methods for Dynamic Analysis of Constrained Mechanical Systems: A Survey," *Computer Aided Analysis and Optimization of Mechanical System Dynamics*, edited by E. J. Haug, Springer-Verlag, Heidelberg, FRG, 1984.

[27]Gear, C. W., and Petzold, L. R., "ODE Methods for the Solution of Differential/Algebraic Systems," *SIAM Journal of Numerical Analysis*, Vol. 21, Aug. 1984, pp. 716–728.

[28]Wehage, R. A., and Haug, E. J., "Generalized Coordinate Partitioning of Dimension in Analysis of Constrained Dynamic Systems," *ASME Journal of Mechanical Design*, Vol. 104, Jan. 1982, pp. 247–255.

[29]Mani, N. K., Haug, E. J., and Atkinson, K. E., "Application of Singular Value Decomposition for Analysis of Mechanical Systems Dynamics," *Proceedings of ASME Design Engineering Technical Conference*, ASME Paper 84-DET-89, 1984.

[30]Singh, R. P., and Likins, P. W., "Singular Value Decomposition for Constrained Dynamic Systems," *ASME Journal of Applied Mechanics*, Vol. 52, No. 4, Dec. 1985, pp. 943–948.

[31]Kim, S. S., and Vanderploeg, M. J., "QR Decomposition for State Space Representation of Constrained Mechanical Dynamic Systems," *ASME Journal of Mechanisms, Transmissions, and Automation in Design*, Vol. 108, June 1986, pp. 183–188.

[32]Rosenberg, R. M., *Analytic Dynamics of Discrete Systems*, Plenum, New York, 1977.

[33]Goldstein, H., *Classical Mechanics*, Addison-Wesley, Reading MA, 1965.

[34]Meirovitch, L., *Methods of Analytic Dynamics*, McGraw-Hill, New York, 1970.

[35]Golub, G. H., and Van Loan, C. F., *Matrix Computations*, John Hopkins Univ. Press, Baltimore, MD, 1983.

[36]Klema, V. C., and Laub, A. J., "The Singular Value Decomposition: Its Computation and Some Applications," *IEEE Transactions on Automatic Control*, Vol. AC-25, Jan. 1980, 164–177.

[37]Lawson, C. L., and Hanson, R. J., *Solving Least Squares Problems*, Prentice-Hall, Englewood Cliffs, NJ, 1974.

[38]Stewart, G. W., *Introduction to Matrix Computations*, Academic, New York, 1973.

[39]Dongarra, J. J., Moler, C. B., Bunch, J. R., and Stewart, G. W., *LINPACK User's Guide*, Society of Industrial & Applied Mechanics, 1979.

[40]Meirovitch, L., *Analytic Methods in Vibrations*, Macmillan, New York, 1971.

[41]Quast, E. A., and Bonnickson, B. A., *TRW Structural Analysis Program Manual*, TRW Space Technology Group, Redondo Beach, CA, 1979.

[42]Golub, G. H., and Reinsch, C., "Singular Value Decomposition and Least Squares Solutions," *Numerical Mathematics*, Vol. 14, March 1970, pp. 403–420.

[43]Chan, T. F., "An Improved Algorithm for Computing the Singular Value Decomposition," *ACM Transactions on Mathematical Software*, Vol. 8, Jan. 1982, pp. 72–83.

Chapter 3

Integrated Structure-Control Optimization of Space Structures

Raphael T. Haftka*
Virginia Polytechnic Institute and State University, Blacksburg, Virginia

Introduction

TRADITIONALLY, the process of designing a structure and a control system for damping vibration in the structure has been sequential. First the structure is optimized to minimize an objective such as cost or weight subject to constraints on response quantities (e.g., displacements, stresses, or vibration frequencies). Then the control system is optimized to minimize some measure of control performance, such as control effort, subject to constraints on other performance measures, such as stability margins or robustness.

This process of sequential design of the structure and its control system can result in suboptimal performance. For large space structures the weight of the system consisting of the structure, the control system, and the power source for the controls is extremely critical because of the high cost of transferring mass to orbit. Therefore, there is merit in considering the simultaneous or integrated design of the structure and control system.

One of the key ingredients in the integrated design procedure is cross-sensitivity information. This includes the sensitivity of structural performance and response to changes in control system parameters and the sensitivity of control system performance to changes in structural parameters. This sensitivity information can be used to assess the potential gain, or synergistic effect, associated with the integrated design. For example, Haftka et al.[1] showed that, for a laboratory structure designed to simulate large space structures, the performance of the control system could be

*Christopher Kraft Professor, Department of Aerospace and Ocean Engineering.

significantly enhanced by minor modifications to the structure. This enhancement of the control system performance was predicted well by cross-sensitivity derivatives. Of course, sensitivity derivatives are also the lifeblood of most optimization procedures. Therefore, the calculation of sensitivity derivatives has received much attention in papers dealing with integrated structure-control design, and this is also the first topic discussed in this chapter. One major issue associated with derivative calculation is the accuracy of these derivatives when they are calculated from a reduced structural model.

A second major issue associated with integrated design is the choice of objective function. Here the problem is how to reconcile the different objectives used in the design of the structure and control system. Several approaches to this problem will be discussed in the next section.

Most of the work on integrated structure-control system design was performed for linear-quadratic Gaussian (LQG) control laws. Although other control laws may be equally appropriate, this historical precedent also guides this chapter, with the discussion focused on LQG control.

Analysis and Sensitivity

The discretized equations of motion for a flexible structure are usually written as

$$M\ddot{\boldsymbol{q}} \quad D\dot{\boldsymbol{q}} + K\boldsymbol{q} = \boldsymbol{L} \tag{1}$$

where M, D, and K are the mass, damping, and stiffness matrices, respectively, $\boldsymbol{L}$ is the load vector, $\boldsymbol{q}$ is the structural response vector, and an overdot denotes differentiation with respect to time. The structural response $\boldsymbol{q}$ can be written in physical coordinates if Eq. (1) is directly obtained from a finite-element model. However, in most applications the order of the finite-element model of the structure is too high for control system design; thus, it is reduced by employing some type of modal reduction. When the natural vibration modes of the structure are employed as reduced-basis vectors, M and K become diagonal matrices. However, here we will assume that they can be fully populated.

Equation (1) can be reduced to a more general first-order form as

$$\dot{\boldsymbol{x}} = A\boldsymbol{x} + f \tag{2}$$

One way to accomplish this reduction to a first-order system is to define $\boldsymbol{x}^T = [\dot{\boldsymbol{q}}, \boldsymbol{q}]$; then,

$$A = \begin{bmatrix} -M^{-1}D & -M^{-1}K \\ I & 0 \end{bmatrix}, \qquad f = \begin{Bmatrix} M^{-1}L \\ 0 \end{Bmatrix} \tag{3}$$

where I denotes the unit matrix. In the consideration of the control problem we limit ourselves to linear control so that f is given as

$$f = B\boldsymbol{u} + \boldsymbol{w} \tag{4}$$

where $\boldsymbol{u}$ is the control vector, and $\boldsymbol{w}$ is a vector of white noise in the control commands assumed to be zero-mean, Gaussian stationary process. The control law is assumed to be linear and time invariant, i.e.,

$$\boldsymbol{u} = -G\boldsymbol{x} \tag{5}$$

where G is a matrix of constant gains and is found by solving an optimal control problem. Substituting Eqs. (4) and (5) into Eq. (2), we get

$$\dot{\boldsymbol{x}} = (A - BG)\boldsymbol{x} + \boldsymbol{w} \tag{6}$$

so that the stability of the closed-loop system is determined by the eigenvalues λ_i of $A - BG$.

The derivatives of these eigenvalues with respect to structural or control parameters are given as

$$\lambda_i' = \boldsymbol{x}_{Li}^T(A' - B'G - BG')\boldsymbol{x}_{Ri} \tag{7}$$

where a prime denotes a derivative with respect to the parameter, and $\boldsymbol{x}_{Li}$ and $\boldsymbol{x}_{Ri}$ are the left and right eigenvectors of the matrix $A - BG$, respectively, corresponding to λ_i and normalized so that

$$\boldsymbol{x}_{Li}^T \boldsymbol{x}_{Ri} = 1 \tag{8}$$

In calculating A' we need to calculate derivatives of K, D, and M. When $\boldsymbol{q}$ represents a vector of modal amplitudes, these matrices are obtained from equations of the form

$$M = \Phi^T \bar{M} \Phi \tag{9}$$

where $\bar{M}$ is a mass matrix obtained from a finite-element analysis, and Φ is a matrix of modes. Differentiating Eq. (9) we get

$$M' = \Phi'^T \bar{M} \Phi + \Phi^T \bar{M}' \Phi + \Phi^T \bar{M} \Phi' \tag{10}$$

so that even though we do not need derivatives of the eigenvectors $\boldsymbol{x}_{Li}$ and $\boldsymbol{x}_{Ri}$ for calculating λ_i', we may need derivatives of the vibration modes. Calculating derivatives of vibration modes is expensive. Fortunately, however, Φ' may often be neglected (see, e.g., Sandridge and Haftka[2]).

Control Design Formulations

With linear-quadratic regulators (LQR) the gain matrix G in Eq. (5) is found by minimizing a quadratic performance index J:

$$J = E(\boldsymbol{x}^T Q \boldsymbol{x} + \boldsymbol{x}^T R \boldsymbol{x}) \tag{11}$$

where E is the expected value operator, and Q and R are positive-definite matrices.

The gain matrix that minimizes the quadratic performance index is given as

$$G = R^{-1}B^TS \tag{12}$$

where S is the solution of a matrix Riccati equation:

$$A^TS + SA - SBR^{-1}B^TS + Q = 0 \tag{13}$$

The sensitivity of the gain matrix can be calculated by differentiating Eq. (12) and using the relation

$$(R^{-1})' = -R^{-1}R'R^{-1} \tag{14}$$

to obtain

$$G' = -R^{-1}R'R^{-1}B^TS + R^{-1}B'^TS + R^{-1}B^TS' \tag{15}$$

where S' is obtained by differentiating Eq. (13) and using Eq. (12) and the symmetry of S:

$$S'(A - BG) + (A - BG)^TS' + SA' + A'^TS + Q' - S(BR^{-1}B^T)'S = 0 \tag{16}$$

Equation (16) is a Liapunov equation that can be solved for S'; then G' is calculated from Eq. (15) and the eigenvalue derivatives are calculated from Eq. (7). The optimum value of J is given as

$$J^* = \mathrm{tr}\{SW\} \tag{17}$$

where W is the noise covariance matrix, i.e.,

$$E[w(t)w^T(\tau)] = W\delta(t - \tau) \tag{18}$$

An interesting property of J^* (see Gilbert[3]) is that its derivatives with respect to changes in the gains are zero.

It is usually impossible to measure all of the elements of $\boldsymbol{x}$. Instead, we have sensors that observe a vector $\boldsymbol{z}$:

$$\boldsymbol{z} = C\boldsymbol{x} + \boldsymbol{v} \tag{19}$$

where $\boldsymbol{v}$ is observation noise with covariance matrix V. We assume that the full vector $\boldsymbol{x}$ is estimated as $\hat{\boldsymbol{x}}$ from $\boldsymbol{z}$ by a Kalman filter that integrates

$$\dot{\hat{\boldsymbol{x}}} = A\hat{\boldsymbol{x}} + B\boldsymbol{u} - F(\boldsymbol{z} - C\boldsymbol{x}) \tag{20}$$

where the filter matrix F is found as the solution to another Riccati matrix equation. For expressions of the sensitivity of the observation solution, see Gilbert.[3] The combination of linear-quadratic regulator with a Kalman filter is called linear-quadratic Gaussian (LQG) control.

LQG control minimizes a quadratic performance index J [Eq. (11)], which is a compromise between vibration levels and control effort. However, the choice of the weighting matrices Q and R is somewhat arbitrary, and these variables are typically adjusted by the designer to achieve satisfactory response characteristics.

The primary measures of the response of the controlled structure are the closed-loop eigenvalues or the damping ratios associated with them. The Q and R matrices are often adjusted to satisfy limits (referred to as stability margins) on the real part of the eigenvalues or their damping ratios. Alternatively, the gain matrix G can be calculated to obtain desired eigenvalues in a process called "pole placement" (the closed-loop eigenvalues are poles of the system transfer matrix).

When the control system is designed subject to stability margin limits, it may be reasonable to minimize only a measure of control effort because the response is limited by the stability margins. One measure of control effort used[4] is the Frobenius norm of the control gains $\|G\|_f$:

$$\|G\|_f = \mathrm{tr}\{G^T R G\} \tag{21}$$

The structural model is often a poor representation of the structure because of either modeling errors or deliberate model truncation. Therefore, it is important to design a robust control system; that is, one that is not sensitive to small variations in the properties of the structure. Control systems are sometimes designed to maximize robustness. One measure of robustness[5] is the condition number of the closed-loop modal matrix. Another method used to maximize robustness is to minimize a norm of the sensitivity of the closed-loop eigenvalues to a set of specified parameters.[6]

Truncation Error in Reduced Models

As noted earlier, the size of the structural model is reduced in order to make the design of the control system tractable. This is typically done by using a subset of the vibration modes of the structure as reduced-basis vectors. Modal cost techniques[7] are often used to select the vibration modes to be included in the reduced model. However, the vibration modes of the structure are smooth functions and thus cannot represent well the response to point actuators that introduce spatial discontinuities. In the presence of such point actuators, a reduced model based on vibration modes can be virtually useless.

The difficulty is demonstrated by a five-span beam example[2] shown in Fig. 1. For this simple example direct-rate feedback controllers are used that can be modeled as dashpots. One controller is a displacement rate controller on the second span, and the second is a rotation-rate controller on the middle span. The controllers are designed to produce damping ratios of 1–10% in the first six modes. Derivatives of the damping ratios are calculated with respect to an added mass at point A on the beam.

Figure 2 shows the error in the fourth damping ratio and its derivative with respect to the mass for a finite-element model of three beam finite

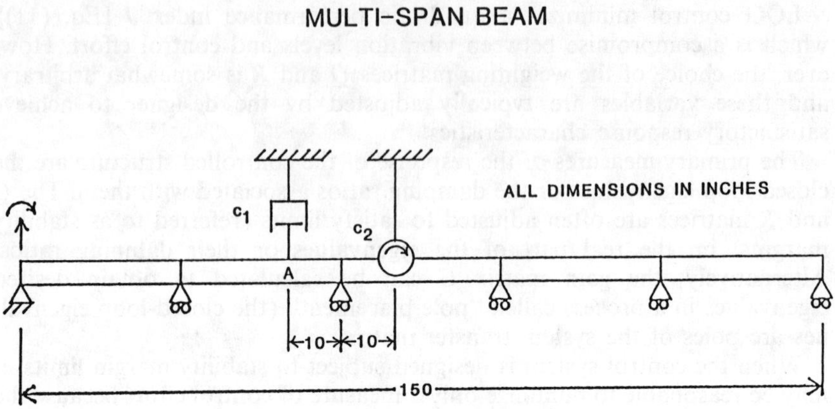

Fig. 1 Multispan beam geometry.

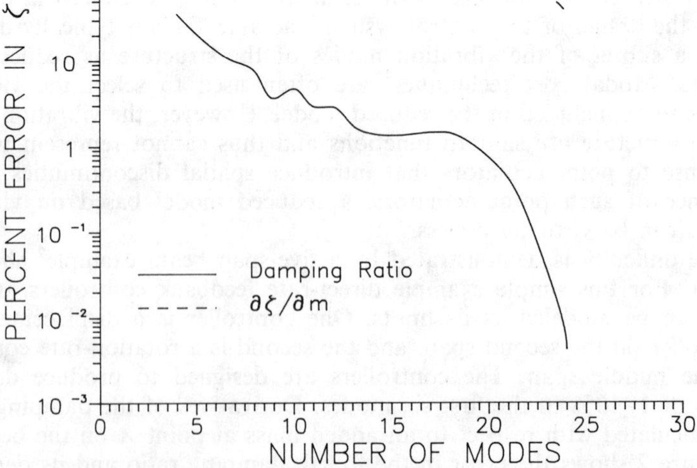

Fig. 2 Error in fourth damping ratio and its derivative for multispan beam.

elements per span. The model has 26 unrestrained degrees of freedom, and it is seen that even for 24 of 26 modes the error in the derivative is very high.

To check the convergence beyond 24 modes, more refined models with 6, 9, and 12 elements per span were also analyzed. Additionally, analytical modes were calculated in closed form from a continuum model of the beam, and these exact modes were also used to reduce the model. We now have three types of models. Two use a reduced model based on vibration modes (one with finite-element modes and one with continuum modes). The third is the full finite-element model. The order of the model for the first two types is equal to the number of modes, whereas for the third it is equal to the number of finite-element degrees of freedom.

The errors in the derivative of the fourth damping ratio for all three types of models are given in Fig. 3 as a function of the order of the model. The top solid line is the error in the derivative calculated using the continuum modes. The dashed lines are the errors in the reduced-model derivatives for modes obtained from the four finite-element models, and the bottom solid line is the error in the finite-element derivative.

From Fig. 3 it is clear that the convergence of the continuum, reduced-model derivative is much slower than that for the finite-element derivative. For the same order the finite-element derivative is two to four orders of

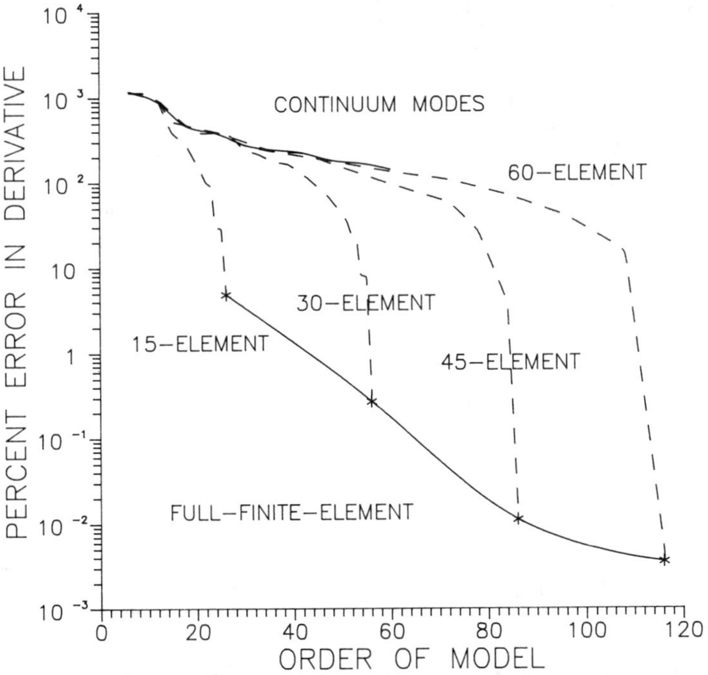

Fig. 3 Errors in fourth-mode derivatives for multispan beam.

magnitude more accurate. It is obvious that, for this example, the approach of using vibration modes to obtain a reduced model is counterproductive in that it requires a higher-order model for the same accuracy.

As expected, finite-element, reduced-order results are close to the continuum results when the number of modes is small compared to the number of finite-element degrees of freedom. As the number of modes approaches the number of finite-element degrees of freedom, the reduced-order results approach the full finite-element results. This produces the paradoxical result that, for a fixed number of modes, the accuracy of the derivative deteriorates as the finite-element model is refined. For example, with 48 modes, the 30-element model (56 DOF) has an error of 60%, the 45-element model (86 DOF) has an error of 158%, and the 60-element model (116 DOF) has an error of 177%.

The solution to this convergence problem is to include in the reduced basis modes that can model the discontinuity associated with the actuator action. One possibility is to indicate the static response vectors due to a unit force by each actuator. These vectors are referred to as Ritz vectors.

Two Ritz vectors corresponding to the static response to a unit load at the translation-rate actuator and a unit torque at the rotation-rate actuator were added to the exact eigenvectors to form the reduced model. Thus, for

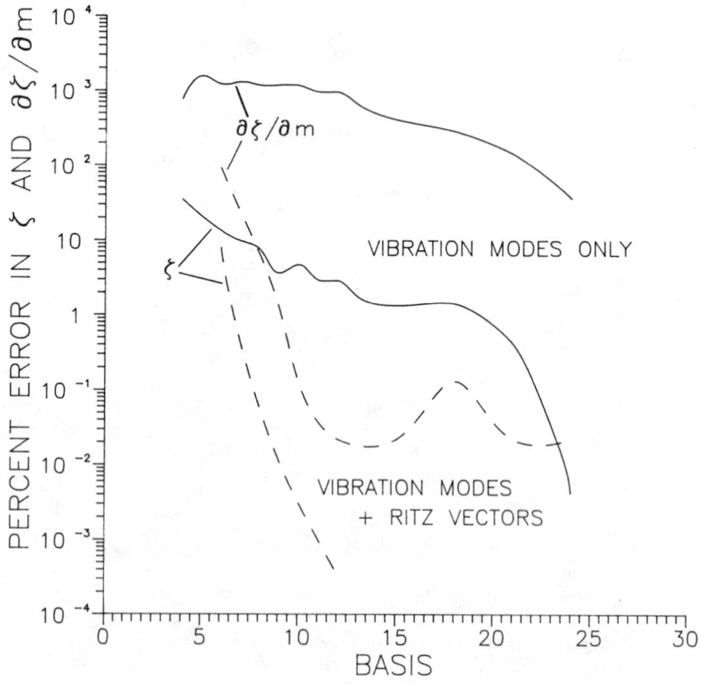

Fig. 4 Effect of Ritz vectors on convergence of fourth damping ratio and its derivative.

a six-mode model, the two Ritz vectors and the first four vibration modes were used to form the reduced model. The calculations for the reduced-model derivatives were the same as before, except that the reduced mass and stiffness matrices were no longer diagonal.

Figure 4 shows the effect of the additional Ritz vectors on the convergence of the fourth damping ratio and its derivative. It can be seen that the convergence is greatly improved with the Ritz vectors for both the damping ratio and derivative. Without the Ritz vectors there is a 34% error in the derivative with 24 of 26 modes used. However, with the Ritz vectors included, the derivative converges to within 10% error with only eight modes.

These results were obtained with a very simple control system. However, the same problems of modal truncation for point actuators were demonstrated for LQG control.[8] Also, the use of a Ritz vector was shown to alleviate such problems for LQG control.

Integrated Design Formulations

The optimization of structural parameters for a space structure can be formulated as

$$\begin{aligned} &\text{minimize} && m(\boldsymbol{v}) \\ &\text{such that} && g_j(\boldsymbol{v}) \geqslant 0, \qquad j = 1, \ldots, n_g \end{aligned} \tag{22}$$

where $\boldsymbol{v}$ is a vector of structural sizes, and m is the structural objective function, typically the mass. The constraint function $g_j(\boldsymbol{v})$ represents limits on stresses, displacements, vibration frequencies, buckling loads, etc.

When we consider the integrated design of the structure and control system, many formulations are possible. It is possible to design first the structure and then permit small changes in structural parameter in order to optimize the control objective function.[9–11]

Similarly, it is possible[12] to use the design of the control system to infer required structural changes, because some parts of the gain matrix are equivalent to modifications in the structural stiffness matrix.

It is also possible to treat the structural variables as primary and assume that for a given structure we use an optimal LQR or LQG control system. Khot and Grandhi[13–16] used this approach with the weight or Frobenius norm of the gains as objective functions with limits on closed-loop eigenvalues.

More truly integrated structure-control design procedures operate simultaneously on a set of structural and control variables to optimize a combined structure-control problem. The mass of the structure can be combined with the quadratic performance index J [Eq. (11)] as

$$m^* = \theta m + (1 - \theta)J, \qquad 0 \leqslant \theta \leqslant 1 \tag{23}$$

to obtain a composite objective m^*. The weighting parameter θ may be chosen so as to emphasize the importance of m or J. This type of an approach was employed by Salama et al.,[17,18] Hale et al.,[19] and Miller and Shim.[20] Alternatively, the quadratic performance index can be optimized for a given structural mass,[21] or the structural objective minimized for a given J.[22] Similar formulations are possible with control design procedures that do not work with the quadratic performance measure but with other performance measures such as stability margins,[1,23] robustness measures,[5,6,24] or actuator force magnitudes.[25]

Instead of combining the structure and control objectives as in Eq. (23), or relegating one of them to a constraint, it is possible to approach the problem in the context of multiobjective optimization[26] or game theory.[27] When a multiobjective optimization is used, we try to optimize all objectives. For example, if our objectives are the structural mass and quadratic performance index, the problem may be formulated as

$$\begin{aligned} &\text{minimize} && m(\boldsymbol{v}), \quad J(G,\boldsymbol{v}) \\ &\text{such that} && g_j(\boldsymbol{v}) \geqslant 0, \qquad j = 1, \ldots, n_g \\ & && h_i(\boldsymbol{v},G) \geqslant 0, \qquad i = 1, \ldots, n_h \end{aligned} \tag{24}$$

where $h_i(\boldsymbol{v},G)$ represents a set of constraints on the closed-loop system such as stability margins and robustness requirements. Multiobjective optimization procedures seek so-called efficient designs. These are designs where one objective cannot be improved without a loss in the other objective. Under some mild conditions all of the efficient designs can be obtained by minimizing m^* of Eq. (23) subject to the constraints of Eq. (24) for all values of θ. Homotopy continuation methods can be used to generate efficiently the optima for the entire range of θ.[5,28] This weighting approach can be extended to more than two objectives, but generating all of the efficient solution becomes much more expensive.

The multiobjective formulation [Eq. (24)] has two shortcomings. First, the effect of the control system on the mass and structural constraints is neglected. This effect can be significant when the magnitude of the control effort determines the mass of the control system and its power supply. Second, the Q and R matrices used in the quadratic performance index are somewhat arbitrary and are typically tuned by the control designer. It is therefore unreasonable to select them a priori and leave them unchanged in the design procedure.

A formulation that attempted to address these two concerns was suggested by Onoda and Haftka.[29] This formulation is summarized in the next section.

Onoda's Formulation

We assume that the mass of the control system m_c is related to the control effort as

$$m_c = \alpha C_E^{\beta} \tag{25}$$

where α and β are constants, and C_E is the quadratic measure of the control effort:

$$C_E = E(\boldsymbol{u}^T R \boldsymbol{u}) \tag{26}$$

The optimization problem is to minimize the total mass of the system consisting of the structural mass m_s and control mass m_c subject to a constraint on the magnitude of the response, as measured by the quadratic measure

$$r = E(\boldsymbol{x}^T Q \boldsymbol{x}) \tag{27}$$

The optimization problem is then formulated as

$$\text{find} \quad \boldsymbol{v} \quad \text{and} \quad G \quad \text{to minimize} \quad m = m_s(\boldsymbol{v}) + m_c(\boldsymbol{v}, G)$$

$$\text{such that} \quad r(\boldsymbol{v}, G) \leqslant \sigma_1 \tag{28}$$

where σ_1 is a response magnitude allowable. In this formulation it may seem that the matrices Q and R are specified, but an alternate formulation of the same problem reveals that the *control system* is designed with some flexibility of Q and R. The alternative formulation is

$$\text{find} \quad \boldsymbol{v} \quad \text{and} \quad k \quad \text{to minimize} \quad m = m_s(\boldsymbol{v}) + m_c(\boldsymbol{v}, G)$$

$$\text{such that} \quad r(\boldsymbol{v}, G) \leqslant \sigma_1 \tag{29}$$

where G is found by minimizing

$$J^* = E[\boldsymbol{x}^T Q \boldsymbol{x} + \boldsymbol{u}^T (kR) \boldsymbol{u}] = r + kC_E \tag{30}$$

It can be shown that the two formulations are equivalent. The variable k plays the role of a scaling factor that permits just enough control effort to meet the response limit. This formulation also has the advantage that the gains G can be obtained by a standard Riccati equation solver that minimizes J^*. This is a two-level optimization solution where the control system is reoptimized for each structural design.

It is possible to add to the problem additional structural constraints that do not depend on G. However, the addition of structural constraints that depend on G (e.g., via the mass of the control system) or of stability margins would preclude the use of the Riccati equation to obtain the gains G.

Reference 29 presented a free-free beam (see Fig. 5) example under Gaussian white noise disturbance. The beam supports a uniform payload and is controlled by two altitude-control units with torque actuators symmetrically installed on the structure. The beam was modeled by finite elements, and its cross-sectional thickness as well as the actuator location were used as structural design variables. Five elements were used to model

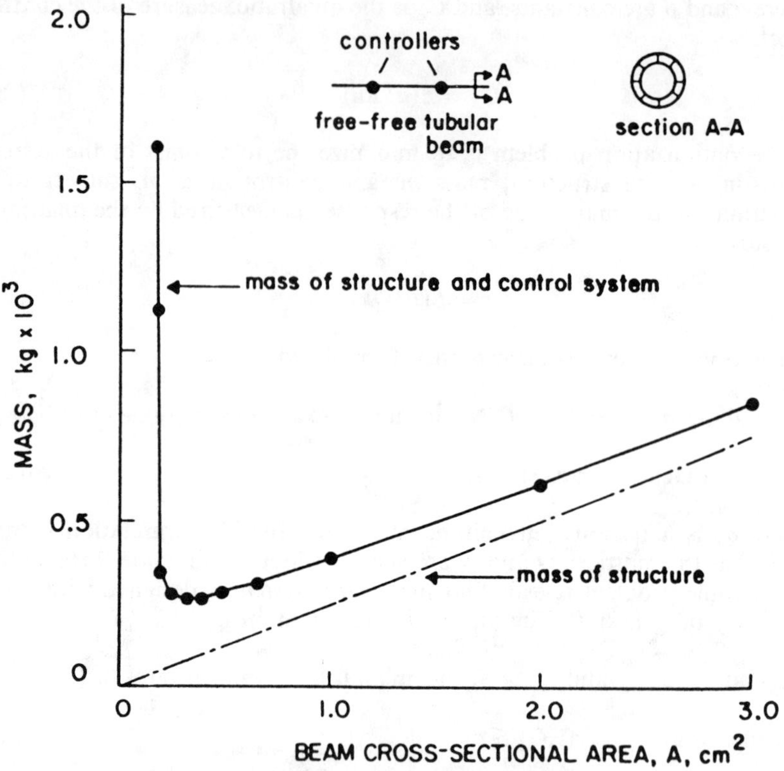

Fig. 5 Combined structure/control optimization of beam structure.

one-half of the beam. In the results presented here it is assumed that the control mass m_c is proportional to the control effort C_E (i.e., $\beta = 1$), that the beam thickness is proportional to the thickness, and that the payload is equal to the structural mass. It is assumed that the entire vector $\boldsymbol{x}$ is sensed; hence, there is no need for a Kalman filter.

Table 1 compares four cases with varying response requirements (σ_1) and control cost (α). As can be seen from the table, the average thickness (proportional to the structural mass m_s) is very sensitive to the mass cost of the control. When that cost is reduced by a factor of 10, the structural mass is reduced by a factor of two to three. Also, both control effort and mass are sensitive to the required response limit σ_1. Thus, it appears to be quite important to do the combined optimization. Another demonstration of this importance is shown in Fig. 5, which shows the total mass as a function of the structural mass for a uniform beam. The figure indicates that even with an optimal control system the control effort is sensitive to the structural stiffness. In fact, if the structure is made too flexible, the total mass becomes very high.

Table 1 Effect of control cost and response limit on combined optimum design for controlled beam

Response limit	σ_1	0.01	0.001	0.01	0.001
Control cost	α	1.0	1.0	0.1	0.1
Normalized beam thickness	ξ_1	0.262	0.822	0.120	0.342
	ξ_2	0.620	1.785	0.184	0.506
	ξ_3	0.811	2.23	0.156	0.408
	ξ_4	0.639	1.14	0.159	0.430
	ξ_5	0.114	0.713	0.242	1.27
Actuator location	$\bar{x}_c$	0.786	0.622	0.386	0.391
Average thickness	$\bar{\xi}$	0.489	1.338	0.172	0.591
Total normalized mass	m_T	1.063	2.179	0.268	0.603

Onoda and Watanabe[30] extended this approach to the case where the entire state vector $\boldsymbol{x}$ is not sensed; hence there is a need for an observer.

References

[1]Haftka, R. T., Martinovic, Z. N., and Hallauer, W. L., "Enhanced Vibration Controllability by Minor Structural Modification," *AIAA Journal*, Vol. 23, Aug. 1985, pp. 1260–1266.

[2]Sandridge, C. A., and Haftka, R. T., "Accuracy of Eigenvalue Derivatives from Reduced-order Structural Model," *Journal of Guidance*, Vol. 12, Nov.–Dec. 1989, pp. 822–829.

[3]Gilbert, M. G., "Results of an Integrated Structure/Control Law Design Sensitivity Analysis," *Recent Advances in Multidisciplinary Analysis and Optimization*, edited by J.-F. M. Barthelemy, NASA CP-3031, 1988, Pt. 2, pp. 727–746.

[4]Khot, N. S., Öz, H., Grandhi, R. V., and Venkayya, V. B., "Optimal Structural Design with Control Gain Norm Constraint," *AIAA Journal*, Vol. 26, May 1988, pp. 604–611.

[5]Junkins, J. L., and Rew, D. W., "Unified Optimization of Structures and Controllers," *Large Space Structures: Dynamics and Control*, edited by S. N. Atluri and A. K. Amos, Springer-Verlag, Berlin 1987.

[6]Lim, K. B., and Junkins, J. L., "Robust Optimization of Structural and Controller Parameters," *Journal of Guidance*, Vol. 12, Jan. 1989, pp. 89–96.

[7]Skelton, R. E., Hughes, P. C., and Hablani, H. B., "Order Reduction for Models of Space Structures Wing Modal Cost Analysis," *Journal of Guidance*, Vol. 5, 1982, pp. 351–357.

[8]Sandridge, C. A., and Haftka, R. T., "Effect of Modal Truncation on Derivatives of Closed-Loop Damping Ratios in Optimal Structural Control," *Computer Aided Optimum Design of Structures: Applications*, edited by C. A. Brebbia and S. Hernandez, Computational Mechanics Publications, Springer-Verlag, Berlin, 1989, pp. 159–168.

[9]Khot, N. S., Venkayya, V. B., and Eastep, F. E., "Structural Modifications to Reduce the LOS-Error in Large Space Structures," AIAA Paper 84-0997-CP, May 1984, Pt. 2, pp. 296–305.

[10]Khot, N. S., Venkayya, V. B., and Eastep, F. E., "Optimal Structural Modifications to Enhance the Active Vibration Control of Flexible Structures," *AIAA Journal*, Vol. 24, Aug. 1986, pp. 1368–1374.

[11]Venkayya, V. B., and Tischler, V. A., "Frequency Control and Its Effects on the Dynamic Response of Flexible Structures," *AIAA Journal*, Vol. 22, Sept. 1984, pp. 1293–1298.

[12]Hanks, B. R., and Skelton, R. E., "Designing Structures for Reduced Response by Modern Control Theory," AIAA Paper 83-0810, May 1983.

[13]Khot, N. S., "Minimum Weight and Optimal Control Design of Space Structures," *Computer Aided Optimal Design: Structural and Mechanical Systems*, edited by C. A. Mota Soares, Springer-Verlag, Berlin, 1987, pp. 389–403.

[14]Khot, N. S., "Structure/Control Optimization to Improve the Dynamic Response of Space Structures," *Computational Mechanics*, Vol. 3, pp. 179–186, 1988.

[15]Grandhi, R. V., "Structures and Control Optimization of Space Structures," *Computers and Structures*, Vol. 31, No. 2, 1989, pp. 139–150.

[16]Khot, N. S., and Grandhi, R. V., "Structural and Control Optimization with Weight and Frobenius Norm as Performance Functions," *Structural Optimization*, edited by G. I. N. Rozvany and B. L. Karihaloo, Kluwer Academic, The Netherlands, pp. 151–158, 1988.

[17]Salama, M., Hamidi, M., and Demestz, L., "Optimization of Controlled Structures," *Proceedings of the JPL Workshop on Identification and Control of Flexible Space Structures*, Jet Propulsion Lab. Publ. 85-29, Vol. II, April 1985, pp. 311–327.

[18]Salama, M., Garba, J., and Demesetz, L., "Simultaneous Optimization of Controlled Structures," *Computational Mechanics*, Vol. 3, 1988, pp. 275–282.

[19]Hale, A. L., Lisowski, R. J., and Dahl, W. G., "Optimum Simultaneous Structural and Control Design of Maneuvering Flexible Spacecraft," *Journal of Guidance and Control and Dynamics*, Vol. 8, No. 1, 1985, pp. 86–93.

[20]Miller, D. F., and Shim, J., "Gradient-Based Combined Structural and Control Optimization," *Journal of Guidance*, Vol. 10, No. 3, 1987, pp. 291–298.

[21]Messac, A., Turner, J., and Soosaar, K., "An Integrated Control and Minimum Mass Structural Optimization Algorithm for Large Space Structures," *Proceedings of the JPL Workshop on Identification and Control of Flexible Space Structures*, Jet Propulsion Lab. Publ. 85-29, Vol. II, April 1985, pp. 231–236.

[22]Dracopoulos, T. N., and Öz, H., "Integrated Aeroelastic Control Optimization," 7th VPI&SU/AIAA Symposium on Dynamics and Control of Large Structures, Blacksburg, VA, May 1989.

[23]Bodden, D. S., and Junkins, J. L., "Eigenvalue Optimization Algorithms for Structure/Controller Design Interactions," *Journal of Guidance and Control and Dynamics*, Vol. 8, No. 6, 1985, pp. 679–706.

[24]Bendsøe, M. P., Olhoff, N., and Taylor, J. E., "On the Design of Structure and Control for Optimal Performance of Actively Controlled Flexible Structures," *Mechanics of Structures & Machines*, Vol. 15, No. 3, 1987, pp. 265–295.

[25]Thomas, H. L., and Schmit, L. A., Jr., "Control Augmented Structural Synthesis with Dynamic Stability Constraints," AIAA Paper 89-1216-CP, April 1989, Pt. I, pp. 521–531.

[26]Rao, S. S., "Combined Structural and Control Optimization of Flexible Structures," *Engineering Optimization*, Vol. 13, 1988, pp. 1–16.

[27]Rao, S. S., Venkayya, V. B., and Khot, N. S., "Game Theory Approach for the Integrated Design of Structures and Controls," *AIAA Journal*, Vol. 26, No. 4, 1988, pp. 463–469.

[28]Milman, M., Scheid, R. E., and Salama, M., "Methods for Combined Control-Structure Optimization," VPI&SU 7th Symposium on Dynamics and Control of Large Structures, Blacksburg, VA, May 8–10, 1989.

[29]Onoda, J., and Haftka, R. T., "An Approach to Simultaneous Structure-Control Optimization of Large Flexible Spacecraft," *AIAA Journal*, Vol. 25, No. 8, 1987, pp. 1133–1138.

[30]Onoda, J., and Watanabe, N., "Integrated Direct Optimization of Structure/Regulator/Observer for Large Flexible Spacecraft," AIAA Paper 89-1313CP, April 1989, Pt. 3, pp. 1336–1344.

Chapter 4

Multibody Dynamics Formulations via Kane's Equations

Ronald L. Huston*
University of Cincinnati, Cincinnati, Ohio

Introduction

THE past decade has seen a rapidly growing interest in multibody dynamics. This interest has been spurred by three factors: 1) a desire to obtain comprehensive analyses of large dynamic systems; 2) advances in modeling and dynamic analysis procedures; and 3) increased availability of computer systems. The growing interest in multibody dynamics has stimulated many research efforts. Indeed, the activity in multibody dynamics research parallels that seen in the early development of finite-element methods.

The desire to obtain comprehensive analyses of large systems stems from the realization that many physical systems of interest and of practical importance (such as chains, cables, robots, mechanisms, space stations, and biodynamic models) can be effectively modeled as multibody systems. The advances in modeling and analysis techniques are based on recent computer-oriented advances in dynamic analyses. Among these advances are the advent of Kane's equations, the development of body configuration arrays, the formulation of differentiation algorithms, the use of Euler parameters, and the use of orthogonal complement arrays. The access to computer systems is continuing to expand, including access to class VI supercomputers. Also, recent advances in computer graphics are leading to a higher level of input and output, including animation of the modeled physical systems.

Surprisingly, however, with all of this activity and despite all of these developments, there is still disagreement among analysts about which

*Professor of Mechanics, Department of Mechanical and Industrial Engineering.

dynamic formulation methods are most suitable for multibody analyses. Indeed, in the development of governing dynamic equations, some analysts use Lagrange's equations, some use Newton-Euler methods, some use virtual work, and recently some have suggested the use of Gibbs-Appell equations. In this writer's opinion, however, the use of Kane's equations is preferable to the use of each of these other methods, particularly for the automated numerical analysis of large multibody systems. It is believed that Kane's method (sometimes called "Lagrange's form of d'Alembert's principle," "Jourdain's principle," or the "principle of virtual power") has the advantages of both Lagrangian and Newton-Euler formulations but without the corresponding disadvantages. That is, with Kane's equations nonworking constraint forces between adjoining bodies are automatically eliminated in an analysis without the introduction of tedious, unwieldy differentiation of scalar energy functions. This writer believes that Kane's equations are both intuitive and easy to develop and also that they lead naturally to computational algorithms for efficient numerical analyses.

The objective of this chapter is therefore to present a summary of multibody dynamics formulations using Kane's equations. The balance of the chapter itself is divided into four parts with the following part presenting some preliminary ideas useful in the sequel. The next part presents the dynamic analysis. The third part discusses numerical methods and computer analyses. The final part contains a discussion and concluding remarks. Additional details and related information may be obtained in earlier writings of this writer and others.[1–8]

Preliminary Considerations

Multibody Systems

A multibody system is a collection of bodies with a given connection configuration. Figure 1 depicts a typical multibody system. Even though the bodies have a "connection configuration" or, alternatively, an "arrangement pattern," there may nevertheless be translation and rotation between adjoining bodies. Also, the bodies of the system may be deformable and may form closed loops.

A multibody system may be subjected to both external and internal force systems. For example, external forces might include gravity and contact forces. Internal forces might involve forces and couples exerted between adjoining bodies. In addition, the system may be subjected to geometric and kinematic constraints, both locally, between adjoining bodies, and globally, affecting the entire system.

In modeling a particular physical system, a multibody system will generally not have all these features. Indeed, many physical systems may be modeled simply as a set of rigid bodies connected with spherical joints, with no closed loops. Such systems are called "open-chain systems" or "open-tree systems."

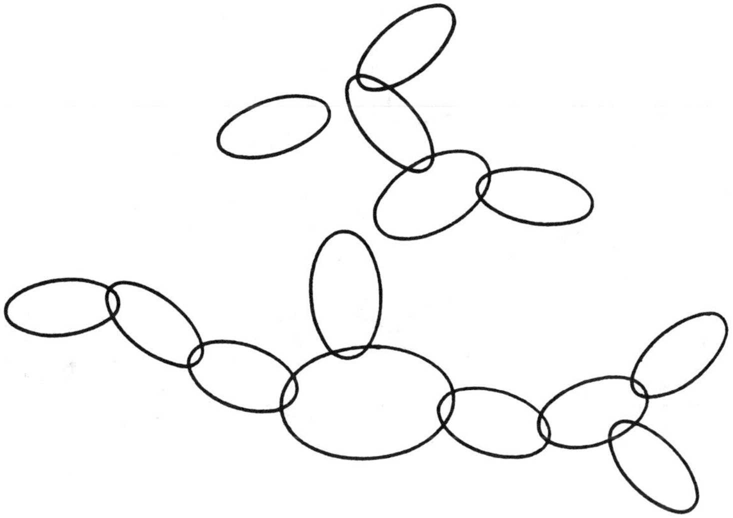

Fig. 1 A multibody system.

Body Connection Arrays

To develop automated numerical analyses of multibody systems, it is convenient to define the connection configuration in terms of a body connection array.[2] To this end an arbitrary body of the system is selected as the reference body and labeled as "body 1," "B_1," or "1." The remaining bodies of the system are then numbered in ascending progression through the branches of the tree system. (If the system has closed loops, the loops are temporarily broken at arbitrary joints.) Figure 2 shows such a numbering procedure for the system of Fig. 1.

When the system is numbered or labeled in this manner, each body has a unique adjoining lower numbered body. The lower numbered body of body 1 is the inertia frame R in which the entire system moves. R is assigned the number 0. The connection configuration of the system can then be defined by an integer array of numbers of lower numbered bodies. This array, names $L(K)$, is called the lower-body array. For the system of Fig. 2 the body array $L^0(K)$ (the serial array of body numbers) and the lower-body array are

$$L^0(K) = (1,2,3,4,5,6,7,8,9,10,11,12,13,14) \tag{1}$$

and

$$L(K) = (0,1,2,3,3,1,6,7,7,9,9,1,12,13) \tag{2}$$

Once the system is numbered as given in Eqs. (1) and (2), the lower-body array, $L(K)$, is uniquely determined. Alternatively, when $L(K)$ is known, the system may be constructed with its connection configuration.

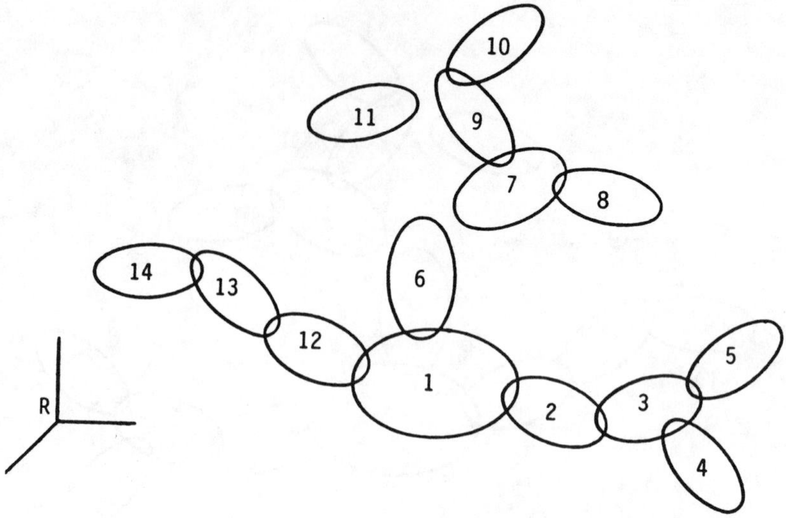

Fig. 2 A numbered multibody system.

In another context $L(K)$ may be thought of as an operator on body number K. Hence, we can form a lower-body array of $L(K)$ by "operating" on $L(K)$; that is, $L(L(K))$ [or $L^2(K)$]. $L^2(K)$ is called the second lower-body array. For the system of Fig. 2 $L^2(K)$ is

$$L^2(K) = (0,0,1,2,2,0,1,6,6,7,7,0,1,12) \tag{3}$$

where $L(0)$ is defined as 0.

By continuing this process we can form $L(L^2(K))$ [or $L^3(K)$], etc. Eventually, all of the entries will be zero. For the system of Fig. 2 the complete set of lower-body arrays is listed in Table 1.

Observe that once $L(K)$ is known the entire set of lower-body arrays can be automatically generated.

In addition to defining the connection configuration of the system, the lower-body array is useful for developing algorithms for an automated

Table 1 Lower-body arrays for the system of Fig. 2

Array	Body number													
$L^0(K)$	1	2	3	4	5	6	7	8	9	10	11	12	13	14
$L^1(L)$	0	1	2	3	3	1	6	7	7	9	9	1	12	13
$L^2(K)$	0	0	1	2	2	0	1	6	6	7	7	0	1	12
$L^3(K)$	0	0	0	1	1	0	0	1	1	6	6	0	0	1
$L^4(K)$	0	0	0	0	0	0	0	0	0	1	1	0	0	0
$L^5(K)$	0	0	0	0	0	0	0	0	0	0	0	0	0	0

(numerical) development of the system kinematics. For example, for the system of Fig. 2 the addition theorem for angular velocity[9] states that the angular velocity of body 11 may be expressed as

$$\omega_{11} = \hat{\omega}_{11} + \hat{\omega}_9 + \hat{\omega}_7 + \hat{\omega}_6 + \hat{\omega}_1 \tag{4}$$

where the "hat" over the symbol designates an angular velocity relative to the adjacent lower numbered body. Observe that the indices of Eq. (4) correspond to the entries in the column of body 11 in Table 1.

Transformation Matrices

Consider two typical adjoining bodies of the system, such as B_j and B_k, as depicted in Fig. 3. Let $\boldsymbol{n}_{ji}$ and $\boldsymbol{n}_{ki}$ $(i = 1,2,3)$ be dextral, orthogonal unit vector sets fixed in B_j and B_k, respectively. The relative orientation of B_j and B_k can be defined in terms of the relative inclination of these unit vector sets.

Let SJK be a transformation matrix between the unit vector sets. Specifically, let the elements of SJK be the projections of the unit vectors from one set to the other. That is,

$$SJK_{mn} = \boldsymbol{n}_{jm} \cdot \boldsymbol{n}_{kn} \tag{5}$$

The transformation properties of SJK are seen from the expressions

$$\boldsymbol{n}_{jm} = SJK_{nm}\boldsymbol{n}_{kn} \qquad \text{and} \qquad \boldsymbol{n}_{kn} = SJK_{mn}\boldsymbol{n}_{jm} \tag{6}$$

Regarding notation, the J and K in SJK and the first subscripts of the unit vectors refer to bodies B_j and B_k. Repeated indices, such as n in Eq. (6), designate a sum over the range of the index.

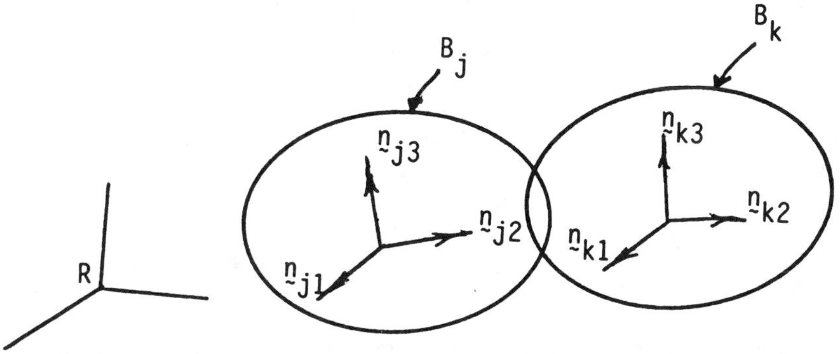

Fig. 3 Two typical adjoining bodies.

The transformation matrices are orthogonal, with the inverse equal to the transpose. They also obey the transitive relation

$$SJL = SJK\,SKL \tag{7}$$

By repeated use of Eq. (6) we can obtain a transformation matrix between the unit vectors of any given body and those of the inertia frame *R*. For example, for body 11 of the system of Fig. 2, we have

$$S011 = S01\,S16\,S67\,S79\,S911 \tag{8}$$

Observe, as in Eq. (4), that the indices of Eq. (8) correspond to the column entries of body 11 in Table 1.

The transformation matrices are useful for obtaining inertia frame components of vectors. For example, if a vector $\boldsymbol{V}$ associated with body k is expressed in the form $V_{kn}\boldsymbol{n}_{kn}$ (no sum on k), then $\boldsymbol{V}$ may also be expressed in terms of unit vectors $\boldsymbol{n}_{om}$ fixed in R as $V_{om}\boldsymbol{n}_{om}$. The relations between the components V_{kn} and V_{om} are

$$V_{om} = SJK_{mn}V_{kn} \qquad \text{and} \qquad V_{kn} = SJK_{mn}V_{om} \tag{9}$$

Transformation matrices may be constructed using a variety of different orientation angles. For example, suppose the unit vector sets $\boldsymbol{n}_{jm}$ and $\boldsymbol{n}_{kn}$ of bodies j and k are mutually aligned. Then B_k may be brought into a general orientation relative to B_j by three successive dextral rotations of B_k about $\boldsymbol{n}_{k1}$, $\boldsymbol{n}_{k2}$, and $\boldsymbol{n}_{k3}$ through the angles α_k, β_k, and γ_k. When this is done SJK may be expressed as[3]

$$SJK = AJK\,BJK\,CJK$$

$$= \begin{bmatrix} 1 & 0 & 0 \\ 0 & c_{\alpha k} & -s_{\alpha k} \\ 0 & s_{\alpha k} & c_{\alpha k} \end{bmatrix} \begin{bmatrix} c_{\beta k} & 0 & s_{\beta k} \\ 0 & 1 & 0 \\ -s_{\beta k} & 0 & c_{\beta k} \end{bmatrix} \begin{bmatrix} c_{\gamma k} & -s_{\gamma k} & 0 \\ s_{\gamma k} & c_{\gamma k} & 0 \\ 0 & 0 & 1 \end{bmatrix}$$

$$= \begin{bmatrix} c_{\beta k}c_{\gamma k} & -c_{\beta k}s_{\gamma k} & s_{\beta k} \\ (s_{\alpha k}s_{\beta k}c_{\gamma k} + s_{\gamma k}c_{\alpha k}) & (-s_{\alpha k}s_{\beta k}s_{\gamma k} + c_{\gamma k}c_{\alpha k}) & -s_{\alpha k}c_{\beta k} \\ (-c_{\alpha k}s_{\beta k}c_{\gamma k} + s_{\gamma k}s_{\alpha k}) & (c_{\alpha k}s_{\beta k}s_{\gamma k} + c_{\gamma k}s_{\alpha k}) & c_{\alpha k}c_{\beta k} \end{bmatrix} \tag{10}$$

where s and c are abbreviations for sine and cosine, respectively.

The angles α_k, β_k, and γ_k are called "dextral" orientation angles, or Bryan angles. Other commonly used orientation angles are Euler angles generated either by an $\boldsymbol{n}_{k1}$, $\boldsymbol{n}_{k3}$, $\boldsymbol{n}_{k1}$, or an $\boldsymbol{n}_{k3}$, $\boldsymbol{n}_{k1}$, $\boldsymbol{n}_{k3}$ rotation sequence. Still others can be developed by different sequences. Kane et al.[10] provide an exhaustive listing of these angles and their associated transformation matrices.

Euler Parameters

Equation (10) shows that the transformation matrix elements are nonlinear functions of the orientation angles. Such nonlinearities occur no matter how the orientation angles are chosen. Moreover, these nonlinearities are passed on to the relative angular velocity vectors (the time rates of change of the relative orientations). Indeed, if $\hat{\boldsymbol{\omega}}_k$ is the angular velocity of $\boldsymbol{B}_k$ in $\boldsymbol{B}_j$, then $\hat{\boldsymbol{\omega}}_k$ may be written as

$$\hat{\boldsymbol{\omega}}_k = (\dot{\alpha}_k + s_{\beta k}\dot{\gamma}_k)\boldsymbol{n}_{j1} + (c_{\alpha k}\dot{\beta}_k - s_{\alpha k}c_{\beta k}\dot{\gamma}_k)\boldsymbol{n}_{j2} + (s_{\alpha k}\dot{\beta}_k + c_{\alpha k}c_{\beta k}\dot{\gamma}_k)\boldsymbol{n}_{j3} \tag{11}$$

If this equation is solved for $\dot{\alpha}_k$, $\dot{\beta}_k$, and $\dot{\gamma}_k$ in terms of the relative angular velocity components, we obtain

$$\dot{\alpha}_k = \hat{\omega}_{k1} + s_{\beta k}(s_{\alpha k}\hat{\omega}_{k2} - c_{\alpha k}\hat{\omega}_{k3})/c_{\beta k} \tag{12a}$$

$$\dot{\beta}_k = c_{\alpha k}\hat{\omega}_{k2} + s_{\alpha}\hat{\omega}_{k3} \tag{12b}$$

$$\dot{\gamma}_k = (c_{\alpha k}\hat{\omega}_{k3} - s_{\alpha k}\hat{\omega}_{k2})/c_{\beta k} \tag{12c}$$

A singularity is seen to occur when β_k is 90 or 270 deg.

These difficulties can be avoided by using Euler parameters to describe the relative orientation of the bodies. Specifically, let $\boldsymbol{B}_k$ be brought into a general orientation relative to $\boldsymbol{B}_j$ by a rotation through an angle θ_k about a line fixed in both $\boldsymbol{B}_j$ and $\boldsymbol{B}_k$. Let $\boldsymbol{\lambda}_k$ be a unit vector parallel to this line. Then Euler parameters, describing the relative orientation, may be defined as[10,11]

$$\epsilon_{ki} = \lambda_{km}\sin(\theta_k/2) \quad (m = 1,2,3), \qquad \epsilon_{k4} = \cos(\theta_k/2) \tag{13}$$

where the λ_{km} are the $\boldsymbol{n}_{jm}$ components of $\boldsymbol{\lambda}_k$.

Four Euler parameters are used to describe the relative orientation of adjoining bodies, as opposed to three orientation angles. Hence, Euler parameters are not independent. Indeed, from Eq. (13) we see that

$$\epsilon_{k1}^2 + \epsilon_{k2}^2 + \epsilon_{k3}^2 + \epsilon_{k4}^2 = 1 \tag{14}$$

Any disadvantage arising from this dependency, or redundancy, is minor compared with the major advantage of having linear kinematic equations. Specifically, with Euler parameters, the $\boldsymbol{n}_{jm}$ components of $\hat{\boldsymbol{\omega}}_k$ become[2]

$$\hat{\omega}_{k1} = 2(\epsilon_{k4}\dot{\epsilon}_{k1} - \epsilon_{k3}\dot{\epsilon}_{k2} + \epsilon_{k2}\dot{\epsilon}_{k3} - \epsilon_{k1}\dot{\epsilon}_{k4}) \tag{15a}$$

$$\hat{\omega}_{k2} = 2(\epsilon_{k3}\dot{\epsilon}_{k1} + \epsilon_{k4}\dot{\epsilon}_{k2} - \epsilon_{k1}\dot{\epsilon}_{k3} - \epsilon_{k2}\dot{\epsilon}_{k4}) \tag{15b}$$

$$\hat{\omega}_{k3} = 2(-\epsilon_{k2}\dot{\epsilon}_{k1} + \epsilon_{k1}\dot{\epsilon}_{k2} + \epsilon_{k4}\dot{\epsilon}_{k3} - \epsilon_{k3}\dot{\epsilon}_{k4}) \tag{15c}$$

Solving for the $\dot{\epsilon}_{ki}$ we obtain

$$\dot{\epsilon}_{k1} = (1/2)(\epsilon_{k4}\hat{\omega}_{k1} + \epsilon_{k3}\hat{\omega}_{k2} - \epsilon_{k2}\hat{\omega}_{k3}) \tag{16a}$$

$$\dot{\epsilon}_{k2} = (1/2)(-\epsilon_{k3}\hat{\omega}_{k1} + \epsilon_{k4}\hat{\omega}_{k2} + \epsilon_{k1}\hat{\omega}_{k3}) \tag{16b}$$

$$\dot{\epsilon}_{k3} = (1/2)(\epsilon_{k2}\hat{\omega}_{k1} - \epsilon_{k1}\hat{\omega}_{k2} + \epsilon_{k4}\hat{\omega}_{k3}) \tag{16c}$$

$$\dot{\epsilon}_{k4} = (1/2)(-\epsilon_{k1}\hat{\omega}_{k1} - \epsilon_{k2}\hat{\omega}_{k2} - \epsilon_{k3}\hat{\omega}_{k3}) \tag{16d}$$

[Compare this result with the expressions of Eqs. (12).]

Finally, in terms of the Euler parameters, the transformation matrix SJK becomes

$$SJK = \begin{bmatrix} (\epsilon_{k1}^2 - \epsilon_{k2}^2 - \epsilon_{k3}^2 + \epsilon_{k4}^2) & 2(\epsilon_{k1}\epsilon_{k2} - \epsilon_{k3}\epsilon_{k4}) & 2(\epsilon_{k1}\epsilon_{k3} + \epsilon_{k2}\epsilon_{k4}) \\ 2(\epsilon_{k1}\epsilon_{k2} + \epsilon_{k3}\epsilon_{k4}) & (-\epsilon_{k1}^2 + \epsilon_{k2}^2 - \epsilon_{k3}^2 + \epsilon_{k4}^2) & 2(\epsilon_{k2}\epsilon_{k3} - \epsilon_{k1}\epsilon_{k4}) \\ 2(\epsilon_{k1}\epsilon_{k3} - \epsilon_{k2}\epsilon_{k4}) & 2(\epsilon_{k2}\epsilon_{k3} + \epsilon_{k1}\epsilon_{k4}) & (-\epsilon_{k1}^2 - \epsilon_{k2}^2 + \epsilon_{k3}^2 + \epsilon_{k4}^2) \end{bmatrix} \tag{17}$$

Differentiation Algorithm

Kane's equations employ vector quantities and the derivatives of vector quantities. Hence, in automated numerical analyses with Kane's equations, it is useful to have algorithms for computing the derivatives of vector quantities. To this end consider again a typical body B_k of a multibody system. Let $\boldsymbol{r}$ be a vector fixed in B_k. Then the derivative of $\boldsymbol{r}$ in inertial frame R may be expressed as[9]

$$\mathrm{d}\boldsymbol{r}/\mathrm{d}t = \boldsymbol{\omega}_k \times \boldsymbol{r} \tag{18}$$

This expression is an algorithm for differentiation in that the derivative is evaluated by multiplication—a useful computational procedure.

Equation (18) forms the basis for a number of differentiation algorithms. For example, for the elements of the transformation matrix SOK we have

$$\begin{aligned} \mathrm{d}(SOK_{mn})/\mathrm{d}t &= \mathrm{d}(\boldsymbol{n}_{om} \cdot \boldsymbol{n}_{kn}) = \boldsymbol{n}_{om} \cdot \mathrm{d}\boldsymbol{n}_{kn}/\mathrm{d}t = \boldsymbol{n}_{om} \cdot \boldsymbol{\omega}_k \times \boldsymbol{n}_{kn} \\ &= \boldsymbol{n}_{om} \cdot \omega_{kr}\boldsymbol{n}_{or} \times SOK_{pn}\boldsymbol{n}_{op} = e_{mrp}\omega_{kr}S_{pn} \end{aligned} \tag{19}$$

where the ω_{kr} ($r = 1,2,3$) are the $\boldsymbol{n}_{or}$ components of $\boldsymbol{\omega}_k$, and the e_{mrp} are the elements of the standard permutation symbol.[12]

In matrix form Eq. (19) can be written as

$$\mathrm{d}(SOK)/\mathrm{d}t = WOK\;SOK \tag{20}$$

where WOK is the 3×3 array with elements

$$WOK_{mp} = e_{mrp}\omega_{kr} \tag{21}$$

Dynamics

Degrees of Freedom, Coordinates, Generalized Speeds

If a multibody system has N bodies, it can have as many as $6N$ degrees of freedom. In the most general case the connecting joints will allow both translational and rotational motion. However, if the joints are more restrictive (such as allowing only rotation), the system will have fewer than $6N$ degrees of freedom. Also, if the system is subjected to constraints, such as closed loops or specified motion, there will be fewer than $6N$ degrees of freedom.

Let n ($\leqslant 6N$) be the number of these degrees of freedom, and let them be described by n parameters (or coordinates) x_ℓ ($\ell = 1, ..., n$). (These coordinates will in general include both translation and rotation variables.) Let n variables y_ℓ ($\ell = 1, ..., n$), called "generalized speeds," be introduced such that the y_ℓ are independent linear combinations of the $\dot{x}_\ell$. For example, the individual y_ℓ could be set equal to the individual $\dot{x}_\ell$, respectively. However, it is often convenient to let the y_ℓ associated with the rotational degrees of freedom be components of the relative angular velocities of the bodies. In this case there are no single x_ℓ that can be differentiated to produce a given y_ℓ. (The analysis is then said to employ "quasicoordinates.") As discussed earlier, to avoid singularities it is advantageous to describe the relative rotations of the bodies in terms of Euler parameters.

Kinematics

Consider again a typical body B_k of a multibody system S. Let G_k be the mass center of B_k. In earlier analyses[2–4] it has been shown that the velocity of G_k in inertia frame R and the angular velocity of B_k in R may be written in the forms

$$\boldsymbol{V}_k = V_{k\ell m} y_\ell \boldsymbol{n}_{om} \qquad \text{and} \qquad \boldsymbol{\omega}_k = \omega_{k\ell m} y_\ell \boldsymbol{n}_{om} \tag{22}$$

where the coefficients $V_{k\ell m}$ and $\omega_{k\ell m}$ ($k = 1, ..., N$; $\ell = 1, ..., n$; $m = 1,2,3$) are components of the "partial velocity" and "partial angular velocity" vectors originally introduced by Kane.[13]

Observe that $\boldsymbol{V}_k$ and $\boldsymbol{\omega}_k$ are linear combinations of the generalized speeds. Hence, the partial velocity and partial angular velocity vectors ($V_{k\ell m}\boldsymbol{n}_{om}$ and $\omega_{k\ell m}\boldsymbol{n}_{om}$) may be considered to be base vectors in the n-dimensional space of the generalized speeds.

By differentiating in Eq. (21) the acceleration of G_k and the angular velocity of B_k in R may be expressed as

$$\boldsymbol{a}_k = (V_{k\ell m}\dot{y}_\ell = \dot{V}_{k\ell m} y_\ell)\boldsymbol{n}_{om} \qquad \text{and} \qquad \boldsymbol{\alpha}_k = (\omega_{k\ell m}\dot{y}_\ell + \dot{\omega}_{k\ell m} y_\ell)\boldsymbol{n}_{om} \tag{23}$$

Computer algorithms have been written producing an automated numerical development of these kinematic quantities.[14,15] Specifically, efficient algorithms for the numerical evaluation of the $V_{k\ell m}$ and the $\omega_{k\ell m}$, and their derivatives, have been developed and validated.[2,3,15,16] These algorithms employ body connection arrays and the differentiation algorithm of Eq. (18).

Kinetics

Suppose a multibody system S is subjected to an externally applied force field. Let this force field be represented on typical body B_k by a single force $\boldsymbol{F}_k$ passing through G_k together with a couple with torque $\boldsymbol{M}_k$. Then, by using the procedures associated with Kane's equations,[9,10] the generalized applied ("active") force on B_k, corresponding to the generalized speed y_ℓ, may be expressed as

$$F_\ell = F_{km} V_{k\ell m} + M_{km} \omega_{k\ell m} \tag{24}$$

where F_{km} and M_{km} are the $\boldsymbol{n}_{om}$ components of $\boldsymbol{F}_k$ and $\boldsymbol{M}_k$.

Similarly, let the inertia forces on B_k be replaced by a single force $\boldsymbol{F}_k^*$ passing through G_k together with a couple with torque $\boldsymbol{M}_k^*$. $\boldsymbol{F}_k^*$ and $\boldsymbol{M}_k^*$ may be expressed as[9]

$$\boldsymbol{F}_k^* = -m_k \boldsymbol{a}_k \quad \text{and} \quad \boldsymbol{M}_k^* = -\boldsymbol{I}_k \cdot \boldsymbol{\alpha}_k - \boldsymbol{\omega}_k \times (\boldsymbol{I}_k \cdot \boldsymbol{\omega}_k) \quad (\text{no sum on } k) \tag{25}$$

where m_k and $\boldsymbol{I}_k$ are the mass and central inertia dyadic of B_k. Then the generalized inertia ("passive") force on B_k, corresponding to the generalized speed y_ℓ, may be expressed as

$$F_\ell^* = F_{km}^* V_{k\ell m} + M_{km}^* \omega_{k\ell m} \tag{26}$$

where F_{km}^* and M_{km}^* are the $\boldsymbol{n}_{om}$ components of $\boldsymbol{F}_k^*$ and $\boldsymbol{M}_k^*$.

Finally, if the system S is subjected to constraints, there will be constraining forces on the bodies of S. As before, let these forces be represented on B_k by a single force $\boldsymbol{F}_k'$ passing through G_k together with a couple with torque $\boldsymbol{M}_k'$. Then the generalized constraint force on B_k, corresponding to the generalized speed y_ℓ, may be expressed as

$$F_\ell' = F_{km}' V_{k\ell m} + M_{km}' \omega_{k\ell m} \tag{27}$$

where F_{km}' and M_{km}' are the $\boldsymbol{n}_{om}$ components of $\boldsymbol{F}_k'$ and $\boldsymbol{M}_k'$.

Dynamic Equations (Kane's Equations)

Using Kane's equations the governing dynamic equations for a multibody system are

$$F_\ell + F_\ell^* + F_\ell' = 0 \tag{28}$$

Using Eqs. (22–27) these equations may be expressed as

$$a_{\ell p}\dot{y}_p = f_\ell + F'_\ell \tag{29}$$

where the coefficients $a_{\ell p}$ and f_ℓ are

$$a_{\ell p} = m_k V_{km} V_{kpm} + I_{kmn}\omega_{k\ell m}\omega_{kpn} \tag{30}$$

and

$$f_\ell = F_\ell - (m_k V_{k\ell m} V_{kpm} y_p + I_{kmn}\omega_{k\ell m}\dot{\omega}_{kpn} y_p + e_{rsm} I_{ksn}\omega_{k\ell m}\omega_{kqr}\omega_{kpn} y_p y_q) \tag{31}$$

where I_{kmn} are the $\boldsymbol{n}_{om}$, $\boldsymbol{n}_{on}$ components of $\boldsymbol{I}_k$.

In matrix form Eqs. (28) and (29) may be expressed as

$$F + F^* + F' = 0 \tag{32}$$

and

$$A\dot{y} = f + F' \tag{33}$$

Constraint Equations

As noted earlier, a multibody system may be constrained in several ways: 1) there may be constraints at the joints; 2) there may be closed loops (geometric constraints); 3) and there may be specified motion for some points or bodies of the system (kinematic constraints). These constraints can frequently be modeled in terms of constraint equations of the form

$$b_{q\ell} y_\ell = g_q, \qquad q = 1, \ldots, m \qquad (m < n) \tag{34}$$

where the coefficients $b_{q\ell}$ and g_q are, in general, functions of x_ℓ (or Euler parameters) and time t. In matrix form these equations may be written as

$$By = g \tag{35}$$

When the system is constrained as in Eq. (34), the y_ℓ are no longer independent. Indeed, only $n - m$ of the y_ℓ are independent; thus, there are $n - m$ degrees of freedom.

When a multibody system is constrained, constraint forces and moments on the bodies occur. If the components of these forces and moments can be arranged into a $m \times 1$ array λ, then the bases for λ and B can be chosen so that the generalized constraint force array F' is related to λ and B as[7]

$$F' = B^T\lambda \tag{36}$$

Solution Methods

From Eqs. (32), (33), and (36) the governing dynamic and constraint equations may be written as

$$A\dot{y} = f + B^T\lambda \tag{37a}$$

and

$$By = g \tag{37b}$$

where, as before, A is an $n \times n$ array and B is an $m \times n$ array. These equations are equivalent to a set of $m + n$ scalar equations for the n y_ℓ and the m elements of the λ array. They form a system of coupled differential/algebraic equations. As such, they are not readily adaptable to a differential equation solver. However, they may be converted into a "solvable" system by eliminating the λ array. To this end, let C be an orthogonal complement of B; that is, let C be an $n \times (n - m)$ array such that

$$BC = 0 \qquad \text{or} \qquad C^T B^T = 0 \tag{38}$$

Then, by premultiplying Eq. (37a) by C^T, the governing equations become

$$C^T A\dot{y} = C^T f \qquad \text{and} \qquad By = g \tag{39}$$

Equations (39) form a set of $(n - m) + m$, or n, equations for the ny_ℓ.

Kane's Triangle

Equation (32) may be viewed as a force triangle in the n-dimensional space of the generalized speeds, as depicted in Fig. 4.

Let "reduced" generalized applied and inertia force arrays, K and K^*, be defined as

$$K = C^T F \qquad \text{and} \qquad K^* = C^T F^* \tag{40}$$

where, as before, C is an orthogonal complement of the constraint matrix B. Then, in view of Eqs. (32) and (36) we have

$$K + K^* = 0 \tag{41}$$

This is the traditional form of Kane's equations as introduced in 1961.[13]

Since C^T is "orthogonal" to the constraint matrix B^T, C^T defines the "freedom directions" of the system. Hence, K and K^* may be interpreted as projections of F and F^* onto these freedom directions as in Fig. 4.

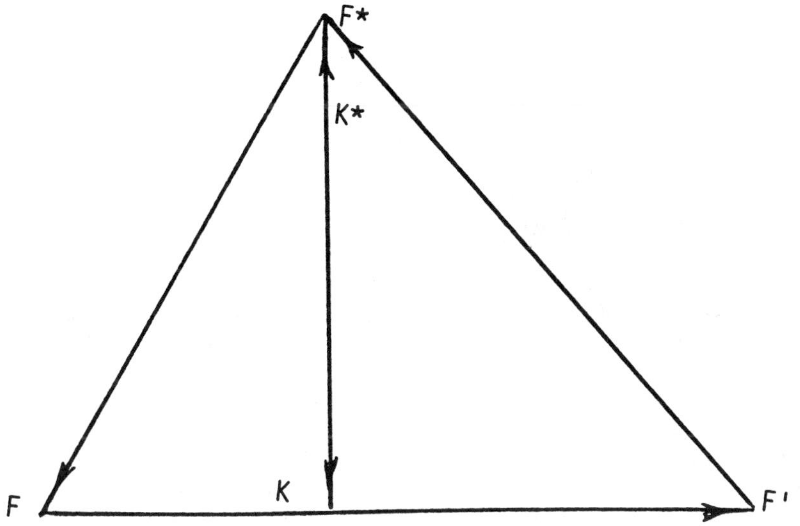

Fig. 4 Kane's force triangle.

Numerical Analyses

A number of computer programs have been written to simulate the dynamics of multibody systems by using the foregoing analysis.[14,15,17–21] The success of these efforts is based on the explicit form of the dynamic equations. Observe in Eqs. (30) and (31) that the elements of the arrays a and f are explicit functions of the arrays $\omega_{k\ell m}$, $\dot{\omega}_{k\ell m}$, $V_{k\ell m}$, and $\dot{V}_{k\ell m}$. Hence, the governing dynamic equations can be generated once these four arrays are determined. Reference 16 contains explicit algorithms for the development of these arrays. The differentiation algorithm [Eqs. (18) and (20)] is used to obtain the array derivatives.

By employing generalized speeds in the analysis, the dynamics equations occur as a system of first-order ordinary differential equations. As such, they are in a form that is readily adaptable for numerical solution. Also, since the dynamic equations involve generalized speeds (relative angular velocity components) and Euler parameters as dependent variables, it is necessary to append Eqs. (16) to the dynamic equations to form the equations to be integrated.

If, in addition, the multibody system is constrained, the system of constraint equations needs to be satisfied and integrated as well. As noted earlier, this requires the use of an orthogonal complement of the constraint array to convert the equations to a consistent and solvable set. There are several ways of obtaining an orthogonal complement array. These include the use of singular value decomposition[23] and the use of the zero eigenvalues of the array B^TB.[5,6,22]

The references include a number of examples of numerical solutions of the governing equations for both constrained and unconstrained systems. It has been this writer's experience that the most stable numerical integrators are those based on a Runge-Kutta (power series) algorithm.

Discussion

The central feature of these procedures is the use of Kane's equations to obtain the dynamic equations. The utility of Kane's equations is enhanced by the auxiliary procedures of the body connection arrays, the differentiation algorithm, partial velocities, generalized speeds, and the use of Euler parameters.

Interestingly, when Kane's equations were introduced in 1961 they were viewed as a Lagrangian method for studying nonholonomic systems. At that time computer analyses were not practicable, and there was little if any interest in multibody dynamics. However, the very features that make Kane's equations useful for nonholonomic systems are the same features that make them effective in multibody system analyses. Specifically, in Eqs. (24–28) it is seen that Kane's equations are a vector approach based on the projection of forces and torques along the freedom directions defined by the partial velocity and partial angular velocity vectors. As noted, the partial velocity vectors are base vectors in the n-dimensional space of the generalized speeds. They determine the coordinate directions of the system. By projecting the applied forces along these directions, the "nonworking" constraint forces are eliminated from the analysis. These projections produce the generalized applied (or "active") forces commonly used in Lagrangian analyses. With Kane's equations, however, the process is carried further to include the projection of the inertia forces as well. The principal benefit gained by this projection is the avoidance of differentiating a kinetic energy function, a tedious and arduous task for large multibody systems. With Kane's equations the differentiation can be obtained via multiplication using the differentiation algorithm.

The simplicity and balance nature of Kane's equations allow the analysis to be developed so that an explicit form of the governing dynamic equations is obtained, as in Eqs. (29–31). Indeed, the coefficients of these equations are directly dependent on the partial velocity and partial angular velocity arrays and their derivatives. Hence, the entire analysis may be automated by using numerical algorithms for determining these arrays and their derivatives.[16]

Finally, as interest in multibody dynamics increases, it is expected that the use of Kane's equations to conduct the analyses will increase as well. It is believed by this writer that this in turn will lead to a deepening understanding of the behavior of multibody systems.

Acknowledgment

Research leading to the development of these concepts was partially supported by National Science Foundation Grant MSM-8612970.

References

[1]Huston, R. L., "Use of Kane's Equations in Multibody Dynamics," *Proceedings, Twenty-first Midwestern Mechanics Conference*, Michigan Technological University, Houghton, MI, 1989.

[2]Huston, R. L., Passerello, C. E., and Harlow, M. W., "Dynamics of Multirigid Body Systems," *Journal of Applied Mechanics*, Vol. 45, 1978, pp. 889–894.

[3]Huston, R. L., and Passerello, C. E., "On Multi-Rigid Body-Systems Dynamics," *Computers and Structures*, Vol. 10, 1979, pp. 439–446.

[4]Huston, R. L., and Passerello, C. E., "Multibody Structural Dynamics Including Translation Between the Bodies," *Computers and Structures*, Vol. 11, 1980, pp. 713–720.

[5]Kamman, J. W., and Huston, R. L., "Constrained Multibody System Dynamics—An Automated Approach," *Computers and Structures*, Vol. 18, 1984, pp. 999–1003.

[6]Kamman, J. W., and Huston, R. L., "Dynamics of Constrained Multibody Systems," *Journal of Applied Mechanics*, Vol. 51, 1984, pp. 899–904.

[7]Wang, J. T., and Huston, R. L., "Kane's Equations with Undetermined Multipliers—Application with Constrained Multibody Systems," *Journal of Applied Mechanics*, Vol. 54, 1987, pp. 424–429.

[8]Huston, R. L., King, T. P., and Kamman, J. W., "UCIN-DYNOCOMBS—Software for the Dynamic Analysis of Constrained Multibody Systems," *Multibody System Handbook*, edited by W. Schiehlen, Springer-Verlag, Berlin, 1989, Chap. 4.

[9]Kane, T. R., and Levinson, D. A., *Dynamics: Theory and Applications*, McGraw-Hill, New York, 1985.

[10]Kane, T. R., Likins, P. W., and Levinson, D. A., *Spacecraft Dynamics*, McGraw-Hill, New York, 1983, pp. 12, 422–431.

[11]Whittaker, E. T., *Analytical Dynamics*, Cambridge Univ. Press, London, 1937.

[12]Brand, L., *Vector and Tensor Analysis*, Wiley, New York, 1947.

[13]Kane, T. R., "Dynamics of Nonholonomic Systems," *Journal of Applied Mechanics*, Vol. 28, 1961, pp. 574–578.

[14]Kamman, J. W., and Huston, R. L., "User's Manual for UCIN DYNOCOMBS-II," National Technical Information Service, Springfield, VA, Rept. PB87-216594/A05, 1987.

[15]Huston, R. L., King, T. P., and Kamman, J. W., "UCIN-DYNOCOMBS—Software for the Dynamic Analysis of Constrained Multibody Systems," *Multibody System Handbook*, edited by W. Schiehlen, Springer-Verlag, Berlin, 1989, Chap. 6.

[16]Huston, R. L., "Computing Angular Velocity in Multibody Systems," *Engineering Computations*, Vol. 3, 1986, pp. 223–230.

[17]Kamman, J. W., and Huston, R. L., "User's Manual for UCIN-CABLE III—A Two-Dimensional Finite Segment Computer Code for Submerged and Partially Submerged Cable Systems," Naval Coastal Systems Center Contractor Rept. 105-84, 1984.

[18]Huston, R. L., Harlow, M. W., and Gausewitz, N. L., "User's Manual for UCIN-EULER, A Multipurpose, Multibody Systems Dynamics Computer Program," National Technical Information Service, Springfield, VA, Rept. AD-A120403, 1982.

[19]Huston, R. L., Passerello, C. E., Harlow, M. W., and Winget, J. M., "User's Manual for UCIN-SUPER: CHAIN SYSTEMS," National Technical Information Service, Springfield, VA, Rept. PB2459, 1975.

[20]Huston, R. L., Hessel, R. E., and Winget, J. M., "Dynamics of a Crash-Victim—A Finite Segment Model," *AIAA Journal*, Vol. 14, Feb. 1976, pp. 173–178.

[21]Huston, R. L., Huston, J. C., and Harlow, M. W., "A Comprehensive, Three-Dimensional Head-Neck Model for Impact and High Acceleration Studies," *Aviation, Space and Environmental Medicine*, Vol. 49, 1978, pp. 205–210.

[22]Walton, W. C. Jr., and Steeves, E. C., "A New Matrix Theorem and Its Application for Establishing Independent Coordinates for Complex Dynamical Systems with Constraints," NASA TR-326, 1969.

[23]Lawson, C. L., and Hanson, R. J., *Solving Least Squares Problems*, Prentice-Hall, Englewood Cliffs, NJ, 1974.

Chapter 5

Input/Output System Identification: Learning from Repeated Experiments

Jer-Nan Juang,* Lucas G. Horta†
NASA Langley Research Center, Hampton, Virginia
and
Richard W. Longman‡
Columbia University, New York, New York

Introduction

CONTROL systems are often used to execute the same command repeatedly, for example, the controller for each joint of a robot performing a task on an assembly line. In tracking control problems, the controller, even under ideal circumstances, will execute only an approximate version of the command given repeatedly. It is natural to use a more sophisticated controller that can learn from past experience in order to do a better job of executing a command each time it is given the command.

Such learning control systems have received considerable attention in the recent literature.[1-10] Learning control differs from adaptive control in that it explicitly takes advantage of the knowledge of the repetitive nature of the process and the command. The controller changes as a function of repetition as opposed to sample time. Thus, the repetition number is thought of as a domain. In indirect learning control[10] a system identification step in the repetition domain is involved in generating the control law. As in adaptive control, it is not necessarily required that the identification step converge in order to have the controller accomplish its goal of zero tracking error. Here

---

*Principal Scientist, Spacecraft Dynamics Branch.
†Aerospace Technologist, Spacecraft Dynamics Branch.
‡Professor of Mechanical Engineering, Columbia University.

we study system identification in the repetition domain, and in order to ensure convergence of the identified parameters, we must change the command from one repetition to the next. The result might be thought of as performing a series of experiments varying the command from one experiment to the next.

In Ref. 11 a learning identification based on an error-correcting training procedure is presented for the case of noise-free data. Reference 12 gives a general formulation for the learning identification of linear discrete systems in the repetition domain with unknown repetitive disturbances possibly present. These two schemes make use of error projection procedures from adaptive control. The fundamental difference between the two is that Ref. 11 deals with the output data directly, whereas Ref. 12 uses differences between successive experiments and is formulated for time-varying systems. This paper presents three approaches, using learning ideas, for the identification of Markov parameters. The first two approaches are direct solutions of the Markov parameters from repetitive experiments. The first solution arranges the input/output data to take advantage of the system equation structure due to causality. Essentially, once a solution is obtained for the first parameter, it is used to solve for the second, and the process continues until all parameters are evaluated. This bootstrapping procedure is computationally efficient but it is prone to large propagation errors. The second approach uses singular value decomposition for the solution of all parameters at once. This form has no propagation errors, but it could require a large matrix decomposition. The third approach is an application of a scheme presented in Ref. 12 where the Markov parameters are identified recursively in the repetition domain. Each time an experiment is repeated, an updated set of parameters is obtained with guaranteed reduction in the errors between the estimated parameters and their true values.

All of the approaches aforementioned solve for as many Markov parameters as there are sample points in the experiment. To describe the output of an nth-order system, $n + 1$ parameters are needed. References 13–19 present procedures for the minimum-order realization of systems from input/output data. Reference 19 combines realization theory with singular value decomposition into a computationally efficient algorithm that has been successfully applied in the identification of large structures in the presence of noise. The procedure uses free decay data in the identification of the system matrices. Extending this algorithm, the Markov parameters for forced response data are evaluated by the procedure developed, and then the realization of the system is performed, as in Ref. 19, to determine the system order and the input/output model description.

Numerical results for all three approaches are presented for a simple discrete time system. For noise-free data the exact Markov parameters are obtained in a predefined number of experiment repetitions. From the numerical simulation in the presence of noise, approach 2, which uses singular value decomposition of the input data, yields the most accurate results in the least number of repetitions. The additional computational requirement of this method is often of no concern because the computations can be performed between experiments.

Problem Formulation

A finite-dimensional, discrete-time, time-invariant dynamic system has the state variable equations

$$x(k+1) = Ax(k) + Bu(k) \tag{1a}$$

$$y(k) = Cx(k) + Du(k) \tag{1b}$$

where x is an n-dimensional state vector, u is an r-dimensional control input, and y is a q-dimensional output or measurement vector. The integer k is the sample indicator. The matrix A is the discrete system matrix, and B is the input influence matrix. The identification problem is to find $[A,B,C,D]$ from experimental data. Once this is accomplished, one can obtain a continuous time realization from $[A,B,C,D]$ if desired.

Repeated substitution of Eqs. (1) into itself for different time steps produces

$$y(k) = \sum_{i=1}^{k} CA^{i-1}Bu(k-i) + CA^k x(0) + Du(k) \tag{2}$$

The Markov parameters for the discrete system in Eqs. (1) are

$$P_0 = D \qquad \text{and} \qquad P_i = CA^{i-1}B \qquad \text{for} \qquad i = 1, 2, ..., \infty \tag{3}$$

where P_0 is dimension $q \times r$. The response of the system in terms of the Markov parameters is

$$y(k) = \sum_{i=0}^{k} P_i u(k-i) + CA^k x(0) \tag{4}$$

including initial conditions. For identification purposes, if the system is not relaxed at the initial time but it is controllable and observable, then there exists an input sequence $u(k)$ for $k = \{-n_0, -n_0+1, -n_0+2, ..., -1\}$ such that the system is relaxed at $k = -n_0$. For time-invariant systems the time reference frame can be shifted without altering the system dynamics. Furthermore, the Markov parameters completely define the system dynamics. If the initial conditions of the system need to be identified, Eq. (1a) could be altered as

$$x(k+1) = Ax(k) + [Bx_0]\begin{bmatrix} u(k) \\ \delta_{-1,k} \end{bmatrix} \tag{5}$$

where now $x(-1) = 0$, $u(-1) = 0$, $\delta_{-1,-1} = 1$ and 0 otherwise, and x_0 is the vector of initial conditions. Note that the system realization of $[A,[B\,x_0],C,D]$ obtained by using Eq. (5) does not necessarily produce a minimum realization of $[A,B,C,D]$, since x_0 may be exciting some portion of the system that is not excited by the input u.

The mathematical formulation for the learning identification problem was presented in Ref. 12 for general time-varying systems. The concept of learning identification used here requires that succesive experiments be performed starting from the same initial state, either known or unknown. Using this idea, a system whose output is given by Eq. (4) can be driven by input $\boldsymbol{u}_j(k)$ in experiment j. The identification problem requires that the set of the input sequences $\boldsymbol{u}(k)$ be sufficiently rich that all the system parameters participate in the output sequence. In the time-invariant case the solution for the Markov parameters can be obtained in several different ways. Let the changes in the output and input sequences for two repetitive experiments be given by

$$\delta_j \boldsymbol{y}(k) = \boldsymbol{y}_j(k) - \boldsymbol{y}_{j-1}(k); \qquad \delta_j \boldsymbol{u}(k) = \boldsymbol{u}_j(k) - \boldsymbol{u}_{j-1}(k) \tag{6}$$

where $\boldsymbol{y}_j(k)$ and $\boldsymbol{u}_j(k)$ are the system response and input at time step k for the jth repetition. Using Eqs. (4) and (6) the system output differences for two repetitions with p samples each are

$$\begin{bmatrix} \delta_j \boldsymbol{y}(0) \\ \delta_j \boldsymbol{y}(1) \\ \vdots \\ \delta_j \boldsymbol{y}(p-1) \end{bmatrix} = \begin{bmatrix} P_0 & 0 & \cdots & 0 \\ P_1 & P_0 & \cdots & 0 \\ \vdots & \vdots & & \vdots \\ P_{p-1} & P_{p-2} & \cdots & P_0 \end{bmatrix} \begin{bmatrix} \delta_j \boldsymbol{u}(0) \\ \delta_j \boldsymbol{u}(1) \\ \vdots \\ \delta_j \boldsymbol{u}(p-1) \end{bmatrix} \tag{7}$$

which is in lower block triangular form due to causality. The matrix of parameters in Eq. (7) is of dimensions $pq \times pr$, which is a rather large matrix depending on the number of sample points taken in each repetition. In the following, three different approaches to obtain the Markov parameters are discussed.

Approach 1

Taking advantage of the lower triangular form of the matrix in Eq. (7), the parameters can be evaluated by bootstrapping the solution, i.e., solving for P_0 using data from the first sample point of each experiment and substituting the result to obtain P_1 from the second sample point, etc. Define the following matrices for s repetitions $Y(0) = [\delta_1 \boldsymbol{y}(0)\ \delta_2 \boldsymbol{y}(0) \ldots \delta_s \boldsymbol{y}(0)]$ and $U(0) = [\delta_1 \boldsymbol{u}(0)\ \delta_2 \boldsymbol{u}(0) \ldots \delta_s \boldsymbol{u}(0)]$ of dimensions $q \times s$ and $r \times s$, respectively, for $s \geqslant r$. The first Markov parameter is given by

$$P_0 = Y(0)U(0)^+ \tag{8}$$

where $U(0)^+$ is the pseudoinverse defined by $U(0)U(0)^+U(0) = U(0)$. For r inputs to the system there can be at most r-independent columns of $U(0)$, for $s \geqslant r$. When the experiments are designed, the input sequences must be chosen such that the first r columns of $U(0)$ have full column rank. In general the kth Markov parameter can be written as

$$P_k = \left\{ Y(k) - [P_0 \quad P_1 \quad \dots \quad P_{k-1}] \begin{bmatrix} U(k) \\ U(k-1) \\ \vdots \\ U(1) \end{bmatrix} \right\} U(0)^+ \tag{9}$$

This approach is attractive because only one decomposition of small dimension is done initially and then used in the solution of all the parameters. Equation (9) evaluates k parameters, where $1 < k < p - 1$, which are not all necessarily independent. Equation (9) has a very simple structure that permits easy computation of as many Markov parameters as there are sample points in the experiment. The disadvantage of this approach is that $P_0, P_1, \dots, P_j$ are obtained based on a reduced data set, and inaccuracies in these results propagate to P_{j+1}.

Approach 2

Improved error properties can be obtained if all Markov parameters are evaluated simultaneously using all data. Let

$$\underline{U}_i = \begin{bmatrix} \delta_i \boldsymbol{u}(0) & \delta_i \boldsymbol{u}(1) & \dots & \delta_i \boldsymbol{u}(p-1) \\ 0 & \delta_i \boldsymbol{u}(0) & \dots & \delta_i \boldsymbol{u}(p-2) \\ \vdots & 0 & & \vdots \\ & \vdots & \ddots & \vdots \\ 0 & 0 & & \delta_i \boldsymbol{u}(0) \end{bmatrix} \tag{10}$$

$$\underline{Y} = [\delta_1 y(0) \quad \delta_1 y(1) \quad \dots \quad \delta_1 y(p-1) | \delta_2 y(0) \quad \delta_2 y(1) \quad \dots \quad \delta_2 y(p-1)| \quad \dots \quad |\delta_s y(0) \quad \delta_s y(1) \quad \dots \quad \delta_s y(p-1)]$$

The equation for the Markov parameters using Eq. (10) is

$$\underline{Y} = [P_0 \quad P_1 \quad \dots \quad P_{p-1}] W \tag{11}$$

where $W = [\underline{U}_1 \ \underline{U}_2 \ \underline{U}_3 \ \dots \ \underline{U}_s]$ is of the dimension $rp \times ps$. Using a pseudoinverse W^+ defined such that $WW^+W = W$, the solution for the Markov parameters is

$$[P_1 \quad P_1 \quad \dots \quad P_{p-1}] = \underline{Y} W^+ \tag{12}$$

The resulting parameters avoid the propagation of errors in Eq. (9) at the expense of a singular value decomposition of a rather large dimensional matrix W.

An alternative to this approach combines approaches 1 and 2 by solving for successive sets of P_j from approach 2 and substituting these results into a generalized form of approach 1. This allows one to adjust the size of the decomposition with a tradeoff in accuracy. For example, one can use approach 2 to solve for P_0 and P_1 simultaneously and then use a generalized approach 1 to obtain P_2 and P_3 simultaneouly with a decomposition whose dimension is doubled.

Approach 3

The first two approaches solve for the parameters in r repetitions. Reference 12 presents schemes to update the parameters recursively. Recursive algorithms are frequently used in adaptive control for identification of the system parameters where the basic idea is to update the parameters in such a way as to make the error vector orthogonal to the parameter plane. A least-squares algorithm presented in Ref. 12 can be used for the identification of the Markov parameters. Let an mth row of the parameter matrix in Eq. (7) with all the identically zero terms removed be written as a column vector:

$$\boldsymbol{R}_m(k) = [\hat{P}_{kq+m,1} \quad \hat{P}_{kq+m,2} \quad \cdots \quad \hat{P}_{kq+m,(k+1)r}]^T \tag{13}$$

The subscripts indicate the position of the element in the parameter matrix, and (^) refers to estimated values. The $(k+1)r \times 1$ input matrix is

$$\delta_j \underline{U}(k) = [\delta_j \boldsymbol{u}^T(0) \quad \delta_j \boldsymbol{u}^T(1) \quad \ldots \quad \delta_j \boldsymbol{u}^T(k)]^T \tag{14}$$

for $0 < k < p - 1$. Using Eqs. (13) and (14) the mth row of Eq. (7) is written using the preceding notation as

$$\delta_j y_m(k) = \boldsymbol{R}_m^T(k) \delta_j \underline{U}(k) \tag{15}$$

The parameter vector is evaluated recursively using

$$\boldsymbol{R}_{m,j+1}(k) = \boldsymbol{R}_{m,j}(k) + \alpha_{j-1}(k)\, \Delta\boldsymbol{R}_{m,j+1}(k) \tag{16}$$

with

$$\Delta\boldsymbol{R}_{m,j+1}(k) = \frac{a_j \delta_{j+1} \underline{U}(k)[\delta_{j+1} y_m(k) - \delta_{j+1} \underline{U}^T(k) \boldsymbol{R}_{m,j}(k)]}{1 + \delta_{j+1} \underline{U}^T(k) \alpha_{j-1}(k) \delta_{j+1} \underline{U}(k)} \tag{17}$$

and

$$\alpha_j(k) = \alpha_{j-1}(k) - \frac{\alpha_{j-1}(k) \delta_{j+1} \underline{U}(k) \delta_{j+1} \underline{U}^T(k) \alpha_{j-1}(k)}{1 + \delta_{j+1} \underline{U}^T(k) \alpha_{j-1}(k) \delta_{j+1} \underline{U}(k)} \tag{18}$$

for $j \geqslant 0$, $0 < a_j < 2$, and $\alpha_{-1} = I_{(k+1)r}$. Initially, large changes in the parameters are expected when the initial guess is far from the true value. To accelerate convergence the $\alpha_j(k)$ matrix can be set to any arbitrarily large positive-definite matrix. The numerical results that will be shown take advantage of this idea.

A limitation of this approach is that output data from only one time step is used to estimate all of the parameters. An improved version of approach 3 is to construct the Auto Regressive Moving Average (ARMA) model involved in the recursive least-squares algorithm from the rows of Eq. (11). Like approach 2, this allows all data to be used in calculating each P_j for improved accuracy.

System Realization

The formulation thus far evaluates p Markov parameters, where p is the number of sample points. Knowledge of the first $n + 1$ Markov parameters (with zero initial conditions) is sufficient to determine the complete output in Eq. (4) for an nth-order system. The problem of minimum-order realization has been the subject of numerous papers. The mathematical formulation was initially presented in Ref. 13 and was later on refined in Refs. 14–19. For system realization the procedure begins by defining a Hankel matrix (of order $qi \times ri$) of the form

$$H(k-1,i) = \begin{bmatrix} P_k & P_{k+1} & \cdots & P_{k+i} \\ P_{k+1} & P_{k+2} & \cdots & P_{k+i+1} \\ \vdots & \vdots & & \vdots \\ P_{k+i} & P_{k+i+1} & & P_{k+2i} \end{bmatrix} \tag{19}$$

for $k + 2i < p$. Let us define a $q \times qi$ matrix $E_q^T = [I_q | \underline{0}]$ where I_q is a $q \times q$ identity matrix and $\underline{0}$ is a null matrix with proper dimensions. Similarly, let $E_r^T = [I_r | \underline{0}]$ be an $r \times ri$ matrix. Using these an equivalent expression for the Markov parameters is,

$$P_k = E_q^T H(k-1,i) E_r, \qquad k = 1, 2, \ldots, \infty \tag{20}$$

Comparing Eqs. (20) and (3), the identification problem now is to extract a minimum order realization of the triple system matrices $[A,B,C]$ with $D = P_0$. Decomposing $H(0,i)$ using singular value decomposition,[20]

$$H(0,i) = K \begin{bmatrix} \Sigma & 0 \\ 0 & 0 \end{bmatrix} L \tag{21}$$

where $\Sigma = \text{diag}(\sigma_1, \sigma_2, \ldots, \sigma_n)$, and σ_j are the positive square roots of the eigenvalues of $H(0,i)^T\, H(0,i)$. The matrices K and L of order $qi \times qi$ and $ri \times ri$, respectively, are isometric, i.e., $K^T K = I_{qi}$ and $L^T L = I_{ri}$. When noise-corrupted data are used, the lower diagonal terms of Eq. (21) are small nonzero values. The order of the system is then determined by looking for a step decrease in the magnitude of the singular values[21] or by more subtle means as in Ref. 22. Using only those columns and rows of K and L that multiply the nonzero singular values, the matrix product can be written as

$$H(0,i) = K_1 \Sigma L_1 = K_1 \Sigma^{1/2} \Sigma^{1/2} L_1 = VU \tag{22}$$

The matrix $V = K_1\Sigma^{1/2}$ is of order $qi \times n$ and $U = \Sigma^{1/2}L_1$ is of order $n \times ri$. With the aid of the following matrices,

$$V^+ = \Sigma^{-1/2}K_1^T; \quad U^+ = L_1^T\Sigma^{-1/2} \tag{23}$$

where $K_1^TK_1 = I_n$ and $L_1L_1^T = I_n$, a discrete-time minimum-order realization of the system is given by

$$A = V^+H(1,i)U^+ \tag{24a}$$

$$B = UE_r^T \tag{24b}$$

$$C = E_qV \tag{24c}$$

$$D = P_0 \tag{24d}$$

This realization can be used to obtain a z transfer function description or to convert to a continuous time model.[19] The procedure described herein has been successfully applied to the identification of large flexible structures in the presence of noise.[19]

Numerical Example

To demonstrate the procedures presented in the preceding sections, a simple numerical example is chosen from Ref. 23. The model consists of a second-order discrete-time system with two inputs ($r = 2$) and two outputs ($q = 2$). The governing difference equation is

$$\boldsymbol{x}(k+1) = \begin{bmatrix} 0 & 1 \\ 1 & 0 \end{bmatrix} \boldsymbol{x}(k) + \begin{bmatrix} 1 & 1 \\ -0.5 & 1 \end{bmatrix} \boldsymbol{u}(k) \tag{25a}$$

$$\boldsymbol{y}(k) = \begin{bmatrix} 1 & 0 \\ 2/3 & 4/3 \end{bmatrix} \boldsymbol{x}(k) + \begin{bmatrix} 0 & 0 \\ 1 & 2 \end{bmatrix} \boldsymbol{u}(k) \tag{25b}$$

The system response is obtained by propagating the response starting from zero initial state using Eq. (4). The system eigenvalues are easily evaluated to be $(\lambda_1,\lambda_2) = (1,-1)$. The first four Markov parameters are obtained using Eqs. (3) and (25):

$$P_0 = \begin{bmatrix} 0 & 0 \\ 1 & 2 \end{bmatrix}, \quad P_1 = \begin{bmatrix} 1 & 1 \\ 0 & 2 \end{bmatrix}, \quad P_2 = \begin{bmatrix} -1/2 & 1 \\ 1 & 2 \end{bmatrix}, \quad P_3 = P_1 \tag{26}$$

A sequence of simulated experiments is performed using different random inputs with normal distribution (zero mean and 1.0 variance). The input/output data are numerically generated and used with each of the three different approaches for parameter determination. For each simulated experiment or repetition, 20 sample points are used to identify 20 Markov parameters. Note that, since the order of the system in the preceding example is two, three parameters are enough to describe the output. This knowledge are not known a priori; therefore, additional parameters should be retained for purposes of determining the system order and smoothing the effects of noise. Initially, noise-free data are used to verify the formulation. With the first two approaches the parameters are identified exactly in two successive experiments. For the recursive approach generically the least number of successive experiments required is k (number of sample points) times r (number of inputs), which for this case is 40. For noise-free data the exact Markov parameters can be identified with any of the three approaches.

To examine noise effects, random signals from a normal distribution with 0 mean and 0.01 variance were added to the output signals of the system. Results using approach 1 with $k = 20$ are shown in Fig. 1. The error convergence histories for the first four Markov parameters as a function of the repetition number are shown. The parameter error matrix e_i $(i = 1, 2, 3)$ is the difference between the estimated parameters and their true values P_i given in Eq. (3). Its spectral norm denoted by $\|e_i\|_2$ is used to measure the accuracy of estimated values.

The bootstrapping (approach 1) has poor error propagation properties because subsequent parameters depend on previous parameters that are in error. The parameters in repetition 10 produce the following system matrices:

$$A = \begin{bmatrix} 1.004 & -0.0024 \\ -0.0005 & -0.9984 \end{bmatrix}, \quad B = \begin{bmatrix} 0.3709 & 1.4721 \\ -0.9499 & 0.0002 \end{bmatrix}$$
$$C = \begin{bmatrix} 0.6780 & -0.7935 \\ 1.3584 & 0.5220 \end{bmatrix}, \quad D = \begin{bmatrix} -0.0014 & 0.0008 \\ 0.9977 & 1.9962 \end{bmatrix} \tag{27}$$

The corresponding singular values with the largest normalized to 1 are $\Sigma = \{1.0, 0.39, 0.004, 0.002, \ldots\}$ and the identified eigenvalues are $(\lambda_1, \lambda_2) = (0.9999, -0.9999)$. Only the first two nonzero values are retained. Note that the triple $[A,B,C]$ in Eq. (27) is different from that in Eq. (25) because the triple $[A,B,C]$ is not uniquely determined. However, the Markov parameters of the identified system and the eigenvalues are in agreement.

The same data are now used with approach 2 and the error histories are depicted in Fig. 2. Note that the error for the parameters shown in Fig. 2 are significantly less than those shown in Fig. 1. This is because approach 2 uses information from all sample points in the estimation of all the parameters. This effectively filters out some of the noise in the data. The converged parameters for repetition 10 produce the system matrices:

$$A = \begin{bmatrix} 0.9999 & -0.0001 \\ 0.0000 & -0.9999 \end{bmatrix}, \quad B = \begin{bmatrix} 0.3685 & 1.4728 \\ -0.9493 & 0.0004 \end{bmatrix}$$
$$C = \begin{bmatrix} 0.6794 & -0.7897 \\ 1.3577 & 0.5268 \end{bmatrix}, \quad D = \begin{bmatrix} -0.0002 & 0.0021 \\ 1.0024 & 1.9986 \end{bmatrix} \tag{28}$$

The system order is correctly identified as in the previous case to be 2.

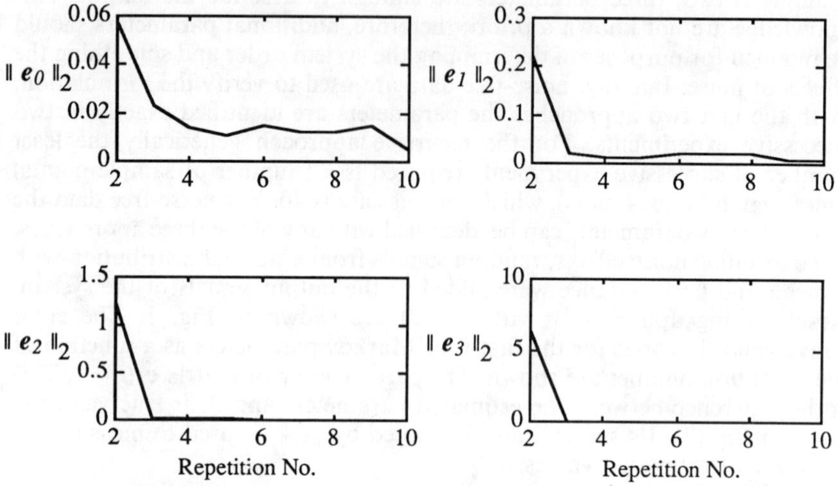

Fig. 1 Error histories using bootstrapping approach to determine first four Markov parameters.

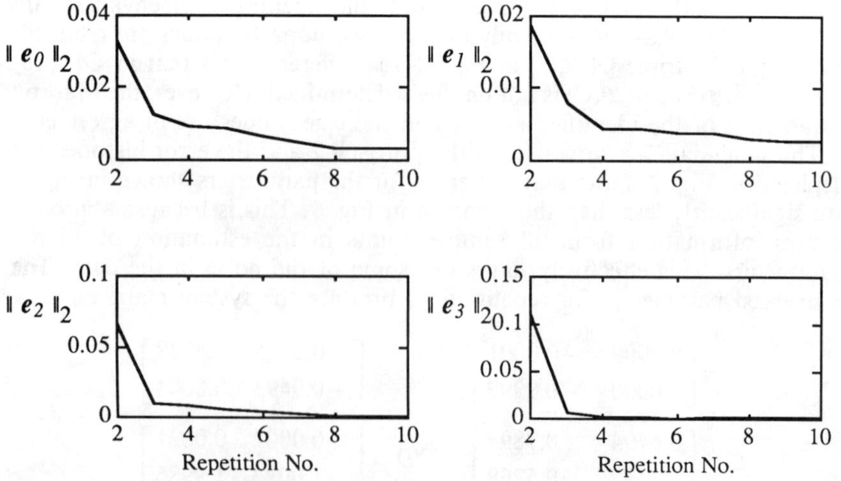

Fig. 2 Error histories using singular value decomposition to determine first four Markov parameters.

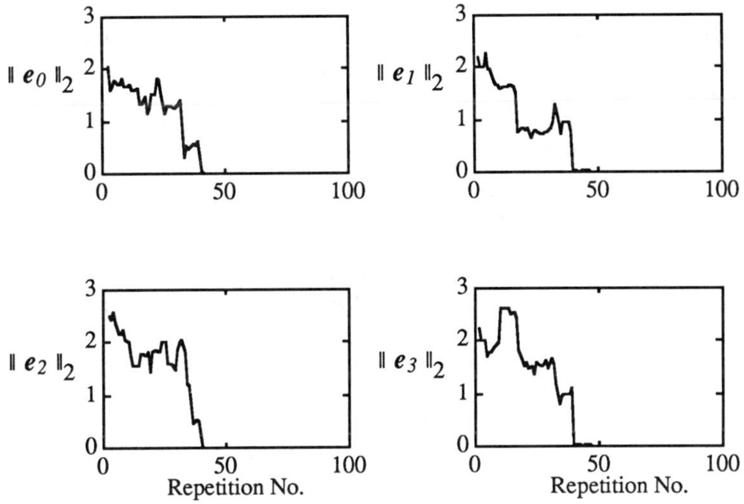

Fig. 3 Error histories for recursive determination of first four Markov parameters.

Figure 3 shows the error histories using approach 3. To accelerate convergence, the initial value α_{-1} is set to $\alpha_{-1} = 1000\ I_{(k+1)r}$. A significant drop in the error in all of the parameter estimates is seen when the number of experiments approached 40. The converged parameters are used for realization and the corresponding system matrices are

$$A = \begin{bmatrix} 1.0000 & 0.0000 \\ 0.0000 & -0.9999 \end{bmatrix}, \quad B = \begin{bmatrix} 0.3677 & 1.4728 \\ -0.9502 & -0.0001 \end{bmatrix}$$
$$C = \begin{bmatrix} 0.6785 & -0.7903 \\ 1.3580 & 0.5275 \end{bmatrix}, \quad D = \begin{bmatrix} -0.0012 & -0.0009 \\ 0.9993 & 2.0021 \end{bmatrix} \tag{29}$$

From the figures it is clear that the converged parameters in all three cases matched very well with the Markov parameters in Eq. (26).

Examination of the results suggest using the processing of sets of P_j suggested in the modified approach 2, as starting values for the improved version of approach 3. This combines the computational convenience of approach 1 with the accuracy of 2 and the recursive properties of 3.

Conclusions

This paper presents three approaches and possible variations for the determination of the Markov parameters for forced response data using general inputs. When the parameters in the solution procedure are bootstrapped, the results can be obtained very efficiently, but errors propagate throughout all parameters. By arranging the data in a different form and singular value decomposition is used, the resulting identified parameters are more accurate, in the least number of successive experiments, at the expense of a large matrix singular value decomposition. When a recursive procedure is used, the calculations can be performed very efficiently, but the number of repetitions of the experiment is much larger for a given accuracy than any of the previous approaches. An alternative formulation is suggested to combine the advantages of each of the approaches.

Acknowledgment

The authors thank Dr. Minh Phan for his suggestions and discussions during the preparation of this manuscript.

References

[1]Aizerman, M. A., "Potential Functions Technique and Extrapolation in Learning System Theory," *Proceedings of the 3rd IFAC Congress*, Butterworths, London, 1966.

[2]Fu, K., "Learning Control Systems—Review and Outlook," *IEEE Transaction on Automatic Control*, Vol. AC-15, April 1970, pp. 210–221.

[3]Arimoto, S., "Mathematical Theory of Learning with Applications to Robot Control," *Proceedings of the Fourth Yale Workshop on Applications of Adaptive Systems Theory*, May 1985.

[4]Arimoto, S., Kawamura, S., Miyazaki, F., and Tamaki, S., "Learning Control Theory for Dynamical Systems," *Proceedings of the 24th IEEE Conference on Decision and Control*, IEEE, New York, 1985.

[5]Kawamura, S., Miyazaki, F., and Arimoto, S., "Applications of Learning Method for Dynamic Control of Robot Manipulators," *Proceedings of the 24th Conference on Decision and Control*, IEEE, New York, 1985.

[6]Harokopos, E. G., "Optimal Learning Control of Mechanical Manipulators in Repetitive Motions," *Proceedings of the IEEE Conference on Robotics*, IEEE, New York, 1986.

[7]Harokopos E. G., and Longman, R. W., "A Learning Controller for Tracking Arbitrary Inputs in Robot Motions," *Tenth IASTED International Symposium on Robotics and Automation*, 1987.

[8]Phan, M., and Longman, R. W., "A Mathematical Theory of Learning Control for Linear Discrete Multivariable Systems," AIAA Paper 88-4313, Aug. 1988.

[9]Phan, M., and Longman R. W., "Liapunov Based Model Reference Learning Control," *Proceedings of the 26th Annual Allerton Conference on Communication, Control, and Computing*, 1988.

[10]Phan, M., and Longman, R. W., "Indirect Learning Control with Guaranteed Stability," *Proceedings of the 1989 Johns Hopkins Conference on Information Sciences and Systems*, Johns Hopkins University, Baltimore, MD, 1989.

[11]Nagumo, J., and Noda, A., "A Learning Method for System Identification," *IEEE Transactions*, Vol. AC-12, No. 3, June 1967, pp. 282–287.

[12]Phan, M., and Longman, R. W., "Learning System Identification," *Modeling and Simulation*, Instrument Society of America, Vol. 20, Part 5, 1989.

[13]Ho, B. L., and Kalman, R. E., "Effective Construction of Linear State-Variable Models from Input/Output Data," *Proceedings of the 3rd Annual Allerton Conference on Circuit and System Theory*, University of Illinois, Urbana-Champaign, IL, 1965, pp. 449–459.

[14]Tether, A. J., "Construction of Minimal Linear State-Variable Models from Finite Input-Output Data,' *IEEE Transactions on Automatic Control*, Vol. AC-15, No. 4, Aug. 1970, pp. 427–436.

[15]Silverman, L. M., "Realization of Linear Dynamical Systems," *IEEE Transactions on Automatic Control*, Vol. AC-16, No. 6, 1971, pp. 554–567.

[16]Rosen, R. H., and Lapidus, L., "Minimum Realization and System Modeling: I. Fundamental Theory and Algorithms," *AIChE Journal*, Vol. 18, No. 4, July 1972, pp. 673–684.

[17]Rosen, R. H., and Lapidus, L., "Minimum Realizations and System Modeling: II. Theoretical and Numerical Extensions," *AIChE Journal*, Vol. 18, No. 5, Sept. 1972, pp. 881–882.

[18]Kung, S., "A New Identification and Model Reduction Algorithm via Singular Value Decomposition," *12th Asilomar Conference on Circuits, Systems, and Computers*, 1978, pp. 705–714.

[19]Juang, J. N., and Pappa, R. S., "An Eigensystem Realization Algorithm for Modal Parameter Identification and Model Reduction," *Journal of Guidance, Control, and Dynamics*, Vol. 8, No. 5, Sept.–Oct. 1985, pp. 620–627.

[20]Golub, G. H., and Reinsch, C., "Singular Value Decomposition and Least Squares Solutions," *Numerical Mathematics*, Vol. 14, 1970, pp. 403–420.

[21]Juang, J. N., and Pappa, R. S., "Effects of Noise on Modal Parameters Identified by the Eigensystem Realization Algorithm," *Journal of Guidance, Control, and Dynamics*, Vol. 9, No. 3, May–June, pp. 294–303.

[22]Bergmann, M., Longman, R. W., and Juang, J. N., "Variance and Bias Computation for Enhanced System Identification," *Proceedings of the 28th IEEE Conference on Decision and Control*. IEEE, New York, 1989.

[23]Chen, C. T., *Linear System Theory and Design*, Holt, Rinehart and Winston, New York, 1984.

Chapter 6

Multibody Dynamics Formulations Using Maggi's Approach

A. J. Kurdila*
Texas A&M University, College Station, Texas

Introduction

IN this chapter a unified numerical theory is presented for a large class of recent multibody dynamics formulations.[1-4] It is shown that, although the derivation and presentation of many of these methods are quite dissimilar, they may all be represented concisely using Maggi's dynamic equations and common projective methods from linear operator theory. These methods all eliminate constraint contributions from the governing equations. It is shown in this chapter that this "reduction" process amounts to projecting the system of governing differential-algebraic equations to a "constraint-free" subspace. The process of equation reduction and constraint force recovery is achieved by generating an orthogonal basis for the null space of the constraint matrix and its orthogonal complement, the range space of the transposed constraint matrix. By explicitly deriving these dual bases for several typical methods in the class, it is shown that the methods differ only in the particular basis selected for the null space and its orthogonal complement.

Several advantages are derived from the theory presented herein:

1) The approach generalizes several of the existing methods so that constraint forces can be calculated after the reduced system is solved.

2) The technique enhances one's ability to contrast and compare available methods. In particular, "singularities" that have been noted in some methods are readily identified in the theory.

3) The method can be extended in principle to more general types of constraints.

---

*Assistant Professor, Department of Aerospace Engineering.

4) The approach suggests many interesting hybrid methods that have not yet appeared in the literature.

5) Most importantly, the approach enables one to derive concurrent solution methods that are applicable to a wide variety of individual methods.

Maggi's Equations of Motion

Maggi's equations were first presented in 1896 and again in 1901. Although the method is conspicuously absent from most texts in the English language, Neimark and Fufaev[5] summarize the essentials of Maggi's approach. Until quite recently, the method has not been amenable to actual calculations and has been of theoretical interest only. However, to obtain practical implementations of the method, it is necessary to extend the original, classical definition of the method using common projective techniques from linear algebra. As will be shown in this section, the advent of digital computers now permits a wide variety of practical numerical implementations of the approach.

Let S be a system of v particles whose position r^p relative to an inertial frame of reference is a function of N generalized system coordinates $\boldsymbol{q}_\kappa$, $k = 1, ..., N$ and the time t. Simply stated, the dynamic principle of virtual work states that the difference in the virtual work of the applied forces and the virtual work of the inertial forces must vanish for all virtual displacements consistent with the constraints. As a consequence of Newton's second law of motion for each particle, the virtual work of the constraint forces must vanish independently:

$$\int \mathrm{d}m\boldsymbol{a}^p \cdot \delta\boldsymbol{r}^p = \int \mathrm{d}\boldsymbol{f}^p \cdot \delta\boldsymbol{r}^p \tag{1}$$

$$\int \mathrm{d}\boldsymbol{R}^p \cdot \delta\boldsymbol{r}^p = 0 \tag{2}$$

where $\mathrm{d}f^p$ is the resultant of all impressed forces acting on a particle p, excluding constraint forces $\mathrm{d}R^p$, and the integrations are carried out over the entire system of particles. The quantities $\mathrm{d}m$, a^p, and δr^p are the mass, acceleration, and virtual displacement of the particle p. Since one can express the virtual variation of r^p as a linear combination of the variations in the generalized coordinates $\delta\boldsymbol{q}_k$,

$$\delta\boldsymbol{r}^p = \boldsymbol{e}_k^p \delta\boldsymbol{q}_k \qquad \text{where} \qquad \boldsymbol{e}_k^p \equiv \frac{\partial \boldsymbol{r}^p}{\partial \boldsymbol{q}_k} \tag{3}$$

Equation (1) can be rewritten in the familiar form

$$\int \{\mathrm{d}m\boldsymbol{a}^p - \mathrm{d}\boldsymbol{f}^p\} \cdot \boldsymbol{e}_k^p \delta\boldsymbol{q}_k = 0 \tag{4}$$

or

$$\{\boldsymbol{E}_k(T) - \boldsymbol{Q}_k\}\delta\boldsymbol{q}_k = 0 \tag{5}$$

where

$$\boldsymbol{E}_k(T) \equiv \frac{\mathrm{d}}{\mathrm{d}t}\left\{\frac{\partial T}{\partial \dot{\boldsymbol{q}}_k}\right\} - \frac{\partial T}{\partial \boldsymbol{q}_k}, \qquad T \equiv \int \frac{1}{2}\,\mathrm{d}mv \cdot v$$

and

$$\boldsymbol{Q}_k \equiv \int \mathrm{d}\boldsymbol{f}^p \cdot \boldsymbol{e}_k^p$$

and the summation convention has been used in Eqs. (3–5). For unconstrained problems the variations δq_k above are independent and their respective coefficients in Eq. (5) must be identically zero for $k = 1, \ldots, N$. In constrained problems, however, the $\delta\boldsymbol{q}_k$ are not independent.

It is possible to create independent virtual displacements, thus resulting in a useful set of equations similar to Eq. (5). For holonomic systems, a large class of nonholonomic systems, constraints on the system can be expressed in terms of D independent equations that are linear in the "kinematically possible" differentials of the generalized coordinated and time

$$a_{lk}\,\mathrm{d}\boldsymbol{q}_k + b_\ell\,\mathrm{d}t = 0 \tag{6}$$

In order that the virtual displacements δr can be consistent with the constraints, it is necessary that the virtual variations in the generalized coordinates δq_k satisfy the relations

$$a_{lk}\delta_{qk} = 0 \tag{7}$$

Equation (7) can be alternately represented in terms of a $D \times N$ matrix a and an N vector $\delta\boldsymbol{q} = [\delta q_1, \ldots, \delta_{qN}]^T$ as

$$\boldsymbol{a}\delta\boldsymbol{q} = 0 \tag{8}$$

Equations (7) and (8) can now be interpreted as requiring that the N vector $\delta\boldsymbol{q}$ lie in the null space of the operator a, or that $\delta\boldsymbol{q}$ be orthogonal to the range of a^T.

$$\delta\boldsymbol{q} \in \text{null space } (\boldsymbol{a}) \tag{9}$$

$$\delta\boldsymbol{q} \perp \text{range } (\boldsymbol{a}^T) \tag{10}$$

The range of a^T is equal to the span of the columns of the matrix a^T and is consequently of dimension D if the constraints are independent. Since the null space of a and range of a^T are orthogonal subspaces, the dimension of the null space of a is $I \equiv N - D$. Deferring for the moment the discussion of their calculation, let $[A_1, ..., A_i]$ be any basis for the null space of a, and let $[p_1, ..., p_D]$ be any basis for the range of a^T. The vector $\delta \boldsymbol{q}$ may lie anywhere within the span of $[A_1, ..., A_I]$. In other words, it is true that

$$\delta \boldsymbol{q}_k = A_{ki} \delta \theta_i \tag{11}$$

$$\delta \boldsymbol{q} = \boldsymbol{A} \delta \boldsymbol{\theta} \tag{12}$$

where $\delta \theta_i$ are arbitrary coefficients, and $\boldsymbol{A} = [\boldsymbol{A}_1, ..., \boldsymbol{A}_I] = [A_{ki}]$ is an $N \times (N - D)$ matrix.

Substituting Eq. (12) into Eq. (5), a form of virtual work emerges that is well suited for constrained systems. Using the fact that each $\delta \theta_i$ is arbitrary, one obtains Maggi's equations,

$$\{\boldsymbol{E}_k(T) - \boldsymbol{Q}_k\} A_{ki} \delta \theta_i = 0 \tag{13}$$

and

$$\{\boldsymbol{E}_k(T) - \boldsymbol{Q}_k\} A_{ki} = 0 \tag{14}$$

In effect, Eq. (13) expresses the fact that the virtual work due to each of the independent virtual displacements obtained from Eqs. (11) and (3),

$$\delta \boldsymbol{r}^p = \boldsymbol{e}_k^p \delta \boldsymbol{q}_k \tag{15}$$

$$\delta \boldsymbol{r}^p = \boldsymbol{e}_k^p A_{ki} \delta \theta_i \tag{16}$$

be zero. It should be noted that Maggi's Eqs. (13) and (14) do not distinguish between genuinely nonintegrable constraints and holonomic ones in differential form. In this respect Maggi's equations are a quite general formulation of dynamics.

For ease of comparison to other methods, it is useful to express Eq. (14) in operator form as

$$\boldsymbol{A}^T\{\boldsymbol{E}(T) - \boldsymbol{Q}\} = 0 \tag{17}$$

The preceding system of $N-D$ equations, together with the D constraints in Eq. (6), comprise a "determinate" system of N equations in terms of N unknowns and are sufficient to solve for the displacements, velocities, and accelerations of the system. An illustrative account of the actual details of implementing Eqs. (6) and (17) in a specific numerical simulation can be found in Ref. 6.

Supposing that Eq. (17) has been solved either analytically or numerically at a particular time t, the decomposition of R^N into span $\{\boldsymbol{A}_I\}$ and $\{\boldsymbol{p}_D\}$ is also useful since it is possible to solve for constraint forces from the base vectors $[\boldsymbol{p}_1, ..., \boldsymbol{p}_D]$. Because the constraint forces perform zero virtual work, most formulations up until now have not explicitly contained the constraint forces $\mathrm{d}R^p$. If one forms the inner product between Newton's second law for each particle and $\boldsymbol{e}_k^p$, and subsequently sums over all of the particles in the system, one obtains the equations

$$\boldsymbol{E}(T) - \boldsymbol{Q} - \boldsymbol{Q}^{(r)} = 0 \tag{18}$$

$$\boldsymbol{Q}^{(r)} = [Q_1^{(r)} \dots Q_N^{(r)}]^T \tag{19}$$

where

$$Q_k^{(r)} = \int \mathrm{d}R^p \cdot e_k^p \tag{20}$$

and $\mathrm{d}\boldsymbol{R}^p \equiv$ total reaction force on particle p, so that $\mathrm{d}ma^p = \mathrm{d}f^p + \mathrm{d}R^p$. Moreover, Eq. (1) requires that the N vector $\boldsymbol{Q}^{(r)}$ must be a member of the range of a^T, the orthogonal complement of the range of A. It follows that, for some D vector γ,

$$\boldsymbol{Q}^{(r)} = \boldsymbol{p}\gamma \tag{21}$$

Making use of the identity in Eq. (21), it is possible to express Eq. (18) in either of two ways:

$$\boldsymbol{E}(T) - \boldsymbol{Q} - \boldsymbol{a}^T\gamma = 0 \tag{22}$$

$$\boldsymbol{E}(T) - \boldsymbol{Q} - \boldsymbol{p}\gamma = 0 \tag{23}$$

These two equations are clearly equivalent, since the columns of p are any basis for the range of a^T. It should be noted that the generalized multipliers γ have no physical meaning in their "raw" form. They are simply the constraint reaction terms with respect to a different basis for the range of the transposed constraint matrix. The two representations coincide when $[\boldsymbol{p}_1 \dots \boldsymbol{p}_D]$ are chosen to be the columns of a^T. The derivations that follow are carried out in the generic basis $\{\boldsymbol{p}_D\}$, a description well suited to numerical techniques that automatically generate a basis for the range (a^T) that does not coincide with the columns of a^T. Assuming that Eq. (17) has been solved and that the original unknowns q and their derivatives $\dot{\boldsymbol{q}}$ and $\ddot{\boldsymbol{q}}$ have been reconstructed, the unknown constraint contributions are simply the unique minimizer of the N vector norm

$$\boldsymbol{Q}^{(r)} = \boldsymbol{p}\gamma = \underset{X \in R^N}{\text{minimizer}} \, \|\boldsymbol{E}(T) - \boldsymbol{Q} - \boldsymbol{X}\| \tag{24}$$

or

$$\gamma = \underset{\xi \in R^D}{\text{minimizer}} \, \|\boldsymbol{E}(T) - \boldsymbol{Q} - \boldsymbol{p}\xi\| \tag{25}$$

Equations (17) and (24) suggest a simple "geometric" interpretation of the problem of solving constrained dynamics problems that has been widely used in applied linear algebra. Equation (17) may be interpreted as a statement of orthogonality. It requires that the N vector $\boldsymbol{E}(T) - \boldsymbol{Q}$ be perpendicular to the null space a. In other words, the projection of $\boldsymbol{E}(T) - \boldsymbol{Q}$ onto the null space of the variational constraints is zero. Likewise, when Eq. (24) requires that $\boldsymbol{Q}^{(r)}$ is the vector that lies in the range of a^T that is closest to $\boldsymbol{E}(T) - \boldsymbol{E}(T) - \boldsymbol{Q}$, it implies that $\boldsymbol{Q}^{(r)}$ is the orthogonal projection of $\boldsymbol{E}(T) - \boldsymbol{Q}$ onto the range of a^T, the complementary space to the null space of a:

$$\boldsymbol{Q}^{(r)} = \boldsymbol{P}_a T\{\boldsymbol{E}(T) - \boldsymbol{Q}\} \tag{26}$$

Carefully note that the orthogonality statement in Eq. (11) requires that the "abstract angle" θ must be zero:

$$\tan\theta = \frac{\boldsymbol{P}_A\{\boldsymbol{E}(T) - \boldsymbol{Q}\}}{\boldsymbol{P}_a T\{\boldsymbol{E}(T) - \boldsymbol{Q}\}} = 0 \tag{27}$$

where P_A and $P_a T$ are the orthogonal projections onto the null space of a and the range of a^T, respectively.

The solution of Eqs. (24) and (25) is given, in principle, by the normal equations

$$\gamma = (\boldsymbol{p}^T\boldsymbol{p})^{-1}\boldsymbol{p}^T\{\boldsymbol{E}(T) - \boldsymbol{Q}\} \tag{28}$$

A strategy similar to this approach in the literature on constrained dynamics is derived via algebraic considerations in Ref. 7. However, it is neither recommended nor usually required to form the "generalized inverse" depicted in Eq. (28) (see Ref. 8). Instead, orthogonal factorizations are employed in solving the minimization problem in Eq. (24). If $\boldsymbol{\theta}$ is an orthogonal transformation, that is, $\boldsymbol{\theta}^T\boldsymbol{\theta} = \boldsymbol{I}$, then $\boldsymbol{\theta}$ is an isometry. In other words, the transformation θ is norm preserving so that

$$\|\boldsymbol{E}(T) - \boldsymbol{Q} - \boldsymbol{Q}^{(r)}\| = \|\boldsymbol{\theta}\{\boldsymbol{E}(T) - \boldsymbol{Q} - \boldsymbol{Q}^{(r)}\}\| \tag{29}$$

A judicious choice of θ can greatly simplify the form of the norm shown in Eq. (29) and lead to a straightforward solution procedure for the constraint contributions. Typical choices of $\boldsymbol{\theta}$ are derived from the QR and singular value decompositions.[8]

Equations (24), (25), and (28) provide a complete description of the solution for the constraint forces. When the calculations are performed via orthogonal transformations, and not the normal equations, they exhibit well-known stability characteristics. However, Eqs. (15) and (16) are too general to permit a detailed comparison of methods that choose different bases in A and $\boldsymbol{p}$. An alternative description of Maggi's equations given in Ref. 9, although not computationally efficient, depicts information regarding the bases selected for $\boldsymbol{p}$ and A in a compact form. In addition, the formalism displays enough information to reconstruct the projection operator $\boldsymbol{P}_{\boldsymbol{a}}T$ explicitly. In fact, when an orthogonal basis is chosen for A and $\boldsymbol{p}$, Maggi's equations generate the same expressions obtained in the least-squares solution of Eq. (24).

Papastavridis[9] generates the governing relations via Maggi's equations by first supplementing the constraint matrix a to form a full rank $N \times N$ matrix $\hat{\boldsymbol{a}}$:

$$\hat{\boldsymbol{a}} = \begin{bmatrix} \boldsymbol{a} \\ \boldsymbol{a}' \end{bmatrix} \tag{30}$$

Next, the inverse $\hat{\boldsymbol{A}} = \hat{\boldsymbol{a}}^{-1}$ can be calculated and partitioned in the following manner:

$$\hat{\boldsymbol{A}} = [A' \quad A] \tag{31}$$

By inspection it is clear that the columns of a^T and A' are dual bases for the range of a^T. Likewise, the columns of a'^T and A are dual bases for the null space of a:

$$\begin{bmatrix} \boldsymbol{a} \\ \boldsymbol{a}' \end{bmatrix} [A' \quad A] = \begin{bmatrix} \boldsymbol{I} & 0 \\ 0 & \boldsymbol{I} \end{bmatrix} \tag{32}$$

By premultiplying Eq. (22) by $\hat{\boldsymbol{A}}^T$, one again obtains Maggi's governing equations with or without reactions

$$(A')^T\{\boldsymbol{E}(T) - \boldsymbol{Q}\} = \boldsymbol{\lambda} \tag{33}$$

$$A^T\{\boldsymbol{E}(T) - \boldsymbol{Q}\} = 0 \tag{34}$$

Another description of this procedure is given in Ref. 68. Upon comparing Eqs. (33) and (26), one can see that the projection onto the range of a^T is easily constructed from the dual basis $(A')^T$:

$$\boldsymbol{P}_{\boldsymbol{a}}T = \boldsymbol{a}^T(A')^T \tag{35}$$

The precise relationship defining the dual basis $[\boldsymbol{A}'_1 \ldots \boldsymbol{A}'_D]$ of $[\boldsymbol{a}_1^T \ldots \boldsymbol{a}_D^T]$ can be extracted from Eq. (32) and is simply

$$\boldsymbol{A}'^T\boldsymbol{a}^T = \boldsymbol{a}\boldsymbol{A}' = \boldsymbol{I} \tag{36}$$

But if we choose $\boldsymbol{p} = a^T$ in Eq. (28), it is clear that A'^T is equal to the coefficient matrix $(aa^T)^{-1}a$ since there is exactly one dual basis corresponding to $[a_1^T \ldots a_D^T]$, and

$$\{(\boldsymbol{a}\boldsymbol{a}^T)^{-1}\boldsymbol{a}\}\boldsymbol{a}^T = (A')^T\boldsymbol{a}^T \tag{37}$$

Thus, in this case the normal equations solution reduces to Eq. (33).

In cases where the dual bases comprising the matrices $\hat{a}$ and $\hat{A}$ can be derived analytically, the notation in Ref. 9 is concise in its representation of the solution. A knowledge of $\hat{a}$ and $\hat{A}$ conveys both the specific basis chosen to reduce the constrained system to minimal order and the projection operator selected to calculate constraint forces. It will be shown later that the theory is sufficiently detailed to provide an excellent means of tabulating computational alternatives that have appeared in recent literature.

Related Computational Methods

Kane's Method

Kane's method is a collection of steps that yields the smallest possible set of reaction less equations for a constrained dynamic system. It will be shown later that Kane's method can be considered as an analytical procedure for choosing the null-space basis $[A_1 \ldots A_I]$ appearing in Eq. (17). The following steps summarize the approach:

1) Choose a set of N generalized coordinates $\boldsymbol{q}_k$, $k = 1, \ldots, N$. These coordinates may be dependent or constrained. As is well known, the system number of degrees of freedom equals the number of independent $\dot{q}_k$, or, in other words, the number of independent nonholonomic constraints.

2) Choose N quantities $\boldsymbol{U}_k$, $k = 1, \ldots, N$, the "generalized speeds," defined in terms of the generalized holonomic velocities $\dot{\boldsymbol{q}}$ according to

$$\boldsymbol{U}_k = Y_{kl}\dot{\boldsymbol{q}}_l + Z_k \tag{38}$$

or as

$$\boldsymbol{U} = \boldsymbol{Y}\dot{\boldsymbol{q}} + \boldsymbol{Z} \tag{39}$$

This definition is valid only if these equations that are linear in the $\dot{\boldsymbol{q}}_k$ are invertible. That is, the matrix Y must be invertible. If $Y^{-1} = y$, one has

$$\dot{\boldsymbol{q}}_1 = u_{lk}\boldsymbol{U}_k z_l \tag{40}$$

or

$$\dot{\boldsymbol{q}} = \boldsymbol{y}\boldsymbol{U} + \boldsymbol{z} \tag{41}$$

3) Express all constraints on the system in terms of D equations having the kinematic form

$$\boldsymbol{U}_d = \boldsymbol{W}_{di}\boldsymbol{U}_i - \boldsymbol{\Gamma}_d \tag{42}$$

$$[\boldsymbol{I} - \boldsymbol{W}]\boldsymbol{U} + \boldsymbol{\Gamma} = 0 \tag{43}$$

$$\boldsymbol{c}\boldsymbol{U} + \boldsymbol{\Gamma} = 0 \tag{44}$$

These constraints arise either because the system is genuinely nonholonomic or by differentiating the holonomic constraints due to the dependent coordinates $\boldsymbol{q}_k$. Another useful representation of these constraints is achieved by partitioning the N vector $\boldsymbol{U}$ into an $I = (N - D)$ vector of "independent" generalized speeds $\boldsymbol{U}^I$ and a D vector of "dependent" generalized speeds $\boldsymbol{U}^D$:

$$\boldsymbol{U} = \begin{bmatrix} \boldsymbol{U}^D \\ \boldsymbol{U}^I \end{bmatrix} = \begin{bmatrix} \boldsymbol{W} \\ \boldsymbol{I} \end{bmatrix} \boldsymbol{U}^I - \begin{bmatrix} \boldsymbol{\Gamma} \\ \boldsymbol{0} \end{bmatrix} \tag{45}$$

$$\boldsymbol{U} = \boldsymbol{C}\boldsymbol{U}^I - \boldsymbol{\Gamma} \tag{46}$$

The matrix W is $D \times I$, the matrix C is $N \times I$, and I is an $I \times I$ identity matrix. Note that $cC = 0$, just as $aA = 0$ in Eq. (32).

4) By virtue of the definitions in Eqs. (38–46), the velocity of every particle in the system can be expressed as

$$\boldsymbol{v}^p = \boldsymbol{v}^p_i U_{i+D} + \boldsymbol{v}^p_t \tag{47}$$

The preceding summation only includes elements of the independent generalized speeds $\boldsymbol{U}^I$. In the absence of constraints the summation runs from 1 to N. The I vectors $\boldsymbol{v}^p_i = \boldsymbol{e}^p_k A_{ki}$ of Eq. (47) are defined to be the "partial velocities" of particle p.

5) Last, Kane postulates the governing dynamic equations

$$F_i + F^*_i = 0 \tag{48}$$

where F_i and F^*_i are defined via integration over all of the particles in the system:

$$F_i = \int \mathrm{d}\boldsymbol{f}^p \cdot \boldsymbol{v}^p_i \tag{49}$$

$$F^*_i = -\int \mathrm{d}m\boldsymbol{a}^p \cdot \boldsymbol{v}^p_i \tag{50}$$

Although Kane's terminology is quite different, his method can be expressed in terms of the virtual work expressions. If Eqs. (40) and (46) are substituted into

$$\boldsymbol{v}^p = \boldsymbol{e}^p_k \dot{q}_k + \frac{\partial \boldsymbol{r}^p}{\partial t} = \boldsymbol{e}^p_k y_{kl} C_{li} U_{i+D} + \boldsymbol{e}^p_k (y_{kl}\Gamma_l + z_k) + \frac{\partial \boldsymbol{r}^p}{\partial t} \tag{51}$$

then the partial velocities may be obtained by inspection as

$$\boldsymbol{v}^p_i = \boldsymbol{e}^p_k y_{kl} C_{li} \tag{52}$$

Upon inserting the preceding partial velocities into Eq. (48) and rearranging the summations, Kane's governing equations are just

$$\int \{\mathrm{d}m\boldsymbol{a}^{p} - \mathrm{d}\boldsymbol{f}^{p}\}\boldsymbol{e}_{k}^{p} y_{kl} C_{li} = 0 \tag{53}$$

or

$$\{\boldsymbol{E}_k(T) - \boldsymbol{Q}_k\} y_{kl} C_{li} = 0 \tag{54}$$

or

$$\boldsymbol{C}^{T}\boldsymbol{y}^{T}\{\boldsymbol{E}(T) - \boldsymbol{Q}\} = 0$$

The identity between Eqs. (55) and (17) is immediately apparent. To prove that Kane's method is a particular a null-space-generating technique, it only remains to be shown that the range of yC is indeed equal to the null space of the constraints. If the constraints are given in Kane's notation, as in Eq. (44), the corresponding constraints are

$$\boldsymbol{c}\boldsymbol{U} + \boldsymbol{\Gamma} = 0 \tag{55}$$

$$\boldsymbol{c}\boldsymbol{Y}\dot{\boldsymbol{q}} + (\boldsymbol{c}\boldsymbol{Z} + \boldsymbol{\Gamma}) = 0 \tag{56}$$

so that

$$\boldsymbol{a} = \boldsymbol{c}\boldsymbol{Y} = [\boldsymbol{I} - \boldsymbol{W}]\boldsymbol{Y} \tag{57}$$

But the columns of the matrix $A = yC$ clearly constitute a basis for the null space of the variational constraints, since

$$\boldsymbol{a}\boldsymbol{A} = \boldsymbol{c}\boldsymbol{Y}\boldsymbol{y}\boldsymbol{C} = \boldsymbol{c}\boldsymbol{C} = 0 \tag{58}$$

Alternatively, the analytical nature of this formulation enables the solution procedure to be represented by the method described earlier. In this case, by combining Eqs. (33) and (55) one obtains the dual bases that may be used to calculate the reduced equations as well as the constraint forces in Eq. (33). These bases are depicted in Table 1. However, it should be pointed out that it is not meant to suggest in Table 1 that the matrices y, Y need to be formed in a practical implementation of the dual basis methods. As will soon be illustrated, the dual bases can be utilized in this case without forming y or Y.

Coordinate Partitioning

Generalized coordinate partitioning[10] is a computational method that uses Gaussian elimination with full pivoting on the $D \times N$ constraint matrix a to select a subset of $I = N - D$ independent generalized

Table 1 Maggi's equations for multibody simulation

Integration method	Augmented constraints $\hat{a}$	Augmented null-space matrix $\hat{A}$
Coordinate partitioning	$\begin{bmatrix} [a^D & a^I] \\ [0 & I] \end{bmatrix}$	$\left[\begin{bmatrix} (a^D)^{-1} \\ 0 \end{bmatrix} \begin{bmatrix} -(a^D)^{-1}a^I \\ I \end{bmatrix}\right]$
Mani's method	$\begin{bmatrix} [a] \\ [U_2^T] \end{bmatrix}$	$\left[\left[U_1 \begin{bmatrix} \frac{1}{\sigma} \end{bmatrix} V^T\right][U_2]\right]$
Null-space updating	$\begin{bmatrix} c \\ \tilde{Q}_2^T \end{bmatrix} Y$	$y[\tilde{Q}_1(c\tilde{Q}_1)^{-1}(I - \tilde{Q}_1(c\tilde{Q}_1)^{-1})\tilde{Q}_2]$
Kane's method	$[c^D \quad c^I]$	$\begin{bmatrix} -(c^D)^{-1}c^I \\ I \end{bmatrix}$
Singh/Huston's method	$\begin{bmatrix} [c] \\ [\hat{U}_2^T] \end{bmatrix} Y$	$y\left[\left[\hat{U}_1 \frac{1}{\hat{\sigma}} \hat{V}^T\right][\hat{U}_2]\right]$
Null-space updating	$\begin{bmatrix} [a] \\ [Q_2^T] \end{bmatrix}$	$[[Q_1(aQ_1)^{-1}][(I - Q_1(aQ_1)^{-1}a)Q_2]]$

velocities $\dot{q}^I$ and D dependent velocities $\dot{q}^D$. If a has full row rank, then it is always possible to partition a into $[a^D a^I]$ such that a^D is a $D \times D$ nonsingular matrix:

$$a\dot{q} + b = [a^D a^I] \begin{bmatrix} \dot{q}^D \\ \dot{q}^I \end{bmatrix} + b \tag{59}$$

$$A = [-(a_I^D)^{-1} a^I] \tag{60}$$

With this choice of independent velocities $\dot{q}^I$, the columns of the matrix A are a basis for the null space of the constraints. Furthermore, premultiplication of Lagrange's equations by A^T as in Eq. (17) yields precisely the governing equations obtained by Wehage in Eq. (26) of Ref. 10. As in Kane's method, the analytical nature of this strategy permits a closed-form expression of the dual bases in Eq. (31), which are displayed in Table 1. It should also be noted that premultiplication by $(A')^T$ for Wehage's method in Table 1 generates identical expressions for the Lagrange multipliers in Eq. (24) of Ref. 10.

Mani's Method

Although Gaussian elimination has been used quite effectively in some simulations utilizing coordinate partitioning,[11] one deficiency in the method has been noted in Ref. 12. The choice of a fixed set of independent coordinates may lead to poorly conditioned matrices that must be factored. On inspection of the null-space matrix A in Eq. (61), one can observe that the columns of A span the null space of the constraint matrix a only as long as $(\boldsymbol{a}^D)^{-1}$ exists. Since the matrix $\boldsymbol{a} = [\boldsymbol{a}^D \boldsymbol{a}^I]$ is time dependent, one cannot in general guarantee that this inverse is well defined for the entire time span of interest.

Mani addresses this problem by utilizing the singualr value decomposition of the constraint matrix a^T,

$$\boldsymbol{a}^T = [U_1\, U_2]\boldsymbol{\Sigma} V^T \tag{61}$$

$$\boldsymbol{\Sigma} = \begin{bmatrix} \boldsymbol{\sigma} \\ 0 \end{bmatrix} \tag{62}$$

where $U = [U_1\, U_2]$ is an $N \times N$ orthogonal matrix, $\boldsymbol{\Sigma}$ is an $N \times D$ diagonal matrix, and V is a $D \times D$ orthogonal matrix. The diagonal terms σ in the matrix $\boldsymbol{\Sigma}$ in Eq. (62) are the singular values of the constraint matrix. It is a property of the singular value decomposition that the $N \times D$ matrix U_1 and the $N \times (N - D)$ matrix U_2 are bases for the range of a^T and null space of a, respectively. In terms of the theory presented earlier, the integration process carried out by Mani is equivalent to choosing the dual bases as depicted in Table 1. It should be noted that the actual algorithm outlined in Ref. 12 is quite complicated in that the singular valued decomposition is carried out on "as needed" basis at certain time steps, and any constraint violation that occurs is corrected by employing Newton-Raphson iteration.

Null-Space Updating

A more recent technique is based on the QR decomposition of the constraint matrix a^T (Ref. 13). In this decomposition the $N \times D$ matrix a^T is expressed as the product of an $N \times N$ orthogonal matrix Q and a $D \times N$ upper triangular matrix R. It is a characteristic of the decomposition that the matrix Q can be partitioned into two matrixes Q_1 and Q_2, where the columns of Q_1 are a basis for the range of a^T and the columns of Q_2 are a basis for the null space of a, respectively. In other words, one can write

$$\boldsymbol{a}^T = \boldsymbol{QR} = [\boldsymbol{Q}_1\, \boldsymbol{Q}_2]\begin{bmatrix} \boldsymbol{R} \\ 0 \end{bmatrix} \tag{63}$$

where Q is an $N \times N$ orthogonal matrix and R is an $N \times D$ upper triangular matrix. In Ref. 13, Q_1 and Q_2 are used to create two other matrices $\hat{\boldsymbol{Q}}_1$ and $\hat{\boldsymbol{Q}}_2$, which, although not orthogonal for all time, still satisfy Maggi's Eqs. (30) and (31), respectively:

$$\hat{\boldsymbol{Q}}_1 = \boldsymbol{Q}_1(\boldsymbol{a}\boldsymbol{Q}_1)^{-1} \tag{64}$$

$$\hat{\boldsymbol{Q}}_2 = (\boldsymbol{I} - \boldsymbol{Q}_1(\boldsymbol{a}\boldsymbol{Q}_1)^{-1}\boldsymbol{a})\boldsymbol{Q}_2 \tag{65}$$

Choosing $\hat{Q}_2$ as the matrix A and premultiplying Lagrange's equation by A^T again produces the reduced governing system in Eq. (33) of Ref. 13.

The calculation of Lagrange multipliers in Ref. 13 is also a special case of the general method in the section on Maggi's equations of motion. Suppose that the QR decomposition has just been calculated for the matrix A^T at the current time step. Choosing the columns of $\hat{Q}_1$ as the basis for the range of a^T, Eq. (28) allows one to calculate the Lagrange multipliers referred to this basis:

$$\gamma = (\hat{Q}_1^T \hat{Q}_1)^{-1} \hat{Q}_1 \{E(T) - Q\} \tag{66}$$

$$\gamma = (aa^T) \hat{Q}_1^T \{E(T) - Q\} \tag{67}$$

One can now change basis from γ to λ by the normal equations

$$\lambda = (aa^T)^{-1} a \hat{Q}_1 \gamma \tag{68}$$

Making this replacement above gives the equations used to calculate the Lagrange multipliers in

$$\lambda = \hat{Q}_1^T \{E(T) - Q\} \tag{69}$$

The dual bases appropriate for this method are shown in Table 1. Again, Eq. (69) can be derived in short order by simply extracting the dual basis A' from Table 1 for Kim's method.

Singh/Huston's Method

As another example of computational applications, Singh and Likins[6] and Huston[14] present a modification of Kane's method based on the singular value decomposition of the constraint matrix that is similar to the method employed by Mani. In this approach Kane's equations are first generated for the unconstrained system, that is, with $C = I$ in Eq. (46). As in Mani's method, the singular value decomposition

$$c^T = \hat{U} \hat{\Sigma} \hat{V}^T \tag{70}$$

where $\hat{U}$, $\hat{\Sigma}$, and $\hat{V}^T$ are defined in a manner analogous to that shown in Eq. (62), is then used to obtain a basis $\hat{U}_2$ for the null space of the coefficient matrix c in Eq. (44). It is subsequently shown that the reduced governing equations are

$$\hat{U}^T \{F + F^*\} = 0 \tag{71}$$

Comparing Eqs. (22), (34), and (55) with the above, one can alternatively express the equations derived by Singh as

$$\hat{U}_2^T y^T \{E(T) - Q - a^T \lambda\} = 0 \tag{72}$$

These equations are identical to those derived in Ref. 15 for a very specific choice of basis for the null space of the constraints. Since the columns of $\hat{\boldsymbol{U}}_2$ are a basis for the null space of c, it is clear that $\boldsymbol{A} = \boldsymbol{y}\hat{\boldsymbol{U}}_2$ is a basis for the null space of a. It is consequently apparent that this method simply chooses a different basis for the null space of a in the method of virtual work. As with all previous methods, this approach may also be succinctly represented in terms of the dual bases it uses for the null space of the variational constraints and its orthogonal complement. These choices are found in Table 1.

Advantages of the Unified Theory

Several benefits can be derived from the theory discussed in the section on Maggi's equations of motion. In one sense all of the advantages of the approach stem from the fact that the description given enhances one's ability to compare and contrast specific techniques. For instance, it is shown in this section that the features of Lagrange's equations that make constraint reaction calculation possible may be found in other reduced formulations, if they are perceived in a "unified" context. It is also demonstrated that certain ill conditioning noted in coordinate partitioning methods is predicted by the unified theory and has been found in some implementations of Kane's method. As a final example, it is illustrated that the method provides a foundation for numerical methods suitable for a wider class of constraints.

Constraint Reaction Calculation

One benefit of the general null-space decomposition theory presented in the first section is that it enables the calculation of Lagrange multipliers from Kane's equations after the reduced system of equations has been solved. Neither the method of Kane and Levinson[16] nor its recent modifications[6,14,15] have suggested an equivalent means of calculating the eliminated constraint forces. Furthermore, the development that follows extends the work in Ref. 17 in two important respects. Whereas the derivation in Ref. 17 is limited to rigid bodies, no such assumption is made in this paper. Also, although Ref. 17 establishes a form of Kane's equations that retain multiplicative "undetermined" coefficients, the relationship between these terms and the multipliers in a Lagrangian formulation is not explored. In the following discussion it is shown that the undetermined multipliers in Huston's formulation are precisely the multipliers obtained in Lagrange's equations using the same generalized coordinates.

From Eqs. (22), (34), and (55), Kane's equations for the "released" or "relaxed" system, that is, with $C = I$, are

$$\boldsymbol{y}^T\{\boldsymbol{E}(T) - \boldsymbol{Q} - \boldsymbol{a}^T\boldsymbol{\lambda}\} = 0 \tag{73}$$

$$\boldsymbol{F} + \boldsymbol{F}^* - \boldsymbol{y}^T\boldsymbol{a}^T\boldsymbol{\lambda} = 0 \tag{74}$$

Consequently, the constraint force contribution can be calculated from

$$\boldsymbol{y}^T\boldsymbol{a}^T\boldsymbol{\lambda} = P_y T_a T\{\boldsymbol{F} + \boldsymbol{F}^*\} \tag{75}$$

In this equation F and F^* are the generalized active forces and generalized inertia forces for the unconstrained system, in other words, with $C = I$. But if one is given the constraints in terms of the generalized speeds as shown in Eq. (56), the solution can be expressed exclusively in terms of matrices available in Kane's formulation as

$$\boldsymbol{\lambda} = \underset{\xi \in R^D}{\text{minimizer}} \; \|\boldsymbol{F} + \boldsymbol{F}^* - \boldsymbol{c}^T\boldsymbol{\xi}\| \tag{76}$$

$$\boldsymbol{c}^T\boldsymbol{\lambda} = \boldsymbol{P}_c T\{\boldsymbol{F} + \boldsymbol{F}^*\} \tag{77}$$

As noted earlier, specific forms for the projection $P_c T$ onto the range of c^T can be derived from the normal equations, or from the dual basis for Kane's method in Table 1.

Following derivations along the same lines, Lagrange multipliers are also calculated in Singh's approach. It is still true that

$$\boldsymbol{\lambda} = \underset{\xi \in R^D}{\text{minimizer}} \; \|\boldsymbol{F} + \boldsymbol{F}^* - \boldsymbol{c}^T\boldsymbol{\xi}\| \tag{78}$$

But in this method the singular value decomposition of c has already been calculated. Using the fact that an orthogonal transformation is an isometry, the solution to Eq. (79) can be calculated via standard techniques in numerical methods[8]:

$$\boldsymbol{\lambda} = \hat{\boldsymbol{V}} \begin{bmatrix} \dfrac{1}{\hat{\boldsymbol{\sigma}}} \end{bmatrix} \hat{\boldsymbol{U}}_1^T\{\boldsymbol{F} + \boldsymbol{F}^*\} \tag{79}$$

It is important to note that all of the information required to evaluate the constraint forces is readily available, and no further matrix decomposition need be performed in Eq. (79). This same result can be derived simply by inspecting the dual bases for Singh's method in Table 1. Equations (78) and (80) illustrate the fact mentioned earlier that the dual bases shown in Table 1 can be utilized in practical implementations without forming the matrices y and Y explicitly.

Explanation of Singularities

A second advantage of the unified theory is the ease with which these methods can be compared. In this example, Singh[6] has noted certain "numerical instabilities" in a general implementation of Kane's method, but offers no quantitative explanation for their origin. However, by comparing the basis implicitly chosen in Kane's method to that selected in the method of Ref. 6, the nature of the singularities is brought to light.

In Kane's method it is assumed that the constraint matrix c has the form

$$c = [I - W] \tag{80}$$

The requirement is equivalent to stipulating that one can partition the constraint matrix $c = [c^D c^I]$ so that c^D is invertible, whence

$$cU + \phi = 0 \tag{81}$$

$$[c^D c^I]U + \phi = 0 \tag{82}$$

$$[I(c^D)^{-1} c^I]U + (c^D)^{-1}\phi = 0 \tag{83}$$

Consequently, $W = -(c^D)^{-1} c^I$, and the basis for the null space of the constraints in Eq. (82) is just

$$A_{\text{Kane}} = y \begin{bmatrix} -(c^D)^{-1} c^I \\ I \end{bmatrix} \tag{84}$$

Recalling for a moment the basis utilized in Ref. 10,

$$A_{\text{Wehage}} = \begin{bmatrix} -(a^D)^{-1} a^I \\ I \end{bmatrix} \tag{85}$$

the remarkable similarity in form in Eqs. (58) and (59) is immediately apparent. As noted earlier, Ref. 12 explains the numerical difficulties associated with Wehage's method in terms of the poor conditioning of $(a^D)^{-1}$. The above makes clear that a general formulation of Kane's method may suffer the same fate, if the condition number of $(c^D)^{-1}$ becomes poor. In fact, this phenomenon is guaranteed to occur if the constraint matrix becomes rank defective, as commonly occurs when mechanical systems assume "singular" configurations.[13,18] The observation that one must be at least cognizant of the potential for ill conditioning of submatrices arising in some implementations of Kane's equations is documented elsewhere, e.g., Ref. 15.

Nonlinear Nonholonomic Constraints

A third advantage derives from one's ability to extend the theory presented to first-order nonlinear nonholonomic constraints, i.e., constraints having the form

$$H_i(q, \dot{q}, t) = 0 \tag{86}$$

where each equation above is nonlinear, independent, and nonintegrable. One should note that these constraints differ from those considered earlier in that they are not required to be linear in the derivatives of the generalized coordinates. The physical and analytical aspects of these constraints have already been examined over the past 75 years (e.g., Ref. 19). As of yet, however, very little research on their computational treatment has appeared in the open literature. Such an extension is straightforward, in principle, within the theory described herein. Following the derivations carried out earlier in this paper and the results outlined in Ref. 19, one need only replace the constraint matrix a from Eq. (6) by the Jacobian matrix of partial derivatives

$$a_{ik} = \frac{\partial H_i}{\partial \dot{q}_k} \tag{87}$$

With this new definition of the constraint matrix a, one can follow the same computational steps outlined in the section on Maggi's equations of motion to obtain the governing equations:

$$A^T\{E(T) - Q\} = 0 \tag{88}$$

$$Q^{(r)} a^T \lambda = P_a T\{E(T) - Q\} \tag{89}$$

where $aA = 0$.

Hybrid Methods

Finally, the theory presented in this paper suggests interesting "hybrid" methods that can be obtained by combining methods shown in Table 1. For example, the method of null-space updating described in Ref. 13 need not be limited to only Lagrangian formulations. By inspection of Table 1, it is clear that a null-space updating formulation of Kane's equations is obtained by choosing the dual bases

$$\hat{a} = \begin{bmatrix} c_T \\ \tilde{Q}_2 \end{bmatrix} Y \tag{90}$$

and

$$\hat{A} = y[\tilde{Q}_1(c\tilde{Q}_1)^{-1}(I - \tilde{Q}_1(c\tilde{Q}_1)^{-1})\tilde{Q}_2] \tag{91}$$

where

$$c^T = \tilde{Q}\tilde{R} = [\tilde{Q}_1 \tilde{Q}_2]\tilde{R} \tag{92}$$

is the QR decomposition of the matrix c^T. Of course, as mentioned earlier, a practical implementation of Eqs. (90–92) does not require actual formation and calculation of Y and y.

It seems that Kim in Ref. 13 is well aware that evaluating the bases at each time step

$$\hat{a} = \begin{bmatrix} a_T \\ Q_2 \end{bmatrix} \tag{93}$$

and

$$\hat{A} = [Q_1(aQ_1)^{-1}Q_2] \tag{94}$$

are perfectly reasonable choices of dual bases, although perhaps not as computationally efficient as null-space updating, where the bases shown in Eqs. (91) and (92) are calculated only at certain time steps. The corresponding bases for a Kane-based algorithm can then be extracted from

$$\hat{a} = \begin{bmatrix} c_T \\ \tilde{Q}_2 \end{bmatrix} Y \tag{95}$$

and

$$\hat{A} = y\tilde{Q}_1(c\tilde{Q}_1^{-1})\tilde{Q}_2 \tag{96}$$

Null-Space and Range-Space Equations

In the last section it was shown that a wide class of formulations of the dynamics of multibody systems can be represented succinctly in terms of Maggi's equations

$$A^T\{E(T) - Q\} = 0 \tag{97}$$

$$(A')^T\{E(T) - Q\} = \lambda \tag{98}$$

where the columns of A form a basis for the null space of the constraint matrix a and the columns of A' form a basis dual to the columns of a^T:

$$a\dot{q} + b = 0 \tag{99}$$

$$aA = 0 \tag{100}$$

$$aA' = I \tag{101}$$

However, the discussion in the last section is in some sense incomplete, since the D reduced Eq. (97) is expressed in terms of $N = I + D$ unknown derivatives $\ddot{q}$. In this final section it is shown that Maggi's equations naturally lead to two "complementary" coordinate reduction techniques. This section presents the null-space equations and range-space equations for dynamics, respectively. It is shown that a primary computational cost of both methods is the calculation of an orthogonal basis at certain time steps.

In the following derivations it is assumed that either

1) the second-order equations derived from Maggi's equations have been reduced to first-order form by a change of variables having the form

$$\omega = \dot{q} \tag{102}$$

$$\dot{\omega} = \ddot{q} \tag{103}$$

2) or a quasicoordinate formulation has been employed that expresses the governing equations in temrs of the quasivelocities ω and the quasi-accelerations $\dot{\omega}$.

In either case it is assumed that the governing equations can be written in a form more general than that considered in the last section:

$$\boldsymbol{M}(\omega,t)\dot{\omega} = \boldsymbol{g}(\omega,t) + \boldsymbol{c}^T\boldsymbol{\lambda} \tag{104}$$

$$\boldsymbol{c\omega} + \boldsymbol{d} = 0 \tag{105}$$

In Eq. (104) M is an $N \times N$ system coefficient matrix, c is a $D \times N$ constraint matrix, $\boldsymbol{\omega}$ and $\boldsymbol{g}$ are N vectors, and $\boldsymbol{\lambda}$ and $\boldsymbol{d}$ are D vectors. If a Lagrangian formulation as in Ref. 20 is utilized, then the vector $\boldsymbol{\omega}$ contains derivatives of generalized coordinates. In the case where Kane's equations [Eq. (53)], Jourdain's principle,[7] or Newton-Euler equations are employed, ω may contain derivatives of quasicoordinates. In the following paragraphs the null-space and dual bases A and A' are defined by using the more general form of the constraints represented in Eq. (105):

$$cA = 0 \tag{106}$$

$$cA' = I \tag{107}$$

Null-Space Equations

Both the null-space and range-space formulations of the governing equations can be derived from Eq. (104) by decomposing the space in which the vector $\dot{\omega}$ is contained. By introducing the definition of the nullspace of the constraint matrix c and the range of c^T,

$$\text{null space}(\boldsymbol{c}) = \{\boldsymbol{x} \in R^N : \boldsymbol{cx} = 0\} \tag{108}$$

$$\text{range}(\boldsymbol{c}^T) = \{\boldsymbol{x} \in R^N : \boldsymbol{y}R^D \quad \text{with} \quad x = \boldsymbol{c}^T y\} \tag{109}$$

The rank nullity theorem guarantees that a direct sum decomposition of R^N is possible:

$$R^N = \text{null space}(\boldsymbol{c}) + \text{range}(\boldsymbol{c}^T) \tag{110}$$

Consequently, the N vector of unknown derivatives $\dot{\omega}$ can be expressed as the sum of two orthogonal components at a specific time t,

$$\dot{\omega} = \boldsymbol{A\xi} + \boldsymbol{c}^T\boldsymbol{\eta} \tag{111}$$

where $\boldsymbol{\xi}$ and $\boldsymbol{\eta}$ are $I = N - D$ and D vectors, respectively, and $\boldsymbol{A}$ is an $N \times (N - D)$ matrix whose columns span the null space of c. By differentiating the constraints in Eq. (99) and utilizing the orthogonal decomposition in Eq. (111), one can find that

$$\dot{\omega} = \boldsymbol{A\xi} - \boldsymbol{c}^T(\boldsymbol{cc}^T)^{-1}(\dot{\boldsymbol{c}}\omega + \dot{\boldsymbol{d}}) \tag{112}$$

The null-space equations for constrained dynamics result from substituting Eq. (112) into the first of Maggi's equations [Eq. (97)] and premultiplying by A^T so that

$$\xi = (A^TMA)^{-1}\{A^Tg + A^TMc^T(cc^T)^{-1}(\dot{c}\omega + \dot{d})\} \tag{113}$$

and

$$\dot{\omega} = A(A^TMA)^{-1}\{A^Tg + A^TMc^T(cc^T)^{-1}(\dot{c}\omega + \dot{d})\} - c^T(cc^T)^{-1}(\dot{c}\omega + \dot{d}) \tag{114}$$

A special case of these equations for a particular choice of null-space basis A can be found in Ref. 6.

It should be noted that the governing equation [Eq. (114)] has been derived by assuming that the constraints are, at most, of the "linear, nonholonomic" type. These equations generalize the null-space equations derived in Refs. 1 and 4, which assume that the constraints are instead of the "holonomic, scleronomic" type. These works derive the null-space equations via a decomposition having the structure

$$\dot{y} = A\dot{\xi} + \dot{A}\xi \tag{115}$$

Although the final null-space equations derived in this fashion are subtly different from those shown in Eq. (114), they have the same character. That is, both representations are functions of the null-space matrix A.

Range-Space Equations

Another way of solving for the vector of derivatives $\dot{\omega}$ in Eq. (106) using the orthogonal decomposition in Eq. (111) leads to the range-space equations. If one premultiplies Eq. (104) by cM^{-1} and substitutes Eq. (112) for the vector $\dot{\omega}$, the undetermined multipliers are then

$$\lambda = -(cM^{-1}c^T)^{-1}\{\dot{c}\omega + d + cM^{-1}g\} \tag{116}$$

The important attribute of the above equations is that they are independent of the quasivelocities $\dot{\omega}$. Hence, the governing equations can be expressed as

$$\dot{\omega} = M^{-1}\{g - c^T(cM^{-1}c^T)^{-1}\{\dot{c}\omega + \dot{d}) + cM^{-1}\}\} \tag{117}$$

Equation (117) is referred to as the range-space equation for constrained multibody dynamics. Various forms of this equation have appeared in dynamics formulations in recent literature.[1,21] As in the case of the null-space formulations, these equations can be derived easily by employing Maggi's equations. By making the specific choice for the dual basis A' in the second of Maggi's Eq. (98),

$$A'^{T} = (cM^{-1}c^{T})^{-1}cM^{-1} \tag{118}$$

the governing equation becomes

$$(cM^{-1}c^{T})c\dot{\omega} = (cM^{-1}c^{T})^{-1}cM^{-1}g + \lambda \tag{119}$$

Substituting the constraint equation [Eq. (105)] for $c\dot{\omega}$ yields again the expression [Eq. (116)] for the multipliers and [Eq. (117)] for the reactionless equations of motion.

Relationship to Dual Principles of Optimization

Another benefit derived from the null-space and range-space equations for multibody dynamics is that the integration procedures closely resemble dual principles of optimization. Since these dual optimization principles have been the focus of much research since the 1970s, much useful information can be extracted from the relationship between the methods. The null-space and range-space equations for optimization arise from the solution of the quadratic minimization problem

$$\underset{\chi \in R^{N}}{\text{minimizer}} \left\{\tfrac{1}{2}x^{T}Mx + x^{T}g\right\} \tag{120}$$

subject to the equality constraints

$$cx = 0 \tag{121}$$

The null-space and range-space methods of optimization solve the constrained minimization problem, respectively, by the equations

$$x = A(A^{T}MA)^{-1}A^{T}g \tag{122}$$

or

$$x = M^{-1}\{g - c^{T}(cM^{-1}c^{T})^{-1}cM^{-1}g\} \tag{123}$$

where $cA = 0$; i.e., A is a basis for the null space of c. The similarity between these methods and the corresponding equations for dynamics derived from Maggi's equations is immediately apparent.

Furthermore, objective criteria have been developed to select either the null-space or range-space equations in optimization:

1) If there are many constraints, in other words, if D is much greater than $I = N - D$, then the system coefficient matrix $A^{T}MA$ is of relatively low dimension and is easily factored. In this case null-space equations are more efficient.

2) If there are relatively few constraints, then the system coefficient matrix $cM^{-1}c^{T}$ is easily factored, and the range-space method is preferred.

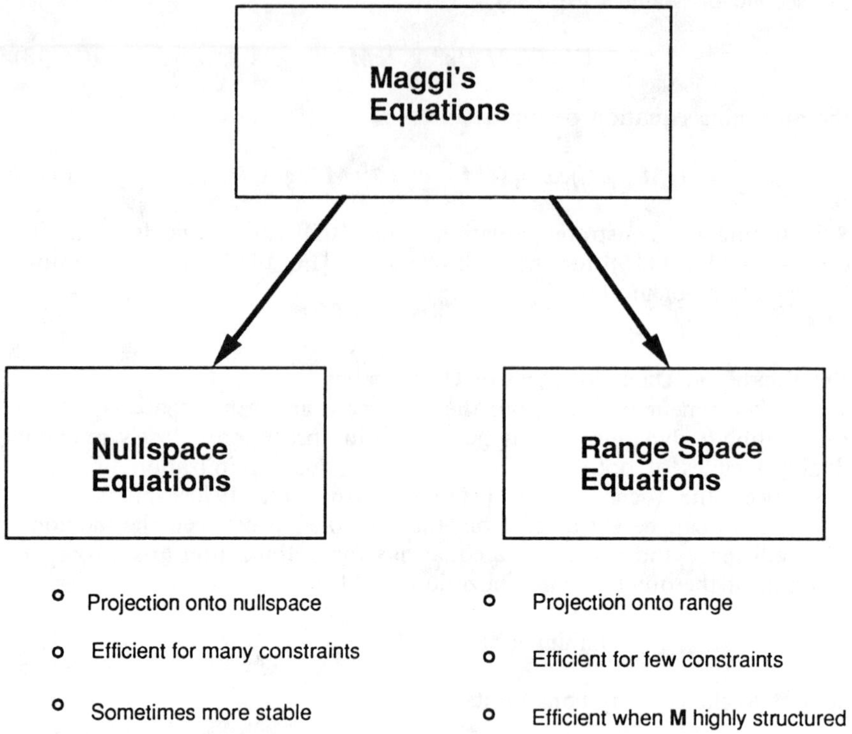

Fig. 1 Coordinate reduction methods for Maggi's equations.

3) If the structure of the matrix M is such that it is easily factored, the range-space method may be advisable.

4) If neither method is clearly superior in terms of computational work, the null-space method is preferred due to its superior stability.

These criteria obviously can be extended to the null-space and range-space methods for dynamics. Moreover, they agree with some empirical results recently collected by Refs. 1 and 4. Figure 1 depicts the relationship of Maggi's equations to the null-space and range-space equations and summarizes the criteria for selecting a particular approach.

Computational Complexity Analysis

When developing a concurrent multiprocessing algorithm for a particular task, it is imperative to identify the steps that constitute the greatest computational cost. Only in this fashion can an efficient parallel method be designed. In this section a brief discussion of the computational cost associated with the null-space and range-space formulations is presented. Surprisingly, despite the fact that the null-space and range-space formulations are in a sense complementary, it is shown in this section that the calculation of an orthogonal basis at specified time steps dominates the overall solution time in both methods.

When some forward, time-marching algorithm is applied to the null-space equations in Eq. (114), the nonlinear right-hand side of the equation must be evaluated at several specific values of t and $\boldsymbol{\omega}$ defined by the integration rule. Assuming that the nonlinear system coefficient matrix M, constraint matrix c, and vector $\boldsymbol{g}$ have been formed at the specified integration time, the solution of Eq. (114) for the derivatives $\dot{\boldsymbol{\omega}}$ entails the following primary (cubic order) computational steps:

1) Calculation of an orthogonal basis for the null space of the constraint matrix. In terms of floating-point operations (flops), this step requires roughly $2N^2D + 11D^3$ flops if a singular valued decomposition is employed, whereas a QR decomposition may also require on the order of D^3 flops.

2) Formation of the reduced system coefficient matrix $\boldsymbol{A}^T\boldsymbol{M}\boldsymbol{A}$. This step requires between NI^2 and $\frac{1}{2}N^2I^2$ flops, depending on the structure of the coefficient matrix M.

3) Factorization of the reduced system coefficient matrix $\boldsymbol{A}^T\boldsymbol{M}\boldsymbol{A}$. The cost of the step is $\frac{1}{3}D^3$ or $\frac{1}{6}D^3$, depending on whether an LU or LDL^T factorization is employed.

The determination of a similar breakdown for the time spent solving the range-space equation [Eq. (117)] for the derivatives $\dot{\boldsymbol{\omega}}$ is a bit more complicated. The difficulty arises when one attempts to factor the coefficient matrix $cM^{-1}c^T$ appearing in Eq. (117). The error induced in factoring this matrix directly is bounded by

$$\{\kappa(\boldsymbol{c})\}^2\kappa(\boldsymbol{M}) \tag{124}$$

where $\kappa(\ )$ denotes the condition number of the matrix in parentheses. When the condition of the constraint matrix is large, as is the case when a system of constraints becomes nearly redundant, the bound in Eq. (124) indicates, heuristically, a relatively large induced error. A means of circumventing this problem is to employ an orthogonal factorization of the constraint matrix c^T:

$$\boldsymbol{c}^T = \boldsymbol{O}\boldsymbol{P} \tag{125}$$

where $\boldsymbol{O}$ is an $N \times D$ orthogonal matrix, and $\boldsymbol{P}$ is a $D \times D$ upper triangular matrix. With this factorization, the calculation of $cM^{-1}c^T$ is achieved by using the identity

$$(\boldsymbol{c}\boldsymbol{M}^{-1}\boldsymbol{c}^T)^{-1} = \boldsymbol{P}^{-T}(\boldsymbol{O}^T\boldsymbol{M}^{-1}\boldsymbol{O})^{-1}\boldsymbol{P}^{-1} \tag{126}$$

The factorization of P and P^T in Eq. (126) is easily achieved due to the upper triangular form of P. In addition, the factorization of $O^TS^{-1}O$ induces an error that is bounded by

$$(\kappa(\boldsymbol{O}))^2\kappa(\boldsymbol{M}) = \kappa(\boldsymbol{M}) \tag{127}$$

since $\boldsymbol{O}$ is orthogonal and $\kappa(\boldsymbol{O}) = 1$. When this procedure is used to calculate the factorization of $\boldsymbol{cM}^{-1}\boldsymbol{c}^T$, the solution of the range-space equations for the derivatives $\dot{\boldsymbol{\omega}}$ entails the following primary computational steps:

1) Calculation of an orthogonal basis for the range space of the constraint matrix c. The computational cost of this step is roughly ND^2 flops if the Gram-Schmidt procedure is employed.

2) Factorization of the system coefficient matrix M^{-1}.

3) Formation of the reduced system coefficient matrix $cM^{-1}c^T$. As in the null-space equations, this step varies from ND^2 to $\frac{1}{2}N^2D^2$ flops, depending on the form of the matrix M.

4) Factorization of the reduced system coefficient matrix $cM^{-1}c^T$, at a cost of $\frac{1}{3}D^3$ for the LU decomposition, or $\frac{1}{6}D^3$ flops for the LDL^T decomposition.

From the preceding tabulations it is clear that the computational cost incurred in solving for the derivatives $\dot{\boldsymbol{\omega}}$ is divided among the following: 1) matrix factorization, 2) orthogonal basis calculation, and 3) matrix products $cM^{-1}c^T$ and A^TMA.

Of the operations required by the null-space and range-space formulations, concurrent strategies for matrix factorization have been explored extensively. Furthermore, the products $cM^{-1}c^T$ and A^TMA are rather straightforward to calculate in a concurrent computing environment. Hence, one topic suggested by this paper is the investigation of novel concurrent orthogonal basis algorithms for multibody simulation.

Computational Examples

By carrying out some simple numerical experiments, one can easily demonstrate that the most computationally demanding phases of the class of formulations described are thus far

1) The calculation of the orthogonal bases required to eliminate constraint force reactions in either the null-space or range-space formulation. This computational cost is dominant when the number of constraints is large relative to the total number of system degrees of freedom.

2) The generation and assembly of the nonlinear equations. This step requires an inordinate amount of computer time. In fact, as the number of flexible degree of freedom increases, this step quickly overtakes the "projection" phase as the most time-consuming.

The relative cost of these steps is illustrated in this section in two examples carried out in a sequential computing environment. In the first test case, the 12-bar system in Fig. 2 comprised of rigid links is considered. This system has been modeled using 72 (quasi-) coordinates and is subject to 48 constraints arising from the 16 spherical joints connecting the system. Tables 2 and 3 summarize the time spent during several phases of the sequential solution processing the null-space and range-space methods. The Givens QR decomposition and the modified Gram-Schmidt procedure are used to calculate the bases for the null-space and range, respectively, of the transposed constraint matrix. Both tables indicate the times in terms of

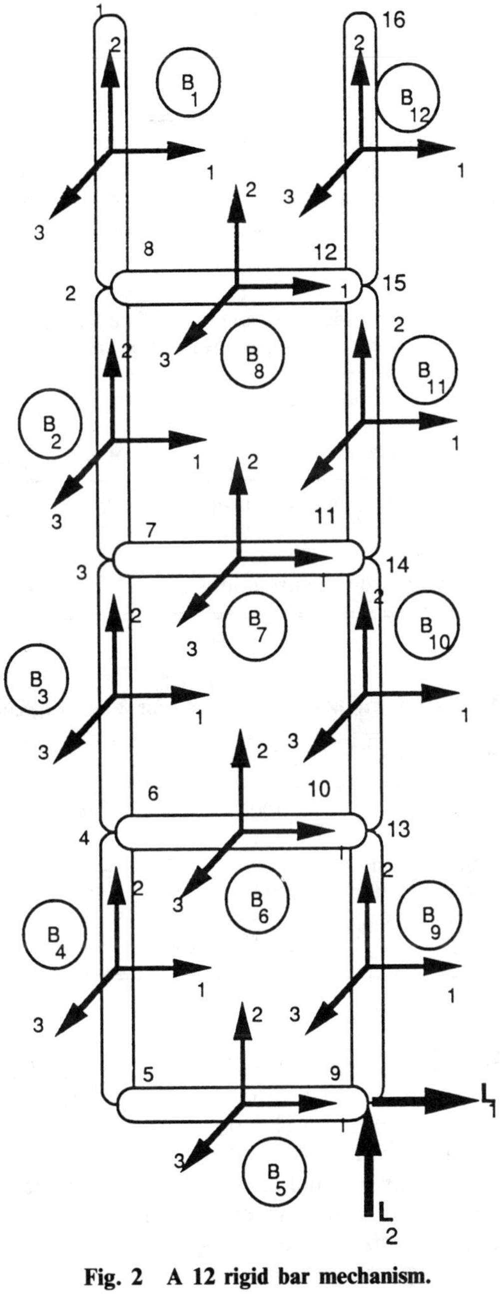

Fig. 2 A 12 rigid bar mechanism.

Table 2 Breakdown of simulation time for 12-bar mechanism: null-space simulation method

Null-space timing results		
Type of calculation	1/50 s	%
Orthogonal basis calculation	914	83
Formation/factorization	143	13
Quadratic order contributions	41	4

Table 3 Breakdown of simulation time for 12-bar mechanism: range-space simulation method

Range-space timing results		
Type of calculation	1/50 s	%
Orthogonal basis calculation	440	40
Formation	497	45
Factorization	136	12
Quadratic order contributions	12	1

Table 4 Breakdown of simulation time for 6-beam mechanism

Computation type	1/50 s	%	1/50 s	%
Gram-Schmidt	100	4	100	10
Formation	511	21	511	51
LU Decomposition	20	1	20	2
Nonlinear Formation	1681	71	296	30
Miscellaneous	69	3	69	7
Total	2381	100	996	100

1/50 of a second for a single evaluation of the right-hand side of the governing equations to calculate the derivative terms $\dot{\omega}$. Clearly, the calculation of an orthonormal basis constitutes a significant portion of the overall solution time in both methods. From the first table the calculation of an orthogonal basis for the null space of the constraints using a Givens QR decomposition is seen to require over 80% of the total solution time. The formation and factorization of the system coefficient matrix A^TMA requires a mere 13% by comparison. When the range-space equtions are employed the calculation of an orthogonal basis via the modified Gram-Schmidt procedure for the range of the transposed constraint matrix takes

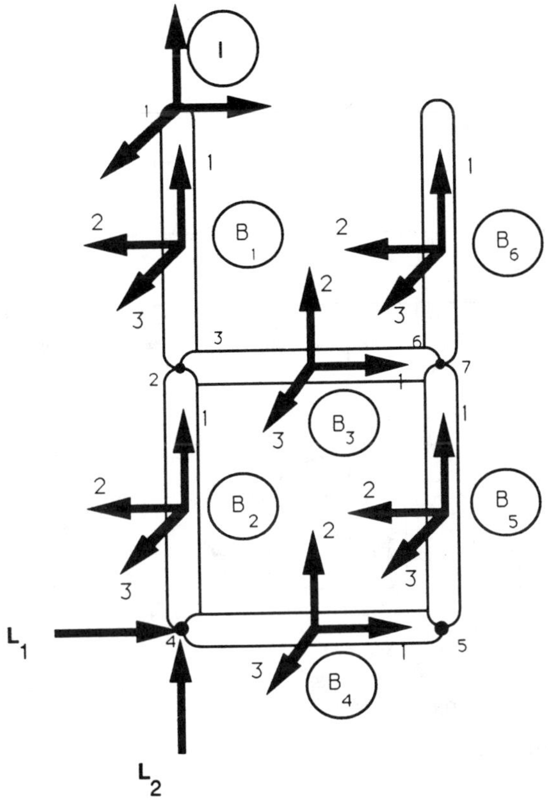

Fig. 3 A 6-beam mechanism.

40% of the total solution time. A six-beam, flexible mechanism is considered in Fig. 3 and in Table 4. In this simulation each beam has six rigid and four flexible degrees of freedom; thus, the system has a total of 60 degrees of freedom. Table 4 illustrates the fact that over 70% of the time spent in calculating the derivatives $\dot{\omega}$ required by the integration scheme is spent forming and assembling the governing equations. In comparison, only 4% of the total solution time is spent calculating an orthogonal basis for constraint elimination.

Survey of Concurrent Orthogonal Basis Methods

In view of the previous numerical examples, this section summarizes some recently developed concurrent orthogonal basis methods that employ graph theoretic representations of system connectivity to assign tasks to independent processors. Although many algorithms are considered, the "columnwise" modified Gram-Schmidt and columnwise QR procedures derived in Ref. 22 consistently appear to be the most efficient. The actual

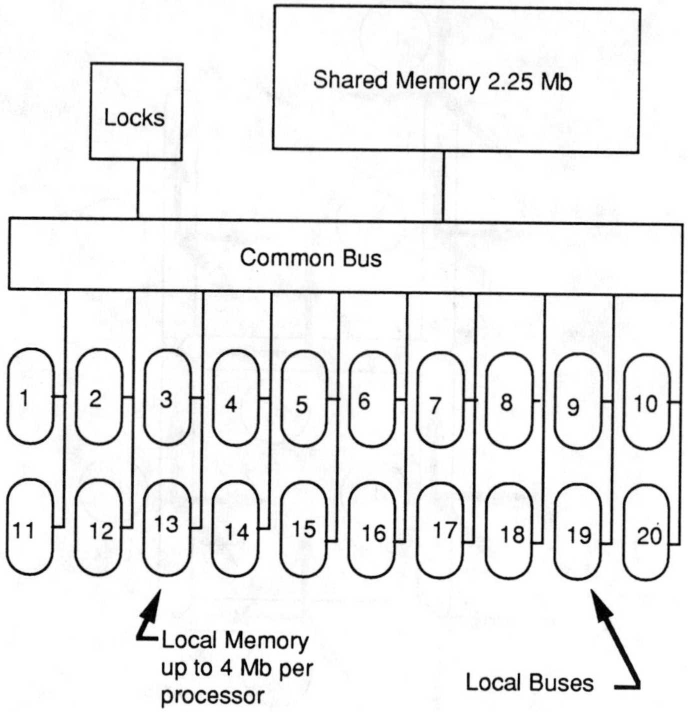

Fig. 4 Flex 32 architecture schematic for 20 processors.

task scheduling that is carried out in these procedures is somewhat complicated and is beyond the scope of this paper. The interested reader is referred to Ref. 22 for details.

The columnwise versions of these orthogonal basis methods have the following advantages over similar approaches: The techniques reduce subdiagonal fill-in to the original block band of the transposed constraint matrix, the methods dramatically decrease the overall number of operations that must be applied to the matrix, and the approaches are automatically load-balanced. It should be noted that, if every column of the matrix has a full, nonzero length equal to the number of rows, then the columnwise QR approaches described in Ref. 22 reduces to the Greedy scheme outlined in Ref. 23. Furthermore, the task of regularly ordering the directed graph before orthogonalization can be likened to node numbering schemes used in finite-element methods to reduce the bandwidth prior to factorization.

Figures 5 and 6 succinctly represent the outcome of the comparison of the various concurrent techniques for calculating a range-space basis in terms of the relative speedups defined earlier. With respect to the execution time for the Greedy QR decomposition, the concurrent modified Gram-Schmidt procedures are by far the most efficient algorithms for calculating an orthogonal basis for the range space of the constraints. The concurrent and

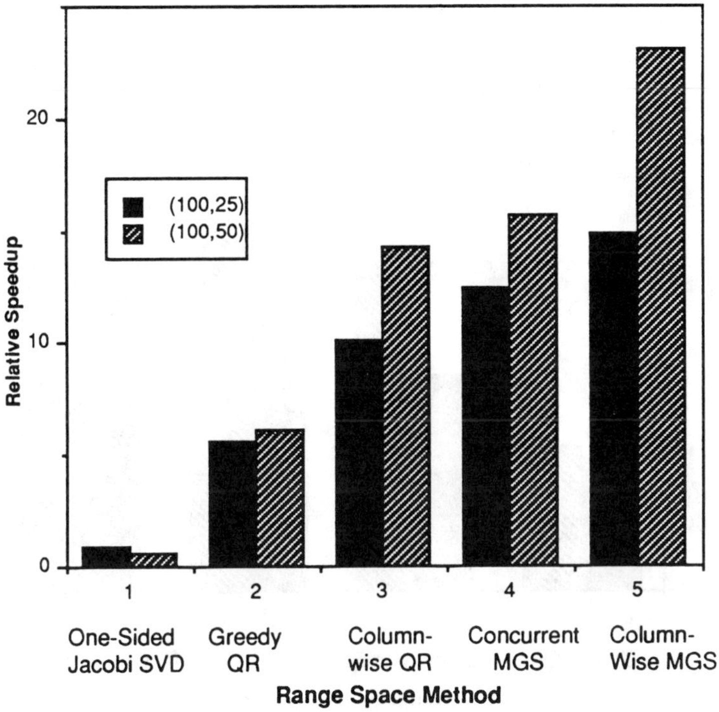

Fig. 5 Relative speedup for range-space basis generation.

columnwise modified Gram-Schmidt procedures have yielded speedups for the 100×50 matrix of approximately 16 and 23, respectively. In other words, it is estimated that on seven processors these concurrent algorithms require around $1/16$ and $1/23$ of the execution time required by the sequential Greedy QR decomposition to calculate an orthonormal basis for the range of the constraint matrix. After the modified Gram-Schmidt methods, the Givens-based concurrent columnwise and concurrent Greedy QR decompositions are seen to have relative speedups near 15 and 6, respectively, for the 100×50 matrices using 7 processors. However, it must be kept in mind when interpreting these results that these encouraging relative speedups are due to several different factors.

The relative speedups are not due solely to parallelism, but are due to the fact that the sequential methods appearing in the literature make no use of the special block upper triangular structure of the constraint matrix. These comparisons are important to those currently employing "off-the-shelf" matrix packages that do not consider the sparsity of the matrix.

It should be noted in Fig. 5 that the Jacobi-based singular value decomposition is not competitive with the other methods investigated. Although regular ordering does improve the performance of the method in some test cases not shown here, the decrease in execution time is not

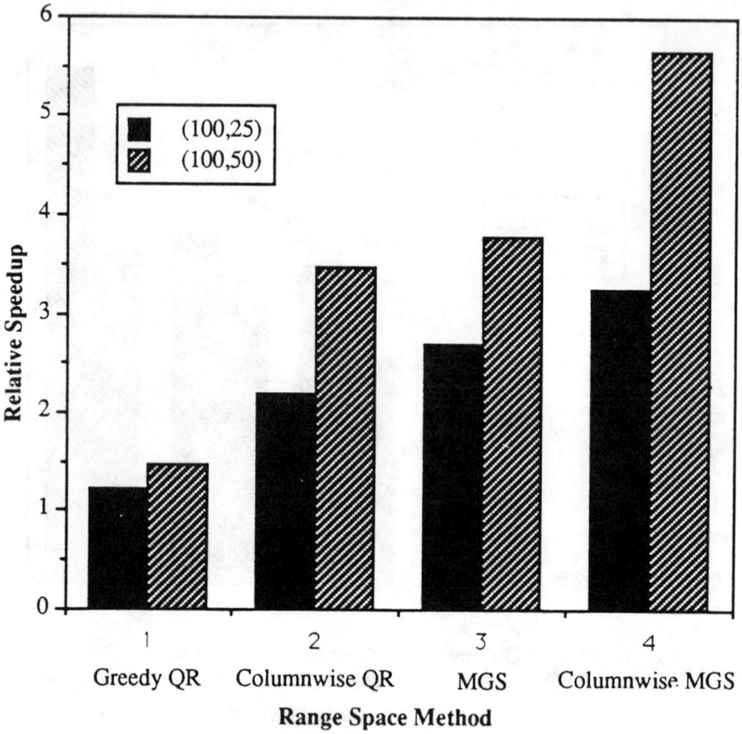

Fig. 6 Relative speedup for range-space basis generation.

significant. The method is not included in the remaining comparisons that follow.

A more pessimistic comparison of the merits of the various algorithms is achieved by normalizing the relative speedups with respect to the sequential modified Gram-Schmidt procedure. Figure 6 illustrates that the columnwise *QR*, concurrent modified Gram-Schmidt, and columnwise modified Gram-Schmidt require about $\frac{1}{4}$, $\frac{1}{4}$, and $\frac{1}{6}$, respectively, of the execution time of the sequential modified Gram-Schmidt procedure.

Figure 7 illustrates the relative speedup of the Greedy *QR*, Householder *QR*, and columnwise *QR* for the calculation of an orthogonal basis for the null space of the constraint matrix employing seven processors. In this graph the execution times of the sequential Householder *QR* method have been used to normalize the relative speedups. Clearly, the columnwise *QR* decomposition is by far the most efficient algorithm. For the largest matrices tested the columnwise *QR* decomposition is an order of magnitude faster than the sequential *QR* decomposition.

Theoretical Speedup Comparison

Figures 8–10 summarize the projected, theoretical speedup as the number of processors is increased beyond the seven available on the FLEX 32

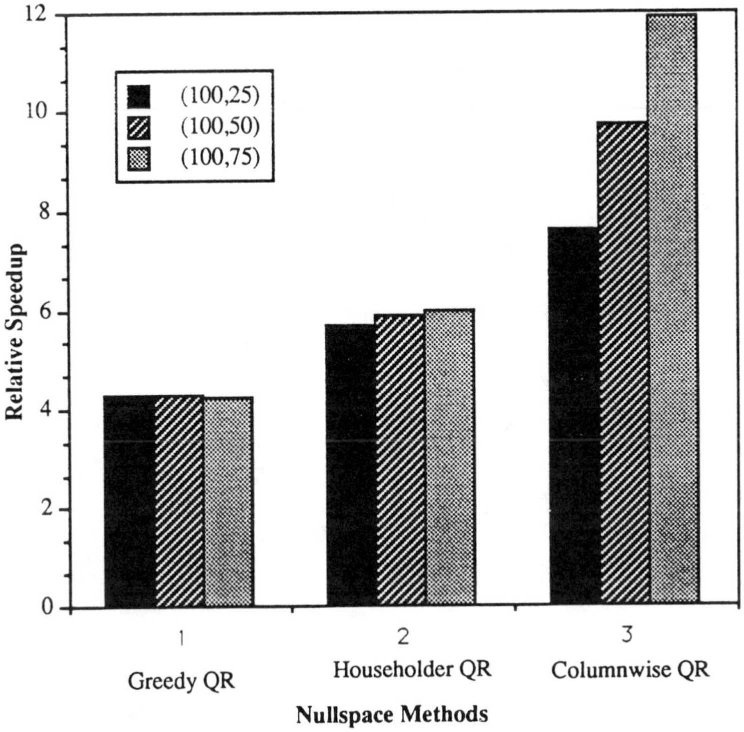

Fig. 7 Relative speedup for null-space basis generation.

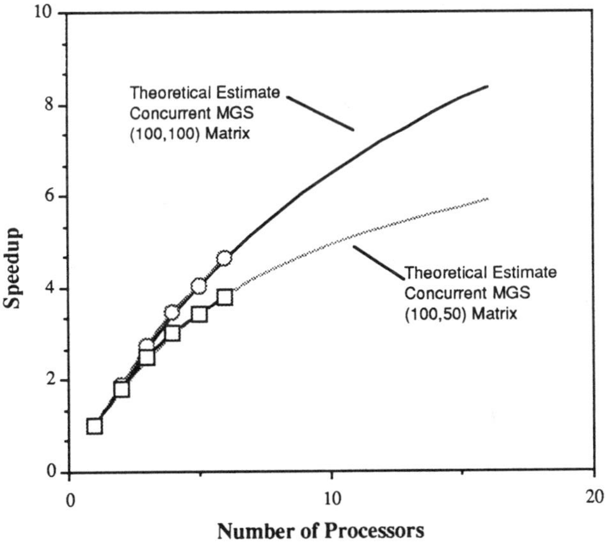

Fig. 8 Theoretical speedup for concurrent modified Gram-Schmidt procedure.

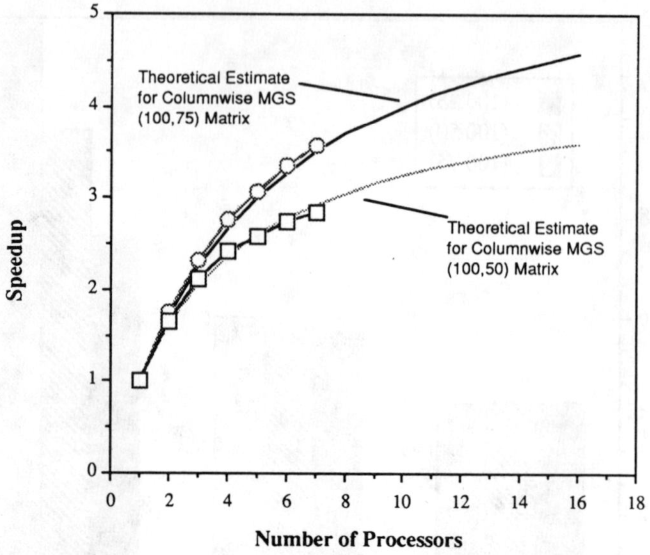

Fig. 9 Theoretical speedup for columnwise concurrent modified Gram-Schmidt procedure.

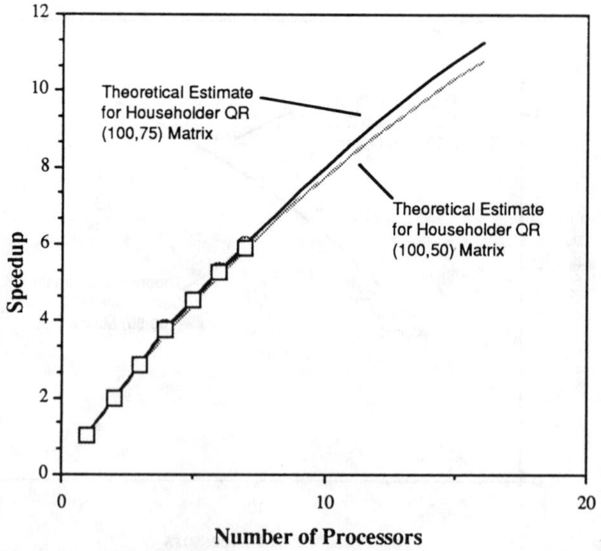

Fig. 10 Theoretical speedup for concurrent Householder *QR* decomposition.

multiprocessor at the Georgia Institute of Technology. In the case of the concurrent modified Gram-Schmidt procedure, excellent correlation has been achieved between the theoretical estimates derived in Ref. 22 and the experimental derived values. It is estimated in Fig. 8 that the use of 16 processors with the concurrent modified Gram-Schmidt (MGS) procedure should yield speedups as high as nine for large problems, matrices of order (100×100). Thus, an order-of-magnitude improvement is a reasonable expectation for the concurrent MGS procedure when applied to "large" problems.

The columnwise version of the MGS procedure, on the other hand, is projected to yield speedups using 16 processors near five for large problems, say, on the order of (75×75). Of course, the fact that the columnwise MGS procedure is projected to have a lower speedup than the MGS *a* procedure is not surprising. The columnwise algorithm has far fewer operations overall and a faster execution time.

Finally, Fig. 10 depicts the correlation between theoretical estimates and numerical estimates for the concurrent Householder QR decomposition. As shown in the figure, excellent agreement is achieved for the test cases considered. When 16 processors are employed to calculate an orthogonal basis for the null space using the concurrent Householder method for large problems, an order-of-magnitude improvement in performance is projected.

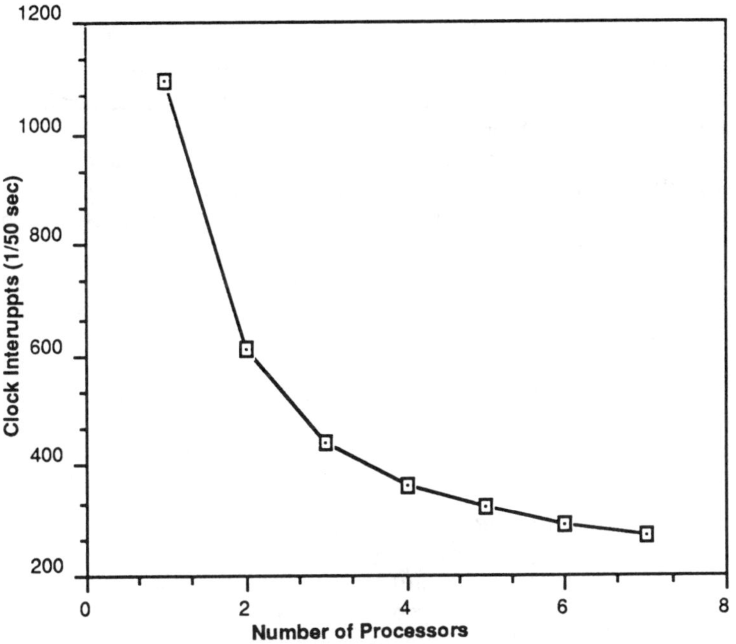

Fig. 11 Execution time for null-space solution of derivatives.

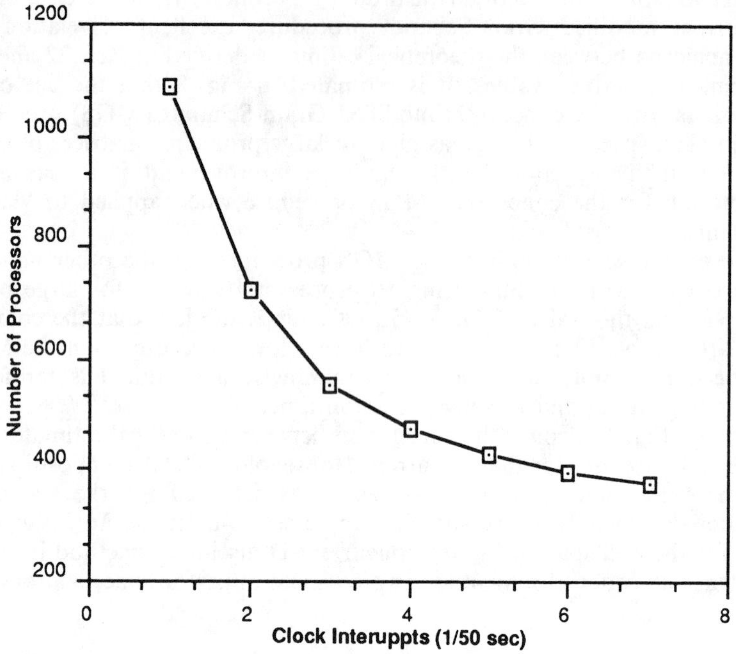

Fig. 12 Execution time for range-space solution of derivatives.

Figures 11 and 12 summarize the performance of the concurrent orthogonal basis methods for the 12 rigid bar system considered earlier in Fig. 2. Each figure depicts the time spent to calculate once the derivatives $\dot{\omega}$ in the governing equations. When seven processors are employed for the null-space equations, the solution time is reduced by a factor of four when the columnwise QR decomposition is used to calculate the null-space basis. Similarly, the solution time for the range-space equations is reduced by a factor of three when the columnwise modified Gram-Schmidt procedure is used to solve for the derivatives $\dot{\omega}$. The improved performance of the null-space equations can be atrributed to the fact that a larger fraction of the overall solution time is spent calculating the null-space basis than the basis for the range space of the constraints. The results for both methods are very good indeed when one remembers that the "cubic order" LU factorization in the solution procedure has been carried out sequentially.

Concurrent Equation Formation and Assembly

Methods for selecting generalized (or quasi-) coordinates for multibody systems and corresponding techniques for calculating the governing equations of motion abound. The approach followed in this paper is similar to that derived in Ref. 24. By employing the formalism discussed in Ref. 24, the equations for each of the uncoupled substructures are block structured,

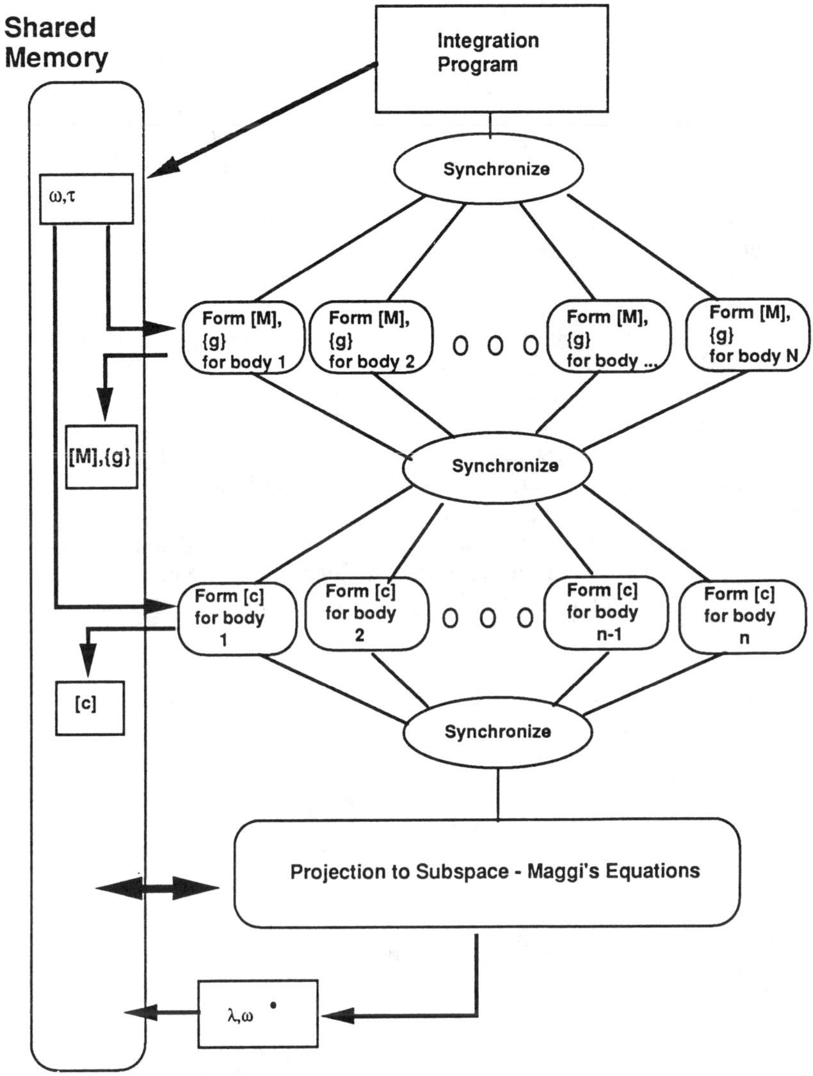

Fig. 13 Schematic of concurrent formation.

and the concurrent integration strategy in Fig. 13 can be employed. At the beginning of each integration step, the integration values $\hat{\omega}$ and $\hat{t}$ are set in shared memory. All processors are then launched, and each processor reads the integration values $\hat{\omega}$ and $\hat{t}$ required to form the contributions to the governing equations of the substructure to which it is assigned. Some processors may be responsible for several individual substructures, so that each processor loops over its collection of substructures until it is finished. After each processor completes the calculation of the contributions due to a particular substructure, it immediately stores that M matrix and $\boldsymbol{g}$ vector in shared memory. Because the governing equations are highly structured, and because the substructure contributions reside in different physical locations in shared memory, this portion of the integration procedure can proceed with no synchronization. Furthermore, each processor requires the physical and material properties of only the substructures for which it is responsible. The overall storage requirements are consequently distributed among the processors' local memories.

After the nonlinear formation and assembly of the matrices M and c and the vector $\boldsymbol{g}$, the projection phase of the solution can proceed. By using any of the methods presented in the section on Maggi's equations of motion, the governing equations can be solved as shown in Fig. 13. The orthogonal basis calculation is carried out directly in shared memory on the matrix $\boldsymbol{c}$, and the solution using the range-space or null-space equations likewise is carried out in shared memory. The procedure ends by storing the calculated values of $\boldsymbol{\lambda}$ and $\dot{\boldsymbol{\omega}}$ in shared memory, and the procedure repeats for all of the required integration values specified in the integration rule.

The results of the simulation of the six-beam mechanism are contained in Figs. 14–19. Figure 14 shows that one can be confident in the accuracy of the results, as the maximum constraint violation ϵ_{max} remained on the order of 10^{-14} following the impulsive load. The location of the center of mass of beam 4 is plotted in Fig. 15. The graph shows both the location of the

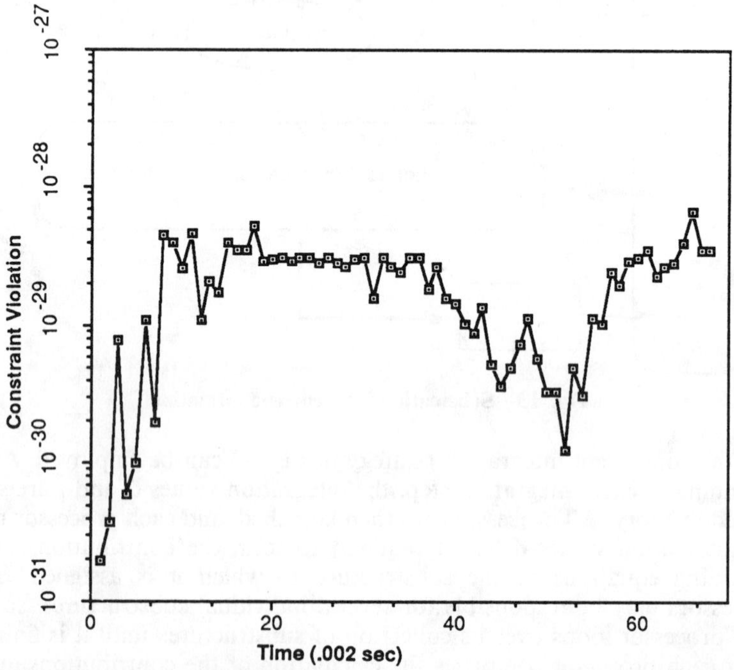

Fig. 14 Constraint violation: 6-beam mechanism, 24 modes, $DT = 0.001$ s.

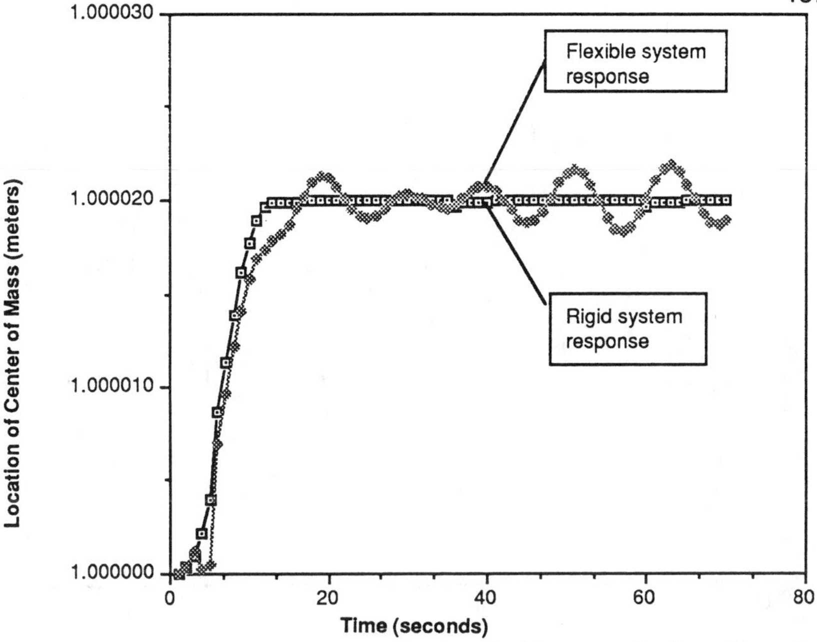

Fig. 15 ***X*-location of center of mass for bar 4; 6-beam mechanism, 24 modes, $DT = 0.001$ s.**

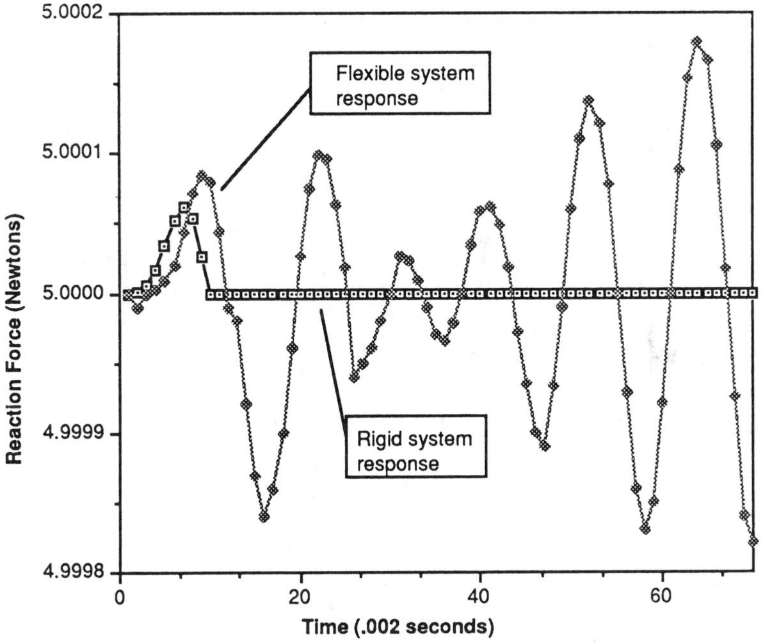

Fig. 16 ***Y*-direction reaction force on body 1, node 2, due to ground; 6-beam mechanism, 24 modes, $DT = 0.001$ s.**

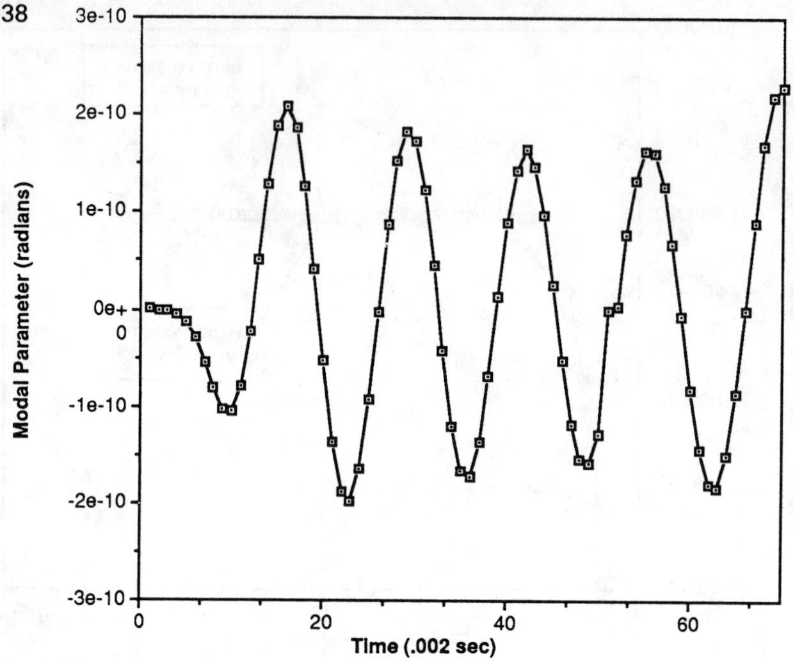

Fig. 17 ***x-y*** **Plane symmetric bending mode; time history, body 4; 6-beam mechanism, 24 modes,** $DT = 0.001$ **s.**

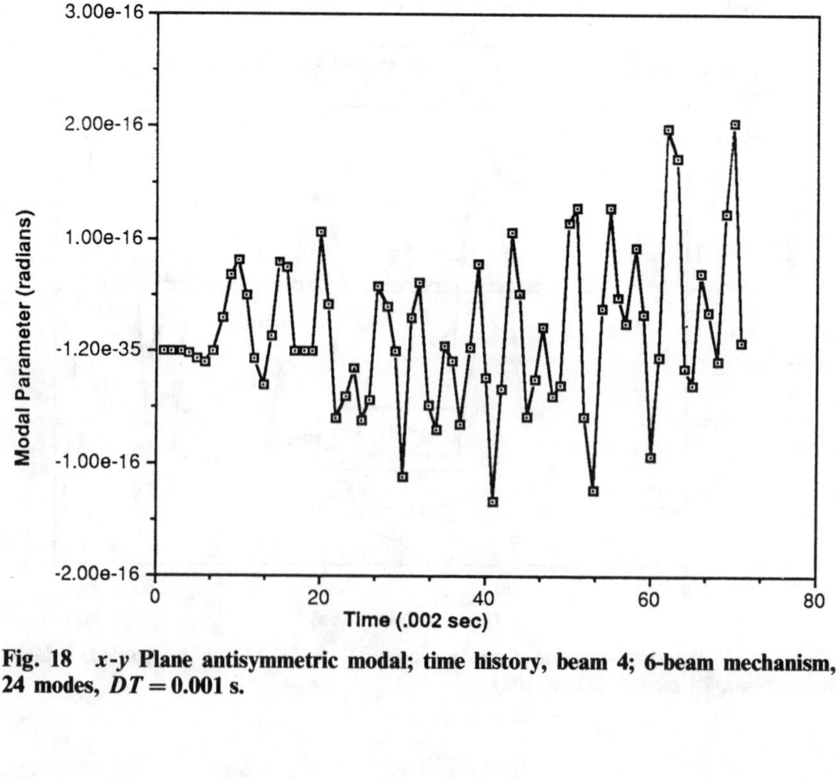

Fig. 18 ***x-y*** **Plane antisymmetric modal; time history, beam 4; 6-beam mechanism, 24 modes,** $DT = 0.001$ **s.**

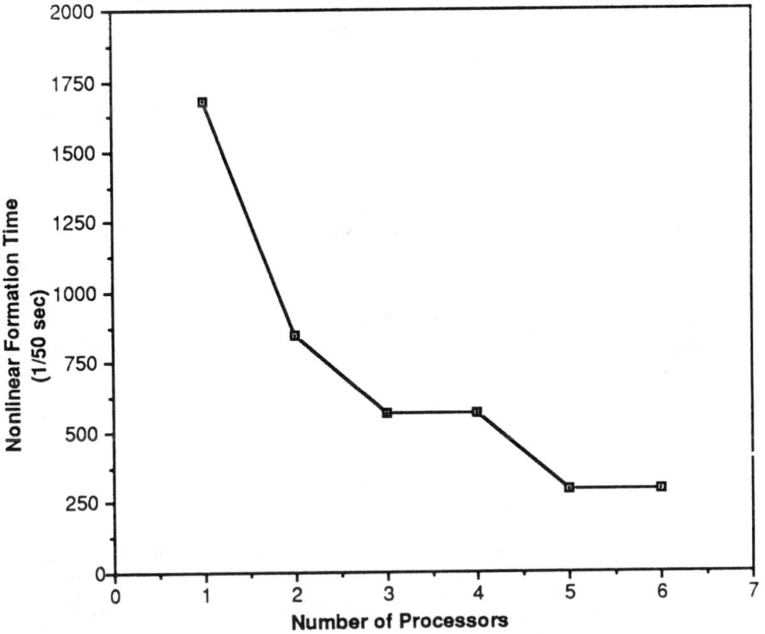

Fig. 19 Time to form the nonlinear equations; 6-beam mechanism, 24 modes, $DT = 0.001$ s.

center of mass for the system comprised of six beams and the corresponding result for six rigid bars. Figure 16 depicts the constraint force acting on one of the bodies immediately following impact. This figure illustrates the importance of flexibility in achieving accurate representations of internal forces for design purposes. Figures 17 and 18 depict the modal parameters corresponding to the x-y plane symmetric and antisymmetric bending modes, respectively. These two figures illustrate why multibody simulation of flexible systems can be so time-consuming. Even though the modal parameters in Figs. 17 and 18 have a very small amplitude, the frequency of oscillation is two orders of magnitude higher than the frequency associated with the "gross motion" of the center of mass of the beams. Very small time steps must be taken to capture these high-frequency modal contributions to the equations of motion.

A breakdown of the time spent in several phases of the simulation is depicted in Table 4. When the concurrent formulation of the governing equations is employed, the execution time in the formation phase is reduced by a factor of 5.7 when six processors are employed. Figure 19 displays how the execution time in formation decreases as a function of the number of processors. As shown in the figure, the greatest decreases occur at divisors of the number of substructures. Even though every other calculation in the simulation procedure has been carried out sequentially, the speedup of the simulation is near 2.5.

Conclusions

Several conclusions can be drawn from this investigation that are of interest to those involved in multibody dynamics research. It has been shown that Maggi's equations can be extended using common projective methods to provide the theoretical foundation for a wide class of recent methods in multibody simulation.

The unified method enhances one's ability to compare and contrast available methods. Moreover, theory explicitly depicts the nature of singularities that can arise in some simulation methods, and it generalizes some simulation procedures to enable constraint force calculation by clearly depicting the projection operator used to calculate the constraint forces. The theory suggests novel hybrid techniques that have not yet appeared in the literature and the theory closely resembles dual principles in optimization. This similarity can be exploited to obtain objective criteria for choosing a particular method and is amenable to concurrent multiprocessing.

References

[1]Agrawal, O. P., "Dynamic Analysis of Multi-Body Systems Using Tangent Coordinates," *The Theory of Machines and Mechanisms*, edited by E. Bautista, J. Garcia-Lomas, and A. Navarro, Pergamon, Oxford, UK, 1987, pp. 533–536.

[2]Amirouche, F. M., and Huston, R. L., "Dynamics of Large Constrained Flexible Structures," *Journal of Dynamic Systems, Measurement and Control*, Vol. 110, March 1988, pp. 78–83.

[3]Amirouche, F. M., and Jia, T., "Automatic Elimination of the Undetermined Multipliers in Kane's Equations Using a Pseudo-Uptriangular Decomposition (putd) Method," *Computers and Structures*, Vol. 27, No. 2, Nov. 1987, pp. 203–210.

[4]Garcia de Jalon, J., and Jimenez, J. M., "Basis for the Nullspace for the Jacobean of Constrained Multibody Systems," *The Theory of Machines and Mechanisms*, edited by E. Bautista, J. Garcia-Lomas, and A. Navarro, Pergamon, Oxford, UK, 1987, pp. 501–504.

[5]Neimark, J. I., and Fufaev, N. A., *Dynamics of Nonholonomic Systems*, American Mathematical Society, Providence, RI, 1972.

[6]Singh, R. P., and Likins, P. W., "Singular Value Decomposition for Constrained Dynamical Systems," *Journal of Applied Mechanics*, Vol. 52, Dec. 1985, pp. 943–948.

[7]Schielen, W. O., "Dynamics of Complex Multibody Systems," *SM Archives*, Vol. 9, July 1984, pp. 159–195.

[8]Golub, G. H., and Vanm Loan, C. F., *Matrix Computations*, Johns Hopkins Univ. Press, Baltimore, MD, 1983.

[9]Kurdila, A. J., Papastavridis, J. G., and Kannt, M. P., "Role of Maggi's Equations in Computational Methods for Constrained Multibody Systems," Journal of Guidance, Control, and Dynamics, Vol. 13, No. 1, Jan. 1990, pp. 113–120.

[10]Wehage, R. A., and Haug, E. J., "Generalized Coordinate Partitioning for Dimension Reduction in Analysis of Constrained Dynamic Systems," *Journal of Mechanical Design*, Vol. 104, Jan. 1982, pp. 247–255.

[11]Agrawal, O. P., and Shabana, A. A., "Automated Visco-Elastic Analysis of Large Scale Inertia-Variant Spatial Vehicles," *Computers and Structures*, Vol. 22, No. 2, 1986, pp. 165–178.

[12]Mani, N. K., *Use of Singular Value Decomposition for Analysis and Optimization of Mechanical System Dynamics*, PhD Dissertation, University of Iowa, Iowa City, IA, 1984.

[13]Kim, S. S., and Vandoerploeg, M. J., "*QR* Decomposition for State Space Representation of Constrained Mechanical Dynamic Systems," *Journal of Mechanisms, Transmissions, and Automation in Design*, Vol. 108, June 1986, pp. 183–188.

[14]Huston, R. L., "Useful Procedures in Multibody Dynamics," *Dynamics of Multibody Systems*, edited by G. Bianchi, Springer-Verlag, Berlin, 1986, pp. 69–78.

[15]Wampler, C., Buffinton, K., and Shu-Hui, J., "Formulation of Equations of Motion for Systems Subject to Constraints," *Journal of Applied Mechanics*, Vol. 52, June 1985, pp. 465–470.

[16]Kane, T. R., and Levinson, D. A., *Dynamics: Theory and Applications*, McGraw-Hill, New York, 1985.

[17]Wang, J. T., and Huston, R. L., "Kane's Equations with Undetermined Multipliers Application to Constrained Multibody Systems," *Transactions of the ASME*, Vol. 54, June 1987, pp. 424–429.

[18]Wittenburg, J., *Dynamics of Systems of Rigid Bodies*, Teubner, Stuttgart, 1977.

[19]Papastavridis, J. G., "On the Nonlinear Appell's Equations and the Determination of Generalized Reaction Forces," *International Journal of Engineering Science*, Vol. 26, No. 6, 1988, pp. 609–625.

[20]Song, J. O., and Haug, E. J., "Dynamic Analysis of Planar Flexible Mechanisms," *Computer Methods in Applied Mechanics and Engineering*, Vol. 24, Dec. 1980, pp. 359–381.

[21]Placek, B., "Contribution to the Solution of the Equations of Motion of the Discrete Dynamical System with Holonomic Constraints," *The Theory of Machines and Mechanisms*, edited by E. Bautista, J. Garcia-Lomas, and A. Navarro, Pergamon, Oxford, UK, 1987, pp. 379–382.

[22]Kurdila, A. J., "Concurrent Multiprocessing Methods for Calculating Nullspace and Range Space Bases for Multibody Simulation," *Proceedings of the 29th Structures, Structural Dynamics and Materials Conference*, April 1988, pp. 1447–1459.

[23]Cosnard, M., and Robert, Y., "Complexity of the Parallel *QR* Decomposition of a Rectangular Matrix," *Parallel Computing 85*, edited by M. Feilmeier, North-Holland, New York, 1986, pp. 123–128.

[24]Singh, R. P., VanderVoort, R. J., and Likins, P. W., "Dynamics of Flexible Bodies in Tree Topology—A Computer Oriented Approach," *Journal of Guidance*, Vol. 8, No. 5, Oct. 1985, pp. 584–590.

Bibliography

Bae, D., and Haug, E. J., "A Recursive Formulation for Constrained Mechanical System Dynamics, Part I, Open Loop Systems," *Mechanics of Structures and Machines*, Vol. 15, No. 359, 1987.

Banerjee, A. K., "Comment on Relationship Between Kane's Equations and the Gibbs-Appell Equations," *Journal of Guidance*, Vol. 10, No. 6, Nov. 1987, pp. 596–597.

Berry, M., and Sameh, A., "Multiprocessor Jacobi Algorithms for Dense Symmetric Eigenvalue and Singular Value Decompositions," Proceedings of the 1986 International Conference on Parallel Processing, Aug. 1986, pp. 433–440.

Boland, P., Samin, J. C., and Willems, P. Y., "Stability Analysis for Interconnected Deformable Bodies in a Topological Tree," *AIAA Journal*, Vol. 12, No. 8, Aug. 1974, pp. 1025–1030.

Boland, P., Samin, J. C., and Willems, P. Y., "Stability Analysis of Interconnected Deformable Bodies with Closed-Loop Configuration," *AIAA Journal*, Vol. 13, No. 7, July 1975, pp. 864–867.

Brauchli, H., and Weber, R. W., "Canonical Approach to Multibody Systems Using Redundant Coordinates," *Dynamics of Multibody Systems*, Springer-Verlag, Berlin, 1986, pp. 31–41.

Buffinton, K. W., and Kane, T. R., "Dynamics of a Beam Moving Over Supports," *International Journal of Solids and Structures*, Vol. 21, No. 7, Oct. 1985, pp. 617–643.

Chace, M. A., "Methods and Experience in Computer Aided Design of Large-Displacement Mechanical Systems," *Computer Aided Analysis and Optimization of Mechanical System Dynamics*, edited by E. J. Haug, Springer-Verlag, Berlin, 1984, pp. 233–256.

Craig, R. R., *Structural Dynamics, An Introduction to Computer Methods*, Wiley, New York, 1981.

De Jalon, J. G., Unda, J., Avello, A., and Jimenez, J. M., "Dynamic Analysis of Three-Dimensional Mechanisms in Natural Coordinates," *Transactions of the ASME*, Vol. 109, Dec. 1987, pp. 460–474.

Desloge, E. A., "A Comparison of Kane's Equations of Motion and the Gibbs-Appell Equations of Motion," *American Journal of Physics*, Vol. 54, No. 5, May 1986, pp. 470–472.

Desloge, E. A., "Relationship Between Kane's Equations and the Gibbs-Appell Equations," *Journal of Guidance, Control, and Dynamics*, Vol. 10, Jan. 1987, pp. 120–122.

Dongarra, J. J., Biench, J. R., Moler, C. B., and Stewart, G. W., *LINPACK User's Guide*, Society of Industrial and Applied Mathematics, 1979.

Farhat, C., "Multiprocessors in Computational Mechanics," PhD Dissertation, University of California, Berkeley, 1986.

Gear, C. W., "Differential-Algebraic Equations," *Computer Aided Analysis and Optimization of Mechanical System Dynamics*, edited by E. J. Haug, Springer-Verlag, Berlin, 1984, pp. 401–414.

Gill, P. E., Murray, W., and Wright, M. H., *Practical Optimization*, Academic, London, 1981.

Gluck, R., "Hope for Simulating Flexible Spacecraft," *Aerospace America*, Nov. 1986, pp. 40–44.

Haug, E. J., "Elements and Methods of Computational Dynamics," *Computer Aided Analysis and Optimization of Mechanical System Dynamics*, edited by E. J. Haug, Springer-Verlag, Berlin, 1984, pp. 3–37.

Heath, M. T., Plemmons, R. J., and Ward, R. C., "Sparse Orthogonal Schemes for Structural Optimization Using the Force Method," *SIAM J. Sci. Stat. Comput.*, Vol. 5, No. 3, Sept. 1984, pp. 514–528.

Hollerbach, J. M., "A Recursive Lagrangian Formulation of Manipulator Dynamics and a Comparative Study of Dynamics Formulation Complexity," *IEEE Transactions on Systems, Man, and Cybernetics*, Vol. 10, Nov. 1980, pp. 730–736.

Huston, R. L., "Multi-Body Dynamics Including the Effects of Flexibilityand Compliance," *Computers and Structures*, Vol. 14, No. 5–6, Feb. 1981, pp. 443–451.

Huston, R. L., "On the Equivalence of Kane's Equations and Gibbs Equations for Multibody Dynamics Formulations," *Mechanics Research Communications*, Vol. 14, No. 2, 1987, pp. 123–131.

Keat, J. E., "Comment on 'Relationship' Between Kane's Equations and the Gibbs-Appell Equations," *Journal of Guidance*, Vol. 10, No. 6, Nov. 1987, pp. 594–595.

Ju, M. S., and Mansour, J. M., "Comparison of Methods for Developing the Dynamics of Rigid-Body Systems Using Macsyma," *19th Midwestern Conference in Developments in Mechanics*, Ohio State University, Columbus, OH, 1985, pp. 185–186.

Kane, T. R., and Levinson, D. A., "The Use of Kane's Dynamical Equations in Robotics," *International Journal of Robotics Research*, Vol. 2, No. 3, Fall 1983, pp. 3–21.

Kane, T. R., and Levinson, D. A., "Large Motions of Unrestrained Space Trusses," *Journal of the Astronautical Sciences*, Jan.–March 1980, pp. 49–88.

Kane, T. R., and Levinson, D. A., "Simulation of Large Motions of Nonuniform Beams in Orbit: Part ii—The Unrestrained Beam," *Journal of the Astronautical Sciences*, Vol. 29, No. 3, July–Sept. 1981, pp. 245–275.

Kim, S. S., and Vanderploeg, M. J., "A General and Efficient Method for Dynamic Analysis of Mechanical Systems, Using Velocity Transformations," *Journal of Mechanisms, Transmissions, and Automation in Design*, Vol. 198, June 1986, pp. 176–182.

Levinson, D. A., "Comment on 'Relationship' Between Kane's Equations and the Gibbs-Appell Equations," *Journal of Guidance and Control*, Vol. 10, No. 6, Nov. 1987, pp. 593.

Levinson, D. A., "Equations of Motion for Multiple Rigid Body Systems via Symbolic Manipulation," *Journal of Spacecraft*, Vol. 8, April 1976, pp. 479–487.

Levinson, D. A., and Kane, T. R., "Simulation of Large Motions of Nonuniform Beams in Orbit: Part 1—The Cantilver Beam," *Journal of the Astronautical Sciences*, Vol. 29, No. 3, July–Sept. 1981, pp. 213–244.

Levinson, D. A., and Kane, T. R., "Spin Stability of a Satellite Equipped with Four Booms," *Journal of Spacecraft*, Vol. 13, No. 4, April 1976, pp. 208–213.

Liang, C. G., and Lance, G. M., "A Differentiable Null Space Method for Constrained Dynamic Analysis," *Journal of Mechanisms, Transmissions, and Automation in Design*, Vol. 109, Sept. 1987, pp. 405–411.

Lilov, L., "Dynamics of Chains of Rigid Bodies and elastic Rods with Revolute and Prismatic Joints," *Dynamics of Multibody Systems*, edited by G. Bianchi, Springer-Verlag, Berlin, 1986, pp. 141–152.

Nikravesh, P. E., "Spatial Kinematic and Dynamic Analysis with Euler Parameters," *Computer Aided Analysis and Optimization of Mechanical System Dynamics*, edited by E. J. Haug, Springer-Verlag, Berlin, 1984, pp. 261–281.

Orin, D. E., and McGee, R. B., "Kinematic and Kinetic Analysis of Open-Chain Linkages Utilizing Newton-Euler Methods," *Mathematical Biosciences*, Vol. 43, Feb. 1979, pp. 106–130.

Passerello, C. E., and Huston, R. L., "Another Look at Nonholonomic Systems," *Journal of Applied Mechanics*, March 1973, pp. 101–104.

Rosenthal, D. E., "Comment on 'Relationship Between Kane's Equations and the Gibbs-Appell Equations'," *Journal of Guidance*, Vol. 10, No. 6, 1987, pp. 595–596.

Ryan, R. R., and Kane, T. R., "Dynamics of a Modular Space Station," *Journal of the Astronautical Sciences*, Vol. 33, No. 4, Oct.–Dec. 1985, pp. 381–400.

Sameh, A. H., and Kuck, D. J., "On Stable Parallel Linear System Solvers," *Journal of the Association for Computing Machinery*, Vol. 25, No. 1, Jan. 1978, pp. 81–91.

Sameh, A. H., and Kuck, D. J., "A Parallel *QR* Algorithm for Symmetric Tridiagonal Matrices," *IEEE Transactions on Computers*, Vol. 26, No. 2, Feb. 1977, pp. 147–152.

Samin, J. C., "A Multibody Approach for Dynamic Investigation of Rolling Systems," *Engineering Archives*, Vol. 54, 1984, pp. 1–15.

Schwertassek, R., and Roberson, R., "A State-Space Dynamical Representation for Multibody Mechanical Systems, Part I: Systems with Tree Configuration," *Acta Mechanica*, Vol. 50, 1984, pp. 141–161.

Schwertassek, R., and Roberson, R. E., "A Perspective on Computer-Oriented Multibody Dynamical Formalisms and their Implementations," *Dynamics of Multibody Systems*, edited by G. Bianchi, Springer-Verlag, Berlin, 1986, pp. 261–274.

Shabana, A. A., and Wehage, R. A., "A Coordinate Reduction Technique for Dynamic Analysis of Spatial Substructures with Large Angular Rotations," *Journal of Structural Mechanics*, Vol. 11, No. 3, 1983, pp. 401–413.

Singh, R., Vandervoort, R., and Likins, P., "Interactive Design for Flexible Multibody Control," *Dynamics of Multibody Systems*, edited by G. Bianchi, Springer-Verlag, Berlin, 1986, pp. 275–286.

Soni, A. H., "Research Needs in Supercomputing," *CIME*, March 1986, pp. 40–47.

Storaasli, O. O., and Bergan, P., "Nonlinear Substructuring Method for Concurrent Processing Computers," *AIAA Journal*, Vol. 25, June 1987, pp. 871–876.

Walker, M. W., and Orin, D. E., "Efficient Dynamic Computer Simulation of Robotic Mechanisms," *Journal of Dynamic Systems, Measurement and Control*, Vol. 104, Sept. 1982, pp. 205–211.

Yoo, W. S., and Haug, E. J., "Dynamics of Articulated Structures (Part I: Theory)," *Journal of Structural Mechanics*, Nov. 1984, pp. 1–31.

Yoo, W. S., and Haug, E. J., "Dynamics of Articulated Structures (Part II: Computer Implementation and Application)," *Journal of Structural Mechanics*, Vol. 1, No. 20, Nov. 1984, pp. 1–20.

Chapter 7

Dynamics and Control of Tethered Spacecraft During Deployment and Retrieval

V. J. Modi,* P. K. Lakshmanan,†
The University of British Columbia, Vancouver, British Columbia, Canada
and
A. K. Misra‡
McGill University, Montreal, Quebec, Canada

Introduction

THE advent of the Space Shuttle and the proposed U.S. space station have presented a wide range of possibilities for space exploration and exploitation. One approach to this end is the concept of tethered satellite system (TSS). The innovative idea is attributed to the Russian scientist K. E. Tsiolkovsky, who explored the possiblity of using an extremely high tower, extending from the equator to the geosynchronous altitude and beyond, as a means of escaping the Earth's gravity. In his paper,[1] aptly called "Day Dreams of Heaven and Earth," he discussed changes in the gravity force an individual would experience as he climbed up the tower, thus uncovering one of the principal novelties of the concept that has now grown to become the TSS. As far as the actual application is concerned, interest in the system was initially associated with the retrieval of a stranded astronaut by throwing a buoy on a tether from the rescue vehicle and reeling in the tether. The first space-based experiments with tether were carried out during the two Gemini missions.[2] The manned space vehicle Gemini XI was tethered to an Agena vehicle and a rotating configuration maintained through the Gemini thruster reaction control system. Later, the Gemini XII flight demonstrated the feasibility of a gravity gradient

*Professor, Department of Mechanical Engineering.
†Graduate Research Assistant, Department of Mechanical Engineering.
‡Associate Professor, Department of Mechanical Engineering.

stabilized tethered configuration. Since then, possible applications of the system have grown to cover a broad spectrum:

1) sophisticated scientific experiments aimed at gravity gradient, magnetic, ionospheric, aerothermodynamic, and radio astronomy experiments;

2) use of the tethered system as a flying wind tunnel;

3) depolyment of payloads into new orbits and retrieval of satellites for servicing;

4) provision of a desired controlled microgravity environment for scientific experiments and space manufacturing;

5) generation of electricity (electrodynamic tether);

6) power and cargo transfer;

and many others. These have been documented in some detail in a two-volume proceedings of a NASA workshop compiled by Cron.[3] The vast potential of tether-connected orbiting systems has led to numerous investigations aimed at their dynamics and control during deployment, operational (stationkeeping), and retrieval phases. The objective here is to briefly touch on some of the important aspects of the studies, significant conclusions based on them, and the course of future efforts that would aid in a better understanding of the system performance.

Remarks on Modeling of Tether Dynamics and Control

Important Parameters

Consider a system of two bodies (in general flexible), connected by an elastic tether of finite mass, in an arbitrary orbit around the center of force E (Fig. 1). The center of mass S of the system can be located with respect to E by the orbital coordinates R and θ. Let A and B be the instantaneous centers of mass of the end bodies, and P_a, P_b represent the points of connections with finite offset from A and B, respectively. In its utmost generality the problem is quite challenging, since the system dynamics is governed by a set of ordinary and partial nonlinear, nonautonomous, and coupled differential equations that account for

1) three-dimensional rigid-body dynamics (librational motion) as well as elastic response of the end bodies;

2) swinging in-plane and out-of-plane motions of the tether, finite mass, and elasticity, with longitudinal and transverse vibrations superposed on them;

3) offsets of the tether attachment points with the problem further complicated by their specified variations in case of the offset control strategy;

4) external environmental forces due to aerodynamic drag in a rotating environment, solar radiation pressure, thermal stresses, etc.; and

5) control forces provided by tether tension, thrusters, etc., and their combinations.

Models

Over the years investigators have attempted to obtain some insight into the complex dynamics of the system using a variety of models that have

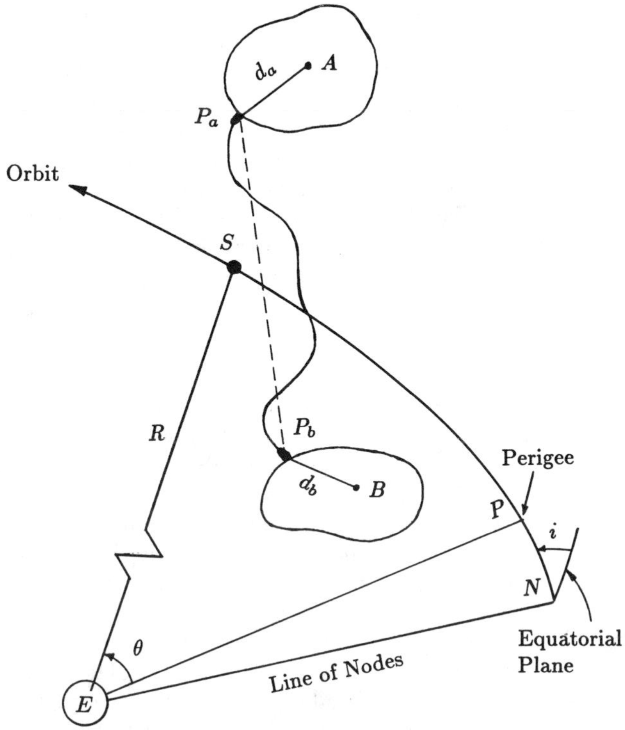

Fig. 1 A schematic diagram of the tether-connected two-body system free to undergo general three-dimensional librational and vibrational motions. Note offsets of the tether attachment points.

been summarized by Misra and Modi.[4,5] In general, the studies show that the dynamics of the system during deployment is stable; however, the retrieval dynamics is basically unstable. The system involves a negative damping approximately proportional to $\bar{l}'/l_{\mathrm{ref}}$, where $\bar{l}$ and l_{ref} are the unstretched and reference lengths, respectively, and the prime denotes differentiation with respect to the true anomaly θ. This suggests a need for an active control strategy, particularly to limit in-plane (α_t) and out-of-plane (γ_t) swing motions of the tether. The pioneering contribution that may help realize this objective is due to Rupp,[6] who introduced a tension control law for the system. Librational motion in the orbital plane was analyzed, and the growth of pitch oscillations during the retrieval phase was noted. The system was further studied in detail by Baker et al.,[7] who took into account the three-dimensional character of the dynamics as well as the aerodynamic drag in a rotating atmosphere. Several more sophisticated models have followed since[8-15]; however, one of the major conclusions of all of the analyses remains essentially the same: even when the various tension control schemes are used, a large-amplitude motion can result under certain conditions, particularly during retrieval, which may not be acceptable.

Tension is directly dependent on the gravity gradient, i.e., on the tether length. Hence, at shorter lengths during retrieval, there exists a possibility of the tether becoming slack, rendering the tension control strategy ineffective. This can be avoided by providing thrusters in the three orthogonal directions at the subsatellite end.[16,17] The strategy has been shown to be quite effective; however, the use of thrusters in close proximity to a space platform is undesirable due to the danger of plume impingement and associated effects. The offset control scheme involving displacement of the tether attachment point as proposed by Lakshmanan et al. overcomes these limitations quite effectively.[18–20]

Environmental Effects

Major external forces on the TSS arise from the gravitational, solar, aerodynamic, and electrodynamic effects. As far as the attitude dynamics of the TSS is concerned, gravitational force may be considered to be that due to a homogeneous spherical Earth with negligible contributions from the sun and the moon. For low-altitude orbits, the force resulting from solar radiation pressure is small compared to gravity gradient and aerodynamic forces and hence has very little influence on the dynamics of the TSS. However, the radiation has a significant effect on the temperature of the tether and hence its longitudinal extension. Alternate heating and cooling of the tether due to solar radiation and the Earth's shadow, respectively, can cause large temperature variations. For example, Kalaghan et al.[9] have shown that a 20-km stainless steel tether has a peak-to-peak temperature excursion of 60 K and a net thermal elongation of 18 m during the period of shadow exit to shadow entrance. The free molecular aerodynamic forces are the most significant for the near-Earth configurations under consideration for future experiments. The aerodynamic force on an element of the system can be approximated as

$$\Delta \bar{F} = -(1/2) C_d \rho \, \Delta A |\bar{V}| \bar{V} \tag{1}$$

where C_d and ρ are the drag coefficient and density of air at the altitude of the element, respectively, and ΔA is the projected area of the element normal to its velocity $\bar{V}$ relative to the atmosphere. The coefficient C_d somewhat varies with the geometry of the element, angle of attack, and atmospheric conditions; however, to make dynamic analysis practicable, C_d is usually assumed to be constant. A crucial quantity to be modeled is the atmospheric density. A simple model assumes that the density of air decreases exponentially with the altitude, i.e.,

$$\rho = \rho_{\text{ref}} \exp[-(h - h_{\text{ref}})/h_0] \equiv \rho_0 \exp(-h/h_0) \tag{2}$$

where h is the altitude of the element above the surface of the Earth represented by an ellipsoid, ρ_{ref} is the density of air at an altitude h_{ref}, and h_0 is a scaling parameter. Note that ρ_0 is a constant equal to $\rho_{\text{ref}} \exp(h_{\text{ref}}/h_0)$. It is clear that, for orbits inclined to the equatorial plane, the oblateness of the Earth results in a periodically varying altitude of the tether element. This can have a significant effect on the dynamics of the system, since the density varies exponentially with altitude. To obtain a reasonable correlation with experimental density data, it becomes necessary to use different parameters for different altitude ranges, i.e., to introduce a set of piecewise exponential functions. A simple approach is to choose ρ_0 and h_0 in order to match the lower portion of the tether accurately. This yields an inaccurate aerodynamic force at the upper portion of the tether, but it is of little consequence since both the actual and approximate aerodynamic drag at the upper end are several orders of magnitude smaller than that at the lower end. Salient features of the system behavior can be assessed using this simple model; however, for accurate results more precise representation of the density and its variation may be necessary.[9]

The relative velocity $\bar{V}$ in Eq. (1) depends on the orbital velocity of the center of mass of the system, the velocity of the element relative to the center of mass of the system, and the velocity of the atmosphere due to its rotation about the Earth's axis. The first component is the largest; the second is small and can be ignored. The velocity of the atmosphere due to its rotation is not insignificant compared to the orbital velocity and should be taken into account. Furthermore, for orbits inclined to the equatorial plane, rotation of atmosphere is responsible for excitation of motion normal to the orbital plane, which may have important bearing on deployment and retrieval dynamics.

In the proposed electrodynamic tether experiment[3] the conducting tether, while traversing the geomagnetic field, will be subjected to electrodynamic forces that, in turn, will affect the tether dynamics. In fact, Columbo et al.[22] have shown that the electrodynamic forces can be used for attitude stabilization.

Equilibrium Orientation

As a result of aerodynamic drag, a subsatellite deployed downward trails the platform space (Space Shuttle, space station); i.e., there is an in-plane rotation, $\alpha_{t,e}$, of the tether in the steady state from the local vertical. It can be shown[10] that $\alpha_{t,e}$ is given approximately by the solution of the transcendental equation

$$\cos\alpha_{t,e}[(3m_s + m_t)\sin\alpha_{t,e} + \tfrac{1}{2}C_d R^2\{\bar{\rho}_t d_t|\cos\alpha_{t,e}| + \rho_s A_s/\bar{l}\}] = 0 \qquad (3)$$

where m_s and m_t are mass of the subsatellite and tether, respectively; R is the radius of the circular orbit; d_t and A_s are the diameter of the tether and area of the subsatellite, respectively; and ρ_s and $\bar{\rho}_t$ are the density of air at the subsatellite altitude and the weighted altitude applicable to the tether, respectively. Even for a tether length as long as 50 km, its deviation from the local vertical is almost imperceptible when deployed from an altitude of 220 km (Fig. 2). However, if $(3m_s + m_t) < \tfrac{1}{2}C_d R^2 \rho_s A_s/\bar{l}$, then $\alpha_{t,e}$ is equal to $-\pi/2$; i.e., the tether attains a horizontal equilibrium configuration. For an elliptic orbit the equilibrium orientation undergoes periodic variation. In

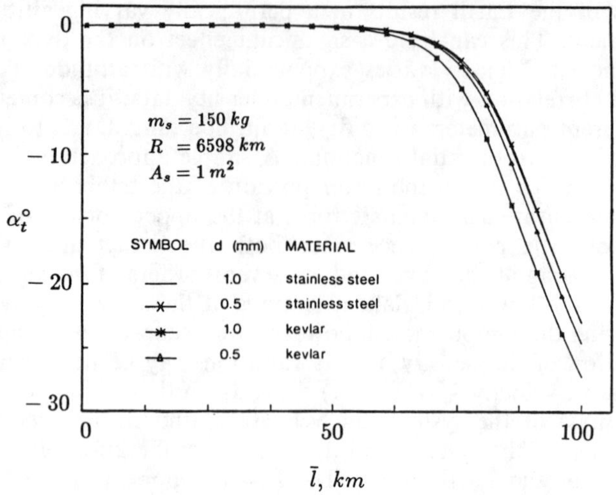

Fig. 2 Variation of the equilibrium in-plane rotation with the tether length for a system in a circular orbit at a height of 220 km. The α_t, measured from the nadir, is considered positive clockwise.

the case of inclined orbits, out-of-plane rotations are present as well, although, relatively, they are quite small. Aerodynamic drag also causes steady-state transverse deflections. In-plane transverse deflections are typically several hundred meters for a 100-km tether deployed from an orbiting platform at a height of 220 km.[5] Out-of-plane deflections are smaller. Hence, the tether configuration does not deviate significantly from a straight line.

Deployment and Retrieval Procedures

Deployment and retrieval of the subsatellite can be carried out by controlling the tension in tether through a reel mechanism as proposed by Rupp.[6] Two of the simplest deployment rates include the following:

1) uniform deployment rate (i.e., $\dot{\bar{l}} = c\bar{l}_i$, where $\bar{l}_i$ is the initial length); clearly,

$$\bar{l} = \bar{l}_i(1 + ct) \tag{4a}$$

2) exponential deployment, where the rate at any instant is proportional to the deployed length (i.e., $\dot{\bar{l}} = c\bar{l}$); integrating,

$$\bar{l} = \bar{l}_i \exp(ct) \tag{4b}$$

The constant c is positive for deployment and negative for retrieval. For elliptic orbits, it is more convenient to express $\bar{l}$ in terms of the true anomaly θ, and Eqs. (4) transform to

$$\bar{l} = \bar{l}_i[1 + (c/\Omega)\{H(\theta) - H(\theta_i)\}] \tag{5a}$$

and

$$\bar{l} = \bar{l}_i \exp[1 + (c/\Omega)\{H(\theta) - H(\theta_i)\}] \tag{5b}$$

where

$$H(\theta) = -\tfrac{1}{2}(1 - e^2)^{1/2}F(\theta) + 2\tan^{-1}[\{(1-e)/(1+e)\}^{1/2}\tan(\theta/2)] \tag{5c}$$

and θ_i is the initial anomaly. It may be noted that, for circular orbits, $H(\theta) = \theta$. Let the length of the tether increase from $\bar{l}_i$ to $\bar{l}_f$ during n orbits. Then, from Eqs. (5), c/Ω, which corresponds to the two deployment procedures, is given by

$$c/\Omega = \{(\bar{l}_f/\bar{l}_i) - 1\}/2\pi n \tag{6a}$$

and

$$c/\Omega = \ln(\bar{l}_f/\bar{l}_i)/2\pi n \tag{6b}$$

In practice, the deployment is likely to involve a combination of the two procedures discussed above. For example, a deployment strategy may be as follows:

$$\dot{\bar{l}} = \begin{cases} c\bar{l}, & \bar{l}_i \leqslant \bar{l} \leqslant \bar{l}_1 \\ c\bar{l}_1, & \bar{l}_1 \leqslant \bar{l} \leqslant \bar{l}_2 \\ c(\bar{l}_1 + \bar{l}_2 - \bar{l}), & \bar{l}_2 \leqslant \bar{l} \leqslant \bar{l}_f \end{cases} \tag{7}$$

The first and the third phases correspond to the exponential rate, and the second stage represents the uniform rate. The length history can be written as

$$\bar{l} = \begin{cases} \bar{l}_i \exp[(c/\Omega)\{H(\theta) - H(\theta_i)\}], & \theta_i \leqslant \theta \leqslant \theta_1 \\ \bar{l}_1[1 + (c/\Omega)\{H(\theta) - H(\theta_1)\}], & \theta_1 \leqslant \theta \leqslant \theta_2 \\ (\bar{l}_1 + \bar{l}_2) - \bar{l}_1 \exp[-(c/\Omega)\{H(\theta) - H(\theta_2)\}], & \theta_2 \leqslant \theta \leqslant \theta_f \end{cases} \tag{8}$$

The average tension in the tether can be calculated since $\bar{l}$ is known as a function of θ.

Outline of the Formulation

A rather general formulation for studying the dynamics and control of tether-connected two-body systems, taking into account three-dimensional librational motion, longitudinal and transverse vibrations of the continuous tether, eccentricity of the orbit, and aerodynamic drag in a rotating, oblate atmosphere, has been presented by Modi and Misra.[10,21] The governing highly nonlinear, nonautonomous, and coupled set of hybrid differential equations are rather lengthy and difficult to handle even with the help of a computer.

To facilitate appreciation of the system dynamics, particularly the effectiveness of various control strategies, it is convenient to consider the case of a rigid system.[20] Figure 3 shows a schematic diagram of a two-body tethered satellite system with an offset of the attachment point. The first body represents a plate-type structure to simulate a large satellite in space, for example, the space station. The second body may be a smaller subsatellite on the end of a tether which is capable of being deployed or retrieved from the space platform. Accordingly, the subscripts p, s, and t refer to the plate, subsatellite, and tether, respectively. The offset is designated by means of orthogonal scalar components of a three-dimensional vector denoted by $D_{x,p}$, $D_{y,p}$, and $D_{z,p}$. These scalar components, measured relative to the center of mass of the platform, can be varied by means of the boom trolley arrangement as indicated. Thus, the magnitude and direction of the vector $\boldsymbol{d}_p$ can be controlled relative to the space platform. The boom

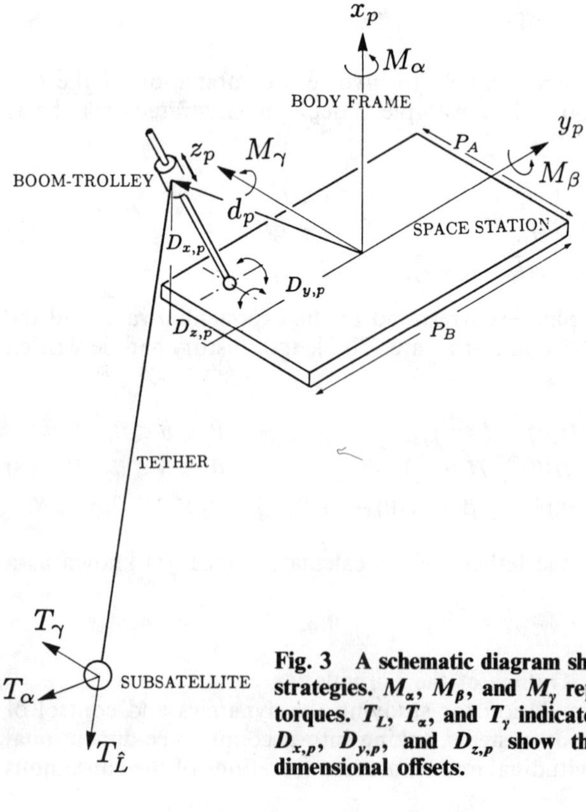

Fig. 3 A schematic diagram showing the various control strategies. M_α, M_β, and M_γ represent momentum wheel torques. T_L, T_α, and T_γ indicate the thruster forces, and $D_{x,p}$, $D_{y,p}$, and $D_{z,p}$ show the time-dependent three-dimensional offsets.

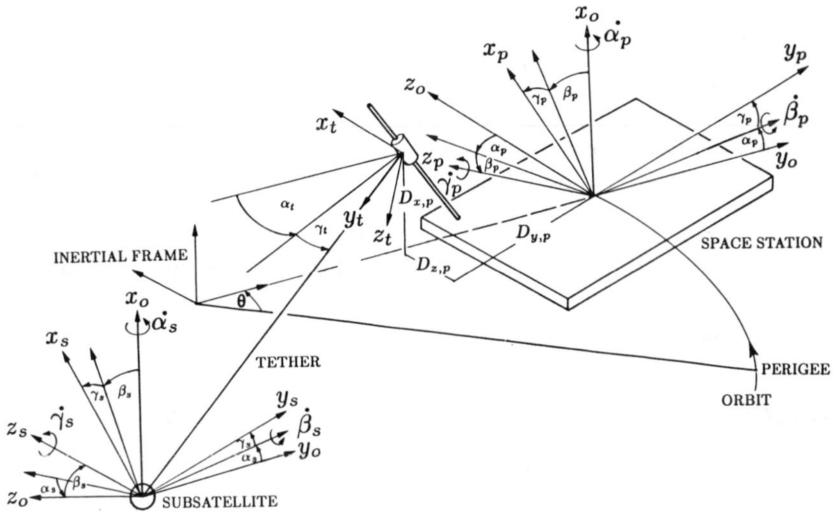

Fig. 4 The space platform based TSS configuration with a time-varying three-dimensional offset of the attachment point. The kinematics of the motion is described using modified Eulerian relations.

and trolley masses, being very small compared to the mass of the station, can thus be neglected. Three reference frames are utilized to specify the motion of the TSS relative to the orbital frame: x_0, y_0, and z_0, as shown in Fig. 4. Here the y_0 axis is along the local vertical, z_0 is along the local horizontal, and x_0 is aligned with the orbit normal. Starting from the orbital frame orientation, a set of modified Eulerian rotations are applied successively to arrive at the final configuration. For example, the plate fixed frame x_p,y_p,z_p is attached to the center of mass of the plate. Its orientation is obtained by first applying a pitch rotation α_p about the x_0 axis, a yaw rotation β_p about the y_1 axis (intermediate frame), and, finally, a roll of γ_p about the z_2 axis. Similar transformations are used to obtain the frame x_s,y_s,z_s located at the center of mass of the subsatellite. In the case of the tether, two rotations α_t (in plane) and γ_t (out of plane) are sufficient to establish orientation of the frame x_t,y_t,z_t. Note that the origin of the frame x_t,y_t,z_t, is considered to be at the point of attachment of the tether with the boom. Since the mass of the platform is likely to be much larger than that of the subsatellite and tether, the overall center of mass may be considered coincident with that of the plate. However, such a simplification must be used with caution and must be guided by relative values of system parameters, since there are situations where it can lead to significant errors in the response analysis.

The system under consideration has nine generalized coordinates: three corresponding to the platform librations α_p (pitch), β_p (yaw), and γ_p (roll); three representing the subsatellite rotations α_s, β_s, and γ_s; and three defining the tether motions α_t (in-plane rotation), γ_t (out-of-plane swing angle), and $\bar{l}$ (deviation of tether length from a nominal value). The Lagrangian formulation procedure may be adopted in deriving the equations of motion. General expressions for the kinetic, potential, and strain energies of the system should be obtained accounting for the librations and vibrations of the system. The governing nonlinear, nonautonomous, and coupled equations of motion, even for the rigid case being extremely lengthy, are omitted here for conciseness. In order to obtain some appreciation of the complex character of the system, a typical equation for the platform pitch, linearized about the equilibrium configuration, is presented as an example:

α_p: *Platform pitch equation*:

$$
\begin{aligned}
&\ddot{\alpha}_p[I_{x,p} + m_{rst}(\bar{d}_{y,p}^2 + \bar{d}_{z,p}^2)] + \alpha_p\Bigg[-\dot{m}_t\dot{\bar{l}}\bar{d}_{y,p} - m_{st}\ddot{\bar{l}}\bar{d}_{y,p} - \frac{\dot{m}_t}{2}\bar{l}\dot{\theta}\bar{d}_{z,p} \\
&\quad - m_{st2}(\dot{\bar{l}}\dot{\theta} + \bar{l}\ddot{\theta})\bar{d}_{z,p} - m_{st}\dot{\bar{l}}\dot{\theta}\bar{d}_{z,p} + m_{st2}\bar{l}\dot{\theta}^2\bar{d}_{y,p} \\
&\quad + \frac{GM}{R^3}\{3m_{rst}(\bar{d}_{y,p}^2 - \bar{d}_{z,p}^2) + m_{st2}2\bar{l}\bar{d}_{y,p} + 3(I_{z,p} - I_{y,p})\}\Bigg] \\
&\quad - \ddot{\beta}_p[m_{rst}\bar{d}_{x,p}\bar{d}_{y,p}] - \dot{\beta}_p[2m_{rst}\dot{\theta}\bar{d}_{x,p}\bar{d}_{z,p}] \\
&\quad + \beta_p\Bigg[\left(2m_{rst}\ddot{\theta}\bar{d}_{x,p}\bar{d}_{z,p} + \dot{m}_t\dot{\bar{l}}\bar{d}_{x,p} + m_{st}\ddot{\bar{l}}\bar{d}_{x,p} - m_{st2}\bar{l}\dot{\theta}^2\bar{d}_{x,p}\right) \\
&\quad + \frac{GM}{R^3}\{-3m_{rst}\bar{d}_{x,p}\bar{d}_{y,p} - m_{st2}2\bar{l}\bar{d}_{x,p}\}\Bigg] \\
&\quad - \ddot{\gamma}_p[m_{rst}\bar{d}_{x,p}\bar{d}_{z,p}] + \dot{\gamma}_p[2m_{rst}\dot{\theta}\bar{d}_{x,p}\bar{d}_{y,p}] \\
&\quad + \gamma_p\Bigg[2m_{rst}\ddot{\theta}\bar{d}_{x,p}\bar{d}_{y,p} + \frac{\dot{m}_t}{2}\bar{l}\dot{\theta}\bar{d}_{x,p} + m_{st2}(\dot{\bar{l}}\dot{\theta} + \bar{l}\ddot{\theta})\bar{d}_{x,p} \\
&\quad + m_{st}\dot{\bar{l}}\dot{\theta}\bar{d}_{x,p} + \frac{GM}{R^3}\{3m_{rst}\bar{d}_{x,p}\bar{d}_{z,p}\}\Bigg] + \ddot{\alpha}_t[m_{st2}\bar{l}\bar{d}_{y,p}] \\
&\quad + \dot{\alpha}_t\Bigg[\frac{\dot{m}_t}{2}\bar{l}\bar{d}_{y,p} + m_{st2}\dot{\bar{l}}\bar{d}_{y,p} + m_{st}\dot{\bar{l}}\bar{d}_{y,p} + m_{st2}2\bar{l}\dot{\theta}\bar{d}_{z,p}\Bigg] \\
&\quad + \alpha_t\Bigg[\frac{\dot{m}_t}{2}\dot{\bar{l}}\bar{d}_{y,p} + \frac{\dot{m}_t}{2}\bar{l}\dot{\theta}\bar{d}_{z,p} + m_{st}\ddot{\bar{l}}\bar{d}_{y,p} + m_{st2}(\dot{\bar{l}}\dot{\theta} + \bar{l}\ddot{\theta})\bar{d}_{z,p} \\
&\quad + m_{st}\dot{\bar{l}}\dot{\theta}\bar{d}_{z,p} - m_{st2}\bar{l}\dot{\theta}^2\bar{d}_{y,p} + \frac{GM}{R^3}m_{st2}\bar{l}\bar{d}_{y,p}\Bigg] - \ddot{\bar{l}}[m_{st}\bar{d}_{z,p}] \\
&\quad + \dot{\bar{l}}\Bigg[m_t\dot{\theta}\bar{d}_{y,p} + m_{st}\dot{\theta}\bar{d}_{y,p} + m_{st2}\dot{\theta}\bar{d}_{y,p} + \frac{\dot{m}_t}{2}2\bar{d}_{z,p}\Bigg]
\end{aligned}
$$

$$+\hat{l}\left[m_{st2}\ddot{\theta}\bar{d}_{y,p}+\frac{m_t}{2}\ddot{\theta}\bar{d}_{y,p}+\dot{m}_t 2\dot{\theta}\bar{d}_{y,p}+m_{st}\dot{\theta}^2\bar{d}_{z,p}\right.$$

$$\left.+\frac{GM}{R^3}\{m_{st2}2\bar{d}_{z,p}+m_t\bar{d}_{z,p}\}\right]-\ddot{\hat{d}}_{y,p}[m_{rst}\bar{d}_{z,p}]+\dot{\hat{d}}_{y,p}[2m_{rst}\dot{\theta}\bar{d}_{y,p}]$$

$$+\hat{d}_{y,p}\left[2m_{rst}\ddot{\theta}\bar{d}_{y,p}+m_{st2}\bar{l}\ddot{\theta}+m_{st}2\dot{\bar{l}}\dot{\theta}+\frac{GM}{R^3}\{3m_{rst}\bar{d}_{z,p}\}\right]$$

$$+\ddot{\hat{d}}_{z,p}[m_{rst}\bar{d}_{y,p}]+\dot{\hat{d}}_{z,p}[2m_{rst}\dot{\theta}\bar{d}_{z,p}]$$

$$+\hat{d}_{z,p}\left[2m_{rst}\ddot{\theta}\bar{d}_{z,p}+m_{st2}\bar{l}\dot{\theta}^2-\dot{m}_t\dot{\bar{l}}-m_{st}\ddot{\bar{l}}+\frac{GM}{R^3}\{3m_{rst}\bar{d}_{y,p}+m_{st2}\bar{l}\}\right]$$

$$+\ddot{\theta}[I_{x,p}+m_{rst}(\bar{d}^2_{y,p}+\bar{d}^2_{z,p})+m_{st2}\bar{l}\bar{d}_{y,p}]$$

$$+\dot{\theta}\left[\frac{\dot{m}_t}{2}\bar{l}\bar{d}_{y,p}+m_{st}\dot{\bar{l}}\bar{d}_{y,p}+m_{st2}\{\dot{\bar{l}}\bar{d}_{y,p}+\bar{l}\dot{\theta}\bar{d}_{z,p}\}\right]+\dot{m}_t\dot{\bar{l}}\bar{d}_{z,p}-m_{st}\ddot{\bar{l}}\bar{d}_{z,p}$$

$$+\frac{GM}{R^3}\{3m_{rst}\bar{d}_{y,p}\bar{d}_{z,p}+m_{st2}2\bar{l}\bar{d}_{z,p}\}=Q_{\alpha p} \tag{9}$$

where

$\bar{d}_{j,p}$, $\hat{d}_{j,p}$	= nominal offset values and their corresponding deviations, respectively, along orthogonal directions, $j = x, y, z$
$\bar{l}$	= instantaneous nominal unstretched tether length
$\hat{l}$	= deviation of tether length from nominal value
m	= mass of the entire system
m_p	= mass of the space platform
m_s	= mass of the subsatellite
m_t	= mass of the tether
m_r	= mass of reel mechanism and undeployed tether
m_{st}	$= m_s + m_t$
m_{st2}	$= m_s + (m_t/2)$
m_{rst}	$= m - m_p = m_s + m_t + m_r$
GM	= gravitational constant
$I_{x,p}$, $I_{y,p}$, $I_{z,p}$	= platform inertias about x_p, y_p, and z_p axes, respectively
$Q_{\alpha p}$	= generalized force associated with α_p
α_p, β_p, γ_p	= pitch, yaw, and roll of platform, respectively
α_s, β_s, γ_s	= pitch, yaw, and roll of subsatellite, respectively

The overdots indicate differentiation with respect to time t. The governing equations thus consider rigid-body dynamics of the tether, platform, and subsatellite; deployment and retrieval of the tether accounting for its mass; and changes in the magnitude and direction of the offset and orbits of varying eccentricities. Note that the equations can be used to study the effectiveness of the tether tension, thruster, and offset control strategies.

Furthermore, since the tether dynamics and its interaction with the platform are of major interest, the subsatellite rotations may be neglected to help appreciate the system response better. Even with these simplifications, numerical integration of the equations presents a challenging task. A particular case of the system equations in the stationkeeping mode while negotiating a circular oribit is shown here in detail. The equations are nondimensionalized with respect to the subsatellite mass m_s; the tether reference length l_{ref}, which is chosen to be the initial length of the tether before retrieval; and the orbital rate $\dot{\theta}$. In other words, the nondimensionalizing parameter for the length equation is $m_s l_{\mathrm{ref}} \dot{\theta}^2$, whereas all other equations are nondimensionalized with respect to $m_s l_{\mathrm{ref}}^2 \dot{\theta}^2$. Now the equations simplify considerably, although they still remain quite formidable:

α_p: *Platform pitch*:

$$
\begin{aligned}
&\{M_{p,s}K_{xp}^2 + (1+M_{t,s})\bar{D}_{xxp}\}\alpha_p'' \\
&+3\left\{\left(1+\frac{\bar{L}M_{t,s}}{2}\right)\bar{D}_{yp} + (1+M_{t,s})(\bar{D}_{zzp}-\bar{D}_{xxp}) + M_{p,s}(K_{xp}^2 - K_{yp}^2)\right\}\alpha_p \\
&-(1+M_{t,s})\bar{D}_{xyp}\beta_p'' - 2(1+M_{t,s})\bar{D}_{xzp}\beta_p' \\
&-3\left\{\left(1+\frac{\bar{L}M_{t,s}}{2}\right)\bar{D}_{xp} + (1+M_{t,s})\bar{D}_{xyp}\right\}\beta_p - (1+M_{t,s})\bar{D}_{xzp}\gamma_p'' \\
&+2(1+M_{t,s})\bar{D}_{xyp}\gamma_p' + 3(1+M_{t,s})\bar{D}_{xzp}\gamma_p + 2\left(1+\frac{\bar{L}M_{t,s}}{2}\right)\bar{D}_{yp}\alpha_t'' \\
&-(1+\bar{L}M_{t,s})\bar{D}_{zp}\hat{L}'' + 2(1+\bar{L}M_{t,s})\bar{D}_{yp}\hat{L}' + 3(1+\bar{L}M_{t,s})\bar{D}_{zp}\hat{L} \\
&-(1+M_{t,s})\bar{D}_{zp}\hat{D}_{y,p}'' + 2(1+M_{t,s})\bar{D}_{yp}\hat{D}_{y,p}' + 3(1+M_{t,s})\bar{D}_{zp}\hat{D}_{y,p} \\
&+(1+M_{t,s})\bar{D}_{yp}\hat{D}_{z,p}'' + 2(1+M_{t,s})\bar{D}_{zp}\hat{D}_{z,p}' \\
&+3\left\{\left(1+\frac{\bar{L}M_{t,s}}{2}\right)\bar{L} + (1+M_{t,s})\bar{D}_{yp}\right\}\hat{D}_{z,p} \\
&+3\left\{\left(1+\frac{\bar{L}M_{t,s}}{2}\right)\bar{D}_{zp} + (1+M_{t,s})\bar{D}_{yzp}\right\} \\
&+3\left\{\left(1+\frac{\bar{L}M_{t,s}}{2}\right)\bar{D}_{zp} + (1+M_{t,s})\bar{D}_{yzp}\right\} = T_\alpha\bar{D}_{yp} - T_{\bar{L}}\bar{D}_{zp} + M_\alpha \qquad \text{(10a)}
\end{aligned}
$$

β_p: *Platform yaw*:

$$
\begin{aligned}
&-(1+M_{t,s})\bar{D}_{xyp}\alpha_p'' + 2(1+M_{t,s})\bar{D}_{xzp}\alpha_p' \\
&\quad - 3\left\{\left(1+\frac{\bar{L}M_{t,s}}{2}\right)\bar{D}_{xp} + (1+M_{t,s})\bar{D}_{xyp}\right\}\alpha_p \\
&\quad + \{M_{p,s}K_{yp}^2 + (1+M_{t,s})\bar{D}_{yyp}\}\beta_p'' + \{M_{p,s}(K_{xp}^2 - K_{yp}^2) \\
&\quad + (1+M_{t,s})(\bar{D}_{xxp} - \bar{D}_{yyp})\}\beta_p - \{(1+M_{t,s})\bar{D}_{yzp}\}\gamma_p'' \\
&\quad + \{M_{p,s}(K_{xp}^2 - K_{yp}^2 - K_{zp}^2) + (1+M_{t,s})(\bar{D}_{xxp} - \bar{D}_{yyp} - \bar{D}_{zzp})\}\gamma_p' \\
&\quad - (1+M_{t,s})\bar{D}_{yzp}\gamma_p - \left(1+\frac{\bar{L}M_{t,s}}{2}\right)\bar{D}_{xp}\alpha_t'' - \left(1+\frac{\bar{L}M_{t,s}}{2}\right)\bar{D}_{zp}\gamma_t'' \\
&\quad - \left(1+\frac{\bar{L}M_{t,s}}{2}\right)\bar{D}_{zp}\gamma_t' - 2\left(1+\frac{\bar{L}M_{t,s}}{2}\right)\bar{D}_{xp}\hat{L}' - (1+M_{t,s})\bar{D}_{xzp}\gamma_t'' \\
&\quad + (1+M_{t,s})\bar{D}_{zp}\hat{D}_{x,p}'' - (1+M_{t,s})\bar{D}_{zp}\hat{D}_{x,p} - 2(1+M_{t,s})\bar{D}_{xp}\hat{D}_{y,p}' \\
&\quad - (1+M_{t,s})\bar{D}_{xp}\hat{D}_{z,p}'' - (1+M_{t,s})\bar{D}_{xp}\hat{D}_{z,p} = T_\gamma\bar{D}_{zp} - T_\alpha\bar{D}_{xp} + M_\beta \qquad \text{(10b)}
\end{aligned}
$$

γ_p: *Platform roll*:

$$
\begin{aligned}
&-(1+M_{t,s})\bar{D}_{xzp}\alpha_p'' - 2(1+M_{t,s})\bar{D}_{xyp}\alpha_p' + 3(1+M_{t,s})\bar{D}_{xzp}\alpha_p \\
&\quad - (1+M_{t,s})\bar{D}_{yzp}\beta_p'' + \{M_{p,s}(K_{zp}^2 + K_{yp}^2 - K_{xp}^2) \\
&\quad + (1+M_{t,s})(\bar{D}_{zzp} + \bar{D}_{yyp} - \bar{D}_{xxp})\}\beta_p' - (1+M_{t,s})\bar{D}_{yzp}\beta_p \\
&\quad + \{M_{p,s}K_{zp}^2 + (1+M_{t,s})\bar{D}_{zzp}\}\gamma_p'' + \Bigg[4\{M_{p,s}(K_{xp}^2 - K_{yp}^2) \\
&\quad + (1+M_{t,s})(\bar{D}_{xxp} - \bar{D}_{yyp})\} + 3\left(1+\frac{\bar{L}M_{t,s}}{2}\right)\bar{D}_{yp}\Bigg]\gamma_p \\
&\quad - 2\left(1+\frac{\bar{L}M_{t,s}}{2}\right)\bar{D}_{xp}\alpha_t' + \left(1+\frac{\bar{L}M_{t,s}}{2}\right)\bar{D}_{yp}\gamma_t \\
&\quad + (1+\bar{L}M_{t,s})\bar{D}_{xp}\hat{L}'' - 3(1+\bar{L}M_{t,s})\bar{D}_{xp}\hat{L} \\
&\quad - (1+M_{t,s})\bar{D}_{yp}\hat{D}_{x,p}'' - \left\{4(1+M_{t,s})\bar{D}_{yp} + 3\left(1+\frac{\bar{L}M_{t,s}}{2}\right)\bar{L}\right\}\hat{D}_{x,p} \\
&\quad + (1+M_{t,s})\bar{D}_{xp}\hat{D}_{y,p}'' - 4(1+M_{t,s})\bar{D}_{xp}\hat{D}_{y,p} - 2(1+M_{t,s})\bar{D}_{xp}\hat{D}_{z,p}' \\
&\quad - \{4(1+\bar{L}M_{t,s})\bar{D}_{xyp} + 3(1+\bar{L}M_{t,s})\bar{D}_{xp}\} = T_\gamma\bar{D}_{yp} + T_{\bar{L}}\bar{D}_{xp} + M_\gamma \qquad \text{(10c)}
\end{aligned}
$$

α_t: *Tether in-plane swing*:

$$\left(1+\frac{\bar{L}M_{t,s}}{2}\right)\bar{D}_{yp}\alpha_p''-2\left(1+\frac{\bar{L}M_{t,s}}{2}\right)\bar{D}_{zp}\alpha_p'-\left(1+\frac{\bar{L}M_{t,s}}{2}\right)\bar{D}_{xp}\beta_p''$$

$$+2\left(1+\frac{\bar{L}M_{t,s}}{2}\right)\bar{D}_{xp}\gamma_p'+\left(1+\frac{\bar{L}M_{t,s}}{3}\right)\alpha_t''$$

$$+3\left\{\left(1+\frac{\bar{L}M_{t,s}}{2}\right)\bar{D}_{yp}+\left(1+\frac{\bar{L}M_{t,s}}{2}\right)\right\}\alpha_t+2\left(1+\frac{\bar{L}M_{t,s}}{2}\right)\hat{L}'$$

$$+2\left(1+\frac{\bar{L}M_{t,s}}{2}\right)\hat{D}_{y,p}'+\left(1+\frac{\bar{L}M_{t,s}}{2}\right)\hat{D}_{z,p}''=T_\alpha\bar{L} \qquad (10d)$$

γ_t: *Tether out-of-plane swing*:

$$-\left(1+\frac{\bar{L}M_{t,s}}{2}\right)\bar{D}_{zp}\beta_p''-\left(1+\frac{\bar{L}M_{t,s}}{2}\right)\bar{D}_{zp}\beta_p+\left(1+\frac{\bar{L}M_{t,s}}{2}\right)\bar{D}_{yp}\gamma_p''$$

$$+\left(1+\frac{\bar{L}M_{t,s}}{2}\right)\bar{D}_{yp}\gamma_p+\left(1+\frac{\bar{L}M_{t,s}}{3}\right)\gamma_t''+\left\{3\left(1+\frac{\bar{L}M_{t,s}}{2}\right)\bar{D}_{yp}\right.$$

$$\left.+4\left(1+\frac{\bar{L}M_{t,s}}{3}\right)\bar{L}\right\}\gamma_t-(1+\bar{L}M_{t,s})\bar{D}_{xp}\hat{L}-\left(1+\frac{\bar{L}M_{t,s}}{2}\right)\bar{D}_{xp}$$

$$-\left(1+\frac{\bar{L}M_{t,s}}{2}\right)\bar{L}\hat{D}_{x,p}''-\left(1+\frac{\bar{L}M_{t,s}}{2}\right)\bar{L}\hat{D}_{x,p}=-T_\gamma\bar{L} \qquad (10e)$$

$\hat{L}$: *Deviation from nominal length*:

$$-(1+\bar{L}M_{t,s})\bar{D}_{zp}\alpha_p''-2(1+\bar{L}M_{t,s})\bar{D}_{yp}\alpha_p'+3(1+\bar{L}M_{t,s})\bar{D}_{zp}\alpha_p$$

$$+2(1+\bar{L}M_{t,s})\bar{D}_{xp}\beta_p'+(1+\bar{L}M_{t,s})\bar{D}_{xp}\gamma_p''-3(1+\bar{L}M_{t,s})\bar{D}_{xp}\gamma_p$$

$$-2\left(1+\frac{\bar{L}M_{t,s}}{2}\right)\alpha_t'-(1+\bar{L}M_{t,s})\bar{D}_{xp}\gamma_t+(1+\bar{L}M_{t,s})\hat{L}''$$

$$-3\left\{\bar{L}M_{t,s}(\bar{D}_{yp}+\tfrac{1}{2})+\left(1+\frac{\bar{L}M_{t,s}}{2}\right)\right\}\hat{L}+(1+\bar{L}M_{t,s})\hat{D}_{y,p}''$$

$$-3(1+\bar{L}M_{t,s})\hat{D}_{y,p}-2(1+\bar{L}M_{t,s})\hat{D}_{z,p}'-T_{\hat{L}}=Q_T \qquad (10f)$$

where

$\bar{D}_{i,p}$, $\hat{D}_{i,p}$ = nondimensionalized nominal and controlled offsets, respectively, $i = x,y,z$

$$\bar{D}_{xx,p} = \bar{D}^2_{y,p} + \bar{D}^2_{z,p}$$

$$\bar{D}_{yy,p} = \bar{D}^2_{x,p} + \bar{D}^2_{z,p}$$

$$\bar{D}_{zz,p} = \bar{D}^2_{y,p} + \bar{D}^2_{x,p}$$

$$\bar{D}_{xy,p} = \bar{D}_{x,p}\bar{D}_{y,p}$$

$$\bar{D}_{yz,p} = \bar{D}_{y,p}\bar{D}_{z,p}$$

$$\bar{D}_{zx,p} = \bar{D}_{z,p}\bar{D}_{x,p}$$

$$\hat{L} = \hat{l}/l_{\mathrm{ref}}$$

$$\bar{L} = \bar{l}/l_{\mathrm{ref}}$$

$M_{p,s}$ = ratio of platform mass to subsatellite mass

$M_{t,s}$ = ratio of tether mass to subsatellite mass

$Q_{\bar{l}}$ = equilibrium tether tension force

T_α, T_γ, $T_{\hat{L}}$ = thruster forces along α_t, γ_t, and y_t directions, respectively, nondimensionalized with respect to $m_s l^2_{\mathrm{ref}}\dot{\theta}^2$

M_α, M_β, M_γ = momentum wheel torques about x_p, y_p, and z_p directions, respectively, nondimensionalized with respect to $m_s l_{\mathrm{ref}}\dot{\theta}^2$

Primes indicate differentiation with respect to θ, the true anomaly.

System Dynamics and Control: End Bodies Treated as Point Masses (No Offset Effects)

Important features of the system dynamics can be illustrated quite conveniently by treating the end bodies as point masses, considering the tether rigid but with finite mass, and retaining the nonlinearities. These equations are presented by Modi et al.[15] The amount of information that can be accumulated through variation of the various system parameters such as orbital eccentricity and inclination, deployment and retrieval modes, control strategies, choice of gains, atmospheric drag effects, etc., is rather extensive. For conciseness, only a few of the typical results useful in establishing trends are recorded here.

Deployment Dynamics

During deployment the subsatellite is positioned at the desired altitude corresponding to its operational state. Obviously, one would like to complete this maneuver as quickly as possible so that the subsatellite can undertake its intended mission. As pointed out earlier, all reported analyses, irrespective of the system model used, suggest that deployment is essentially a stable operation unless the rate of deployment exceeds a critical limit[8]:

$$|\dot{\bar{l}}/\omega\bar{l}| < 0.75$$

where ω is the mean orbital angular rate.

Figure 5 shows in-plane and out-of-plane responses of the tether during downward deployment of a 170-kg subsatellite from the Shuttle in a 200-km equatorial circular orbit. Physical properties of the tether, subsatellite, atmosphere as well as the three-stage deployment procedure are taken to be the same as those used by Baker et al.,[7] i.e., a combination of

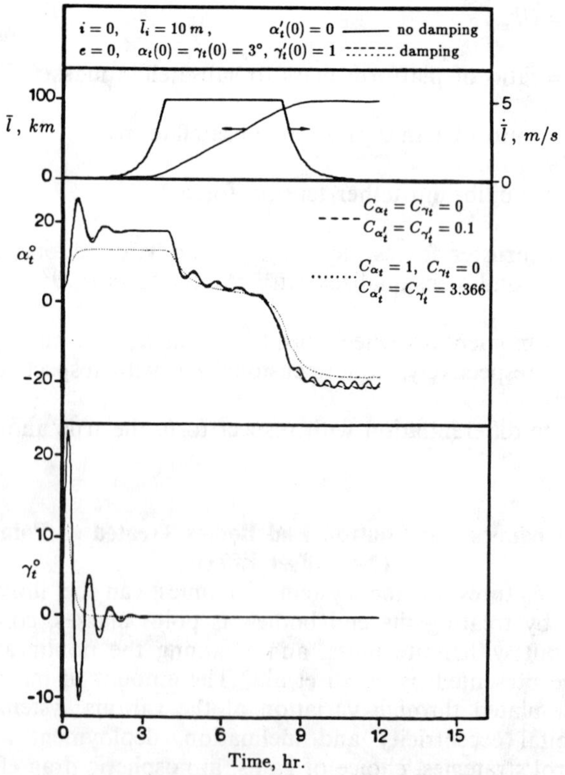

Fig. 5 In-plane and out-of-plane responses of the Space Shuttle based 100-km tether while in a circular equatorial orbit. Note the effect of damping.

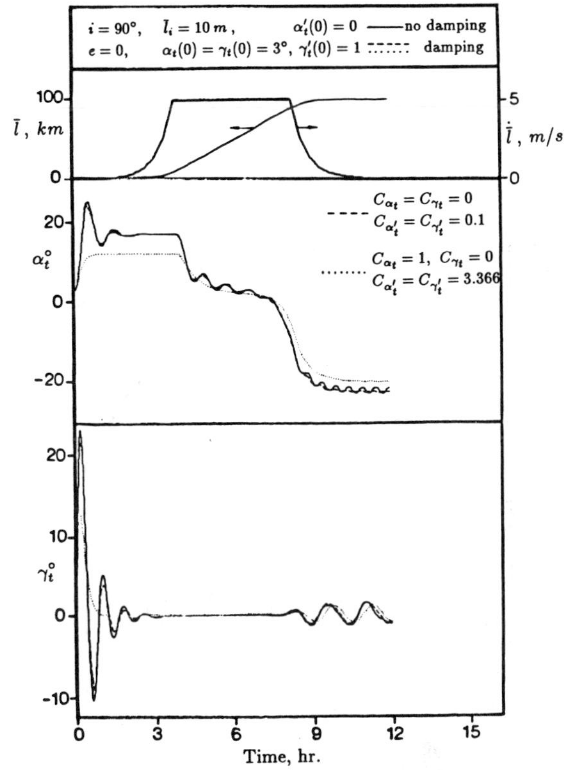

Fig. 6 Influence of the rotating atmosphere-induced periodic disturbance on the deployment dynamics during the Shuttle in a circular polar orbit.

exponential and uniform $\dot{L}$. For a relatively small in-plane disturbance of $\alpha_t(0) = 3$ deg, a large-amplitude motion sets in that gradually drifts toward the equilibrium orientation of the tether at around -22 deg when fully deployed. Modulations at the natural frequency near the terminal phase are attributed to atmospheric effects, which become significant as the tether length approaches around 90 km. On the other hand, in spite of rather severe out-of-plane initial conditions [$\gamma_t(0) = 3$ deg, $\gamma'_t(0) = 1$] the roll degree of freedom returns to the equilibrium position quite quickly. The dotted lines show the effect of a relatively large damping (C corresponds to the damping ratio). Figure 6 considers essentially the same deployment condition as before, but now the satellite is in a polar orbit ($i = 90$ deg). Note the forced out-of-plane oscillations at the orbital frequency, caused by the rotating atmosphere, which the damper is unable to suppress. The effect of eccentricity of the orbit ($e = 0.00076$) is to accentuate this behavior both in pitch and roll motions as shown by the larger-amplitude limit cycles (Fig. 7), which are at the orbital frequency, as expected. The tether

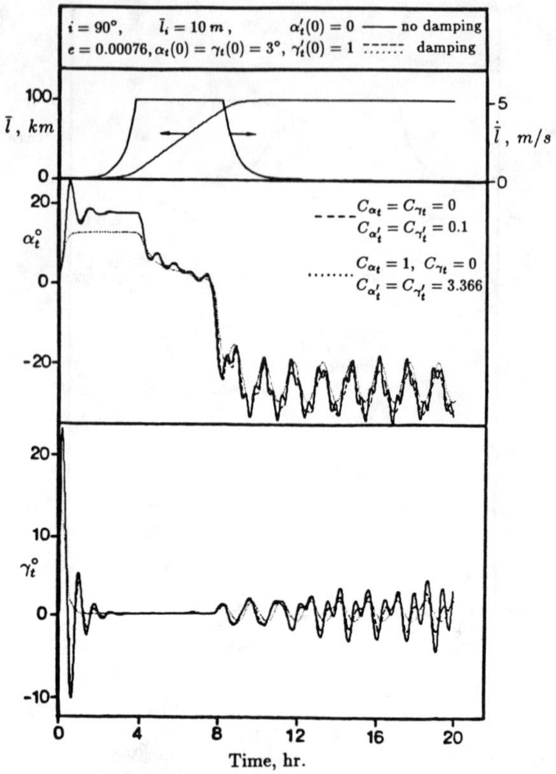

Fig. 7 Effect of the orbital eccentricity on the deployment dynamics of the Space Shuttle based 100-km tether in a polar orbit.

librational amplitude of around 20 deg, though undesirable, may not prove to be critical.

On the other hand, the situation during retrieval using similar exponential strategy is disastrous. Both the in-plane and out-of-plane tether librations become unstable even in the presence of dissipation ($C_{\alpha_t'} = C_{\gamma_t'} = 0.1$) due to the negative damping introduced by the terms proportional to $\dot{\bar{l}}/\bar{l}$ (Fig. 8). Depending on the retrieval rate used, one can easily calculate the damping level required to guarantee stability; however, it may not always be possible to provide the required level of damping. This clearly suggests a need for the control of the type proposed by Rupp[6] and others.[4,5]

Tension Control

Figure 9 shows tether dynamics during retrieval using the tension control strategy as proposed by Kissel (Baker et al.[7]). Effects of aerodynamic drag and damping are also included that were not considered by Kissel. Even in the absence of any damping, the control appears to be quite effective;

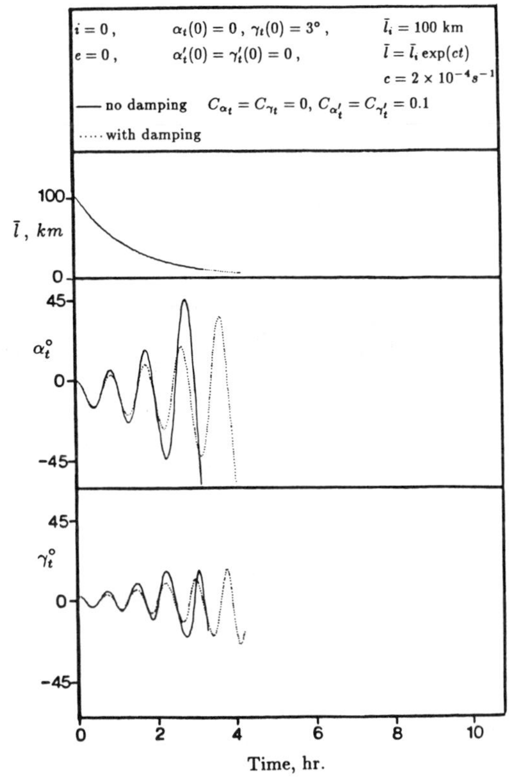

Fig. 8 Unstable tether librations due to a negative damping during retrieval.

however, this is misleading because limit cycles develop, in both pitch and of a large amplitude (50 deg) in the out-of-plane roll motion, during retrieval of the last 1 km of tether length. Although the damping is able to eliminate the limit cycles, the dissipation level used is unrealistically large ($C_{\alpha_t'} = C_{\gamma_t'} = 3.366$). Thus, it is necessary for an alternate control strategy to evolve that may lead to an acceptable level of oscillations using realistic values of damping.

Bainum and Kumar[12] developed an optimal control law based on an application of the linear regulator problem. The control was by modulating the tether tension level T as a function of the nondimensional difference between the actual and the commanded tether length l, length rate l', in-plane swing angle α_t, and in-plane swing angle rate α_t':

$$T = K_l l + K_{l'} l' + K_{\alpha_t} \alpha_t + K_{\alpha_t'} \alpha_t' \tag{11}$$

Figure 10 shows the controller to be effective in rendering the response stable; however, the steady-state amplitude, particularly for the out-of-plane limit cycle, can be quite large (approximately ± 45 deg) in the

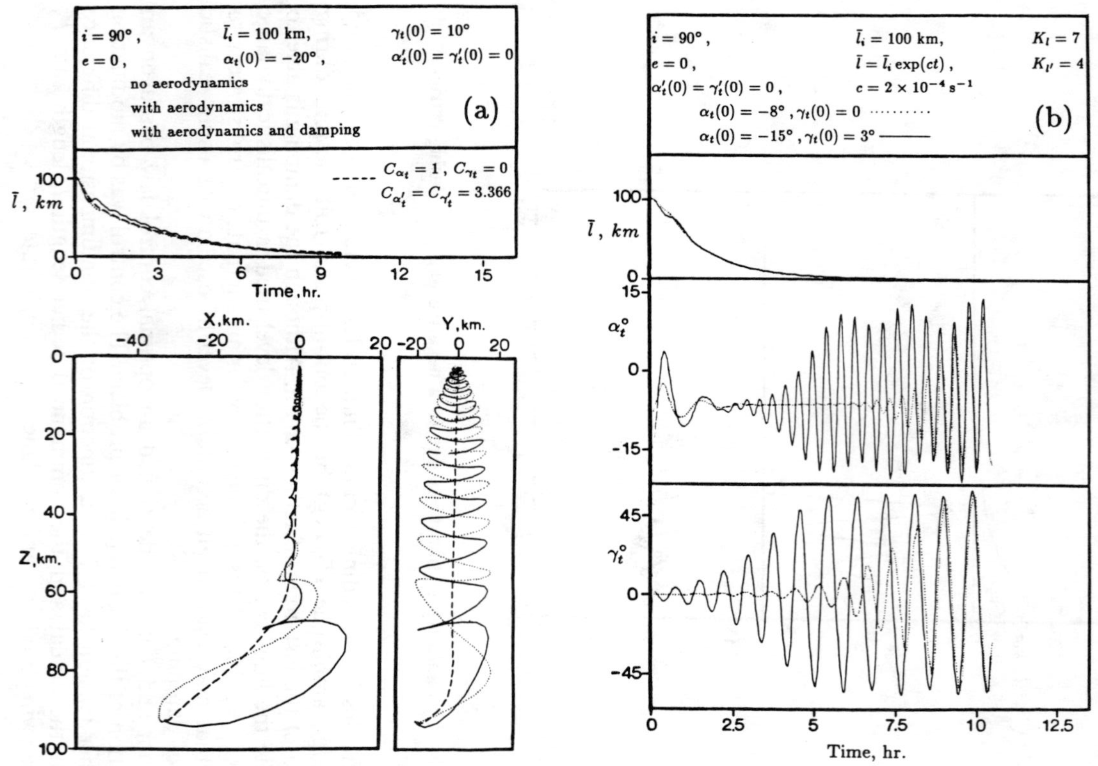

Fig. 9 Effect of the Kissel-type tether tension control on the librational dynamics of the system during retrieval: a) control appears effective; b) a large-amplitude limit cycle in roll appears at smaller lengths which is not apparent in part a.

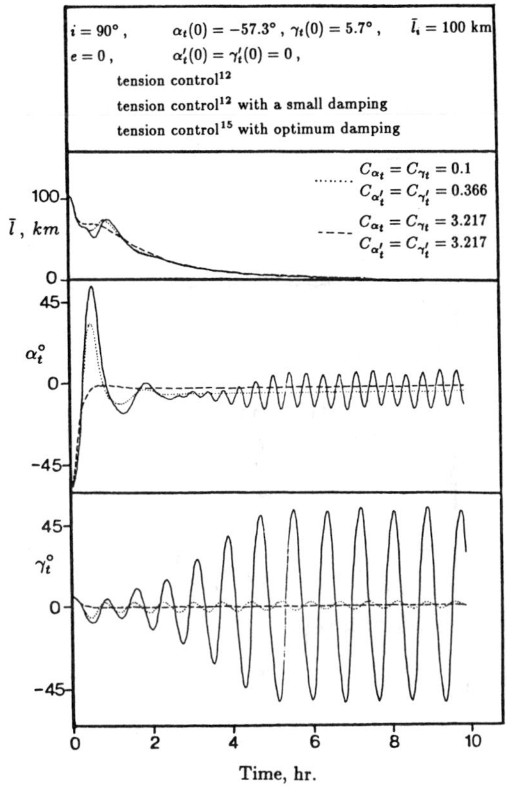

Fig. 10 Stable but large-amplitude response of the tether during retrieval using the optimum linear control strategy as proposed by Bainum and Kumar.[12]

presence of atmospheric effects that were neglected in the analysis of Bainum and Kumar. The figure also indicates the influence of a change in initial conditions. As can be expected, this only alters the time taken to attain the limit cycle amplitude.

Under such a situation, one is tempted to assess the effectiveness of nonlinear control strategies, such as

$$\begin{aligned} T &= T(l,l',|\gamma_t|,|\gamma_t'|) \\ &= T(l,l',\alpha_t,\alpha_t',\gamma_t'^2) \\ &= T(l,l',\alpha_t,\alpha_t',|\gamma_t'|) \\ &= T(l,l',\alpha_{t'},\gamma_t'^2) \end{aligned}$$

and others. A procedure where the tether tension is controlled as

$$T = K_l l + K_{l'} l' + K_{\gamma_t'} \gamma_t'^2 \tag{12}$$

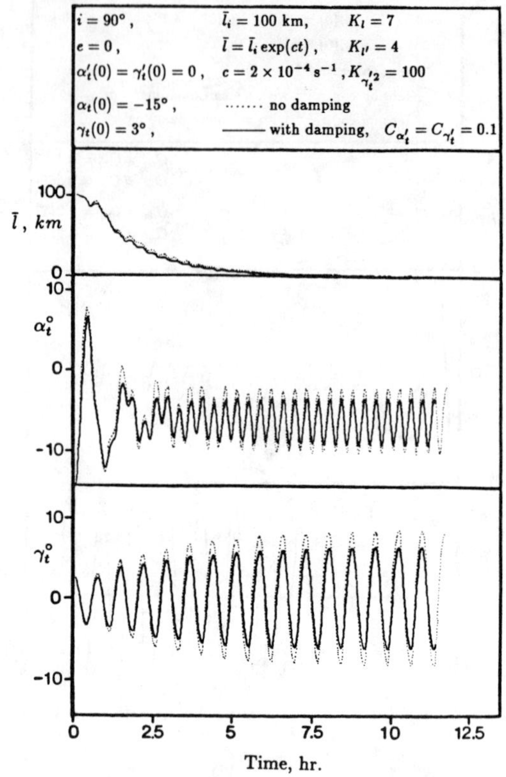

Fig. 11 Retrieval dynamics in the presence of a nonlinear control strategy as proposed by Modi et al.[15] showing dramatic reduction in the limit cycle amplitudes.

appeared to be the most promising. Figure 11 presents representative results for the tether response during retrieval from 100 km to 500 m using the control strategy stated in Eq. (12). Note that the limit cycle amplitudes in both pitch and roll are substantially reduced. The controller continued to remain effective over a wide range of situations involving different initial conditions, orbital elements, physical properties of the system, retrieval rate, etc.

In actual practice the tether material is indeed elastic, causing a longitudinal stretch (ζ) of the tether (besides transverse deflections). Any realistic analysis must account for it. Hence, it was essential to check the controller's effectiveness when the tether is elastic. To this end the appropriate stretch equation was added to the governing set. Figure 12 summarizes controlled retrieval dynamics of such a tethered subsatellite system for the Space Shuttle in a circular polar orbit at a height of 220 km. The high-frequency longitudinal oscillations of the tethered system made the numerical integration of the nonlinear coupled equations quite expensive. Clearly, the nonlinear control strategy remains effective even for an elastic

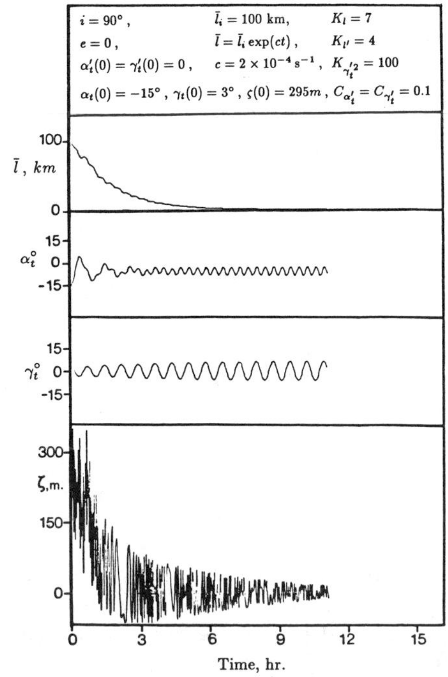

Fig. 12 Plots showing effectiveness of the nonlinear control strategy accounting for longitudinal flexibility of the tether.

tether during the longitudinal stretch for this specific model of point end masses.

However, the question of transverse oscillations of the tether in addition to the longitudinal vibrations and rigid-body rotations still remains. This has been investigated at length by Modi et al.[15,17] The problem appears to be that of maintaining sufficient tension even at shorter tether lengths, to render the system stable. They have considered the case of a 100-km Space Shuttle-based tether supporting a 170-kg subsatellite. The effects of the aerodynamic environment are accounted for. The subsatellite is provided with an orthogonal set of thrusters with a rating of ± 5 N. Employing exponential followed by uniform retrieval rates and feedback of transverse vibrations in conjunction with a length rate control law with thrusters, the authors showed complete control of the system (vibrational as well as librational motions) even in the presence of severe disturbances. Typically, the retrieval time was around 2 h, with a demand on thruster impulse of about 10,000 Ns.

Finite End Bodies with the Point of Attachment Offset

Three strategies have been reported in the literature[18–20] that use linear regulator theory to obtain optimal state variable feedback control. Each strategy takes advantage of a particular aspect of TSS dynamics and is accordingly classified as tension, thruster, or offset control. In each case the minimum fuel regulator performance index was chosen:

$$J = \int_0^{\infty} (\boldsymbol{x}^T Q \boldsymbol{x} + \boldsymbol{u}^T R \boldsymbol{u})\, \mathrm{d}t$$

The optimal control $\boldsymbol{u}$ that minimizes this criterion is given by

$$\boldsymbol{u} = -R^{-1} B^T K \boldsymbol{x} = -G \boldsymbol{x}$$

where Q is the positive-definite state penalty matrix; R is the positive-definite control penalty matrix; and K is the solution to the steady-state Ricatti matrix equation,

$$-KA - A^T K + KBR^{-1} B^T K - Q = 0$$

Tension Control

It is apparent that the instantaneous length of the tether affects its tension and librational dynamics through coupling. Hence, as seen before, tether tension can be utilized to control the system. It would be of interest to assess its effectiveness in the presence of platform rotations as well as tether offset during stationkeeping and retrieval phases. The station is provided with momentum wheels. M_α, M_β, and M_γ represent the nondimensionalized momentum wheel torques (Fig. 3). Here, $\boldsymbol{u}$ is a 4×1 control vector consisting of M_α, M_β, M_γ, and T_e:

$$\boldsymbol{u} = \begin{Bmatrix} M_\alpha \\ M_\beta \\ M_\gamma \\ T_e \end{Bmatrix}$$

where T_e represents tension in the tether line nondimensionalized with respect to $m_s l_{\mathrm{ref}} \dot{\theta}^2$.

Thruster Control

The effectiveness of thrusters during retrieval of the tether has been demonstrated by several authors.[16,17] However, in these investigations librational dynamics of the station and offset of the tether attachment point were ignored. Lakshmanan et al.[18–20] have studied the system control accounting for the end body librations and the attachment point offsets using a set of orthogonal thrusters. The nondimensionalized thruster force $T_{\hat{L}}$ acts along the tether line, whereas T_α and T_γ are directed normal to the tether line in the orbital plane and along the orbit normal, respectively. Here $\boldsymbol{u}$ is a 6×1 control vector consisting of M_α, M_β, M_γ, T_α, T_γ, and $T_{\hat{L}}$ as shown by

$$\boldsymbol{u} = \begin{Bmatrix} M_\alpha \\ M_\beta \\ M_\gamma \\ T_\alpha \\ T_\gamma \\ T_{\hat{L}} \end{Bmatrix}$$

Offset Control

When the out-of-plane offset $\bar{D}_{x,p}$ is set to zero, the in-plane degrees of freedom (α_p, α_t, and $\hat{L}$) completely decouple from the out-of-plane degrees of freedom (β_p, γ_p, and γ_t) in the linear case. This decoupling is illustrated in the tether equations shown by:

α_t: *Tether in-plane swing*:

$$\left(1+\frac{\bar{L}M_{t,s}}{2}\right)\bar{D}_{yp}\alpha_p'' - 2\left(1+\frac{\bar{L}M_{t,s}}{2}\right)\bar{D}_{zp}\alpha_p' + \left(1+\frac{\bar{L}M_{t,s}}{3}\right)\alpha_t''$$
$$+3\left\{\left(1+\frac{\bar{L}M_{t,s}}{2}\right)\bar{D}_{yp} + \left(1+\frac{\bar{L}M_{t,s}}{2}\right)\right\}\alpha_t + 2\left(1+\frac{\bar{L}M_{t,s}}{2}\right)\hat{L}'$$
$$+2\left(1+\frac{\bar{L}M_{t,s}}{2}\right)\hat{D}_{y,p}' + \left(1+\frac{\bar{L}M_{t,s}}{2}\right)\hat{D}_{z,p}'' = T_\alpha \bar{L}$$

γ_t: *Tether out-of-plane swing*:

$$-\left(1+\frac{\bar{L}M_{t,s}}{2}\right)\bar{D}_{zp}\beta_p'' - \left(1+\frac{\bar{L}M_{t,s}}{2}\right)\bar{D}_{zp}\beta_p + \left(1+\frac{\bar{L}M_{t,s}}{2}\right)\bar{D}_{yp}\gamma_p''$$
$$+\left(1+\frac{\bar{L}M_{t,s}}{2}\right)\bar{D}_{yp}\gamma_p + \left(1+\frac{\bar{L}M_{t,s}}{3}\right)\gamma_t''$$
$$+\left\{3\left(1+\frac{\bar{L}M_{t,s}}{2}\right)\bar{D}_{yp} + 4\left(1+\frac{\bar{L}M_{t,s}}{3}\right)\bar{L}\right\}\gamma_t$$
$$-\left(1+\frac{\bar{L}M_{t,s}}{2}\right)\bar{L}\hat{D}_{x,p}'' - \left(1+\frac{\bar{L}M_{t,s}}{2}\right)\bar{L}\hat{D}_{x,p} = -T_\gamma \bar{L}$$

$\hat{L}$: *Deviation from nominal length*:

$$-(1+\bar{L}M_{t,s})\bar{D}_{zp}\alpha_p'' - 2(1+\bar{L}M_{t,s})\bar{D}_{yp}\alpha_p' + 3(1+\bar{L}M_{t,s})\bar{D}_{zp}\alpha_p$$
$$-2\left(1+\frac{\bar{L}M_{t,s}}{2}\right)\alpha_t' + (1+\bar{L}M_{t,s})\hat{L}''$$
$$-3\left\{\bar{L}M_{t,s}(\bar{D}_{yp}+\tfrac{1}{2}) + \left(1+\frac{\bar{L}M_{t,s}}{2}\right)\right\}\hat{L} + (1+\bar{L}M_{t,s})\hat{D}_{y,p}''$$
$$-3(1+\bar{L}M_{t,s})\hat{D}_{y,p} - 2(1+\bar{L}M_{t,s})\hat{D}_{z,p}' - T_{\hat{L}} = Q_{\bar{l}}$$

Thus, control of the out-of-plane motion of the tether using a length control strategy alone would not be possible. Each of the three offsets couples the dynamics of the tether with that of the platform in a unique way. For example, if the in-plane offset $\bar{D}_{y,p}$, nominally along the local vertical, is ignored, the platform roll motion γ_p and the tether out-of-plane motion γ_t become decoupled. Three-dimensional offset of the tether attachment point leads to an increased influence on the system dynamics, thus providing a mechanism for improved control. However, it must be recognized that this will also change the system equilibrium configuration.

It is apparent from the librational equations that the offset perturbations denoted by $\hat{d}_{j,p}$ and their derivatives affect all of the degrees of freedom. In order to control the system one may choose any set of three input variables (e.g., the offset positions, velocities, or accelerations). If one selected $\hat{D}_{x,p}$, $\hat{D}_{y,p}$, and $\hat{D}_{z,p}$ as the offset control variables (M_α, M_β, and M_γ, would make up the rest), then

$$\boldsymbol{D} = \boldsymbol{D}(t) = \begin{Bmatrix} \hat{D}_{x,p}(t) \\ \hat{D}_{y,p}(t) \\ \hat{D}_{z,p}(t) \end{Bmatrix}$$

From the feedback law,

$$\boldsymbol{D} = -\boldsymbol{K}\boldsymbol{x}$$

where $\boldsymbol{x}$ is the state vector. In order to evaluate the offset positions $D(t)$ and $\dot{D}(t)$, the state vector $\boldsymbol{x}$ would have to be differentiated twice. On the other hand, choosing the three offset accelerations as control inputs ($\ddot{\hat{D}}_{x,p}$, $\ddot{\hat{D}}_{y,p}$, $\ddot{\hat{D}}_{z,p}$) and integrating them to obtain the offset velocities and positions present a number of advantages:

1) The linearized closed-loop equations indicate that the three offset accelerations are the only set of variables that directly influence all the degrees of freedom. The offset velocities have no effect on the tether out-of-plane swing, whereas the offset positions have no effect on the tether in-plane motion when the equations are linearized.

2) From a numerical standpoint integration schemes are more accurate than differentiation algorithms.

3) From a practical standpoint the relationship between motor torques (proportional to accelerations) and the motor input voltage is direct; hence, a servo loop is not necessary.

4) An offset position specification scheme may result in velocities and accelerations that are impossible to implement through conventional motors. On the other hand, the maximum offset accelerations generated for control could be easily lowered by increasing the elements of the penalty matrix R associated with them.

Accordingly, the control vector $\boldsymbol{u}$ is of order 6×1 and consists of M_α, M_β, M_γ, $\ddot{\hat{D}}_{x,p}$, $\ddot{\hat{D}}_{y,p}$, and $\ddot{\hat{D}}_{z,p}$. It should be noted that the offset velocities and positions become specified quantities in this case and must be accounted for appropriately in the equations during numerical simulations.

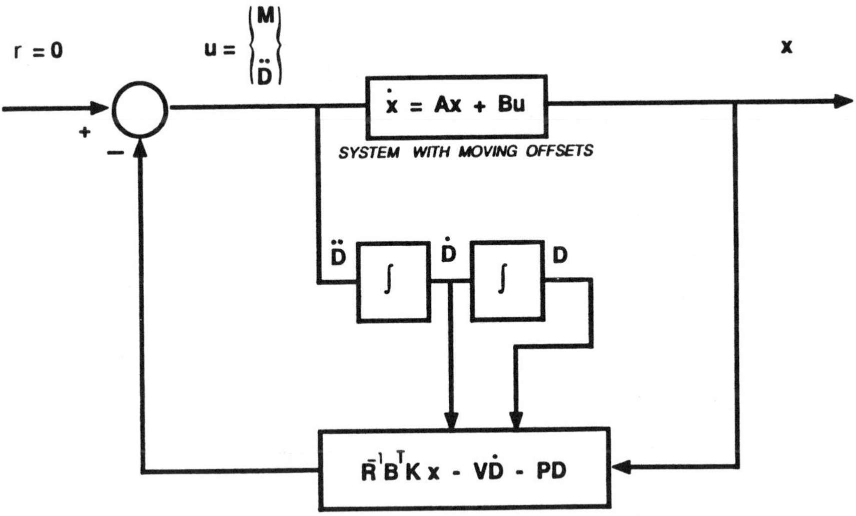

Fig. 13 Modified linear regulator based offset controller with a feedforward strategy.

Furthermore, they must be restricted and returned to rest at the end of the control, preferably at the starting position. This can be achieved through feedforward of the offset state. The system control block diagram can be represented as indicated in Fig. 13.

Results and Discussion

The authors[18–20] have carried out extensive parametric study involving platform dimensions, the ratio of the platform to tether mass, tether length, offset, and initial conditions. Only one typical set of response results is presented here for the case:

Orbital altitude	= 500 km
Platform dimensions	= 100 × 120 m
Platform mass	= 10,000 kg
Platform inertias	
$I_{x,p}$	$= 20.33 \times 10^6$ kg-m^2
$I_{y,p}$	$= 8.33 \times 10^6$ kg-m^2
$I_{z,p}$	$= 12.00 \times 10^6$ kg-m^2
Subsatellite mass	= 100 kg
Tether mass	= 10 kg
Tether length	= 100 m, 1000 m
Maximum offset in each direction	= 20 m
Initial disturbance, $\alpha_t(0)$	= 0.5 rad

A relatively large in-plane tether disturbance was purposely chosen to establish effectiveness of the various strategies. Inherent coupling in the

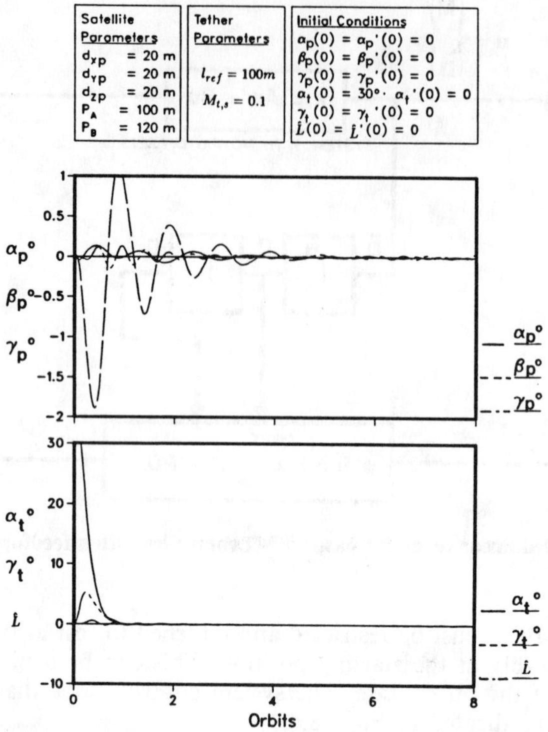

Fig. 14 Control of the plate-type platform and tether configuration using tension variation in the tether line. The controller was found to damp tether motions in the shortest time.

system causes all of the degrees of freedom to be excited through this disturbance.

Stationkeeping Phase

Figure 14 attempts to evaluate performance of the tension control strategy when the nominal tether length is held fixed at 100 m. It was found to utilize the maximum energy among the three control strategies studied here. It requires close to six orbits to damp the platform motion but only one orbit to arrest the tether dynamics. The maximum platform amplitude is around 2 deg. Figure 15 shows the tether tension force profile during the application of this control strategy. Note that there is an equilibrium tension of 5.7 N corresponding to the tether and subsatellite masses. The maximum tension force does not exceed 5.74 N.

Figure 16 shows the platform and tether response for the case of thruster control. Note that the platform motion dies out in around 6 orbits, whereas the tether returns to the equilibrium configuration in slightly more than 1.75 orbits. The peak platform rotation in the pitch degree is observed to

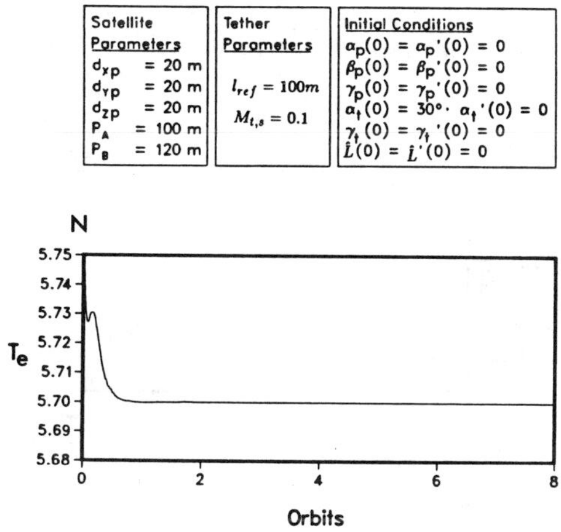

Fig. 15 Variation of tether tension during the tension control. Impulse requirement for this controller turned out to be the highest (55.31 Ns).

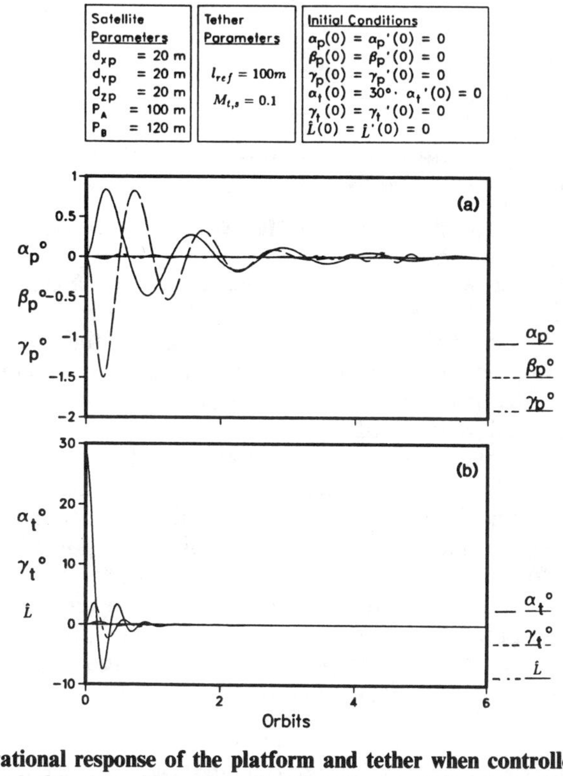

Fig. 16 Librational response of the platform and tether when controlled by momentum wheels and thrusters. Note that the thruster control scheme is successful in limiting the motion to a very small value.

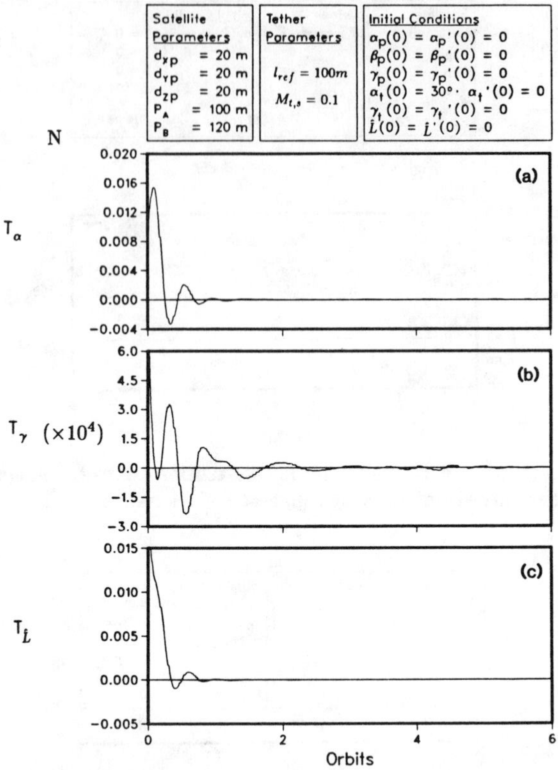

Fig. 17 Thrust time histories. Impulse required is 36.21 Ns.

be less than 1.5 deg, whereas the peak tether libration is around 8 deg. Both values are within an acceptable limit considering a rather severe initial disturbance. Figure 17 shows the thruster profiles in the three orthogonal directions during the controlled motion. For the chosen system parameters and the initial disturbance, the thrust required to control the motion is confined to a small value (0.02 N).

The response of the system with the offset control is depicted in Fig. 18. The controller utilizes minimum energy; however, as can be expected, it requires more than 10 orbits to stabilize the platform. It should be noted here that the platform motion tends to the equilibrium position that is itself changing due to the offset control. Consequently, the regulator returns the system to the equilibrium state corresponding to the terminal position of the attachment point. The offset control forces are proportional to the accelerations of the attachment point. Their magnitudes are found to be very small ($<10^{-3}$ N), as shown in Fig. 19.

Retrieval Phase

The thruster and tension control strategies were also tested during the exponential retrieval of the tether. In this phase of simulations the nominal

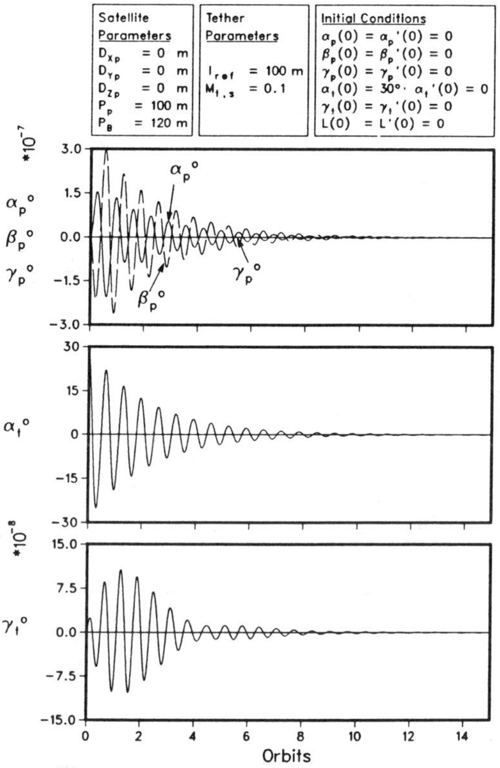

Fig. 18 Offset-controlled response of the station-tether-subsatellite configuration in the stationkeeping mode.

length of the tether was varied continuously with time to a final length of 10 m. The responses of the system with starting lengths of 1000 and 100 m were compared. The retrieval scheme chosen was of exponentially decaying type with $c/\dot{\theta}$ taken to be $-3 \times 10^{-4}\ \mathrm{s}^{-1}$. This leads to retrieval of a 100-m tether in 1.4 orbits and a 1000-m tether in 3 orbits. The scheme results in a maximum retrieval velocity of 0.302 m/s. It is clear that the equilibrium configuration changes with the length during retrieval and must therefore be considered quasistatic. The changing length would necessitate continuous determination of optimal gains during retrieval. In the results presented here, evaluation of the gains was carried out at an interval $\theta = 0.1 - 1$ rad, depending on the rate of retrieval. The solution of the steady-state Ricatti equations requires determination of eigenvalues of a matrix of order $2n$ where n is the order of the system.

All of the three control strategies were found to be effective during retrieval of 100-m as well as 1000-m tethers. As can be expected, the control impulse demanded increased with the length almost linearly for the tension and thruster control strategies. Surprising as it may seem, the

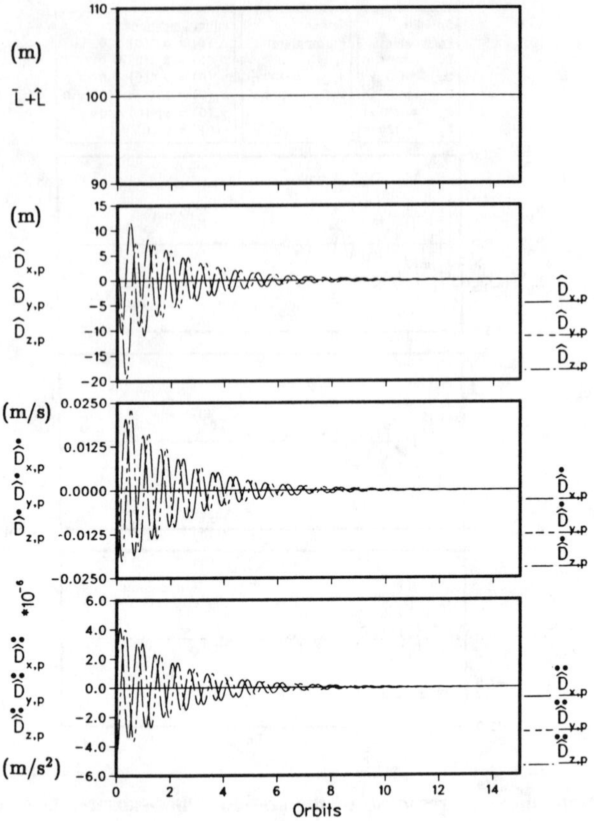

Fig. 19 Time history of the offset accelerations, velocities, and positions. This control strategy resulted in a minimum impulse of 28.05 Ns.

thruster controller damped the system in shorter time but also required smaller impulse for the tether lengths considered. This may be attributed to the progressive deterioration of the tension control at shorter tether lengths. The authors have presented extensive simulation results for these two control strategies.[18–20]

Of particular interest is the offset control strategy, which proved to be quite effective with a significantly lower impulse demand, particularly at shorter tether lengths. Ironically, this is the state where tension control loses its effectiveness while the thruster control cannot be employed due to safety considerations and the plume impingement problem, as mentioned earlier.

The offset control technique is similar to balancing a rod on the palm of one's hand. As can be expected, for a given angular disturbance, the motion required at the point of attachment of the tether would grow proportionally to the length of the tether. Figures 20 and 21 demonstrate

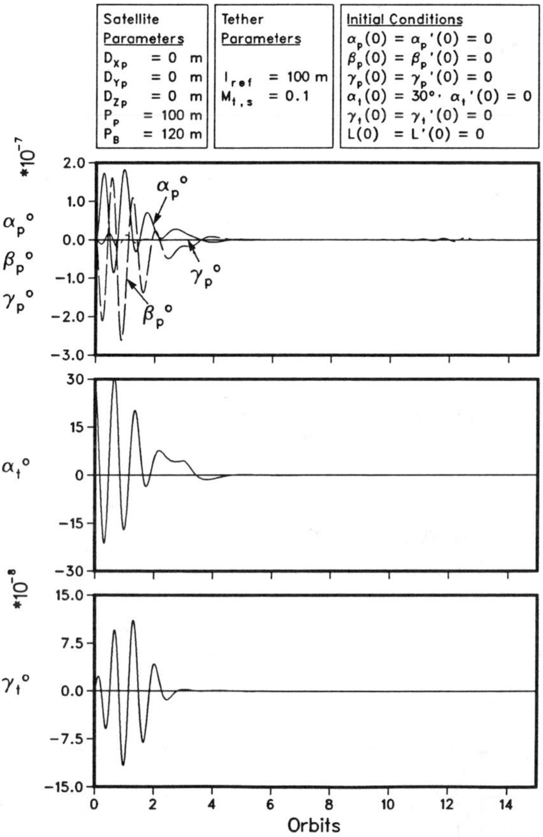

Fig. 20 Offset-controlled response of the platform-based tethered system during exponential retrieval from 100 m to 10 m in three orbits for a tether in-plane disturbance $\alpha_t(0)$ of 30 deg.

the effectiveness of the offset controller during the retrieval phase of a 100-m tether. Note that even a disturbance as large as $\alpha_t(0) = 30$ deg is damped in less than six orbits, with the platform remaining almost insensitive to the disturbance as well as the offset motion. In fact, the strategy can be used rather effectively to damp the station librations induced by astronauts' activities, manipulator motion, Shuttle docking, satellite launch, and other disturbances. Such essentially passive control of the space station appears quite attractive from an economic standpoint.

Table 1 compares impulse requirements for the three control strategies during stationkeeping and retrieval phases. It suggests that tension and thruster control strategies, implemented individually or together, in conjunction with the offset control at a shorter length of the tether are likely to be the most effective procedures from energy as well as safety considerations. Such a hybrid control procedure should receive more attention in the future.

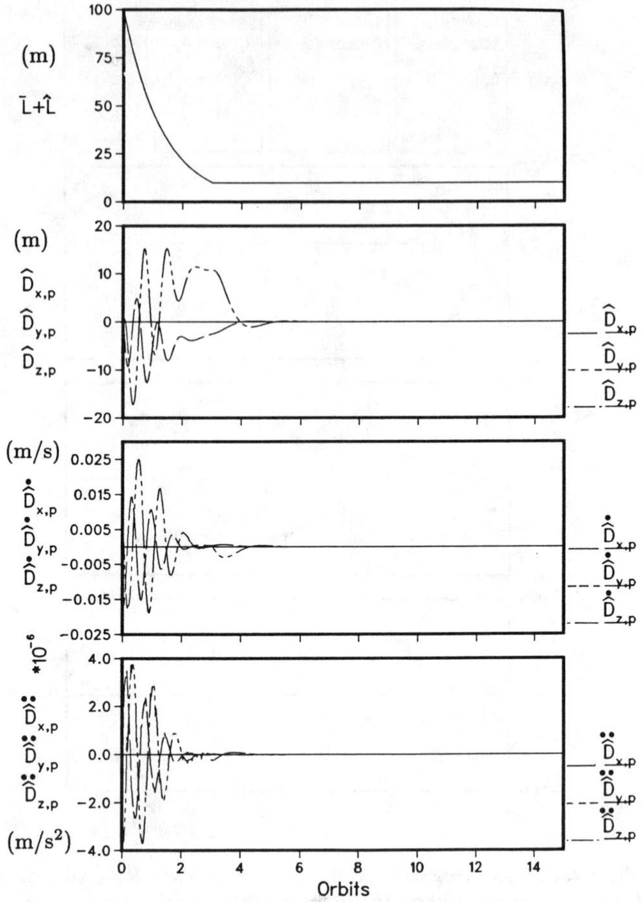

Fig. 21 Time histories of the offsets and their derivatives during exponential retrieval of the platform-based tether for an in-plane disturbance of 30 deg. Note that the offset motions do not exceed 20 m.

Direction of Future Efforts

It is apparent that there is a rich body of literature in the area of dynamics and control of the space platform supported tethered systems. Both rotational dynamics as well as vibrations during the deployment, stationkeeping, and retrieval phases have received attention from a growing community of researchers in this fascinating area. However, some basic questions pertaining to efficient and successful retrieval of flexible continuum tether remain unanswered. Concerted efforts in this direction are necessary to arrive at optimum control strategies based on tension, thruster, and offset algorithms, and particularly their judicious combinations. Of course, eventually we will have to plan an on-orbit demonstration

Table 1 Impulse demand for the three control strategies during stationkeeping and retrieval phases

Mode	Control	Tether length, m	Disturbance, α_t, deg	Control impulse, Ns
Stationkeeping	Tension	100	30	55.31
	Thruster	100	30	36.21
		100	30	39.23
	Offset	100	6	22.12
Retrieval	Tension	100	30	59.11
		1000	30	531.91
	Thruster	100	30	42.23
		1000	30	374.27
	Offset	100	30	30.76
		100	6	17.57

of the control algorithms in order to check their validity and modify them appropriately. The proposed TSS-1 mission is a welcome step in that direction. However, there is a considerable scope for ground-based and smaller space-based experiments focused at specific aspects pertaining to material properties, damping, deployment mechanism, sensor-actuator design and operation, reduction of observed information and state estimation, control strategies, etc. Ground-based and on-flight verification of existing dynamic models and control schemes is fundamental to the success of a variety of visionary concepts involving application of tethers in space.

With reference to the proposed space station Freedom, all of the participating agencies have shown interest in the tether-supported facilities aimed at a variety of missions. When the highly flexible character of the station is recognized, the problem of dynamics and control of the space station-based tethered facility attains the challenge of a higher order than that encountered or studied so far. Lakshmanan et al.[18–20] have initiated investigations in that area; however, eventually the models will have to be made more sophisticated to account for station, tether, and payload flexibility. The influence of free molecular reaction forces, solar-radiation-induced heating, and electromagnetic forces for conducting tethers will have to be incorporated realistically. This class of problems has remained virtually untouched so far.

A new class of problems involving tether-supported systems of three, four, or more bodies is also receiving some attention lately. The objectives are quite varied, ranging from simultaneous sounding of the environment at several altitudes to communication antenna, payload transfer, and microgravity control. This represents a fertile field of considerable practical significance and promise.

Orbital transfer using a tether depends strongly on the attitude dynamics of the tethered system. A parametric study aimed at interactions between the orbital dynamics and swing motions to optimize the transfer would lay a sound foundation for further evolution of that subject. Our present knowledge in the area consists of isolated studies with specific idealized configurations of limited value.

The awareness of scientists of some of the previously stated problems is indicated by studies presented at a recent workshop and a conference on the subject.[23] However, like any process of inquiry, philosophical or scientific, no matter how far we advance we will always be on the shores of an unexplored ocean of knowledge. But then a journey fulfills itself in every step.

References

[1]Tsiolkovsky, K. E., "Speculations between Earth and Sky," *Science Fiction Works*, izd-vo AN-SSSR, Moscow, 1859, p. 35 (reprinted 1959).

[2]Lang, D. L., and Nolting, R. K., "Operations with Tethered Space Vehicles," NASA SP-138, Feb. 1967.

[3]Cron, A. C., "Application of Tethers in Space," NASA CP-2365, June 1983.

[4]Misra, A. K., and Modi, V. J., "Dynamics and Control of Tether Connected Two-Body Systems—A Brief Review," 33rd Congress of the International Astronautical Federation, Paris, Sept. 1982, Paper IAF-82-316; also *Space 2000*, edited by L. G. Napolitano, AIAA, New York, 1983, pp. 473–514.

[5]Misra, A. K., and Modi, V. J., "A Survey on the Dynamics and Control of Tethered Satellite Systems," NASA/AIAA/PSN International Conference on Tethers, Arlington, VA, Sept. 1986, Paper AAS-86-246; also *Advances in Astronautical Sciences*, Vol. 62, edited by P. M. Bainum et al., American Astronautical Society, Washington, DC, pp. 667–719.

[6]Rupp, C. C., "A Tether Tension Control Law for Tethered Subsatellites Deployed Along Local Vertical," NASA TM-X-64963, Sept. 1975.

[7]Baker, P. W., et al., "Tethered Subsatellite Study," NASA TM-X73314, March 1976.

[8]Kulla, P., "Dynamics of Tethered Satellites," *Proceedings of the Symposium on Dynamics and Control of Non-Rigid Spacecraft*, European Space Agency, Frascati, Italy, May 1976, pp. 349–354.

[9]Kalaghan, P. N., et al., "Study of the Dynamics of a Tethered Satellite System (Skyhook), Smithsonian Institution, Astrophysical Observatory, Cambridge, MA, Final Rept., Contract NAS8-32199, March 1978.

[10]Modi, V. J., and Misra, A. K., "On the Deployment Dynamics of Tether Connected Two-Body Systems," *Acta Astronautica*, Vol. 6, No. 9, 1979, pp. 1183–1197.

[11]Misra, A. K., and Modi, V. J., "A General Dynamical Model for the Space Shuttle Based Tethered Subsatellite System," *Advances in the Astronautical Science*, American Astronautical Society, Washington, DC, Vol. 40, Pt. II, 1979, pp. 537–557.

[12]Bainum, P. M., and Kumar V. K., "Optimal Control of the Shuttle-Tethered System" *Acta Astronautica*, Vol. 7, No. 12, 1980, pp. 1333–1348.

[13]Xu, D. M., Misra A. K., and Modi, V. J., "Three Dimensional Control of the Shuttle Supported Tethered Satellite Systems During Deployment and Retrieval," edited by L. Meirovitch, *Proceedings of the 3rd VPI&SU/AIAA Symposium on Dynamics and Control of Large Flexible Spacecraft*, 1981, pp. 453–469.

[14]Xu, D. M., Misra A. K., and Modi, V. J., "On Vibration Control of Tethered Satellite Systems," *NASA/JPL Workshop on Applications of Distributed Systems Theory to the Control of Large Space Structures*, 1982, pp. 317–327.

[15]Modi, V. J., Gung, C. F., and Misra, A. K., "Effect of Damping on the Control Dynamics of the Space Shuttle Based Tethered System," *Journal of the Astronautical Sciences*, Vol. 21, No. 1, 1983, pp. 135–149.

[16]Banerjee, A. K., and Kane, T. R., "Tethered Satellite Retrieval with Thruster Augmented Control," *Journal of Guidance, Control, and Dynamics*, Vol. 7, No. 1, 1984, pp. 45–50.

[17]Xu, D. M., Misra A. K., and Modi, V. J., "Thruster Augmented Active Control of a Tethered Satellite During Its Retrieval," *Journal of Guidance, Control, and Dynamics*, Vol. 9, No. 6, Nov.–Dec. 1986, pp. 663–672.

[18]Lakshmanan, P. K., Modi, V. J., and Misra, A. K., "Dynamics and Control of the Tethered Satellite System in the Presence of Offsets," *Acta Astronautica*, Vol. 15, No. 12, 1987, pp. 1053–1057.

[19]Lakshmanan, P. K., Modi, V. J., and Misra, A. K., "Dynamics and Control of the Space Station Based Tethered Payload," AAS/AIAA Astrodynamics Conference, Kalispell, MT, Paper AAS-87-434, Aug. 1987; also *Advances in the Astronautical Sciences*, Vol. 65, Pt. 1, American Astronautical Society, Washington, DC, pp. 425–440.

[20]Lakshmanan, P. K., Modi, V. J., and Misra, A. K., "Dynamics and Control of the Tethered Satellite System in the Presence of Offsets," *Acta Astronautica*, Vol. 19, No. 2, 1989, pp. 145–160.

[21]Modi, V. J., and Misra, A. K., "Deployment Dynamics of Tethered Satellite Systems," AIAA/AAS Astrodynamics Conference, Palo Alto, CA, 1978, Paper AAS-78-1398.

[22]Columbo, G., et al., "Investigation of Electrodynamic Stabilization and Control of Long Orbiting Tethers," Smithsonian Institution, Astrophysical Observatory, Cambridge, MA, 1980.

[23]*Proceedings of the Third International Conference on Tethers in Space*, AIAA, Washington, DC, 1989.

[illegible]

Chapter 8

Staggered Solution Procedures for Multibody Dynamics Simulation

K. C. Park,* J. C. Chiou,† and J. D. Downer†
University of Colorado, Boulder, Colorado

Introduction

SIMULATION of multibody dynamics systems, such as robotic manipulators, automobile maneuvering, and satellite deployment, remains a challenge to the dynamicist due to its increasing role in design improvements, control, and safe operation. Because of substantial progress made during the past three decades in formulation,[1–19] constraint treatment and solution techniques,[21–36] and the availability of multibody dynamics simulation packages,[37–42] it has now become almost a routine practice to perform realistic modeling and assessment of some practical problems such as mechanical linkages and manipulations of robotic arms if multibody components consist mostly of rigid bodies, discrete springs, and dampers (see, e.g., Haug[15]). However, substantial advances in modeling, formulation, and computational methods are necessary in order to develop a real-time simulation capability for ground vehicle maneuvering dynamics, robotic manipulation, and space structure deployment/assembly.

Specifically, improved modeling of flexibility for localized motions and geometric nonlinearities, material nonlinearities, and contact/friction phenomena, robust and accurate treatment of the system constraint conditions, and efficient use of emerging computer hardware/software technology continue to offer intense research opportunities. Thus, the development of a real-time multibody dynamics simulation capability requires a concerted integration of various modeling, formulation, and computational aspects.

*Professor, Department of Aerospace Engineering Sciences and Center for Space Structures and Controls.

†Research Assistant, Department of Aerospace Engineering Sciences and Center for Space Structures and Controls.

These include selection of a data structure for describing the system topology, computerized generation of the governing equations of motion, implementation of suitable solution algorithms, incorporation of constraint conditions, and easy interpretation of the simulation results. Of these, this chapter is concerned with three computational aspects of multibody dynamics simulation: direct time integration of the governing equations of motion, stabilization of the constraint solution process, and their computer implementation aspects.

From the computational viewpoint, multibody dynamics (MBD) problems are distinct from the structural dynamics problems in that the solution of MBD problems must also satisfy, at each time integration step, the attendant kinematic and equilibrium constraints. This has motivated many dynamicists to develop various techniques, in addition to direct integration algorithms, for accurately and efficiently handling the system constraints. Hence, the reliability and cost of existing MBD simulation packages have been strongly affected by how efficiently and accurately the constraints are preserved during the numerical solution stage.

In general, there have been two types of direct time integration algorithms for the transient response analysis of dynamic systems: explicit and implicit algorithms (see, e.g., Hughes and Belytschko,[43] Park,[44] and Belytschko, et al.[45]). Currently, implicit algorithms appear to be favored by many MBD specialists when both the generalized coordinates and the constraint forces are treated as the unknowns. In this case the corresponding formulations incorporate the system constraints by the Lagrange multipliers method. It has been well known that the resulting Newton-like solution matrix is stiff. This has led to implicit time discretization of the constraint-augmented equations and simultaneous solution of both the generalized coordinates and the Lagrange multipliers. This approach has been extensively investigated by Gear,[21] Baumgarte,[22,29] Orlandea et al.,[23] Petzold,[27] Nikravesh,[31] and others. Because these methods solve both the generalized coordinates and the constraint forces simultaneously, they will be called the *simultaneous solution methods* in this chapter.

On the other hand, if the constraints are eliminated so that the number of unknowns is reduced, it is possible for one to employ either implicit or explicit algorithms. For this situation one may invoke either a geometric or algebraic procedure to streamline the resulting equations of motion if the system topology is an open tree. In essence, geometric procedures have utilized an open-tree topology such as the use of the incidence matrix by Wittenburg[10] and the body array matrix by Huston.[19] Some of the proposed algebraic procedures include the singular decomposition by Walton and Steeves,[20] the use of the generalized speed of Kane and Levinson,[13] the coordinate partitioning technique by Wehage and Haug,[28] the selection of independent coordinates through the natural-coordinate formulation of Garcia de Jalon et al.,[33] and the so-called order-N procedures of Armstrong,[11] Hollerbach,[12] Schwertassek and Roberson,[17] Orin et al.,[25] and others.

As the complexity of MBD systems increases, the simultaneous solution methods have become less attractive. This is due to matrix ill-conditioning, especially for the so-called index two and higher index problems (see, e.g., Petzold[27] and Brenan et al.[46] for the definition of "index" for constraint characterization), divergence of the solution away from the constraint conditions, and, ultimately, to a large number of equations that must be handled. As an alternative to the simultaneous solution methods, a series of computational methods that employ a "divide-and-conquer" strategy have been developed, which are termed partitioned solution procedures presented in Park,[47] Felippa and Park,[48] and Park and Felippa.[49] As an example, partitioned solution procedures allow one to analyze fluid-structure interaction problems with two separate single-field analysis packages, namely, the structural dynamics module and the fluid dynamics analyzer. At each time integration step, one may advance the solution of structural equations of motion by treating the fluid coupling term as an external force. Once the structural coordinates are advanced, the fluid state variables can be advanced by treating the structural coupling terms as a source term. However, a naïve partitioned procedure can suffer from a loss of accuracy as well as computational stability. Thus, a combination of equation augmentation and stabilization should be devised to recover the accuracy loss and maintain unconditional stability. Such a solution procedure is in contrast to a practice of embedding both the structural and fluid dynamics attributes into a combined analysis program.

The numerical solution procedure for MBD systems that we advocate in this chapter is termed a "staggered MBD solution procedure," which solves the generalized coordinates in a separate module from that for the constraint force. This requires a reformulation of the constraint conditions so that the constraint forces can also be integrated in time. A major advantage of such a partitioned solution procedure is that additional analysis capabilities such as active controller and design optimization molecules can be easily interfaced without being embedded into a monolithic program. To this end, the rest of the chapter is organized as follows.

After the basic equations of motion for MBD systems are introduced in the next section, the section on constraint-handling techniques briefly reviews some constraint-handling techniques and introduces the staggered stabilized technique[34,35] for the solution of the constraint forces as independent variables.

The numerical direct time integration of the equations of motion is described in the section on solution algorithms for generalized coordinates. Since accurate damping treatment is important for the dynamics of space structures, we have employed the central-difference method and the midpoint form of the trapezoidal rule since they engender no numerical damping. This is in contrast to the current practice in dynamic simulations of ground vehicles by employing a set of backward-difference formulas.[46] First, the equations of motion are partitioned according to the translational and the rotational coordinates. This sets the stage for an efficient treatment of the rotational motions via the singularity-free Euler parameters. The resulting partitioned equations of motion are then integrated via a two-stage explicit stabilized algorithm for updating both the translational coordinates and angular velocities.[34] Once the angular velocities are obtained, the angular orientations are updated via the midpoint implicit formula employing the Euler parameters.

When the two algorithms, namely, the two-stage explicit algorithm for the generalized coordinates and the implicit staggered procedure for the constraint Lagrange multipliers, are brought together in a staggered manner, they constitute a staggered explicit-implicit procedure that is summarized in the section on implementation. The numerical examples present some example problems, and discussions concerning several salient features of the staggered MBD solution procedure are offered in the last section.

Governing Equations of Motion

The Lagrangian equations of motion for mechanical systems that are free from any constraint can be written, for the generalized coordinate component u_i, as

$$\frac{\mathrm{d}}{\mathrm{d}t}\frac{\partial L}{\partial \dot{u}_i} - \frac{\partial L}{\partial u_i} = Q_i, \qquad i = 1, \ldots, n \tag{1}$$

where L is the system Lagrangian, t is time, $(\dot{\cdot})$ denotes time differentiation, and Q_i is the generalized applied force. It is well known that, if there are m-constraint conditions imposed on $\{u_i, i = 1, \ldots, n\}$, Eq. (1) must be modified as

$$\frac{\mathrm{d}}{\mathrm{d}t}\frac{\partial L}{\partial \dot{u}_i} - \frac{\partial L}{\partial u_i} = Q_i + \sum_{k=1}^{m} \lambda_k B_{ki}, \qquad i = 1, \ldots, n \tag{2}$$

where λ is the Lagrange multiplier, and B_{ki} is the ith gradient component of the kth constraint equation, viz., for configuration constraints,

$$\Phi_k(\boldsymbol{u}) = 0, \qquad B_{ki} = \frac{\partial \Phi_k}{\partial u_i}, \qquad k = 1, \ldots, m \tag{3}$$

and for motion constraints,

$$\Phi_k(\boldsymbol{u},\dot{\boldsymbol{u}}) = 0, \qquad B_{ki} = \frac{\partial \Phi_k}{\partial \dot{u}_i}, \qquad k = 1, \ldots, m \tag{4}$$

Therefore, regardless of the nature of constraints, one may express the equations of motion with constraints in the following form:

$$\begin{bmatrix} M & B^T \\ B & 0 \end{bmatrix} \begin{pmatrix} \ddot{u} \\ \lambda \end{pmatrix} = \begin{pmatrix} Q \\ c \end{pmatrix} \qquad \begin{matrix} (5a) \\ (5b) \end{matrix}$$

where M is a positive-definite matrix, and c depends on the nature of constraints. For example, for configuration constraints we have

$$c = -\frac{\partial}{\partial u}\left(\frac{\partial \Phi}{\partial u}\dot{u}\right) - 2\frac{\partial}{\partial t}\left(\frac{\partial \Phi}{\partial u}\dot{u}\right) - \frac{\partial^2 \Phi}{\partial t^2} \tag{6}$$

and for motion constraints,

$$c = -\frac{\partial \Phi}{\partial t} \tag{7}$$

An implicit time integration formula to solve Eq. (5) may be written as

$$\dot{\boldsymbol{u}}^n = \delta \ddot{\boldsymbol{u}}^n + \boldsymbol{h}^n_{\dot{u}} \tag{8a}$$

$$\boldsymbol{u}^n = \delta \dot{\boldsymbol{u}}^n + \boldsymbol{h}^n_{u} \tag{8b}$$

where δ is a step size that is dependent on the choice of formula, and $\boldsymbol{h}^n_{\dot{u}}$ and $\boldsymbol{h}^n_u$ are formula-dependent historical vectors that consist of past-step solution components.[48,50] As an example, the trapezoidal rule has the following δ and historical vectors:

$$\delta = h/2 \tag{9a}$$

$$\boldsymbol{h}^n_{\dot{u}} = \dot{\boldsymbol{u}}^{n-1} + \delta \ddot{\boldsymbol{u}}^{n-1} \tag{9b}$$

$$\boldsymbol{h}^n_u = \boldsymbol{u}^{n-1} + \delta \dot{\boldsymbol{u}}^{n-1} \tag{9c}$$

where h is the time-step increment.

Substitution of Eqs. (8) into Eq. (5) yields

$$\begin{bmatrix} M & \delta^2 B^T \\ B & 0 \end{bmatrix} \begin{pmatrix} \boldsymbol{u}^n \\ \lambda^n \end{pmatrix} = \begin{pmatrix} r^n_u \\ r^n_\lambda \end{pmatrix} = \begin{pmatrix} Mh^n + \delta^2 Q^n \\ Bh^n + \delta^2 c \end{pmatrix} \qquad \begin{matrix} (10a) \\ (10b) \end{matrix}$$

In practice, in order to avoid pivoting and to maintain high accuracy, the solution of the preceding difference equations is carried out as follows. First, since M is nonsingular for properly formulated dynamic problems, one computes

$$\boldsymbol{u}_u = M^{-1}\boldsymbol{r}^n_u, \qquad C = M^{-1}B^T, \qquad A = BC \tag{11}$$

and factors A. Second, one obtains λ^n by solving

$$\lambda^n = A^{-1}(B\boldsymbol{u}_u - \boldsymbol{r}^n_\lambda)/\delta^2 \tag{12}$$

Finally, $\boldsymbol{u}^n$ is obtained from

$$\boldsymbol{u}^n = \boldsymbol{u}_u - \delta^2 C \lambda^n \tag{13}$$

It should be noted that the accuracy loss associated with the factoring of an ill-conditioned matrix $BA^{-1}B^T$ and the subsequent backsubstitutions can severely influence the solution accuracy of not only the Lagrange multipliers but also the generalized coordinates as seen from Eqs. (12) and

(13). This has motivated many numerical analysts to undertake the development of methods for differential-algebraic systems as a recent monograph[46] and references therein attest to their rich numerical properties. It is generally agreed that the present status of differential-algebraic methods yields robust solutions for problems of *index one*, but can suffer from inaccurate solutions of the Lagrange multipliers for higher index problems. Observe that many practical multibody dynamics problems are characterized by index greater than one. Hence, the need to compute accurately the constraint forces remains a challenge. For instance, for lock-up mechanisms that are activated when truss structures are fully deployed in space often introduce stiff responses with nearly singular state of $BM^{-1}B^T$. More robust constraint computation algorithms are required for these problems.

One way to improve the accuracy of constraint force computations is to adopt index reduction strategies as discussed in Ref. 46. However, such an index reduction inevitably introduces additional system degrees of freedom in the resulting differential-algebraic equations, thus destroying the matrix sparsity of Eq. (5) in addition to increasing the size of the matrix B. In the following we present an alternative approach based on a parabolic regularization of the equations for the Lagrange multipliers, which preserves the first row of Eq. (5) and enables us to solve λ from the parabolic differential equations.

Constraint-Handling Techniques

As alluded to in the Introduction, techniques for handling the system constraints constitute a major part of solution procedures for the numerical simulation of multibody dynamics systems. In this section we will first review the coordinate partitioning technique, Baumgarte's technique, and the penalty technique. The staggered stabilization procedure that we advocate will then be described in detail. A distinct feature of the staggered stabilization procedure is that it can be implemented in a stand-alone module; thus, it can be interfaced not only with the equation solver for rigid-body systems but with that for flexible-body systems as well.

Coordinate Partitioning Technique

In the coordinate partitioning[28,33] or singular decomposition technique,[20] one selects a rank sufficient part of $\boldsymbol{B}$ and partitions it as

$$B = \lfloor B_i \quad B_e \rfloor, \qquad \boldsymbol{u} = \lfloor \boldsymbol{u}_i \quad \boldsymbol{u}_e \rfloor \tag{14}$$

where the rank of $B_i (m \times m)$ is m, and the subscripts (i,e) refer to internal and external variables, respectively. First, we express $\boldsymbol{u}_i$ in terms of $\boldsymbol{u}_e$ as

$$\boldsymbol{u}_i^n = B_i^{-1}(\boldsymbol{r}_\lambda^n - B_e \boldsymbol{u}_e^n) \tag{15}$$

Since we have

$$\lfloor -B_e^T B_i^{-T} \quad I \rfloor \begin{Bmatrix} B_i^T \\ B_e^T \end{Bmatrix} = 0 \tag{16}$$

the first row of Eq. (10) reduces to

$$(M_e + T^T M_i T)\boldsymbol{u}_e^n = \boldsymbol{r}_e^n \tag{17}$$

where

$$T = B_i^{-1} B_e \tag{18a}$$

$$M = \begin{bmatrix} M_i & 0 \\ 0 & M_e \end{bmatrix} \tag{18b}$$

$$\boldsymbol{r}_u^n = \begin{Bmatrix} \boldsymbol{r}_{u_i}^n \\ \boldsymbol{r}_{u_e}^n \end{Bmatrix} \tag{18c}$$

and

$$\boldsymbol{r}_e^n = \boldsymbol{r}_{u_e}^n - T^T \boldsymbol{r}_{u_i}^n + T^T M_i B_i^{-1} \boldsymbol{r}_\lambda^n \tag{19}$$

Once one obtains $\boldsymbol{u}_e^n$, one can obtain $\boldsymbol{u}_e^n$ from Eq. (15) and similarly λ from Eq. (12). Note that, even though Eq. (17) has a smaller dimension than Eq. (10a), its left-hand-side matrix is in general full since T given by Eq. (18a) is in general full. Hence, unless T is a constant matrix, one must refactor the solution matrix in Eq. (17) whenever a new T is formed.

Baumgarte's Technique

Baumgarte's technique[22,29] is based on the observation that the errors committed in computing the constraint conditions given by Eqs. (3) or (4) can either be critically damped out or exponentially decreased as the integration process continues. Mathematically, this can be stated for the configuration constraint [Eq. (3)] as

$$\ddot{\Phi} + 2\alpha\dot{\Phi} + \beta\Phi = 0 \tag{20}$$

or the motion constraint [Eq. (4)] as

$$\dot{\Phi} + \gamma\Phi = 0 \tag{21}$$

In terms of the general constraint equation augmentation as given by Eq. (5b), the preceding stabilization is equivalent to modifying c in Eq. (5b) accordingly. Hence, the technique can be implemented within the standard augmented form of the equations of motion [Eqs. (5)]. However, if $BM^{-1}B^T$ is ill-conditioned, which can happen since B is in general state-dependent, the accuracy of the generalized constraint force λ can be considerably degraded. This can occur if any two rows of B are physically similar (i.e., when two members form a straight line) or numerically close during three-dimensional orientations.

Penalty Technique

In the two constraint-handling techniques outlined so far, the objective was to satisfy the constraint condition

$$\Phi = 0 \tag{22}$$

whose differentiated forms were augmented to the equations of motion. In the penalty procedure one adopts

$$\lambda = \frac{1}{\varepsilon}\Phi, \qquad \varepsilon \to 0 \tag{23}$$

as the basic constraint equations instead of the twice-differentiated form adopted in Eq. (5).

It is noted that the penalty formulation tacitly assumes that there will be violations of the constraint condition in actual computations as discussed in Lanczos.[51] If one substitutes Eq. (23) into the governing equations of motion, the resulting equation becomes

$$M\ddot{u} + \frac{1}{\varepsilon} B^T \Phi = Q \tag{24}$$

A major drawback of the preceding penalty procedure is that, once an error is committed in computing λ, there is no compensation scheme by which the drifting of the numerical solution can be corrected. This has led to the development of a staggered stabilized procedure as described later.

Staggered Stabilization Procedure

To illustrate this procedure, we will consider the case of nonholonomic constraints. Instead of substituting the penalty expression directly into the governing equations of motion, first we differentiate Eq. (23) once to obtain

$$\dot{\lambda} = \frac{1}{\varepsilon}\left(B\dot{u} + \frac{\partial \Phi}{\partial t}\right) \tag{25}$$

where we assume the penalty parameter ε to be constant.

Second, we obtain $\ddot{u}$ from Eq. (5a) in the form

$$\ddot{u} = M^{-1}(\bar{Q} - B^T\lambda) \tag{26}$$

and substitute it into Eq. (25) to yield

$$\varepsilon\dot{\boldsymbol{\lambda}} + \boldsymbol{B}\boldsymbol{M}^{-1}\boldsymbol{B}^T\boldsymbol{\lambda} = \boldsymbol{r}_\lambda = \boldsymbol{B}\boldsymbol{M}^{-1}\bar{\boldsymbol{Q}} + \frac{\partial\Phi}{\varphi t} \tag{27}$$

Notice that the homogeneous part of the preceding stabilized equation in terms of the generalized constraint forces $\boldsymbol{\lambda}$ has the following companion eigenvalue problem:

$$(\lambda + \boldsymbol{B}\boldsymbol{M}^{-1}\boldsymbol{B}^T/\varepsilon)y = 0 \tag{28}$$

where $\{\lambda_k,\ k = 1, ..., m\}$ are the eigenvalues of the homogeneous operator for the new stabilized constraint equation [Eq. (27)]. Since γ_k also dictates how the errors in the constraint forces will diminish with time, the errors committed in the constraint conditions will decay with their corresponding different response time constants. This physically oriented stabilization property of the present technique is in contrast to that of Baumgarte's technique, wherein all of the error components diminish according to a single time constant.

Third, this technique enables one to solve for $\boldsymbol{\lambda}$ from the stabilized differential equation [Eq. (27)]. Specifically, one now has two coupled equations, one set for the generalized coordinates $\boldsymbol{u}$ and the other for the generalized constraint forces λ, which are recalled here from Eqs. (5a) and (27) for the case of nonholonomic constraints:

$$\begin{bmatrix} M & 0 \\ 0 & \varepsilon \end{bmatrix}\begin{Bmatrix} \ddot{\boldsymbol{u}} \\ \dot{\boldsymbol{\lambda}} \end{Bmatrix} + \begin{bmatrix} 0 & \boldsymbol{B}^T \\ 0 & \boldsymbol{B}\boldsymbol{M}^{-1}\boldsymbol{B}^T \end{bmatrix}\begin{Bmatrix} \dot{\boldsymbol{u}} \\ \boldsymbol{\lambda} \end{Bmatrix} = \begin{Bmatrix} \bar{\boldsymbol{Q}} \\ \boldsymbol{r}_\lambda \end{Bmatrix} \tag{29}$$

Note that the preceding coupled equations directly provide the desired differential equations for a pair of $\lfloor \boldsymbol{u}\ \ \boldsymbol{\lambda} \rfloor$.

For holonomic constraints, one has several stabilization possibilities. The one we have chosen is to integrate the governing equations of motion once to obtain

$$\dot{\boldsymbol{u}}^n = \delta\boldsymbol{M}^{-1}(\bar{\boldsymbol{Q}}^n - \boldsymbol{B}^T\boldsymbol{\lambda}^n) + \boldsymbol{h}^n_{\dot{u}} \tag{30}$$

which is substituted into

$$\dot{\boldsymbol{\lambda}} = \frac{1}{\varepsilon}\left(\boldsymbol{B}\dot{\boldsymbol{u}} + \frac{\partial\Phi}{\partial t}\right) \tag{31}$$

to yield

$$\varepsilon\dot{\boldsymbol{\lambda}}^n + \delta\boldsymbol{B}\boldsymbol{M}^{-1}\boldsymbol{B}^T\boldsymbol{\lambda}^n = \boldsymbol{B}(\delta\boldsymbol{M}^{-1}\bar{\boldsymbol{Q}}^n + \boldsymbol{h}^n_{\dot{u}}) + \frac{\partial\Phi}{\partial t} \tag{32}$$

It is observed that, even if $\boldsymbol{B}\boldsymbol{M}^{-1}\boldsymbol{B}^T$ is almost singular, this stabilization technique as derived in Eqs. (27) and (32) would not cause numerical difficulty in computing $\boldsymbol{\lambda}$ since the solution iteration matrix becomes $(\varepsilon + \delta \boldsymbol{BM}^{-1}\boldsymbol{B}^T)$ for nonholonomic cases and $(\varepsilon + \delta^2 \boldsymbol{BM}^{-1}\boldsymbol{B}^T)$ for holonomic cases. It is noted that one must choose ε in such a way to maintain a robust solution when $\boldsymbol{BM}^{-1}\boldsymbol{B}^T$ becomes ill-conditioned by choosing $\varepsilon \sim c/|(\boldsymbol{BM}^{-1}\boldsymbol{B}^T)^{-1}| \cdot |\boldsymbol{BM}^{-1}\boldsymbol{B}^T|$ where c is the solution accuracy desired for λ.

Integration of the preceding equation by the midpoint implicit rule yields the following difference equation:

$$\left(\varepsilon \boldsymbol{I} + \frac{h}{4}\boldsymbol{BM}^{-1}\boldsymbol{B}^T\right)\boldsymbol{\lambda}^{n+1/4} = \frac{h}{4}(\boldsymbol{r}_\lambda^{n+1/2} + \boldsymbol{r}_\lambda^{n}) + \varepsilon\boldsymbol{\lambda}^n \tag{33a}$$

$$\lambda^{n+1/2} = 2\boldsymbol{\lambda}^{n+1/4} - \boldsymbol{\lambda}^n \tag{33b}$$

It has been shown that the staggered stabilized procedure for the solution of the constraints offers not only a modular software package to treat the constraints but also has been found to yield more robust solutions compared to the techniques proposed by Baumgarte as reported in Park and Chiou.[35] In particular, even when $\boldsymbol{BM}^{-1}\boldsymbol{B}^T$ becomes nearly singular, the staggered stabilized procedure [Eqs. (33)] gives stable and acceptable solutions, whereas the constraint forces computed by Baumgarte's technique diverge.

Solution Algorithms for Generalized Coordinates

In addition to the choice of implicit and explicit formulas, the recognition that the equations of motion for multibody systems with constraints are not ordinary differential equations (ODEs) (see, e.g., Petzold[27]) has placed a unique requirement in the selection of solution algorithms for multibody dynamics problems. From the user's viewpoint one has the option of either employing one of the available ODE packages (see Enright[32] for existing ODE packages) or building a special solution module. It should be noted that, since the integration of angular velocity vector does not lead to angular orientations, one must solve a set of kinematic equations to obtain the desired angular orientations.

In this section we describe an explicit-implicit transient analysis algorithm that exploits the special kinematic relationships of the generalized rotational coordinates vs the angular velocity, namely, the Euler parameters.[34] The integration of the translational coordinates and the angular velocity is accomplished by the central difference formula. It should be mentioned that the use of the central-difference formula does impose a step-size restriction due to its stability limit ($\omega_{\max} h \leqslant 2$) where $\omega_{\max}$ is the highest angular velocity of the system components for rigid-body systems or the highest frequency of the entire flexible members for flexible-body systems. The simplicity of the formula's programming effort and the robustness of its solution results can often become compelling enough to adopt an explicit formula, which is the view taken here.

In conventional structural dynamics analysis, explicit time integration of the equations of motion by the central-difference formula involves the following two updates per step:

$$\dot{\boldsymbol{u}}^{n+1/2} = \dot{\boldsymbol{u}}^{n-1/2} + h\ddot{\boldsymbol{u}}^{n} \tag{34a}$$

$$\boldsymbol{u}^{n+1} = \boldsymbol{u}^{n} + h\dot{\boldsymbol{u}}^{n+1/2} \tag{34b}$$

Unfortunately, this simplistic procedure is not directly applicable to the rotational part of the equations of motion, since ω is not directly integrable, except for some special kinematic configurations. This motivates us to partition $\dot{\boldsymbol{q}}$ into the translational velocity vector $\dot{\boldsymbol{d}}$, which is directly integrable and the angular velocity vector $\boldsymbol{\omega}$, which is not, and treat them differently, viz.,

$$\ddot{\boldsymbol{u}} = \begin{Bmatrix} \ddot{\boldsymbol{d}} \\ \dot{\boldsymbol{\omega}} \end{Bmatrix}, \qquad \dot{\boldsymbol{u}} = \begin{Bmatrix} \dot{\boldsymbol{d}} \\ \boldsymbol{\omega} \end{Bmatrix} \tag{35}$$

The equation of motion [Eq. (5a)] can be partitioned according to the preceding partitioning:

$$\begin{bmatrix} M_d & 0 \\ 0 & M_\omega \end{bmatrix} \begin{Bmatrix} \ddot{\boldsymbol{d}} \\ \dot{\boldsymbol{\omega}} \end{Bmatrix} = \begin{Bmatrix} Q_d \\ Q_\omega \end{Bmatrix} \tag{36}$$

where

$$\begin{Bmatrix} Q_d \\ Q_\omega \end{Bmatrix} = \begin{Bmatrix} \boldsymbol{f}_d - \boldsymbol{D}_d(\dot{\boldsymbol{d}}) - \boldsymbol{S}_d(\boldsymbol{d}, \boldsymbol{e}) - \boldsymbol{B}_d^T\boldsymbol{\lambda} \\ \boldsymbol{f}_\omega - \boldsymbol{D}_\omega(\boldsymbol{\omega}) - \boldsymbol{S}_\omega(\boldsymbol{d}, \boldsymbol{e}) - \boldsymbol{B}_\omega^T\boldsymbol{\lambda} \end{Bmatrix} \tag{37}$$

in which the subscripts $(\boldsymbol{d}, \boldsymbol{\omega})$ refer to the translational and the rotational motions, respectively, $\boldsymbol{f}$ is the external force vector, $\boldsymbol{D}$ is the generalized damping force including the centrifugal force, $\boldsymbol{S}$ is the internal force vector including member flexibility, $\boldsymbol{q}$ is the angular orientation parameters, $\boldsymbol{B}_d$ and $\boldsymbol{B}_\omega$ are the partition of the combined gradient matrices of the constraint conditions [Eq. (3) or (4)] that are symbolically expressed as

$$\boldsymbol{B} = \boldsymbol{B}_N + \boldsymbol{B}_H, \qquad \boldsymbol{\lambda} = \boldsymbol{\lambda}_N + \boldsymbol{\lambda}_H \tag{38}$$

To effect the body-by-body integration for the rotational degrees of freedom, we partition $\dot{\boldsymbol{\omega}}$ further into

$$\dot{\boldsymbol{\omega}} = \lfloor \dot{\boldsymbol{\omega}}^1, \dot{\boldsymbol{\omega}}^2, \ldots, \dot{\boldsymbol{\omega}}^P \rfloor^T \tag{39}$$

where $\dot{\boldsymbol{\omega}}^{(j)}$ is a (3×1) angular acceleration vector for the jth body,

$$\dot{\boldsymbol{\omega}}^{(j)} = \lfloor \omega_1^{(j)}, \omega_2^{(j)}, \omega_3^{(j)} \rfloor \tag{40}$$

Next we present the update algorithm for both translational and rotational coordinates.

Update of Translational and Angular Velocity

First, assume that $\boldsymbol{d}^{n+1/2}$ and $\boldsymbol{q}^{n+1/2}$ are already computed so that we can compute $\ddot{\boldsymbol{d}}^{n+1/2}$ and $\dot{\boldsymbol{\omega}}^{n+1/2}$ by Eq. (36), namely,

$$\begin{Bmatrix} \ddot{\boldsymbol{d}}^{n+1/2} \\ \dot{\boldsymbol{\omega}}^{n+1/2} \end{Bmatrix} = -M^{-1} \begin{Bmatrix} \boldsymbol{D}_d^{n+1/2} + \boldsymbol{S}_d^{n+1/2} - \boldsymbol{B}_d^T \boldsymbol{\lambda}^{n+1/2} \\ \boldsymbol{D}_\omega^{n+1/2} + \boldsymbol{S}_\omega^{n+1/2} - \boldsymbol{B}_\omega^T \boldsymbol{\lambda}^{n+1/2} \end{Bmatrix} \tag{41}$$

Second, we update the translational velocity and the angular velocity vectors at the step $(n + 1)$ by

$$\begin{cases} \dot{\boldsymbol{d}}^{n+1} = \dot{\boldsymbol{d}}^n + h\ddot{\boldsymbol{d}}^{n+1/2} \\ \boldsymbol{\omega}^{n+1} = \boldsymbol{\omega}^n + h\dot{\boldsymbol{\omega}}^{n+1/2} \end{cases} \tag{42}$$

Third, we update the translational displacement $\boldsymbol{d}$ by

$$\boldsymbol{d}^{n+3/2} = \boldsymbol{d}^{n+1/2} + h\dot{\boldsymbol{d}}^{n+1} \tag{43}$$

However, the updating of the angular orientation requires somewhat involved computations. To this end, we will employ the Euler parameters and update them accordingly.

Update of Euler Parameters and Angular Velocity

As mentioned in conjunction with a direct use of Eqs. (34) for integrating the rotational equations of motion, it is necessary to introduce a set of generalized coordinates whose time rate can be related to the angular velocity. To this end, we employ the four-parameter Euler representation of the angular velocity for each body as (see, e.g., Wittenburg[10])

$$\dot{\boldsymbol{q}} = \frac{1}{2} \begin{bmatrix} 0 & -\omega^T \\ \omega & -\tilde{\omega} \end{bmatrix} \boldsymbol{q} = A(\omega)\boldsymbol{q}, \qquad \boldsymbol{q} = \lfloor q_0 \quad q_1 \quad q_2 \quad q_3 \rfloor^T \tag{44}$$

that is subject to the constraint

$$\boldsymbol{q}^T \boldsymbol{q} = 1 \tag{45}$$

where

$$\tilde{\boldsymbol{\omega}} = \begin{bmatrix} 0 & -\omega_3 & \omega_2 \\ \omega_3 & 0 & -\omega_1 \\ -\omega_2 & \omega_1 & 0 \end{bmatrix}, \qquad \boldsymbol{\omega} = \lfloor \omega_1 \quad \omega_2 \quad \omega_3 \rfloor^T \tag{46}$$

and the nodal-designation superscript is omitted for notational simplicity. We adopt the midpoint implicit procedure to integrate the Euler parameters:

$$\dot{\boldsymbol{q}}^{n+1} = \boldsymbol{A}(\omega^{n+1}) \cdot \boldsymbol{q}^{n+1} \tag{47a}$$

$$\boldsymbol{q}^{n+1} = \boldsymbol{q}^{n+1/2} + (h/2)\dot{\boldsymbol{q}}^{n+1} \tag{47b}$$

$$\boldsymbol{q}^{n+3/2} = 2\boldsymbol{q}^{n+1} - \boldsymbol{q}^{n+1/2} \tag{47c}$$

$$(\boldsymbol{q}^{n+3/2})^T \cdot \boldsymbol{q}^{n+3/2} = 1 \tag{47d}$$

It should be noted that the midpoint implicit update is no more costly than any explicit as the solution matrix inversion can be explicitly obtained.

Finally, once $\boldsymbol{q}^{n+3/2}$ is computed from Eq. (47), it is often required to compute the body-fixed basis vector, $\boldsymbol{b} = \lfloor \boldsymbol{b}_1\ \boldsymbol{b}_2\ \boldsymbol{b}_3 \rfloor^T$ in terms of the inertial basis vectors, $\boldsymbol{e} = \lfloor \boldsymbol{e}_1\ \boldsymbol{e}_2\ \boldsymbol{e}_3 \rfloor^T$. These two vectors are related by

$$\boldsymbol{b} = \boldsymbol{R}\boldsymbol{e} \tag{48}$$

where

$$\boldsymbol{R} = \begin{bmatrix} 2(q_0^2 + q_1^2) - 1 & 2(q_1q_2 + q_0q_3) & 2(q_1q_3 - q_0q_2) \\ 2(q_1q_2 - q_0q_3) & 2(q_0^2 + q_2^2) - 1 & 2(q_2q_3 + q_0q_1) \\ 2(q_1q_3 + q_0q_2) & 2(q_2q_3 - q_0q_1) & 2(q_0^2 + q_3^2) - 1 \end{bmatrix} \tag{49}$$

Update of $\dot{d}$, ω, d, and q at the $(n+2)$ Step

Thus far we have advanced from the step $(n+1)$ to the step $(n+\frac{3}{2})$. In other words, we have advanced only half of the total step. For the next step, viz., the step $(n+2)$ from $(n+3/2)$, we employ the following sequence of computations:

$$\begin{Bmatrix} \ddot{\boldsymbol{d}}^{n+1} \\ \dot{\boldsymbol{\omega}}^{n+1} \end{Bmatrix} = -\boldsymbol{M}^{-1} \begin{Bmatrix} \boldsymbol{D}_d^{n+1} + \boldsymbol{S}_d^{n+1} - \boldsymbol{B}_d^T \boldsymbol{\lambda}^{n+1} \\ \boldsymbol{D}_\omega^{n+1} + \boldsymbol{S}_\omega^{n+1} - \boldsymbol{B}_\omega^T \boldsymbol{\lambda}^{n+1} \end{Bmatrix} \tag{50}$$

$$\dot{\boldsymbol{d}}^{n+3/2} = \dot{\boldsymbol{d}}^{n+1/2} + h\ddot{\boldsymbol{d}}^{n+1} \tag{51a}$$

$$\boldsymbol{\omega}^{n+3/2} = \boldsymbol{\omega}^{n+1/2} + h\dot{\boldsymbol{\omega}}^{n+1} \tag{51b}$$

$$\boldsymbol{d}^{n+2} = \boldsymbol{d}^{n+1} + h\dot{\boldsymbol{d}}^{n+3/2} \tag{52a}$$

$$\dot{\boldsymbol{q}}^{n+3/2} = \boldsymbol{A}(\omega^{n+3/2})\boldsymbol{q}^{n+3/2} \tag{52b}$$

$$\boldsymbol{q}^{n+3/2} = \boldsymbol{q}^{n+1} + \frac{h}{2}\dot{\boldsymbol{q}}^{n+3/2} \tag{52c}$$

$$\boldsymbol{q}^{n+2} = 2\boldsymbol{q}^{n+3/2} - \boldsymbol{q}^{n+1} \tag{52d}$$

$$(\boldsymbol{q}^{n+2})^T\boldsymbol{q}^{n+2} = 1 \tag{52e}$$

Note that we do not use $\boldsymbol{d}^{n+3/2}$ and $\boldsymbol{q}^{n+3/2}$ in advancing from the step $(n+3/2)$ to the present step $(n+2)$ in computing $\boldsymbol{d}^{n+2}$ and $\boldsymbol{q}^{n+2}$. Instead, we employ $\boldsymbol{d}^{n+1}$ and $\boldsymbol{q}^{n+1}$, hence the name "two-stage staggered explicit procedure."[34] The net result is that, even though we take a full step (h instead of $h/2$), we only advance half the step at a time. In other words, we evaluate the acceleration and the angular acceleration vectors twice for each full step.

Implementation

We will now outline the implementation aspects of the partitioned MBD solution procedure. The procedure is implemented into two separate integration modules: generalized-coordinate integrator (CINT) and Lagrange multiplier solver (LINT). The generalized-coordinate integrator employs a two-stage modified form of the central-difference method for updating the angular velocity vector and the midpoint implicit rule for updating the angular orientations via the Euler parameters. The Lagrange multipliers solver adopts a staggered form of the midpoint implicit method.

Generalized-Coordinate Integrator (CINT)

The modular receives $f_\lambda^n = \boldsymbol{B}^T\boldsymbol{\lambda}^n$ from LINT and advances the solution of the MBD equation [Eq. (1)] from time t^n to t^{n+1}. At each integration step, CINT performs the following computations.

Given

$$\boldsymbol{p}^n = (\dot{\boldsymbol{d}}^{n-1/2}, \boldsymbol{d}^n, \boldsymbol{\omega}^{n-1/2}, \boldsymbol{q}^n) \qquad \text{and} \qquad \boldsymbol{g}^n = (\boldsymbol{\omega}^n, f_\lambda^n = \boldsymbol{B}^T\boldsymbol{\lambda}^n)$$

Compute $\ddot{\boldsymbol{d}}^n$ and $\dot{\boldsymbol{\omega}}^n$ by Eq. (41).

Advance

$$\dot{\boldsymbol{d}}^{n+1/2} = \dot{\boldsymbol{d}}^{n-1/2} + h\ddot{\boldsymbol{d}}^n \tag{53a}$$

$$\boldsymbol{d}^{n+1} = \boldsymbol{d}^n + h\dot{\boldsymbol{d}}^{n+1/2} \tag{53b}$$

$$\boldsymbol{\omega}^{n+1/2} = \boldsymbol{\omega}^{n-1/2} + h\dot{\boldsymbol{\omega}}^n \tag{54a}$$

$$\tilde{\boldsymbol{q}}^{n+1/2} = \frac{1}{\Delta}[I + (h/2)A(\boldsymbol{\omega}^{n+1/2})] \cdot \boldsymbol{q}^n,$$

$$\Delta = 1 + (h^2/4)(\omega_1^2 + \omega_2^2 + \omega_3^2) \tag{54b}$$

$$\boldsymbol{q}^{n+1} = 2\tilde{\boldsymbol{q}}^{n+1/2} - \boldsymbol{q}^n, \qquad (\boldsymbol{q}^{n+1})^T \cdot \boldsymbol{q}^{n+1} = 1 \tag{54c}$$

Output:

$$\boldsymbol{p}^{n+1} = (\dot{\boldsymbol{d}}^{n+1/2}, \boldsymbol{d}^{n+1}, \boldsymbol{\omega}^{n+1/2}, \boldsymbol{q}^{n+1})$$

Module invoke:

$$\text{Call CINT}(\boldsymbol{p}^n, \boldsymbol{g}^n, h, \boldsymbol{p}^{n+1})$$

where h is the step size, and $\boldsymbol{A}(\omega)$ is given by

$$\boldsymbol{A}(\omega) = \frac{1}{2}\begin{bmatrix} 0 & -\omega_1 & -\omega_2 & -\omega_3 \\ \omega_1 & 0 & \omega_3 & -\omega_2 \\ \omega_2 & -\omega_3 & 0 & \omega_1 \\ \omega_3 & \omega_2 & -\omega_1 & 0 \end{bmatrix} \tag{55}$$

and $\tilde{\boldsymbol{q}}^{n+1/2}$ is an intermediate vector, and Eq. (54c) must be solved to obtain $\boldsymbol{q}^{n+1}$ in order to satisfy the linear dependency constraint, $\boldsymbol{q}^T\boldsymbol{q} = 1$.

Lagrange Multiplier Solver (LINT)

This module receives $(\dot{\boldsymbol{d}}, \boldsymbol{d}, \boldsymbol{\omega}, \boldsymbol{q})$ from CINT and performs the following computations:

Given

$$\ell^{n+1/2} = (\dot{\boldsymbol{d}}^{n+1/2}, \boldsymbol{d}^{n+1/2}, \boldsymbol{\omega}^{n+1/2}, \boldsymbol{q}^{n+1/2}, \boldsymbol{\lambda}^n)$$

Compute

$$\boldsymbol{B}^{n+1/2}, \qquad \boldsymbol{B}\boldsymbol{M}^{-1}\boldsymbol{B}^T \qquad \text{and} \qquad \boldsymbol{r}_\lambda^{n+1/2}$$

by Eqs. (3) and (4).

Advance

$$\bar{\boldsymbol{\lambda}}^{n+1/4} = [\varepsilon\boldsymbol{I} + (h/4)\boldsymbol{B}\boldsymbol{M}^{-1}\boldsymbol{B}^T]^{-1}[\varepsilon\boldsymbol{\lambda}^n + (\boldsymbol{h}/4)(\boldsymbol{r}_\lambda^n + \boldsymbol{r}_\lambda^{n+1/2})] \tag{56a}$$

$$\boldsymbol{\lambda}^{n+1/2} = 2\bar{\boldsymbol{\lambda}}^{n+1/4} - \boldsymbol{\lambda}^n \tag{56b}$$

$$\boldsymbol{f}_\lambda^{n+1/2} = (\boldsymbol{B}^{n+1/2})^T \cdot \boldsymbol{\lambda}^{n+1/2} \tag{56c}$$

Output:

$$\boldsymbol{\lambda}^{n+1/2}, \boldsymbol{f}_\lambda^{n+1/2}$$

Module invoke:

$$\text{Call LINT}(\ell^{n+1/2}, h, \boldsymbol{\lambda}^{n+1/2}, \boldsymbol{f}_\lambda^{n+1/2})$$

Two-Stage Explicit-Implicit Staggered Procedure

In order to evaluate $\dot{\omega}^{n+1}$, ω^{n+1} must be known. Notice from the preceding section that only $\dot{\omega}^{n+1/2}$ is available. Because inaccurate treatments of the gyroscopic damping and the centrifugal force terms can quickly lead to computational instability in computing $\dot{\omega}^{n+1}$, it is not advisable to obtain ω^{n+1} by extrapolating with $\omega^{n+1/2}$ and $\omega^{n-1/2}$. To mitigate this difficulty, we advance only to the next half step, at each CINT and LINT call. This is illustrated as follows:

$$t = t^n$$

$$\text{Call CINT}(\boldsymbol{p}^n, \boldsymbol{g}^n, h, \boldsymbol{p}^{n+1})$$

$$\text{Call LINT}(\ell^{n+1/2}, h, \boldsymbol{\lambda}^{n+1/2}, \boldsymbol{f}_\lambda^{n+1/2})$$

$$t = t^n + h/2 (n \leftarrow n + 1/2)$$

$$\text{Call CINT}(\boldsymbol{p}^{n+1/2}, \boldsymbol{g}^{n+1/2}, h, \boldsymbol{p}^{n+3/2})$$

$$\text{Call LINT}(\ell^{n+1}, h, \boldsymbol{\lambda}^{n+1}, \boldsymbol{f}_\lambda^{n+1})$$

$$t = t^n + h$$

Note that

$$\boldsymbol{g}^{n+1/2} = (\boldsymbol{\omega}^{n+1/2}, \boldsymbol{f}_\lambda^{n+1/2})$$

together with

$$\boldsymbol{p}^{n+1/2} = (\dot{\boldsymbol{d}}^n, \boldsymbol{d}^{n+1/2}, \boldsymbol{\omega}^n, \boldsymbol{q}^{n+1/2})$$

provides the necessary input data to compute $\ddot{\boldsymbol{d}}^{n+1/2}$ and $\dot{\boldsymbol{\omega}}^{n+1/2}$ in the second call of CINT in the preceding calling sequence. In summary, the present procedure requires two function evaluations and two λ solutions per each full step, hence the name "two-stage explicit-implicit staggered procedure."

Numerical Examples

The two modules, the generalized coordinate integrator (CINT) and the Lagrange multipliers solver (LINT), have been implemented in FORTRAN 77. In solving the following three example problems, we have incorporated the constraint conditions through the use of Lagrange multipliers instead of eliminating the constraints. It is therefore necessary to solve the governing equations of motion in a way that satisfies the constraint equations. Hence, efficient and accurate solutions of these problems will confirm not only the viability of the present integration procedure for the solution of the multibody equations of motion with or without constraints but also the constraint stabilization procedure in their combined totality.

Plane Three-Link Manipulator

The first problem tested is a simplified version of the seven-link manipulator deployment problem.[52] The three links are initially folded and, for modeling simplicity, between the two joints is a coil spring that resists a constant deploying force at the tip of the third link. Also, the left-hand end of the first link is fixed through the same coil spring to the wall. These three coil springs are to be "locked up" once the links are deployed straight. The deployment sequence of the manipulator is illustrated in Fig. 1. The time-discretized difference equations both for Baumgarte's technique and the staggered stabilization technique have been solved at each time increment by a Newton-type iterative procedure to meet a specified accuracy level. Hence, the performance of the two techniques can be assessed by the average number of iterations taken per time increment. This is presented in Fig. 2 for the accuracy of 10^{-4}. Notice that the staggered stabilization technique requires on the average about 4.5 iterations per step, whereas Baumgarte's technique requires about 22 iterations per step.

Note that Baumgarte's technique fails to converge for time, $t \approx 1.1$ as manifested in Fig. 2, because the rows in $\boldsymbol{B}$ become numerically dependent on one another when the links are in a straight configuration. This corroborates the theoretical prediction of nonconvergence whenever the

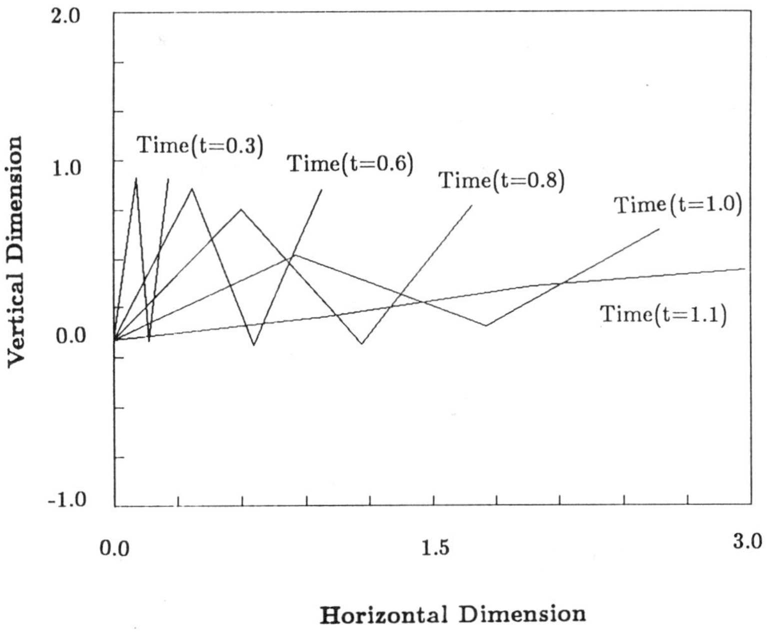

Fig. 1 Deployment of three-link remote manipulator.

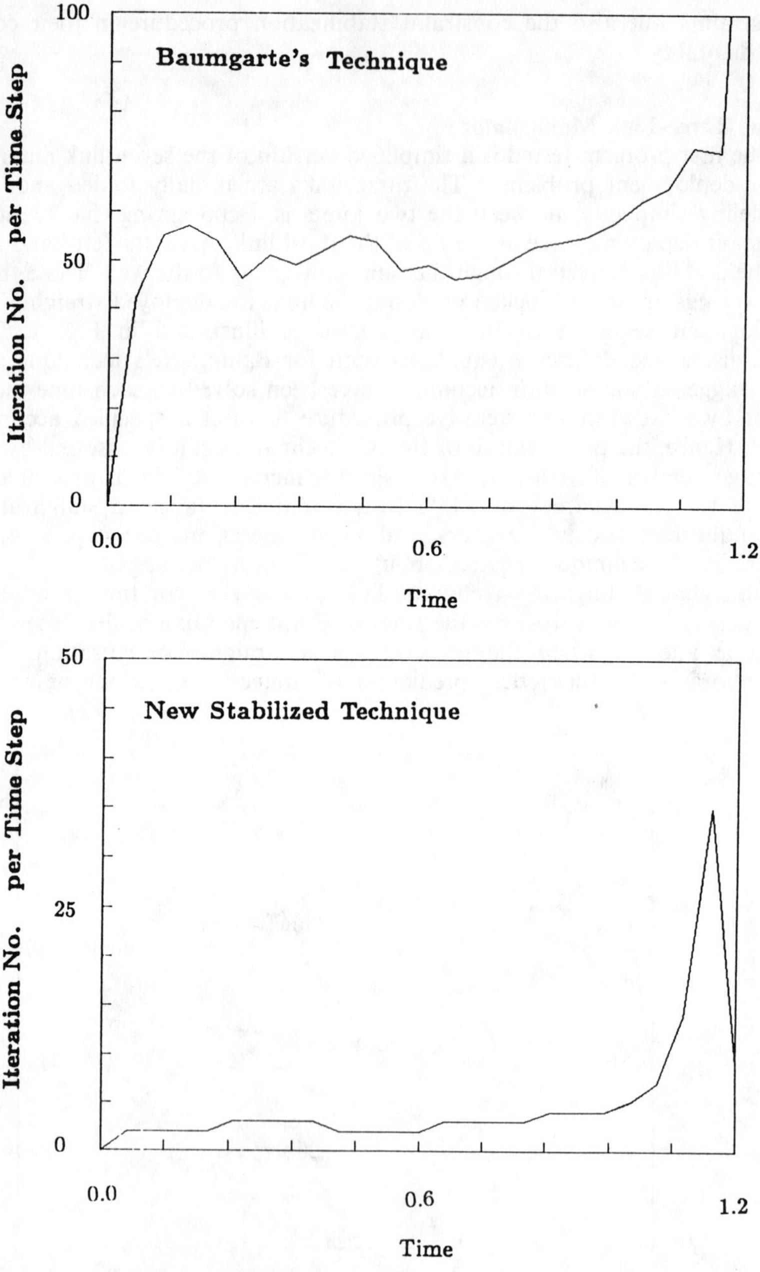

Fig. 2 Performance of two stabilization techniques for three-link remote manipulator (solution accuracy = 10^{-6}).

solution matrix, $BM^{-1}B^T$, for Baumgarte's technique [see Eqs. (5b), (20), and (21)] becomes singular. On the other hand, the staggered stabilization technique still converges within 30 iterations, because it overcomes this singularity difficulty, since $\dot{\lambda}$ still exists, as can be seen from Eqs. (27) and (32).

It should be noted that, in order to avoid such ill-conditioning, one must differentiate the constraint equations once or twice more and recast the resulting higher-order constraint equations in terms of first-order equations with an increased number of equations. This process is known as an index reduction strategy.[46] Thus, one must restructure the augmented equations of motion [Eqs. (5)] with the net result of increased solution variables. Other techniques involve singular value decompositions, e.g., as advocated by Führer and Leimkuhler.[53] On the other hand, the present staggered stabilization technique overcomes the ill-conditioning difficulty without restructuring the governing equations of motion. Instead, the constraint equations are enforced in a separate module by the parabolically regularized equations for the Lagrange multipliers as derived in Eqs. (27) and (32).

Although not reported here, the same relative performance has been observed for different accuracy levels, i.e., for the accuracy of 10^{-5} and 10^{-6}.

From this test problem we conclude that the staggered stabilization technique yields both improved accuracy over and greater computational robustness than the Baumgarte technique. In addition, the staggered stabilization technique offers software modularity in that the solution of the constraint force λ can be carried out separately from that of the generalized displacement q. The only data each solution module needs to exchange with the other is a set of vectors, plus a common module to generate the gradient matrix of the constraints, B. However, one should be cautioned not to extrapolate blindly to complex problems the results of the present simple examples. Further judicious experiments are needed in applying the present staggered stabilization technique to complex production-level problems before it can be adopted for general applications in multibody dynamic simulations.

Three-Dimensional Double Pendulum

The second problem with which we have tested the present procedure is a spatially moving double pendulum as shown in Fig. 3. The governing equations of motion become those of two separate rigid bars, except they are connected by two spherical joints. From Fig. 3 we have the following quantities:

$$\Phi^i = \dot{d}^i - \tfrac{1}{2}\omega^i \times z^i = 0, \qquad i = 1, 2 \tag{57}$$

$$M = \text{diag}\{m^1, J^1, m^2, J^2\} \tag{58}$$

$$B = \begin{bmatrix} I & \tfrac{1}{2}z^1\times & 0 & 0 \\ I & -\tfrac{1}{2}\tilde{z}^1\times & -\mathrm{I} & -\tfrac{1}{2}\tilde{z}^2\times \end{bmatrix} \tag{59}$$

$$F_\omega = \begin{Bmatrix} f^1 \\ f^2 \end{Bmatrix}, \quad f^i = -\begin{bmatrix} 0 \\ 0 \\ 0 \\ \omega_2\omega_3(J_2 - J_3) \\ \omega_3\omega_1(J_3 - J_1) \\ \omega_1\omega_2(J_1 - J_2) \end{bmatrix}^i, \quad i = 1, 2 \tag{60}$$

$$\ddot{u}^i = \{\ddot{d}, \dot{\omega}\}^i, \quad \ddot{d}^i = [\ddot{x}, \ddot{y}, \ddot{z}]^T, \quad \dot{\omega}^i = [\dot{\omega}_1, \dot{\omega}_2, \dot{\omega}_3]^T \tag{61}$$

$$\lambda = [\lambda_1, \lambda_2, \lambda_3, \lambda_4, \lambda_5, \lambda_6]^T \tag{62}$$

where $\frac{1}{2}z$ is the vectorial distance from the center of the bar to the spherical joint constraints, m and J the three translational and rotatory inertia matrics, $\tilde{z}$ the skew symmetric matrix formed by the three components of z, $\times$ implies a vector cross multiplication, and the superscript designates the ith bar.

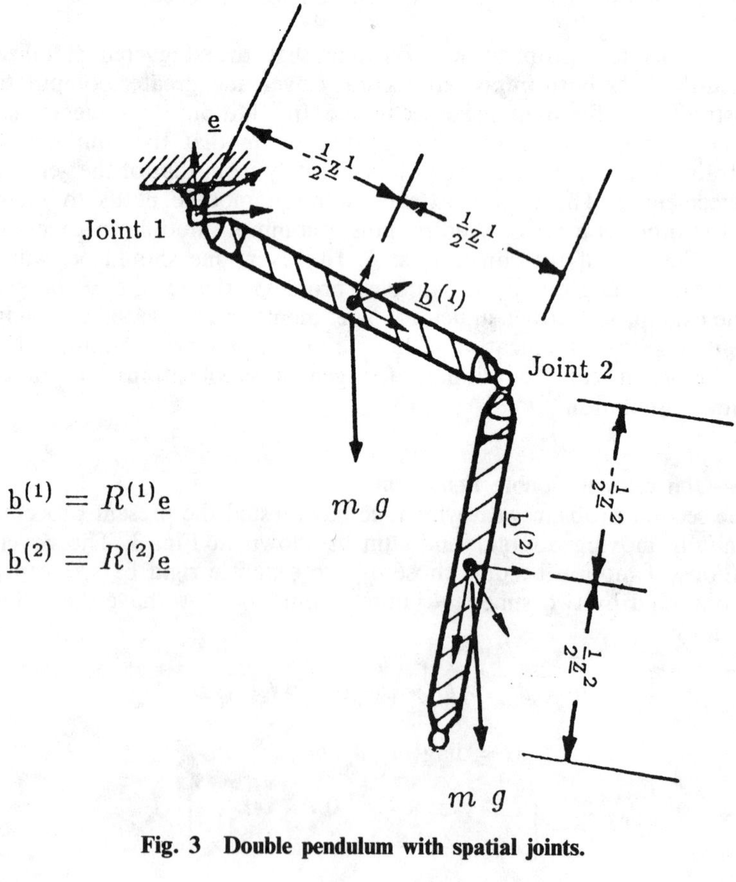

Fig. 3 Double pendulum with spatial joints.

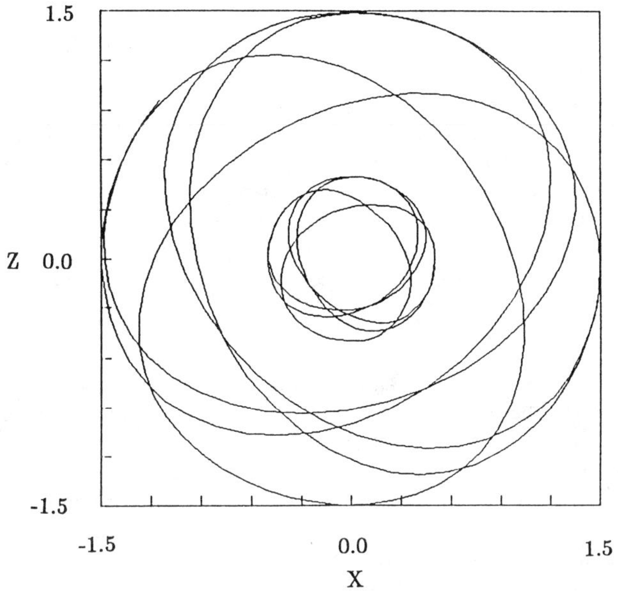

Fig. 4a Trajectories of double pendulum on X-Z plane.

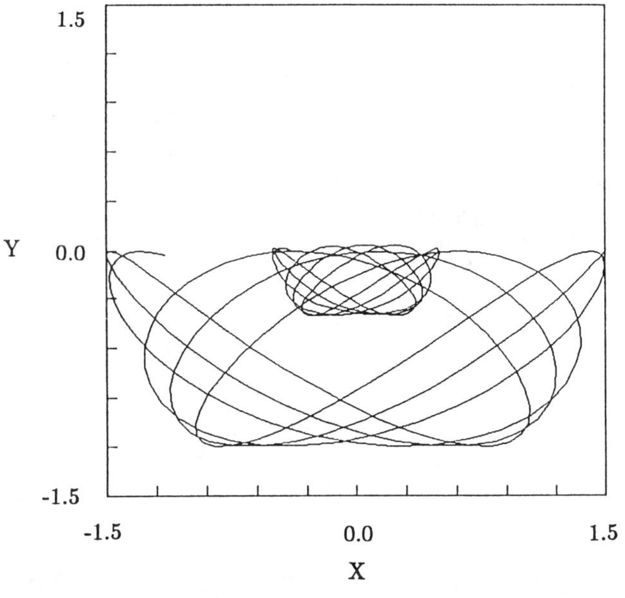

Fig. 4b Trajectories of double pendulum on X-Y plane.

The pendulum is originally positioned in a gravity field with initial horizontal angular velocities ($\omega_z^{(1)} = \omega_z^{(2)} = 1$). Figure 4 shows the spatial trajectories of the two mass centers as projected on the horizontal surface and on the vertical plane. It is noted that the two trajectories form a similar pattern. The constraint forces and angular velocities, although not reported herein, exhibit patterns that are analogous in their characteristics for the two joints and two mass centers, respectively.

We have performed convergence studies by using different step sizes h. Numerical evaluations indicate, as with the rigid-link problem, that when the step-size samples more than 20 per period, the present procedure yields both good accuracy and stability.

Open-Loop Torque for Three-Link Manipulator

The third problem is a three-link manipulator maneuvering under a specified nonholonomic tip velocity constraint. For this problem, both rigid links and flexible links with four beam elements per link have been investigated. The flexible beam was modeled with a constant-strain Timoshenko beam element that allows large rotations. The three joints are modeled as spherical ones, and the Lagrange multipliers have been introduced to enforce the joint constraints as well as the nonholonomic constraint at the manipulator tip. The trajectories of the manipulator and the tip velocity specification are shown in Figs. 5 and 6. The corresponding

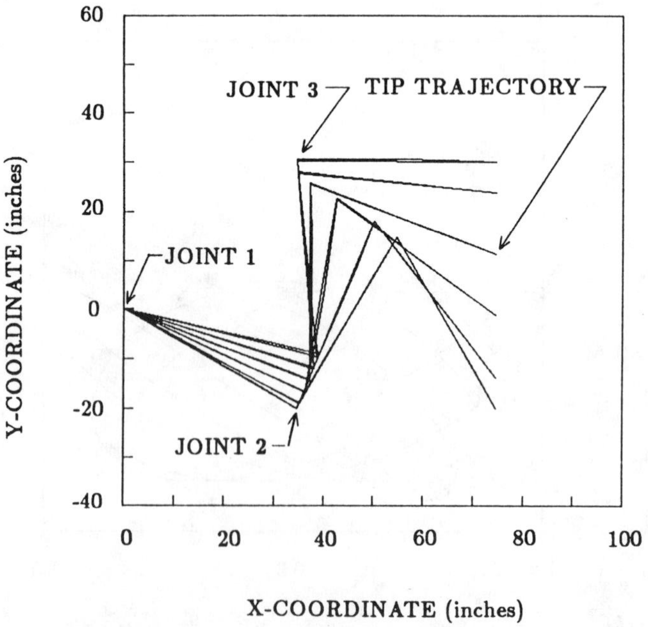

Fig. 5 Crane tip trajectory of rigid and flexible arms.

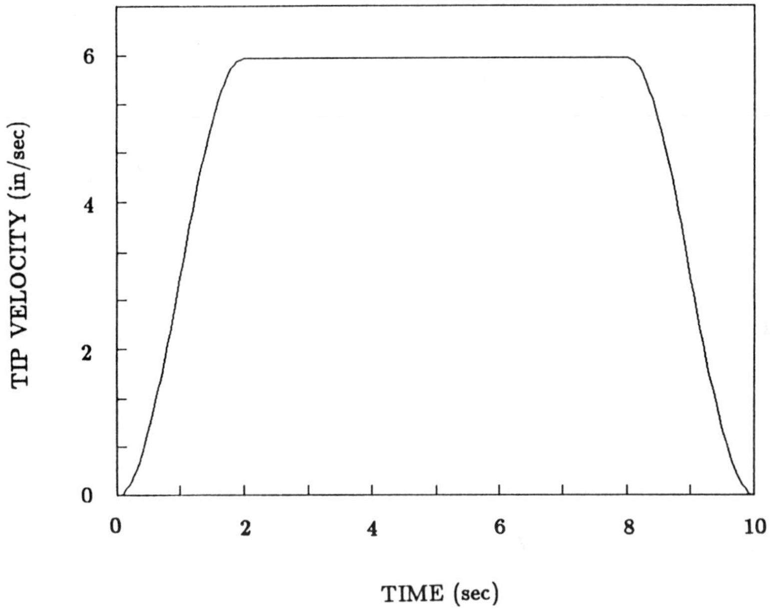

Fig. 6 Crane tip velocity of rigid and flexible arms.

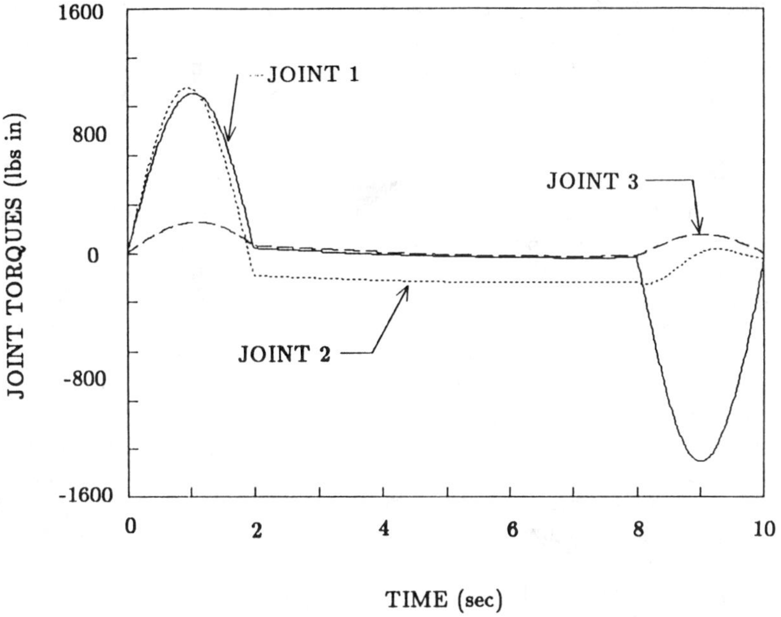

Fig. 7. Crane joint torques vs time for rigid arms.

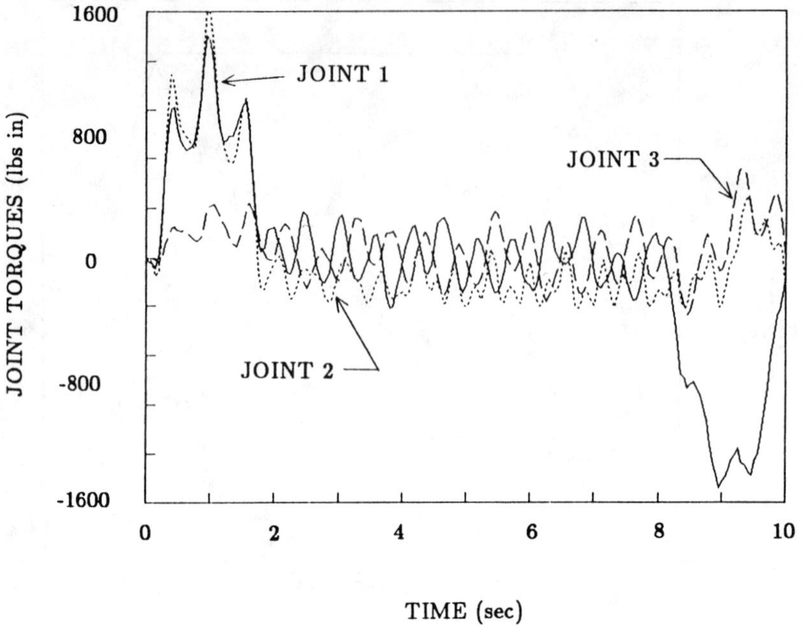

Fig. 8. Crane joint torques vs time for flexible arms.

joint torques for the rigid and flexible links are also shown in Figs. 7 and 8, respectively. Note that, even though there exists little difference in the two trajectories of the rigid and flexible cases, there are significant differences in the open-loop joint torques, these will play an important role in the design of controller for vibration suppression in the manipulator arms.

Discussion

In this chapter we have presented a computational procedure for direct integration of the MBD equations with constraints.

Because of its step-advancing nature, the procedure is labeled as a two-stage staggered explicit-implicit algorithm: explicit for solving the generalized coordinates (CINT) and implicit for Lagrange multipliers to incorporate constraints (LINT). Our numerical experiments indicate that it is essential to enforce the linear dependency constraint condition on the Euler parameters at each integration step.

Numerical experiments reported herein and additional applications conducted so far indicate that the present procedure yields robust solutions if the step size gives more than 20 samples for the period of the apparent highest response frequency of a given multibody system. Hence, the present procedure appears to have accomplished the following:

1) For closed-loop multibody systems and/or problems with complex topology wherein it is practically inadvisable to eliminate the constraints, the present procedure facilitates a straightforward construction of the

governing equations of motion with appropriate constraints. The generalized coordinates and the system open- and closed-loop Lagrange multipliers can then be solved by the present procedure in a partitioned manner.

2) For problems that involve lock-up mechanisms or similar discontinuities, the present procedure appears to overcome numerical difficulties encountered in using the Baumgarte stabilization. This may be an important impetus for applying the present procedure for the simulation of deployment dynamics of space structures.

3) The angular velocity is obtained by an adaptation of the central-difference algorithm in a two-stage form, and the update of angular orientations is based on the Euler parameters by adopting the midpoint implicit formula. Both of the integrators conserve the system energy, which is important when the multibody simulation package is to be interfaced with an active control synthesis module. This is because stability margins of active control systems are sensitive to the system damping characteristics, either physical or numerical.

4) The present MBD solution procedure is implemented into two separate modules: the generalized coordinates solver (CINT) and the constraint Lagrange multiplier solver (LINT). Hence, the task for interfacing of the present MBD solution modules with additional capabilities such as active controller, observer, and other analysis and design software modules becomes relatively straightforward. Such software architecture is in contrast to most of the existing programming practice wherein several analysis capabilities are embedded into a single monolithic program.

Applications of the present procedure to flexible multibody systems are currently being carried out, and preliminary results are quite encouraging. We hope to report on the results of flexible-body dynamics as well as on large-scale multibody problems in the near future.

Acknowledgments

The authors wish to thank Dr. Jerry Housner for his keen interest and encouragement during the couse of this work. This work was supported by the NASA/Langley Research Center under Grant NAG-1-756.

References

[1]Hooker, W., and Margulies, G., "The Dynamical Attitude Equations for an N-body Satellite," *Journal of the Astronautical Sciences*, Vol. 12, 1965, pp. 123–128.

[2]Roberson, R., and Wittenburg, J., "A Dynamical Formalism for an Arbitrary Number of Interconnected Rigid Bodies with Reference to the Problem of Satellite Attitude Control," *Proceedings of the Third International Congress of Automatic Control*, Butterworths, London, 1965.

[3]Roberson, R., "A Form of the Translational Dynamical Equations for Relative Motion in Systems of Many Non-Rigid Bodies," *Acta Mechanica*, Vol. 14, 1972, pp. 297–308.

[4]Huston, R. L., and Passerello, C. E., "On Lagrange's Form of d'Alembert's Principle," *Matrix and Tensor Quarterly*, Vol. 23, 1973, pp. 109–112.

[5]Boland, P., Samin, J., and Willems, P., "Stability Analysis of Interconnected Deformable Bodies in a Topological Tree," *AIAA Journal*, Vol. 12, 1974, pp. 1025–1030.

[6]Likins, P., "Analytical Dynamics and Nonrigid Spacecraft Simulation," Jet Propulsion Lab., Pasadena, CA, TR-32-1593, 1974.

[7]De Veubeke, B. F., "The Dynamics of Flexible Bodies," *International Journal of Engineering Sciences*, Vol. 14, 1976, pp. 895–913.

[8]Jerkovsky, W., "The Transformation Operator Approach to Multisystem Dynamics, Part I: The General Approach," *Matrix and Tensor Quarterly*, Vol. 27, Dec. 1976, pp. 48–59.

[9]Ho, J. Y. L., "Direct Path Method for Flexible Multibody Spacecraft Dynamics," *Journal of Spacecraft and Rockets*, Vol. 14, No. 2, 1977, pp. 102–110.

[10]Wittenburg, J., *Dynamics of Systems of Rigid Bodies*, Teubner, Stuttgart, FRG, 1977.

[11]Armstrong, W. W., "Recursive Solution to the Equations of Motion of an N-Link Manipulator," *Proceedings of the 5th World Congress, Theory of Machines, Mechanisms*, Vol. 2, 1979, pp. 1343–1346.

[12]Hollerbach, J. M., "A Recursive Lagrangian Formulation of Manipulator Dynamics and a Comparative Study of Dynamics Formulation Complexity," *IEEE Transactions on Systems, Man, and Cybernetics*, SMC-10, IEEE, New York, 1980, pp. 730–736.

[13]Kane, T., and Levinson, D., "Formulation of Equations of Motion for Complex Spacecraft," *Journal of Guidance and Control*, Vol. 3, March–April 1980, pp. 99–112.

[14]Keat, J. E., "Dynamical Equations of Body Systems with Applications Space Structure Deployment," Ph.D. Dissertation, Massachusetts Institute of Technology, Cambridge, MA, 1983.

[15]Haug, E. J. (ed.), *Computer Aided Analysis and Optimization of Mechanical System Dynamics*, Springer-Verlag, Berlin, 1984.

[16]Roberson, R. E., and Schwertassek, R., *Dynamics of Multibody Systems*, Springer-Verlag, New York, 1984.

[17]Schwertassek, R., and Roberson, R. E., "A State-Space Dynamical Representation for Multibody Mechanical Systems, Part II," *Acta Mechanica*, Vol. 51, 1984, pp. 15–29.

[18]Bianchi, G., and Schielen, W. (eds.), *Dynamics and Multibody Systems*, Springer-Verlag, Berlin, 1986.

[19]Huston, R. L., *Lecture Notes on Dynamics*, Preprint, Univ. of Cincinnati, Cincinnati, OH, 1988.

[20]Walton, W. C., and Steeves, E. C., "A New Matrix Theorem and Its Application for Establishing Independent Coordinates for Complex Dynamical Systems with Constraints," NASA TR-R326, 1969.

[21]Gear, C. W., "Simultaneous Numerical Solution of Differential/Algebraic Equations," *IEEE Transactions on Circuit Theory*, CT-18, IEEE, New York, 1971, pp. 89–95.

[22]Baumgarte, J. W., "Stabilization of Constraints and Integrals of Motion in Dynamical Systems," *Computer Methods in Applied Mechanics and Engineering*, Vol. 1, 1972, pp. 1–16.

[23]Orlandea, N., Chase, M. A., and Calahan, D. A., "A Sparsity-Oriented Approach to the Dynamic Analysis and Design of Mechanical Systems—Parts I

and II," *Transactions of ASME, Journal of Engineering for Industry, Series B*, Vol. 99, Aug. 1977, pp. 773–784.

[24]Lötstedt, P., "On a Penalty Function Method for the Simulation of Mechanical Systems Subject to Constraints," Royal Inst. of Technology, TRITA-NA-7919, Stockholm, 1979.

[25]Orin, D. E., McGhee, R. B., Vukobratovic, M., and Hartoch, G., "Kinematic and Kinetic Analysis of Open-chain Linkages Utilizing Newton-Euler Methods," *Mathematical Biosciences*, Vol. 43, 1979, pp. 106–130.

[26]Huston, R. L., and Kamman, J. W., "A Discussion on Constraint Equations in Multibody Dynamics," *Mechanics Research Communications*, Vol. 9, 1982, pp. 251–256.

[27]Petzold, L., "Differential/Algebraic Equations are not ODEs," *SIAM Journal on Scientific and Statistical Computing*, Vol. 3, 1982, pp. 367–384.

[28]Wehage, R. A., and Haug, E. J., "Generalized Coordinate Partitioning for Dimension Reduction in Analysis of Constrained Dynamic Systems," *ASME Journal of Mechanical Design*, Vol. 104, 1982, pp. 247–255.

[29]Baumgarte, J. W., "A New Method of Stabilization for Holonomic Constraints," *Journal of Applied Mechanics*, Vol. 50, Dec. 1983, pp. 869–870.

[30]Führer, C., and Wallrapp, O., "A Computer-Oriented Method for Reducing Linearized Multibody System Equations by Incorporating Constraints," *Computer Methods in Applied Mechanics and Engineering*, Vol. 46, Oct. 1984, pp. 169–175.

[31]Nikravesh, P. E., "Some Methods for Dynamic Analysis of Constrained Mechanical Systems: A Survey," *Computer Aided Analysis and Optimization of Mechanical System Dynamics*, edited by E. J. Haug, NATO ASI series, F9, Springer-Verlag, Berlin, 1984, pp. 351–367.

[32]Enright, W. H., "Numerical Methods for Systems of Initial Value Problems—The State of the Art," *Computer Aided Analysis and Optimization of Mechanical System Dynamics*, edited by E. J. Haug, NATO ASI series, F9, Springer-Verlag, Berlin, 1984, pp. 309–322.

[33]Garcia de Jalon, J., Unda, J., Avello, A., and Jimenez, J. M., "Dynamic Analysis of Three-Dimensional Mechanisms in Natural Coordinates," *Journal of Mechanisms, Transmissions and Automation in Design*, Vol. 109, Dec. 1987, pp. 460–465.

[34]Park, K. C., Chiou, J. C., and J. D. Downer, "A Computational Procedure for Large Rotational Motions in Multibody Dynamics," AIAA Paper 88-2416, 1988, pp. 1593–1601; also *Journal of Guidance, Control and Dynamics* (to be published).

[35]Park, K. C., and Chiou, J. C., "Stabilization of Computational Procedures for Constrained Dynamical Systems," *Journal of Guidance, Control, and Dynamics*, Vol. 11, July–Aug. 1988, pp. 365–370.

[36]Geradin, M., and Cardona, A., "Kinematic and Dynamics of Rigid and Flexible Mechanisms Using Finite Elements and Quaternion Algebra," *Computational Mechanics*, Vol. 4, 1989, pp. 115–135.

[37]Bodley, C. S., Devers, A. D., Park, A. C., and Frish, H. P., "A Digital Computer Program for the Dynamic Interaction Simulation of Controls and Structures (DISCOS)," NASA TP-1219, 1978.

[38]*The ADAMS User's Guide*, Mechanical Dynamics, Inc., Ann Arbor, MI, 1979.

[39]Schwertassek, R., "Der Roberson/Wittenburg Formalismus and das Programmsystem MULTIBODY zur Rechnersimulation von Mehrköpersystemen," German Aerospace Research Establishment, Cologne, FRG, Rept. DFVLR-FB-78/08, 1978.

[40]Huston, R. L., Harlow, M. W., and Gausewitz, N. L., "User's Mannual for UCIN-EULER—A Multipurpose, Multibody Systems Dynamics Computer Program," National Technical Information Service Rept. AD-A120403, 1982.

[41]Haug, E. J., Lance, G. M., Nikravesh, P. E., Vanderploeg, M. J., and Wehage, R. A., *DADS* (*Dynamic Analysis and Design Systems*), Computer Aided Design Software, Inc., Oakdale, IA, 1985.

[42]Housner, J. M., McGowan, P. E., Abrahamson, A. L., and Powel, M. G., "The LATDYN User's Manual," NASA TM-87635, NASA/Langley Research Center, Jan. 1986.

[43]Hughes, T. J. R., and Belytschko, T., "A Prècis of Developments in Computational Methods for Transient Analysis," *Journal of Applied Mechanics*, Vol. 50, 1983, pp. 1033–1041.

[44]Park, K. C., "Transient Analysis Methods in Computational Methods," *Finite Elements: Theory and Applications*, edited by D. L. Dwoyer, M. Y. Hussaini, and R. G. Voigt, Springer-Verlag, Berlin, 1988, pp. 240–267.

[45]Belytschko, T., Englemann, B. E., and Liu, W. K., "A Review of Recent Developments in Time Integration," *State-of-the-Art Surveys on Computational Mechanics*, edited by A. K. Noor and J. T. Oden, American Society of Mechanical Engineers, New York, 1989, pp. 185–200.

[46]Brenan, K. E., Campbell, S. L., and Petzold, L. R., *The Numerical Solution of Initial Value Problems in Ordinary Differential-Algebraic Equations*, Elsevier, 1989.

[47]Park, K. C., "Partitioned Analysis Procedures for Coupled-Field Problems: Stability Analysis," *Journal of Applied Mechanics*, Vol. 47, June 1980, pp. 370–378.

[48]Felippa, C. A., and Park, K. C., "Staggered Transient Analysis Procedures for Coupled Mechanical Systems," *Computer Methods in Applied Mechanics and Engineering*, Vol. 24, Oct. 1980, pp. 61–111.

[49]Park, K. C., and Felippa, C. A., "Partitioned Analysis of Coupled Systems," *Computational Methods for Transient Analysis*, edited by T. Belytschko and T. J. R. Hughes, Elsevier, 1983, pp. 157–219.

[50]Felippa, C. A., and Park, K. C., "Computational Aspects of Time Integration Procedures in Structural Dynamics, Part 1: Implementation," *Journal of Applied Mechanics*, Vol. 45, Sept. 1978, pp. 595–602.

[51]Lanczos, L., *The Variational Principles of Mechanics*, 4th ed., Univ. of Toronto Press, Toronto, Canada, 1970, pp. 141–147.

[52]Housner, J. M., "Convected Transient Analysis for Large Space Structure Maneuver and Deployment," AIAA Paper 84-1023-CP, May 1984, Pt. 2, pp. 616–619.

[53]Führer, C., and Leimkuhler, B., "Formulation and Numerical Solution of the Equations of Constrained Mechanical Motion," German Aerospace Research Establishment, Cologne, FRG, Rept. DFVLR-FB 89-08, D-5000 90, 1989.

Chapter 9

Precision Pointing of Large Antennas by Static Shape Estimation

R. E. Scheid* and G. Rodriguez†
California Institute of Technology, Pasadena, California

Introduction

THIS chapter reviews the status of a methodology that the authors have been developing for some time[1–5] to solve the problem of estimating the static shape of the surface of a space structure. We combine an a priori spatially random model for the structure with information from a set of measurements of the surface deflections. In particular, this chapter focuses on a concrete application of this methodology to solve the problem of compensating for static deflections in achieving precise pointing of the large ground antennas that are part of the NASA Deep Space Network (DSN)[6] used to track and communicate with planetary spacecraft.

A number of important emerging applications of the technology developed for large space structures can be appropriately analyzed as statically loaded structural systems. This approach is applicable to systems where the time-dependent effects are negligible and where the relevant forces do not significantly excite dynamic behavior in the system. More generally, one can consider problem settings where the stiffness properties dominate the mass properties in determining the energy balances. Then a static deformation of the structure defines a reference configuration about which dynamic motions can be resolved with greater accuracy. In this context one can include applications where the time-varying effects of the model are changing slowly with respect to the scale on which operations must be carried

*Member Technical Staff, Jet Propulsion Laboratory.
†Senior Member Technical Staff, Jet Propulsion Laboratory.

out. These quasistatic disturbances may, for example, include gravitational and thermal effects.

The introduction of statistical model errors allows the smearing of effects that have been ignored in the modeling or that occur on too fine a scale to be modeled as deterministic. This treatment of the system error distinguishes this approach from purely dynamic treatments of the problem. As a result, the observational data can be statistically referenced to a plant, and questions regarding the validity of the estimates can be treated in a statistical framework that includes modeling errors as well as observational errors. The approach developed here leads to a framework that is suitable for combining a variety of models, both geometric and structural, together with a variety of sensing systems, both mechanical and electro-optical.

Of particular interest are the many large space antenna systems such as the DSN antennas. Performance requirements in communications and radiometry call for antennas that have increasingly large diameters and surface error requirements far beyond current capabilities.

Current pointing system designs for antennas do not typically incorporate effects beyond some reference plane that separates the control system from the antenna tipping structure (which includes the primary reflector, the secondary reflector, the quadripod, and the feed cone). For example, an antenna pointing system may determine the main azimuth and elevation pointing servo-drive error signals from a two-axis autocollimator mounted on an intermediate reference structure (IRS)[7] (see Fig. 1). By projecting a light beam onto a precision mirror mounted on the master equatorial (ME) and measuring the angular deviation of the reflected beam from the nominal orientation, the autocollimator determines an error signal that results from the nonparallelism of the two surfaces.

Because the tipping structure is outside of the pointing control loop, uncompensated boresight shifts will typically result from the distortion of the structure due to environmental loads. Self-compensating effects in the structural design of the antenna limit the net peak pointing offset to approximately 100 mdeg, even though structural deformations of the components of the antenna system can individually produce equivalent pointing shifts of nearly a degree due to unmodeled environmental loads.

The schematic diagram in Fig. 2 illustrates the effect of structural distortion on pointing capability. According to design specifications, the axes of the main reflector paraboloid and the subreflector hyperboloid lie on an axis normal to the IRS, which passes through the vertex of the paraboloid. As indicated by the broken line, environmental forces that distort the structure cause rays reflected from different segments of the antenna not to be focused onto the IRS reference axis. The design objective of a pointing-error compensation system is to bias the pointing command so that the centroid of energy with respect to the target falls along the IRS reference axis. Other antennas also employ compensator designs that do not correct for the effect of distortions due to unmodeled environmental loads on the tipping structure.

The principal environmental effects acting on the antenna are gravity, wind, and thermal effects. Gravity, the largest but most predictable load on

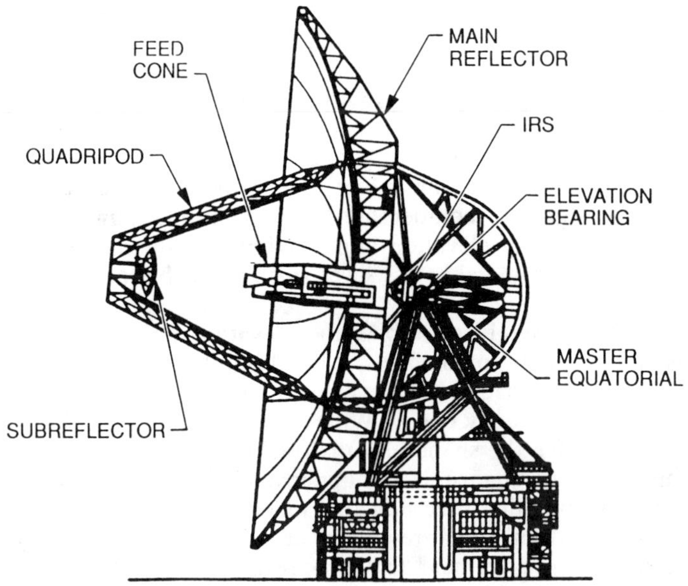

Fig. 1 70-Meter deep space network antenna.

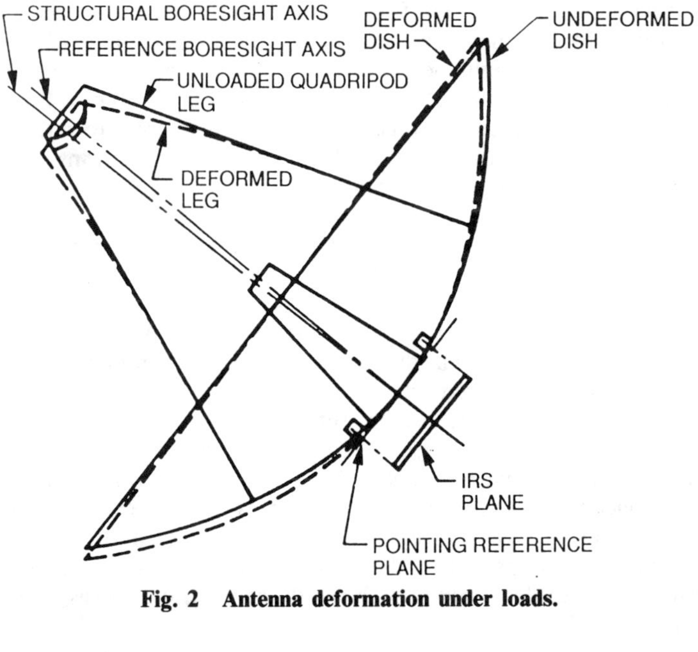

Fig. 2 Antenna deformation under loads.

the structure, causes sagging in the primary reflector and varies as a function of elevation angle. In addition, the gravitational effects lead to a displacement of the quadripod structure and a resultant displacement of the subreflector. The effects of gravity can be modeled by finite-element models that are available for antennas, and, since the effects are predictable, the resulting boresight errors can be offset through a calibration process. The resulting look-up tables then provide an elevation pointing command bias signal to compensate for gravitationally induced pointing errors as a function of elevation angle. Wind and thermal effects on the structure lead to considerably less severe distortions but are also much less predictable.

By using look-up tables to bias the elevation pointing command to compensate for gravitational effects, a blind pointing capability of approximately 10 mdeg is currently achievable in weather conditions that range from benign to moderately windy. With the recently completed upgrades to enlarge and improve the shape of large tracking antennas, in combination with future plans to quadruple the upper operating frequency, it is desirable to improve pointing precision to 1 mdeg.

The following discussion presents a description and an analysis of a real-time optical measurement and processing concept that delivers real-time pointing system bias commands to compansate for the effects of environmentally induced structural distortions. The concept requires optical measurements of the displacements of selected points on the primary and secondary reflectors relative to some reference coordinate system that lies at the base of the feed cone (e.g., the pointing reference plane in Fig. 2).

The spatial high-accuracy position encoding sensor (SHAPES) system under development at the Jet Propulsion Laboratory[8] was taken as the baseline sensor for the measurement system, since it is capable of delivering multiple simultaneous high-speed (10 frames/s) ranging measurements with accuracies at the submillimeter level. SHAPES is a time-of-flight laser ranging sensor that measures ranges from multiple sources to retroreflector targets that must be placed at optically unobstructed locations. By means of a system of fiber-optic connections, each SHAPES sensor head can accommodate up to 24 targets, whose actual locations can be physically remote from the head. By means of triangularization, the range measurements taken from sensor heads fixed with respect to a known reference coordinate system can be combined to give two- and three-dimensional displacement coordinates for a given retroreflector. A discussion of the effects on the SHAPES system due to anticipated environmental uncertainties in applications has recently been undertaken.[9]

This discussion deals only with the effects of unmodeled structural distortions beyond the reference coordinate plane (i.e., primary reflector, secondary reflector, quadripod, and feed cone) and assumes an a priori spatial knowledge of a reference coordinate system, the determination of which is a subject of current investigations in this technology. Although other authors have approached the problem of real-time compensation by combining SHAPES technology with purely geometric analysis,[10] the treatment here is based on the solution of an underlying structural estimation problem combined with a geometric synthesis. It is the framework of the estimation problem that provides for a meaningful quantitative evaluation of a given compensator design.

The next section begins with a general description of the proposed compensator design and continues with a more detailed discussion of the resulting structural and geometric problems. The section on model description and simulation results presents simulation results based on a reduced-order finite-element model of an antenna. The final sections outline future work and conclusions.

Problem Formulation

Compensator Design

The current compensator design is based on a calibration process for the effects of external loads such as gravity, which can be predicted by finite-element modeling analysis (Fig. 3). The result is a tabulation of compensation signals that depends on the current attitude of the antenna tipping structure.

As a starting point for the calibration process, the modeled external forces are taken as static load inputs to a high-order finite-element model, and this results in a predicted estimate for the shape of the main reflector as well as location estimates for the subreflector and the feed. From this a parabolic-fit routine is used to obtain a least-squares fit of the main reflector to a parabola of revolution. From this routine the effective

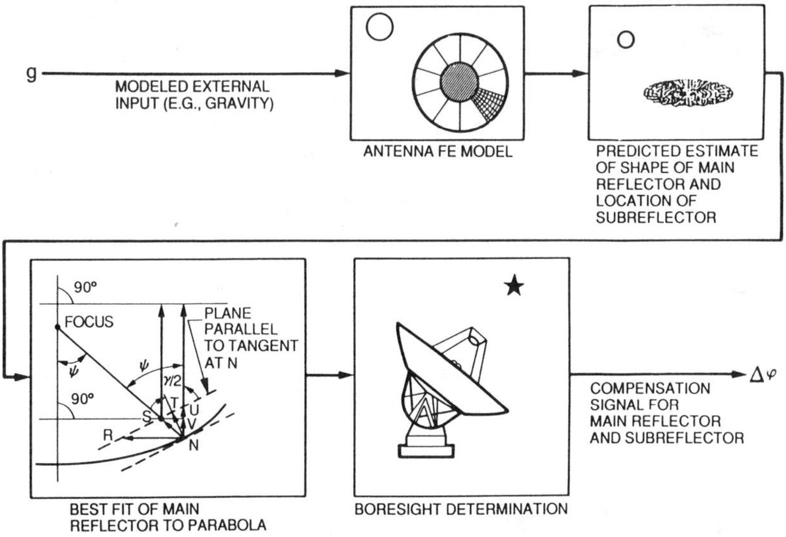

Fig. 3 Current compensator design.

boresight axis can be determined as a linear combination of terms that depend on the parameters of the best-fit parabola and the position of the other rigidly modeled structural components of the system.[11] More precisely, the boresight error is determined as a linear combination of the vertex shift and the rotation of the best-fit parabola, the lateral translation and rotation of the subreflector, and the lateral translation of the feed.

Such a control process is called open loop in that there is no feedback from the state of the system. The development proposed here is a real-time structural-optical compensator (Fig. 4) that incorporates measurement data into the process of estimating the state of the structural system. The blocks of the design which indicate the parabolic best fit and the boresight determination can be used as before. The crucial block of this real-time compensator design is the structural-optical estimator, which combines modeled responses to static loads with real-time measurement data (Fig. 5).

This block has the structure of a predictor-corrector system whereby the difference between the observed data and that which can be predicted by a priori modeling leads to a correction that is added to the prediction and that is derived before measurements are taken. Thus, the dotted box in Fig. 5 denotes the structural predictor, which leads to the estimate that conceptually determines the current structural compensator (Fig. 4). The key ingredient in the structural-optical estimator is the filter gain, which is determined by the solution to an optimization problem that is posed in the finite-element space.

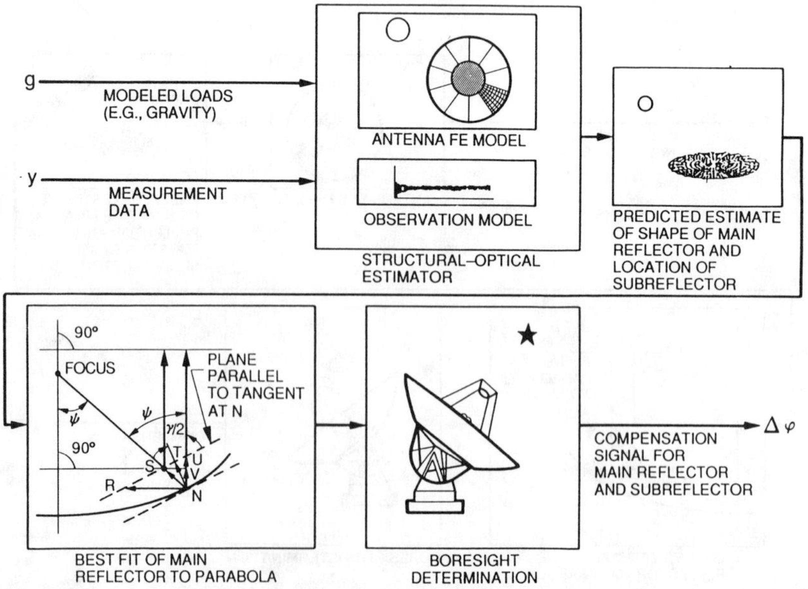

Fig. 4 Structural-optical compensator.

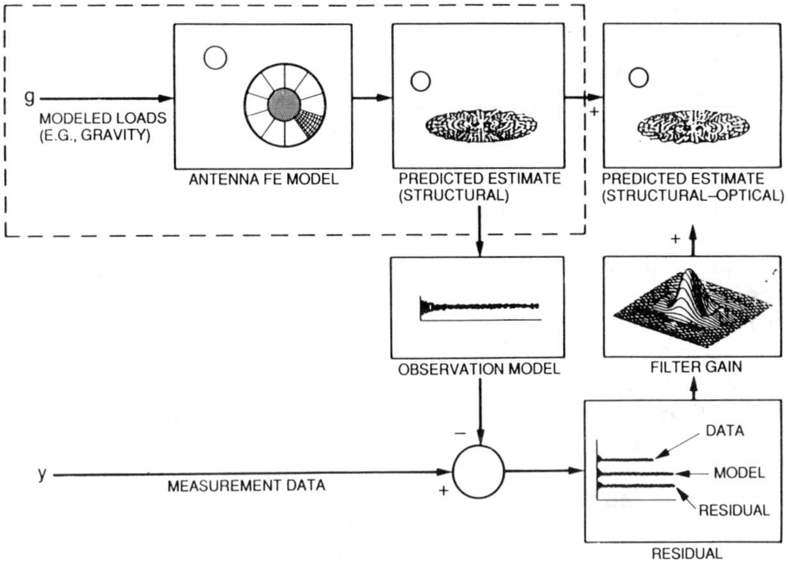

Fig. 5 Structural-optical estimator.

In the next subsections key components of the structural-optical compensator design are discussed in more detail. The structural-optical estimator is derived by a shape determination problem posed in a finite-element space. Next, the parabolic-fit analysis is discussed and specialized to the two-dimensional case, which was the version used in the simulations subsequently undertaken.

Static Shape Determination

In this section the linear system model that supports the static shape determination algorithms is summarized. This approach to shape estimation is part of an integrated methodology that combines the techniques of modeling, identification, and estimation for static control of distributed systems. A more complete discussion of this framework is given elsewhere.[1–5]

Let the state variable be given by $\boldsymbol{u}$. The models considered have the general form

$$A\boldsymbol{u} = Cf + B\omega \tag{1}$$

$$y = H\boldsymbol{u} + F\eta \tag{2}$$

where A is an operator that represents the system model. The preliminary analysis can begin with coarse geometric models that are suitable for resolving overall features. More detailed resolutions are then obtained by taking A to be the stiffness matrix. This type of fine-scale resolution can be idealized by taking A to be an invertible self-adjoint elliptic differential operator defined over some spatial domain. The operator A is assumed to be invertible with inverse Φ. This gives

$$\Phi A = I \tag{3}$$

where I is the identity. The two terms on the right-hand side of Eq. (1) represent the modeled and unmodeled loads on the system, respectively. In particular, C is an operator that characterizes the relevant deterministic forces, and B is an operator that characterizes the process errors that influence the state.

The vector $\boldsymbol{y}$ represents an observation of the system (i.e., a set of measurements). H is an operator that characterizes the state-to-observation map. It is assumed that the observation space has dimension N_s that corresponds to a finite-dimensional sensing scheme. The second term on the right-hand side of Eq. (2) represents the measurement errors present in the observation $\boldsymbol{y}$. The integrated form of the observation equation is given by

$$\boldsymbol{y} = H\Phi C\boldsymbol{f} + H\Phi B\boldsymbol{\omega} + F\boldsymbol{\eta} \tag{4}$$

There are three inner-product spaces of primary interest: the input space S_1, to which the process error $B\boldsymbol{\omega}$ and the deterministic input $C\boldsymbol{f}$ belong; the state space S_2 containing the state $\boldsymbol{u}$; and the measurement space S_3, where the data $\boldsymbol{y}$ and the observation error $F\eta$ belong. The inner product between two arbitrary elements u and v in the space S_i is denoted by $\langle u,v\rangle_i$ or by the simpler notation $u * v = \langle u,v\rangle_i$. Similarly, uv^* denotes an outer product. The corresponding norms are given by $\|\ \|_i$ or, when the context is clear, by the simpler notation $\|\ \|$.

The appropriately dimensioned operators B and F model the statistical influence of the process error and the measurement error; these errors, ω and η, form the model error vector

$$\xi = (\omega,\eta) \tag{5}$$

which represents spatial white noise and is characterized by the covariance operator

$$E(\xi\xi^*) = I \tag{6}$$

where $E(\ )$ denotes the expectation operator. More precisely, for $x_1 \varepsilon S_1$ and $x_3 \varepsilon S_3$, one has

$$\begin{aligned} E[\boldsymbol{\omega}] &= 0, \qquad E[\langle \boldsymbol{x}_1,\boldsymbol{\omega}\rangle_1 \boldsymbol{\omega}] = \boldsymbol{x}_1 \\ E[\boldsymbol{\eta}] &= 0, \qquad E[\langle \boldsymbol{x}_3,\boldsymbol{\eta}\rangle_3 \boldsymbol{\eta}] = \boldsymbol{x}_3 \end{aligned} \tag{7}$$

The limiting cases, $\|B\| \to 0$ and $\|F\| \to 0$, represent the assumptions of perfect modeling and perfect measurements, respectively.

The abstract quantities in Eqs. (1) and (2) take on relevant physical meanings in the context of the application under consideration. Here $\boldsymbol{u}$, the state, represents the elastic deformation of the antenna tipping structure as modeled by A, a finite-element stiffness matrix. The modeled forces Cf are restricted to gravity only, whereas the process error term $B\boldsymbol{\omega}$ represents all unmodeled environmental loads as well as inaccuracies in the finite-element model. The vector of observations $\boldsymbol{y}$ results from the sampling of the state by the measurement system (i.e., SHAPES), which is characterized by the state-to-observation map H and the measurement error term $F\boldsymbol{\eta}$.

The preceding assumptions lead to an appropriate framework for the analysis of minimum variance estimators of the state. Here the expected observation m is characterized by

$$\boldsymbol{m} = H\Phi Cf$$

$$E[\boldsymbol{y}] = \boldsymbol{m}, \qquad E[(\boldsymbol{y} - \boldsymbol{m})(\boldsymbol{y} - \boldsymbol{m})^*] = HR_\omega H^* + R_\eta \tag{8}$$

and the expected process and measurement covariances R_ω and R_η are given by

$$R_\omega = \Phi BB^*\Phi^*, \qquad R_\eta = FF^* \tag{9}$$

The resulting formulas are similar to the typical Kalman gain formulations used in the analysis of dynamic systems.[12] The minimum-variance estimate u_{est} has the form

$$\boldsymbol{u}_{\text{est}} = \boldsymbol{u}_0 + g(\boldsymbol{y} - \boldsymbol{m}) \tag{10a}$$

$$\boldsymbol{u}_0 = Cf \tag{10b}$$

where the gain g is given by

$$g = R_\omega H^*(R_\eta + HR_\omega H^*)^{-1} \tag{11}$$

Thus, as illustrated in Fig. 5, the structural-optical estimator acts as a predictor-corrector scheme where the prediction comes from the modeled value of the state and where the correction is determined by the discrepancy between the modeled state and the actual measurements.

The covariance operator associated with the estimation error is given by

$$P = R_\omega - R_\omega H^*(R_\eta + HR_\omega H^*)^{-1}HR_\omega \tag{12}$$

The covariance can then be used to determine the expected error before and after measurements:

$$E[\|u - u_0\|^2] = \text{tr}[R_\omega] \tag{13a}$$

$$E[\|\boldsymbol{u} - \boldsymbol{u}_{\text{est}}\|^2] = \text{tr}[P] \tag{13b}$$

From Eqs. (9), (12), and (13) it follows that

$$E[\|\boldsymbol{u} - \boldsymbol{u}_{\text{est}}\|^2] < E[\|\boldsymbol{u} - \boldsymbol{u}_0\|^2] \tag{14}$$

That is, the incorporation of measurements reduces the expected error in the estimate of the state (i.e., the shape of the antenna tipping structure).

In the case where A represents a finite-element stiffness matrix, a number of static loading problems must be solved in order to approximate the gain and the covariance. For example, when the state-to-observation map H returns point values of the state, one needs to compute fundamental solutions that correspond to unit static loads at the observed points.

One can also consider possibilities for recursive batch processing of data sets obtained by means of different sensing strategies. For example, Eq. (2) can be replaced by two sets of observations:

$$y_1 = H_1\boldsymbol{u} + F_1\eta_1 \tag{15a}$$

$$y_2 = \boldsymbol{H}_2\boldsymbol{u} + \boldsymbol{F}_2\boldsymbol{\eta}_2 \tag{15b}$$

An analysis of the second-order statistics of the related estimation error can also be conducted. Such an approach has been implemented to combine optical measurements with data taken from radiation patterns to obtain optimal estimates for antenna surface deformations.[3]

Parabolic-Fit Analysis

The problem of fitting a reflector deformation model with many degrees of freedom to a parabola of revolution has been treated elsewhere in detail.[13,14] The problem is generally treated in a least-squares formulation about the undisturbed dish, and the minimization is carried out with respect to the sums of squares of the path length errors associated with each point in the discretization. It can be shown that the path length error at a point can be trigonometrically related to the normal component of the surface error [cf. Eq. (30)]. Here a summary is given of the analysis when specialized to two spatial dimensions. The relevant equation is

$$z = y^2/(4f), \qquad -1 \leqslant y \leqslant 1 \tag{16}$$

where the coordinates y and z and the focal parameter f have been normalized by the radius of the dish.

The grid $G = \{(y_i, z_i)\}$ is chosen so that the deformation can be represented sufficiently accurately by these point values. For this application the y_i values are the nodal locations of the finite-element model, and the corresponding z_i values are the displacements at the nodes. The differential distortion of the parabola is represented by the so-called[13,14] homology parameters $\boldsymbol{h} = (h^{(1)}, h^{(2)}, h^{(3)}, h^{(4)})^T$, defined as

$h^{(1)}$ = shift in vertex (y coordinate)
$h^{(2)}$ = shift in vertex (z coordinate)
$h^{(3)}$ = rotation about origin in the y/z plane
$h^{(4)}$ = relative change in focal length ($\Delta f/f$)

The analysis is very similar in the three-dimensional case, where there are two additional homology parameters (one for the shift in vertex, one for the rotation).

Retaining only first-order terms, one can compute the differential distortion at the point $\binom{y}{z}$:

$$\begin{pmatrix} h^{(1)} \\ h^{(2)} \end{pmatrix} + \begin{bmatrix} 1 & -h^{(3)} \\ h^{(3)} & 1 \end{bmatrix} \begin{pmatrix} y \\ z \end{pmatrix} + \begin{pmatrix} 0 \\ -zh^{(4)} \end{pmatrix} \tag{17}$$

where geometrically these three terms represent the translation, rotation, and focal shift of the parabola, respectively. Thus, the differential displacement of the parabolic point is given by

$$\hat{C}(y,z)\boldsymbol{h} \tag{18}$$

where the 2×4 matrix $\hat{C}(y,z)$ is defined by

$$\hat{C}(y,z) = \begin{pmatrix} 1 & 0 & -z & 0 \\ 0 & 1 & y & -z \end{pmatrix} \tag{19}$$

The normal to the surface at $\binom{y}{z}$ is given by

$$\boldsymbol{n}(y,z) = \frac{1}{\sqrt{\dfrac{z}{f}+1}} \begin{pmatrix} \dfrac{-y}{2f} \\ 1 \end{pmatrix} \tag{20}$$

Then projecting the deflection onto the normal gives

$$\hat{B}(y,z) = \boldsymbol{n}(y,z)^T \hat{C}(y,z) = \frac{1}{\sqrt{\dfrac{z}{f}+1}} \left[\frac{-y}{2f}, 1, \left(\frac{yz}{2f} + y \right), -z \right] \tag{21}$$

The deflection at each point is also projected onto the normal direction:

$$\rho(y,z)\boldsymbol{n}(y,z) \tag{22}$$

where $\rho(y,z)$ is a scalar.

It is now possible to form the normal equations on the grid G by defining the $N \times 4$ matrix:

$$\hat{A} = \begin{bmatrix} \hat{B}(y_1,z_1) \\ \hat{B}(y_2,z_2) \\ \vdots \\ \hat{B}(y_N,z_N) \end{bmatrix} \tag{23}$$

and the N-dimensional vector:

$$\boldsymbol{b} = \begin{pmatrix} \rho(y_1, z_1) \\ \vdots \\ \rho(y_N, z_N) \end{pmatrix} \tag{24}$$

This allows one to form the N-dimensional residual vector for the normal deflections:

$$\boldsymbol{r} = \hat{A}\boldsymbol{h} - \boldsymbol{b} \tag{25}$$

The best-fit optimization problem is now posed as a minimization of the residual with respect to the norm:

$$\|\boldsymbol{r}\|_W^2 = \boldsymbol{r}^T W \boldsymbol{r} \tag{26}$$

where W is an $N \times N$ symmetric positive definite matrix.[15] The resulting normal equations are given by

$$(\hat{A}^T W \hat{A})\boldsymbol{h} = \hat{A}^T W \boldsymbol{b} \tag{27}$$

or, equivalently,

$$(\tilde{A}^T \tilde{A})\boldsymbol{h} = \tilde{A}^T \tilde{\boldsymbol{b}} \tag{28a}$$

$$\tilde{A} = W^{1/2} \hat{A} \tag{28b}$$

$$\tilde{\boldsymbol{b}} = W^{1/2} \boldsymbol{b} \tag{28c}$$

W can be chosen as a diagonal matrix:

$$W = \mathrm{diag}[w_1, w_2, \ldots, w_N] \tag{29}$$

where w_i has the form

$$w_i = a_i \left(\frac{f}{f + z_i} \right) \tag{30}$$

and a_i is the length (or area) associated with each point. The additional factor results in a minimization with respect to path length.

The solution of Eq. (27) is then given by

$$\boldsymbol{h} = \hat{T}\boldsymbol{b} \tag{31a}$$

$$\hat{T} = (\hat{A}^T W \hat{A})^{-1} \hat{A}^T W \tag{31b}$$

Here the $4 \times N$ matrix $\hat{T}$ is often called the pseudoinverse. Practical computational issues for the calculation in Eq. (31) are addressed in standard references.[15]

Model Description and Simulation Results

The functional block diagram in Fig. 6 outlines the conceptual framework in which the boresight determination problem is treated in this paper. The relevant synthesis problem focuses on the determination of three antenna boresights: 1) a gravitational boresight derived from a model for gravitational deformations combined with a parabolic fit; 2) a true boresight derived from a model for gravitational plus unmodeled deformations combined with a parabolic fit; and 3) an estimated boresight derived from an estimator that integrates structural finite-element models, gravitational deformation models, sampled measurements of the true deformation vector, and a parabolic fit.

As has been previously noted, current boresight pointing is based on a calibration with respect to forces whose effects can be adequately predicted (e.g., gravity); however, this calibration process is less successful when the disturbance forces cannot be adequately modeled (e.g., thermal effects, wind). The subsequent simulations were intended to determine the requirements for a boresight estimator that, under nominal uncertainty conditions, would recover the true boresight to within a millidegree, a requirement that is beyond the capabilities of the current calibration process. The design parameters include the properties of the disturbance forces (modeled and unmodeled), as well as the requirements for the sensing configuration (number, placement, and accuracy). In order to make the simulations more tractable, simplifications were introduced in the structural, geometric, and statistical analysis.

For the structural analysis a reduced two-dimensional antenna deformation model was derived from a high-order finite-element model of a 34-m antenna. More precisely, two ribs were designated as the north and south ribs, and all deformations calculated from the three-dimensional model were projected onto this pair (see Fig. 7). The total number of nodal values retained was 26, and these were evenly divided between the two ribs.

For the geometric analysis of the parabolic fit, the two-dimensional least-squares formulation given in the subsection on static shape determination was implemented. In addition, only the third component of the homology vector ($h^{(3)}$) was used in the estimate of the boresight. This component, the rotation about the origin in the y/z plane, is the geometric boresight of the best-fit parabola and corresponds to the largest term in the computation of the structural boresight.[10] Thus, in the spirit of the analysis of the subsection on static shape determination, this work focuses on the boresight errors associated with the structural deformations of the primary reflector, and attention is restricted to the dominant contribution. The other relevant components of the structural boresight noted in the subsection on compensator design (vertex shift of best-fit parabola, lateral translational and rotation of the subreflector, and the lateral translation of

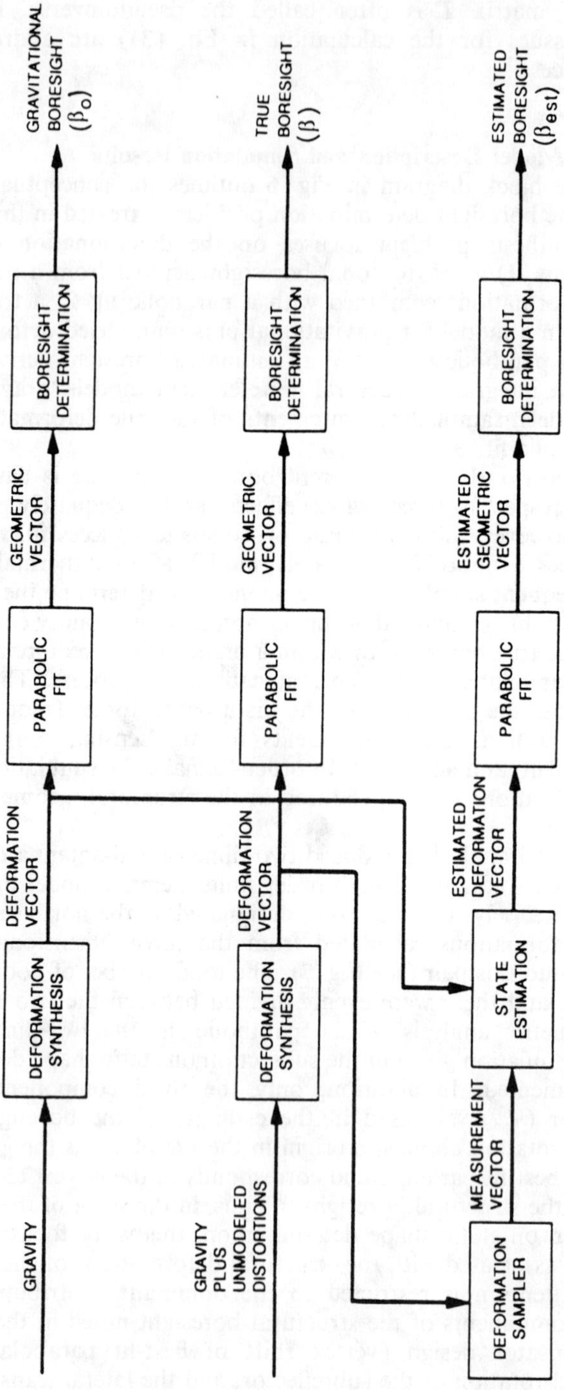

Fig. 6 Analysis of boresight synthesis problem.

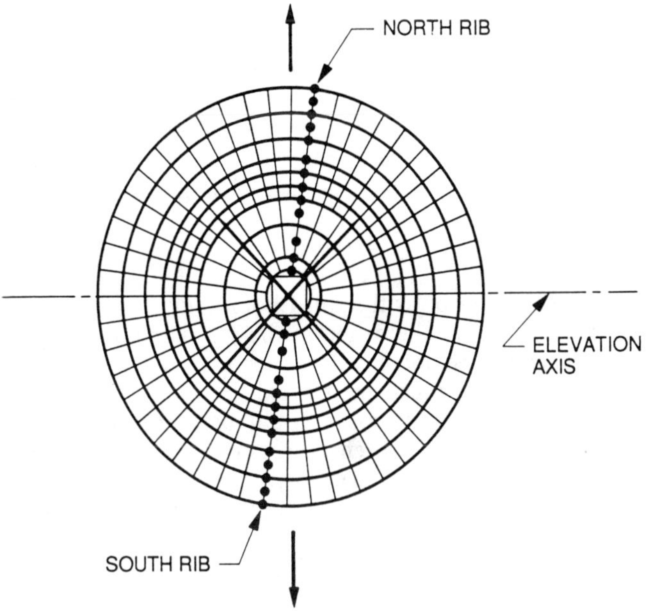

Fig. 7 Reduced-order finite-element model of main reflector.

the feed) could also have been included in the analysis, making the simulations somewhat more complex. Instead, the boresight errors associated with estimating these contributions were acceptably bounded. In particular, it was determined that the SHAPES system, when combined with a charge-coupled device (CCD) angle-measuring camera, is capable of measuring the subreflector position sufficiently accurately so that the resulting uncertainty in the boresight is smaller than the errors considered here.

For the statistical analysis the weighting matrices for the process error and the measurement error were taken to be scalar:

$$F = \sigma_\eta I \tag{32a}$$

$$B = \sigma_\omega I \tag{32b}$$

where I is the identity. More general forms for F and B could be used to characterize such features as a structural finite-element model that is less reliable at some nodes than at others or a measurement system with some components that are more accurate than others.

The preceding assumptions lead to the following version of the formulations given in the preceding section:

$$A\boldsymbol{u} = C\boldsymbol{f} + \sigma_\omega \boldsymbol{\omega} \tag{33}$$

$$\boldsymbol{y} = H\boldsymbol{u} + \sigma_\eta \boldsymbol{\eta} \tag{34}$$

$$\beta = \hat{Q}\hat{T}\hat{N}\boldsymbol{u} \tag{35}$$

where Eqs. (33) and (34) are simply Eqs. (1) and (2) combined with the simplification given by Eqs. (32). In Eq. (35) $\hat{N}$ is the mapping that projects the state (i.e., the deflection of the structure) onto the normal component. Then $\hat{T}$ is the pseudoinverse, given in Eqs. (31), which maps the normal deflections onto the vector of homology parameters $\boldsymbol{h}$. Finally, $\hat{Q}$ maps the homology parameters onto the boresight value. In accordance with the simplifying assumptions we have outlined, $\hat{Q}$ is given by

$$\hat{Q} = [0,0,1,0] \tag{36}$$

that is, $\hat{Q}$ is a projection onto the third homology parameter ($h^{(3)}$), which denotes the rotation about the origin in the y/z plane. More generally, the dimension of β could be increased to account for the other rotational degree of freedom that must be treated in the full three-dimensional problem. And also, the input space could be augmented to include the measured estimates of the subreflector and feed cone positions. By a principle of optimality for linear systems,[12] the minimum-variance estimate for β can be written in terms of the minimum-variance estimate for $\boldsymbol{u}$:

$$\beta_{\text{est}} = \hat{Q}\hat{T}\hat{N}\boldsymbol{u}_{\text{est}} \tag{37}$$

The other statistical qualities are similarly transformed. Thus, with the definitions

$$\beta_0 = \hat{Q}\hat{T}\hat{N}\boldsymbol{u}_0 \tag{38}$$

$$(\sigma_\beta^0)^2 = E[(\beta - \beta_0)^2] \tag{39}$$

$$(\sigma_\beta^1)^2 = E[(\beta - \beta_{\text{est}})^2] \tag{40}$$

one can show the precise relationship between the state covariances [cf. Eqs. (9) and (12)] and the boresight covariances:

$$(\sigma_\beta^0)^2 = (\hat{Q}\hat{T}\hat{N})R_\omega(\hat{Q}\hat{T}\hat{N})^* \tag{41}$$

$$(\sigma_\beta^1)^2 = (\hat{Q}\hat{T}\hat{N})P(\hat{Q}\hat{T}\hat{N})^* \tag{42}$$

These quantities can be identified with the outputs of the block diagram in Fig. 6. Thus, the quantities β, β_0, and β_{est} represent the true boresight, the modeled boresight, and the estimated boresight, respectively. The expected boresight errors before and after measurements are given by σ_β^0 and σ_β^1, respectively. To these one can add a fourth boresight value:

$$\tilde{\beta}_{\text{est}} = \beta_0 + \hat{W}\hat{T}_{\text{fit}}\hat{N}_{\text{fit}}(\boldsymbol{y} - H\boldsymbol{u}_0) \tag{43}$$

where β_0, $\boldsymbol{y}$, $\boldsymbol{u}_0$, and H are as previously given, the matrix $\hat{N}_{\text{fit}}$ gives the normal projection at the observed points only, and the matrix $\hat{T}_{\text{fit}}$ is the pseudoinverse [cf. Eqs. (31)], which carries out the parabolic fit with respect to the values at the observed point only. The quantity $\tilde{\beta}_{est}$ represents the boresight estimate, which is obtainable by purely geometric analysis. In correspondence to Eq. (40), one also has an expression for the covariance:

$$
\begin{aligned}
(\tilde{\sigma}_\beta^1)^2 &= E[(\tilde{\beta}_{est} - \beta)^2] \\
&= (\sigma_\omega)^2(\hat{Q}\hat{T}\hat{N})R_\omega(\hat{Q}\hat{T}\hat{N})^* \\
&\quad + (\sigma_\omega)^2(\hat{Q}\hat{T}_{fit}\hat{N}_{fit})R_\omega(\hat{Q}\hat{T}_{fit}\hat{N}_{fit})^* \\
&\quad + (\sigma_\eta)^2(\hat{Q}\hat{T}_{fit}\hat{N}_{fit})(\hat{Q}\hat{T}_{fit}\hat{N}_{fit})^* \\
&\quad - (\sigma_\omega)^2(\hat{Q}\hat{T}\hat{N})R_\omega(\hat{Q}\hat{T}_{fit}\hat{N}_{fit})^* \\
&\quad - (\sigma_\omega)^2(\hat{Q}\hat{T}_{fit}\hat{N}_{fit})R_\omega(\hat{Q}\hat{T}\hat{N})^*
\end{aligned}
\tag{44}
$$

The simulations focused on the covariance analysis of the estimates. Thus, given the expected boresight error before measurements (σ_β^0), the number and placement of targets where measurements are taken (H), and the accuracy of the measurements (σ_η), one can calculate the expected boresight error of the estimates ($\sigma_\beta^1, \tilde{\sigma}_\beta^1$). The following specifications were made for the results presented in Fig. 8. First, the model error parameter σ_ω was chosen so that the expected error before measurements σ_β^0 would be 10 mdeg. The observation error parameter σ_η was taken as 50 μ, a value that is consistent with the capability of the SHAPES system.[9] The only additional specifications required are the number and placement of targets where the displacement values are measured. Thus, in Fig. 8 the quantities $\sigma_\beta^1(j)$ and $\tilde{\sigma}_\beta^1(j)$, where j is an even number, refer to the expected boresight errors with $j/2$ optimally located targets on each of the two ribs in the reduced-order antenna model. Here the optimization was carried out by evaluating all possible distinct target locations and choosing the set that gave the lowest values for the expected boresight error.

Thus, with a total of six optimally placed measurement points, the structural-optical estimated achieves a 10:1 reduction in the expected error of the estimated boresight. One should note that the structurally based estimators are superior to the purely geometric estimators, but they are not substantially better. This observation may be interpreted as being supportive of compensator designs based on real-time optical measurements. That is, the sophistication of the underlying estimator and the fidelity of the model may not be too important as long as measurements are available. The simulation results presented in Fig. 8 are the main quantitative outcome of this work; they demonstrate that under nominal uncertainty conditions a 10:1 reduction in the boresight error can be achieved by integrating structural models, geometric fitting analysis, and a real-time optical measurement system. Even though the underlying structural finite-element model may have a large dimension, the real-time computational

j NUMBER OF OPTIMALLY LOCATED TARGETS	$\sigma_\beta^1(j)$ EXPECTED BORESIGHT ERROR (MILLIDEGREES): STRUCTURAL-OPTICAL ESTIMATOR	$\tilde{\sigma}_\beta^1(j)$ EXPECTED BORESIGHT ERROR (MILLIDEGREES): SIMPLE GEOMETRIC FIT
2	1.4	–
4	1.1	1.6
6	0.9	1.3

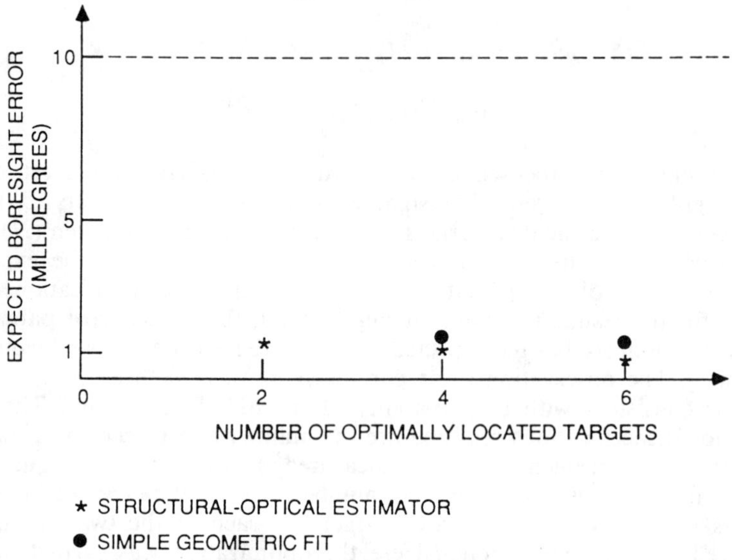

Fig. 8 Expected boresight error ($\sigma_B^0 = 10$ mdeg).

requirements involve only matrix and vector operations where the relevant dimensions are determined by the number of measurements taken.

The results of a test case derived from the gravity model at horizon (elevation angle equals zero) are given in Fig. 9. To test the estimator's ability to recover the boresight from the measurement values only, the modeled gravitational boresight before measurements was taken to be zero ($\beta_0 = 0$). For this trial the deflection was scaled to correspond to a 10-mdeg offset in the boresight, and the measurement locations were evenly spaced on each of the two ribs in the model. (The tips of the ribs were always included as measurement locations.) The measurement error was taken to be zero ($\sigma_\eta = 0$). Thus, $\beta_{est}(j)$ and $\tilde{\beta}_{est}(j)$ are the estimated boresight offsets with j evenly spaced measurement locations.

This may be thought of as a worst-case scenario for the structurally based estimator since the deflection is spatially biased (i.e., distorted in a

j NUMBER OF EVENLY SPACED TARGETS	$\|\beta-\beta_{est}(j)\|$ ERROR IN BORESIGHT ESTIMATE (MILLIDEGREES): STRUCTURAL-OPTICAL ESTIMATOR	$\|\beta-\tilde{\beta}_{est}(j)\|$ ERROR IN BORESIGHT ESTIMATE (MILLIDEGREES): SIMPLE GEOMETRIC FIT
2	12.8	–
4	7.5	0.8
6	0.2	0.3

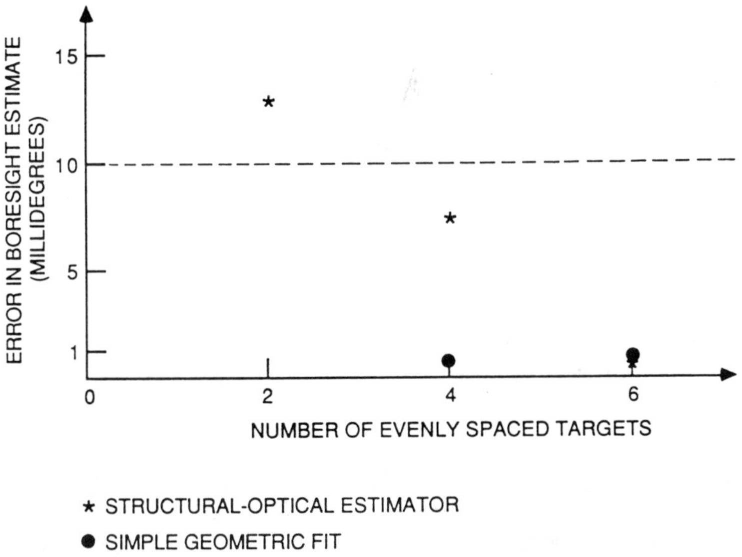

Fig. 9 Boresight error in simulations with gravity model ($\beta = 10$ mdeg, $\beta_0 = 0$ mdeg).

preferential direction), and the estimators were designed to perform best with all distortion directions equally likely.[9,10] Nevertheless, a 10:1 reduction in the error is achieved. Again, the geometric estimators perform well with no structural information incorporated into the design.

Since the simple geometric fit recovered the spatially biased component so well, this suggests that the two designs might be used in combination. First, a simple geometric fit can be performed to recover the spatially biased component, and then a version of the structural-optical estimator can be implemented to obtain a finer resolution about this set point.

Although additional simulations with the full three-dimensional model will be necessary to advance the design process, it is already possible to estimate the number of measurements required to achieve a 10 : 1 reduction in the pointing uncertainty. For example, two adequately resolved cross

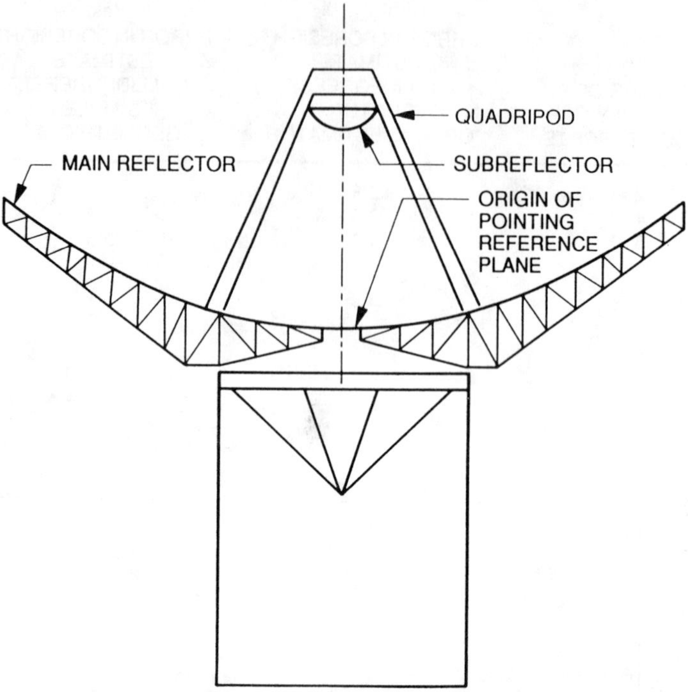

Fig. 10 Key locations on antenna for measurement system.

sections would determine all of the relevant parameters of a three-dimensional parabola of revolution (paraboloid). This would lead to a requirement of 12 measurements (6 per cross section). Another proposed baseline would require 18 targets to similarly cover 3 cross sections. This last design is motivated by the observation of three cross-sectional nodal lines on the main reflector of some antennas; these nodal lines apparently lie on or near the best-fit paraboloid that is derived from the gravitational models.[16]

In Fig. 10 the key locations on the antenna for the measurement system are noted. The design would include approximately three SHAPES sensor heads mounted on the quadripod near the subreflector. These sensor heads would return range measurements from the targets located on the main reflector, and triangularization methods would be used to recover the displacement values relative to the sensor heads. The connection from the sensor heads to the reference plane would be made by taking additional range measurements at targets located near the origin of the reference coordinate system (base of the feed cone). Calculated error estimates indicate that an additional angle-measuring camera (CCD array) at the base of the feed cone would adequately measure the angular displacement of the subreflector and the sensor heads from their nominal positions.

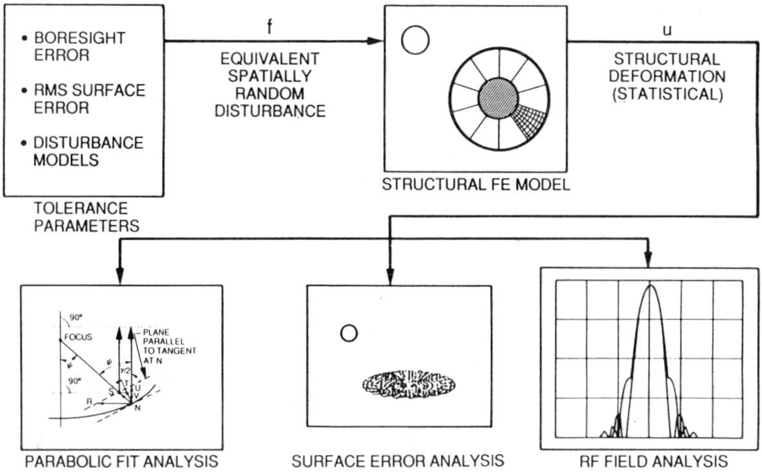

Fig. 11 Antenna estimator design by means of spatially random disturbances.

Future Work: Short-Wavelength Spatial Deflections

This work represents one method of improving the pointing capabilities of antennas by using the techniques of the spatially random analysis of static systems. As illustrated in Fig. 11, this approach is suitable for investigating many aspects of antenna analysis related to surface deformations. Although this work mainly concerns algorithms that estimate the structural boresight of an antenna, other relevant problems concerning the analysis of rms surface errors and the synthesis of antenna radiation patterns can also be studied in this framework.

One of the key issues for future analytical and experimental research is that of estimating the antenna distortions occurring at short spatial wavelengths. These are distortions that are governed by manufacturing tolerances and surface roughness. The static shape estimation approaches outlined in this paper are smoothing solutions in the sense that they recover a smooth interpolated estimate of the antenna shape from relatively few measurements. The estimation algorithms can be viewed as "low-pass" spatial filters that accentuate the long spatial wavelengths describing the large-scale features and filter out the short wavelengths that govern the fine scale structure. Of course, these smooth estimates are necessary to determine the global deflections of the antenna in response to disturbances. The problem of estimating short spatial-wavelength distortions presents two additional challenges: the need for sensing the surface distortions at a finer granularity and the need to extend the estimation algorithms to arrive at fine-granularity solutions.

The formulation for this new class of "high-pass" algorithms begins by denoting the short-wavelength displacement of the antenna by the symbol $\boldsymbol{v}$. This deflection is assumed to be orthogonal to the solutions $\boldsymbol{u}$ of the long-wavelength model in Eq. (1), i.e.,

$$\boldsymbol{v} * \boldsymbol{u} = 0; \qquad A\boldsymbol{u} = C\boldsymbol{f} + B\boldsymbol{\omega} \tag{45}$$

Since the solutions $\boldsymbol{u}$ of Eq. (1) are smoothed solutions, the orthogonal displacement will have a short-wavelength content. The objective of the high-pass estimator, then, is to estimate this orthogonal component $\boldsymbol{v}$ from the measurement

$$\boldsymbol{y} = H\boldsymbol{v} + F\boldsymbol{\eta} \tag{46}$$

This measurement equation assumes that sensor data has enough spatial resolution to make the short-wavelength signal $\boldsymbol{v}$ observable in the measurements. This would be the case, for example, if direct RF field or interferometric measurements were taken. The solution to this estimation problem, and the corresponding experiments with numerical models, are currently under investigation and will be reported on by the authors in the near future.

Conclusions

The concept proposed and analyzed in this chapter combines a real-time measurement system, a structural estimator, and a parabolic fitting algorithm to determine antenna pointing offsets that result from unmodeled structural distortions. This pointing offset data can be used to generate real-time command biases to correct for pointing errors. Simulations with a reduced-order antenna finite-element model have demonstrated that a 10:1 reduction in boresight uncertainty can be achieved. The concept requires optical measurements of the displacements of selected points on the primary and secondary reflectors relative to some reference coordinate system at the base of the feed cone.

An application of the techniques of static shape estimation has been applied to a model representative of large ground antenna system pointing control. Further applications of these techniques are under investigation to determine RMS surface tolerances from electro-optical and RF measurement data.

Acknowledgments

Acknowledgments are given to C.-T. Chian of JPL for his essential contributions to the finite-element simulations and to R. Levy and J. Cucchissi, also of Jet Propulsion Laboratory, for many interesting discussions on the subject of this work.

References

[1]Rodriguez, G., and Scheid, R. E., "Modeling, Estimation and Identification Methods for Static Shape Control of Flexible Structures," *AAS/AIAA Astrodynamics Specialists Conference*, Vail, CO, Aug. 1985.

[2]Rodriguez, G., and Scheid, R. E., "An Integrated Approach to Modeling, Identification, Estimation and Control for Static Distributed Systems," *Proceedings of the Fourth IFAC Symposium on the Control of Distributed Parameter Systems*, Los Angeles, CA, June 1987.

[3]Rodriguez, G., and Scheid, R. E., "Kalman-Like Estimation for Static Distributed Systems: Antenna Shape from Radiation Measurements," *Proceedings of the Fifth VPI & SU/AIAA Symposium on Dynamics and Control of Large Structures*, Blacksburg, VA, June 1985.

[4]Rodriguez, G., and Scheid, R. E., "A Residuals Approach to Filtering, Smoothing and Identification for Static Distributed Systems," *Proceedings of the Workshop on Identification and Control of Flexible Space Structures*, Jet Propulsion Lab. Publ. 85-29, April 1985.

[5]Scheid, R. E., and Rodriguez, G., "Numerical Experimentation with Maximum Likelihood Identification in Static Distributed Systems," *Proceedings of the Workshop on Identification and Control of Flexible Space Structures*, Jet Propulsion Lab. Publ. 85-29, April 1985.

[6]Scheid, R. E., "Precision Pointing Compensation for DSN Antennas with Optical Measuring Sensors," Jet Propulsion Lab., Pasadena, CA, TDA Progress Rept. 42-97, May 15, 1989, pp. 127–136.

[7]Nelson, M. D., Schroeder, J. R., and Tubbs, E. F., "Optical Links in the Angle-Data Assembly of the 70-meter Antennas," Jet Propulsion Lab., Pasadena, CA, TDA Progress Rept. 42-92, Oct.–Dec. 1987, Feb. 15, 1988, pp. 154–165.

[8]McLauchlan, J., Goss, W., and Tubbs, E., "SHAPES: A Spatial, High Accuracy, Position Encoding Sensor for Space System Control Applications," American Astronautical Society Publ. 82-032, Jan. 20, 1983.

[9]Nerheim, N., "Effect of Atmospheric Turbulence on Precision Optical Measurements Used for Antenna Pointing Compensation," Jet Propulsion Lab., Pasadena, CA, TDA Progress Rept. 42-97, May 15, 1989, pp. 141–148.

[10]Schumacher, L. L., and Vivian, H. C., "Antenna Pointing Compensation Based on Precision Optical Measurement Techniques," Jet Propulsion Lab., Pasadena, CA, TDA Progress Rept. 42-94, Vol. April–June 1988, pp. 135–144.

[11]Levy, R., "Optimization of Antenna Structure Design," *Eighth Conference on Electronic Computation*, 1983.

[12]Bryson, A. E., and Ho, Y.-C., *Applied Optimal Control*, Hemisphere, Washington, DC, 1975.

[13]Katow, M. S., "Antenna Structures: Evaluation of Reflector Surface Distortions," JPL TR-32-1526, Vol. 1, Jet Propulsion Lab., Pasadena, CA, 1981, pp. 76–80.

[14]Katow, M. S., and Schmele, L. W., "Antenna Structures: Evolution Techniques of Reflector Distortions," *Supporting Research and Advanced Development*, Jet Propulsion Lab., Pasadena, CA, Space Programs Summary 37-40, Vol. IV, Aug. 31, 1966, pp. 176–184.

[15]Golub, G. H., and Van Loan, C. F., *Matrix Computations*, Johns Hopkins Univ. Press, Baltimore, MD, 1983.

[16]Tracking and Data Acquisition Organization (Technical Staff), "The NASA/JPL 64-Meter-Diameter Antenna at Goldstone, California Project Report," Jet Propulsion Lab., Pasadena, CA, TM-33-671, July 15, 1974.

Part 2. Control, Stability Analysis, and Optimization

Chapter 10

Optimal Projection Approach to Robust Fixed-Structure Control Design

Dennis S. Bernstein* and David C. Hyland†
Harris Corporation, Melbourne, Florida

Overview of the Optimal Projection Approach to Control Design

THE challenge of modern control theory research is to develop effective methodologies for multivariable control system design. Both the importance and difficulty of the task should not be underestimated. Although three decades of intensive effort have now been devoted to this task, its attainment continues to elude us. Meanwhile, the challenges of high-performance control system design for complex systems continue to mount. As sensor/processor/actuator technologies continue to improve, there arises a corresponding burden on the control system designer to effectively utilize the available hardware; to minimize design, implementation, testing, and maintenance costs; and to guarantee the highest possible levels of performance and reliability.

Optimal projection theory is one approach to this challenge. The optimal projection approach was originally developed to address the problem of fixed-order linear-quadratic dynamic compensator design to obtain low-order, high-performance controllers.[1–14] The motivation for restricting the controller order arises from limitations on computer processing capability typical of applications such as vibration suppression in large space structures. Subsequently, a decade of effort has extended the optimal projection approach to address a broad range of issues in multivariable control system design, including the following:

1) distributed-parameter systems[15–17];
2) discrete-time and sampled-data systems[18,19];

---

*Staff Engineer, Structural Controls Group.
†Senior Principal Engineer and Group Leader, Structural Controls Group.

3) decentralized control[20,21];
4) model reduction[22–24];
5) H_2/H_∞ control[25–30];
6) parameter-robust control[31–56];
7) fixed-structure design[57,58]; and
8) pole placement.[59]

The key feature of optimal projection theory is that all of the preceding issues are addressed within a *common* mathematical framework that effectively permits simultaneous treatment of multiple design goals. For example, results on reduced-order design and parameter-robust design can be merged to produce parameter-robust, reduced-order designs.

As its name implies, optimal projection theory is an optimal control methodology. However, it is a *constrained* optimal control methodology. That is, optimal projection theory does not seek to optimize a performance measure per se, but rather seeks to optimize performance within a class of controllers satisfying specified side constraints such as robust stability or fixed controller order. Unconstrained optimal controllers, although globally optimal in a narrow sense, may be quite unacceptable in meeting additional specifications. For example, for linear plants with Gaussian white noise disturbances, linear-quadratic Gaussian (LQG) controllers provide the best nominal quadratic performance compared to all (even nonlinear) feedback controllers, yet the order of an LQG controller is equal to the order of the plant and hence may be impossible or impractical to implement if the plant is of high order. Furthermore, LQG controllers may be arbitrarily sensitive to plant parameter variations and unmodeled dynamics. Optimal projection theory overcomes these defects by generalizing LQG theory to include these and other design aspects. Precisely how these design aspects are incorporated as side constraints is the subject of subsequent sections of this chapter.

How the Optimal Projection Approach Differs from Other Fixed-Structure Design Methods

The most basic constraint incorporated within optimal projection theory involves fixing the order of the feedback compensator to be less than that of the plant. This is the fixed-order dynamic compensation problem. The fundamental nature of this problem is reflected by the extensive literature devoted to it.[60–83] In the standard approach to this problem, a quadratic criterion is minimized over the class of compensator gain matrices of given dimension. By setting the cost gradients to zero, the first-order necessary conditions for the problem consist of a collection of complicated nonlinear matrix equations without apparent structure. Because the equations are complicated, they are difficult to analyze, insight into their solution structure is lacking, and search methods are often the only viable numerical algorithms.

Optimal projection theory, however, makes no attempt to solve these complicated nonlinear matrix equations. Rather, its point of departure involves transforming these complicated nonlinear matrix equations into a highly structured form that is consistent with LQG theory. The key to this transformation is the recognition that the optimality conditions give rise to an idempotent matrix, that is, a nonorthogonal projection. The transformation yields a highly structured system of four coupled algebraic matrix equations consisting of two modified Riccati equations and two modified Liapunov equations. The coupling terms, which depend on the projection, clearly show that there is no estimator/regulator separation in the reduced-order case. However, if the controller order is chosen to be equal to the plant order, then the projection becomes the identity, the coupling terms are zero, the four equations reduce to two standard Riccati equations, and the LQG controller is recovered.

These four optimal projection equations present new challenges and opportunities. Mathematical challenges arise in analyzing these equations. Since the Riccati and Liapunov equations are nonstandard, their analysis requires extensions of existing techniques as well as the development of new methods. In addition, the coupling between the equations presents additional difficulties. Hence, the equations present nontrivial mathematical challenges relevant to a problem of practical interest.

On the other hand, practical benefits lie in the structured form of the equations. We already have a crucial insight, namely, the mechanism responsible for the breakdown of estimator/regulator separation, i.e., the coupling terms. As will be seen in the section on relationships between optimal projection theory and model/controller reduction techniques, it is precisely these coupling terms that are responsible for the suboptimality of controller reduction techniques. The main practical benefit of these equations is the opportunity to develop novel computational methods. By exploiting the Riccati and Liapunov structure, a new class of algorithms can be constructed as described in the section on numerical solution of the optimal projection equations. Furthermore, the optimal projection equations provide insight into the multiplicity of solutions and the guaranteed stability of the closed-loop system. To illustrate these points, we now turn to a brief review of basic optimal projection theory.

Brief Review of Basic Optimal Projection Theory

Consider the system

$$\dot{x}(t) = Ax(t) + Bu(t) + D_1 w(t) \tag{1}$$

$$y(t) = Cx(t) + D_2 w(t) \tag{2}$$

where $x \in \mathbb{R}^n$, $u \in \mathbb{R}^m$, $y \in \mathbb{R}^\ell$, and $w \in \mathbb{R}^p$ is white noise with unit intensity I_p. The intensities of $D_1 w(t)$ and $D_2 w(t)$ are thus given, respectively, by $V_1 \triangleq D_1 D_1^T \geq 0$ and $V_2 \triangleq D_2 D_2^T > 0$. For convenience we assume that $V_{12} \triangleq D_1 D_2^T = 0$; i.e., the plant disturbance and measurement noise are uncorrelated. The goal of the fixed-order dynamic compensation problem is to determine an n_cth-order optimal dynamic compensator

$$\dot{x}_c(t) = A_c x_c(t) + B_c y(t) \tag{3}$$

$$u(t) = C_c x_c(t) \tag{4}$$

that minimizes the steady-state performance criterion

$$J(A_c,B_c,C_c) \triangleq \lim_{t\to\infty} \mathbb{E}[x^T(t)R_1 x(t) + u^T(t)R_2 u(t)] \tag{5}$$

where $x_c \in \mathbb{R}^{n_c}$, $n_c \leqslant n$, $R_1 = R_1^T \geq 0$, and $R_2 = R_2^T > 0$.

The $\tilde{n}$-dimensional closed-loop system corresponding to Eqs. (1) and (2) ($\tilde{n} \triangleq n + n_c$) can be expressed as

$$\dot{\tilde{x}}(t) = \tilde{A}\tilde{x}(t) + \tilde{D}w(t) \tag{6}$$

where

$$\tilde{x}(t) \triangleq \begin{bmatrix} x(t) \\ x_c(t) \end{bmatrix}, \qquad \tilde{A} \triangleq \begin{bmatrix} A & BC_c \\ B_c C & A_c \end{bmatrix}, \qquad \tilde{D} = \begin{bmatrix} D_1 \\ B_c D_2 \end{bmatrix} \tag{7}$$

In addition, the performance (5) can be expressed as

$$J(A_c,B_c,C_c) = \lim_{t\to\infty} \mathbb{E}[\tilde{x}^T(t)\tilde{R}\tilde{x}(t)] \tag{8}$$

where

$$\tilde{R} \triangleq \begin{bmatrix} R_1 & 0 \\ 0 & C_c^T R_2 C_c \end{bmatrix} \tag{9}$$

To guarantee that the cost J is finite and independent of initial conditions, we restrict our attention to the set of n_cth-order stabilizing compensators

$$\mathcal{S}_{n_c} \triangleq \{(A_c,B_c,C_c)\colon \tilde{A} \text{ is asymptotically stable}\} \tag{10}$$

We require no explicit characterization of $\mathcal{S}_{n_c}$; its role is theoretical only. However, we shall *assume* that $\mathcal{S}_{n_c}$ is not empty.

Now let $(A_c,B_c,C_c) \in \mathcal{S}_{n_c}$ so that $\tilde{A}$ is asymptotically stable, and let $\tilde{Q} \in \mathbb{R}^{\tilde{n}\times\tilde{n}}$ and $\tilde{P} \in \mathbb{R}^{\tilde{n}\times\tilde{n}}$ be the closed-loop steady-state covariance and its dual, i.e.,

$$0 = \tilde{A}\tilde{Q} + \tilde{Q}\tilde{A}^T + \tilde{V} \tag{11}$$

$$0 = \tilde{A}^T\tilde{P} + \tilde{P}\tilde{A} + \tilde{R} \tag{12}$$

where

$$\tilde{V} \triangleq \tilde{D}\tilde{D}^T = \begin{bmatrix} V_1 & 0 \\ 0 & B_c V_2 B_c^T \end{bmatrix} \tag{13}$$

Then the performance (5) can be expressed as

$$J(A_c,B_c,C_c) = \operatorname{tr} \tilde{Q}\tilde{R} = \operatorname{tr} \tilde{P}\tilde{V} \tag{14}$$

Also, if $\tilde{Q}$ and $\tilde{P}$ are partitioned according to

$$\tilde{Q} = \begin{bmatrix} Q_1 & Q_{12} \\ Q_{12}^T & Q_2 \end{bmatrix}, \qquad Q_1 \in \mathbb{R}^{n \times n}, \qquad Q_2 \in \mathbb{R}^{n_c \times n_c} \tag{15}$$

$$\tilde{P} = \begin{bmatrix} P_1 & P_{12} \\ P_{12}^T & P_2 \end{bmatrix}, \qquad P_1 \in \mathbb{R}^{n \times n}, \qquad P_2 \in \mathbb{R}^{n_c \times n_c} \tag{16}$$

then Q_1 is the covariance of the plant states, Q_2 is the covariance of the compensator states, and Q_{12} is the cross covariance of the plant and controller states.

Since the value of J is independent of the internal realization of the compensator, in the following we will further restrict our attention to minimal compensators. Thus, we define the admissible set as

$$\mathscr{S}_{n_c}^+ = \{(A_c,B_c,C_c) \in \mathscr{S}_{n_c}\colon (A_c,B_c) \text{ is controllable, } (A_c,C_c) \text{ is observable}\} \tag{17}$$

Note that $\mathscr{S}_{n_c}^+$ is an open set.

Before we present the main results we review a lemma that is relevant to the solution of the optimal fixed-order compensation problem (see Refs. 7 and 56). This result follows from *Theorem* 6.2.5 of Ref. 84.

Lemma 1. Suppose $\hat{Q} \in \mathbb{R}^{n \times n}$ and $\hat{P} \in \mathbb{R}^{n \times n}$ are symmetric and nonnegative-definite and assume rank $\hat{Q}\hat{P} = n_c$. Then the following statements hold:

(*i*) There exists invertible $W \in \mathbb{R}^{n \times n}$ such that $W^{-1}\hat{Q}W^{-T}$ and $W^T\hat{P}W$ are both diagonal.

(*ii*) $\hat{Q}\hat{P}$ is diagonalizable and has nonnegative eigenvalues.

(*iii*) The $n \times n$ matrix

$$\tau \triangleq \hat{Q}\hat{P}(\hat{Q}\hat{P})^{\#} \tag{18}$$

where $(\ )^{\#}$ denotes the group or Drazin generalized inverse,[85] is idempotent; i.e., τ is an oblique projection, and

$$\operatorname{rank} \tau = n_c \tag{19}$$

Furthermore, $\tau_\perp \triangleq I_n - \tau$ satisfies

$$\operatorname{rank} \tau_\perp = n - n_c \tag{20}$$

(*iv*) Let $\sigma_1 \geq \sigma_2 \geq \ldots \geq \sigma_{n_c} > 0$ denote the positive eigenvalues of $\hat{Q}\hat{P}$, and let invertible $W \in \mathbb{R}^{n \times n}$ be such that

$$\hat{Q}\hat{P} = W \begin{bmatrix} \sigma_1 & & & & \\ & \ddots & & & \\ & & \sigma_{n_c} & & \\ & & & 0 & \\ & & & & \ddots & \\ & & & & & 0 \end{bmatrix} W^{-1} \tag{21}$$

and define the ith eigenprojection $\Pi_i[\hat{Q}\hat{P}]$ of $\hat{Q}\hat{P}$ by

$$\Pi_i[\hat{Q}\hat{P}] \triangleq W e_i e_i^T W^{-1} \tag{22}$$

where e_i is the ith column of I_n. Then

$$\hat{Q}\hat{P} = \sum_{i=1}^{n_c} \sigma_i \Pi_i[\hat{Q}\hat{P}] \tag{23}$$

and

$$\tau = \sum_{i=1}^{n_c} \Pi_i[\hat{Q}\hat{P}] = W \begin{bmatrix} I_{n_c} & 0 \\ 0 & 0 \end{bmatrix} W^{-1} \tag{24}$$

(*v*) There exist $G, \Gamma \in \mathbb{R}^{n_c \times n}$ and nonsingular $M \in \mathbb{R}^{n_c \times n_c}$ such that

$$\hat{Q}\hat{P} = G^T M \Gamma \tag{25}$$

$$\Gamma G^T = I_{n_c} \tag{26}$$

(*vi*) If G, Γ, and M are as in (*v*), then

$$\operatorname{rank} G = \operatorname{rank} \Gamma = \operatorname{rank} M = n_c \tag{27}$$

$$(\hat{Q}\hat{P})^{\#} = G^T M^{-1} \Gamma \tag{28}$$

$$\tau = G^T \Gamma \tag{29}$$

$$\tau G^T = G^T, \qquad \Gamma \tau = \Gamma \tag{30}$$

(*vii*) The matrices G, Γ, and M given in (*v*) are unique except for a change of basis in $\mathbb{R}^{n_c}$; i.e., if G', Γ', and M' also satisfy property (*v*), then there exists nonsingular $S \in \mathbb{R}^{n_c \times n_c}$ such that $G' = S^T G$, $\Gamma' = S^{-1}\Gamma$, and $M' = S^{-1}MS$. Furthermore, all such M are diagonalizable with positive eigenvalues.

(*viii*) If G, Γ, and M are as in (*v*), then M is given by

$$M = \Gamma\hat{Q}\Gamma^T G\hat{P}G^T \tag{31}$$

and the $n_c \times n_c$ matrices $\Gamma\hat{Q}\Gamma^T$ and $G\hat{P}G^T$ are positive-definite.

(*ix*) If rank $\hat{Q}$ = rank $\hat{P}$ = rank $\hat{Q}\hat{P} = n_c$, then there exists a (nonunique) invertible $W \in \mathbb{R}^{n\times n}$ such that

$$\hat{Q} = W\begin{bmatrix}\Lambda & 0\\ 0 & 0\end{bmatrix}W^T \tag{32}$$

$$\hat{P} = W^{-T}\begin{bmatrix}\Omega & 0\\ 0 & 0\end{bmatrix}W^{-1} \tag{33}$$

where $\Lambda \in \mathbb{R}^{n_c\times n_c}$ and $\Omega \in \mathbb{R}^{n_c\times n_c}$ are diagonal and nonsingular. In addition,

$$\hat{Q} = \tau\hat{Q} = \hat{Q}\tau^T = \tau\hat{Q}\tau^T \tag{34}$$

$$\hat{P} = \tau^T\hat{P} = \hat{P}\tau = \tau^T\hat{P}\tau \tag{35}$$

Remark 1. The transformation W in statement (*ix*) meets the requirements of statement (*iv*).

Definition 1. A triple (G,M,Γ) satisfying statement (*v*) of *Lemma* 1 is called a "projective factorization" of $\hat{Q}\hat{P}$.

To optimize Eq. (14) subject to the constraint (11), form the Lagrangian

$$\mathscr{L}(A_c,B_c,C_c,\tilde{Q},\tilde{P},\lambda_0) \triangleq \text{tr}[\lambda_0\tilde{Q}\tilde{R} + (\tilde{A}\tilde{Q} + \tilde{Q}\tilde{A}^T + \tilde{V})\tilde{P}] \tag{36}$$

where the Lagrange multipliers, scalar $\lambda_0 \geq 0$ and $\tilde{P} \in \mathbb{R}^{(n+n_c)\times(n+n_c)}$, are not both zero. It can be shown[7] that without loss of generality we can set $\lambda_0 = 1$. The stationary conditions are then given by

$$\frac{\partial\mathscr{L}}{\partial\tilde{P}} = 0, \qquad \frac{\partial\mathscr{L}}{\partial\tilde{Q}} = 0 \tag{37}$$

$$\frac{\partial\mathscr{L}}{\partial A_c} = 0, \qquad \frac{\partial\mathscr{L}}{\partial B_c} = 0, \qquad \frac{\partial\mathscr{L}}{\partial C_c} = 0 \tag{38}$$

Definition 2. A compensator (A_c,B_c,C_c) is an *extremal* of the optimal fixed-order dynamic compensation problem if it satisfies the stationary conditions [Eqs. (37) and (38)].

Definition 3. A compensator (A_c,B_c,C_c) is an *admissible extremal* of the optimal fixed-order dynamic compensation problem if it is an extremal and is also in $\mathscr{S}^+_{n_c}$.

Remark 2. The optimal (admissible) fixed-order dynamic compensator (if it exists) can be found by computing all admissible extremals.

We can now state the basic result of optimal projection theory, namely, the necessary conditions for characterizing admissible extremals of the optimal fixed-order dynamic compensation problem. For convenience define

$$\Sigma \triangleq BR_2^{-1}B^T, \qquad \bar{\Sigma} \triangleq C^TV_2^{-1}C \tag{39}$$

Theorem 1. Suppose (A_c,B_c,C_c) is an admissible extremal of the fixed-order dynamic compensation problem. Then there exist non-negative-definite matrices Q, P, $\hat{Q}$, and $\hat{P}$ such that A_c, B_c, and C_c are given by

$$A_c = \Gamma(A - Q\bar{\Sigma} - \Sigma P)G^T \tag{40}$$

$$B_c = \Gamma QC^TV_2^{-1} \tag{41}$$

$$C_c = -R_2^{-1}B^TPG^T \tag{42}$$

for some projective factorization (G,M,Γ) of $\hat{Q}\hat{P}$ and such that the following conditions are satisfied:

$$0 = AQ + QA^T + V_1 - Q\bar{\Sigma}Q + \tau_\perp Q\bar{\Sigma}Q\tau_\perp^T \tag{43}$$

$$0 = A^TP + PA + R_1 - P\Sigma P + \tau_\perp^T P\Sigma P\tau_\perp \tag{44}$$

$$0 = (A - \Sigma P)\hat{Q} + \hat{Q}(A - \Sigma P)^T + Q\bar{\Sigma}Q - \tau_\perp Q\bar{\Sigma}Q\tau_\perp^T \tag{45}$$

$$0 = (A - Q\bar{\Sigma})^T\hat{P} + \hat{P}(A - Q\bar{\Sigma}) + P\Sigma P - \tau_\perp^T P\Sigma P\tau_\perp \tag{46}$$

$$\operatorname{rank} \hat{Q} = \operatorname{rank} \hat{P} = \operatorname{rank} \hat{Q}\hat{P} = n_c \tag{47}$$

$$\tau = \hat{Q}\hat{P}(\hat{Q}\hat{P})^{\#} \tag{48}$$

Furthermore, the extremal cost is given by

$$J(A_c,B_c,C_c) = \operatorname{tr}[QR_1 + P(Q\bar{\Sigma}Q - \tau_\perp Q\bar{\Sigma}Q\tau_\perp^T)] \tag{49}$$

or, equivalently,

$$J(A_c,B_c,C_c) = \operatorname{tr}[PV_1 + Q(P\Sigma P - \tau_\perp^T P\Sigma P\tau_\perp)] \tag{50}$$

Remark 3. Notice that in the full-order case $n_c = n$ it follows that $\tau = I_n$; hence, without loss of generality one can choose $G = \Gamma = I_n$. In this case Eqs. (43) and (44) reduce to the standard observer and regulator Riccati equations, and Eqs. (40–42) yield the usual LQG expressions. Finally, in the full-order case $\tau_\perp = 0$, and Eqs. (49) and (50) reduce to Eq. (5.77) of Ref. 62.

Remark 4. The alternative projective factorization (S^TG, $S^{-1}MS$, $S^{-1}\Gamma$) of $\hat{Q}\hat{P}$ corresponds to a change of basis of the compensator, i.e., replacing (A_c,B_c,C_c) by ($S^{-1}A_cS$, $S^{-1}B_c$, C_cS). Hence, all compensator realizations equivalent to Eqs. (40–42) are effectively characterized by the optimal projection equations.

Remark 5. The plant state covariance Q_1, the controller state covariance Q_2, and the cross covariance of the plant and controller states Q_{12} are given by

$$Q_1 = Q + \hat{Q}, \qquad Q_{12} = \hat{Q}\Gamma^T, \qquad Q_2 = \Gamma\hat{Q}\Gamma^T \tag{51}$$

Similarly, the corresponding partitions of $\tilde{P}$ are given by

$$P_1 = P + \hat{P}, \qquad P_{12} = -\hat{P}G^T, \qquad P_2 = G\hat{P}G^T \tag{52}$$

The optimal projection equations [Eqs. (43–46)] are coupled by the terms $\tau_\perp Q\Sigma Q\tau_\perp^T$ and $\tau_\perp^T P\Sigma P\tau_\perp$. If these coupling terms were not present, then Eqs. (43) and (44) could be solved independently of Eqs. (45) and (46). The "strength" of these coupling terms is dependent on R_2 and V_2, which essentially govern regulator and estimator authority relative to the state weighting R_1 and plant disturbance intensity V_1. Hence, if $\|R_1\| \gg \|R_2\|$ and $\|V_1\| \gg \|V_2\|$, then one would expect the terms $\tau_\perp Q\bar{\Sigma}Q\tau_\perp^T$ and $\tau_\perp^T P\Sigma P\tau_\perp$ to lead to strong coupling and hence play a significant role in determining the optimal compensator gains. The strength of these coupling terms is also affected by the choice of n_c. If, for example, $n_c \approx n$, then rank $\tau_\perp = n - n_c \approx 0$ so that $\tau_\perp Q\bar{\Sigma}Q\tau_\perp^T$ is reduced relative to $Q\bar{\Sigma}Q$. However, if $n_c \ll n$, then rank $\tau_\perp = n - n_c \approx n$ so that the term $\tau_\perp Q\bar{\Sigma}Q\tau_\perp^T$ is comparable to $Q\bar{\Sigma}Q$. Note also that the plus sign associated with this term shows that it counteracts the usual quadratic term $Q\bar{\Sigma}Q$ in accordance with the excluded portion of the compensator.

The next result is a restatement of *Theorem* 1 in the form of sufficient conditions. Furthermore, conditions are given to guarantee closed-loop stability.

Theorem 2. Suppose there exist nonnegative-definite matrices Q, P, $\hat{Q}$, and $\hat{P}$ satisfying Eqs. (43–48) and let A_c, B_c, and C_c be given by Eqs. (40–42). Then the compensator (A_c,B_c,C_c) is an extremal of the optimal fixed-order dynamic compensation problem. Furthermore, the following are equivalent:

$$\tilde{A} \text{ is stable} \tag{53}$$

$$(\tilde{A},\tilde{D}) \text{ is stabilizable} \tag{54}$$

$$(\tilde{A},\tilde{R}^{1/2}) \text{ is detectable} \tag{55}$$

In addition,

$$(A_c,B_c) \text{ is controllable} \Leftrightarrow A_c + B_cCG^T \text{ is stable} \tag{56}$$

$$(A_c,C_c) \text{ is observable} \Leftrightarrow A_c - \Gamma BC_c \text{ is stable} \tag{57}$$

Proof. That the compensator (A_c, B_c, C_c) is an extremal follows immediately from the proof of *Theorem* 1 presented in Refs. 7 and 56. It is also shown in Refs. 7 and 56 that, if nonnegative-definite Q, P, $\hat{Q}$, and $\hat{P}$ satisfy Eqs. (43–48) and the compensator (A_c, B_c, C_c) is given by Eqs. (40–42), then (independent of the stability of $\tilde{A}$) there exist $(n+n_c) \times (n+n_c)$ real matrices $\tilde{P}$ and $\tilde{Q}$ satisfying Eqs. (11) and (12) and having partitioned forms [Eqs. (15) and (16)] with the partitions satisfying Eqs. (51) and (52). It then follows that $\tilde{Q}$ and $\tilde{P}$ can be expressed as

$$\tilde{Q} = \begin{bmatrix} Q & 0 \\ 0 & 0 \end{bmatrix} + \begin{bmatrix} I_n \\ \Gamma \end{bmatrix} \hat{Q} [I_n \quad \Gamma^T] \tag{58}$$

$$\tilde{P} = \begin{bmatrix} P & 0 \\ 0 & 0 \end{bmatrix} + \begin{bmatrix} -I_n \\ G \end{bmatrix} \hat{P} [-I_n \quad G^T] \tag{59}$$

Thus,

$$\tilde{Q} \geqslant 0, \qquad \tilde{P} \geqslant 0 \tag{60}$$

Obviously Eq. (53) implies Eqs. (54) and (55). To show that Eqs. (55) and (56) imply Eq. (53), use Eqs. (11) and (12) and *Lemma* 12.2 of Ref. 86.

Next it follows from (*vii*) of *Lemma* 1 that the (2,2) block Q_2 of $\tilde{Q}$ satisfies

$$Q_2 = \Gamma \hat{Q} \Gamma^T > 0 \tag{61}$$

and it is shown in Ref. 7 that Q_2 satisfies

$$0 = (A_c + B_c C G^T) Q_2 + Q_2 (A_c + B_c C G^T)^T + B_c V_2 B_c^T \tag{62}$$

The equivalence [Eq. (56)] then follows from Eqs. (61), (62), and *Lemmas* 2.1 and 12.2 of Ref. 86. The equivalence [Eq. (57)] follows in similar fashion by noting that the (2,2) block P_2 of $\tilde{P}$ satisfies

$$P_2 > 0 \tag{63}$$

and

$$0 = (A_c - \Gamma B C_c)^T P_2 + P_2 (A_c - \Gamma B C_c) + C_c^T R_2 C_c \tag{64}$$

Theorem 1 states conditions that all *admissible* extremals must satisfy. *Theorem* 2 includes a converse result which shows that each compensator (A_c, B_c, C_c) that satisfies the conditions of *Theorem* 1 is an extremal. Since it is possible that multiple extrema exist, the optimal projection equations may have multiple solutions. However, it is now conjectured that in many practical situations only one solution exists to the optimal projection equations.

Numerical Solution of the Optimal Projection Equations

Despite the analytical insights gained from the optimal projection equations, this approach to controller design would be little more than a theoretical curiosity without concrete, practical benefits. That such benefits are forthcoming will be demonstrated in the next section, where comparisons are made to suboptimal controller reduction methods. However, before doing so we will provide a brief overview of numerical algorithms for solving the optimal projection equations.

Essentially, two distinct approaches have been developed for solving the optimal projection equations: iterative refinement of the projection and homotopic continuation. We shall make no attempt to provide an in-depth description of these algorithms since to do so would require a lengthy development. Rather, a high-level summary will be given to convey the nature of the approaches.

The iterative refinement method, as its name implies, attempts to satisfy the four coupled equations by progressively refining the projection matrix until a solution is achieved.[4,6,11] The basic idea is to initialize $\tau = I_n$ so that the starting solution is the full-order LQG solution. Once $\hat{Q}$ and $\hat{P}$ are computed, the key step is to select τ to be a sum of eigenprojections of $\hat{Q}\hat{P}$ [see Eq. (22)]. Note that at this stage $\hat{Q}$ and $\hat{P}$ will generally have full rank so that the rank conditions [Eq. (47)] are not satisfied. There are essentially two methods for selecting eigenprojections: balancing or component cost analysis. Balancing selects the eigenprojections of $\hat{Q}\hat{P}$ in accordance with the largest eigenvalues of $\hat{Q}\hat{P}$. Roughly speaking, this choice truncates from the compensator those states that are least controllable and observable. The second approach to eigenprojection selection is via component cost analysis. This approach is somewhat more relevant to the optimization problem since it deletes those eigenprojections that contribute most greatly to the cost. Pertinent references to both balancing and component cost analysis will be given in the next section in connection with controller reduction methods.

The second numerical approach to solving the optimal projection equations is based on homotopic continuation methods. To illustrate such methods, suppose we wish to solve $f(x) = 0$ where $f : \mathbb{R}^n \to \mathbb{R}^n$. The idea is to replace this problem with a simpler problem whose solution is known. For example, let $x_0 \in \mathbb{R}^n$ be arbitrary, let $\alpha \in [0,1]$, and define the homotopy function

$$H(x(\alpha),\alpha) \triangleq \alpha f(x(\alpha)) + (1-\alpha)[f(x(\alpha)) - f(x_0)] \tag{66}$$

For $\alpha = 0$ the problem $H(x(0),0) = 0$ has a known solution, namely, $x(0) = x_0$. However, for $\alpha = 1$, solutions to the problem $H(x(1),1) = 0$ are precisely solutions to the original problem $f(x) = 0$ with $x = x(1)$. Once a homotopy function has been chosen the objective is to follow a path from the initial solution of $H(x(0),0) = 0$ to the desired solution of $H(x(1),1) = 0$. This can be done by integrating Davidenko's equation

$$\frac{\mathrm{d}x(\alpha)}{\mathrm{d}\alpha} = [H_x(x(\alpha),\alpha)]^{-1} H_\alpha(x(\alpha),\alpha), \qquad \alpha \in [0,1] \tag{67}$$

with the initial condition given by the known solution $x(0)$ of $H(x(0),0) = 0$. An alternative but related approach called a "discrete homotopy" involves defining a sequence of points

$$0 = \alpha_0 < \alpha_1 < ... < \alpha_N = 1$$

and solving $H(x(\alpha_{k+1}),\alpha_{k+1}) = 0$ given $x(\alpha_k)$ using a local iteration scheme. Although this approach avoids direct integration of Davidenko's equation, it does, nevertheless, follow the homotopy path from $x(0)$ to $x(1)$. References 88–101 describe the theoretical basis for homotopic continuation as well as a variety of applications.

Below we briefly describe a homotopy algorithm for solving the optimal projection equations developed in Refs. 10 and 13. This algorithm uses both discrete homotopy and continuous homotopy techniques. The overall algorithm is a discrete homotopy algorithm. However, the local iteration scheme that advances the discrete homotopy uses continuous homotopies to solve certain sets of nonlinear algebraic equations that are subsets of the optimal projection equations. We begin by introducing some additional notation and definitions.

For the homotopy parameter $\alpha \in [0,1]$ define

$$A(\alpha) \triangleq A_0 + f_1(\alpha)[A - A_0] \tag{68}$$

$$B(\alpha) \triangleq B_0 + f_2(\alpha)[B - B_0] \tag{69}$$

$$C(\alpha) \triangleq C_0 + f_3(\alpha)[C - C_0] \tag{70}$$

$$R_1(\alpha) \triangleq (R_1)_0 + f_4(\alpha)[R_1 - (R_1)_0] \tag{71}$$

$$R_2^{-1}(\alpha) \triangleq (R_2^{-1})_0 + f_5(\alpha)[R_2^{-1} - (R_2^{-1})_0] \tag{72}$$

$$V_1(\alpha) \triangleq (V_1)_0 + f_6(\alpha)[V_1 - (V_1)_0] \tag{73}$$

$$V_2^{-1}(\alpha) \triangleq (V_2^{-1})_0 + f_7(\alpha)[V_2^{-1} - (V_2^{-1})_0] \tag{74}$$

$$\Sigma(\alpha) \triangleq B(\alpha)R_2^{-1}(\alpha)B^T(\alpha) \tag{75}$$

$$\bar{\Sigma}(\alpha) \triangleq C^T(\alpha)V_2^{-1}(\alpha)C(\alpha) \tag{76}$$

where the functions $f_j(\;)$ are monotonically increasing functions on [0,1] satisfying

$$f_j(0) = 0 \qquad \text{and} \qquad f_j(1) = 1, \qquad j = 1,2, ..., 7 \tag{77}$$

and the initial matrices [i.e., the matrices expressed by the notation $(\;)_0$] are chosen according to the following general guidelines:

1) A_0, B_0, C_0, $(R_1)_0$, $(R_2^{-1})_0$, $(V_1)_0$, and $(V_2^{-1})_0$ should be chosen to yield a low authority controller;

2) V_0, C_0, $(R_2^{-1})_0$, and $(V_2^{-1})_0$ should also be chosen such that, for each $\alpha \in [0,1]$,

$$\mathscr{R}(\Sigma(\alpha)) = \mathscr{R}(\Sigma), \qquad \mathscr{R}(\bar{\Sigma}(\alpha)) = \mathscr{R}(\bar{\Sigma}) \tag{78}$$

where $\mathscr{R}(\ )$ denotes range.

Now, for some positive integer N let $\{\alpha_k\}_{k=0}^{N}$ be a sequence with the property

$$0 = \alpha_0 < \alpha_1 < \ldots < \alpha_{N-1} < \alpha_N = 1 \tag{79}$$

Equations (68–77) and (79) will be used to define a discrete homotopy that is the basis of the homotopy algorithm for solving the optimal projection equations.

Remark 6. The choice $f_j(\alpha) = \alpha$ for $j \in \{1,2,\ldots,7\}$ gives monotonically increasing functions satisfying the conditions (77).

For $i \in \{0,1,2,\ldots\}$ we will subsequently consider the equations

$$0 = A(\alpha)Q^{(i+1)} + Q^{(i+1)}A^T(\alpha) + V_1(\alpha) - Q^{(i+1)}\bar{\Sigma}(\alpha)Q^{(i+1)} + \tau_\perp^{(i)}Q^{(i+1)}\bar{\Sigma}(\alpha)Q^{(i+1)}(\tau_\perp^{(i)})^T \tag{80}$$

$$0 = A^T(\alpha)P^{(i+1)} + P^{(i+1)}A(\alpha) + R_1(\alpha) - P^{(i+1)}\Sigma(\alpha)P^{(i+1)} + (\tau_\perp^{(i)})^T P^{(i+1)}\Sigma(\alpha)P^{(i+1)}\tau_\perp^{(i)} \tag{81}$$

$$0 = [A(\alpha) - \Sigma(\alpha)P^{(i+1)}]\hat{Q}^{(i+1)} + \hat{Q}^{(i+1)}[A(\alpha) - \Sigma(\alpha)P^{(i+1)}]^T + Q^{(i+1)}\bar{\Sigma}(\alpha)Q^{(i+1)} - \tau_\perp^{(i+1)}Q^{(i+1)}\bar{\Sigma}(\alpha)Q^{(i+1)}(\tau_\perp^{(i+1)})^T \tag{82}$$

$$0 = [A(\alpha) - Q^{(i+1)}\bar{\Sigma}(\alpha)]^T\hat{P}^{(i+1)} + \hat{P}^{(i+1)}[A(\alpha) - Q^{(i+1)}\bar{\Sigma}(\alpha)] + P^{(i+1)}\Sigma(\alpha)P^{(i+1)} - (\tau_\perp^{(i+1)})^T P^{(i+1)}\Sigma(\alpha)P^{(i+1)}\tau_\perp^{(i+1)} \tag{83}$$

$$\operatorname{rank}\hat{Q}^{(i+1)} = \operatorname{rank}\hat{P}^{(i+1)} = \operatorname{rank}\hat{Q}^{(i+1)}\hat{P}^{(i+1)} = n_c \tag{84}$$

$$\tau^{(i+1)} = \hat{Q}^{(i+1)}\hat{P}^{(i+1)}[\hat{Q}^{(i+1)}\hat{P}^{(i+1)}]^{\#} \tag{85}$$

Also, define

$$\Delta_Q^{(i)}(\alpha) \triangleq A(\alpha)Q^{(i)} + Q^{(i)}A^T(\alpha) + V_1(\alpha) - Q^{(i)}\bar{\Sigma}(\alpha)Q^{(i)} + \tau_\perp^{(i)}Q^{(i)}\bar{\Sigma}(\alpha)Q^{(i)}(\tau_\perp^{(i)})^T \tag{86}$$

$$\Delta_P^{(i)}(\alpha) \triangleq A^T(\alpha)P^{(i)} + P^{(i)}A(\alpha) + R_1(\alpha) - P^{(i)}\Sigma(\alpha)P^{(i)} + (\tau_\perp^{(i)})^T P^{(i)}\Sigma(\alpha)P^{(i)}\tau_\perp^{(i)} \tag{87}$$

$$\Delta_{\hat{Q}}^{(i)}(\alpha) \triangleq [A(\alpha) - \Sigma(\alpha)P^{(i)}]\hat{Q}^{(i)} + \hat{Q}^{(i)}[A(\alpha) - \Sigma(\alpha)P^{(i)}]^T + Q^{(i)}\bar{\Sigma}(\alpha)Q^{(i)} - \tau_\perp^{(i)}Q^{(i)}\bar{\Sigma}(\alpha)Q^{(i)}(\tau_\perp^{(i)})^T \tag{88}$$

$$\Delta_{\hat{P}}^{(i)}(\alpha) \triangleq [A(\alpha) - Q^{(i)}\bar{\Sigma}(\alpha)]^T\hat{P}^{(i)} + \hat{P}^{(i)}[A(\alpha) - Q^{(i)}\bar{\Sigma}(\alpha)] + P^{(i)}\Sigma(\alpha)P^{(i)} - (\tau_\perp^{(i)})^T P^{(i)}\Sigma(\alpha)P^{(i)}\tau_\perp^{(i)} \tag{89}$$

Equations (80) and (81) are identical in form to Eqs. (43) and (44), and Eqs. (82–85) are identical in form to Eqs. (45–48).

In the homotopy algorithm the matrix functions $\Delta_Q^{(i)}(\alpha)$, $\Delta_P^{(i)}(\alpha)$, $\Delta_{\hat{Q}}^{(i)}(\alpha)$, and $\Delta_{\hat{P}}^{(i)}(\alpha)$, given by Eqs. (86), (87), (88), and (89), respectively, are considered equation errors. The normalized equation errors norms are defined by

$$e_Q^{(i)}(\alpha) \triangleq \|\Delta_Q^{(i)}(\alpha)\|/\|V_1(\alpha)\| \tag{90}$$

$$e_P^{(i)}(\alpha) \triangleq \|\Delta_P^{(i)}(\alpha)\|/\|R_1(\alpha)\| \tag{91}$$

$$e_{\hat{Q}}^{(i)}(\alpha) \triangleq \|\Delta_{\hat{Q}}^{(i)}(\alpha)\|/\|Q^{(i)}\bar{\Sigma}(\alpha)Q^{(i)}\| \tag{92}$$

$$e_{\hat{P}}^{(i)}(\alpha) \triangleq \|\Delta_{\hat{P}}^{(i)}(\alpha)\|/\|P^{(i)}\Sigma(\alpha)P^{(i)}\| \tag{93}$$

where $\| \;\|$ denotes the maximum absolute value of the matrix elements. The maximum normalized error norm $e_{\max}^{(i)}(\alpha)$ is defined by

$$e_{\max}^{(i)}(\alpha) \triangleq \max\{e_Q^{(i)}(\alpha), e_P^{(i)}(\alpha), e_{\hat{Q}}^{(i)}(\alpha), e_{\hat{P}}^{(i)}(\alpha)\} \tag{94}$$

Figure 1 presents a general flowchart of the homotopy algorithm. The outer (k) loop corresponds to a discrete homotopy that is advanced by the iteration scheme described by the inner (i) loop. The inner loop requires the solution of Eqs. (80) and (81) and then the simultaneous solution of Eqs. (82–85). These solutions can be obtained by using continuous homotopies.[13] The sequence $\{\epsilon_k\}_{k=0}^N$ of "small" positive numbers determines how closely the algorithm actually tracks the homotopy curve. The details of this algorithm are presented in Ref. 13.

We now present some remarks concerning the choice of matrices $(\;)_0$ used to initialize the homotopy algorithm. In practice we have found it sufficient to choose $B_0 = B$ and $C_0 = C$ [i.e., $B(\alpha) = B$ and $C(\alpha) = C$ for $\alpha \in [0,1]$], whereas for systems with lightly damped modes we have sometimes used $A_0 = A - \sigma D$ for some positive diagonal D and $\sigma \geqslant 0$. The

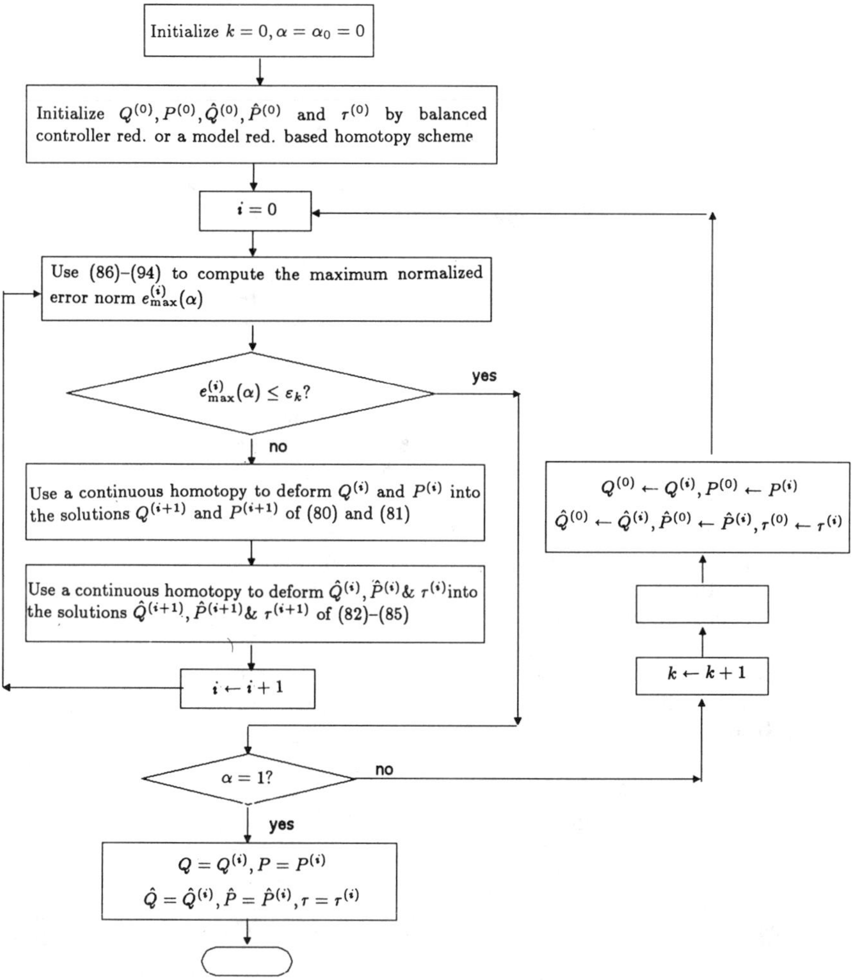

Fig. 1 The homotopy algorithm for solving the optimal projection equations uses both discrete homotopy and continuous homotopy techniques.

initial weighting matrices are usually selected by one of the following two schemes:

Scheme 1. Choose $(R_2^{-1})_0 = \rho_R R_2^{-1}$ and $(V_2^{-1})_0 = \rho_V V_2^{-1}$ for some $\rho_R < 1$ and $\rho_V < 1$, and let $(R_1)_0 = R_1$ and $(V_1)_0 = V_1$ (i.e., $\boldsymbol{R}_1(\alpha) = R_1$ and $\boldsymbol{V}_1(\alpha) = V_1$ for $\alpha \in [0,1]$).

Scheme 2. Choose $(R_1)_0 = \gamma_R R_1$ and $(V_1)_0 = \gamma_V V_1$ for some $\gamma_R < 1$ and $\gamma_V < 1$, and let $(R_2^{-1})_0 = R_2^{-1}$ and $(V_2^{-1})_0 = V_2^{-1}$ (i.e., $\boldsymbol{R}_2^{-1}(\alpha) = R_2^{-1}$ and $\boldsymbol{V}_2^{-1}(\alpha) = V_2^{-1}$ for $\alpha \in [0,1]$).

Relationships Between the Optimal Projection Approach and Model/Controller Reduction Techniques

Because of widespread interest in designing low-order controllers for high-order systems, numerous reduction methods have been proposed.[102–146] These methods usually involve either plant approximation prior to controller design or controller reduction once a high-order controller has been obtained (Fig. 2). The purpose of this section is to shed some light on the relationship between optimal projection theory and controller reduction methods.

In many applications, such as finite-element modeling of flexible structures, the models possess far too many degrees of freedom to be tractable for realistic control design. Hence, at least some form of model reduction is usually required to reduce the plant model to a tractable, perhaps "Riccati solvable" dimension. However, it is important to stress that reduction of the open-loop plant model will generally result in the loss of modeling information. Hence, it is desirable that this reduction step be limited as much as possible. Once a model of tractable order is obtained, it is still often necessary to design a controller of significantly lower order than the model. Controller reduction methods perform this step by first designing a high-order controller and then applying model reduction techniques to the controller itself. The principal drawback of this two-step approach is that guarantees of stability, optimality, robustness, and other desirable design objectives are often lacking.

However, it is possible to argue that such *indirect* controller reduction methods are still useful in practice. For example, reduction methods may be computationally simpler than direct methods and may give the user insight into the reduction procedure. However, such features are not surprising, since, as will be seen, at least some controller reduction methods are, in fact, suboptimal approximations to the optimal projection equations.

To demonstrate these connections, we shall consider three indirect methods of reduced-order controller design (for details see Ref. 6). The first method is the component cost analysis (CCA) approach.[134] In the notation of the section containing a brief overview of basic optimal projection theory, CCA can be summarized by the equations

$$0 = AQ + QA^T + V_1 - Q\bar{\Sigma}Q \tag{95}$$

$$0 = A^TP + PA + R_1 - P\Sigma P \tag{96}$$

$$0 = (A - \Sigma P)^T\hat{Q} + \hat{Q}(A - \Sigma P) + Q\bar{\Sigma}Q \tag{97}$$

$$\tau = \sum_{i=1}^{n_c} \Pi_i[\hat{Q}P\Sigma P] \tag{98}$$

As in optimal projection theory, the reduced-order controller gains are

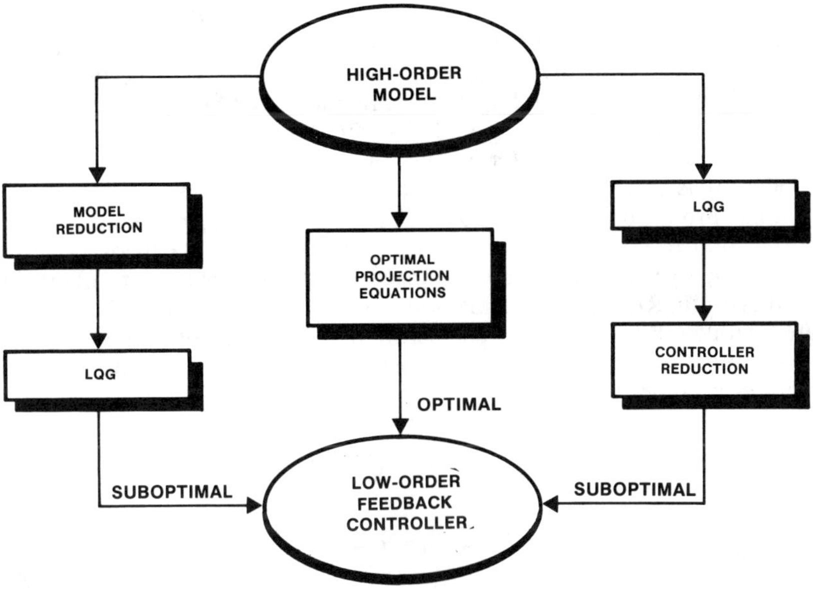

Fig. 2 There are three principal approaches to reduced-order controller design: model reduction followed by low-order controller design, high-order controller design followed by controller reduction, and direct optimization of the fixed-order compensator gains.

formed by factoring the projection τ by means of $\tau = G^T\Gamma$, $\Gamma G^T = I_{n_c}$, and forming the gains [Eqs. (40–42)].

Within this common framework it can immediately be seen that CCA lacks the fourth equation for $\hat{P}$ present in optimal projection theory; furthermore, the remaining three equations lack coupling terms involving τ. Note also that CCA forms the projection by using $P\Sigma P$ in place of $\hat{P}$. This is not altogether unreasonable since $P\Sigma P$ is essentially the nonhomogeneous driving term in the $\hat{P}$ equation [Eq. (46)].

Next we turn to the balanced controller reduction algorithm (BCRA), which corresponds to the application of internal balancing model reduction[122] to an LQG controller. Again, in the previous notation, this method can be written as

$$0 = AQ + QA^T + V_1 - Q\bar{\Sigma}Q \tag{99}$$

$$0 = A^TP + PA + R_1 - P\Sigma P \tag{100}$$

$$0 = (A - Q\bar{\Sigma} - \Sigma P)\hat{Q} + \hat{Q}(A - Q\bar{\Sigma} - \Sigma P)^T + Q\bar{\Sigma}Q \tag{101}$$

$$0 = (A - Q\bar{\Sigma} - \Sigma P)^T\hat{P} + \hat{P}(A - Q\bar{\Sigma} - \Sigma P) + P\Sigma P \tag{102}$$

$$\tau = \sum_{i=1}^{n_c} \Pi_i[\hat{Q}\hat{P}] \tag{103}$$

where τ is again factored as $\tau = G^T\Gamma$, $\Gamma G^T = I_{n_c}$, and the reduced-order controller gains are given by Eqs. (40–42). Note that this method, in contrast to CCA, involves four equations as in optimal projection theory. Again, however, the coupling terms present in optimal projection theory are absent. Note also that $\hat{Q}$ and $\hat{P}$ in BCRA are essentially the controllability and observability Gramians for the LQG compensator since this coefficient is $A + BC_c - B_cC$. In contrast, the coefficients of $\hat{Q}$ and $\hat{P}$ in optimal projection theory are $A - B_cC$ and $A + BC_c$, respectively, which correspond to estimator and regulator dynamics in LQG theory.

Finally, we consider the balanced controller reduction algorithm-modified (BCRAM).[140] This method corrects an obvious drawback associated with BCRA in that the coefficient $A - Q\bar{\Sigma} - \Sigma P$ of $\hat{Q}$ and $\hat{P}$ in Eqs. (101) and (102) may not be asymptotically stable. In our notation this method becomes

$$0 = AQ + QA^T + V_1 - Q\bar{\Sigma}Q \tag{104}$$

$$0 = A^TP + PA + R_1 - P\Sigma P \tag{105}$$

$$0 = (A - \Sigma P)\hat{Q} + \hat{Q}(A - \Sigma P)^T + Q\bar{\Sigma}Q \tag{106}$$

$$0 = (A - Q\bar{\Sigma})^T\hat{P} + \hat{P}(A - Q\bar{\Sigma}) + P\Sigma P \tag{107}$$

$$\tau = \sum_{i=1}^{n_c} \Pi_i[\hat{Q}\hat{P}] \tag{108}$$

with the usual factorization of τ and gain expressions. Note that the coefficients of $\hat{Q}$ and $\hat{P}$, namely, $A - \Sigma P$ and $A - Q\bar{\Sigma}$, are now guaranteed to be asymptotically stable since they correspond to regulator and observer dynamics of LQG theory. Clearly, this method is most similar to optimal projection theory since it lacks only the coupling terms. Indeed, if the optimal projection equations are solved by means of an iterative substitution method with $\tau = I_n$ as the starting projection (see the previous section), then the first iteration of this algorithm corresponds to BCRAM. Of course, the result of this first iteration cannot be expected to be optimal, or even stabilizing. However, further iterations can lead to progressive refinement of this initial BCRAM design.

Hence, we have seen that three controller reduction methods can be viewed as suboptimal approximations to the optimal projection equations. Each method involves the construction of a projection matrix whose factorization leads to a projection of the LQG controller. However, it should be stressed that the optimal projection controller is not a projection of the full-order controller but rather is obtained by solving a system of four matrix equations with additional coupling terms not present in the suboptimal methods.

An alternative indirect approach to controller reduction is the LQG-balancing method.[124,133] This method was discussed in Ref. 140, where it was noted that it may fail to yield a minimal order compensator when one exists. Although this property was not, in fact, claimed in Refs. 124 and 133, this fact serves as a further illustration of the drawbacks inherent in indirect methods. For optimal projection theory, *Theorem* 2 shows that the rank conditions [Eq. (47)] on $\hat{Q}$ and $\hat{P}$ are directly related to the minimality of (A_c,B_c,C_c). To illustrate this point, consider the example of Ref. 140:

$$A=\begin{bmatrix}0 & -1\\ 1 & 0\end{bmatrix},\qquad B=\begin{bmatrix}1\\ 0\end{bmatrix},\qquad C=[1\quad -1]$$

$$R_1=\begin{bmatrix}4 & 0\\ 0 & 0\end{bmatrix},\qquad R_2=1,\qquad V_1=\begin{bmatrix}1 & -1\\ -1 & 1\end{bmatrix},\qquad V_2=1$$

Then the LQG controller is found to be

$$A_c=\begin{bmatrix}-3 & 0\\ 5 & -4\end{bmatrix},\qquad B_c=\begin{bmatrix}1\\ -4\end{bmatrix},\qquad C_c=[-2\quad 0]$$

However, this realization is not observable, and a minimal realization is given by

$$A_c=-3,\qquad B_c=1,\qquad C_c=-2$$

This nonminimality is not surprising since letting $n_c=2$ and $\tau=I_2$ yields

$$Q=\begin{bmatrix}1 & 0\\ 0 & 4\end{bmatrix},\qquad P=\begin{bmatrix}2 & 0\\ 0 & 2\end{bmatrix}$$

$$\hat{Q}=\begin{bmatrix}\frac{17}{4} & -8\\ -8 & \frac{65}{4}\end{bmatrix},\qquad \hat{P}=\begin{bmatrix}2 & 0\\ 0 & 0\end{bmatrix}$$

Note that since rank $\hat{P}=1$ the rank conditions [Eq. (47)] are not satisfied. However, if $n_c=1$, then

$$Q=\begin{bmatrix}1 & 0\\ 0 & \alpha\end{bmatrix},\qquad P=\begin{bmatrix}2 & 0\\ 0 & 2\end{bmatrix}$$

$$\hat{Q}=\frac{17}{4}\begin{bmatrix}1 & \beta\\ \beta & \beta^2\end{bmatrix},\qquad \hat{P}\begin{bmatrix}2 & 0\\ 0 & 0\end{bmatrix}$$

$$\tau=\begin{bmatrix}1 & 0\\ \beta & 0\end{bmatrix}$$

where

$$\alpha = \frac{17}{4} + \frac{16}{17}, \qquad \beta = \frac{-32}{17}$$

Now rank $\hat{Q} = \text{rank}\hat{P} = \text{rank}\hat{Q}\hat{P} = 1$ as required by Eq. (47).

Numerical Comparisons

We now consider some numerical comparisons for several additional LQG reduction methods. A convenient starting point for such comparisons is the study carried out in Ref. 144. The problem considered in Ref. 144, which originated in Ref. 147, involves multiple disks mounted on a flexible shaft with a noncollocated sensor/actuator pair. Because of noncollocation, the system has a complex pair of nonminimum phase zeros. The problem data are

$$A = \begin{bmatrix} -0.161 & 1 & 0 & 0 & 0 & 0 & 0 & 0 \\ -6.004 & 0 & 1 & 0 & 0 & 0 & 0 & 0 \\ -0.5822 & 0 & 0 & 1 & 0 & 0 & 0 & 0 \\ -9.9835 & 0 & 0 & 0 & 1 & 0 & 0 & 0 \\ -0.4073 & 0 & 0 & 0 & 0 & 1 & 0 & 0 \\ -3.982 & 0 & 0 & 0 & 0 & 0 & 1 & 0 \\ 0 & 0 & 0 & 0 & 0 & 0 & 0 & 1 \\ 0 & 0 & 0 & 0 & 0 & 0 & 0 & 0 \end{bmatrix} \qquad B = \begin{bmatrix} 0 \\ 0 \\ 0.0064 \\ 0.00235 \\ 0.0713 \\ 1.0002 \\ 0.1045 \\ 0.9955 \end{bmatrix}$$

$$C = [1 \quad 0 \quad 0 \quad 0 \quad 0 \quad 0 \quad 0 \quad 0]$$

$$R_1 = (1.0 \times 10^{-6}) E_1^T E_1,$$

$$E_1 = [0 \quad 0 \quad 0 \quad 0 \quad 0.55 \quad 11 \quad 1.32 \quad 18.0],$$

$$R_2 = 1$$

$$V_1 = q_2 B B^T, \qquad V_2 = 1$$

The parameter q_2 is a scale factor for the plant disturbance noise used to adjust controller authority. The methods compared in Ref. 144 include the following:

1) Enns: a frequency-weighted balancing technique[137];
2) Glover: an optimal Hankel norm approximation method[138];
3) Davis and Skelton: an extension of balancing to address unstable controllers[139];
4) Yousuff and Skelton: the BCRAM method discussed earlier[140]; and
5) Liu and Anderson: an approximation technique applied to the component parts of a fractional representation of the controller.[144]

Table 1 summarizes the results reported in Ref. 144 along with the results of an optimal projection study performed in Ref. 11. Clearly, all of the LQG reduction methods experience increasing difficulty as controller order decreases and controller authority increases (Table 2). This is not unexpected, since, as discussed in the overview, the coupling terms involving $\tau_\perp$ play an increasingly important role. However, these terms have no counterpart in indirect methods.

Note that, although the indirect methods failed to reliably stabilize the plant with reduced-order controllers, the optimal projection approach succeeded in every case considered. This is not surprising since *Theorem* 2 *guarantees* closed-loop stability in the presence of a generically satisfied disturbability condition. Alternatively, stability is guaranteed by the fact that the numerical solution of the optimal projection equations effectively corresponds to the construction of the Liapunov function $V(\tilde{x}) = \tilde{x}^T\tilde{P}\tilde{x}$, where $\tilde{x}$ is the closed-loop state and $\tilde{P}$ satisfies Eq. (12). In addition, optimal projection theory chooses compensator gains to minimize the quadratic cost subject to the satisfaction of the Liapunov constraint [Eq. (12)]. These observations thus clarify the guarantees inherent in optimal projection theory vs the lack thereof in indirect methods.

More recently, an additional improved indirect method involving controller canonical correlation coefficients C^4 was proposed in Ref. 145. This method was also applied to the example of Ref. 147 considered in Ref. 144 with q_2 now as large as 10^6. This new method proved to be more reliable with regard to stability even with larger values of q_2. A corresponding study in Ref. 12 yielded uniform improvement in quadratic costs—not a surprising result since optimal projection theory is based on quadratic optimality (see Table 3).

The differences between optimal projection theory and the C^4 method can be seen more clearly by considering the results for the lightly damped beam example addressed in Ref. 145. The data for this problem are

$$A = \text{block diag}_{i=1,\ldots,5}\begin{bmatrix} 0 & 1 \\ -\omega_i^2 & -2\zeta\omega_i \end{bmatrix}$$

$$\zeta = 0.005, \qquad \omega_i = i^2, \qquad i = 1, \ldots, 5$$

$$B = [0 \quad 0.9877 \quad 0 \quad -0.3090 \quad 0 \quad -0.8910 \quad 0 \quad -0.5878 \quad 0 \quad 0.7071]^T$$

$$C = [0.9877 \quad 0 \quad 0.3090 \quad 0 \quad -0.8910 \quad 0 \quad -0.5878 \quad 0 \quad 0.7071 \quad 0]$$

$$R_1 = C^TC, \qquad R_2 = \rho, \qquad V_1 = BB^T, \qquad V_2 = 0.1$$

where ρ is the design parameter. Results for both the C^4 method and optimal projection theory are shown in Fig. 3. Note that at higher authority levels the suboptimal C^4 method exhibits degraded performance

Table 1 LQG reduction methods applied to example problems to determine the stability of the resulting closed-loop system

		$q_2 =$						
Method	n_c	0.01	0.1	1	10	100	1000	2000
Enns	7	S	S	S	S	S	S	S
	6	S	S	S	S	S	S	S
	5	S	S	S	S	S	S	S
	4	S	S	S	S	S	S	U
	3	S	S	S	S	S	S	S
	2	S	S	S	S	U	U	U
Glover	7	S	S	S	S	S	U	S
	6	S	S	S	S	U	U	U
	5	S	S	S	S	U	U	U
	4	S	S	S	S	U	U	U
	3	S	S	U	S	U	U	U
	2	S	U	S	U	S	U	U
Davis and Skelton	7	S	U	U	S	S	S	S
	6	S	S	S	S	S	S	S
	5	S	U	S	S	S	U	U
	4	S	S	U	S	S	U	U
	3	U	U	U	U	U	U	U
	2	S	U	S	U	U	U	U
Yousuff and Skelton	7	S	S	S	S	U	U	U
	6	S	S	S	S	U	U	U
	5	S	S	S	U	U	U	U
	4	S	S	S	U	U	U	U
	3	S	U	U	U	U	U	U
	2	S	S	S	U	U	U	U
Liu and Anderson	7	S	S	S	S	S	S	U
	6	S	S	S	S	S	S	U
	5	S	S	S	S	S	S	S
	4	S	S	S	S	S	S	S
	3	S	S	S	S	S	U	U
	2	S	S	S	S	S	S	S
Optimal projection	7	S	S	S	S	S	S	S
	6	S	S	S	S	S	S	S
	5	S	S	S	S	S	S	S
	4	S	S	S	S	S	S	S
	3	S	S	S	S	S	S	S
	2	S	S	S	S	S	S	S

S, the closed-loop system is stable; U, the closed-loop system is unstable.
Only the optimal projection equations yielded stabilizing controllers in every case, even at high authority.

Table 2 Percentage totals of the controller reduction results reveal increasing instability at high authority and low controller order, whereas optimal projection theory guarantees closed-loop asymptotic stability

Method	$q_2 =$ 0.01	0.1	1	10	100	1000	2000	Total % for all cases
Enns	100	100	100	100	83.3	83.3	66.7	90.5
Glover	100	83.3	83.3	83.3	33.3	0	16.7	57.1
Davis and Skelton	83.3	33.3	50.0	66.7	66.7	33.3	33.3	52.4
Yousuff and Skelton	100	83.3	83.3	33.3	0	0	0	42.9
Liu and Anderson	100	100	100	100	100	83.3	50.0	90.5
Optimal projection	100	100	100	100	100	100	100	100.0

Table 3 Optimal projection theory yielded improved performance compared to the C^4 method

q_2	Controller order	Method	State cost (rms)	Control cost (rms)	Total cost (rms)
2000	8	LQG	8.45	12.7	15.3
	4	C^4	15.0	14.7	21.0
		OP	9.28	13.3	16.2
	2	C^4	15.1	14.7	21.0
		OP	9.83	13.5	16.8
10^6	8	LQG	174	277	327
	6	C^4	214	288	359
		OP	195	288	348
	5	C^4	363	376	522
		OP	203	295	358
	4	C^4	355	355	502
		OP	209	297	362
	3	C^4	986	857	1310
		OP	214	299	368
	2	C^4	349	343	489
		OP	220	303	375

tending toward instability. However, the optimal projection designs yield monotonically improving performance at higher authority levels. Note that for $n_c = 2,4$ the optimal projection designs exhibit an upper bound on allowable controller effort. Hence, these plots do not extend indefinitely to the right as in the $n_c = 6$ and full-order cases. These results thus reveal an inherent limitation on the ability of the controller to further improve the state cost as imposed by the controller order. This is precisely the source of difficulty encountered by the C^4 method: By attempting to reduce a high-authority LQG controller, it cannot appropriately curtail the controller effort in accordance with the order-imposed bounds.

Finally, it should be noted that comparisons between direct and indirect design approaches have also been carried out experimentally, for a truss structure with numerous flexible modes. LQG truncation was compared to optimal projection theory.[148,149] The experimental results obtained confirm qualitatively the differences discussed in this section between indirect methods and optimal projection theory.

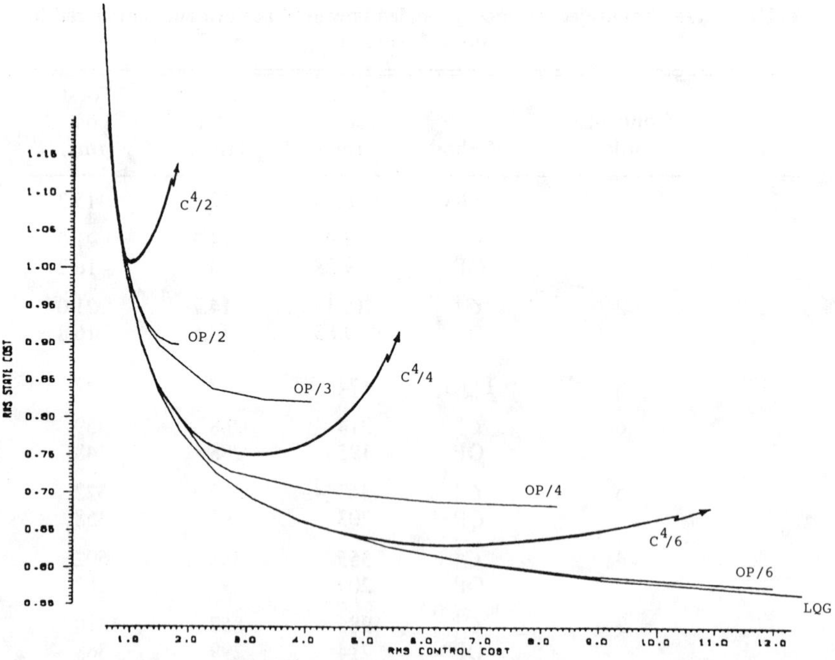

Fig. 3 A study of rms state error vs control effort reveals the tendency of C^4 controllers to destabilize at high authority, whereas optimal projection controllers exhibit monotonically improving performance.

Optimal Projection/H_∞ Theory: Frequency Domain Extensions

Although the development of LQG theory in the 1960s promised a truly multivariable design theory, it eventually became clear that LQG theory had numerous deficiencies in practical application. For example, LQG controllers lacked rudimentary forms of robustness such as gain and phase margins[150] and could be arbitrarily sensitive to parameter variations.[151,152] However, such deficiencies were not to be unexpected, since an LQG controller is constructed to optimize a single, narrowly defined performance measure, namely, a quadratic cost. Although a quadratic performance measure is physically relevant to many applications such as regulation with rms error specifications in the presence of persistent disturbances, the deficiencies of LQG design largely precluded its practical application. Hence, classical single-loop synthesis techniques (e.g., root locus, Bode, Nyquist, and Nichols) were in no danger of being dislodged by the newfangled quadratic machinery.

However, some relief was offered when it was shown that a quadratically optimal full-state feedback (LQR) controller has 6-dB/60 deg gain and phase margins.[153] However, full-state feedback controllers are only implementable in the case $C = I_n$. Thus, such regulators are inapplicable even to single-input single-output systems.

However, such restrictions notwithstanding, these LQR features suggest a remedy to the LQG defects: speed up the regulator and/or estimator dynamics to "recover" the LQR properties.[154–156] This approach is a specific application of a more general class of "loop shaping" procedures ultimately aimed at allowing performance/robustness tradeoffs.[157–159] By minimizing sensitivity at low frequency and complementary sensitivity at high frequency, it becomes possible to achieve both performance specifications and robustness to unmodeled dynamics.

However, such loop shaping procedures were predicated on a quadratic performance criterion. Thus, there were two drawbacks: 1) since the quadratic criterion concerns only the H_2 norm of the frequency response, the designer has only limited control over the loop shape; and 2) since the description of plant uncertainty is not consistent with the performance metric, it is difficult to derive robust performance bounds. Both difficulties were eventually overcome with the development of H_∞ control theory.[160–164] Since the H_∞ performance measure corresponds to worst-case frequency attenuation, it permits precise loop shaping and uncertainty characterization by means of dynamic weights.

Although H_∞ theory addressed "classical" control concerns and filled the gaps of H_2/loop shaping, it had one major drawback: it lacked the elegance of LQG synthesis via algebraic matrix Riccati equations. Until now.

Optimal projection theory was unified with H_∞ theory in Refs. 25 and 26. The technique by which this was done is quite simple and can be described as follows: Optimal projection design is based on minimizing

$$J(A_c,B_c,C_c) = \operatorname{tr} \tilde{Q}\tilde{R} \tag{109}$$

where $\tilde{Q}$ satisfies

$$0 = \tilde{A}\tilde{Q} + \tilde{Q}\tilde{A}^T + \tilde{V} \tag{110}$$

But suppose Eq. (110) is replaced by

$$0 = \tilde{A}\mathcal{Q} + \mathcal{Q}\tilde{A}^T + \gamma^{-2}\mathcal{Q}\tilde{R}_\infty\mathcal{Q} + \tilde{V} \tag{111}$$

where $\gamma > 0$ is a given constant, and $\tilde{R}_\infty \geqslant 0$. Then it is easy to show that

$$\tilde{Q} \leqslant \mathcal{Q} \tag{112}$$

Thus,

$$J(A_c,B_c,C_c) \leqslant \mathcal{J}(A_c,B_c,C_c) \tag{113}$$

where

$$\mathcal{J}(A_c,B_c,C_c) \triangleq \operatorname{tr}\mathcal{Q}\tilde{R} \tag{114}$$

The motivation for replacing Eq. (110) by Eq. (111) stems from the fact that, if Eq. (111) has a solution, then

$$\|\tilde{G}(s)\|_\infty \leqslant \gamma \tag{115}$$

where

$$\tilde{G}(s) \triangleq \tilde{E}_\infty(sI_{\tilde{n}} - \tilde{A})^{-1}\tilde{D}, \qquad \tilde{E}_\infty^T\tilde{E}_\infty = \tilde{R}_\infty, \qquad \tilde{D}\tilde{D}^T = \tilde{V} \tag{116}$$

The constraint (115) can then be used to enforce loop shaping, robustness, or frequency domain performance specifications.

To enforce Eq. (115), we thus need only rederive optimal projection theory with Eqs. (111) and (114) in place of Eqs. (110) and (109). In place of Eqs. (43–49) one then obtains[25,26]

$$0 = AQ + QA^T + V_1 + \gamma^{-2}QR_{1\infty}Q - Q\bar{\Sigma}Q + \tau_\perp Q\bar{\Sigma}Q\tau_\perp^T \tag{117}$$

$$\begin{aligned}0 = (A + \gamma^{-2}[Q + \hat{Q}]R_{1\infty})^TP + P(A + \gamma^{-2}[Q + \hat{Q}]R_{1\infty}) + R_1 \\ - S^TP\Sigma PS + \tau_\perp^TS^TP\Sigma PS\tau_\perp\end{aligned} \tag{118}$$

$$\begin{aligned}0 = (A - \Sigma PS + \gamma^{-2}QR_{1\infty})\hat{Q} + \hat{Q}(A - \Sigma PS + \gamma^{-2}QR_{1\infty})^T \\ + \gamma^{-2}\hat{Q}(R_{1\infty} + \beta^2S^TP\Sigma PS)\hat{Q} + Q\bar{\Sigma}Q - \tau_\perp Q\bar{\Sigma}Q\tau_\perp^T\end{aligned} \tag{119}$$

$$0 = (A - Q\bar{\Sigma} + \gamma^{-2}QR_{1\infty})^T\hat{P} + \hat{P}(A - Q\bar{\Sigma} + \gamma^{-2}QR_{1\infty}) + S^TP\Sigma PS - \tau_\perp^T S^TP\Sigma PS\tau_\perp \tag{120}$$

$$\text{rank }\hat{Q} = \text{rank }\hat{P} = \text{rank }\hat{Q}\hat{P} = n_c \tag{121}$$

$$\tau = \hat{Q}\hat{P}(\hat{Q}\hat{P})^{\#} \tag{122}$$

where the compensator gains are given by

$$A_c = \Gamma(A - Q\bar{\Sigma} - \Sigma PS + \gamma^{-2}QR_{1\infty})G^T \tag{123}$$

$$B_c = \Gamma QC^TV_2^{-1} \tag{124}$$

$$C_c = -R_2^{-1}B^TPSG^T \tag{125}$$

the H_2 performance is bounded by

$$J(A_c,B_c,C_c) \le \text{tr}[(Q + \hat{Q})R_1 + \hat{Q}S^TP\Sigma PS] \tag{126}$$

and

$$R_{1\infty} \triangleq E_{1\infty}^T E_{1\infty}, \qquad R_{2\infty} \triangleq E_{2\infty}^T E_{2\infty} = \beta^2 R_2, \qquad \tilde{E} \triangleq E_{1\infty} + E_{2\infty}C_c \tag{127}$$

$$S \triangleq (I_n + \beta^2\gamma^{-2}\hat{Q}P)^{-1} \tag{128}$$

The additional terms in Eqs. (117–120) guarantee that, if these equations have a solution, then the transfer function $\tilde{G}(s)$ from disturbances $w(t)$ to H_∞ performance variables $z_\infty(t) = E_{1\infty}x(t) + E_{2\infty}u(t)$ has H_∞ norm bounded by γ. Note formally that, if $\gamma \to \infty$, then all terms multiplied by γ^{-2} drop out and the usual H_2 optimal projection equations are recovered. Also, note that this result addresses a mixed-norm problem, since both H_2 and H_∞ objectives are treated.

In the full-order case $n_c = n$ the four optimal projection/H_∞ equations specialize to the three equations:

$$0 = AQ + QA^T + V_1 + \gamma^{-2}QR_{1\infty}Q - Q\bar{\Sigma}Q \tag{129}$$

$$0 = (A + \gamma^{-2}[Q + \tilde{Q}]R_{1\infty})^TP + P(A + \gamma^{-2}[Q + \hat{Q}]R_{1\infty}) + R_1 - S^TP\Sigma PS \tag{130}$$

$$0 = (A - \Sigma PS + \gamma^{-2}QR_{1\infty})\hat{Q} + \hat{Q}(A - \Sigma PS + \gamma^{-2}QR_{1\infty})^T + \gamma^{-2}\hat{Q}(R_{1\infty} + \beta^2S^TP\Sigma PS)\hat{Q} + Q\bar{\Sigma}Q \tag{131}$$

with gains

$$A_c = A - Q\bar{\Sigma} - \Sigma PS + \gamma^{-2}QR_{1\infty} \tag{132}$$

$$B_c = QC^TV_2^{-1} \tag{133}$$

$$C_c = -R_2^{-1}B^TPS \tag{134}$$

Specializing even further to the equalized weight case $R_1 = R_{1\infty}$, $R_2 = R_{2\infty}$, and transforming to a new variable Y_∞ yields the pair of equations:

$$0 = AQ + QA^T + V_1 + \gamma^{-2}QR_{1\infty}Q - Q\bar{\Sigma}Q \tag{135}$$

$$0 = A^TY_\infty + Y_\infty A + R_{1\infty} + \gamma^{-2}Y_\infty V_1 Y_\infty - Y_\infty \Sigma Y_\infty \tag{136}$$

where Q and Y_∞ satisfy

$$\rho(QY_\infty) < \gamma^2 \tag{137}$$

and with gains given by

$$A_c = A - Q\bar{\Sigma} - \Sigma(Y_\infty^{-1} - \gamma^{-2}Q)^{-1} + \gamma^{-2}QR_{1\infty} \tag{138}$$

$$B_c = QC^TV_2^{-1} \tag{139}$$

$$C_c = -R_{2\infty}^{-1}B^T(Y_\infty^{-1} - \gamma^{-2}Q)^{-1} \tag{140}$$

The special case of full-state feedback $u = Kx$ is given by

$$0 = A^TP + PA + R_{1\infty} + \gamma^{-2}PV_1P - P\Sigma P \tag{141}$$

with gain

$$K = -R_{2\infty}^{-1}B^TP \tag{142}$$

The Riccati equation approach to H_∞ control can be traced to Refs. 165 and 166. Synthesis results for full-state feedback first appeared in Ref. 167 with extensions in Refs. 168 and 169. The full-order dynamic-compensation problem with equalized weights was treated in Refs. 170 and 171. We also note interesting connections with the exponential-of-quadratic cost problem,[172–174] whose solution involves essentially the same Riccati equations. Finally, connections between the H_2 cost bound $\mathcal{J}(A_c,B_c,C_c)$ and a frequency domain entropy functional are explored in Refs. 30 and 175–177.

The unification of optimal projection theory and H_∞ theory represents a potentially useful and flexible combination of approaches to controller synthesis. Clearly, although optimal projection theory alone lacks the means to meet frequency domain specifications, it brings to H_∞ theory the means to design low-order H_∞ controllers *directly*. Besides the plant order itself, low-order design is desirable within H_∞ theory to counteract the net increase in plant dimension due to the inclusion of dynamic weights for performing loop shaping and for bounding plant uncertainties.

Optimal Projection/Guaranteed Cost Theory: Parameter-Robust Extensions

H_∞ theory provides the means for guaranteeing robust stability and performance in the presence of unstructured modeling uncertainty, that is, uncertainty modeled by bounds on transfer function perturbations. However, if the uncertainty is more naturally expressed as matrix perturbations to a state space plant realization, then alternative techniques are required. In this section we discuss extensions of optimal projection theory to this problem involving Liapunov functions.

To illustrate the basis for our approach, consider the system

$$\dot{x}(t) = (A + \Delta A)x(t) + Dw(t), \qquad t \in [0,\infty) \tag{143}$$

$$z(t) = Ex(t) \tag{144}$$

where $x(t)$ is an n vector, A is an $n \times n$ matrix denoting the nominal dynamics matrix, ΔA denotes an uncertain perturbation of A belonging to a specified set $\mathcal{U}$, $Dw(t)$ is a white noise signal of intensity $V \triangleq DD^T$, and $z(t)$ is a vector of outputs. The initial condition $x(0)$ is unspecified, since it plays no role in the analysis of steady-state performance criteria.

For the system (143) the performance measure involves the steady-state second moment of the variables $z(t)$. In practice, the diagonal elements of the second moment are measures of the ability of the external disturbances $Dw(t)$ to excite specified states. In the presence of uncertainties ΔA, it is of interest to determine the *worst-case* steady-state values of the second moments of selected states. Thus, we define the scalar performance criterion

$$J(\mathcal{U}) \triangleq \sup_{\Delta A \in \mathcal{U}} \lim_{t \to \infty} \mathbb{E}[z^T(t)z(t)] \tag{145}$$

To evaluate Eq. (145), define the second-moment matrix

$$Q_{\Delta A}(t) \triangleq \mathbb{E}[x(t)x^T(t)] \tag{146}$$

which satisfies the Liapunov differential equation

$$\dot{Q}_{\Delta A}(t) = (A + \Delta A)Q_{\Delta A}(t) + Q_{\Delta A}(t)(A + \Delta A)^T + V \tag{147}$$

so that Eq. (145) becomes

$$J(\mathscr{U}) = \sup_{\Delta A \in \mathscr{U}} \lim_{t \to \infty} \operatorname{tr} Q_{\Delta A}(t) R \tag{148}$$

where $R \triangleq E^T E$. To guarantee both robust stability and performance, we consider modified algebraic Liapunov equations of the form

$$0 = AQ + QA^T + \Omega(Q) + V \tag{149}$$

where $\Omega(\ )$ is a matrix operator satisfying

$$\Delta A Q + Q \Delta A^T \leqslant \Omega(Q) \tag{150}$$

for all $\Delta A \in \mathscr{U}$ and all nonnegative-definite matrices Q. The ordering in Eq. (150) is defined with respect to the cone of nonnegative-definite matrices. Our results are based on the following robust stability and performance result. (For convenience, assume that V is positive-definite.) If there exists a positive-definite solution Q to Eq. (149), where $\Omega(\ )$ satisfies Eq. (150), then $A + \Delta A$ is asymptotically stable for all $\Delta A \in \mathscr{U}$. Furthermore, the worst-case performance can be bounded according to

$$J(\mathscr{U}) \leqslant \operatorname{tr} QR \tag{151}$$

The robust stability result is a direct consequence of Liapunov theory, whereas the performance bound (151) follows from the fact that since $A + \Delta A$ is asymptotically stable, $Q_{\Delta A} \triangleq \lim_{t \to \infty} Q_{\Delta A}(t)$ exists, is independent of $Q_{\Delta A}(0)$, and satisfies

$$0 = (A + \Delta A) Q_{\Delta A} + Q_{\Delta A}(A + \Delta A)^T + V \tag{152}$$

Now subtracting Eq. (152) from Eq. (149) yields

$$\begin{aligned} 0 = {} & (A + \Delta A)(Q - Q_{\Delta A}) + (Q - Q_{\Delta A})(A + \Delta A)^T \\ & + \Omega(Q) - (\Delta A Q + Q \Delta A^T) + V \end{aligned} \tag{153}$$

which, by Eq. (150) and the fact that $A + \Delta A$ is stable, implies

$$Q_{\Delta A} \leqslant Q \tag{154}$$

Now Eqs. (148) and (154) yield the bound (151).

Since the ordering induced by the cone of nonnegative-definite matrices in only a partial ordering, it should not be expected that there exists an operator $\Omega(\ )$ satisfying Eq. (150) that is a least upper bound. Indeed, there are many alternative definitions for the bound $\Omega(\ )$. To illustrate some of these alternatives, assume for convenience that ΔA is of the form

$$\Delta A = \sigma_1 A_1, \qquad |\sigma_1| \leqslant \Delta_1 \tag{155}$$

where σ_1 is an uncertain real scalar parameter assumed only to satisfy the stated bounds, and A_1 is a known matrix denoting the structure of the parametric uncertainty. The bound $\Omega(\ )$ utilized in Refs. 179 and 180 for full-state-feedback design was chosen to be

$$\Omega(Q) = \Delta_1 |A_1 Q + Q A_1^T| \tag{156}$$

where $|\ |$ denotes the nonnegative-definite matrix obtained by replacing each eigenvalue by its absolute value. More recently, the *quadratic* (in Q) bound

$$\Omega(Q) = \Delta_1 [A_L A_L^T + Q A_R^T A_R Q] \tag{157}$$

has been considered where A_L, A_R are a factorization of A_1 of the form $A_1 = A_L A_R$. A third bound that has been considered is the *linear* (in Q) bound

$$\Omega(Q) = \Delta_1 [\alpha Q + \alpha^{-1} A_1 Q A_1^T] \tag{158}$$

where α is an arbitrary positive scalar. A thorough development of bounds (156–158) as well as additional bounds for robustness analysis can be found in Ref. 55.

To make connections with the standard literature, we note that the procedure outlined earlier is closely related to the usual quadratic Liapunov function

$$V(x) = x^T P x \tag{159}$$

Since the matrix P can be characterized by means of an equation of the form

$$0 = A^T P + PA + R \tag{160}$$

it follows that the covariance Q can be viewed as providing a Liapunov function

$$\hat{V}(x) = x^T Q x \tag{161}$$

for the dual system, i.e., with A replaced by A^T. The reason for working with the dual setting is that it leads more directly to robust performance bounds.

Guaranteed Cost Synthesis

The development of Liapunov techniques for robust analysis and synthesis is quite extensive.[178–196] The bounding technique described earlier is useful for developing a robust optimal projection theory for two main reasons. First, the bounding technique yields an upper bound (guaranteed cost) for the worst-case performance and guarantees robust stability over the given uncertainty set $\mathcal{U}$. Second, the uncertainty bound can be directly incorporated into the design procedure by predicating the design optimization on the modified Liapunov equation [Eq. (149)] in place of the standard Liapunov equation.

Although the guaranteed cost approach originated in Refs. 178 and 180, it turns out that the bound (156) utilized there is unsuitable for optimal projection synthesis for the simple reason that $\Omega(\ )$ is not differentiable. However, both the linear and the quadratic bounds can be incorporated directly within optimal projection theory to characterize robust fixed-order controllers. The key to this development is to derive the optimality conditions for the cost bound (151). Controllers that minimize this "auxiliary cost" thus minimize a bound on worst-case performance and guarantee robust stability. We now briefly review this development for both the linear and quadratic bounds.

Linear Bound

The origins of the linear bound can be traced back to the stochastic control literature. To see the connections, suppose that the uncertainty ΔA is modeled as $v_1(t)A_1$, where $v_1(t)$ is a scalar white noise process. Then $v_1(t)$ multiplies x and hence is a multiplicative rather than additive noise input disturbance. The idea of representing an uncertain parameter by multiplicative noise is certainly heuristic since uncertain parameters rarely behave in this manner. However, we shall show that such an approach is indeed sound and, consequently, that such a heuristic, nonphysical model is, in a sense, equivalent to Liapunov bounds.

If $\Delta A = v_1(t)A_1$, where $v_1(t)$ is white noise, then it can be shown that under a second-moment stability condition the steady-state covariance equation is given by[197]

$$0 = AQ + QA^T + A_1QA_1^T + V \tag{162}$$

where V is the intensity of the additive white noise. To bridge the gap to robust stability and performance, we make one additional modification to Eq. (162); namely, we replace A by the right-shifted dynamics $A + \frac{1}{2}I_n$ to obtain

$$0 = (A + \tfrac{1}{2}I_n)Q + Q(A + \tfrac{1}{2}I_n)^T + A_1QA_1^T + V \tag{163}$$

It is now easy to show that the existence of a solution to Eq. (163) guarantees *deterministic* robust stability of $A + \Delta A$ for all ΔA of the form $\sigma_1 A_1$, where $\sigma_1 \in [-1,1]$ is an uncertain constant. Hence, the stochastic result provides a critical part of a deterministic bound that corresponds to a quadratic Liapunov function so that the stochastic interpretation of the bound can be disregarded in favor of a physically meaningful deterministic model. Consequently, this provides a rigorous robustness interpretation for controller synthesis methods based on multiplicative noise.[198–207] An extensive discussion of these issues can be found in Refs. 44 and 48. The linear bound was also considered in Ref. 195.

Optimal projection/guaranteed cost theory with the linear bound can now be summarized. In place of the plant and measurement model (1), (2), consider the perturbed model

$$\dot{x}(t) = (A + \Delta A)x(t) + (B + \Delta B)u(t) + D_1 w(t), \qquad t \in [0,\infty) \tag{164}$$

$$y(t) = (C + \Delta C)x(t) + Du(t) + D_2 w(t) \tag{165}$$

with the worst-case performance criterion

$$J(A_c,B_c,C_c) \triangleq \sup_{(\Delta A,\Delta B,\Delta C) \in \mathscr{U}} \lim_{t \to \infty} \mathbb{E}[x^T(t)R_1 x(t) + 2x^T(t)R_{12}u(t) + u^T(t)R_2 u(t)] \tag{166}$$

For the linear bound the uncertainty set is given by

$$\mathscr{U}_1 = \left\{(\Delta A,\Delta B,\Delta C)\colon \Delta A = \sum_{i=1}^{p} \sigma_i A_i, \quad \Delta B = \sum_{i=1}^{p} \sigma_i B_i, \quad \Delta C = \sum_{i=1}^{p} \sigma_i C_i, \quad \sum_{i=1}^{p} \sigma_i^2/\alpha_i^2 \leqslant 1\right\} \tag{167}$$

where, for $i = 1, \ldots, p$: $A_i \in \mathbb{R}^{n \times n}$, $B_i \in \mathbb{R}^{n \times m}$, and $C_i \in \mathbb{R}^{\ell \times n}$ are fixed matrices denoting the structure of the parametric uncertainty; α_i is a given positive number; and σ_i is an uncertain real parameter. Note that the uncertain parameters σ_i are assumed to lie in a specified ellipsoidal region in $\mathbb{R}^p$. Note also that the uncertainties $\Delta A,\Delta B,\Delta C$ can be correlated in the sense that a given uncertain parameter σ_i may appear in, say, both ΔA and ΔB. Of course, one can set $A_i = 0$, $B_i = 0$, or $C_i = 0$ if there is no correlation.

With this uncertainty model the closed-loop system is given by

$$\dot{\tilde{x}}(t) = (\tilde{A} + \Delta\tilde{A})\tilde{x}(t) + \tilde{D}w(t), \qquad t \in [0,\infty) \tag{168}$$

where

$$\tilde{x}(t) \triangleq \begin{bmatrix} x(t) \\ x_c(t) \end{bmatrix}, \quad \tilde{A} \triangleq \begin{bmatrix} A & BC_c \\ B_c C & A_c + B_c D C_c \end{bmatrix}, \quad \tilde{D} \triangleq \begin{bmatrix} D_1 \\ B_c D_2 \end{bmatrix} \tag{169}$$

$$\Delta\tilde{A} \triangleq \begin{bmatrix} \Delta A & \Delta B C_c \\ B_c\, \Delta C & 0 \end{bmatrix} = \sum_{i=1}^{p} \sigma_i \tilde{A}_i, \quad \tilde{A}_i = \begin{bmatrix} A_i & B_i C_c \\ B_c C_i & 0 \end{bmatrix}, \quad i = 1, \ldots, p \tag{170}$$

and where the closed-loop disturbance $\tilde{D}w(t)$ has the intensity

$$\tilde{V} \triangleq \begin{bmatrix} V_1 & B_c V_{12} \\ V_{12}^T B_c^T & B_c V_2 B_c^T \end{bmatrix} \tag{171}$$

To state the optimal projection/guaranteed cost synthesis result for the linear bound, let $\alpha > 0$ and define the notation

$$\gamma_i \triangleq \alpha_i^2/\alpha, \qquad A_\alpha \triangleq A + \frac{\alpha}{2} I_n \tag{172}$$

$$V_{2s} \triangleq V_2 + \sum_{i=1}^{p} \gamma_i C_i (Q + \hat{Q}) C_i^T, \qquad R_{2s} \triangleq R_2 + \sum_{i=1}^{p} \gamma_i B_i^T (P + \hat{P}) B_i \tag{173}$$

$$Q_s \triangleq QC^T + V_{12} + \sum_{i=1}^{p} \gamma_i A_i (Q + \hat{Q}) C_i^T,$$

$$P_s \triangleq B^T P + R_{12}^T + \sum_{i=1}^{p} \gamma_i B_i^T (P + \hat{P}) A_i \tag{174}$$

Theorem 3. Let $\alpha > 0$ and suppose there exist $Q, P, \hat{Q}, \hat{P} \geq 0$ satisfying

$$0 = A_\alpha Q + QA_\alpha^T + V_1 + \sum_{i=1}^{p} \gamma_i [A_i Q A_i^T + (A_i - B_i R_{2s}^{-1} P_s) \times \hat{Q}(A_i - B_i R_{2s}^{-1} P_s)^T] - Q_s V_{2s}^{-1} Q_s^T + \tau_\perp Q_s V_{2s}^{-1} Q_s^T \tau_\perp^T \tag{175}$$

$$0 = A_\alpha^T P + PA_\alpha + R_1 + \sum_{i=1}^{p} \gamma_i [A_i^T P A_i + (A_i - Q_s V_{2s}^{-1} C_i)^T \times \hat{P}(A_i - Q_s V_{2s}^{-1} C_i)] - P_s^T R_{2s}^{-1} P_s + \tau_\perp^T P_s^T R_{2s}^{-1} P_s \tau_\perp \tag{176}$$

$$0 = (A_\alpha - BR_{2s}^{-1} P_s)\hat{Q} + \hat{Q}(A_\alpha - BR_{2s}^{-1} P_s)^T + Q_s V_{2s}^{-1} Q_s^T - \tau_\perp Q_s V_{2s}^{-1} Q_s^T \tau_\perp^T \tag{177}$$

$$0 = (A_\alpha - Q_s V_{2s}^{-1} C)^T \hat{P} + \hat{P}(A_\alpha - Q_s V_{2s}^{-1} C) + P_s^T R_{2s}^{-1} P_s - \tau_\perp^T P_s^T R_{2s}^{-1} P_s \tau_\perp \tag{178}$$

$$\operatorname{rank} \hat{Q} = \operatorname{rank} \hat{P} = \operatorname{rank} \hat{Q}\hat{P} = n_c \tag{179}$$

$$\tau = \hat{Q}\hat{P}(\hat{Q}\hat{P})^{\#} \tag{180}$$

and let A_c, B_c, C_c be given by

$$A_c = \Gamma(A - BR_{2s}^{-1} P_s - Q_s V_{2s}^{-1} C + Q_s V_{2s}^{-1} D R_{2s}^{-1} P_s) G^T \tag{181}$$

$$B_c = \Gamma Q_s V_{2s}^{-1} \tag{182}$$

$$C_c = -R_{2s}^{-1} P_s G^T \tag{183}$$

Then $(\tilde{A} + \Delta\tilde{A},\tilde{D})$ is stabilizable for all $(\Delta A,\Delta B,\Delta C) \in \mathscr{U}_1$ if and only if $\tilde{A} + \Delta\tilde{A}$ is asymptotically stable for all $(\Delta A,\Delta B,\Delta C) \in \mathscr{U}_1$. In this case the worst-case closed-loop performance [Eq. (166)] satisfies the guaranteed cost bound

$$J(A_c,B_c,C_c) \leqslant \mathrm{tr}[(Q + \hat{Q})R_1 - 2R_{12}R_{2s}^{-1}P_s\hat{Q} + P_s^T R_{2s}^{-1}R_2R_{2s}^{-1}P_s\hat{Q}] \tag{184}$$

It is easy to see that Eqs. (175–178) differ from the usual optimal projection equations [Eqs. (43–46)] precisely because of the bounding terms. If $\gamma_i = 0$, $i = 1, ..., p$, then these terms are absent and Eqs. (43–46) are recovered with a slight generalization due to R_{12} and V_{12}. It is also interesting to specialize Eqs. (175–178) to the full-order case. Although the projection no longer plays a role in coupling the equations, there is still coupling due to the uncertainty bounds since $\hat{Q}$ and $\hat{P}$ appear in Eqs. (175) and (176), respectively. Thus, we see that the guaranteed cost generalization of LQG theory now involves four equations as in optimal projection theory.

Finally, we note that the form of the uncertainty terms in Eqs. (175) and (176) can be interpreted in terms of their effect on control authority. For example, the uncertainty bounds corresponding to ΔB effectively lead to the replacement of R_2 by R_{2s}. Since $R_{2s} \geqslant R_2$, this replacement implies a corresponding *decrease* in regulator authority due to uncertainty ΔB. On the other hand, Eq. (175) for Q now involves a term $B_iR_{2s}^{-1}P_s\hat{Q}(B_iR_{2s}^{-1}P_s)^T$, which corresponds to an augmentation of V_1 and hence to a net *increase* in estimator authority. These phenomena are quite reasonable in the face of uncertainty in the control input matrix B. A dual situation clearly arises when uncertainty ΔC is present. However, uncertainty in the A matrix leads to additional nonnegative terms in both the Q and P equations and hence to both higher regulator and estimator authority. In the next section we shall consider a variant of the linear bound that treats the uncertainty in A somewhat differently.

Quadratic Bound

The quadratic bound appears to have originated more recently than the linear bound.[49,185,188,191,194] The interesting feature of this bound is its similarity to the terms that enforce the H_∞ norm constraint. Indeed, it is this similarity that led to the pivotal paper of Petersen[167] (P^3) on the synthesis of H_∞ controllers.

For the quadratic bound the uncertainty model is given by

$$\mathscr{U}_2 = \left\{(\Delta A,\Delta C):\quad \Delta A = \sum_{i=1}^{p} D_iM_iN_iE_i, \quad \Delta C = \sum_{i=1}^{p} F_iM_iN_iE_i \right.$$

$$\left. M_iM_i^T \leqslant \bar{M}_i, \quad N_i^TN_i \leqslant \bar{N}_i, \quad i = 1, \ldots, p \right\} \tag{185}$$

where, for $i = 1, ..., p$: $D_i \in \mathbb{R}^{n \times r_i}$, $E_i \in \mathbb{R}^{t_i \times n}$, and $F_i \in \mathbb{R}^{l \times r_i}$ are fixed matrices denoting the structure of the uncertainty; $\bar{M}_i \geqslant 0$ and $\bar{N}_i \geqslant 0$ are given uncertainty bounds; and $M_i \in \mathbb{R}^{r_i \times s_i}$ and $N_i \in \mathbb{R}^{s_i \times t_i}$ are uncertain matrices. The closed-loop system thus has structured uncertainty of the form

$$\Delta\tilde{A} = \sum_{i=1}^{p} \tilde{D}_i M_i N_i \tilde{E}_i \tag{186}$$

where

$$\tilde{D}_i \triangleq \begin{bmatrix} D_i \\ B_c F_i \end{bmatrix}, \qquad \tilde{E} \triangleq [E_i \quad 0] \tag{187}$$

The case $\Delta B \neq 0$ and $\Delta C = 0$ can be handled in a dual fashion, whereas the general case $\Delta B \neq 0$, $\Delta C \neq 0$ can also be handled but with increased complexity. To state the optimal projection/guaranteed cost synthesis result for the quadratic bound define the notation

$$D \triangleq \sum_{i=1}^{p} D_i \bar{M}_i D_i^T, \qquad E \triangleq \sum_{i=1}^{p} E_i^T \bar{N}_i E_i \tag{188}$$

$$P_a \triangleq B^T P + R_{12}^T, \qquad Q_a \triangleq QC^T + V_{12} + \sum_{i=1}^{p} D_i \bar{M}_i F_i^T \tag{189}$$

$$A_P \triangleq A - BR_2^{-1}P_a, \qquad A_Q \triangleq A - Q_a V_{2a}^{-1} C, \qquad V_{2a} \triangleq V_2 + \sum_{i=1}^{p} F_i \bar{M}_i F_i^T \tag{190}$$

Theorem 4. Suppose there exist $Q, P, \hat{Q}, \hat{P} \geqslant 0$ satisfying

$$0 = AQ + QA^T + V_1 + D + QEQ - Q_a V_{2a}^{-1} Q_a^T + \tau_\perp Q_a V_{2a}^{-1} Q_a^T \tau_\perp^T \tag{191}$$

$$0 = [A + (Q + \hat{Q})E]^T P + P[A + (Q + \hat{Q})E] + R_1 - P_a^T R_2^{-1} P_a + \tau_\perp^T P_a^T R_2^{-1} P_a \tau_\perp \tag{192}$$

$$0 = (A_P + QE)\hat{Q} + \hat{Q}(A_P - QE)^T + \hat{Q}E\hat{Q} + Q_a V_{2a}^{-1} Q_a^T - \tau_\perp Q_a V_{2a}^{-1} Q_a^T \tau_\perp^T \tag{193}$$

$$0 = (A_Q + QE)^T \hat{P} + \hat{P}(A_Q + QE) + P_a^T R_2^{-1} P_a - \tau_\perp^T P_a^T R_2^{-1} P_a \tau_\perp \tag{194}$$

$$\operatorname{rank}\hat{Q} = \operatorname{rank}\hat{P} = \operatorname{rank}\hat{Q}\hat{P} = n_c \tag{195}$$

$$\tau = \hat{Q}\hat{P}(\hat{Q}\hat{P})^{\#} \tag{196}$$

and let

$$A_c = \Gamma(A - BR_2^{-1}P_a - Q_a V_{2a}^{-1} C + QE)G^T \tag{197}$$

$$B_c = \Gamma Q_a V_{2a}^{-1} \tag{198}$$

$$C_c = -R_2^{-1} P_a G^T \tag{199}$$

Then $(\tilde{A} + \Delta\tilde{A}, \tilde{D})$ is stabilizable for all $(\Delta A, \Delta B, \Delta C) \in \mathcal{U}_2$ if and only if $\tilde{A} + \Delta\tilde{A}$ is asymptotically stable for all $(\Delta A, \Delta B, \Delta C) \in \mathcal{U}_2$. In this case the closed-loop performance satisfies the guaranteed cost bound

$$J(A_c, B_c, C_c) \leqslant \mathrm{tr}[(Q + \hat{Q})R_1 + \hat{Q}PBR_2^{-1}P_a - R_{12}R_2^{-1}P_a\hat{Q}] \tag{200}$$

Finally, we show how optimal projection/guaranteed cost theory can be merged with optimal projection/H_∞ theory to form optimal projection/guaranteed cost/H_∞ theory. To illustrate such a combined theory, consider the uncertainty set $\mathcal{U}_1$, where for simplicity we assume $\Delta B = 0$ (i.e., $B_i = 0$, $i = 1, \ldots, p$). The principal feature of this result is that the H_∞ bound on the transfer function $\tilde{G}_{\Delta\tilde{A}}(s)$ holds for *all* uncertainties in $\mathcal{U}_1$. The following results can be found in Ref. 27; for a related approach see Ref. 208.

Theorem 5. Suppose there exist $Q, P, \hat{Q}, \hat{P} \geqslant 0$ satisfying

$$\begin{aligned} 0 = A_\alpha Q + QA_\alpha^T + \gamma^{-2} QR_{1\infty}Q + V_1 + \sum_{i=1}^{p} \gamma_i A_i(Q + \hat{Q})A_i^T \\ - Q_s V_{2s}^{-1} Q_s^T + \tau_\perp Q_s V_{2s}^{-1} Q_s^T \tau_\perp^T \end{aligned} \tag{201}$$

$$\begin{aligned} 0 = (A_\alpha + \gamma^{-2}[Q + \hat{Q}]R_{1\infty})^T P + P(A_\alpha + \gamma^{-2}[Q + \hat{Q}]R_{1\infty}) + R_1 \\ + \sum_{i=1}^{p} \gamma_i [A_i^T P A_i + (A_i - Q_s V_{2s}^{-1} C_i)^T \hat{P}(A_i - Q_s V_{2s}^{-1} C_i)] \\ - S^T P_s R_2^{-1} P^s + \tau_\perp^T S^T P_s R_2^{-1} P_s S \tau_\perp \end{aligned} \tag{202}$$

$$\begin{aligned} 0 = (A_\alpha - BR_2^{-1}P_s S + \gamma^{-2} QR_{1\infty})\hat{Q} + \hat{Q}(A_\alpha - BR_2^{-1}P_s S \\ + \gamma^{-2} QR_{1\infty})^T + \gamma^{-2} \hat{Q}(R_{1\infty} + \beta^2 S^T P_s R_2^{-1} P_s S)\hat{Q} \\ + Q_s V_{2s}^{-1} Q_s^T - \tau_\perp Q_s V_{2s}^{-1} Q_s^T \tau_\perp^T \end{aligned} \tag{203}$$

$$\begin{aligned} 0 = (A_\alpha - Q_s V_{2s}^{-1} C + \gamma^{-2} QR_{1\infty})^T \hat{P} + \hat{P}(A_\alpha - Q_s V_{2s}^{-1} C \\ + \gamma^{-2} QR_{1\infty}) + S^T P_s R_2^{-1} P_s S - \tau_\perp^T S^T P_s R_2^{-1} P_s S \tau_\perp \end{aligned} \tag{204}$$

$$\mathrm{rank}\, \hat{Q} = \mathrm{rank}\, \hat{P} = \mathrm{rank}\, \hat{Q}\hat{P} = n_c \tag{205}$$

$$\tau = \hat{Q}\hat{P}(\hat{Q}\hat{P})^{\#} \tag{206}$$

and let A_c, B_c, C_c be given by

$$A_c = \Gamma(A - BR_2^{-1}P_sS - Q_sV_{2s}^{-1}C + \gamma^{-2}QR_{1\infty})G^T \tag{207}$$

$$B_c = \Gamma Q_s V_{2s}^{-1} \tag{208}$$

$$C_c = -R_2^{-1}P_sSG^T \tag{209}$$

Then $(\tilde{A} + \Delta\tilde{A}, \tilde{D})$ is stabilizable for all $(\Delta A, \Delta C) \in \mathscr{U}_1$ if and only if $\tilde{A} + \Delta\tilde{A}$ is asymptotically stable for all $(\Delta A, \Delta C) \in \mathscr{U}_1$. In this case, the closed-loop transfer function $G_{\Delta\tilde{A}}(s)$ satisfies the H_∞ disturbance attenuation constraint

$$\|G_{\Delta\tilde{A}}(s)\|_\infty \leqslant \gamma \qquad \text{for all} \qquad (\Delta A, \Delta C) \in \mathscr{U}_1 \tag{210}$$

and the worst-case H_2 performance criterion satisfies the bound

$$J(A_c, B_c, C_c) \leqslant \mathrm{tr}[(Q + \hat{Q})R_1 + \hat{Q}S^TP_sR_2^{-1}P_sS] \tag{211}$$

Optimal Projection/Positive Real Theory

We now turn to an extension of optimal projection theory that was developed specifically for controlling flexible structures with numerous lightly damped modes. Although this extension is similar in form to the optimal projection/guaranteed cost theory with the linear bound, it operates via wholly different principles. Specifically, optimal projection/positive real theory does not achieve robustness by bounding the worst-case value of the performance functional over a class of uncertainties. Rather, it seeks to configure the controller to exploit the dissipative nature of the structural system. In classical terminology we are thus seeking a theory of phase stabilization in contrast to the more standard approach of gain stabilization. The goal of this section is to demonstrate how optimal projection theory can be extended to achieve this goal.

To begin, we consider a damped oscillator with force input and velocity measurement. For such a system the transfer function is known to be positive real, i.e., to have phase shift less than 90 deg. In a negative feedback configuration, a controller that is strictly positive real cannot destabilize the system since the loop transfer function has phase less than 180 deg. Hence, such a control system will be unconditionally robust to natural frequency and damping uncertainties. In controlling structural vibrations a common application of this principle is the rate feedback damper with a colocated sensor/actuator pair. Of course, these observations assume perfect sensors and actuators in order to avoid introducing additional phase shift. If the sensors and actuators do have significant dynamics, then the feedback law must be chosen so that the transfer function from sensor input to actuator is strictly positive real. If in practice positive realness can only be enforced over a limited frequency band, then loop gain rolloff is required when unknown phase shifts become significant.

Optimal projection/positive real theory seeks to embed these design principles within a compensator synthesis theory for lightly damped structures. This extension was developed in a series of papers predating optimal projection theory itself.[31–34] The original motivation arose from a branch of structural analysis known as statistical energy analysis (SEA).[209–214] SEA is a structural analysis theory developed to analyze systems with numerous lightly damped modes. It has found considerable practical applications in analyzing acoustic vibrations in cases in which very little information is available concerning modal frequency and mode shapes. Often the only available knowledge is the modal density, i.e., the average number of modes per frequency band.

The basic concept of SEA is power flow among coupled modes. Estimates of modal density provide the means for computing the energy distribution of modes in steady-state excitation. An important phenomenon in this regard is the *equipartition* of energy, that is, the tendency of modal energy (rms vibration levels) to become equalized among modes due to power flow. Equipartitioning leads to *isotropy*, the state of uniform energy distribution.

For control system analysis and design the modal phenomena of SEA can be reinterpreted in terms of model uncertainty. Specifically, the covariance matrix of a modal system may represent mean square response due to both external excitation and modal uncertainty. Highly uncertain lightly damped modes will thus give rise to a diagonal covariance matrix with equal diagonal elements denoting equipartition in the isotropic state. Off-diagonal elements of the covariance matrix are essentially zero due to the absence of correlation (called "decorrelation") caused by modal uncertainty. At slightly lower frequencies, the diagonal elements may be different because of greater modeling fidelity, although diagonal dominance may still occur. At low frequencies, a reasonably accurate model will, of course, result in a fully dense covariance matrix.

The contribution of Refs. 31–41 was the construction of an SEA covariance equation that captures the effects of modal uncertainty. As we shall see, optimal projection synthesis with the SEA covariance equation induces a robust controller with positive real characteristics that robustify against modal uncertainty. The SEA covariance equation is given by

$$0 = A_s Q + Q A_s^T + \sum_{i=1}^{r} A_i Q A_i^T + V \tag{212}$$

where

$$A = \text{block diag}\left(\begin{bmatrix} -\zeta_1 & -\omega_1 \\ \omega_1 & -\zeta_1 \end{bmatrix}, \ldots, \begin{bmatrix} -\zeta_r & -\omega_r \\ -\omega_r & -\zeta_r \end{bmatrix} \right) \tag{213}$$

$$A_i = \text{block diag}\left(0, \ldots, 0, \begin{bmatrix} 0 & -\beta_i \\ \beta_i & 0 \end{bmatrix}, 0, \ldots, 0 \right) \tag{214}$$

$$A_s \triangleq A + \frac{1}{2} \sum_{i=1}^{r} A_i^2 \tag{215}$$

Several observations are in order. First, note that the SEA covariance equation is a modified Liapunov equation of the form of Eq. (150) with an uncertainty operator $\Omega_s(Q)$ defined by

$$\Omega_s(Q) = \sum_{i=1}^{r} [\tfrac{1}{2}A_i^2 Q + A_i Q A_i^T + \tfrac{1}{2}Q A_i^{2T}] \tag{216}$$

Although this operator has the term $A_i Q A_i^T$ in common with the linear bound, it differs in one crucial respect. Whereas the linear bound (158) has a uniform right-shift term $A + (\alpha/2)I_n$, the operator (216) invokes the term

$$A + \frac{1}{2}\sum_{i=1}^{r} A_i^2$$

which is a variable left shift. To see this, note that

$$\tfrac{1}{2}A_i^2 = \text{block diag}(0, ..., 0, -\tfrac{1}{2}\beta_i^2 I_2, 0, ..., 0) \tag{217}$$

which effectively shifts the ith mode to the left by an amount $-\tfrac{1}{2}\beta_i^2$. Thus, β_i can be increased at a higher frequency to represent greater uncertainty.

An interesting observation concerning $\Omega_s(Q)$ is that its form coincides precisely with the covariance equation arising from a multiplicative noise model interpreted in the sense of Stratonovich.[215–217] The "Stratonovich correction" term

$$\frac{1}{2}\sum_{i=1}^{r} A_i^2$$

arises due to the conversion from Stratonovich stochastic calculus to the more common Ito calculus. Justification for the Stratonovich interpretation was originally given in terms of maximum entropy principles.[218–221]

Next note that the uncertainty matrices A_i correspond to uncertainty in the modal frequencies. In contrast to damping uncertainty, modal frequency uncertainty is nondestabilizing. However, it does have a significant effect on the state covariance. To see this, let $r = 2$ in Eq. (213) for convenience and note that $\Omega_s(Q)$ is given by

$$\Omega_s(Q) =$$

$$\begin{bmatrix} \beta_1^2[Q_{22} - Q_{11}] & -2\beta_1^2 Q_{12} & -\tfrac{1}{2}(\beta_1^2+\beta_2^2)Q_{13} & -\tfrac{1}{2}(\beta_1^2+\beta_2^2)Q_{14} \\ -2\beta_1^2 Q_{21} & \beta_1^2[Q_{11} - Q_{22}] & -\tfrac{1}{2}(\beta_1^2+\beta_2^2)Q_{23} & -\tfrac{1}{2}(\beta_1^2+\beta_2^2)Q_{24} \\ -\tfrac{1}{2}(\beta_1^2+\beta_2^2)Q_{31} & -\tfrac{1}{2}(\beta_1^2+\beta_2^2)Q_{32} & \beta_2^2[Q_{44} - Q_{33}] & -2\beta_2^2 Q_{34} \\ -\tfrac{1}{2}(\beta_1^2+\beta_2^2)Q_{41} & -\tfrac{1}{2}(\beta_1^2+\beta_2^2)Q_{42} & -2\beta_2^2 Q_{43} & \beta_2^2[Q_{33} - Q_{44}] \end{bmatrix} \tag{218}$$

Brief inspection of $\Omega_s(Q)$ reveals two things. First, the off-diagonal terms of $\Omega_s(Q)$ are proportional to the corresponding terms of Q with a negative constant of proportionality. By viewing the covariance equation as a differential equation, it follows that these terms cause the off-diagonal terms to decay (i.e., each is of the form $\dot{x} = -\beta x$). This is the phenomenon of decorrelation due to frequency uncertainty. Next, note that the diagonal terms of $\Omega_s(Q)$ are sign variable. If, for example, $Q_{11} > Q_{22}$, then the (1,1) term of $\Omega_s(Q)$ is negative and thus causes a corresponding decrease in Q_{11}, whereas the (2,2) term is positive and causes Q_{22} to increase. However, if $Q_{11} < Q_{22}$, then the (2,2) element is negative and the situation is reversed. The overall effect of such modal interaction is *equilibration* among modes. Thus, the SEA/Stratonovich model captures the phenomena of decorrelation and equilibration.

However, the uncertainty operator $\Omega_s(Q)$ is not a bound on the effects of frequency uncertainty. In fact, it can be shown that $\Omega_s(Q)$ is actually indefinite. Hence, the role of $\Omega_s(Q)$ cannot be simply explained via guaranteed cost theory. To examine the control-design ramifications of $\Omega_s(Q)$ with optimal projection theory, we can derive the appropriate design equations[41,42]:

$$0 = A_s(Q) + QA_s^T + V_1 + \sum_{i=1}^{r} A_i(Q + \hat{Q})A_i^T - Q\bar{\Sigma}Q + \tau_\perp Q\bar{\Sigma}Q\tau_\perp^T \tag{219}$$

$$0 = A_s^T P + PA_s + R_1 + \sum_{i=1}^{r} A_i^T(P + \hat{P})A_i - P\Sigma P + \tau_\perp^T P\Sigma P\tau_\perp \tag{220}$$

$$0 = (A_s - \Sigma P)\hat{Q} + \hat{Q}(A_s - \Sigma P)^T + Q\bar{\Sigma}Q - \tau_\perp Q\bar{\Sigma}Q\tau_\perp^T \tag{221}$$

$$0 = (A_s - Q\bar{\Sigma})^T\hat{P} + \hat{P}(A_s - Q\bar{\Sigma}) + P\Sigma P - \tau_\perp^T P\Sigma P\tau_\perp \tag{222}$$

$$\operatorname{rank} \hat{Q} = \operatorname{rank} \hat{P} = \operatorname{rank} \hat{Q}\hat{P} = n_c \tag{223}$$

$$\tau = \hat{Q}\hat{P}(\hat{Q}\hat{P})^{\#} \tag{224}$$

with gains

$$A_c = \Gamma(A_s - Q\bar{\Sigma} - \Sigma P)G^T \tag{225}$$

$$B_c = \Gamma QC^T V_2^{-1} \tag{226}$$

$$C_c = -R_2^{-1}B^T PG^T \tag{227}$$

To illustrate the consequences of the SEA/Stratonovich model, we consider a simply supported beam with a *single* collocated force actuator/velocity sensor pair. The modal frequencies are assumed to be uncertain with uncertainty increasing linearly with mode number. A design study carried out in Refs. 36 and 40 yielded the results shown in Fig. 4. The principal feature of this result is the tendency of the controller to become

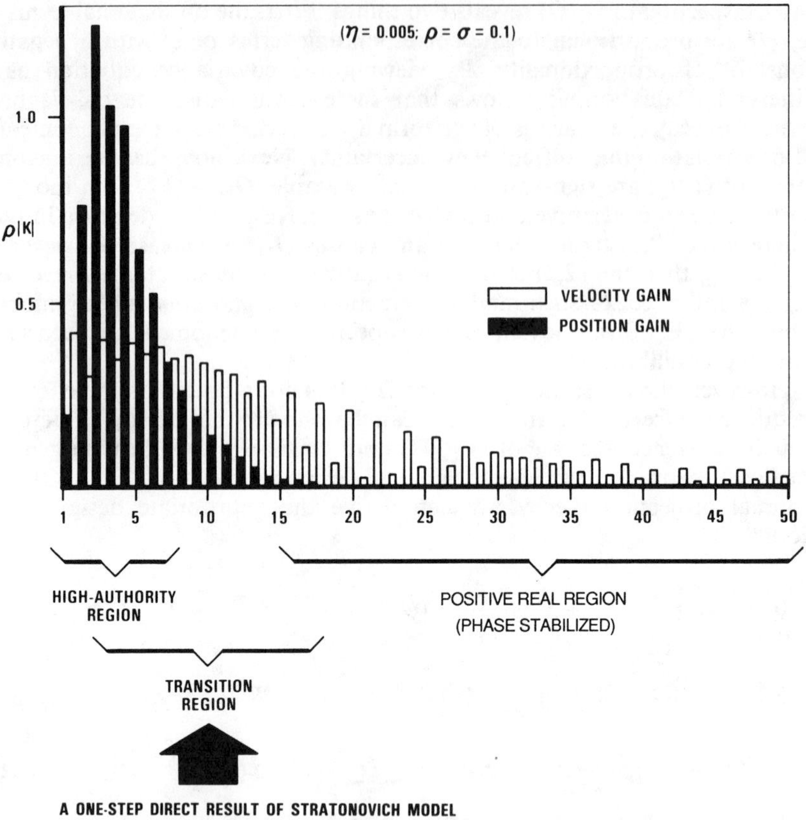

Fig. 4 The optimal projection/positive real controllers achieve robust phase stabilization by means of dissipative rate feedback for highly uncertain modes.

increasingly positive real at higher frequency. That is, in the low-frequency region, the controller assigns high gains to both position and velocity, whereas at high frequency only velocity is fed back, indicating a phase stabilizing, rate dissipative controller. Related results obtained using this approach were reported in Refs. 222–224.

Open Problems and Extensions

Open problems represent a source of vitality and motivation for future development. We close by describing a collection of research topics that are, at the same time, practically motivated and theoretically meaningful. Although all of these problems relate in some way to optimal projection theory, many have much broader significance.

Stabilizability

Optimal projection theory notwithstanding, one of the most basic unsolved problems in control theory is the problem of output feedback stabilizability: Given A,B,C determine necessary and sufficient conditions under which there exists K such that $A + BKC$ is asymptotically stable. If either B or C is the identity matrix, then the solution is well known, whereas in the general case partial results are given in Refs. 225 and 226. The decentralized generalization of this problem is also of interest: Given $A,B_1, ..., B_r,C_1, ..., C_r$, determine necessary and sufficient conditions under which there exist $K_1, ..., K_r$ such that

$$A + \sum_{i=1}^{r} B_i K_i C_i$$

is asymptotically stable. This problem is of interest since it includes stabilizability by reduced-order compensators as a special case. To see this, write the closed-loop dynamics as

$$\begin{bmatrix} A & BC_c \\ B_c C & A_c \end{bmatrix} = \begin{bmatrix} A & 0 \\ 0 & 0 \end{bmatrix} + \begin{bmatrix} 0 \\ I \end{bmatrix} A_c [0 \quad I] + \begin{bmatrix} 0 \\ I \end{bmatrix} B_c [C \quad 0] + \begin{bmatrix} B \\ 0 \end{bmatrix} C_c [0 \quad I]$$

which has decentralized structure with $r = 3$. If $n_c = n$, then necessary and sufficient conditions for stabilizability are well known from LQG theory. The reduced-order case $n_c < n$ is unsolved. Finally, a further extension is to require repetition of certain feedback gains; i.e., $K_j = K_\ell$ for selected indices. It can be shown using the techniques of Refs. 57 and 58 that this feedback structure encompasses all affinely parameterized static and dynamic feedback structures.

Existence Theory

For each optimization problem considered within optimal projection theory, it is of interest to know whether a global minimum exists. In practice such questions are largely academic, since necessary conditions serve as useful tools for computing extremals when they do exist. Nevertheless, in many cases existence theory is related to the achievability of design goals. For example, in the H_∞ problem it is of interest to know whether optimal projection/H_∞ theory is effective in characterizing H_∞-constrained controllers when they exist. Such questions relate directly to the existence of solutions to the constrained optimization problem. Historically, existence problems in nonconvex optimization theory have proven to be difficult to treat mathematically. For fixed-structure optimization problems, it appears that a combination of compactness and monotonicity arguments may be effective.

Analysis of the Design Equations

Analysis of the standard matrix Riccati equation has been ongoing for more than three decades. Such efforts have been extremely fruitful with ramifications in mathematics and other branches of engineering. However, there can be no doubt that the systems of coupled modified Riccati and Liapunov equations arising in optimal projection theory lie beyond the scope of existing Riccati equation theory. Investigation is needed to address questions of existence, multiplicity, monotonicity, stabilizability, and optimality. For example, we would expect that conditions guaranteeing the existence of solutions to the optimal projection equations [Eqs. (44–46)] would be closely related to stabilizability by means of fixed-order compensators. A thorough analysis of these equations represents a significant mathematical challenge for the next decade.

Quadratic Liapunov Bounds

Numerous questions remain unanswered concerning the linear and quadratic Liapunov bounds. For example, the conservatism of such bounds requires further examination. In addition, the gap between robust stability and quadratic robust stability (robustness guaranteed via a fixed quadratic Liapunov function) is not well understood. In this direction it can be seen that the four-polynomial robust stability test of Kharitonov must correspond to a four-Liapunov function robust stability test for an interval matrix in companion form. This observation suggests that a reduction in conservatism may be possible by adopting a multi-Liapunov function approach in place of the single Liapunov function technique.

Positive Real Theory

In the previous section it was argued heuristically and demonstrated numerically that the SEA/Stratonovich uncertainty operator induces robustness by exploiting positive real properties. On the one hand, it would appear to be a straightforward task to prove this mathematically by applying positive real system criteria.[227–232] However, in fact, the situation is much more subtle since the SEA/Stratonovich operator induces positive realness only to the extent that phase shift is needed to robustify with respect to a particular level. Hence, a much finer characterization of phase properties appears to be necessary. The H_∞ approach to positive real design[233] may be a useful starting point in this regard.

Optimal Projection/μ Theory

Structured singular value theory[234,235] was developed to address problems in which direct application of H_∞ bounds leads to overly conservative results. For example, if unstructured uncertainty appears at multiple points within the plant, then the overall uncertainty has a block-diagonal structure. The same situation arises in bounding the worst-case H_∞ performance in the presence of unstructured uncertainty. To address such problems the μ function was developed to generalize the small gain theorem to block-diagonal uncertainties. The extension of optimal projection theory to the μ-synthesis problem[236,237] remains a potentially fruitful research area. Alternative approaches to unstructured uncertainty such as the all-pass characterization[238–240] are also promising in this regard.

Acknowledgments

This chapter describes results obtained over the past decade. We dedicate this paper to Mr. Stobo Wright in honor of his retirement as Group Leader for Control Systems Engineering at Lincoln Laboratory and in sincere appreciation for his encouragement for this research. For supporting this research we are indebted to DARPA/ISTO (1980–1981), Massachusetts Institute of Technology's Lincoln Laboratory (1979–1984), the Air Force Office of Scientific Research (1984–present), and Harris Corporation (1983–present).

References

[1]Hyland, D. C., "Optimality Conditions for Fixed-Order Dynamic Compensation of Flexible Spacecraft with Uncertain Parameters," AIAA Paper 82-0312, Jan. 1982.

[2]Hyland, D. C., and Madiwale, A. N., "Fixed-Order Dynamic Compensation Through Optimal Projection," *Proceedings of the Workshop on Applications of Distributed System Theory to the Control of Large Space Structures*, edited by G. Rodrigues, Jet Propulsion Lab., Pasadena, CA, July 1982, pp. 409–427.

[3]Hyland, D. C., "Mean-Square Optimal Fixed-Order Compensation-Beyond Spillover Suppression," AIAA Paper 82-1403, Aug. 1982.

[4]Hyland, D. C., "The Optimal Projection Approach to Fixed-Order Compensation: Numerical Methods and Illustrative Results," AIAA Paper 83-0303, Jan. 1983.

[5]Hyland, D. C., and Bernstein, D. S., "Explicit Optimality Conditions for Fixed-Order Dynamic Compensation," *Proceedings of the IEEE Conference on Decision and Control*, IEEE, New York, 1983, pp. 161–165.

[6]Hyland, D. C., "Comparison of Various Controller-Reduction Methods: Suboptimal Versus Optimal Projection," *Proceedings of the AIAA Dynamics Specialists Conference*, AIAA, New York, 1984, pp. 381–389.

[7]Hyland, D. C., and Bernstein, D. S., "The Optimal Projection Equations for Fixed-Order Dynamic Compensation," *IEEE Transactions on Automatic Control*, Vol. AC-29, 1984, pp. 1034–1037.

[8]Richter, S., "A Homotopy Algorithm for Solving the Optimal Projection Equations for Fixed-Order Dynamic Compensation: Existence, Convergence and Global Optimality," *Proceedings of the American Control Conference*, 1987, pp. 1527–1531.

[9]Bernstein, D. S., "The Optimal Projection Equations for Static and Dynamic Output Feedback: The Singular Case," *IEEE Transactions on Automatic Control*, Vol. AC-32, 1987, pp. 1139–1143.

[10]Greeley, S. W., and Hyland, D. C., "Reduced-Order Compensation: LQG Reduction Versus Optimal Projection Using a Homotopic Continuation Method," *Proceedings of the IEEE Conference on Decision and Control*, IEEE, New York, 1987, pp. 742–747.

[11]Greeley, S. W., and Hyland, D. C., "Reduced-Order Compensation: Linear-Quadratic Reduction Versus Optimal Projection," *AIAA Journal of Guidance, Control, and Dynamics*, Vol. 11, No. 4, 1988, pp. 328–335.

[12]Hyland, D. C., and Richter, S., "On Direct Versus Indirect Methods for Reduced-Order Controller Design," *IEEE Transactions on Automatic Control*, Vol. 35, 1990.

[13]Richter, S., and Collins, E. G., Jr., "A Homotopy Algorithm for Reduced-

Order Controller Design Using the Optimal Projection Equations," *Proceedings of the IEEE Conference on Decision and Control*, IEEE, New York, 1989.

[14]Halevi, Y., Haddad, W. M., and Bernstein, D. S., "A Riccati Equation Approach to the Singular LQG Problem," *Proceedings of the American Control Conference*, 1989, pp. 2077–2078.

[15]Bernstein, D. S., and Hyland, D. C., "The Optimal Projection Equations for Finite-Dimensional Fixed-Order Dynamic Compensation of Infinite-Dimensional Systems," *SIAM Journal on Control and Optimization*, Vol. 24, 1986, pp. 122–151.

[16]Bernstein, D. S., and Rosen, I. G., "An Approximation Technique for Computing Optimal Fixed-Order Controllers for Infinite-Dimensional Systems," *Proceedings of the IEEE Conference on Decision and Control*, IEEE, New York, 1988, pp. 2023–2028.

[17]Bernstein, D. S., and Rosen, I. G., "Finite-Dimensional Approximation for Optimal Fixed-Order Compensation of Distributed Parameter Systems," *Optimal Control Applications and Methods*, Vol. 11, 1990, pp. 1–20.

[18]Bernstein, D. S., Davis, L. D., and Hyland, D. C., "The Optimal Projection Equations for Reduced-Order, Discrete-Time Modelling, Estimation and Control," *AIAA Journal of Guidance, Control, and Dynamics*, Vol. 9, No. 3, 1986, pp. 288–293.

[19]Bernstein, D. S., Davis, L. D., and Greeley, S. W., "The Optimal Projection Equations for Reduced-Order, Sampled-Data Dynamic Compensation with Computation Delay," *IEEE Transactions on Automatic Control*, Vol. AC-31, 1986, pp. 859–862.

[20]Bernstein, D. S., "Sequential Design of Decentralized Dynamic Compensators Using the Optimal Projection Equations," *International Journal of Control*, Vol. 46, 1987, pp. 1569–1577.

[21]Bernstein, D. S., and Haddad, W. M., "Robust Decentralized Static Output Feedback," *Systems and Control Letters*, Vol. 12, 1989, pp. 309–318.

[22]Hyland, D. C., and Bernstein, D. S., "The Optimal Projection Equations for Model Reduction and the Relationships Among the Methods of Wilson, Skelton and Moore," *IEEE Transactions on Automatic Control*, Vol. AC-30, 1985, pp. 1201–1211.

[23]Haddad, W. M., and Bernstein, D. S., "Robust, Reduced-Order Modeling via the Optimal Projection Equations with Petersen-Hollot Bounds," *IEEE Transactions on Automatic Control*, Vol. 33, 1988, pp. 692–695.

[24]Haddad, W. M., and Bernstein, D. S., "Combined L_2/H_∞ Model Reduction," *International Journal of Control*, Vol. 49, 1989, pp. 1523–1535.

[25]Bernstein, D. S., and Haddad, W. M., "LQG Control with an H_∞ Performance Bound: A Riccati Equation Approach," *Proceedings of the American Control Conference*, 1988, pp. 796–802.

[26]Bernstein, D. S., and Haddad, W. M., "LQG Control With An H_∞ Performance Bound: A Riccati Equation Approach," *IEEE Transactions on Automatic Control*, Vol. 34, 1989, pp. 293–305.

[27]Madiwale, A. N., Haddad, W. M., and Bernstein, D. S., "Robust H_∞ Control Design for Systems with Parameter Uncertainty," *Systems and Control Letters*, Vol. 12, 1989, pp. 393–407.

[28]Haddad, W. M., and Bernstein, D. S., "Complete Solution to the Nonsingular H_2/H_∞ Four Block Problem," *Proceedings of the American Control Conference*, 1989, pp. 187–192.

[29]Haddad, W. M., and Bernstein, D. S., "Generalized Riccati Equations for the Full- and Reduced-Order Mixed-Norm H_2/H_∞ Standard Problem," *Systems and Control Letters*, Vol. 14, 1990.

[30]Haddad, W. M., and Bernstein, D. S., "On the Gap Between H_2 and Entropy Performance Measures in H_∞ Control," Systems and Control Letters, Vol. 14, 1990, pp. 113–120.

[31]Hyland, D. C., "The Modal Coordinate/Radiative Transfer Formulation of Structural Dynamics-Implications for Vibration Suppression in Large Space Platforms," Massachusetts Institute of Technology, Lincoln Lab., TR-27, 14 March 1979.

[32]Hyland, D. C., "Optimal Regulation of Structural Systems With Uncertain Parameters," Massachusetts Institute of Technology, Lincoln Lab., TR-551, 2 Feb. 1981, DDC No. ADA-099111/7.

[33]Hyland, D. C., "Active Control of Large Flexible Spacecraft: A New Design Approach Based on Minimum Information Modelling of Parameter Uncertainties," *Proceedings of the VPI&SU/AIAA Symposium*, Blacksburg, VA, June 1988, pp. 631–646.

[34]Hyland, D. C., "Optimal Regulator Design Using Minimum Information Modelling of Parameter Uncertainties: Ramifications of the New Design Approach," *Proceedings of the VPI&SU/AIAA Symposium*, Blacksburg, VA, June 1981, pp. 701–716.

[35]Hyland, D. C., and Madiwale, A. N., "Minimum Information Approach to Regulator Design: Numerical Methods and Illustrative Results," AIAA Paper 81-1830, June 1981, pp. 101–118.

[36]Hyland, D. C., and Madiwale, A. N., "A Stochastic Design Approach for Full-Order Compensation of Structural Systems with Uncertain Parameters," *Proceedings of the American Control Conference*, 1981, pp. 324–332.

[37]Hyland, D. C., "Minimum Information Stochastic Modelling of Linear Systems with a Class of Parameter Uncertainties," *Proceedings of the American Control Conference*, 1982, pp. 620–627.

[38]Hyland, D. C., "Maximum Entropy Stochastic Approach to Controller Design for Uncertain Structural Systems," *Proceedings of the American Control Conference*, 1982, pp. 680–699.

[39]Hyland, D. C., "Minimum Information Modelling of Structural Systems with Uncertain Parameters," *Proceedings of the Workshop on Applications of Distributed System Theory to the Control of Large Space Structures*, edited by G. Rodriguez, Jet Propulsion Lab., Pasadena, CA, July 1982, pp. 409–427.

[40]Hyland, D. C., "Mean-Square Optimal, Full-Order Compensation of Structural Systems with Uncertain Parameters," Massachusetts Institute of Technology, Lincoln Lab., TR-626, 1 June 1983.

[41]Bernstein, D. S., and Hyland, D. C., "Optimal Projection/Maximum Entropy Stochastic Modelling and Reduced-Order Design Synthesis," *Proceedings of the IFAC Workshop on Model Error Concepts and Compensation*, edited by R. E. Skelton and D. H. Owens, Pergamon, Oxford, UK, 1986, pp. 47–54.

[42]Bernstein, D. S., and Hyland, D. C., "The Optimal Projection/Maximum Entropy Approach to Designing Low-Order, Robust Controllers for Flexible Structures," *Proceedings of the IEEE Conference on Decision and Control*, IEEE, New York, 1985, pp. 745–752.

[43]Bernstein, D. S., Davis, L. D., Greeley, S. W., and Hyland, D. C., "Numerical Solution of the Optimal Projection/Maximum Entropy Design Equations for Low-Order, Robust Controller Design," *Proceedings of the IEEE Conference on Decision and Control*, IEEE, New York, 1985, pp. 1795–1798.

[44]Bernstein, D. S., and Greeley, S. W., "Robust Output-Feedback Stabilization: Deterministic and Stochastic Perspectives," *Proceedings of the American Control Conference*, 1986, pp. 1818–1826.

[45]Bernstein, D. S., and Greeley, S. W., "Robust Controller Synthesis Using the Maximum Entropy Design Equations," *IEEE Transactions on Automatic Control*, Vol. AC-31, 1986, pp. 362–364.

[46]Haddad, W. M. "Robust Optimal Projection Control-System Synthesis," Ph.D. Dissertation, Dept. of Mechanical Engineering, Florida Institute of Technology, Melbourne, FL, March 1987.

[47]Bernstein, D. S., and Haddad, W. M., "Optimal Projection Equations for Discrete-Time Fixed-Order Dynamic Compensation of Linear Systems with Multiplicative White Noise," *International Journal of Control*, Vol. 46, 1987, pp. 65–73.

[48]Bernstein, D. S., "Robust Static and Dynamic Output-Feedback Stabilization: Deterministic and Stochastic Perspectives," *IEEE Transactions on Automatic Control*, Vol. AC-32, 1987, pp. 1076–1084.

[49]Bernstein, D. S., and Haddad, W. M., "The Optimal Projection Equations with Petersen-Hollot Bounds: Robust Stability and Performance via Fixed-Order Dynamic Compensation for Systems with Structured Real-Valued Parameter Uncertainty," *IEEE Transactions on Automatic Control*, Vol. 33, 1988, pp. 578–582.

[50]Bernstein, D. S., and Hyland, D. C., "Optimal Projection for Uncertain Systems (OPUS): A Unified Theory of Reduced-Order, Robust Control Design," *Large Space Structures: Dynamics and Control*, edited by S. N. Atluri, and A. K. Amos, Springer-Verlag, New York, 1988, pp. 263–302.

[51]Bernstein, D. S., and Hyland, D. C., "Optimal Projection Equations for Reduced-Order Modelling, Estimation and Control of Linear Systems with Multiplicative White Noise," *Journal of Optimization Theory and Applications*, Vol. 58, 1988, pp. 387–409.

[52]Bernstein, D. S., "Robust Stability and Performance via Fixed-Order Dynamic Compensation," *SIAM Journal on Control and Optimization*, Vol. 27, 1989, pp. 389–406.

[53]Bernstein, D. S., and Haddad, W. M., "Robust Stability and Performance Analysis for Linear Dynamic Systems," *IEEE Transactions on Automatic Control*, Vol. 34, 1989, pp. 751–758.

[54]Bernstein, D. S., and Hollot, C. V., "Robust Stability for Sampled-Data Control Systems," Systems and Control Letters, Vol. 13, 1989, pp. 217–226.

[55]Bernstein, D. S., and Haddad, W. M., "Robust Stability and Performance Analysis for State Space Systems via Quadratic Lyapunov Bounds," *SIAM Journal of Matrix Analysis and Applications*, Vol. 11, 1990, pp. 239–271.

[56]Bernstein, D. S., and Haddad, W. M., "Robust Stability and Performance via Fixed-Order Dynamic Compensation with Guaranteed Cost Bounds," *Mathematics of Control, Signal, and Systems*, Vol. 3, 1990, pp. 139–163.

[57]Nett, C. N., Bernstein, D. S., and Haddad, W. M., "Minimal Complexity Control Law Synthesis, Part 1: Problem Formulation and Reduction to Optimal Static Output Feedback," *Proceedings of the American Control Conference*, 1989, pp. 2056–2064.

[58]Bernstein, D. S., Haddad, W. M., and Nett, C. N., "Minimal Complexity Control Law Synthesis, Part 2: Problem Solution via H_2/H_∞ Optimal Static Output Feedback," *Proceedings of the American Control Conference*, 1989, pp. 2506–2511.

[59]Haddad, W. M., and Bernstein, D. S., "Regional Pole Placement via Optimal Static and Dynamic Output Feedback," *Proceedings of the American Control Conference*, San Diego, CA, May 1990.

[60]Johnson, T. L., and Athans, M., "On the Design of Optimal Constrained Dynamic Compensators for Linear Constant Systems," *IEEE Transactions on Automatic Control*, Vol. AC-15, 1970, pp. 658–660.

[61]Levine, W. S., Johnson, T. L., and Athans, M., "Optimal Limited State Variable Feedback Controllers for Linear Systems," *IEEE Transactions on Automatic Control*, Vol. AC-16, 1971, pp. 785–793.

[62]Kwakernaak, K., and Sivan, R., *Linear Optimal Control Systems*, Wiley, New York, 1972.

[63]Rom, D. B., and Sarachik, P. E., "The Design of Optimal Compensators for Linear Constant Systems with Inaccessible States," *IEEE Transactions on Automatic Control*, Vol. AC-18, 1973, pp. 509–512.

[64]Sidar, M., and Kurtaran, B. Z., "Optimal Low-Order Controllers for Linear Stochastic Systems," *International Journal of Control*, Vol. 22, 1975, pp. 377–387.

[65]Mendel, J. M., and Feather, J., "On the Design of Optimal Time-Invariant Compensators for Linear Stochastic Time-Invariant Systems," *IEEE Transactions on Automatic Control*, Vol. AC-20, 1975, pp. 653–657.

[66]Basuthakur, S., and Knapp, C. H., "Optimal Constant Controllers for Stochastic Linear Systems," *IEEE Transactions on Automatic Control*, Vol. AC-20, 1975, pp. 664–666.

[67]Asher, R. B., and Durrett, J. C., "Linear Discrete Stochastic Control with a Reduced-Order Dynamic Compensator," *IEEE Transactions on Automatic Control*, Vol. AC-21, 1976, pp. 626–627.

[68]Naeije, W. J., and Bosgra, O. H., "The Design of Dynamic Compensators for Linear Multivariable Systems," *Proceedings of the 1977 IFAC*, 1977, pp. 205–212.

[69]Sirisena, H. R., and Choi, S. S., "Design of Optimal Constrained Dynamic Compensators for Non-Stationary Linear Systems," *International Journal of Control*, Vol. 25, 1977, pp. 513–524.

[70]Blanvillain, P. J., and Johnson, T. L., "Invariants of Optimal Minimal-Order Observer Based Compensators," *IEEE Transactions on Automatic Control*, Vol. AC-23, 1978, pp. 473–474.

[71]Wenk, C. J., and Knapp, C. H., "Parameter Optimization in Linear Systems with Arbitrarily Constrained Controller Structure," *IEEE Transactions on Automatic Control*, Vol. AC-25, 1980, pp. 496–500.

[72]O'Reilly, J., "Optimal Low-Order Feedback Controllers for Linear Discrete-Time Systems," *Control and Dynamic Systems*, Vol. 16, edited by C. T. Leondes, Academic, New York, 1980.

[73]Grimble, M. J., "Reduced-Order Optimal Controller for Discrete-Time Stochastic Systems," *Proceedings of the IEE*, Vol. 127, 1980, pp. 55–63.

[74]Looze, D. P., and Sandell, N. R., Jr., "Gradient Calculations for Linear Quadratic Fixed Control Structure Problems," *IEEE Transactions on Automatic Control*, Vol. AC-25, 1980, pp. 285–288.

[75]Martin, G. D., and Bryson, A. E., Jr., "Attitude Control of a Flexible Spacecraft," *Journal of Guidance and Control*, Vol. 3, No. 1, 1980, pp. 37–41.

[76]Mukhopadhyay, V., Newsom, J. R., and Abel, I., "Reduced-Order Optimal Feedback Control Law Synthesis for Flutter Suppression," *Journal of Guidance and Control*, Vol. 5, No. 4, 1982, pp. 389–395.

[77]Kabamba, P. T., and Longman, R. W., "An Integrated Approach to Reduced-Order Control Theory," *Optimal Control Applications and Methods*, Vol. 4, 1983, pp. 405–415.

[78]Polak, E., Mayne, D. Q., and Stimler, D. M., "Control System Design via Semi-Infinite Optimization: A Review," *IEEE Proceedings*, Vol. 72, 1984, pp. 1777–1794.

[79]Ly, U., Bryson, A. E., and Cannon, R. H., "Design of Low-Order Compensators Using Parameter Optimization," *Automatica*, Vol. 21, 1985, pp. 315–318.

[80]Mukhopadhyay, V., "Digital Active Control Law Synthesis for Large Order Space Systems Using Parameter Optimization," *Proceedings of the AIAA Guidance, Navigation, and Control Conference*, Monterey, CA, Aug. 1987, pp. 1414–1419.

[81]Makila, P. M., and Toivonen, H. T., "Computational Methods for Parametric LQ Problems—A Survey," *IEEE Transactions on Automatic Control*, Vol. AC-32, 1987, pp. 658–671.

[82]Polak, E., "On the Mathematical Foundations of Nondifferential Optimization in Engineering Design," *SIAM Review*, Vol. 29, 1987, pp. 21–90.

[83]Kramer, F. S., and Calise, A. J., "Fixed-Order Dynamic Compensation for Multivariable Linear Systems," *AIAA Journal of Guidance, Control, and Dynamics*, Vol. 11, 1988, pp. 80–85.

[84]Rao, C. R., and Mitra, S. K., *Generalized Inverse of Matrices and Its Applications*, Wiley, New York, 1971.

[85]Campbell, S. L., and Meyer, C. D., Jr., *Generalized Inverses of Linear Transformations*, Pitman, London, 1979.

[86]Wonham, W. M., *Linear Multivariable Control: A Geometric Approach*, Springer-Verlag, New York, 1979.

[87]Lloyd, N. G., *Degree Theory*, Cambridge University Press, London, 1978.

[88]Ortega, J. M., and Rheinboldt, W. C., *Iterative Solution of Nonlinear Equations in Several Variables*, Academic, New York, 1970.

[89]Avila, J. H., "The Feasibility of Continuation Methods for Nonlinear Equations," *SIAM Journal on Numerical Analysis*, Vol. 11, 1974, pp. 102–122.

[90]Wacker, H., *Continuation Methods*, Academic, New York, 1978.

[91]Alexander, J. C., and Yorke, J. A., "The Homotopy Continuation Method: Numerically Implementable Topological Procedures," *Transactions of the American Mathematical Society*, Vol. 242, 1978, pp. 271–284.

[92]Garcia, C. B., and Zangwill, W. I., *Pathways To Solutions, Fixed Points and Equilibria*, Prentice-Hall, Englewood Cliffs, NJ, 1981.

[93]Eaves, B. C., Gould, F. J., Peitgen, H. O., and Todd, M. J., *Homotopy Methods and Global Convergence*, Plenum, New York, 1983.

[94]Richter, S. L., and DeCarlo, R. A., "Continuation Methods: Theory and Applications, *IEEE Transactions on Circuits and Systems*, Vol. CAS-30, 1983, pp. 347–352.

[95]Mariton, M., and Bertrand, R., "A Homotopy Algorithm for Solving Coupled Riccati Equations," *Optimal Control Applications and Methods*, Vol. 6, 1985, pp. 351–357.

[96]Watson, L. T., "Numerical Linear Algebra Aspects of Globally Convergent Homotopy Methods," *SIAM Review*, Vol. 28, 1986, pp. 529–545.

[97]Kabamba, P. T., Longman, R. W., and Jian-Guo, S., "A Homotopy Approach to the Feedback Stabilization of Linear Systems," *Journal of Guidance, Control, and Dynamics*, Vol. 10, No. 5, 1987, pp. 422–432.

[98]Watson, L. T., "ALGORITHM 652 HOMPACK: A Suite of Codes for Globally Convergent Homotopy Algorithms," *ACM Transactions on Mathematical Software*, Vol. 13, 1987, pp. 281–310.

[99]Shin, Y. S., Haftka, R. T., Watson, L. T., and Plautt, R. H., "Tracking Structural Optima as a Function of Available Resources by a Homotopy Method," *Computer Methods in Applied Mechanics and Engineering*, Vol. 70, 1988, pp. 151–164.

[100]Watson, L. T., "Globally Convergent Homotopy Methods: A Tutorial," *Applied Mathematics and Computation*, 1989.

[101]Wang, S. S., and Pan, C. T., "Homotopy Method for Solving Algebraic Riccati Equations," *Proceedings of the IEE*, Vol. 136, 1989, pp. 68–72.

[102]Guyan, R. J., "Reduction of Stiffness and Mass Matrices," *AIAA Journal*, Vol. 3, Feb. 1965, p. 380.

[103]Irons, B., "Structural Eigenvalue Problem: Elimination of Unwanted Variables," *AIAA Journal*, Vol. 3, May 1965, pp. 961–962.

[104]Davison, E. J., "A Method for Simplifying Linear Dynamic Systems," *IEEE Transactions on Automatic Control*, Vol. AC-11, 1966, pp. 93–101.

[105]Aoki, M., "Control of Large-Scale Dynamic Systems by Aggregation," *IEEE Transactions on Automatic Control*, Vol. AC-13, 1968, pp. 246–253.

[106]Ramsden, J. N., and Stoker, J. R., "Mass Condensation: A Semi-Automatic Method for Reducing the Size of Vibration Problems," *International Journal for Numerical Methods in Engineering*, Vol. 1, 1969, pp. 333–349.

[107]Mitra, D., "Analytical Results on the Use of Reduced Models in the Control of Linear Dynamical Systems," *Proceedings of the IEE*, Vol. 116, 1969, pp. 1439–1444.

[108]Wilson, D. A., "Optimum Solution of Model-Reduction Problem," *Proceedings of the IEE*, Vol. 117, 1970, pp. 1161–1165.

[109]Kidder, R. L., "Reduction of Structural Frequency Equations," *AIAA Journal*, Vol. 11, June 1973, p. 892.

[110]Sinha, N. K., and De Bruin, H., "Near-Optimal Control of High-Order Systems Using Low-Order Models," *International Journal of Control*, Vol. 17, 1973, pp. 257–262.

[111]Applevich, J. D., "Approximation of Discrete Linear Systems," *International Journal of Control*, Vol. 17, 1973, pp. 565–575.

[112]Wilson, D. A., "Model Reduction for Multivariable Systems," *International Journal of Control*, Vol. 20, 1974, pp. 57–64.

[113]Genesio, R., and Milanese, M., "A Note on the Derivation and Use of Reduced-Order Models," *IEEE Transactions on Automatic Control*, Vol. AC-21, 1976, pp. 118–122.

[114]Hickin, J., and Sinha, N. K., "Application of Projective Reduction Methods to Estimation and Control," *Journal of Cybernetics*, Vol. 8, 1978, pp. 159–181.

[115]Tse, E. C. Y., Medinac, J. V., and Perkins, W. R., "Generalized Hessenberg Transformation for Reduced-Order Modelling of Large-Scale Systems," *International Journal of Control*, Vol. 27, 1978, pp. 493–512.

[116]Hickin, J., and Sinha, N., "Model Reduction for Linear Multivariable Systems," *IEEE Transactions on Automatic Control*, Vol. AC-25, 1980, pp. 1121–1127.

[117]Skelton, R. E., and Hughes, P. C., "Modal Cost Analysis for Linear Matrix Second-Order Systems," *ASME Transactions, Journal for Dynamic Systems, Measurement, and Control*, Vol. 102, 1980, pp. 151–158.

[118]Mishra, R. N., and Wilson, D. A., "A New Algorithm for Optimal Reduction of Multivariable Systems," *International Journal of Control*, Vol. 31, 1980, pp. 443–466.

[119]Kabamba, P. T., "Model Reduction by Euclidean Methods," *Journal of Guidance and Control*, Vol. 3, 1980, pp. 555–562.

[120]Skelton, R. E., "Cost Decomposition of Linear Systems with Application to Model Reduction," *International Journal of Control*, Vol. 32, No. 6, 1980, pp. 1031–1055.

[121]Mishra, R. N., and Wilson, D. A., "A New Algorithm for Optimal Reduction of Multivariable Systems," *International Journal of Control*, Vol. 31, 1980, pp. 443–466.

[122]Moore, B. C., "Principal Component Analysis in Linear Systems: Controllability, Observability, and Model Reduction," *IEEE Transactions on Automatic Control*, Vol. AC-26, 1981, pp. 17–32.

[123]Eitelberg, E., "Model Reduction by Minimizing the Weighted Equation Error," *International Journal of Control*, Vol. 34, 1981, pp. 1113–1123.

[124]Verriest, E. I., "Suboptimal LQG-Design and Balanced Realizations," *Proceedings of the 20th IEEE Conference on Decision and Control*, IEEE, New York, 1981, pp. 686–687.

[125]Pernebo, L., and Silverman, L. M., "Model Reduction via Balanced State Space Representations," *IEEE Transactions on Automatic Control*, Vol. AC-27, 1982, pp. 382–387.

[126]Sims, C. S., "Reduced-Order Modelling and Filtering," *Control and Dynamic Systems*, Vol. 18, edited by C. T. Leondes, Academic Press, Orlando, FL, 1982, pp. 55–103.

[127]Skelton, R. E., Hughes, P. C., and Hablani, H. B., "Order Reduction for Models of Space Structures Using Modal Cost Analysis," *Journal of Guidance, Control, and Dynamics*, Vol. 5, No. 4, 1982, pp. 351–357.

[128]Skelton, R. E., and Yousuff, A., "Component Cost Analysis of Large Scale Systems," *International Journal of Control*, Vol. 37, 1983, pp. 285–304.

[129]Doran, A. L., and Mingori, D. L., "Model Reduction by Cost Decomposition: Implications of Coordinate Selection," *Journal of the Astronautical Sciences*, Vol. 31, 1983, pp. 415–428.

[130]Fernando, K. V., and Nicholson, H., "On the Structure of Balanced and Other Principal Representations of SISO Systems," *IEEE Transactions on Automatic Control*, Vol. AC-28, 1983, pp. 228–231.

[131]Shokoohi, S., Silverman, L. M., and Van Dooren, P. M., "Linear Time-Variable Systems: Balancing and Model Reduction," *IEEE Transactions on Automatic Control*, Vol. AC-28, 1983, pp. 810–822.

[132]Verriest, E. I., and Kailath, T., "On Generalized Balanced Realizations," *IEEE Transactions on Automatic Control*, Vol. AC-28, 1983, pp. 833–844.

[133]Jonckheere, E. A., and Silverman, L. M., "A New Set of Invariants for Linear Systems—Application to Reduced-Order Compensator Design," *IEEE Transactions on Automatic Control*, Vol. AC-28, 1983, pp. 953–964.

[134]Yousuff, A., and Skelton, R. E., "Controller Reduction by Component Cost Analysis," *IEEE Transactions on Automatic Control*, Vol. AC-29, 1984, pp. 520–530.

[135]Gregory, C. Z., Jr., "Reduction of Large Spacecraft Models Using Internal Balancing Theory," *Journal of Guidance, Control, and Dynamics*, Vol. 7, No. 6, 1984, pp. 725–732.

[136]Jonckheere, E. A., "Principal Component Analysis of Flexible Systems—Open Loop Case," *IEEE Transactions on Automatic Control*, Vol. AC-29, 1984, pp. 1095–1097.

[137]Enns, D., "Model Reduction for Control System Design," Ph.D. Dissertation, Stanford Univ., Palo Alto, CA, 1984.

[138]Glover, G., "All Optimal Hankel-Norm Approximations of Linear Multivariable Systems and Their L_∞-Error Bounds," *International Journal of Control*, Vol. 39, 1984, pp. 1115–1193.

[139]Davis, J. A., and Skelton, R. E., "Another Balanced Controller Reduction Algorithm," *Systems and Control Letters*, Vol. 4, 1984, pp. 79–83.

[140]Yousuff, A., and Skelton, R. E., "A Note on Balanced Controller Reduction," *IEEE Transactions on Automatic Control*, Vol. 29, 1984, pp. 254–257.

[141]Vaccaro, R. J., "Deterministic Balancing and Stochastic Model Reduction," *IEEE Transactions on Automatic Control*, Vol. AC-30, 1985, pp. 921–923.

[142]Lu, W.-S., and Lee, E. B., "Model Reduction via a Quasi-Kalman Decomposition," *IEEE Transactions on Automatic Control*, Vol. AC-30, 1985, pp. 786–790.

[143]Kabamba, P. T., "Balanced Gains and Their Significance for Balanced Model Reduction," *IEEE Transactions on Automatic Control*, Vol. AC-30, 1985, pp. 690–693.

[144]Liu, Y., and Anderson, B. D. O., "Controller Reduction via Stable Factorization and Balancing," *International Journal of Control*, Vol. 44, 1986, pp. 507–531.

[145]De Villemagne, D., and Skelton, R. E., "Controller Reduction Using Canonical Interactions," *IEEE Transactions on Automatic Control*, Vol. 33, 1988, pp. 740–750.

[146]Anderson, B. D. O., and Liu, Y., "Controller Reduction: Concepts and Approaches, *IEEE Transactions on Automatic Control*, Vol. 34, 1987, pp. 802–812.

[147]Cannon, R. H., Jr., and Rosenthal, D. E., "Experiments in Control of Flexible Structures with Noncolocated Sensors and Actuators," *AIAA Journal of Guidance, Control, and Dynamics*, Vol. 7, 1984, pp. 546–553.

[148]Peterson, L. D., Allen, J. J., Lauffer, J. P., and Miller, A. K., "An Experimental and Analytical Synthesis of Controlled Structure Design," AIAA Paper 89-1170, April 1989.

[149]Peterson, L. D., "An Optimal Projection Controller for an Experimental Truss Structure," *Proceedings of the AIAA Guidance, Navigation, and Control Conference*, Boston, MA, Aug. 1989, pp. 77–88.

[150]Doyle, J. C., "Guaranteed Margins for LQG Regulators," *IEEE Transactions on Automatic Control*, Vol. AC-23, 1978, pp. 756–757.

[151]Soroka, E., and Shaked, U., "On the Robustness of LQ Regulators," *IEEE Transactions on Automatic Control*, Vol. AC-29, 1984, pp. 664–665.

[152]Shaked, U., and Soroka, E., "On the Stability Robustness of the Continuous-Time LQG Optimal Control," *IEEE Transactions on Automatic Control*, Vol. AC-30, 1985, pp. 1039–1043.

[153]Safonov, M. G., and Athans, M., "Gain and Phase Margin for Multiloop LQG Regulators," *IEEE Transactions on Automatic Control*, Vol. AC-22, 1977, pp. 173–179.

[154]Doyle, J. C., and Stein, G., "Robustness with Observers," *IEEE Transactions on Automatic Control*, Vol. AC-24, 1979, pp. 607–611.

[155]Doyle, J. C., and Stein, G., "Multivariable Feedback Design: Concepts for a Classical/Modern Synthesis," *IEEE Transactions on Automatic Control*, Vol. AC-26, 1981, pp. 4–6.

[156]Stein, G., and Athans, M., "The LQG/LTR Procedure for Multivariable Feedback Control Design," *IEEE Transactions on Automatic Control*, Vol. AC-32, 1987, pp. 105–114.

[157]Gupta, N. K., "Frequency-Shaped Cost Functionals: Extension of Linear-Quadratic-Gaussian Design Methods," *Journal of Guidance, Control, and Dynamics*, Vol. 3, No. 6, 1980, pp. 529–535.

[158]Safonov, M. G., Laub, A. J., and Hartman, G. L., "Feedback Properties of Multivariable Systems: The Role and Use of the Return Difference Matrix," *IEEE Transactions on Automatic Control*, Vol. 26, 1981, pp. 47–65.

[159]Anderson, B. D. O., and Mingori, D. L., "Use of Frequency Dependence in Linear Quadratic Control Problems to Frequency-Shape Robustness," *Journal of Guidance, Control, and Dynamics*, Vol. 8, 1985, pp. 397–401.

[160]Zames, G., "Feedback and Optimal Sensitivity: Model Reference Transformations, Multiplicative Seminorms, and Approximate Inverses," *IEEE Transactions on Automatic Control*, Vol. AC-26, 1981, pp. 301–320.

[161]Zames, G., and Francis, B. A., "Feedback, Minimax Sensitivity, and Optimal Robustness," *IEEE Transactions on Automatic Control*, Vol. AC-28, 1983, pp. 585–601.

[162]Francis, B. A., and Doyle, J. C., "Linear Control Theory with an H_∞ Optimality Criterion," *SIAM Journal on Control and Optimization*, Vol. 25, 1987, pp. 815–844.

[163]Francis, B. A., *A Course in H_∞ Control Theory*, Springer-Verlag, New York, 1987.

[164]Safonov, M. G., and Chiang, R. Y., "CACSD Using the State-Space L^∞ Theory—A Design Example," *IEEE Transactions on Automatic Control*, Vol. 33, 1988, pp. 477–479.

[165]Brockett, R. W., *Finite Dimensional Linear Systems*, Wiley, New York, 1970.

[166]Willems, J. C., "Least Squares Stationary Optimal Control and the Algebraic Riccati Equation," *IEEE Transactions on Automatic Control*, Vol. AC-16, 1971, pp. 621–634.

[167]Petersen, I. R., "Disturbance Attenuation and H^∞ Optimization: A Design Method Based on the Algebraic Riccati Equation," *IEEE Transactions on Automatic Control*, Vol. AC-32, 1987, pp. 427–429.

[168]Khargonekar, P. P., Petersen, I. R., and Rotea, M. A., "H^∞ Optimization: A Design Method with State-Feedback," *IEEE Transactions on Automatic Control*, Vol. 33, 1988, pp. 786–788.

[169]Zhou, K., and Khargonekar, P. P., "An Algebraic Riccati Equation Approach to H^∞ Optimization," *Systems and Control Letters*, Vol. 11, 1988, pp. 85–91.

[170]Glover, K., and Doyle, J. C., "State-Space Formulae for All Stabilizing Controllers that Satisfy an H_∞-Norm Bound and Relations to Risk Sensitivity," *Systems and Control Letters*, Vol. 11, 1988, pp. 167–172.

[171]Doyle, J. C., Khargonekar, P. P., and Francis, B. A., "State-Space Solutions to Standard H_2 and H_∞ Control Problems," *IEEE Transactions on Automatic Control*, Vol. 34, 1989, pp. 831–847.

[172]Jacobson, D. H., *Extensions of Linear-Quadratic Control, Optimization and Matrix Theory*, Academic, New York, 1977.

[173]Whittle, P., "Risk-Sensitive Linear/Quadratic/Gaussian Control," *Advances in Applied Probability*, Vol. 13, 1981, pp. 764–777.

[174]Bensoussan, A., and van Schuppen, J. H., "Optimal Control of Partially Observable Stochastic Systems with an Exponential-of-Integral Performance Index," *SIAM Journal on Control and Optimization*, Vol. 23, 1985, pp. 599–613.

[175]Mustafa, D., and Glover, K., "Controllers Which Satisfy a Closed-Loop H_∞ Norm Bound and Maximize an Entropy Integral," *Proceedings of the IEEE Conference on Decision and Control*, IEEE, New York, 1988, pp. 959–964.

[176]Mustafa, D., "Relations between Maximum Entropy/H_∞ Control and Combined H_∞/LQG Control," *Systems and Control Letters*, Vol. 12, 1989, pp. 193–203.

[177]Mustafa, D., "Minimum Entropy H_∞ Control," Ph.D. Dissertation, Cambridge University Press, London, 1989.

[178]Chang, S. S. L., and Peng, T. K. C., "Adaptive Guaranteed Cost Control of Systems with Uncertain Parameters," *IEEE Transactions on Automatic Control*, Vol. AC-17, 1972, pp. 474–483.

[179]Menga, G., and Dorato, P., "Observer-Feedback Design for Linear Systems with Large Parameter Uncertainties," *Proceedings of the IEEE Conference on Decision and Control*, IEEE, New York, 1974, pp. 872–878.

[180]Vinkler, A., and Wood, L. J., "Multistep Guaranteed Cost Control of Linear Systems with Uncertain Parameters," *Journal of Guidance and Control*, Vol. 2, No. 6, 1979, pp. 449–456.

[181]Leitmann, G., "Guaranteed Asymptotic Stability for a Class of Uncertain Linear Dynamical Systems," *Journal of Optimization Theory and Applications*, Vol. 27, 1979, pp. 99–106.

[182]Eslami, M., and Russell, D. L., "On Stability with Large Parameter Variations: Stemming from the Direct Method of Lyapunov," *IEEE Transactions on Automatic Control*, Vol. AC-25, 1980, pp. 1231–1234.

[183]Corless, M., and Leitmann, G., "Continuous State Feedback Guaranteeing Uniform Ultimate Boundedness for Uncertain Dynamical Systems," *IEEE Transactions on Automatic Control*, Vol. AC-26, 1981, pp. 1139–1144.

[184]Thorp, J. S., and Barmish, B. R., "On Guaranteed Stability of Uncertain Linear Systems via Linear Control," *Journal of Optimization Theory and Applications*, Vol. 35, 1981, pp. 559–579.

[185]Noldus, E., "Design of Robust State Feedback Laws," *International Journal of Control*, Vol. 35, 1982, pp. 935–944.

[186]Barmish, B. R., Corless, M., and Leitmann, G., "A New Class of Stabilizing Controllers for Uncertain Dynamical Systems," *SIAM Journal on Control and Optimization*, Vol. 21, 1983, pp. 246–255.

[187]Barmish, B. R., Petersen, I. R., and Feuer, A., "Linear Ultimate Boundedness Control of Uncertain Dynamic Systems," *Automatica*, Vol. 19, 1983, pp. 523–532.

[188]Petersen, I. R., "A Riccati Equation Approach to the Design of Stabilizing Controllers and Observers for a Class of Uncertain Linear Systems," *IEEE Transactions on Automatic Control*, Vol. AC-30, 1985, pp. 904–907.

[189]Barmish, B. R., "Necessary and Sufficient Conditions for Quadratic Stabilizability of an Uncertain Linear System," *Journal of Optimization Theory and Applications*, Vol. 46, 1985, pp. 399–408.

[190]Yedavalli, R. K., Banda, S. S., and Ridgely, D. B., "Time-Domain Stability Robustness Measures for Linear Regulators," *Journal of Guidance, Control, and Dynamics*, Vol. 8, No. 4, 1985, pp. 520–524.

[191]Petersen, I. R., and Hollot, C. V., "A Riccati Equation Approach to the Stabilization of Uncertain Systems," *Automatica*, Vol. 22, 1986, pp. 397–411.

[192]Galimidi, A. R., and Barmish, B. R., "The Constrained Lyapunov Problem and Its Application to Robust Output Feedback Stabilization," *IEEE Transactions on Automatic Control*, Vol. AC-31, 1986, pp. 410–419.

[193]Zhou, K., and Khargonekar, P. P., "Stability Robustness Bounds for Linear State-Space Models with Structured Uncertainty," *IEEE Transactions on Automatic Control*, Vol. AC-32, 1987, pp. 621–623.

[194]Hinrichsen, D., and Pritchard, A. J., "Stability Radius for Structured Perturbations and the Algebraic Riccati Equation," *Systems and Control Letters*, Vol. 8, 1987, pp. 105–113.

[195]Kosmidou, O. I., and Bertrand, P., "Robust-Controller Design for Systems with Large Parameter Variations," *International Journal of Control*, Vol. 45, 1987, pp. 927–938.

[196]Keel, L. H., Bhattacharyya, S. P., and Howze, J. W., "Robust Control with Structured Perturbations," *IEEE Transactions on Automatic Control*, Vol. 33, 1988, pp. 68–78.

[197]Arnold, L., *Stochastic Differential Equations: Theory and Applications*, Wiley, New York, 1974.

[198]Wonham, W. M., "Optimal Stationary Control of a Linear System with State-Dependent Noise," *SIAM Journal on Control*, Vol. 5, 1967, pp. 486–500.

[199]Wonham, W. M., "On a Matrix Riccati Equation of Stochastic Control," *SIAM Journal on Control*, Vol. 6, 1968, pp. 681–697.

[200]Kleinman, D., "Optimal Stationary Control of Linear Systems with Control-Dependent Noise," *IEEE Transactions on Automatic Control*, Vol. AC-14, 1969, pp. 673–677.

[201]McLane, P., "Optimal Stochastic Control of Linear Systems with State and Control-Dependent Disturbances," *IEEE Transactions on Automatic Control*, Vol. AC-16,1971, pp. 793–798.

[202]Haussmann, U., "Optimal Stationary Control with State and Control-Dependent Noise," *SIAM Journal on Control and Optimization*, Vol. 9, 1971, pp. 419–444.

[203]Bismut, J., "Linear-Quadratic Optimal Stochastic Control and Random Coefficients," *SIAM Journal on Control and Optimization*, Vol. 14, 1976, pp. 419–444.

[204]Merriam, C. W. III, *Automated Design of Control Systems*, Gordon & Breach, New York, 1974.

[205]Gustafson, D. E., and Speyer, J. L., "Design of Linear Regulators for Nonlinear Systems," *Journal of Spacecraft and Rockets*, Vol. 12, June 1975, pp. 351–358.

[206]Mil'shtein, G. N., "Design of Stabilizing Controller with Incomplete State Data for Linear Stochastic System with Multiplicative Noise," *Automation and Remote Control*, Vol. 43, 1982, pp. 653–659.

[207]Phillis, Y. A., "Controller Design of Systems with Multiplicative Noise," *IEEE Transactions on Automatic Control*, Vol. AC-30, 1985, pp. 1017–1019.

[208]Yeh, H.-H., Banda, S. S., and Heise, S. A., "Robust Design of Multivariable Feedback Systems with Real Parameter Uncertainty and Unmodelled Dynamics," *Proceedings of the American Control Conference*, 1989, pp. 662–670.

[209]Lyon, R. H., and Maidanik, G., "Power Flow Between Linearly Coupled Oscillators," *Journal of the Acoustical Society of America*, Vol. 34, 1962, pp. 623–639.

[210]Ungar, E. E., "Statistical Energy Analysis of Vibrating Systems," *Transactions of the American Society of Mechanical Engineers*, Vol. 89, 1967, pp. 626–632.

[211]Scharton, T. D., and Lyon, R. H., "Power Flow and Energy Sharing in Random Vibration," *Journal of the Acoustical Society of America*, Vol. 43, 1968, pp. 1332–1343.

[212]Lyon, R. H., *Statistical Energy Analysis of Dynamical Systems: Theory and Applications*, MIT Press, Cambridge, MA, 1975.

[213]Woodhouse, J., "An Approach to the Theoretical Background of Statistical Energy Analysis Applied to Structural Vibration," *Journal of the Acoustical Society of America*, Vol. 69, 1981, pp. 1695–1709.

[214]Dowell, E. H., and Kubota, Y., "Asymptotic Modal Analysis and Statistical Energy Analysis of Dynamical Systems," *Journal of Applied Mechanical Engineering*, Vol. 52, 1985, pp. 949–957.

[215]Stratonovich, R. L., "A New Representation for Stochastic Integrals," *SIAM Journal on Control*, Vol. 4, 1966, pp. 362–371.

[216]Wong, E., and Zakai, M., "On the Relation between Ordinary and Stochastic Differential Equations," *International Journal of Engineering Science*, Vol. 2, 1965, pp. 213–229.

[217]Ikeda, N., and Watanabe, S., *Stochastic Differential Equations and Diffusion Processes*, North-Holland, Amsterdam, 1981.

[218]Jaynes, E. T., "Information Theory and Statistical Mechanics," *Physical Review*, Vol. 106, 1957, pp. 620–630.

[219]Jaynes, E. T., "New Engineering Applications of Information Theory," *Proceedings of the First Symposium on Engineering Applications of Random Function Theory and Probability*, edited by J. L. Bogdanoff and F. Kozin, Wiley, New York, 1963, pp. 163–203.

[220]Jaynes, E. T., "Prior Probabilities," *IEEE Transactions on Systems Science and Cybernetics*, Vol. SSC-4, 1968, pp. 227–241.

[221]Jaynes, E. T., "Where Do We Stand on Maximum Entropy?" *The Maximum Entropy Formalism*, edited by D. Levine and M. Tribus, MIT Press, Cambridge, MA, 1979, pp. 15–118.

[222]Gruzen, A., "Robust Reduced Order Control of Flexible Structures," C. S. Draper Lab. Rept. CSDL-T-900, April 1986.

[223]Gruzen, A., and Vander Velde, W. E., "Robust Reduced-Order Control of Flexible Structures Using the Optimal Projection/Maximum Entropy Design Methodology," *AIAA Guidance, Navigation, and Control Conference*, Williamsburg, VA, Aug. 1978.

[224]Cheung, M., and Yurkovich, S., "On the Robustness of MEOP Design Versus Asymptotic LQG Synthesis," *IEEE Transactions on Automatic Control*, Vol. 33, 1988, pp. 1061–1065.

[225]Anderson, B. D. O., Bose, N., and Jury, E., "Output Feedback Stabilization and Related Problems-Solutions via Decision Methods," *IEEE Transactions on Automatic Control*, Vol. AC-20, 1975, pp. 53–66.

[226]Byrnes, C. I., and Anderson, B. D. O., "Output Feedback and Generic Stabilizability," *SIAM Journal on Control and Optimization*, Vol. 11, 1984, pp. 362–380.

[227]Anderson, B. D. O., "A System Theory Criterion of Positive Real Matrices," *SIAM Journal on Control and Optimization*, Vol. 5, 1967, pp. 171–182.

[228]Balas, M. J., "Direct Velocity Feedback Control of Large Space Structures," *Journal of Guidance and Control*, Vol. 2, 1967, pp. 252–253.

[229]Dickinson, B. W., "Analysis of the Lyapunov Equation Using Generalized Positive Real Matrices," *IEEE Transactions on Automatic Control*, Vol. AC-25, 1980, pp. 560–563.

[230]Benhabib, R. J., Iwens, R. P., and Jackson, R. L., "Stability of Large Space Structure Control Systems Using Positivity Concepts," *Journal of Guidance and Control*, Vol. 4, No. 5, 1981, pp. 487–494.

[231]Joshi, S. M., "Robustness Properties of Collocated Controllers for Flexible Spacecraft," *Journal of Guidance, Control, and Dynamics*, Vol. 9, No. 1, 1986, pp. 85–91.

[232]McLaren, M. D., and Slater, G. L., "Robust Multivariable Control of Large Space Structures Using Positivity," *Journal of Guidance, Control, and Dynamics*, Vol. 10, No. 4, 1987, pp. 393–400.

[233]Safonov, M. G., Jonckheere, E. A., Verma, M., and Limebeer, D. J. N., "Synthesis of Positive Real Multivariable Feedback Systems," *International Journal of Control*, Vol. 45, 1987, pp. 817–842.

[234]Doyle, J. C., "Analysis of Feedback Systems with Structured Uncertainties," *Proceedings of the IEE-D*, Vol. 129, 1982, pp. 242–250.

[235]Safonov, M. G., "Stability Margins of Diagonally Perturbed Multivariable Feedback Systems," *Proceedings of the IEE-D*, Vol. 129, 1982, pp. 251–256.

[236]Madiwale, A. N., "Reduction of Conservatism in Worst Case Designs: Systems with Block Diagonal Uncertainty Structure," *Proceedings of the American Control Conference*, 1989, pp. 770–771.

[237]Madiwale, A. N., "Design of LQG/H_∞ Dynamic Compensators for Time Varying Nonlinear Feedback Systems: Circle Criterion and Guaranteed Gain and Phase Margin," *Proceedings of the American Control Conference*, 1989, pp. 2049–2055.

[238]Boyd, S., "A Note on Parametric and Nonparametric Uncertainties in Control Systems," *Proceedings of the American Control Conference*, 1986, pp. 1847–1849.

[239]Hollot, C. V., Looze, D. P., and Bartlett, A. C., "Parametric Uncertainty and Unmodelled Dynamics: Analysis via Parameter Space Methods," *Automatica*, 1990.

[240]Iftar, A., and Ozguner, U., "Modeling of Uncertain Dynamics for Robust Controller Design in State Space" (Preprint).

Chapter 11

Low-Order Control of Linear Finite-Element Models of Large Flexible Structures Using Second-Order Parallel Architecture

Mark J. Balas*
University of Colorado, Boulder, Colorado

Introduction to the Problem

ACTIVE control of large flexible structures (LFSs) would be used to improve performance, e.g., shape (figure) fidelity, line-of-sight pointing accuracy, vibration, and disturbance suppression. Large-scale models of such structures are appropriate for dynamic simulation but cannot be used as a basis for control algorithms, which must be computer-implemented in real time. Therefore, control algorithms are based on reduced-order models (ROMs) of the structure dynamics. Whenever such a controller operates in closed loop with the actual structure, unwanted controller/structure interaction (CSI) occurs due to unmodeled dynamics and can cause performance degradation and even instability.[4,5]

Using distributed-parameter system (DPS) theory, we have shown[1-3] that CSI instabilities can be relieved by using very low-order residual mode filters (RMFs) in conjunction with the original ROM controller to produce stable closed-loop control of general linear DPSs. In this chapter we will develop advancd compensation techniques for CSI problems in large space structures, using these RMF ideas on finite-element structure models.

The closed-loop format will consist of a ROM-based controller designed for performance and CSI compensation adjoined to produce stability without substantial loss of performance. The results will follow from the

*Professor, Center for Space Structures and Controls and Department of Aerospace Engineering Sciences.

development of *second-order* observer-based controllers for second-order LFS models.

Structure Models

A DPS model of a structure would look like

$$m(x)\frac{\partial^2 q(x,t)}{\partial t^2} + D_0\frac{\partial q(x,t)}{\partial t} + A_0 q(x,t) = F(x,t) \tag{1a}$$

$$F(x,t) = \sum_{i=1}^{M} b_i(x)\boldsymbol{u}_i(t) + \text{disturbances} \tag{1b}$$

$$y_j(t) = \int_\Omega \left[c_0^j(x)q(x,t) + e_0^j(x)\frac{\partial q(x,t)}{\partial t} \right] \mathrm{d}x, \qquad j = 1, \ldots, P \tag{1c}$$

where $q(x,t)$ represents distributed displacements of the structure Ω from its equilibrium at location x and at time t. The damping D_0 and stiffness A_0 operators would be appropriate differential operators describing these structural properties. The input and output vectors $\boldsymbol{u} = [\boldsymbol{u}_1, \ldots, \boldsymbol{u}_M]^T$ and $\boldsymbol{y} = [y_1, \ldots, y_P]^T$ represent M linearly independent actuator inputs and P linearly independent sensor outputs, respectively.

Typical large-scale approximations of Eqs. (1) can be obtained by using finite-element methods (FEMs):

$$q(x,t) \approx \sum_{k=1}^{L} q_k(l)\theta_k(x) \tag{2}$$

where $\theta_k(x)$ are known local spline approximations of the structure (or finite elements). The Galerkin method yields the large-scale dynamic approximation of the structure in physical coordinates:

$$M_0\ddot{q} + D_0\dot{q} + K_0 q = B_0\boldsymbol{u} \tag{3a}$$

$$\boldsymbol{y} = C_0 q + E_0\dot{q} \tag{3b}$$

where

$$q \equiv [q_1, \ldots, q_L]^T$$

Mass matrix:

$$M_0 \equiv \left[\int_\Omega m(x)\theta_K(x)\theta_1(x)\,\mathrm{d}x] \right]$$

Damping matrix:

$$D_0 \equiv \left[\int_\Omega \theta_k(x)D_0\theta_1(x)\,\mathrm{d}x \right]$$

Stiffness matrix:

$$K_0 \equiv \left[\int_\Omega \theta_k(x) A_0 \theta_1(x) \, dx \right]$$

Input matrix:

$$B_0 \equiv \left[\int_\Omega \theta_k(x) b_i(x) \, dx \right]$$

Output matrices:

$$C_0 \equiv \left[\int_\Omega \theta_k(x) c_0^j(x) \, dx \right]$$

$$E_0 \equiv \left[\int_\Omega \theta_k(x) e_0^j(x) \, dx \right]$$

This is also called the configuration space model. For real structures it is more likely that lumped models such as Eqs. (3) based on FEMs are available through structural dynamics numerical methods (e.g., NASTRAN); the DPS model [Eqs. (1)] may be more difficult to obtain (although it does yield additional theoretical benefits if it is available).

Since the damping in a space structure is rather poorly known and is expected to be quite small, it is usually assumed that $D_0 = 0$. When the effects of damping are needed in the D_0 model, Rayleigh damping is usually introduced by structural dynamists:

$$D_0 = \alpha_1 M_0 + \alpha_2 K_0 \tag{4}$$

This is a convenient damping model for analysis and simulation but not necessarily a correct model for aerospace structures. Whenever damping is essential to our development, we will assume a convenient model, such as Eq. (4).

The mode shapes ϕ_k and mode frequenices ω_k are calculated from the generalized eigenproblem associated with Eqs. (3):

$$(K_0 - \omega_k^2 M_0)\phi_k = 0 \tag{5}$$

where K_0, M_0 are both symmetric and positive-semidefinite (and M_0 is positive-definite). Modal coordinates z are obtained by the transformation

$$q = \Phi z \tag{6}$$

where $\Phi = [\phi_1, ..., \phi_L]$. Thus,

$$\Phi^T M_0 \Phi = I \tag{7a}$$

$$\Phi K_0 \Phi = \Lambda = \text{diag}[\lambda_1, ..., \lambda_L] \tag{7b}$$

where $\lambda_k \equiv \omega_k^2$. Note that $\Phi\Phi^T = M_0^{-1}$ and $\Phi^{-1} = \Phi^T M_0$. The numerical calculation of modal data ω_k, ϕ_k takes great advantage of the symmetry and sparsity of M_0 and K_0 in Eq. (5). It is typical that, even though there are L modes in the FEM approximation [Eqs. (3)], the number L_m of modes computed is much smaller, e.g., $L \sim 1000$ and $L_m \sim 100$.

The most current algorithms that compute modal data are based on the Lanczos method. They give good accuracy when $L \gg L_m$; i.e., the number L_m of usable modes is far less than the number L of degrees of freedom in the physical model [Eqs. (3)]. This is due largely to the accumulation of round-off error in the calculation. Hence, we are restricted to consideration of the system in modal coordinates:

$$\ddot{z} + \Lambda z = \tilde{B}_0 \boldsymbol{u} \tag{8a}$$

$$y = \tilde{C}_0 z + \tilde{E}_0 \dot{z} \tag{8b}$$

where $\dim z = L_m \ll L = \dim q$, and

$$\Phi \equiv [\phi_1, ..., \phi_{L_m}]$$

$$\Lambda \equiv \text{diag}[\lambda_1, ..., \lambda_{L_m}], \qquad \tilde{B}_0 \equiv \Phi^T B_0, \qquad \tilde{C}_0 \equiv C_0 \Phi, \qquad \text{and} \qquad \tilde{E}_0 = E_0 \Phi$$

Model Reduction and Controller Design

Although the modal model [Eqs. (8)] is reduced in dimension from the physical model [Eqs. (3)], it is still too large to be used as the basis for a feedback controller. Further model reduction must be done to obtain a low-order structure controller. We will illustrate this procedure using the modal model [Eqs. (8)] of the structure, although nonmodal model reductions are possible.

Let

$$x \equiv \begin{bmatrix} z \\ \dot{z} \end{bmatrix}$$

and the system in state-space form becomes

$$\dot{x} = Ax + \boldsymbol{B}\boldsymbol{u} \tag{9a}$$

$$y = Cx \tag{9b}$$

where

$$A \equiv \begin{bmatrix} 0 & I \\ -\Lambda & 0 \end{bmatrix}, \qquad B \equiv \begin{bmatrix} 0 \\ \tilde{B}_0 \end{bmatrix}$$

$$C \equiv [\tilde{C}_0 \quad \tilde{E}_0]$$

In modal coordinates each mode is decoupled from all others (as long as no feedback control is applied); hence, it is relatively easy to select a modal

ROM by retaining modes that are most critical to the desired structural performance:

$$\dot{x}_N = A_N x_N + B_N \boldsymbol{u} \tag{10a}$$

$$y_N = C_N x_N \tag{10b}$$

where x_N contains the mode amplitudes and velocities of the critical modes ($\dim x_N = 2N$). The open-loop system [Eqs. (9)] becomes

$$\dot{x}_N = A_N x_N + B_N \boldsymbol{u} \tag{11a}$$

$$\dot{x}_R = A_R x_R + B_R \boldsymbol{u} \tag{11b}$$

$$y = y_N + y_R = C_N x_N + C_R x_R \tag{11c}$$

where x_R contains the mode amplitudes and velocities of the residual modes ($\dim x_R = 2R$ where $N + R = L_m$).

Controller design is based entirely on the ROM [Eqs. (10)]. It is easy to assess controllability and observability of the ROM since it is equivalent to that of $(\Lambda, \tilde{B}_0, \tilde{C}_0)$ when no velocity sensors are used ($E_0 = 0$). This is a simple test because Λ is diagonal; hence, each row of $\tilde{B}_0$ and each column of $\tilde{C}_0$ must have at least one nonzero entry.[4] These tests determine basic actuator-sensor placement. The ROM-based controller will have the form

$$\boldsymbol{u} = L_{11} y + L_{12} \zeta \tag{12a}$$

$$\dot{\zeta} = L_{21} y + L_{22} \zeta \tag{12b}$$

where $\dim \zeta = \alpha \leqslant 2N$, and the controller gains L_{ij} are determined from a design based entirely on the ROM data (A_N, B_N, C_N). For example, a particular design using a state estimator for x_N will have $z = \hat{z}_N$, $L_{11} = 0$, $L_{12} = G_N$, $L_{21} = K_N$, and $L_{22} = A_N + B_N G_N - K_N C_N$; this design is based on gains G_N and K_N chosen so that $A_N + B_N G_N$ and $A_N - K_N C_N$ have appropriate (stable) pole locations.

Closed-Loop Stability and CSI Compensation

Since the controller is based on the ROM equation [Eqs. (10)], it interacts with unmodeled structure dynamics, and this is called CSI. This interaction occurs with the residual modes of Eqs. (11) through the terms $B_R \boldsymbol{u}$ and $y_R = C_R x_R$; CSI can cause deterioration of the designed performance, reduced stability margin, and even instability.[4,5]

If no CSI compensation is used, the closed-loop system [Eqs. (11) and (12)] becomes

$$\frac{\mathrm{d}}{\mathrm{d}t}\begin{bmatrix} x_N \\ \zeta \\ x_R \end{bmatrix} = A_c \begin{bmatrix} x_N \\ \zeta \\ x_R \end{bmatrix} \tag{13}$$

where

$$A_c = \begin{bmatrix} A_N + B_N L_{11} C_N & B_N L_{12} & B_N L_{11} C_R \\ L_{21} C_N & L_{22} & L_{21} C_R \\ B_R L_{11} C_N & B_R L_{12} & A_R + B_R L_{11} C_R \end{bmatrix}$$

It is difficult (usually impossible) to decide whether A_c is stable by simple inspection. Various perturbation techniques have been proposed (e.g., Ref. 4), where

$$A_c = A_0 + \Delta A \tag{14}$$

with A_0 stable and ΔA a perturbation. Then, if ΔA is sufficiently small, i.e., if the CSI terms $(B_N L_{11} C_R, L_{11} L_{21} C_R, B_R L_{11} C_N, B_R L_{12})$ are small enough, A_c is stable. These may be useful for controller/structure redesign; i.e., it may be possible to make small refinements in structure design and/or controller design that will eliminate instability generated by CSI without further complication of the control algorithm. When this is not possible, we must consider CSI compensation:

$$y_c = H_{11} y + H_{12} u + H_{13} v \tag{15a}$$

$$\dot{v} = H_{21} y + H_{22} u + H_{23} v \tag{15b}$$

where $\dim v \equiv \beta$, and the controller becomes (with y replaced by y_c):

$$u = L_{11} y_c + L_{12} \zeta \tag{16a}$$

$$\dot{\zeta} = L_{21} y_c + L_{22} \zeta \tag{16b}$$

Although Eqs. (15) and (16) can be combined into one controller-compensator of dimension α and β, we believe they should be separately designed. The benefits from this include the following:

1) The original ROM-based controller gains remain unchanged.

2) The CSI compensator [Eq. (15)] may be very simple to implement (as it is for residual mode filters).

3) The controller is designed for performance and the CSI compensator is designed for closed-loop stability.

Observer-Based Controllers for LFS

Henceforth, we will assume a *modal model* of the LFS equation [Eq. (3)]:

$$\ddot{z} + 2\zeta \Lambda^{1/2} \dot{z} + \Lambda z = \tilde{B}_0 u \tag{17a}$$

$$y = \tilde{C}_0 z + \tilde{E}_0 \dot{z} \tag{17b}$$

where z, Λ, $\tilde{B}_0$, $\tilde{C}_0$, and $\tilde{E}_0$ are as in Eqs. (8), and

$$\zeta \equiv \text{diag}[\zeta_1, \ldots, \zeta_L], \qquad \Lambda^{1/2} \equiv \text{diag}[\omega_1, \ldots, \omega_L]$$

Thus, we have assumed that the physical damping D_0 in Eqs. (3) is diagonalized by the modal matrix Φ:

$$\Phi^T D_0 \Phi = 2\zeta\Lambda^{1/2} \tag{18}$$

This is a convenience, as was discussed in the section on structure models.

The simplest control approach for LFS models of this form would be to convert to *state-space* or *first-order* form as in Eqs. (9), where

$$A = \begin{bmatrix} 0 & I \\ -\Lambda & -2\zeta\Lambda^{1/2} \end{bmatrix}$$

and B, C are given as before, with

$$x = \begin{bmatrix} z \\ \dot{z} \end{bmatrix}$$

When (A,B,C) is controllable and observable, the *observer-based controller*

$$\boldsymbol{u} = G\hat{x} \tag{19a}$$

$$\dot{\hat{x}} = A\hat{x} + B\boldsymbol{u} + K(y - \hat{y}) \tag{19b}$$

$$\hat{y} = C\hat{x} \tag{19c}$$

produces a closed-loop system whose poles are those of $A + BG$ and $A - KC$, which can be chosen arbitrarily using the gain matrices G and K. This is standard state-space theory where Eqs. (19b) and (19c) are called a state estimator or "observer." Two serious problems arise for implementation of the controller equation [Eqs. (19)]:

1) The controller is in first-order form even though the LFS model is second order.

2) The controller dimension is twice the dimension (L) of the LFS model.

We address these concerns in the next sections.

Second-Order Architecture for LFS Observer-Based Controllers

Since the LFS models [Eqs. (17)] are second order, it seems "natural" (we enjoy the use of this word in the aerospace structures literature because it implies that others are doing unnatural things and perhaps should be arrested) that the observer-based controllers would be also; i.e., the observer in Eqs. (19) could be rewritten as a *natural* second-order controller:

$$\ddot{\hat{z}} + 2\zeta\Lambda^{1/2}\dot{\hat{z}} + \Lambda\hat{z} = \tilde{B}_0 \boldsymbol{u} + K_3(\boldsymbol{y} - \hat{\boldsymbol{y}}) \tag{20a}$$

$$\hat{\boldsymbol{y}} = \tilde{C}_0\hat{z} + E_0\dot{\hat{z}} \tag{20b}$$

Then we would expect that by proper choice of K_3, whenever (A,C) observable,

$$\begin{bmatrix} \hat{z} \\ \dot{\hat{z}} \end{bmatrix}$$

converges to x arbitrarily fast as $t \to \infty$. Except for the example of a single mass and spring with position measurement alone ($M_0 = 1$, $K_0 = 1$, $C_0 = 1$, $E_0 = 0$, $B_0 = 1$), such a natural second-order observer will not converge even though the corresponding (A,C) is observable. Nevertheless, in this section we will find a realizable second-order architecture for the observer-based controller that is nearly natural.

Let

$$\hat{x} = \begin{bmatrix} \hat{z} \\ \hat{\dot{z}} \end{bmatrix}$$

and

$$G = [G_1 \quad G_2], \qquad K = \begin{bmatrix} K_1 \\ K_2 \end{bmatrix}$$

in Eqs. (19):

$$\boldsymbol{u} = G_1\hat{z} + G_2\hat{\dot{z}} \tag{21a}$$

$$\dot{\hat{z}} = \hat{\dot{z}} + K_1(\boldsymbol{y} - \hat{\boldsymbol{y}}) \tag{21b}$$

$$\dot{\hat{\dot{z}}} = -\Lambda\hat{z} - 2\zeta\Lambda^{1/2}\hat{\dot{z}} + \tilde{B}_0\boldsymbol{u} + K_2(\boldsymbol{y} - \hat{\boldsymbol{y}}) \tag{21c}$$

$$\hat{\boldsymbol{y}} = \tilde{C}_0\hat{z} + \tilde{E}_0\hat{\dot{z}} \tag{21d}$$

Note that, in general, $\dot{\hat{z}} \neq \hat{\dot{z}}$; i.e., the derivative of the modal amplitude estimate is not the modal velocity estimate. However, if $K_1 = 0$, then $\dot{\hat{z}} \neq \hat{\dot{z}}$, and indeed Eqs. (21b–21d) become the natural second-order observer [Eqs. (20)] with $K_3 \equiv K_2$. Now it is clear why requiring (A,C) to be observable is not enough to make Eqs. (20) converge arbitrarily fast: the observer gain K is restricted to the form

$$\begin{bmatrix} 0 \\ K_2 \end{bmatrix}$$

Take the derivative of Eq. (21b)

$$\dot{\hat{\dot{z}}} = \ddot{\hat{z}} + K_1 \frac{\mathrm{d}}{\mathrm{d}t}(y - \hat{y}) \tag{22a}$$

Now rewrite Eq. (21c) using Eqs. (22a) and (21b):

$$\ddot{\hat{z}} + 2\zeta\Lambda^{1/2}\hat{\dot{z}} + \Lambda\hat{z} = \tilde{B}_0 u + \tilde{K}_2(y - \hat{y}) + K_1 \frac{\mathrm{d}}{\mathrm{d}t}(y - \hat{y}) \tag{22b}$$

where $\tilde{K}_2 \equiv K_2 + 2\zeta\Lambda^{1/2}K_1$. This looks almost like the natural observer [Eqs. (20)], except for the *unrealizable* term

$$K_1 \frac{\mathrm{d}}{\mathrm{d}t}(y - \hat{y})$$

which is present whenever $K_1 \neq 0$. However, we must accommodate the nonzero K_1 situation in our second-order architecture.

Our solution is the following *nonnatural* (notice that we are careful not to do anything unnatural here) *second-order controller architecture* for Eqs. (21):

Control law:

$$u = G_1\hat{z} + G_2\hat{\dot{z}} \tag{23a}$$

Signal processor:

$$\hat{z} = z + L_1\eta \tag{23b}$$

$$\hat{\dot{z}} = \dot{z} + L_2\eta \tag{23c}$$

η Filter:

$$\dot{\eta} = F_\eta \eta + (y - y_z); \qquad F_\eta \quad \text{any stable} \quad P \times P \quad \text{matrix} \tag{23d}$$

Second-order observer:

$$\ddot{z} + 2\xi\Lambda^{1/2}\dot{z} + \Lambda z = \tilde{B}_0 u + K_3(y - y_z) + \gamma_c \tag{23e}$$

$$y_z = \tilde{C}_0 z + \tilde{E}_0 \dot{z} \tag{23f}$$

Output estimate:

$$\hat{y} = y_z + L_3\eta \tag{23g}$$

Corrector:

$$\gamma_C = L_4\eta \tag{23h}$$

where L_1, L_2, L_3, L_4 are constant matrices of appropriate dimensions. Now we have the following result:

Theorem 1. Equations (23) are exactly the same as Eqs. (19) when

$$L_1 = K_1 \tag{24a}$$

$$L_2 = K_1 F \tag{24b}$$

$$L_3 = C_0 K_1 + E_0 K_1 F \tag{24c}$$

$$L_4 = -[(\Lambda + K_2 C_0)K_1 + (K_2 E_0 + 2\zeta \Lambda^{1/2}) K_1 F F_\eta] \tag{24d}$$

and

$$K_3 = K_2 - K_1 F$$

where

$$F_\eta \equiv (I - E_0 K_1)F - C_0 K_1$$

is stable.

Proof of Theorem 1.

From Eqs. (22) and (23),

$$\ddot{z} + 2\zeta\Lambda^{1/2}\dot{z} + \Lambda z = (\ddot{\hat{z}} - L_1\ddot{\eta}) + 2\zeta\Lambda^{1/2}(\dot{\hat{z}} - L_1\dot{\eta}) + \Lambda(\hat{z} - L_1\eta)$$

$$= \tilde{B}_0 \boldsymbol{u} + \tilde{K}_2(\boldsymbol{y} - \hat{\boldsymbol{y}}) + K_1 \frac{\mathrm{d}}{\mathrm{d}t}(\boldsymbol{y} - \hat{\boldsymbol{y}}) - [L_1\ddot{\eta} + 2\zeta\Lambda^{1/2}L_1\dot{\eta} + \Lambda L_1\eta]$$

$$= \tilde{B}_0 \boldsymbol{u} + K_2(\boldsymbol{y} - \hat{\boldsymbol{y}}) - L_1 \frac{\mathrm{d}}{\mathrm{d}t}[F_\eta \eta + \boldsymbol{y} - (\hat{\boldsymbol{y}} - L_3\eta)]$$

$$- [2\zeta\Lambda^{1/2}L_1\dot{\eta} + \Lambda L_1 \eta] + K_1 \frac{\mathrm{d}}{\mathrm{d}t}(\boldsymbol{y} - \hat{\boldsymbol{y}})$$

$$= \tilde{B}_0 \boldsymbol{u} + \tilde{K}_2(\boldsymbol{y} - \hat{\boldsymbol{y}}) + (K_1 - L_1)\frac{\mathrm{d}}{\mathrm{d}t}(\boldsymbol{y} - \hat{\boldsymbol{y}})$$

$$- [L_1 F_\eta + L_1 L_3 + 2\zeta\Lambda^{1/2}L_1]\dot{\eta} - \Lambda L_1 \eta$$

$$= \tilde{B}_0 \boldsymbol{u} + K_2(\boldsymbol{y} - \hat{\boldsymbol{y}}) + (K_1 - L_1)\frac{\mathrm{d}}{\mathrm{d}t}(\boldsymbol{y} - \hat{\boldsymbol{y}})$$

$$- [(L_1 F_\eta + L_1 L_3 + 2\zeta\Lambda^{1/2}L_1)F_\eta + \Lambda L_1]\eta$$

$$- (L_1 F_\eta + L_1 L_3 + 2\zeta\Lambda^{1/2}L_1)(\boldsymbol{y} - \boldsymbol{y}_z)$$

$$= \tilde{B}_0 \boldsymbol{u} + (K_1 - L_1)\frac{\mathrm{d}}{\mathrm{d}t}(\boldsymbol{y} - \hat{\boldsymbol{y}}) + [\tilde{K}_2 - (L_1 F_\eta + L_1 L_3 + 2\zeta\Lambda^{1/2}L_1)](\boldsymbol{y} - \boldsymbol{y}_z)$$

$$- [\tilde{K}_2 L_3 + (L_1 F_\eta + L_1 L_3 + 2\zeta\Lambda^{1/2}L_1)F_\eta + \Lambda L_1]\eta$$

Now take $L_1 = K_1$:

$$K_3 = \tilde{K}_2 - (K_1 F_\eta + K_1 L_3 + 2\zeta \Lambda^{1/2} K_1) = K_2 - K_1(F_\eta + L_3)$$

and

$$L_4 = -[\tilde{K}_2 L_3 + (K_1 F_\eta + K_1 L_3 + 2\zeta \Lambda^{1/2} K_1) F_\eta + \Lambda K_1]$$

Using Eqs. (20b), (23b), (23c), (23f), and (23g), consider that

$$\hat{y} = C_0 \hat{z} + E_0 \hat{\dot{z}} = C_0(z + L_1 \eta) + E_0(\dot{z} + L_2 \eta)$$

$$= y_z + [C_0 L_1 + E_0 L_2]\eta = y_z + L_3 \eta$$

$$\therefore \quad L_3 = C_0 L_1 + E_0 L_2 = C_0 K_1 + E_0 L_2$$

Furthermore, from Eqs. (21b), (23a), (23d), (23g), and (23c), we have

$$\hat{\dot{z}} = \dot{\hat{z}} - K_1(y - \hat{y}) = \frac{d}{dt}(z + L_1 \eta) - K_1(y - y_z) + K_1 L_3 \eta$$

$$= \dot{z} + (L_1 - K_1)\dot{\eta} + [K_1 F_\eta + K_1 L_3]\eta$$

$$= \dot{z} + K_1(L_3 + F_\eta)\eta = \dot{z} + L_2 \eta$$

$$\therefore \quad L_2 = K_1(L_3 + F_\eta)$$

Therefore,

$$L_2 = K_1(L_3 + F_\eta) = K_1[(C_0 K_1 + E_0 L_2) + F_\eta]$$

Define F so that

$$F_\eta = (I - E_0 K_1)F - C_0 K_1$$

i.e.,

$$F = (I - E_0 K_1)^{-1}(F_\eta - C_0 K_1)$$

$$\therefore \quad L_2 = K_1[E_0 L_2 + (I - E_0 K_1)F]$$

$$= K_1[E_0(L_2 - K_1 F) + F]$$

Note that the "magic choice" (yes, we admit it took a long time and many mistakes to come up with this; alas, it is not considered sophisticated to reveal the tortuous route we followed) is

$$L_2 = K_1 F$$

Also note that

$$F_\eta = F - L_3$$

$$\therefore \quad L_4 = -[\Lambda K_1 + \tilde{K}_2 L_3 + 2\zeta\Lambda^{1/2} K_1 F_\eta + K_1(F_\eta + L_3)F_\eta]$$

$$= -[\Lambda K_1 + (\tilde{K}_2 - 2\zeta\Lambda^{1/2} K_1)L_3 + 2\zeta\Lambda^{1/2} K_1 F + K_1 F F_\eta]$$

$$= -[\Lambda K_1 + K_2(C_0 K_1 + E_0 K_1 F) + 2\zeta\Lambda^{1/2} K_1 F + K_1 F F_\eta]$$

$$= -[(\Lambda + K_2 C_0)K_1 + (K_2 E_0 + 2\zeta\Lambda^{1/2})K_1 F + K_1 F F_\eta]$$

Also,

$$K_3 = K_2 - K_1(F_\eta + L_3) = K_2 - K_1 F$$

The interpretation of this result is that the original state-space controller [Eqs. (19)] can be rewritten in terms of the architecture [Eqs. (23)]. This architecture consists of

1) the original control law in $\hat{z}$ and $\hat{\dot{z}}$;

2) a signal processor (no dynamics) to obtain the estimates $\hat{z}$ and $\hat{\dot{z}}$;

3) a P-dimensional first-order filter that produces a signal η from the innovations term $\boldsymbol{y} - \boldsymbol{y}_z$; and

4) an *unnatural* second-order observer for the original modal LFS model [Eqs. (17)], which is corrected by a linear feedback signal from the filter in Eqs. (3).

Since this architecture is exactly reproducing the original controller, the convergence rate of the observer is the same and so also is the closed-loop stability; however, the second-order architecture seems easier to implement and looks more like the LFS it controls.

Clearly, when $K_1 = 0$, we have the natural second-order observer [Eqs. (20)], which has no need for the η filter or the corrector signal $\gamma_c = L_4\eta$. It is interesting to note that if an F can be found so that both

$$F_\eta = (I - E_0 K_1)F - C_0 K_1 \quad \text{stable} \tag{25a}$$

and

$$(\Lambda + K_2 C_0)K_1 + (K_2 E_0 + 2\zeta\Lambda^{1/2})K_1 F + K_1 F F_\eta = 0 \tag{25b}$$

then we have a "natural" second-order observer plus the η filter; we call these "pseudonatural" second-order observers.

This section has resolved the form of the controller into a nonnatural second-order observer containing a full modal model of the LFS, a low-order η filter, and some signal processing (no dynamics). Although the observer is a parallel bank of second-order modal transfer functions of the form

$$T_k(s) = \frac{1}{s^2 + 2\zeta_K \omega_K s + \omega_K^2}$$

interconnected by the corrector term γ_c and the innovations feedback term $K_3(y - y_z)$, there is still a very large number (L) of these modal systems to be integrated. The next section resolves the problem of reducing this dimension L.

Synthesis of Reduced Second-Order Controllers for LFS

The dimension of the controller [Eqs. (23)] is $L + P$, where L is the number of degrees of freedom in the LFS model; hence, L is *very* large and we must reduce the dimension of the controller. We partition z into three parts:

$$z = \begin{bmatrix} z_N \\ z_Q \\ z_R \end{bmatrix}$$

$$\ddot{z}_N + 2\zeta_N \Lambda_N^{1/2} \dot{z}_N + \Lambda_N z_N = \tilde{\boldsymbol{B}}_N^0 \boldsymbol{u} \tag{26a}$$

$$\ddot{z}_Q + 2\zeta_Q \Lambda_Q^{1/2} \dot{z}_Q + \Lambda_Q z_Q = \tilde{\boldsymbol{B}}_Q^0 \boldsymbol{u} \tag{26b}$$

$$\ddot{z}_R + 2\zeta_R \Lambda_R^{1/2} \dot{z}_R + \Lambda_R z_R = \tilde{\boldsymbol{B}}_R^0 \boldsymbol{u} \tag{26c}$$

$$\boldsymbol{y} = \boldsymbol{y}_N + \boldsymbol{y}_Q + \boldsymbol{y}_R \tag{26d}$$

$$\boldsymbol{y}_N = \tilde{C}_N^0 z_N + \tilde{E}_N^0 \dot{z}_N \tag{26e}$$

$$\boldsymbol{y}_Q = \tilde{C}_Q^0 z_Q + \tilde{E}_Q^0 \dot{z}_Q \tag{26f}$$

$$\boldsymbol{y}_R = \tilde{C}_R^0 z_R + \tilde{E}_R^0 \dot{z}_Q \tag{26g}$$

where

$$\Lambda = \begin{bmatrix} \Lambda_N & 0 & 0 \\ 0 & \Lambda_Q & 0 \\ 0 & 0 & \Lambda_R \end{bmatrix}, \qquad \text{etc.}$$

For now, we concentrate on the ROM:

$$\ddot{z}_N + 2\zeta_N \Lambda_N^{1/2} \dot{z}_N + \Lambda_N z_N = \tilde{\boldsymbol{B}}_N^0 \boldsymbol{u} \tag{27a}$$

$$\boldsymbol{y}_N = \tilde{C}_N^0 z_N + \tilde{E}_N^0 \dot{z}_N \tag{27b}$$

Henceforth, "stable" will mean that a system has all of its poles (eigenvalues) in the set $C\sigma_0 \equiv \{h \text{ complex number}/\text{Re}\, h < -\sigma_0\}$ for some previously selected $\sigma_0 > 0$. We will assume that all instabilities of the LFS model [Eqs. (17)] are contained in the ROM [Eqs. (27)]; i.e., the ROM is the only part of the LFS that does not have the stability margin σ_0. The ROM can be selected by many techniques, but it usually consists of modes that need control to achieve some performance goal; of course, stability is the most fundamental aspect of any performance goal.

If the ROM (A_N, B_N, C_N), where

$$A_N \equiv \begin{bmatrix} 0 & I_N \\ -\Lambda_N & -2\zeta_N \Lambda_N^{1/2} \end{bmatrix}, \qquad B_N = \begin{bmatrix} 0 \\ \tilde{B}_N^0 \end{bmatrix}$$

$$C_N \equiv [\tilde{C}_N^0 \quad \tilde{E}_N^0]$$

is controllable and observable (easy to assess,[4] since this is a modal model), we follow the same approach as in the previous section, treating the ROM as the total system. We obtain the ROM controller in the following manner:

Control law:

$$u = G_1^N \hat{z}_N + G_2^N \hat{\dot{z}}_N \tag{28a}$$

Signal processor:

$$\hat{z}_N = z_N + L_1^N \eta_N \tag{28b}$$

$$\hat{\dot{z}}_N = \dot{z}_N + L_2^N \eta_N \tag{28c}$$

η_N Filter:

$$\dot{\eta}_N = F_\eta N_N + (y - y_{z_N}); \qquad F_N \quad \text{stable} \quad P \times P \quad \text{matrix} \tag{28d}$$

Second-order observer:

$$\ddot{z}_N + 2\zeta_N \Lambda_N^{1/2} z_N + \Lambda_N z_N = \tilde{B}_N^0 u + K_3^N (y - y_{z_N}) + \gamma_C^N \tag{28e}$$

$$y_{z_N} = \tilde{C}_N^0 z_N + \tilde{E}_N^0 \dot{z}_N \tag{28f}$$

Output estimate:

$$\hat{y}_N = y_{z_N} + L_3^N \eta_N \tag{28g}$$

Corrector:

$$\gamma_C^N = L_4^N \eta_N \tag{28h}$$

where

$$L_1^N = K_1^N, \qquad \text{etc.}$$

The design of control gains

$$G_N = [G_1^N \quad G_2^N]$$

and observer gains

$$K_N = \begin{bmatrix} K_1^N \\ K_2^N \end{bmatrix}$$

can be done to achieve performance in $A_N + B_N G_N$ and rate of convergence in $A_N - K_N C_N$. Of course, we make sure that $A_N + B_N G_N$ and $A_N - K_N C$ are stable as part of this design. If $\boldsymbol{y} = \boldsymbol{y}_N$, then closed-loop stability would follow from the same arguments as in the two previous sections. But $\boldsymbol{y} = \boldsymbol{y}_N + \boldsymbol{y}_Q + \boldsymbol{y}_R$, and it is the interaction with these unmodeled dynamics that causes stability problems. In the next section we present a basic plan for CSI compensation to obtain closed-loop stability without redesigning the ROM controller.

CSI Compensation: Second-Order Residual Mode Filters

Since the ROM controller [Eqs. (28)] operates in closed loop with the full dynamics of the LFS as modeled by Eqs. (17) or Eqs. (26), it may produce instabilities. We will assume that any closed-loop unstable modes are included in the z_Q system [Eq. (26b)]. Thus, we define the z_Q system as those modes that, when deleted from the open-loop system [Eqs. (26)], produce closed-loop stability; i.e., set $z_Q \equiv 0$ and the ROM controller plus the remaining open-loop modes z_N, z_R form a stable closed-loop system. Of course, in some cases there will be no instabilities and hence no z_Q modes. However, when there are z_Q modes, we would like to add compensation to counteract the instability without redesigning the ROM controller.

The following CSI compensation can be added to the controller to remove the instability; we call it the second-order RMF:

$$\hat{\boldsymbol{y}}_Q = \boldsymbol{y}_{z_Q} = \tilde{C}_N^0 z_Q + \tilde{E}_N^0 \dot{z}_Q \tag{29a}$$

$$\ddot{z}_Q + 2\zeta_Q \Lambda_Q^{1/2} \dot{z}_Q + \Lambda_Q z_Q = \tilde{B}_Q^0 \boldsymbol{u} \tag{29b}$$

This is a parallel bank of Q second-order filters driven by the ROM controller output $\boldsymbol{u}$ and interconnected to produce $\hat{\boldsymbol{y}}_Q$.

Now we modify the ROM controller [Eqs. (28)] using the RMF output $\hat{\boldsymbol{y}}_Q$ by replacing $\boldsymbol{y}$ with $\boldsymbol{y}_c \equiv \boldsymbol{y} - \hat{\boldsymbol{y}}_Q$. We have the following closed-loop stability result:

Theorem 2. The combination of the RMF [Eqs. (29)] and the modified ROM controller, i.e., Eqs. (28) with $\boldsymbol{y}_c \equiv \boldsymbol{y} - \hat{\boldsymbol{y}}_Q$ replacing $\boldsymbol{y}$, in closed loop with the LFS modal model [Eqs. (26)] is stable.

Proof of Theorem 2. By *Theorem* 1, Eqs. (26) are exactly the same as

$$\boldsymbol{u} = G_N \hat{x}_N \tag{30a}$$

$$\dot{\hat{x}}_N = A_N \hat{x}_N + B_N \boldsymbol{u} + K_N(\boldsymbol{y} - \hat{\boldsymbol{y}}_N) \tag{30b}$$

$$\hat{\boldsymbol{y}}_N = C_N \hat{x}_N \tag{30c}$$

where (A_N, B_N, C_N) are defined by the ROM [Eqs. (27)] with

$$\hat{x}_N \equiv \begin{bmatrix} \hat{z}_N \\ \dot{\hat{z}}_N \end{bmatrix}, \qquad G_N \equiv [G_N^1 G_N^2], \qquad \text{and} \qquad K_N \equiv \begin{bmatrix} K_N^1 \\ K_N^2 \end{bmatrix}$$

Furthermore, by taking

$$\hat{x}_Q \equiv \begin{bmatrix} z_Q \\ \dot{z}_Q \end{bmatrix}$$

we see that Eqs. (29) are the same as

$$\dot{\hat{x}}_Q = A_Q \hat{x}_Q + B_Q \boldsymbol{u} \tag{31a}$$

$$\hat{\boldsymbol{y}}_Q = C_Q \hat{x}_Q \tag{31b}$$

where

$$A_Q \equiv \begin{bmatrix} 0 & I_Q \\ -\Lambda_Q & -2\zeta_Q \Lambda_Q^{1/2} \end{bmatrix}$$

$$B_Q \equiv \begin{bmatrix} 0 \\ \tilde{B}_Q^0 \end{bmatrix} \qquad \text{and} \qquad C_Q \equiv [\tilde{C}_Q^0 \tilde{E}_Q^0]$$

Also note that

$$\hat{x}_Q = \begin{bmatrix} \hat{z}_Q \\ \dot{\hat{z}}_Q \end{bmatrix} \tag{32}$$

When we modify the ROM controller [Eqs. (28)] by replacing $\boldsymbol{y}$ with

$$\boldsymbol{y}_c = \boldsymbol{y} - \hat{\boldsymbol{y}}_Q,$$

we produce

$$\boldsymbol{u} = G_N \hat{x}_N \tag{33a}$$

$$\dot{\hat{x}}_N = A_N \hat{x}_N + B_N \boldsymbol{u} + K_N(\boldsymbol{y}_c - \hat{\boldsymbol{y}}_N) \tag{33b}$$

$$\hat{\boldsymbol{y}}_N = C_N \hat{x}_N \tag{33c}$$

Define

$$e_N \equiv \hat{x}_N - x_N \qquad \text{and} \qquad e_Q \equiv \hat{x}_Q - x_Q$$

It follows that

$$\dot{e}_N = (A_N - K_N C_N) e_N + K_N C_R x_R - K_N C_Q e_Q \tag{34a}$$

$$\dot{e}_Q = A_Q e_Q \tag{34b}$$

Take

$$x_c \equiv \begin{bmatrix} x_N \\ e_N \\ x_R \\ x_Q \\ e_Q \end{bmatrix}$$

The closed-loop system becomes

$$\dot{x}_C = A_C x_C \tag{35}$$

where

$$A_C = \begin{bmatrix} A_N + B_N G_N & B_N G_N & 0 & 0 & 0 \\ 0 & A_N - K_N C_N & K_N C_R & 0 & -K_N C_Q \\ B_R G_N & B_R G_N & A_R & 0 & 0 \\ B_Q G_N & B_Q G_N & 0 & A_Q & 0 \\ 0 & 0 & 0 & 0 & A_Q \end{bmatrix}$$

This is stable when both A_Q and

$$A_{11}^C \equiv \begin{bmatrix} A_N + B_N B_N & B_N B_N & 0 \\ 0 & A_N - K_N C_N & K_N C_R \\ B_R G_N & B_R G_N & A_R \end{bmatrix}$$

are stable. However, A_Q is stable since all open-loop instabilities are contained in the ROM A_N, and A_{11}^C is stable because it is the closed-loop system consisting of Eqs. (26) with $z_Q \equiv 0$ and the original ROM controller [Eqs. (26)], which is the same as Eqs. (30).

Conclusions

We have taken a linear FEM model [Eqs. (3)] of a LFS that has a very large number (L) of degrees of freedom. We have transformed it to modal coordinates [Eqs. (17)] and selected, by *any* of the available techniques based on performance, a ROM [Eqs. (27)]. We have produced a ROM controller [Eq. (28)] that has essentially the same second-order architecture as the LFS model. This architecture consists of an unnatural second-order observer based on the ROM [Eqs. (28)] with a linear corrector signal derived from a P-dimensional (P is the number of sensors) first-order filter; signal processors (no additional dynamics) create the control signals from the above.

From *Theorem* 1 the architecture is exactly the same as the usual first-order or state space ROM controller, but it is more easily implemented via parallel connections of modal transfer functions:

$$\frac{1}{s^2 + 2\zeta_k \omega_k s + \omega_k^2}$$

Since the controller is based on a ROM, there will be controller interactions with unmodeled dynamics, or CSI. The second-order RMF [Eqs. (29)] is a parallel bank of Q second-order filters, where Q is the number of residual mode instabilities produced by the ROM controller due to CSI. The RMF modifies the ROM controller so that closed-loop stability results without redesigning the ROM controller (*Theorem* 2). Second-order parallel architecture of ROM controller and RMF is maintained throughout.

Acknowledgments

Several long discussions with Brian Reisenauer, a graduate student in the Department of Aerospace Engineering Sciences of the University of Colorado-Boulder, helped me to develop this chapter. I thank him for his insights and take full responsibility for any muddling of them that has occurred here.

References

[1]Balas, M., "Exponentially Stabilizing Finite-Dimensional Controllers for Linear Distributed Parameter Systems (DPS): Galerkin Approximation of Infinite-Dimensional Controllers," *Journal of Mathematical Analysis and Applications*, Vol. 117, 1986, pp. 358–384.

[2]Balas, M., "Finite-Dimensional Controllers for Linear DPS: Exponential Stability Using Residual Mode Filters," *Journal of Mathematical Analysis and Applications*, Vol. 133, 1988, pp. 283–296.

[3]Ouyang, J., "Adaptive Residual Mode Filter Control of DPS for Large Space Structure Applications," Ph.D. Dissertation, Electrical Computer and Systems Engineering Dept., Rensselaer Polytechnic Institute, Troy, NY, May 1987.

[4]Balas, M., "Feedback Control of Flexible Systems," *IEEE Transactions on Automatic Control*, Vol. AC-23, 1978, pp. 673–679.

[5]Balas, M., and Ginter, S., "Attitude Stabilization of Large Flexible Spacecraft," *Journal of Guidance and Control*, Vol. 4, 1981, pp. 561–564.

Chapter 12

Boundary-Element Method for Shape Control of Distributed-Parameter Elastostatic Systems

Goong Chen* and Jianxin Zhou†
Texas A&M University, College Station, Texas

Introduction

IN this chapter we examine the problems of shape control for elastic systems. Such problems appear in contemporary space technology and manufacturing processes. The object of shape control is to achieve a desirable shape for a given elastic body by the application of distributed control forces.

For a control system in elasticity there are finitely many sensors located on the elastic body to measure the deformation at the sensor locations, and the control devices are distributed over the domain and/or on the boundary. In this chapter we limit our study to the case where the control devices are distributed only on the boundary. This simplifies the mathematical treatment considerably yet allows ready extension to the case where there are internally distributed actuators. We will study two models: 1) two-dimensional thin elastostatic Kirchhoff-Timoshenko plate and 2) three-dimensional isotropic homogeneous Hookean solid. Only linear elastostatic models are treated here. We also define a quadratic cost performance index, which constitutes a linear-quadratic distributed-parameter control problem.

In space applications the computers for the digital controller usually have only limited memory and power. Also, the control scheme must be relatively simple to ensure fast response. The recent rapid development of the boundary-element method (BEM) seems ideally suited for such technological applications because of its numerical efficiency and small memory requirement. This is our major adopted mathematical approach here, and

---

*Professor, Department of Mathematics.
†Assistant Professor, Department of Mathematics.

in our opinion, it is perhaps superior to several other variational methods developed earlier in the literature.

Numerical results of shape are illustrated by computer graphics.

Kirchhoff-Timoshenko Plate Equation and Shape Control

The Kirchhoff-Timoshenko model of a thin, elastostatic, isotropic, and homogeneous plate subject to pure bending is an important engineering model in structural mechanics. Let Ω be a bounded domain in the two-dimensional Euclidean space $\mathbb{R}^2$ with boundary $\partial\Omega$. We use $x = (x_1,x_2)$ to denote an arbitrary point in $\mathbb{R}^2$. Let $w(x)$ denote the vertical displacement of the plate at x, and assume that there is no external loading. According to Kirchhoff's theory, the strain energy of the plate is given by

$$\mathrm{E}(w) = \frac{D}{2}\int_\Omega \{|\Delta w(x)|^2 + 2(1-\nu)(w^2_{x_1x_2})(x) - w_{x_1x_1}(x)w_{x_2x_2}(x)]\}\,\mathrm{d}x \quad (1)$$

where

$$\begin{aligned} \mathrm{d}x &= \mathrm{d}x_1\,\mathrm{d}x_2 \\ D > 0 &= \text{flexural rigidity} \\ \nu, 0 < \nu < \tfrac{1}{2} &= \text{Poisson ratio} \\ \Delta &= (\partial^2/\partial x_1^2) + (\partial^2/\partial x_2^2) \end{aligned}$$

It is easy to see that

$$\mathrm{E}(w) = \mathrm{E}(w + a_0 + a_1x_1 + a_2x_2)$$

Therefore, the strain energy of a plate is the same up to a three-degree-of-freedom rigid-body motion $a_0 + a_1x_1 + a_2x_2$. In all of the developments below, we assume that the coefficient D has been normalized to 1. By the principle of virtual work, we have

$$0 = \delta\mathrm{E}(w)$$

$$0 = \int_\Omega \{\Delta w(x)\cdot\Delta\delta w(x) + (1-\nu)[2w_{x_1x_2}(x)\delta w_{x_1x_2}(x) - w_{x_1x_1}(x)\delta w_{x_2x_2}(x) - w_{x_2x_2}(x)\delta w_{x_1x_1}(x)]\}\,\mathrm{d}x$$

Integration by parts of the preceding equation (see Timoshenko and Woinowsky-Krieger[1]) yields

$$\int_\Omega \Delta^2 w(x)\cdot\delta w(x)\,\mathrm{d}x = \int_{\partial\Omega}\left[B_1 w(x)\cdot\delta w(x) - B_2 w(x)\frac{\partial}{\partial n}\delta w(x)\right]\mathrm{d}\sigma = 0 \quad (2)$$

where $n = (n_1, n_2)$ is the unit exterior normal at $\partial\Omega$, and

$$B_1 w \equiv \frac{\partial}{\partial n} \Delta w - (1-\nu) \frac{\partial}{\partial s} \left[n_1 n_2 \left(\frac{\partial^2}{\partial x_1^2} w - \frac{\partial^2}{\partial x_2^2} w \right) - (n_1^2 - n_2^2) \frac{\partial^2}{\partial x_1 \, \partial x_2} w \right]$$

is the boundary term containing the shearing force and the twisting moment. We call B_1 the shear operator, and

$$B_2 w \equiv \nu \, \Delta w + (1-\nu) \left[n_1^2 \frac{\partial^2}{\partial x_1^2} w + n_2^2 \frac{\partial^2}{\partial x_2^2} w + 2 n_1 n_2 \frac{\partial^2}{\partial x_1 \, \partial x_2} w \right]$$

is the boundary bending-moment term.

We call B_2 the bending-moment operator.

$$\Delta^2 = \left(\frac{\partial^2}{\partial x_1^2} + \frac{\partial^2}{\partial x_2^2} \right)^2$$

is the biharmonic operator, and

$$\frac{\partial}{\partial s} = -n_2 \frac{\partial}{\partial x_1} + n_1 \frac{\partial}{\partial x_2}$$

is the counterclockwise tangential derivative. From the variational equation [Eq. (2)], we obtain the following four types of biharmonic boundary-value problems:

Type I (displacement-angle):

$$\begin{cases} \Delta^2 w(x) = 0, & x \in \Omega \qquad \text{(3a)} \\ w(x) = g_1(x), & x \in \partial\Omega \qquad \text{(3b)} \\ \dfrac{\partial w}{\partial n}(x) = g_2(x), & x \in \partial\Omega \qquad \text{(3c)} \end{cases}$$

Type II (shear-moment):

$$\begin{cases} \Delta^2 w(x) = 0, & x \in \Omega \qquad \text{(4a)} \\ B_1 w(x) = g_1(x), & x \in \partial\Omega \qquad \text{(4b)} \\ B_2 w(x) = g_2(x), & x \in \partial\Omega \qquad \text{(4c)} \end{cases}$$

Type III (displacement-moment):

$$\begin{cases} \Delta^2 w(x) = 0, & x \in \Omega \qquad \text{(5a)} \\ w(x) = g_1(x), & x \in \partial\Omega \qquad \text{(5b)} \\ B_2 w(x) = g_2(x), & x \in \partial\Omega \qquad \text{(5c)} \end{cases}$$

Type IV (angle-shear):

$$\begin{cases} \Delta^2 w(x) = 0, & x \in \Omega \quad \text{(6a)} \\ \dfrac{\partial w}{\partial n}(x) = g_1(x), & x \in \partial\Omega \quad \text{(6b)} \\ B_1 w(x) = g_2(x), & x \in \partial\Omega \quad \text{(6c)} \end{cases}$$

As will be seen later on [cf. the selected control models of Eqs. (16), (17), (33), and (34)], types II and III boundary conditions are the most relevant to the shape control of a plate.

Boundary Integral Equations and Multilayer Potential Methods for the Plate Equation

The BEM is based on an integral equation approach, called boundary integral equations (BIEs), for solving a system of linear partial differential equations. We briefly present the ideas in the following. The interested reader can consult Chen and Zhou.[2] Let

$$E(x,\xi) = -\frac{1}{8\pi} |x - \xi|^2 \, \ell n |x - \xi| \tag{7}$$

be the fundamental solution of the biharmonic equation; i.e., $E(x,\xi)$ satisfies

$$\Delta_\xi^2 E(x,\xi) = -\delta(x - \xi), \qquad x,\xi \in \mathbb{R}^2 \tag{8}$$

where δ is the Dirac delta function, and Δ_ξ^2 is the biharmonic partial differential operator with respect to the ξ variables. We represent a solution $w(x)$ to problems (3), (4), (5), or (6) as

$$w(x) = \int_{\partial\Omega} \left[E(x,\xi) f_1(\xi) + \frac{\partial E(x,\xi)}{\partial n_\xi} f_2(\xi) \right] d\sigma_\xi + a_0 + a_1 x_1 + a_2 x_2, \qquad x \in \Omega \tag{9}$$

where f_1 and f_2 are two unknown functions called, respectively, the simple and double-layer densities; and a_0, a_1, and a_2 are three unknown constants related to three degrees of freedom of a rigid motion of the plate. We must determine f_1, f_2, a_0, a_1, and a_2 from the two given boundary conditions and three other auxiliary boundary conditions.

Since the shape control problem in this section is primarily related to biharmonic boundary-value problems of types II and III, we will limit our attention to these two cases only. The proofs of the following two theorems can be found in Chen and Zhou.[2]

Theorem 2.1 (simple and double layers representation for biharmonic problem of type II). Let A_0, A_1, and A_2 be three given real constants, and let g_1 and g_2 be, respectively, the boundary shear and bending moment in Eqs. (4). Then the following system of boundary integral equations

$$\left\{\begin{aligned} &\int_{\partial\Omega} f_1(\xi)\, \mathrm{d}\sigma = A_0 \\ &\int_{\partial\Omega} [\xi_1 f_1(\xi) + n_1(\xi) f_2(\xi)]\, \mathrm{d}\sigma = A_1 \\ &\int_{\partial\Omega} [\xi_2 f_1(\xi) + n_2(\xi) f_2(\xi)]\, \mathrm{d}\sigma = A_2 \\ &\int_{\partial\Omega} \left[B_{1x} E(x,\xi) \cdot f_1(\xi) + B_{1x} \frac{\partial}{\partial n_\xi} E(x,\xi) \cdot f_2(\xi) \right] \mathrm{d}\sigma_\xi + \tfrac{1}{2} f_1(x) = g_1(x) \\ &\int_{\partial\Omega} \left[B_{2x} E(x,\xi) \cdot f_1(\xi) + B_{2x} \frac{\partial}{\partial n_\xi} E(x,\xi) \cdot f_2(\xi) \right] \mathrm{d}\sigma_\xi - \tfrac{1}{2} f_2(x) = g_2(x) \end{aligned}\right. \tag{10}$$

have unique solutions f_1 and f_2, provided that g_1 and g_2 satisfy the compatibility conditions

$$\left\{\begin{aligned} &\int_{\partial\Omega} g_1(\xi)\, \mathrm{d}\sigma = 0 \\ &\int_{\partial\Omega} [\xi_1 g_1(\xi) - n_1(\xi) g_2(\xi)]\, \mathrm{d}\sigma = 0 \\ &\int_{\partial\Omega} [\xi_2 g_1(\xi) - n_2(\xi) g_2(\xi)]\, \mathrm{d}\sigma = 0 \end{aligned}\right. \tag{11}$$

The solutions $w(x)$ for the biharmonic boundary-value problem [Eqs. (4)] are given by Eq. (9) for three arbitrary constants a_0, a_1, and a_2.

Theorem 2.2 (simple and double layers representation for biharmonic problem of type III). Let A_0, A_1, and A_2 be three given real constants, and let g_1 and g_2 be, respectively, the boundary displacement and bending moment in Eqs. (5). Then the following system of boundary integral equations,

$$
\begin{cases}
\int_{\partial\Omega} f_1(\xi)\, d\sigma = A_0 \\
\int_{\partial\Omega} [\xi_1 f_1(\xi) + n_1(\xi) f_2(\xi)]\, d\sigma = A_1 \\
\int_{\partial\Omega} [\xi_2 f_1(\xi) + n_2(\xi) f_2(\xi)]\, d\sigma = A_2 \\
\int_{\partial\Omega} \left[E(x,\xi) f_1(\xi) + \frac{\partial}{\partial n_\xi} E(x,\xi) \cdot f_2(\xi) \right] d\sigma_\xi + a_0 + a_1 x_1 + a_2 x_2 = g_1(x) \\
\int_{\partial\Omega} \left[B_{2x} E(x,\xi) \cdot f_1(\xi) + B_{2x} \frac{\partial}{\partial n_\xi} E(x,\xi) \cdot f_2(\xi) \right] d\sigma_\xi - \tfrac{1}{2} f_2(x) = g_2(x)
\end{cases}
\tag{12}
$$

have unique solutions f_1, f_2, a_0, a_1, and a_2. The biharmonic boundary-value problem [Eqs. (5)] has a unique solution $w(x)$ given by Eq. (9). [Note that A_0, A_1, and A_2 in Eq. (12) can be chosen arbitrarily, but the resulting f_1, f_2, a_0, a_1, and a_2 as solutions of Eq. (12) vary according to the choice of A_0, A_1, and A_2.]

Remark 2.3 For solutions to partial differential or integral equations, there is the associated question of *regularity* of solutions, namely, how *smooth* will the solution functions f_1 and f_2 in Eq. (10) or (12) be, given the smoothness of the boundary data g_1 and g_2 in Eqs. (4) or (5)? The regularity question can be addressed using *Sobolev spaces* (see Chen and Zhou[2]).

Remark 2.4 The solutions $w(x)$ [cf. Eq. (9)] to the biharmonic boundary-value problem of type II [Eqs. (4)] contain three arbitrary constants a_0, a_1, and a_2 according to *Theorem* 2.1 thus are undetermined. In practice we replace the first three equations in Eq. (10) by three other auxiliary conditions, such as

$$
\begin{cases}
w(x_0) = \alpha_0, \qquad x_0 \in \partial\Omega \\
\frac{\partial w}{\partial n}(x_0) = \alpha_1 \\
\frac{\partial w}{\partial s}(x_0) = \alpha_2
\end{cases}
\tag{13}
$$

or

$$
\begin{cases}
w(x_1) = \beta_1, \qquad x_1 \in \bar{\Omega} \\
w(x_2) = \beta_2, \qquad x_2 \in \bar{\Omega} \\
w(x_3) = \beta_3, \qquad x_3 \in \Omega
\end{cases}
$$

or a mixture of three conditions of the preceding types, where α_0, α_1, α_2, β_1, β_2, and β_3 are given constants, and x_0, x_1, x_2, and x_3 are given boundary or interior points. Using these three conditions, we can uniquely determine the three constants a_0, a_1, and a_2.

Remark 2.5 In *Theorems* 2.1 and 2.2 we have solved the biharmonic boundary-value problems by a combination of simple and double-layer potentials [Eq. (9)], wherein

$$\int_{\partial\Omega} E(x,\xi) f_1(\xi)\, d\sigma_\xi$$

and

$$\int_{\partial\Omega} \frac{\partial}{\partial n_\xi} E(x,\xi) \cdot f_2(\xi)\, d\sigma_\xi$$

are, respectively, the simple and double layers. This combination yields boundary integral equations with the least severe singularities in their kernels [such as $B_{1x}E(x,\xi)$ or $B_{2x}E(x,\xi)$ in Eqs. (10) and (12)]. Other combinations of layer potentials such as simple and triple layers, double and quadruple layers, and triple and quadruple layers are also feasible, but they lead to more complicated numerical computations and larger errors. The basic approach in this section using a combination of layers is called the multilayer potential method.

Formulation of the Shape Control Problem as a Linear-Quadratic Distributed-Parameter Boundary Control Problem

Some recent applications in space technology, large astronomical telescope control and design, and solar energy utilization suggest the following shape control problem for study. Consider a round plate made of elastic rubber or other material with finitely many internally distributed point sensors that can measure the deformation of the plate at those points. On the boundary of the plate there are distributed actuators that can apply bending and/or shearing forces to the plate boundary to warp the plate to a desirable shape. (This concept or device is called a "deformable mirror" in contemporary space technology.) In most applications the (adjustable) desirable shape is a concave mirror with a certain *curvature* so that the mirror can focus communication signals, laser beams, or sunlight over a distance.

To illustrate the mathematical formulation and approach, let us furnish a concrete example. To be specific, let us assume that the domain Ω is a disk with radius one-half centered at the origin. The desirable shape that we wish to achieve is given by

$$\zeta(x) = \frac{1}{16\pi}\{10|x-\eta_1|^2 \ln|x-\eta_1| + 4|x-\eta_2|^2 \ln|x-\eta_2|$$

$$+ 3|x-\eta_3|^2 \ln|x-\eta_3| + 2|x-\eta_4|^2 \ln|x-\eta_4|\}, \quad x \in \mathbb{R}^2, \quad |x| \leqslant 1/2 \quad (14)$$

where

$$\eta_1 = \left(-\frac{3}{2}, \frac{3}{2}\right), \quad \eta_2 = (2,2), \quad \eta_3 = (3,-3), \quad \eta_4 = \left(-\frac{5}{2}, -\frac{5}{2}\right)$$

The $\zeta(x)$ has a concave mirror shape as shown in Fig. 1.

Let us assume that located at

$$y_0 = (0,0), \quad y_1 = \left(\frac{1}{3}, 0\right), \quad y_2 = \left(-\frac{1}{6}, \frac{\sqrt{3}}{3}\right), \quad y_3 = \left(-\frac{1}{6}, -\frac{\sqrt{3}}{3}\right) \tag{15}$$

on the plate are four symmetrically placed sensors (cf. Fig. 2).

We formulate a linear-quadratic optimal control problem as follows: The governing equations are

$$\begin{cases} \Delta^2 w(x) = 0, & x \in \Omega \\ B_1 w(x) = u_1(x), & x \in \partial\Omega, \quad \nu = 1/8 \\ B_2 w(x) = u_2(x), & x \in \partial\Omega \end{cases} \tag{16}$$

where the control forces are shear u_1 and moment u_2 applied at the

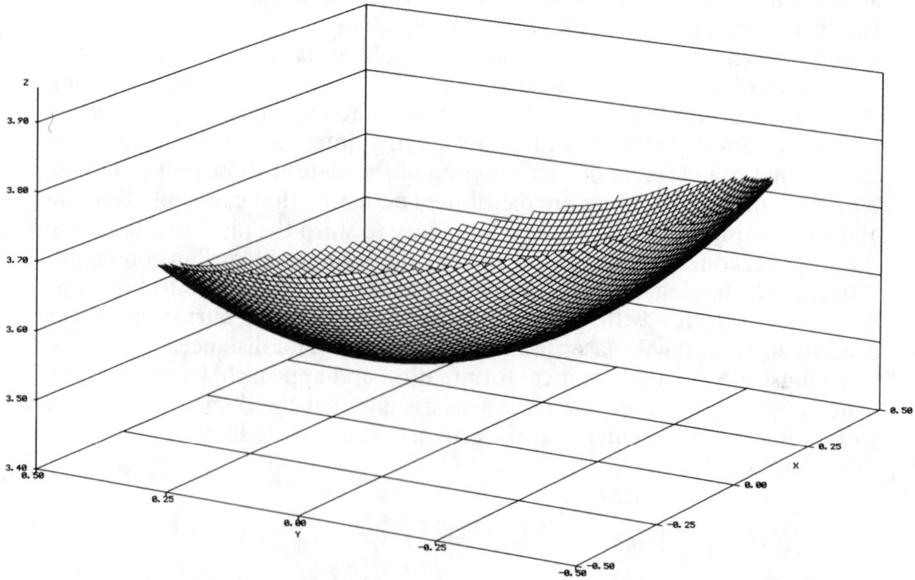

Fig. 1 The concave mirrow shape of $\zeta(x)$ as given by Eq. (14).

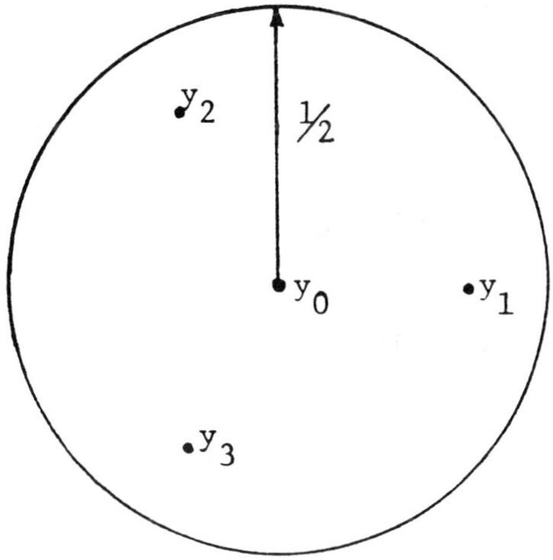

Fig. 2 A circular plate with radius one-half and four symmetrically placed point sensors.

boundary. We choose to minimize a quadratic functional

$$J = \sum_{i=0}^{3} |w(y_i) - \zeta(y_i)|^2 + \int_{\partial\Omega} |w(x) - \zeta(x)|^2 \, d\sigma$$
$$+ \gamma_1 \int_{\partial\Omega} |u_1(x)|^2 \, d\sigma + \gamma_2 \int_{\partial\Omega} |u_2(x)|^2 \, d\sigma, \qquad \gamma_1, \gamma_2 > 0 \tag{17}$$

whose choice is based on the following concerns: At the sensor locations y_0, y_1, y_2, and y_3, we want the deformation w to be close to the objective sensor data $\zeta(y_i)$, $i = 0, 1, 2, 3$. The first integral in J says that the boundary contour of w should be close to that of ζ. The appearance of the last two terms in Eq. (17),

$$\gamma_1 \int_{\partial\Omega} |u_1(x)|^2 \, d\sigma, \qquad \gamma_2 \int_{\partial\Omega} |u_2(x)|^2 \, d\sigma$$

signifies control constraints. Mathematically, these terms give strict convexity and enhance numerical stability. We intend to make the control scheme simple; therefore, we do not adopt cost functionals J containing more complicated Sobolev space norms.

Let us invoke *Theorem* 2.1 to obtain a boundary integral equation solution for the governing system [Eq. (16)]. The solution is represented by

$$w(x) = \int_{\partial\Omega} \left[E(x,\xi) f_1(\xi) + \frac{\partial}{\partial n_\xi} E(x,\xi) f_2(\xi) \right] d\sigma_\xi + a_0 + a_1 x_1 + a_2 x_2, \qquad x \in \Omega \tag{18}$$

where a_0, a_1, $a_2 \in \mathbb{R}$ and f_1, f_2 satisfy

$$\begin{cases} \int_{\partial\Omega} f_1(\xi)\, d\sigma = A_0 \\ \int_{\partial\Omega} [\xi_1 f_1(\xi) + n_1(\xi) f_2(\xi)]\, d\sigma = A_1 \\ \int_{\partial\Omega} [\xi_2 f_1(\xi) + n_2(\xi) f_2(\xi)]\, d\sigma = A_2 \\ \int_{\partial\Omega} \left[B_{1x} E(x,\xi) f_1(\xi) + B_{1x} \frac{\partial}{\partial n_\xi} E(x,\xi) \cdot f_2(\xi) \right] d\sigma_\xi + \tfrac{1}{2} f_1(x) = u_1(x) \\ \int_{\partial\Omega} \left[B_{2x} E(x,\xi) f_1(\xi) + B_{2x} \frac{\partial}{\partial n_\xi} E(x,\xi) \cdot f_2(\xi) \right] d\sigma_\xi - \tfrac{1}{2} f_2(x) = u_2(x) \end{cases} \tag{19}$$

Note that in Eq. (19) we have replaced the first three equations in Eq. (12) by the last three equations in Eq. (16).

To numerically simulate and solve the control system, we proceed to discretize the control system

$$\inf_{u_1, u_2 \in L^2(\partial\Omega)} J, \qquad \text{subject to Eq. (19)} \tag{20}$$

by boundary elements. Any system of sufficiently smooth splines can be used. Here we choose

$$\hat{S}_h = \left\{ \phi_j^h(x) \middle| 1 \leqslant j \leqslant n(h), \quad h = \frac{\pi}{n(h)}, \quad n(h) \quad \text{is a positive integer} \right\}$$

where

$$\phi_j^h(x) = \frac{4}{3} \begin{cases} (x - x_{j-2})^2 / 2h^2, & x_{j-2} \leqslant x \leqslant x_{j-1} \\ (x - x_{j-1})(x_j - x)/h^2 + \tfrac{1}{2}, & x_{j-1} \leqslant x \leqslant x_j \\ (x_{j+1} - x)^2 / 2h^2, & x_j \leqslant x \leqslant x_{j+1} \\ 0 & \text{otherwise} \end{cases} \tag{21}$$

$$x_j = jh \quad \text{for} \quad j = 0,1,2,\ldots, n(h) - 1$$

$$x_{-1} \equiv x_{n(h)-1}, \qquad x_{n(h)} \equiv x_0, \qquad x_{n(h)+1} \equiv x_1$$

We let the unknown potential densities f_1 and f_2 in Eq. (18) be approximated by

$$f_1^h(x) = \sum_{j=1}^{n(h)} \alpha_j \phi_j^h(x), \qquad f_2^h(x) = \sum_{j=1}^{n(h)} \beta_j \phi_j^h(x) \tag{22}$$

Equation (19) is then approximated by *point collocation*:

$$\begin{cases} \sum_{j=1}^{n(h)} \alpha_j \int_{\partial\Omega} \phi_j^h(\xi)\, d\sigma = A_0 \\ \sum_{j=1}^{n(h)} \alpha_j \int_{\partial\Omega} \xi_1 \phi_j^h(\xi)\, d\sigma_\xi + \sum_{j=1}^{n(h)} \beta_j \int_{\partial\Omega} n_1(\xi)\phi_j^h(\xi)\, d\sigma_\xi = A_1 \\ \sum_{j=1}^{n(h)} \alpha_j \int_{\partial\Omega} \xi_2 \phi_j^h(\xi)\, d\sigma_\xi + \sum_{j=1}^{n(h)} \beta_j \int_{\partial\Omega} n_2(\xi)\phi_j^h(\xi)\, d\sigma_\xi = A_2 \\ \sum_{j=1}^{n(h)} \alpha_j \int_{\partial\Omega} B_{1x} E(z_k,\xi)\phi_j^h(\xi)\, d\sigma_\xi + \sum_{j=1}^{n(h)} \beta_j \int_{\partial\Omega} B_{1x} \frac{\partial}{\partial n_\xi} E(z_k,\xi)\phi_j^h(\xi)\, d\sigma_\xi \\ \qquad + \frac{1}{2} \sum_{j=1}^{n(h)} \alpha_j \phi_j^h(z_k) = u_1(z_k), \qquad 1 \leqslant k \leqslant n(h) \\ \sum_{j=1}^{n(h)} \alpha_j \int_{\partial\Omega} B_{2x} E(z_k,\xi)\phi_j^h(\xi)\, d\sigma_\xi + \sum_{j=1}^{n(h)} \beta_j \int_{\partial\Omega} B_{2x} \frac{\partial}{\partial n_\xi} E(z_k,\xi)\phi_j^h(\xi)\, d\sigma_\xi \\ \qquad - \frac{1}{2} \sum_{j=1}^{n(h)} \beta_j \phi_j^h(z_k) = u_2(z_k), \qquad 1 \leqslant k \leqslant n(h) \end{cases} \tag{23}$$

where

$$z_k = (k-1)h + \frac{h}{2}, \qquad k = 1,2, \ldots, n(h) \tag{24}$$

is the node of ϕ_k^h such that $\phi_k^h(z_k) = 1$.

According to *Theorem* 2.1 for Eq. (16) [and consequently Eq. (23)] to be solvable, u_1 and u_2 must satisfy the compatibility conditions [Eq. (11)] with (g_1,g_2) replaced by (u_1,u_2) in Eq. (11). However, u_1 and u_2 are still unknown. It is not obvious how to handle these conditions. As it turns out, they can be disregarded according to the same ideas as in Chen and Tsai[3] (see *Remark* 2.7).

We write Eq. (23) as a linear matrix system

$$A_h F_h = G_h \tag{25}$$

where

$$F_h = (a_0,a_1,a_2,\alpha_1,\alpha_2, \ldots, \alpha_{n(h)},\beta_1,\beta_2, \ldots, \beta_{n(h)}) \in \mathbb{R}^{2n(h)+3} \tag{26}$$

$$G_h = A_0,A_1,A_2,u_1(z_1), \ldots, u_1(z_{n(h)}) \tag{27a}$$

$$u_2(z_1), \ldots, u_2(z_{n(h)}) \in \mathbb{R}^{2n(h)+3} \tag{27b}$$

Similarly, the cost function J in Eq. (17) can be discretized to

$$J_h(F_h,G_h)$$

where $J_h(F_h,G_h)$ is a quadratic functional of F_h and G_h. We solve the linear-quadratic mathematical programming problem

$$\begin{cases} \min J_h(F_h,G_h) \\ F_h,G_h \in \mathbb{R}^{2n(h)+3} \\ \text{subject to} \qquad A_hF_h = G_h \end{cases} \tag{28}$$

using the method of Lagrange multipliers

$$\max_{\lambda_h \in \mathbb{R}^{2n(h)+3}} \min_{F_h \in \mathbb{R}^{2n(h)+3}} [J_h(F_h,G_h) + \langle \lambda_h, A_hF_h - G_h \rangle_{\mathbb{R}^{2n(h)+3}}] \tag{29}$$

The solution of Eq. (29) will yield the optimal $\hat{F}_h,\hat{G}_h$ for Eq. (28). Therefore, the numerically optimal approximate state $\hat{w}_h(x)$ and controls $\hat{u}_{1,h}(x),\hat{u}_{2,h}(x)$ are given by

$$\hat{w}_h(x) = \int_{\partial\Omega} \left\{ E(x,\xi)\left[\sum_{i=1}^{n(h)} \hat{\alpha}_i \phi_i^h(\xi)\right] + \frac{\partial}{\partial n_\xi} E(x,\xi)\left[\sum_{i=1}^{n(h)} \hat{\beta}_i \phi_i^h(\xi)\right]\right\} d\sigma_\xi$$
$$+ \hat{a}_0 + \hat{a}_1 x_1 + \hat{a}_2 x_2, \qquad x \in \Omega \tag{30}$$

$$\hat{u}_{1,h}(x) = \int_{\partial\Omega} \left\{ B_{1x} E(x,\xi) \cdot \left[\sum_{i=1}^{n(h)} \hat{\alpha}_i \phi_i^h(\xi)\right] + B_{1x} \frac{\partial}{\partial n_\xi} E(x,\xi)\right.$$
$$\left. \cdot \left[\sum_{i=1}^{n(h)} \hat{\beta}_i \phi_i^h(\xi)\right]\right\} d\sigma_\xi + \frac{1}{2} \sum_{i=1}^{n(h)} \hat{\alpha}_i \phi_i^h(x), \qquad x \in \partial\Omega \tag{31}$$

$$\hat{u}_{2,h}(x) = \int_{\partial\Omega} \left\{ B_{2x} E(x,\xi) \cdot \left[\sum_{i=1}^{n(h)} \hat{\alpha}_i \phi_i^h(\xi)\right] + B_{2x} \frac{\partial}{\partial n_\xi} E(x,\xi)\right.$$
$$\left. \cdot \left[\sum_{i=1}^{n(h)} \hat{\beta}_i \phi_i^h(\xi)\right]\right\} d\sigma_\xi - \frac{1}{2} \sum_{i=1}^{n(h)} \hat{\beta}_i \phi_i^h(x), \qquad x \in \partial\Omega \tag{32}$$

Remark 2.6. In Eqs. (23), (31), and (32) we need to evaluate *divergent integrals*,

$$\int_{\partial\Omega} B_{1x} E(x,\xi) \cdot \phi_i^h(\xi)\, d\sigma_\xi, \qquad \int_{\partial\Omega} B_{1x} \frac{\partial}{\partial n_\xi} E(x,\xi) \cdot \phi_i^h(\xi)\, d\sigma_\xi$$

$$\int B_{2x} \frac{\partial}{\partial n_\xi} E(x,\xi) \cdot \phi_i^h(\xi)\, d\sigma_\xi$$

The technique can be found in Chen and Zhou.[2]

Remark 2.7. The computational approach [Eqs. (23–32)] will completely bypass the compatibility conditions [Eq. (11)] [with (u_1,u_2) replacing (g_1,g_2)]. The numerical optimal controls $\hat{u}_{1,h}$ and $\hat{u}_{2,h}$ will automatically satisfy Eq. (11).

A sequence of $\hat{w}_h(x)$, $\hat{u}_{1,h}(x)$, and $\hat{u}_{2,h}(x)$ have been plotted by varying γ_1 and γ_2 in Eq. (17) (cf. Figs. 3–10). Numerical values of $\hat{w}_h(x)$, $\hat{u}_{1,h}(x)$, and $\hat{u}_{2,h}(x)$ at selected points can be found in Table 1.

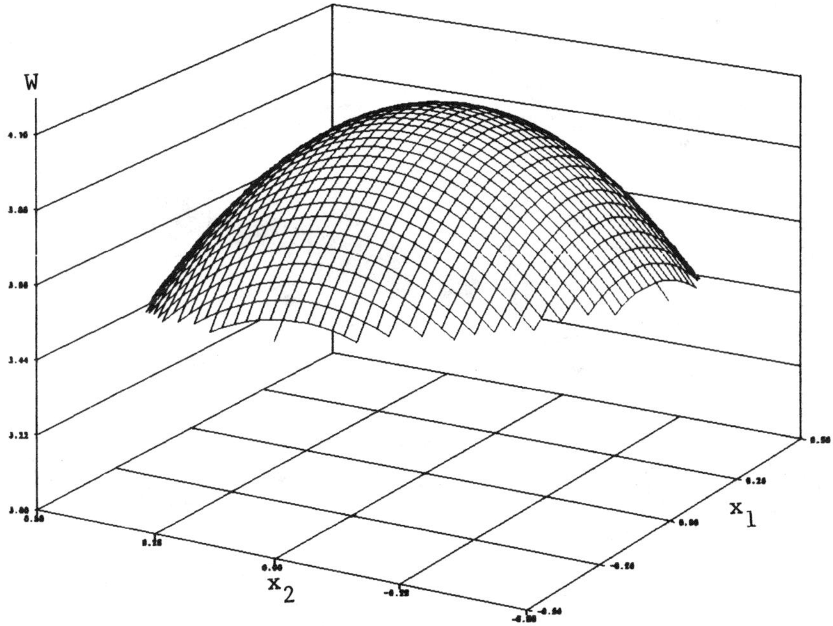

Fig. 3 Deformation with shear and bending-moment controls, $\gamma_1 = 1.0 \times 10^{-1}$, $\gamma_2 = 1.0 \times 10^{-2}$.

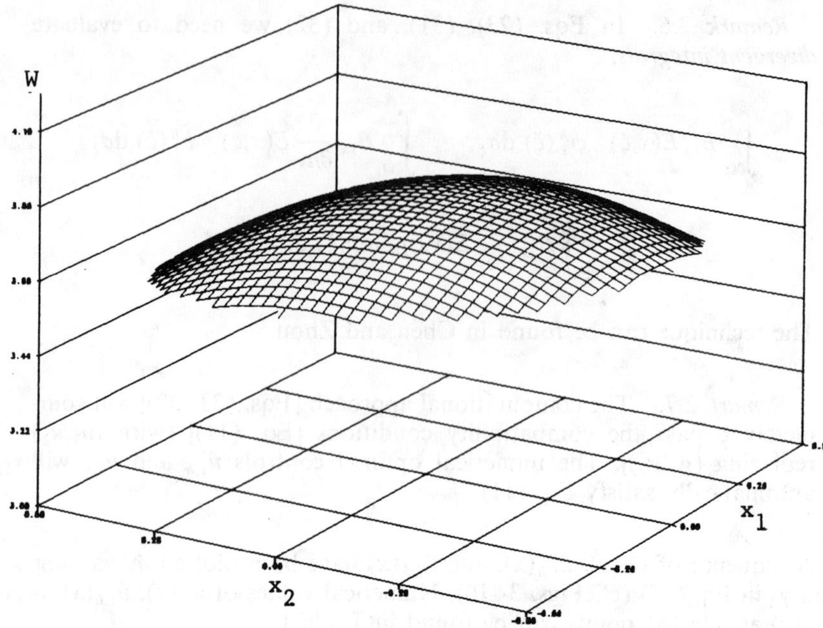

Fig. 4 Deformation with shear and bending-moment controls, $\gamma_1 = 5.0 \times 10^{-2}$, $\gamma_2 = 3.0 \times 10^{-3}$.

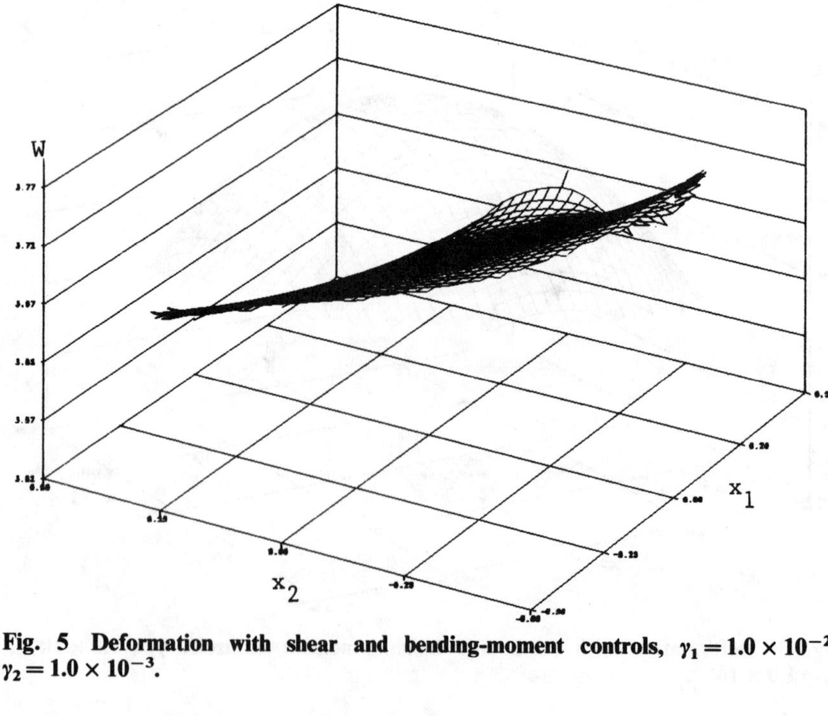

Fig. 5 Deformation with shear and bending-moment controls, $\gamma_1 = 1.0 \times 10^{-2}$, $\gamma_2 = 1.0 \times 10^{-3}$.

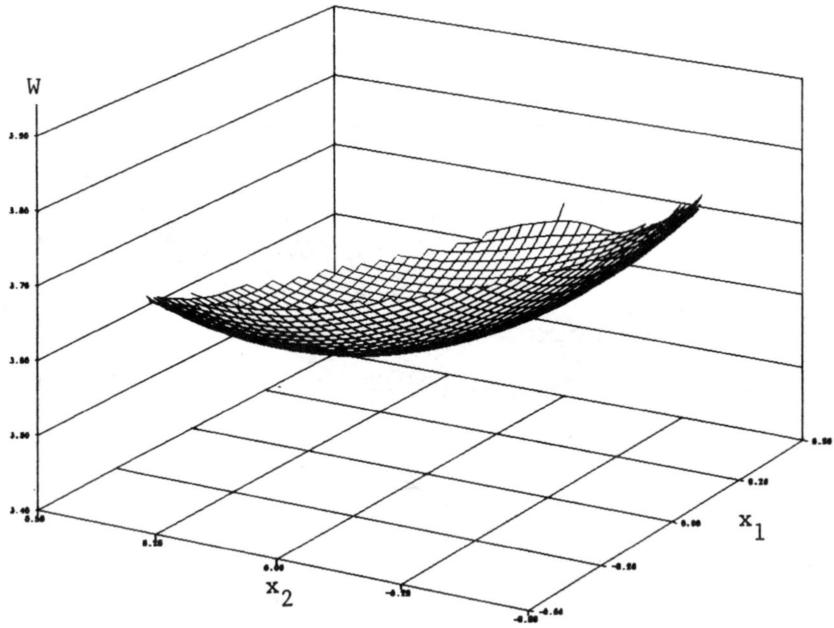

Fig. 6 **Deformation with shear and bending-moment controls,** $\gamma_1 = 5.0 \times 10^{-3}$, $\gamma_2 = 1.0 \times 10^{-4}$.

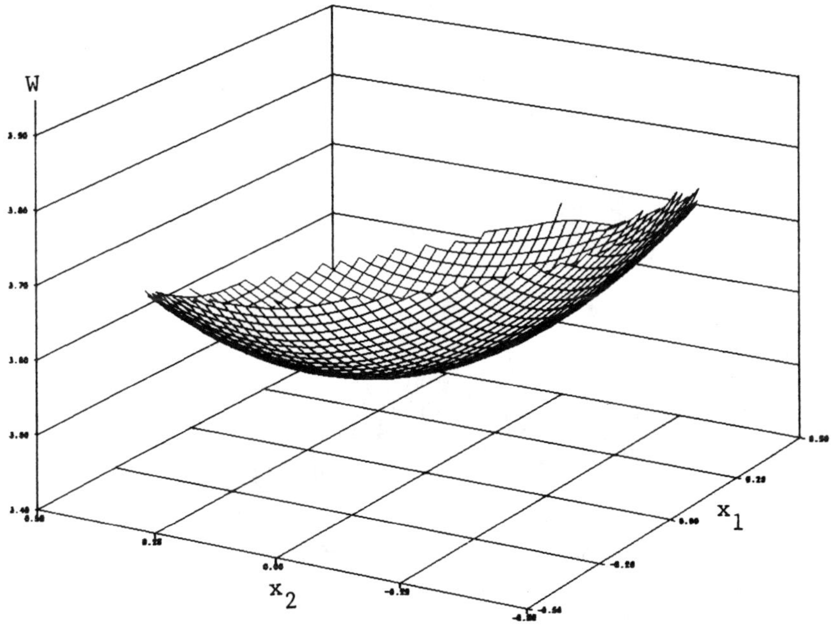

Fig. 7 **Deformation with shear and bending-moment controls,** $\gamma_1 = 1.0 \times 10^{-3}$, $\gamma_2 = 1.0 \times 10^{-4}$.

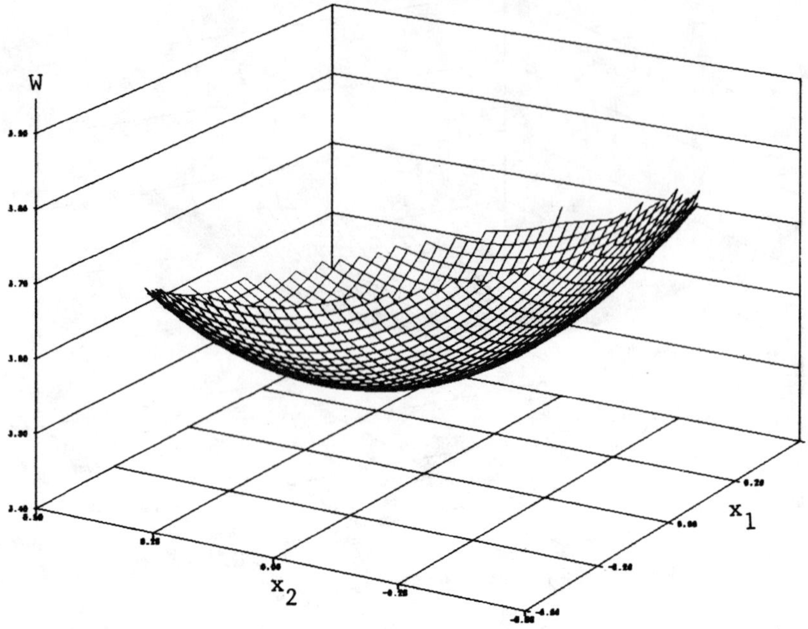

Fig. 8 Deformation with shear and bending-moment controls, $\gamma_1 = 1.0 \times 10^{-4}$, $\gamma_2 = 1.0 \times 10^{-5}$.

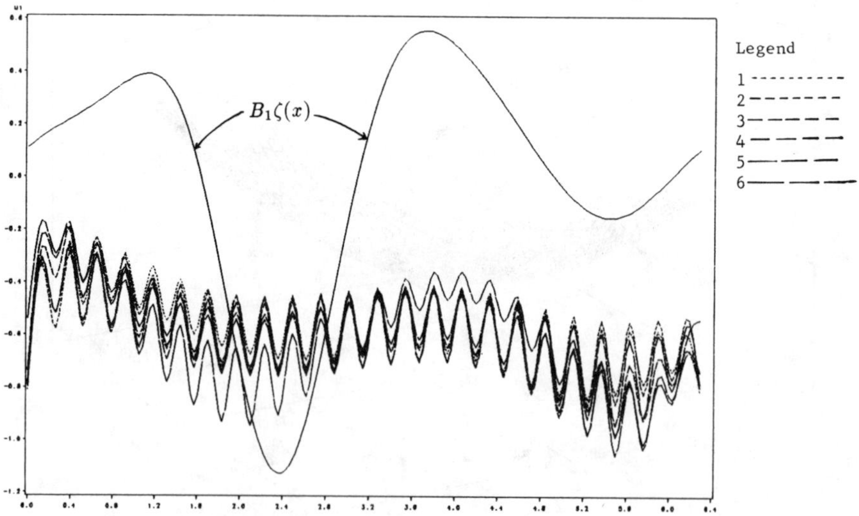

Fig. 9 Numerical solutions of the shear control $\hat{u}^{(i)}_{1,h}(x)$, $i = 1, 2, 3, 4, 5, 6$ and $h = \pi/48$.

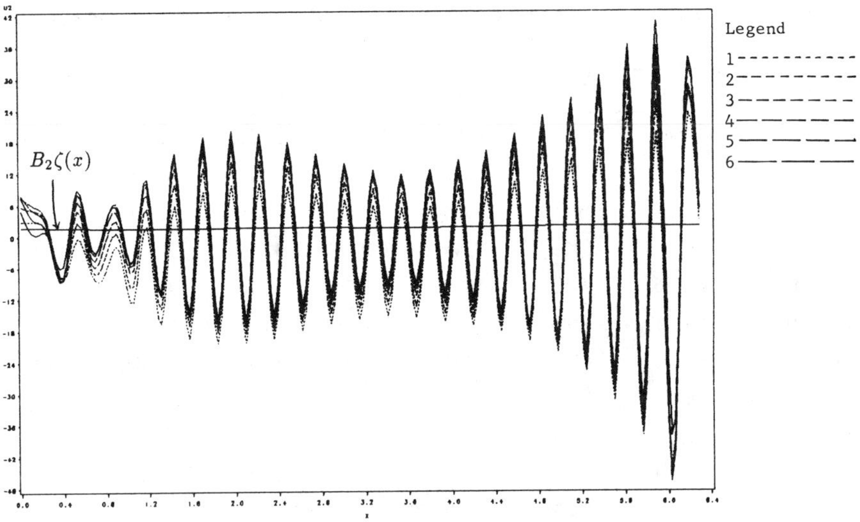

Fig. 10 Numerical solutions of the bending-moment control $\hat{u}^{(i)}_{2,h}(x)$, $i = 1, 2, 3, 4, 5, 6$ and $h = \pi/48$.

Figures 3 and 4 actually show that, when γ_1 and γ_2 are not very small, the plate is deformed *contrary* to the desirable direction. The reason is that the larger weights for the control costs in J severely limit the control performance. When $\gamma_1 = 1.0 \times 10^{-2}$ and $\gamma_2 = 1.0 \times 10^{-3}$, the plate in Fig. 5 begins to bend inside out. After the values of γ_1 and γ_2 decrease further in magnitude, the plate bends in the desirable direction, with increasing curvature (cf. Figs. 6 and 7). The shape of the plate in Fig. 8 corresponding to $\gamma_1 = 1.0 \times 10^{-4}$, $\gamma_2 = 1.0 \times 10^{-4}$ is fairly close to the desirable shape as shown in Fig. 8. Checking the values of $\hat{w}_h(x)$ against those of $\zeta(x)$ at sensor locations y_i, $i = 0, 1, 2, 3$ in Table 1, we have also found that they

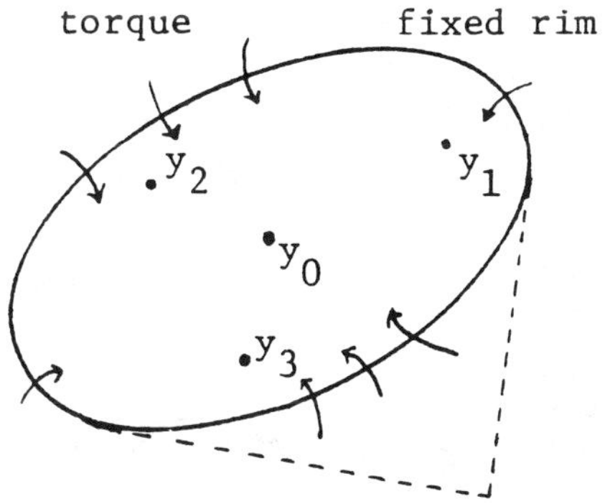

Fig. 11 Torque applied distributively at the boundary of a plate with fixed rim.

Table 1 Numerical values of the state and controls at selected points for the control problem [Eq. (17)] subject to Eq. (16)

x	$\hat{w}_h^{(1)}(x)$	$\hat{w}_h^{(2)}(x)$	$\hat{w}_h^{(3)}(x)$	$\hat{w}_h^{(4)}(x)$	$\hat{w}_h^{(5)}(x)$	$\hat{w}_h^{(6)}(x)$	$\zeta(x)$
y_0	4.0640	3.8554	3.6726	3.5707	3.5355	3.5182	3.5158
y_1	3.8220	3.7327	3.6439	3.5941	3.5776	3.5710	3.5703
y_2	3.7736	3.7069	3.6308	3.5862	3.5703	3.5632	3.5642
y_3	3.8783	3.8123	3.7348	3.6894	3.6736	3.6650	3.6628
$(1/2, \pi/3)$	3.4236	3.4912	3.5371	3.5654	3.5760	3.5796	3.5803
$(1/2, 2\pi/3)$	3.4321	3.5508	3.6092	3.6361	3.6455	3.6500	3.6531
$(1/2, \pi)$	3.5785	3.6636	3.7094	3.7306	3.7373	3.7406	3.7429
$(1/2, 4\pi/3)$	3.6091	3.7050	3.7522	3.7778	3.7876	3.7926	3.7956
$(1/2, 5\pi/3)$	3.4859	3.6518	3.7335	3.7684	3.7758	3.7760	3.7818
$(1/2, 2\pi)$	3.5849	3.6427	3.6665	3.6739	3.6739	3.6692	3.6616

$x = (1/2, \theta)$ θ:	$\hat{u}_{1,h}^{(1)}(x)$	$\hat{u}_{1,h}^{(2)}(x)$	$\hat{u}_{1,h}^{(3)}(x)$	$\hat{u}_{1,h}^{(4)}(x)$	$\hat{u}_{1,h}^{(5)}(x)$	$\hat{u}_{1,h}^{(6)}(x)$	$B_1\zeta(x)$
$\pi/3$	−0.4999	−0.6131	−0.5252	−0.6481	−0.5723	−0.6763	0.3876
$\pi/2$	−0.5994	−0.6940	−0.6675	−0.7267	−0.6982	−0.8692	0.0938
$2\pi/3$	−0.6756	−0.7355	−0.7224	−0.7574	−0.7498	−0.9448	−0.8917
π	−0.6719	−0.7188	−0.6849	−0.7253	−0.6635	−0.6532	0.1006
$4\pi/3$	−0.6970	−0.7206	−0.6755	−0.7214	−0.6455	−0.4779	0.4172
$3\pi/2$	−0.7884	−0.8017	−0.8208	−0.8069	−0.7933	−0.6570	0.1189
$5\pi/3$	−0.8415	−0.8673	−0.9778	−0.9168	−0.9804	−0.8847	−0.1259
2π	−0.8107	−0.7952	−0.7954	−0.7810	−0.7380	−0.5394	0.1121

$x = (1/2, \theta)$ θ:	$\hat{u}_{2,h}^{(1)}(x)$	$\hat{u}_{2,h}^{(2)}(x)$	$\hat{u}_{2,h}^{(3)}(x)$	$\hat{u}_{2,h}^{(4)}(x)$	$\hat{u}_{2,h}^{(5)}(x)$	$\hat{u}_{2,h}^{(6)}(x)$	$B_2\zeta(x)$
$\pi/3$	−12.4998	−9.3855	−6.4040	−4.6125	−4.3462	−5.1655	1.5231
$\pi/2$	−19.4199	−17.0106	−15.2607	−14.7662	−14.8336	−15.1387	1.5333
$2\pi/3$	−20.2135	−18.4864	−17.3312	−16.6061	−16.2533	−16.3032	1.6677
π	−15.7829	−13.7659	−12.0152	−11.0577	−10.5861	−10.0415	1.5807
$4\pi/3$	−16.9920	−15.6086	−14.0623	−12.7873	−12.2267	−12.0243	1.6686
$3\pi/2$	−19.9549	−18.4697	−17.8921	−17.8530	−17.7624	−18.1289	1.8031
$5\pi/3$	−25.7124	−23.3660	−22.0807	−22.3852	−23.7311	−25.6695	1.8866
2π	2.9693	5.7237	7.5113	8.0074	7.5970	5.0503	1.7557

$J^{(1)} = 7.8814$, $J^{(2)} = 2.8086$, $J^{(3)} = 1.0306$, $J^{(4)} = 0.3287$, $J^{(5)} = 0.1113$, $J^{(6)} = 0.0116$.
$n(h) = 48$ is used.
Superscripts $i = 1, 2, 3, 4, 5, 6$, correspond, respectively, to the sequential order of γ_1, γ_2 in Figs. 3, 4, 5, 6, 7, 8.

are extremely close to each other, with fractional deviations no larger than 0.1% for Fig. 8.

How do the digital controllers $\hat{u}_{1,h}$ and $\hat{u}_{2,h}$ behave? This can be found in Figs. 9 and 10. In these figures both $\hat{u}_{1,h}$ and $\hat{u}_{2,h}$ oscillate a great deal as a function of x, with 24 peaks. [That is obviously due to the fact that $n(h) = 48$ is used, because we have found that the number of peaks will be just equal to $n(h)/2$ if we vary $n(h)$.] At present, it is not clear what is the best strategy to eliminate such oscillations. Defining a cost functional J with different Sobolev space norms for u_1 and u_2 in J should help, but feedback control laws for such a cost functional would be harder to design. Perhaps a simple numerical filtering (to eliminate high frequencies above a certain level) could serve the purpose.

How much do the shear u_1 and moment u_2, respectively, contribute to the control of the shape? This can be largely answered by comparing the graphs of $\hat{u}_{1,h}$ (cf. Fig. 9) and $\hat{u}_{2,h}$ (cf. Fig. 10) with the shear and moment of the exact shape, $B_1\zeta(x)$ and $B_2\zeta(x)$. We observe that, as γ_1 and γ_2 vary and tend to the ideal choices $\gamma_1 = 1.0 \times 10^{-4}$, $\gamma_2 = 1.0 \times 10^{-5}$, $\hat{u}_{1,h}(x)$ does not oscillate around $B_1\zeta(x)$, the shear of the exact shape, whereas $\hat{u}_{2,h}(x)$ always oscillates around $B_2\zeta(x)$, the moment of the exact shape. This gives a strong indication that *bending moment* $B_2 w(x) = u_2(x)$ *contributes significantly to shape control.*

The preceding numerical simulation has suggested an improved control-theoretic formulation and modeling approach. We consider

$$\inf_{u \in L^2(\partial\Omega)} J(u) = \sum_{i=0}^{3} |w(y_i) - \zeta(y_i)|^2 + \gamma \int_{\partial\Omega} |u(x)|^2 \, d\sigma, \qquad \gamma > 0 \tag{33}$$

subject to

$$\begin{cases} \Delta^2 w(x) = 0 \\ w(x) = \zeta(x) & \text{on} \quad \partial\Omega \\ B_2 w(x) = u(x) & \text{on} \quad \partial\Omega \end{cases} \tag{34}$$

In this model the rim of the plate is fixed with contour $\zeta(x)$, and only bending-moment u is applied at the boundary as control, as illustrated in Fig. 11. Again, in setting up the cost functional J in Eq. (33), we require the deformation w at sensor locations y_i, $i = 0, 1, 2, 3$ to be close to desirable values $\zeta(y_i)$, $i = 0, 1, 2, 3$. The second integral term as control cost in J is added for the same reasons as before.

To solve the boundary-value problem [Eq. (34)], we adopt the integral equation (12) in *Theorem* 2.2. We use $\hat{S}_h$ and $n(h)$ the same as in Eq. (21) and write f_1 and f_2 as in Eq. (22). We set $A_0 = A_1 = A_2 = 0$ in Eq. (12), and replace (g_1, g_2) by (ζ, u). The discretized equations are

$$\left\{\begin{array}{l}
\sum_{i=1}^{n(h)} \alpha_i \int_{\partial\Omega} \phi_j^h(\xi)\, \mathrm{d}\sigma = 0 \\
\sum_{i=1}^{n(h)} \alpha_j \int_{\partial\Omega} \xi_1 \phi_i^h(\xi)\, \mathrm{d}\sigma + \sum_{i=1}^{n(h)} \beta_i \int_{\partial\Omega} n_1(\xi)\phi_i^h(\xi)\, \mathrm{d}\sigma = 0 \\
\sum_{i=1}^{n(h)} \alpha_i \int_{\partial\Omega} \xi_2 \phi_i^h(\xi)\, \mathrm{d}\sigma + \sum_{i=1}^{n(h)} \beta_i \int_{\partial\Omega} n_2(\xi)\phi_i^h(\xi)\, \mathrm{d}\sigma = 0 \\
\sum_{i=1}^{n(h)} \alpha_i \int_{\partial\Omega} E(z_j,\xi)\phi_i^h(\xi)\, \mathrm{d}\sigma + \sum_{i=1}^{n(h)} \beta_i \int_{\partial\Omega} \frac{\partial}{\partial n_\xi} E(z_j,\xi)\phi_i^h(\xi)\, \mathrm{d}\sigma \\
\qquad + a_0 + a_1 z_{j,1} + a_2 z_{j,2} = \zeta(z_j), \qquad 1 \leqslant j \leqslant n(h) \\
\sum_{i=1}^{n(h)} \alpha_i \int_{\partial\Omega} B_{2x} E(z_j,\xi) \cdot \phi_i^h(\xi)\, \mathrm{d}\sigma + \sum_{i=1}^{n(h)} \beta_i \int_{\partial\Omega} B_{2x} \frac{\partial}{\partial n_\xi} E(z_j,\xi) \cdot \phi_i^h(\xi)\, \mathrm{d}\sigma \\
\qquad - \frac{1}{2} \sum_{i=1}^{n(h)} \beta_i \phi_i^h(z_j) = u(z_j), \qquad 1 \leqslant j \leqslant n(h)
\end{array}\right. \tag{35}$$

where z_j is the same as in Eq. (24). Equation (35) can be assembled into a matrix equation

$$\tilde{A}_h \tilde{F}_h = \tilde{G}_h \tag{36}$$

where $\tilde{F}_h$ is the same as F_h in Eq. (26), and

$$\tilde{G}_h = (0,0,0,\zeta(z_1),\zeta(z_2), \ldots, \zeta(z_{n(h)}),u(z_1),u(z_2), \ldots, u(z_{n(h)})) \tag{37}$$

The cost functional equation [Eq. (33)] is also discretized to a quadratic functional

$$J_h(\tilde{F}_h,\tilde{G}_h) \tag{38}$$

The structure of the minimization problem [Eq. (38)] subject to Eq. (36) suggests that a direct substitution $\tilde{G}_h = \tilde{A}_h\tilde{F}_h$ to eliminate $\tilde{G}_h$ is best:

$$\tilde{J}_h(\tilde{F}_h) = J_h(\tilde{F}_h,\tilde{A}_h\tilde{F}_h)$$

The minimizer $\hat{F}_h$ is obtained by

$$\frac{\partial}{\partial \tilde{F}_h} \tilde{J}_h(\tilde{F}_h)|_{\tilde{F}_h = \hat{F}_h} = 0$$

by the convexity of $\tilde{J}_h$, which yields a $[2n(h) + 3] \times [2n(h) + 3]$ nonsingular matrix linear system. The numerically optimal state and control are given by

$$\hat{w}_h(x) = \int_{\partial\Omega} \left\{ E(x,\xi) \left[\sum_{i=1}^{n(h)} \hat{\tilde{\alpha}}_i \phi_i^h(\xi) \right] + \frac{\partial}{\partial n_\xi} E(x,\xi) \left[\sum_{i=1}^{n(h)} \hat{\tilde{\beta}}_i \phi_i^h(\xi) \right] \right\} d\sigma_\xi$$

$$+ \hat{a}_0 + \hat{a}_1 x_1 + \hat{a}_2 x_2, \qquad x \in \Omega \tag{39}$$

$$\hat{u}_h(x) = \int_{\partial\Omega} \left\{ B_{2x} E(x,\xi) \left[\sum_{i=1}^{n(h)} \hat{\tilde{\alpha}}_i \phi_i^h(\xi) \right] + \left[B_{2x} \frac{\partial}{\partial n_\xi} E(x,\xi) \right] \left[\sum_{i=1}^{n(h)} \hat{\tilde{\beta}}_i \phi_i^h(\xi) \right] \right\} d\sigma_\xi$$

$$- \frac{1}{2} \sum_{i=1}^{n(h)} \hat{\tilde{\beta}}_i \phi_i^h(x), \qquad x \in \partial\Omega \tag{40}$$

We have plotted a sequence of graphs for the state $\hat{w}_h(x)$ and the control $\hat{u}_h(x)$ by varying the weight γ in Eq. (33) [cf. Figs. 12–17). We see that the curvature of the plate is steadily increasing, and the final shape is extremely close to the desirable one. The control $\hat{u}_h(x)$ also oscillates closer to $B_2\zeta(x)$, the "exact" bending moment of the desirable shape function $\zeta(x)$ (cf. Fig. 17).

Numerical values of $\hat{w}_h(x)$ and $\hat{u}_h(x)$ at selected sample points can be found in Table 2.

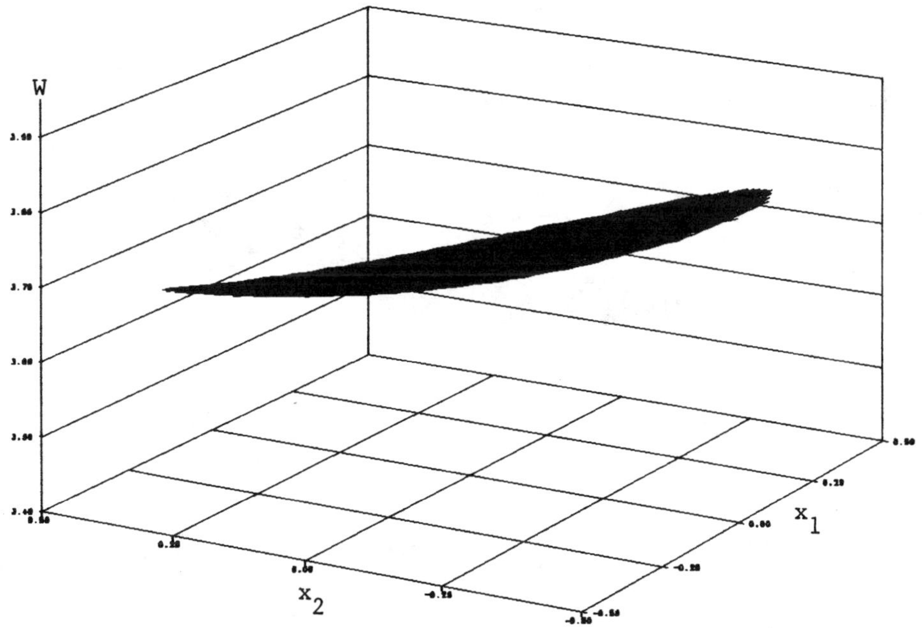

Fig. 12 Deformation with bending-moment control, $\gamma = 1.0$.

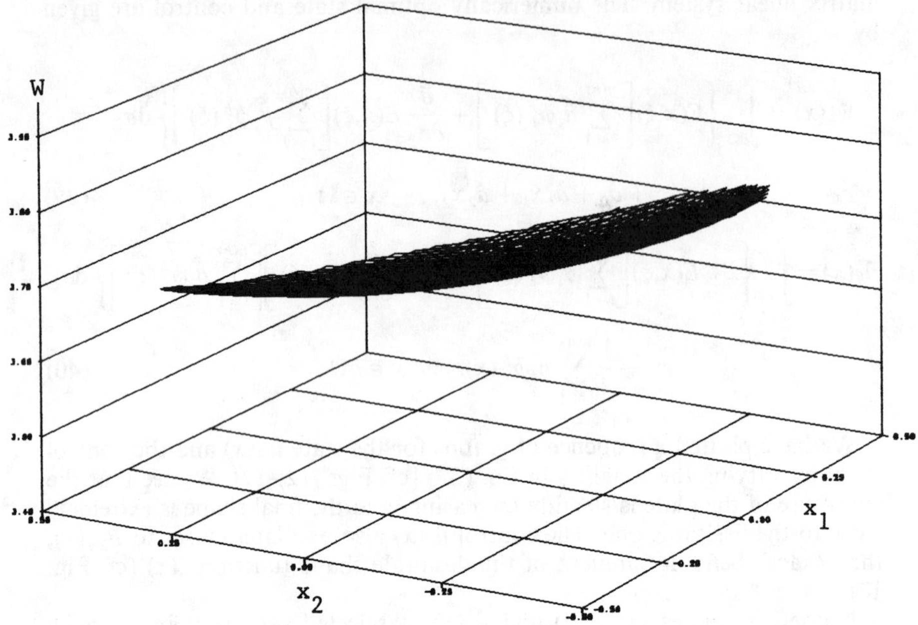

Fig. 13 Deformation with bending-moment control, $\gamma = 0.05$.

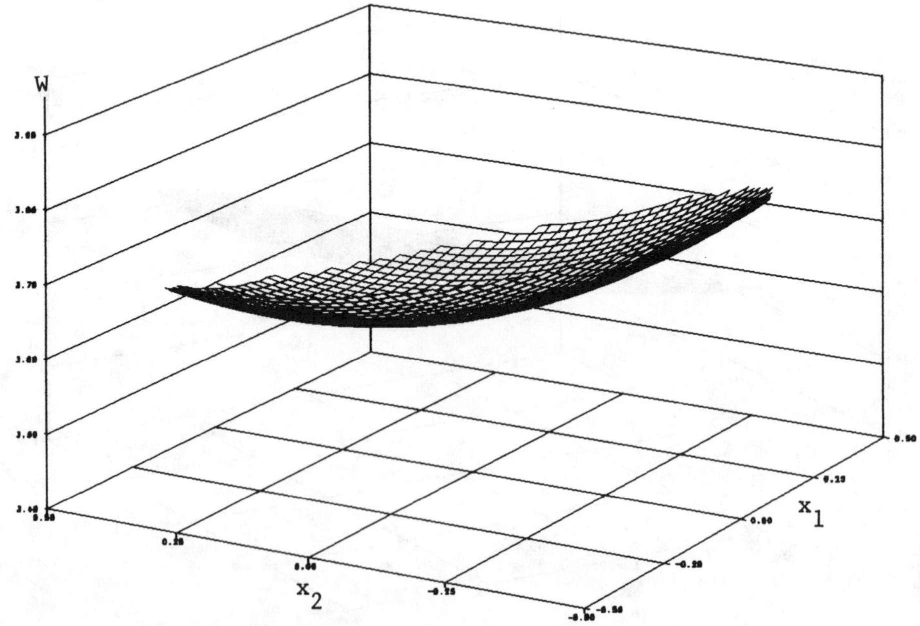

Fig. 14 Deformation with bending-moment control, $\gamma = 0.01$.

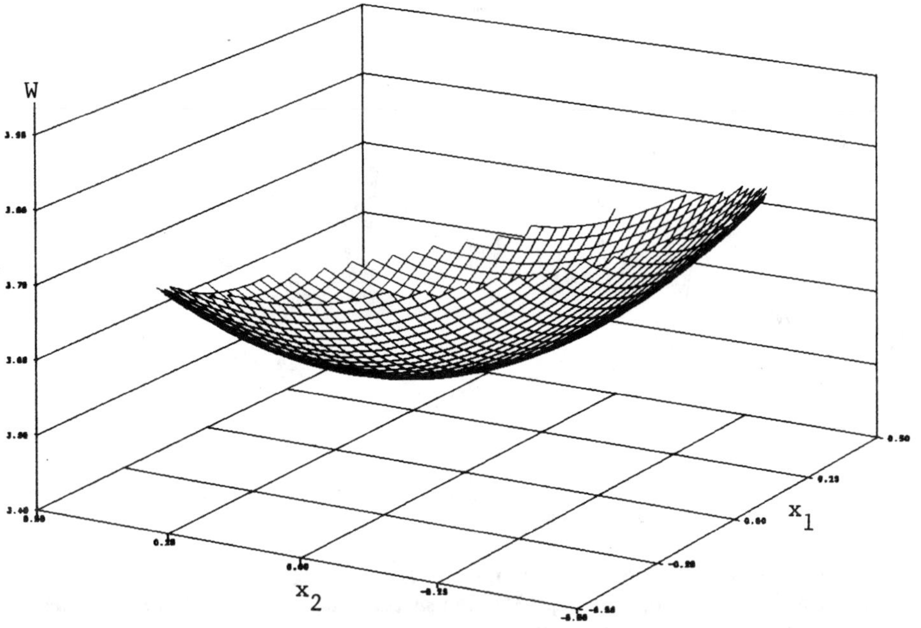

Fig. 15 Deformation with bending-moment control, $\gamma = 0.001$.

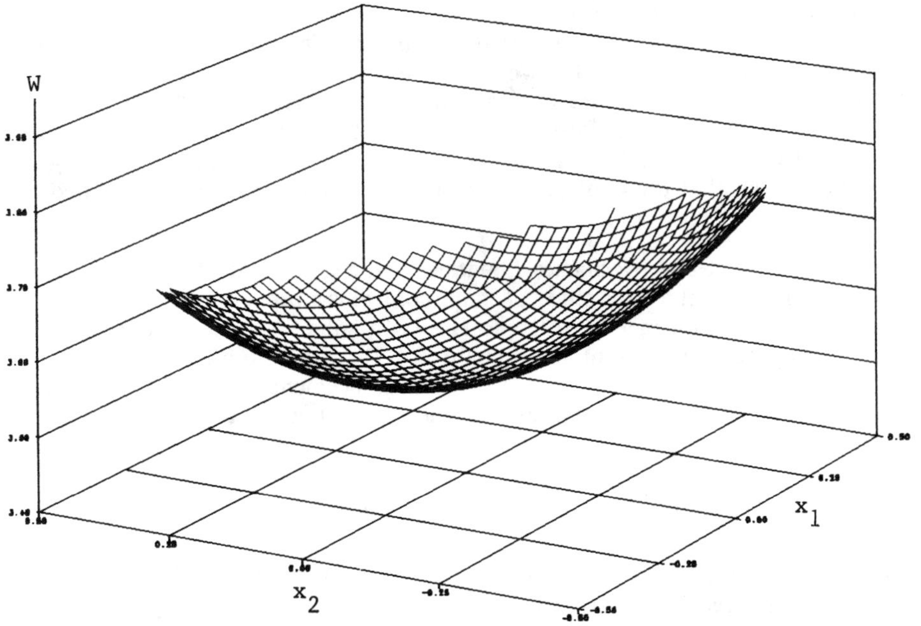

Fig. 16 Deformation with bending-moment control, $\gamma = 0.00001$.

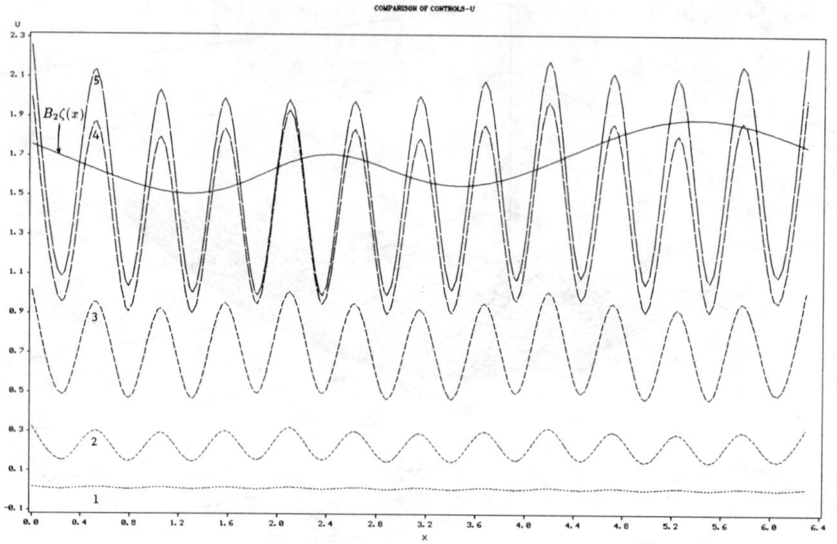

Fig. 17 Numerical solutions of $\hat{u}_h^{(i)}(x)$, $i = 1, 2, 3, 4, 5$ and $h = \pi/24$, for the control problem [Eq. (33)] subject to Eq. (34). The superscripts $i = 1, 2, 3, 4, 5$ correspond, in sequential order, to the γ in Figs. 12–16.

By comparison, we see that the control model [Eqs. (33) and (34)] is much superior to Eqs. (16) and (17).

To close this section, we note that the biharmonic Kirchhoff-Timoshenko plate model applies only to two-dimensional elastic structures with *small* curvature. If the curvature becomes large, one must use the *shell* model, whose boundry-element formulation is not yet available. In our numerical control examples we have purposefully used elastic structures with large curvature so that the control effects can be more dramatically illustrated.

Solid Mechanics Equations and Shape Control

Our next object for the study of shape control is a three-dimensional linear elastostatic homogeneous isotropic Hookean solid. Assume that the solid originally occupies a bounded domain $\Omega \subseteq \mathbb{R}^3$ with smooth boundary $\partial\Omega$. Let $w(x) = (w_1(x), w_2(x), w_3(x))$ be the displacement vector at position $x = (x_1, x_2, x_3) \in \Omega$. From continuum mechanics the linear strain tensor is given by

$$\epsilon_{ij}(x) = \frac{1}{2}\left[\frac{\partial w_i(x)}{\partial x_j} + \frac{\partial w_j(x)}{\partial x_j}\right], \qquad 1 \leqslant i, j \leqslant 3, \qquad x \in \Omega \tag{41}$$

Let $\sigma_{ij}(x)$, $1 \leqslant i,\ j \leqslant 3$ be the stress tensor at point $x \in \Omega$. Then the stress-strain relation is given by

$$\sigma_{ij} = \lambda(\nabla \cdot w)\delta_{ij} + 2\mu\epsilon_{ij}, \qquad 1 \leqslant i, j \leqslant 3 \tag{42}$$

Table 2 Numerical values of the state and control at selected points for the control problem [Eq. (33)] subject to Eq. (34)

x	$\hat{w}_h^{(1)}(x)$	$\hat{w}_h^{(2)}(x)$	$\hat{w}_h^{(3)}(x)$	$\hat{w}_h^{(4)}(x)$	$\hat{w}_h^{(5)}(x)$	$\zeta(x)$
y_0	3.6890	3.6650	3.6101	3.5343	3.5161	3.5158
y_1	3.6675	3.6541	3.6235	3.5811	3.5704	3.5703
y_2	3.6590	3.6457	3.6151	3.5732	3.5642	3.5642
y_3	3.7594	3.7460	3.7154	3.6732	3.6629	3.6628
$(2/5, \pi/3)$	3.6017	3.5931	3.5733	3.5462	3.5397	3.5406
$(2/5, 2\pi/3)$	3.6560	3.6474	3.6275	3.6003	3.5947	3.5947
$(2/5, \pi)$	3.7302	3.7216	3.7019	3.6748	3.6685	3.6688
$(2/5, 4\pi/3)$	3.7736	3.7649	3.7451	3.7176	3.7109	3.7111
$(2/5, 5\pi/3)$	3.7594	3.7508	3.7310	3.7037	3.6965	3.6951
$(2/5, 2\pi)$	3.6647	3.6560	3.6361	3.6086	3.6015	3.6016
$x = (1/2, \theta)$ θ:	$\hat{u}_h^{(1)}(x)$	$\hat{u}_h^{(2)}(x)$	$\hat{u}_h^{(3)}(x)$	$\hat{u}_h^{(4)}(x)$	$\hat{u}_h^{(5)}(x)$	$B_2\zeta(x)$
$\pi/3$	0.0170	0.2932	0.9254	1.7963	2.0344	1.5231
$\pi/2$	0.0176	0.3025	0.9543	1.8390	1.9932	1.5333
$2\pi/3$	0.0186	0.3206	1.0106	1.9354	1.9869	1.6677
π	0.0170	0.2930	0.9247	1.7905	2.0090	1.5807
$4\pi/3$	0.0187	0.3218	1.0163	1.9811	2.1883	1.6686
$3\pi/2$	0.0176	0.3033	0.9581	1.8696	2.1286	1.8031
$5\pi/3$	0.0170	0.2936	0.9274	1.8121	2.1042	1.8866
2π	0.0187	0.3222	1.0183	1.9977	2.2617	1.7557

$n(h) = 24$ is used.
The superscripts $i = 1, 2, 3, 4, 5$ correspond, respectively, to $\gamma = 1.0, 0.05, 0.01, 0.001, 0.00001$ in Eq. (33).

where $\nabla \cdot \boldsymbol{w} = \text{div}\boldsymbol{w}$, δ_{ij} is the Kronecker delta, and λ,μ are Lamé constants. Assume further that there is no body force, and the solid is in equilibrium. Then

$$\sum_{i=1}^{3} \frac{\partial}{\partial x_i} \sigma_{ij}(x) = 0, \qquad j = 1,2,3, \qquad x \in \Omega \tag{43}$$

Equivalently,

$$\mu \Delta \boldsymbol{w}(x) + (\lambda + \mu)\nabla[\nabla \cdot \boldsymbol{w}(x)] = 0 \tag{44}$$

by Eq. (42). The strain energy of the solid is given by

$$\epsilon(\boldsymbol{w}) = \frac{1}{2}\int_\Omega \left[\lambda(\nabla \cdot \boldsymbol{w})^2 + \frac{\mu}{2}\sum_{j,k=1}^{3}\left(\frac{\partial w_j}{\partial x_k} + \frac{\partial w_k}{\partial x_j}\right)^2\right] \mathrm{d}x \tag{45}$$

Let $m_1(x), m_2(x), \ldots, m_6(x)$ be the column vectors of the matrix $M(x)$:

$$[m_1(x), m_2(x), \ldots, m_6(x)] \equiv M(x) \equiv \begin{bmatrix} 1 & 0 & 0 & 0 & x_3 & -x_2 \\ 0 & 1 & 0 & -x_3 & 0 & x_1 \\ 0 & 0 & 1 & x_2 & -x_1 & 0 \end{bmatrix} \tag{46}$$

It is easy to check that

$$\epsilon(\boldsymbol{w}) = \epsilon\left[\boldsymbol{w} + \sum_{i=1}^{6} \alpha_i m_i(x)\right] \tag{47}$$

Hence, the strain energy of a solid is the same up to a six-degree-of-freedom rigid-body motion

$$\sum_{i=1}^{6} \alpha_i m_i(x)$$

At a given boundary point $x \in \partial\Omega$, the traction vector $\boldsymbol{\tau}(w)(x)$ signifies the surface force and is given by

$$\boldsymbol{\tau}(w)(x) = (\tau_1(x), \tau_2(x), \tau_3(x)) \tag{48a}$$

$$\tau_j(x) = \sum_{i=1}^{3} \sigma_{ij}(x) n_i(x) \tag{48b}$$

$$j = 1,2,3, \qquad \boldsymbol{n} = (n_1, n_2, n_3) \tag{48c}$$

is the unit exterior normal.

Then

$$\boldsymbol{\tau}(w) = \lambda(\nabla \cdot \boldsymbol{w})\boldsymbol{n} + 2\mu \frac{\partial \boldsymbol{w}}{\partial n} + \mu \boldsymbol{n} \times (\nabla \times \boldsymbol{w}) \tag{49}$$

where $\boldsymbol{n} \times (\nabla \times \boldsymbol{w})$ is the cross product of $\boldsymbol{n}$ with $\nabla \times \boldsymbol{w}$, the curl of $\boldsymbol{w}$.

From the principle of virtual work,

$$\begin{aligned} 0 = \delta\epsilon(\boldsymbol{w}) &= \int_\Omega \Bigg[\lambda(\nabla \cdot \boldsymbol{w})(\nabla \cdot \delta\boldsymbol{w}) \\ &\quad + \frac{\mu}{2} \sum_{j,k=1}^{3} \left(\frac{\partial w_j}{\partial x_k} + \frac{\partial w_k}{\partial x_j}\right)\left(\frac{\partial}{\partial x_k}\delta w_j + \frac{\partial}{\partial x_j}\delta w_k\right)\Bigg] \mathrm{d}x \\ &= -\int_\Omega [\mu\Delta\boldsymbol{w} + (\lambda + \mu)\nabla(\nabla \cdot \boldsymbol{w})]\delta\boldsymbol{w}\, \mathrm{d}x + \int_{\partial\Omega} \boldsymbol{\tau}(\boldsymbol{w}) \cdot \delta\boldsymbol{w}\, \mathrm{d}\sigma \end{aligned} \tag{50}$$

Thus, we obtain two basic variational types of boundary conditions:

Dirichlet type:

$$w(x) = g(x) \tag{51}$$

is given on $\partial\Omega$, or

Neumann type:

$$\tau(w)(x) = g(x) \tag{52}$$

is given on $\partial\Omega$, for $w(x)$ satisfying the partial differential equation [Eq. (44)].

Boundary Integral Equations for Linear Elastostatics

We are primarily concerned with the Neumann problem

$$\begin{cases} \mu \, \Delta w(x) + (\lambda + \mu)\nabla[\nabla \cdot w(x)] = 0 & \text{on} \quad \Omega \\ \tau(w)(x) = g(x) & \text{on} \quad \partial\Omega \end{cases} \tag{53}$$

since it is the most relevant to the boundary shape control of solids. To formulate boundary integral equations for Eq. (53), we take the approach of a direct representation using Green's formula. It is known (Kupradze[4]) that the fundamental soluton matrix $E(x,\xi)$ for the equations of linear elastostatics satisfying

$$\mu\Delta_\xi E(x,\xi) + (\lambda + \mu)\nabla_\xi[\nabla_\xi \cdot E(x,\xi)] = \delta(x - \xi)I_3 \tag{54}$$

is given by

$$E(x,\xi) = \frac{\lambda + 3\mu}{8\pi(\lambda + 2\mu)} \frac{1}{r_{x\xi}} \left\{ I_3 + \frac{\lambda + \mu}{\lambda + 3\mu} \frac{1}{r_{x\xi}^2} (x - \xi)(x - \xi)^T \right\} \tag{55}$$

where, in Eqs. (54) and (55),

I_3 = the 3 × 3 identity matrix

$r_{x\xi} = |x - \xi|$

$(x - \xi)^T$ = a row vector, is the transpose of the column vector $x - \xi$

For $x \in \Omega$, Green's theorem gives

$$\begin{aligned} w(x) &= \int_\Omega (\{\mu\Delta_\xi E(x,\xi) + (\lambda + \mu)\nabla_\xi[\nabla_\xi \cdot E(x,\xi)]\} w(\xi) \\ &\quad - E(x,\xi)\{\mu\Delta w(\xi) + (\lambda + \mu)\nabla[\nabla \cdot w(\xi)]\})\, d\xi \\ &= \int_{\partial\Omega} [\tau_\xi E(x,\xi) \cdot w(\xi) - E(x,\xi)\tau(w)(\xi)]\, d\sigma_\xi \end{aligned} \tag{56}$$

where

$$\tau_\xi E(x,\xi) = \frac{\mu}{4\pi(\lambda + 2\mu)}\left(\left[I_3 + \frac{3(\lambda+\mu)}{\mu r_{x\xi}^2}(x-\xi)(x-\xi)^T\right]\frac{\partial}{\partial n_\xi}\frac{1}{r_{x\xi}} - \frac{1}{r_{x\xi}^3}\{(x-\xi)n(\xi)^T - [(x-\xi)n(\xi)^T]^T\}\right)$$

In Eq. (56), letting x tend nontangentially to a boundary point on $\partial\Omega$, we have obtained the boundary integral equation

$$\frac{1}{2}\boldsymbol{w}(x) - \int_{\partial\Omega} \tau_\xi E(x,\xi)\cdot \boldsymbol{w}(\xi)\,d\sigma_\xi = -\int_{\partial\Omega} E(x,\xi)\boldsymbol{\tau}(\boldsymbol{w})(\xi)\,d\sigma_\xi, \qquad x \in \partial\Omega \tag{57}$$

or, in operator form,

$$H\boldsymbol{w} = G\boldsymbol{\tau} \tag{58}$$

where H and G represent, respectively, the integral operators on the left and the right-hand sides of Eq. (58). Note again that the operator H involves a divergent integral.

The following theorems were originally due to Kupradze.[4]

Theorem 3.1 (properties of operators H and G).

1) The operator equation

$$G\boldsymbol{h} = \boldsymbol{\psi} \tag{59}$$

is always solvable for given $\boldsymbol{\psi}$ sufficiently smooth; i.e., G is an invertible operator. We write

$$\boldsymbol{h} = G^{-1}\boldsymbol{\psi}$$

as the solution to Eq. (59).

2) The operator equation

$$H\boldsymbol{f} = \boldsymbol{\psi} \tag{60}$$

is solvable for given ψ sufficiently smooth if and only if the six compatibility conditions

$$\int_{\partial\Omega} \boldsymbol{\psi}(x)\cdot m_i(x)\,d\sigma = 0, \qquad i = 1,2,\ldots,6 \tag{61}$$

are satisfied. If $\boldsymbol{f}_0$ satisfies

$$H\boldsymbol{f}_0 = \boldsymbol{\psi}$$

then a general solution $\boldsymbol{f}$ to Eq. (60) is given by

$$\boldsymbol{f}(x) = \boldsymbol{f}_0(x) + \sum_{i=1}^{6} a_i \boldsymbol{m}_i(x) \tag{62}$$

for arbitrarily given $a_1, a_2, \ldots, a_6 \in \mathbb{R}$.

Theorem 3.2 (boundary integral solutions to the Neumann problem). Given $\boldsymbol{h}(x)$ on $\partial\Omega$, satisfying

$$\int_{\partial\Omega} \boldsymbol{h}(x) \cdot \boldsymbol{m}_i(x) \, d\sigma = 0, \qquad i = 1,2, \ldots, 6 \tag{63}$$

the Neumann problem

$$\begin{cases} \mu \Delta \boldsymbol{w}(x) + (\lambda + \mu)\nabla[\nabla \cdot \boldsymbol{w}(x)] = 0 & \text{on} \quad \Omega \\ \boldsymbol{\tau}(\boldsymbol{w})(x) = \boldsymbol{h}(x) \end{cases} \tag{64}$$

has solutions

$$\boldsymbol{w}(x) = \boldsymbol{w}_0(x) + \sum_{i=1}^{6} a_i \boldsymbol{m}_i(x) \quad \text{for arbitrary} \quad a_i \in \mathbb{R}, \quad i = 1,2, \ldots, 6 \tag{65}$$

where

$$\boldsymbol{w}_0(x) = \int_{\partial\Omega} [\tau_\xi E(x,\xi) \cdot \boldsymbol{f}_0(\xi) - E(x,\xi)\boldsymbol{h}(\xi)] \, d\sigma_\xi, \qquad x \in \Omega$$

is a particular solution of Eq. (64), with $\boldsymbol{f}_0$ a particular solution of the boundary integral equation

$$H\boldsymbol{f}_0 = G\boldsymbol{h} \tag{66}$$

Theorem 3.3 (boundary integral solution to the Dirichlet problem). Given $\boldsymbol{g}(x)$ on $\partial\Omega$, there exists a unique solution $\boldsymbol{w}(x)$ on Ω satisfying

$$\begin{cases} \mu \Delta \boldsymbol{w}(x) + (\lambda + \mu)\nabla(\nabla \cdot \boldsymbol{w}(x)) = 0 & \text{on} \quad \Omega \\ \boldsymbol{w}(x) = \boldsymbol{g}(x) \quad \text{on} \quad \partial\Omega \end{cases} \tag{67}$$

The solution $\boldsymbol{w}(x)$ is given by

$$\boldsymbol{w}(x) = \int_{\partial\Omega} \{\tau_\xi E(x,\xi) \cdot \boldsymbol{g}(\xi) - E(x,\xi) \cdot [(G^{-1}H)g](\xi)\} \, d\sigma_\xi, \quad x \in \Omega \tag{68}$$

Shape Control of a Solid by the Application of Traction Force on the Boundary Surface

As in the previous section, we assume that there are finitely many sensors distributed over the solid. Here we choose the sensor locations to be on the boundary surface because they allow easy access, although it is also done very often in practice that the sensors are placed in the interior of the solid. Thus, we assume a quadratic cost functional

$$J(u) = \sum_{i=1}^{n} |\boldsymbol{w}(y_i) - \zeta_i|^2 + \gamma \int_{\partial\Omega} |\boldsymbol{u}(x)|^2 \, d\sigma, \qquad \gamma > 0 \tag{69}$$

where y_i, $i = 1,2, \ldots, n$ are sensor locations on $\partial\Omega$, and $\zeta_i \in \mathbb{R}^3$, $i = 1,2, \ldots, n$ are target deformation vectors at those points, and an integral term of $\boldsymbol{u}$, the traction control on $\partial\Omega$, is included in the overall objective cost functional. The solid is governed by

$$\begin{cases} \mu\Delta\boldsymbol{w}(x) + (\lambda + \mu)\nabla(\nabla \cdot \boldsymbol{w}(x)) = 0 & \text{on} \quad \Omega \\ \tau(\boldsymbol{w})(x) = \boldsymbol{u}(x), & x \in \partial\Omega \end{cases} \tag{70}$$

Because solutions $\boldsymbol{w}(x)$ to Eq. (70) have six degrees of freedom, we must impose six additional auxiliary conditions to Eq. (70). They are commonly given in the form

$$w_{i_0}(y_{j_0}) = c_{i_0 j_0} \tag{71}$$

for some $y_{j_0} \in \partial\Omega$, $1 \leqslant j_0 \leqslant 6$, for some i_0, $1 \leqslant i_0 \leqslant 3$. There are a total of six of them in Eq. (71). We let $\partial\Omega$ be approximated by piecewise polygonal planes or surfaces:

$$\partial\Omega \approx \bigcup_{i=1}^{n(h)} \Gamma_i^h \tag{72}$$

where h signifies a mesh size parameter. The piecewise surfaces Γ_i^h, $1 \leqslant i \leqslant n(h)$ at most overlap on their boundaries. The surface approximation [Eq. (72)] must be "uniform"; i.e., there exist $B_1 > 0$, $B_2 > 0$ such that

$$0 < B_1 \leqslant \frac{\text{area } \Gamma_i^h}{\text{area } \Gamma_j^h} \leqslant B_2 \qquad \text{for all} \qquad 1 \leqslant i, j \leqslant n(h)$$

and the approximation surfaces must be convergent:

$$\lim_{h \downarrow 0} \bigcup_{i=1}^{n(h)} \Gamma_i^h = \partial\Omega$$

We are now in a position to discretize the shape control problem:

$$\inf_{\boldsymbol{u} \in L^2(\partial\Omega)} \text{ Eq. (69) subject to Eq. (70)}$$

We first utilize the fact that on $\partial\Omega$ we have

$$\boldsymbol{u} = G^{-1}H\boldsymbol{w} \tag{73}$$

because G is invertible according to *Theorem* 3.1. We substitute Eq. (73) into Eq. (70) to obtain

$$I(\boldsymbol{w}) \equiv J(G^{-1}H\boldsymbol{w}) \tag{74}$$

By doing so, we have eliminated the traction control $\boldsymbol{u}$ and obtained a minimization problem in terms of the displacement $\boldsymbol{w}$ on the boundary.

We approximate $\boldsymbol{w}$ in Eq. (74) by piecewise constant (vector-valued) functions

$$\boldsymbol{w}_h(x) = \sum_{\ell=1}^{3} \sum_{j=1}^{n(h)} \alpha_{j\ell} \boldsymbol{\phi}^h_{j,\ell}(x) \tag{75}$$

where

$$\boldsymbol{\phi}^h_{j,\ell}(x) = \begin{cases} (1,0,0) & & \ell = 1 \\ (0,1,0) & \text{if} & \ell = 2 \quad \text{for} \quad x \in \Gamma^h_j \\ (0,0,1) & & \ell = 3 \\ (0,0,0) & \text{for} & x \notin \Gamma^h_j \end{cases} \tag{76}$$

Discretizing the functional I in Eq. (74) by

$$I_h(\boldsymbol{w}_h) \equiv J(G_h^{-1} H_h \boldsymbol{w}_h)$$

where G_h, H_h are discretized operators of G and H relative to the choice of boundary-element basis [Eq. (76)]. The $3 \cdot n(h)$ unknowns in Eq. (75) are then determined by

$$\frac{\partial}{\partial \alpha_{j\ell}} I_h(\boldsymbol{w}_h)\big|_{\alpha_{j\ell} = \hat{\alpha}_{j\ell}} = 0, \qquad j = 1,2, \ldots, n(h),\ \ell = 1,2,3$$

The numerically optimal control $\hat{\boldsymbol{u}}_h$ is then determined by

$$\boldsymbol{u}_h = G^{-1}H\boldsymbol{w}_h = \sum_{\ell=1}^{3} \sum_{j=1}^{n(h)} \hat{\alpha}_{j\ell} G^{-1} H(\boldsymbol{\phi}^h_{j,\ell})$$

and the numerically optimal state (on Ω), by Eq. (56), is

$$\boldsymbol{w}^h(x) = \int_{\partial\Omega} [\tau_\xi E(x,\xi) \cdot \boldsymbol{w}_h(\xi) - E(x,\xi) \cdot \boldsymbol{u}_h(\xi)]\, d\sigma_\xi, \qquad x \in \Omega$$

Note that once again the compatibility conditions [Eq. (63)] will be automatically satisfied by $\boldsymbol{u}_h$ and thus do not need any special attention.

Table 3 Comparison of the controlled displacement w with objective displacement ζ_i at sensor locations y_i, $1 \leqslant i \leqslant 14$

i	y_i			ζ_i			$w(y_i)$		
1	1	0	0	$0.000E+00$	$0.000E+00$	$-0.872E+00$	$-0.169E-06$	$0.668E-10$	$-0.871E+00$
2	1	$\pi/4$	0	$0.227E-04$	$0.813E-01$	$-0.411E+00$	$0.228E-04$	$0.813E-01$	$-0.410E+00$
3	1	$\pi/4$	$\pi/2$	$-0.817E-01$	$0.459E-03$	$-0.410E+00$	$-0.818E-01$	$0.459E-03$	$-0.410E+00$
4	1	$\pi/4$	π	$-0.902E-03$	$-0.813E-01$	$-0.411E+00$	$-0.902E-03$	$-0.813E-01$	$-0.410E+00$
5	1	$\pi/4$	$3\pi/2$	$0.808E-01$	$-0.459E-03$	$-0.410E+00$	$-0.809E-01$	$-0.459E-03$	$-0.40E+00$
6	1	$\pi/2$	0	$0.000E+00$	$0.152E+00$	$0.000E+00$	$-0.604E-07$	$0.152E+00$	$-0.760E-17$
7	1	$\pi/2$	$\pi/2$	$-0.152E+00$	$0.000E+00$	$0.000E+00$	$-0.152E+00$	$0.238E-07$	$-0.103E-16$
8	1	$\pi/2$	π	$0.000E+00$	$-0.152E+00$	$0.000E+00$	$-0.109E-06$	$-0.152E+00$	$-0.716E-17$
9	1	$\pi/2$	$3\pi/2$	$0.152E+00$	$0.000E+00$	$0.000E+00$	$0.152E+00$	$-0.240E-07$	$-0.865E-17$
10	1	$3\pi/4$	0	$0.227E-04$	$-0.813E-01$	$0.411E+00$	$0.228E-04$	$0.813E-01$	$0.410E+00$
11	1	$3\pi/4$	$\pi/2$	$-0.817E-01$	$0.459E-03$	$0.410E+00$	$-0.818E-01$	$0.459E-03$	$0.410E+00$
12	1	$3\pi/4$	π	$-0.902E-03$	$-0.813E-01$	$0.411E+00$	$-0.902E-03$	$-0.813E-01$	$0.410E+00$
13	1	$3\pi/4$	$3\pi/2$	$0.808E-01$	$-0.459E-03$	$0.410E+00$	$0.809E-01$	$-0.459E-03$	$0.410E+00$
14	1	π	0	$0.000E+00$	$0.000E+00$	$0.872E+00$	$-0.169E-06$	$0.668E-10$	$0.871E+00$

To furnish a concrete example, let Ω be the unit sphere in $\mathbb{R}^3$ and (r,θ,ϕ) be the spherical coordinates for (x_1,x_2,x_3) in $\mathbb{R}^3$:

$$r > 0, \qquad 0 \leqslant \phi \leqslant \pi, \qquad 0 \leqslant \theta \leqslant 2\pi$$

Then

$$\partial\Omega = \{(1,\phi,\theta)|0 \leqslant \phi \leqslant \pi, 0 \leqslant \theta \leqslant 2\pi\}$$

Suppose that in Eq. (69) 14 sensors at y_i, $1 \leqslant i \leqslant 14$ are used, and ζ_i, $1 \leqslant i \leqslant 14$ are the target deformation vectors. Their data are given in Table 3, where for y_j, $1 \leqslant j \leqslant 14$ spherical coordinates are used, and for ζ_j, $1 \leqslant j \leqslant 14$ rectangular coordinates are used. Use

$$\gamma = 10^{-5}, \qquad \lambda = 2/3, \qquad \mu = 1$$

in Eqs. (69) and (70). The sphere is compressed into the shape shown in Figs. 18 and 19. At these sensor locations, the displacement vectors are computed and listed on the three rightmost columns of Table 3 (in rectangular coordinates). The reader can find that these vector values agree very well with ζ_i, $1 \leqslant i \leqslant 14$, respectively.

The discretization of the boundary by boundary elements of this computation can be seen in Fig. 18, where $n(h) = 11 \times 24 + 2 = 266$ boundary elements are used.

Again, we note that the linear elastostatic model in this section applies only to the case of small deformation. In Figs. 18 and 19 we give large deformation in order to illustrate the dramatic effects of shape control. In reality, *finite elasticity* models must be used in lieu of linear elasticity ones when there is large deformation.

Our boundary-element numerical algorithms have been carefully tested. The computer solutions have agreed extremely well with known closed-form benchmarks. The treatment of divergent integrals (cf. also *Remark* 2.7) is the key to success here.

Concluding Remarks

The deformation and shape control are very sensitive to the choice of sensor locations. In the numerical examples given here, the desirable shapes and sensor locations all have good symmetry and thus yield fairly satisfactory solutions. In general, if the desirable shape lacks certain symmetry, then symmetric placement of sensors will not yield good shape control, and more sensors are required to achieve a satisfactory shape.

This paper has shown us

1) the application of the boundry-element method in numerical simulation and control model selection; and

2) the combination of the boundary-element method, boundary control, and computer graphics to treat and to understand shape control problems.

The controllers obtained here are all open loop. We are presently also exploring the application of the boundary-element method to the feedback boundary shape control of dynamic distributed-parameter systems.

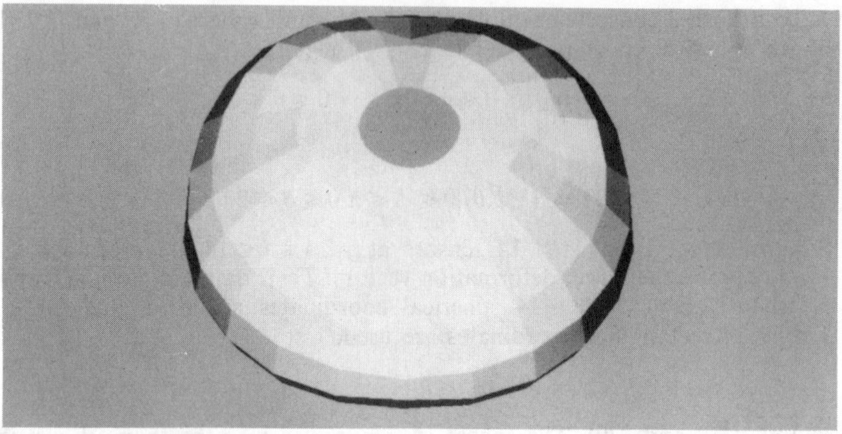

Fig. 18 An apple-like compressed sphere obtained by linear-quadratic shape control method (top view).

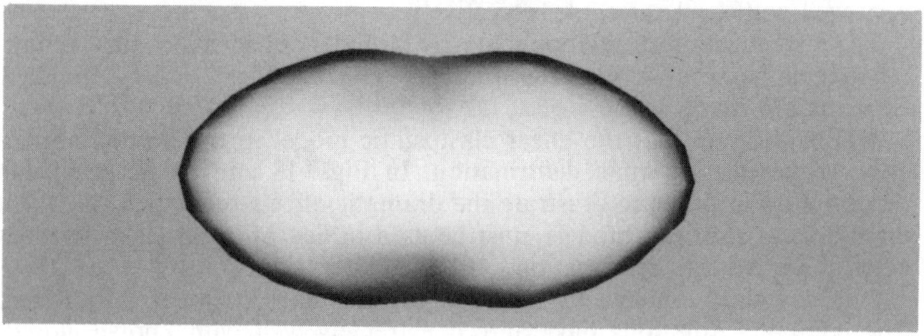

Fig. 19 Same solid as in Fig. 18 (front view).

Acknowledgment

This work was supported in part by AFOSR Grants 87-0334 and 88-0091 and by National Science Foundation Grant DMS 87-18510.

References

[1]Timoshenko, S. P., and Woinowsky-Krieger, S., *Theory of Plates and Shells*, 2nd ed., McGraw-Hill, New York, 1959, pp. 88–91.

[2]Chen, G., and Zhou, J., "Boundary Integral Equations and Boundary Element Computations for the Biharmonic Timoshenko Plate Equation with Applications to Boundary Shape Control" (Preprint).

[3]Chen, G., and Tsai, Y. L., "The Boundary Element Method for Two Dimensional Linear Quadratic Elliptic Problems (I): Neumann Control," *Math. Comp.*, Vol. 49, 1987, pp. 479–498.

[4]Kupradze, V. D., *Potential Methods in the Theory of Elasticity*, Israel Program Scientific Translation, Jerusalem, 1965, Chap. 2.

Chapter 13

Stability of Time-Varying Structural Dynamic Systems

S. Pradeep*
Indian Institute of Technology, Kanpur, India

Introduction

TIME-VARYING systems (with one notable exception—that of periodic systems) are of recent origin in the field of structural dynamics. Presently, most studies in structural dynamics deal with constant coefficient systems. This is due to the fact that, in the time-varying systems that were encountered in the past, the changes in the structural properties were so small that it was possible either to ignore them altogether or to average them over a reasonable period of time and treat the system as a constant coefficient.

The exception of periodically time-varying systems has been of interest for a long time and has been studied extensively. Problems such as the motion of the lunar perigee and the vibrations of a stretched elliptic membrane initiated the studies on periodic systems. Periodically time-varying systems continue to hold the interest of engineers, especially that of structural dynamicists, through current problems such as the dynamics of spacecraft in elliptic orbits and of helicopter rotor blades, which ultimately lead to differential equations with periodic coefficients.

Arbitrarily time-varying systems were introduced by recent developments in spacecraft dynamics and multibody dynamics. The increasing size and complexity of present-day satellites, compounded with the limited size of launch vehicle fairings, make it impossible to carry the spacecraft in the desired form to the final orbit. Therefore, the current trend is to pack spacecraft, or parts of a spacecraft within the launch vehicle, to be deployed and assembled in the final orbit. In the case of tethered satellites

*Assistant Professor, Department of Aerospace Engineering.

the operations of deployment and retrieval are an integral part of the mission. These are typical problems that lead to equations with arbitrarily time-varying coefficients.

Most time-varying systems are not amenable to a closed-form solution. We are therefore forced to take recourse to numerical solutions or to a qualitative study of the solutions. Although numerical methods are widely used, they have well-known drawbacks:

1) The qualitative behavior of the solutions cannot be predicted by numerical integration schemes.

2) In order to determine the effect of variation of initial conditions or changes in parameters, the entire equation must be integrated again with the new values.

3) Errors increase monotonically with the duration of integration.

4) Computer memory, and most often time, are limited.

On the other hand, qualitative methods ascertain the properties of solutions without explicitly solving the equation in question. Of the properties of solutions, stability is perhaps the most important. This stems from three reasons: first, in any physical problem, it is the stable solutions that are of interest. As engineers, we wish to be informed about the unstable solutions only so that we may elude them in design. Second, unstable solutions are extremely sensitive to errors. Compounded with the fact that no physical law is exact, this will render such solutions useless as they will turn out to be poor approximations of observed natural phenomena. Finally, although the unstable states are mathematically rigorous solutions of the differential equations, they cannot be predicted in practice because of the small disturbances that exist in the initial conditions and in the forces acting on the system. These factors have to be disregarded in the formulation of the problem either because of the unwieldy equations that result or because of our sheer ignorance of the way they manifest themselves.

In typical problems of space structures and multibody dynamics, the dimension of the system is rather high. Most often, the equations derived are second order. However, most of the existing results on stability, particularly those for nonlinear systems, are for first-order systems. Because of the large dimension of the system, it is difficult in most cases to convert the second-order system into first order. Therefore, some of the recent research deals with second-order systems directly. In this chapter a review of results on the stability of time-varying systems that are of interest to structural dynamicists is presented.

Stability

Stability has no universally accepted definition. A rough characterization is given by the following: Given a solution to a differential equation, do neighboring solutions remain close as time goes to infinity? (The solution is then stable.) Or do they depart from each other? (The solution is then unstable.) The modern theory of stability was founded by Liapunov.[1] We shall later define some of the more commonly used concepts of stability.

Let the physical system be governed by the following first-order, vector differential equation:

$$y' = g(t; y) \tag{1}$$

where y and g are $(n \times 1)$ vectors, with y representing the state vector and g a nonlinear function. The time t is the independent variable, and the prime denotes differentiation with respect to time. All of the vectors in this chapter are assumed to belong to $\boldsymbol{R}^n$, and the norms associated with them are Euclidean norms. Although the assumptions can be relaxed and the results can be generalized, it is not necessary in the present context, because these assumptions would suffice to describe engineering systems.

Let $y = \eta(t)$ be a particular solution of Eq. (1). In order to study the properties of solutions of Eq. (1) in the neighborhood of $\eta(t)$, we make the substitution

$$x = y - \eta(t) \tag{2}$$

where $x = 0$ is the unperturbed trajectory. The transformed equation is

$$x' = f(t;x) \tag{3}$$

where f is a $(n \times 1)$ nonlinear function. Equation (3) is called the equation of perturbed motion. Henceforth, we shall treat Eq. (3) directly without reference to Eq. (1). The solution $y = \eta(t)$ now corresponds to the solution $x = 0$ and is called the trivial solution or the equilibrium of Eq. (3). The function f has the property

$$f(t;0) = 0, \qquad t \geqslant t_0$$

We assume the existence and uniqueness of solutions with the initial condition x_0 in some neighborhood of the origin for all $t \geqslant t_0$.

Definition 1. The trivial solution $x = 0$ of Eq. (3) is said to be stable in the sense of Liapunov (ISL) if for every positive real number ε there exists another positive real number $\delta = \delta(\varepsilon, t_0)$ such that

$$\|x(t;t_0,x_0)\| < \varepsilon \qquad \text{for} \qquad t \geqslant t_0$$

whenever

$$\|x(t_0)\| < \delta$$

Definition 2. The trivial solution $x = 0$ of Eq. (3) is said to be attractive if there exists a positive real number δ such that

$$\lim_{t \to \infty} x(t;t_0,x_0) = 0 \qquad \text{whenever} \qquad \|x_0\| < \delta$$

It is to be noted that the concepts of stability and attractiveness are independent of each other; a system that is not stable can be attractive, and vice versa.

Definition 3. The trivial solution $x = 0$ of Eq. (3) is said to be asymptotically stable ISL if it is both stable ISL and attractive.

Definition 4. The trivial solution $x = 0$ of Eq. (3) is said to be globally asymptotically stable ISL if it is stable ISL and attractive for every initial value x_0.

Definition 5. The trivial solution $x = 0$ of Eq. (3) is said to be unstable if it is not stable.

Instability implies that there exists a positive real number ε for which no δ can be found complying to the definition. Attractive solutions can be unstable.

Definition 6. Given a system described by

$$x' = f(t;x)$$

a solution $x(t;t_0,x_0)$ is said to be bounded if $\|x\| \leqslant M$ for all $t \geqslant t_0$ and some finite, positive M.

Boundedness and stability are independent concepts; however, for linear systems they are connected.

(*i*) For homogeneous systems,

$$x' = A(t)x \tag{4}$$

where the elements of the matrix $A(t)$ are continuous in $[t_0,\infty)$, the solutions $x(t)$ are all stable if and only if they are all bounded.

(*ii*) For homogeneous linear systems,

$$x' = A(t)x + b(t) \tag{5}$$

where the elements of $A(t)$ and $b(t)$ are continuous in $[t_0,\infty)$ if the solutions are all bounded, then they are stable; if they are all stable and one is known to be bounded, then they are all bounded.

A solution $x(t)$ of Eq. (5) is stable (or asymptotically stable) if and only if the null solution of the corresponding homogeneous equation [Eq. (4)] is stable (or asymptotically stable).

This statement frees the concept of stability from the particular solution considered and of the nonhomogeneous term $b(t)$, enabling us to discuss the stability of a linear differential equation instead of the stability of a particular solution.

Equations of Motion

In this section the equations of motion of structural dynamic systems, obtained by using different formulations, are described to show that they cannot always be analyzed by the existing stability criteria.

Several techniques exist for the formulation of the equations of motion for mechanical systems. Among these, some of the popular ones are 1) the Newton-Euler method, 2) the Lagrange method, 3) the quasicoordinate formulation, 4) the Hamilton canonical equations, and 5) the Kane equations. Depending on the method used, the ensuing equations after linearization are either of the first-order form

$$A(t)x'(t) = B(t)x(t) + b(t)\boldsymbol{v}(t) \tag{6}$$

or of the second-order form

$$M(t)\boldsymbol{q}''(t) + [D(t) + G(t)]\boldsymbol{q}'(t) + [K(t) + A(t)]\boldsymbol{q}(t) = H(t)\boldsymbol{v}(t) \tag{7}$$

In Eq. (7) primes denote differentiation with respect to time t, the independent variable; $\boldsymbol{q}$ is an $(n \times 1)$ vector, denoting the generalized coordinates; M, D, G, K, and A are $(n \times n)$ matrices, whose elements are functions of t and differentiable (M is symmetric and positive-definite, D and K are symmetric, and G and A are skew-symmetric); H is an $(n \times m)$ matrix, whose elements are functions of t; $\boldsymbol{v}$ is an $(m \times 1)$ vector; and n is the dimension of the system, which is often large. For mechanical systems M denotes the inertia matrix, D the damping matrix, G the gyroscopic matrix, K the stiffness matrix and A the circulatory matrix. To simplify notation, Eq. (7) is often written as

$$M(t)\boldsymbol{q}''(t) + B(t)\boldsymbol{q}'(t) + C(t)\boldsymbol{q}(t) = u(t)$$

or, simply, as

$$M\boldsymbol{q}'' + B\boldsymbol{q}' + C\boldsymbol{q} = u \tag{8}$$

where the explicit time dependence of the elements of these matrices and vectors is understood.

Most stability criteria cited in standard mathematical texts or derived in current journals deal with the standard first-order system

$$x'(t) = C(t)x(t) + w(t) \tag{9}$$

Equation (9) can be obtained from Eq. (6) by premultiplying with $A(t)^{-1}$. From Eq. (7) we can obtain Eq. (9) by defining

$$x(t) = [\boldsymbol{q}'(t) \quad \boldsymbol{q}(t)]^T \qquad \text{and} \qquad U(t) = [H(t)\boldsymbol{v}(t) \quad 0]^T$$

Then

$$x'(t) = \begin{bmatrix} -M^{-1}B & -M^{-1}C \\ I & 0 \end{bmatrix} x(t) + U(t)$$

Both involve the processes of inversion and multiplication of matrices. Therefore, the task of conversion of the equation from one form to the other, in addition to being laborious (because of the large dimension of the system), is also computationally and analytically unwieldy for the following reasons:

1) The travail of matrix inversion of even small-order constant matrices is so well known that one need hardly overemphasize the effort required to invert matrices of higher order, and that too, in literal form.

2) Through inversion and multiplication of matrices, the physical feel that one has when working directly with the second-order system is rapidly lost. A designer, who needs to know how much damping (or inertia, or stiffness) he must provide to his system, finds little utility in stability conditions stated in terms of $M^{-1}B$ and $M^{-1}C$. To be of help to the designer, it is essential that the ultimate stability conditions be in terms of M, B, and C.

By means of a suitable transformation, we can convert Eq. (7) to Eq. (6) such that the process of transformation does not involve inversion and multiplication of matrices. Four such transformations are listed below. With x and U as defined earlier,

$$\begin{bmatrix} M & 0 \\ 0 & Y \end{bmatrix} x' = \begin{bmatrix} -B & -C \\ Y & 0 \end{bmatrix} x + U$$

where Y can be any of the matrices I, M, B, or C.

The analysis of Eq. (7) can therefore be sometimes simplified by converting it into one of the preceding forms, or equivalently, to Eq. (6). In view of these difficulties, it is necessary in structural systems to treat Eqs. (6) and (7) in addition to Eq. (9).

Nonlinearities and Stability in the First Approximation

Differential equations governing physical systems are mostly nonlinear. Yet, in the preceding section, our discussion was confined to linear systems. This is because, in most problems of structural dynamics, we are concerned with small motions about the equilibrium position; consequently, linear terms lead to sufficiently accurate results.

In many cases, it is possible to simplify the nonlinear problem by expanding the nonlinear function in a power series in a neighborhood of the origin, truncating the series to include only linear terms, and investigating the resulting linear equation. If the governing equation is

$$x' = f(t;x)$$

and it takes the form

$$x' = A(t)x + g(t;x)$$

after the series expansion of f, then

$$x' = A(t)x$$

is called the equation in the first approximation.

The following theorem[2] relates the stability of the nonlinear equation to the stability of the equation in the first approximation:

Theorem 1. If A is a constant matrix whose eigenvalues all have negative real parts, if $\|g(t,x)\| = 0(\|x\|)$ uniformly in $[0,\infty)$, then every solution $x(t)$ of the system

$$x' = Ax + g(t;x) \tag{10}$$

with $\|x(0)\|$ sufficiently small exists, and the solution $x = 0$ is asymptotically stable.

The solution $x = 0$ of Eq. (10) is unstable if at least one of the eigenvalues of A has a positive real part.

A similar theorem can be proved for the case where A is a periodic matrix.

Theorem 2. If $A(t)$ is a periodic matrix whose characteristic exponents all have negative real parts, if $\|g(t,x)\| = 0(\|x\|)$ uniformly in $[0,\infty)$, then every solution $x(t)$ of the system

$$x' = A(t)x + g(t;x)$$

with $\|x(0)\|$ sufficiently small, exists in $[0,\infty)$, and the solution $x = 0$ of the preceding equation is asymptotically stable. The solution $x = 0$ is unstable if at least one of the characteristic exponents of A has a positive real part.

The following theorem of Bellman[3] relates boundedness of solutions of linear and nonlinear equations:

Theorem 3. Consider the linear equation

$$y' = A(t)y + \phi(t) \tag{11}$$

where the elements of the matrix $A(t)$ are continuous functions of time, $\phi(t)$ is an arbitrary function, and the nonlinear equation

$$x' = A(t)x + \Psi(t;x) \tag{12}$$

where $\Psi(t;x)$ is a continuous function in (t,x).

A necessary condition for all solutions of Eq. (12) to be bounded for all vector functions $\Psi(t;x)$ with either

$$\|\Psi(t;x)\| \leqslant M \qquad \text{for} \qquad t \geqslant 0 \tag{12a}$$

or

$$\|\Psi(t;x)\| \leqslant f_1(t), \qquad \int_0^\infty f_1(t)\,\mathrm{d}t < \infty \tag{12b}$$

or

$$\|\Psi(t;x)\| \leqslant f_2(t), \qquad \int_0^\infty f_2^2(t)\,\mathrm{d}t < \infty \tag{12c}$$

is that the linear system in Eq. (11) has only bounded solutions for all vector functions $\phi(t)$ satisfying, respectively, the condition

$$\|\phi(t)\| \leqslant N \qquad \text{for} \qquad t \geqslant 0 \tag{12d}$$

or

$$\int_0^\infty \|\phi(t)\|\,\mathrm{d}t < \infty \tag{12e}$$

or

$$\int_0^\infty \|\phi(t)\|^2\,\mathrm{d}t < \infty \tag{12f}$$

Conversely, if condition (12d), or (12e), or (12f) is satisfied, and the vector function $\Psi(t;x)$ satisfies, respectively, the further condition

$$\|\Psi(t;x)\| \leqslant c_1 \|x\| \tag{12g}$$

or

$$\|\Psi(t;x)\| \leqslant c_1 f_3(t) \|x\|, \qquad \int_0^\infty f_3(t)\,\mathrm{d}t < \infty \tag{12h}$$

or

$$\|\Psi(t;x)\| \leqslant c_1 f_4(t) \|x\|, \qquad \int_0^\infty f_4^2(t)\,\mathrm{d}t < \infty \tag{12i}$$

for all $t \geqslant 0$ and c_1 sufficiently small, whenever $\|x\| \leqslant c_2$, c_2 is sufficiently small, then every solution of Eq. (12) for which $\|x(0)\|$ is sufficiently small is bounded in $[0, \infty)$. In particular, under conditions (12d) and (12g), $\|x\| \to 0$ as $t \to \infty$.

The rest of our discussion is therefore solely confined to linear equations.

Periodic Systems

Many problems in physics and engineering lead to linear ordinary differential equations with periodic coefficients. Such equations appear in spacecraft dynamics,[4] helicopter rotor blades,[5] columns under periodic loading,[6] and periodic laminar flows.[7] They constitute an important, special class of time-varying systems. Some of the fundamental results on periodic systems are presented in this section.

Floquet Theory

Consider the system of first-order equations

$$x' = A(t)x \tag{13}$$

where A is a periodic matrix with period T; i.e., $A(t+T) = A(t)$. The basic result governing the behavior of the solution of such systems is due to Floquet,[8] who showed that the fundamental matrix of Eq. (13) can be expressed as the product of a matrix exponential and a periodic matrix with the same period T as that of the matrix A.

Theorem 4.

(*i*) If Φ is a fundamental matrix for Eq. (13), then there exists a periodic nonsingular matrix P with period T and a constant matrix R such that

$$\Phi(t) = P(t)\, e^{tR} \tag{14}$$

(*ii*) The matrix Ψ, defined by

$$\Psi(t) = \Phi(t+T)$$

is also a fundamental matrix.

Thus, it may be seen that Floquet's theorem provides valuable information about the nature of the solution, although not the solution itself. The following is a brief description of the stability analysis of linear periodic systems using Floquet's theorem.

With $\Phi(t)$ being a fundamental matrix for Eq. (13),

$$\Phi'(t) = A(t)\Phi(t)$$

$$\Psi'(t) = \Phi'(t+T) = A(t+T)\Phi(t+T) = A(t)\Psi(t)$$

Now

$$\det\Psi(t) = \det\Phi(t+T) \neq 0$$

for all t, and hence Ψ is a fundamental matrix for Eq. (13).

The variables $\Phi(t)$ and $\Phi(t+T)$, being fundamental matrices, are related by

$$\Phi(t+T) = \Phi(t)C$$

where C is a constant nonsingular matrix. The matrix C can be written as

$$C = e^{TR}$$

where R is a constant matrix:

$$\Phi(t+T) = \Phi(t)\, e^{TR}$$

Define P by

$$P(t) = \Phi(t)\, e^{-tR}$$

It can be shown that P is a periodic matrix and is nonsingular for all t. This proves that $\Phi(t)$ can be expressed in the form of Eq. (14). Thus, if the fundamental matrix is determined over one period, say, $[0,T)$, then it is determined at once over $[0,\infty)$, provided the matrices $P(t)$ and R are known. For stability analysis of the system, it is not necessary to compute the matrix $P(t)$; clearly, the stability of the system is governed by the eigenvalues of the matrix R. It will now be shown how this is computed.

The expression $\Phi(T)$ is first determined. C is then given by

$$C = \Phi^{-1}(0)\Phi(T)$$

The expression $\Phi(0)$ is usually chosen to be the identity matrix so that the computation is simplified. The matrix $\Phi(T)$ that satisfies $\Phi(0) = I$ is called the Floquet transition matrix (FTM). For the large systems usually encountered in engineering, the computational effort required for evaluating the FTM is considerable. We shall return to this topic at the end of this section.

The matrix R can be determined from the relation

$$R = \frac{1}{T}\,\ell n C$$

The logarithm is a multiple-valued function. Therefore, R is not determined uniquely. However, this is in no way a handicap regarding stability analysis, because of the following result.

If Φ_1 is any fundamental matrix other than Φ,

$$\Phi = \Phi_1 K$$

We have seen that

$$\Phi(t+T) = \Phi(t)\, e^{TR}$$

Hence,

$$\Phi_1(t+T) = \Phi_1(t)(K e^{TR} K^{-1})$$

Or any fundamental matrix Φ_1 determines a matrix $K e^{TR} K^{-1}$, which is similar to e^{TR}. Therefore, the set of all fundamental matrices of the equation (or, in other words, the matrix A) uniquely determines all quantities that are common to the set of all matrices that are similar to e^{TR}. In particular, A determines uniquely $\lambda_1, ..., \lambda_n$, the eigenvalues of e^{TR}, which are called the multipliers associated with A. The eigenvalues of R are called characteristic exponents. The stability condition for Eq. (13) can now be stated in the following manner:

All solutions $x(t)$ of Eq. (13) approach zero as $t \to \infty$ if and only if $|\lambda_i| < 1$, $i = 1, ..., n$. All solutions $x(t)$ are bounded if and only if $|\lambda_i| \leqslant 1$, $i = 1, ..., n$; the system has a periodic solution of period T if and only if there is at least one root $\lambda_i = 1$.

Efficient Computation of the FTM

Efficient computation of the FTM has been the object of much research. There are two methods for computing the FTM: the n-pass scheme and the single-pass scheme.

In the n-pass scheme[9] the differential equation is numerically integrated over one period, and the process is repeated n times. Each integration over a period is called a pass, and each pass generates one column of $\Phi(T)$. In a given pass an initial condition of zero is chosen for all dependent variables except the one corresponding to the column of $\Phi(T)$ being generated.

In the single-pass scheme[10,11] the $(n \times n)$ FTM is computed in one sweep as an $(n^2 \times 1)$ vector. The FTM is solved directly from the matrix equation

$$\Phi' = A(t)\Phi$$

rather than from Eq. (13). For this the n original equations are reformulated before the integrations are begun. It has been demonstrated that the single-pass scheme is computationally more efficient than the n-pass scheme. Reference 12 provides a comparison of the CPU time taken by different numerical methods for the single- and the n-pass schemes.

An alternate method is to approximate the equation over one period. Several approximation schemes were recently developed by Hsu.[13] The basic idea is to divide the period T into a number of small and equal intervals and to approximate the periodic system by a constant coefficient matrix in each interval. The method is very convenient to program on a digital computer and is numerically efficient.

Computation of the Logarithm of the Matrix C

The logarithm of C can, in principle, be calculated by a matrix infinite series. However, in practical computations the use of the Cayley-Hamilton theorem is preferred. It permits a matrix function to be written as

$$f(C) = Mf(\Lambda)M^{-1}$$

where f is some function of the matrix C, C is an $(n \times n)$ matrix with distinct eigenvalues, Λ is the spectral matrix of C, and M is the matrix whose columns contain the n linearly independent eigenvectors of C.

The formula for logarithm becomes

$$\ell n(C) = M\, \ell n(\Lambda)M^{-1}$$

Thus, the logarithm is determined by computing the logarithms of the scalars that constitute Λ and by two operations of matrix multiplication. Multiple roots make the computation slightly more complex as a result of the involvement of Jordan canonical forms.

Estimates for the Characteristic Exponents

An alternate route to obtaining information regarding stability is to find estimates for the characteristic exponents. A brief description of the method of determination of these estimates is given below. A complete discussion may be found in Article 4, Chaper II, of Ref. 14.

Let $G(t)$ be an $(n \times n)$ Hermitian, periodic (with period T), differentiable, and positive-definite matrix. Define

$$Q = G' + GA + A^*G \qquad (Q^* = Q)$$

and let $q_1(t)$ and $q_2(t)$ denote the smallest and the largest roots of the equation

$$\det(Q - qG) = 0$$

Roots $q_1(t)$ and $q_2(t)$ are real periodic (with period T) functions. If α is a characteristic exponent of Eq. (13), then

$$\frac{1}{2T}\int_0^T q_1(t)\, dt \leqslant \mathrm{Re}(\alpha) \leqslant \frac{1}{2T}\int_0^T q_2(t)\, dt$$

By a suitable choice of $G(t)$, this estimate can be made arbitrarily sharp.

In problems of structural dynamics characteristic exponents are always computed rather than estimated. This is because the process of computation is relatively straightforward and provides exact values (subject to numerical errors), whereas the process of estimation is rather arbitrary.

Reducibility of Periodic Systems

Time-invariant systems are much easier to analyze than time-variant systems. Therefore, most attempts to solve time-variant systems have centered around either transforming the time-variant system to a time-invariant system, or determining connections between the time variant and a corresponding time-invariant system. The concept of reducibility, which belongs to the first category, is explained in this section.

Consider the equation

$$x' = A(t)x \tag{15}$$

Assume that the elements of the matrix $A(t)$ are continuous and bounded functions of t. The transformation

$$x = L(t)y \tag{16}$$

changes Eq. (15) into

$$y' = B(t)y \tag{17}$$

We will express B in terms of A and L. On differentiating Eq. (16),

$$x' = L'(t)y + L(t)y' = A(t)L(t)y$$

Therefore,

$$y' = (L^{-1}AL - L^{-1}L')y$$

which gives

$$B = L^{-1}AL - L^{-1}L'$$

Definition 7. Equation (16) is said to be a Liapunov transformation if it satisfies the following conditions:

(*i*) $L(t)$ has a continuous derivative $\mathrm{d}L/\mathrm{d}t$ in $[0,\infty)$.
(*ii*) $L(t)$ and $\mathrm{d}L/\mathrm{d}t$ are bounded in $[0,\infty)$.
(*iii*) $\det L(t) \neq 0$.

The matrix L is said to be a Liapunov matrix.

It may be verified that Liapunov transformations form a group. They derive their utility from the possession of the following very important property:

Theorem 5. If the zero solution of Eq. (17) is stable, asymptotically stable, or unstable ISL, then the zero solution of the original equation [Eq. (15)] has the same property.

This enables one to simplify the original system of equations. The simplest differential equations are of the linear constant coefficient type. Therefore, the systems that can be carried over to linear constant coefficient equations deserve special attention. Following Liapunov, they are called reducible.

Definition 8. Equation (15) is said to be reducible if the Liapunov transformation [Eq. (16)] converts it into the time-invariant equation

$$y' = By$$

Periodic systems have been shown to be reducible.

Constant Coefficient Systems Perturbed by Small Periodic Terms

An important class of problems is represented by the equation

$$x' = [A + \varepsilon B(t)]x \tag{18}$$

where A is a constant $(n \times n)$ matrix, ε is a small constant, and $B(t)$ is an $(n \times n)$ periodic matrix of period T. Systems of this kind occur when constant coefficient systems undergo small periodic disturbances. It may be expected that the solutions of Eq. (18) closely mimic the behavior of solutions of the constant coefficient system

$$x' = Ax$$

An important theorem[2] of this type is given by the following.

Theorem 6. If A is a constant matrix with characteristic roots λ_j and $\text{Re}(\lambda_j) < 0$, $j = 1, ..., n$, and if $B(t)$ is a continuous periodic matrix of a given period T, then for some $\varepsilon_0 > 0$ and all ε such that $|\varepsilon| < \varepsilon_0$, all solutions of Eq. (18) approach zero as $t \to \infty$.

If $\text{Re}(\lambda_j) > 0$ for some j, then for some $\varepsilon_0 > 0$ and all $|\varepsilon| < \varepsilon_0$, infinitely many solutions of Eq. (18) are unbounded in $[0, \infty)$.

Almost Constant Coefficient Systems

We shall begin with the following fundamental result:

Theorem 7. Let $\Phi(t)$ be a fundamental matrix for the equation

$$x' = A(t)x \tag{19}$$

Then Eq. (19) is

(*i*) stable if and only if there exists a positive constant M such that

$$\|\Phi(t)\| \leqslant M \qquad \text{for} \qquad t \geqslant t_0 \qquad \text{and}$$

(*ii*) asymptotically stable if and only if

$$\|\Phi(t)\| \to 0 \quad \text{as} \quad t \to \infty$$

However, unlike in constant coefficient and periodic systems, determination of the fundamental matrix for a general equation of the form of Eq.

(19) is not an easy task. Therefore, *Theorem* 7 by itself is restricted in applicability.

We begin with systems that are almost constant coefficient in some sense. A number of important results have been obtained in this line because, in such systems, the time-varying terms are assumed to play increasingly smaller roles with the increase in time.

Definition 9. The coefficient matrix $A(t)$ of the equation

$$x' = A(t)x$$

is called almost constant if

$$\lim_{t\to\infty} A(t) = B$$

a constant matrix (see Ref. 15).

Results on almost constant coefficient systems prove that, under certain given hypotheses, the solutions of the time-varying system

$$x' = A(t)x$$

behave (or do not behave) as $t \to \infty$, as the solutions of the corresponding constant coefficient equation

$$y' = By$$

We shall later state a few typical results.[2]

Theorem 8. Let B be a constant $(n \times n)$ matrix and $D(t)$ be an $(n \times n)$ matrix whose elements are continuous functions of t. If all solutions of the equation

$$x' = Bx$$

approach zero as $t \to \infty$ at the rate

$$\|\Phi(t)\| \leqslant m\, e^{-ut}$$

where $\Phi(t)$ is the state transition matrix for the preceding equation, and if

$$\|D(t)\| < d \qquad \text{and} \qquad md < u$$

then all solutions of the system

$$x' = [B + D(t)]x$$

approach zero as $t \to \infty$.

The result does not hold if we replace the phrases "approach zero as $t \to \infty$" by "are bounded as $t \to \infty$." A modification of this result[16,17] is given by the following:

Theorem 9. Let A and B be constant $(n \times n)$ matrices and $C(t)$ and $D(t)$ be $(n \times n)$ matrices whose elements are continuous functions of t. If all solutions of the equation

$$Ax' = Bx$$

where A is nonsingular approach zero as $t \to \infty$ at the rate

$$\|\Phi(t)\| \leqslant m\, e^{-ut}$$

where $\Phi(t)$ is the state transition matrix for the preceding equation, and if

$$\|A^{-1}\| = a, \qquad \|B\| = b, \qquad \|C\| \leqslant c < 1/a, \qquad \|C'(t)\| \leqslant f$$

and these numbers are related by

$$am(f + bc + d) < u(1 - ac)$$

then all solutions of the system

$$[A + C(t)]x' = [B + D(t)]x$$

approach zero as $t \to \infty$.

We have the following result of Dini-Hukuhara[2] for boundedness of solutions:

Theorem 10. If B is a constant matrix, $D(t)$ is a matrix whose coefficients are measurable functions[18] of t and

$$\int_0^\infty \|D(t)\|\,\mathrm{d}t < \infty$$

and if all solutions of the equation $x' = Bx$ are bounded, then all solutions of the equation $x' = [B + D(t)]x$ are also bounded as $t \to \infty$.

A modification of this result[16,17] is given by the following:

Theorem 11. Let A and B be constant matrices and $C(t)$ and $D(t)$ be matrices whose coefficients are measurable functions of t, and

$$\|C(t)\| \leqslant c < 1/\|A^{-1}\|, \qquad \int_0^\infty \|C(t)\|\,\mathrm{d}t < \infty, \qquad \int_0^\infty \|D(t)\|\,\mathrm{d}t < \infty$$

if all solutions of the equation $Ax' = Bx$, where A is nonsingular, are bounded, then all solutions of the equation $[A + C(t)]x' = [B + D(t)]x$ are also bounded as $t \to \infty$.

Another result[2] in this direction is given by the following:

Theorem 12. If the constant matrix A has all eigenvalues λ_j with $\text{Re}(\lambda_j) \leqslant 0$ and the eigenvalues λ_j with $\text{Re}(\lambda_j) = 0$ are simple; if the matrix $B(t)$ has elements of bounded variation[18] in $[0,\infty)$ and $b_{ij}(t) \to 0$ as $t \to \infty$, and the generalized eigenvalues $\lambda_j(t)$ of the matrix $[A + B(t)]$ have real parts $Re[\lambda_j(t)] \leqslant 0$ for all t in $[0,\infty)$; if the matrix $C(t)$ has elements absolutely integrable in $[0,\infty)$, i.e.,

$$\int_0^\infty \|C(t)\| \, \mathrm{d}t < \infty$$

then the equation

$$x' = [A + B(t) + C(t)]x$$

has all of its solutions bounded in $[0,\infty)$.

A generalization of the Dini-Hukuhara theorem is given by the following:

Theorem 13. If all solutions of $y' = A(t)y$ are bounded, and

$$\text{Re} \int_0^t \text{tr}A(t) \, \mathrm{d}t \geqslant -d > -\infty \qquad \text{for all} \qquad t \geqslant 0$$

and

$$\int_0^\infty \|C(t)\| \, \mathrm{d}t < \infty$$

then all solutions of $x' = [A(t) + C(t)]x$ are bounded.
Note: The second condition holds if $\text{tr}(A) = 0$.

General Second-Order Linear Systems with Variable Coefficients

The literature on second-order linear systems with arbitrarily time-varying coefficients is scanty. We begin with the Sonin-Polya theorem,[19] an important result, although it deals with only a single-degree-of-freedom system. This theorem is quite appealing to the user because of the simple structure of the conditions.

Theorem 14. Let $p(x) > 0$ and $q(x) \neq 0$ be continuously differentiable on an interval I, and suppose $p(x)q(x)$ is nonincreasing (nondecreasing) on I. Then the absolute values of the relative maxima and minima of every nontrivial solution of the equation

$$[p(x)y']' + q(x)y = 0$$

is nondecreasing (nonincreasing) as x increases.

Sellappan and Bainum[20] have provided a very interesting application of the Sonin-Polya theorem to space structures. They derived the following equations of motion for a spinning spacecraft during deployment of rigid appendages along one of the transverse axes:

$$h_1'' - \frac{a_2'(t)}{a_2(t)} h_1' + [a_1(t)a_2(t)]h_1 = 0$$

$$h_2'' - \frac{a_1'(t)}{a_1(t)} h_2' + [a_1(t)a_2(t)]h_2 = 0$$

where h_1 and h_2 are the transverse components of angular momentum, and $a_1(t)$ and $a_2(t)$ are functions of the instantaneous values of the principal moments of inertia. The equations being decoupled, the Sonin-Polya theorem is applicable.

An equivalent theorem for multidimensional systems has been proved by Hartman[21] and is given by the following:

Theorem 15. For the system

$$[P(t)x(t)']' + Q(t)x(t) = 0 \tag{20}$$

let $P(t)$, $Q(t)$ be commutative Hermitian matrices whose elements are continuously differentiable functions of t for $t > 0$. Moreover, let their product

$$R(t) = P(t)Q(t) = Q(t)P(t)$$

be positive and monotonic [i.e., $R(t) - R(s)$ is always nonnegative or always nonpositive for $t > s$].

(*i*) If $R(t)$ is a nondecreasing function, then for any solution $x(t)$,

$$E = x^*x + x'^*PQ^{-1}x'$$

is a nonincreasing function. There exists a nontrivial solution for which $E \to 0$ if and only if $\det R \to \infty$ as $t \to \infty$.

(*ii*) If $R(t)$ is a nonincreasing function, then along any solution E is a nondecreasing function. If in addition, $\det R \to 0$ as $t \to \infty$, then there exists a solution for which $E \to \infty$.

The following corollary is important:

Corollary 15.1. If R is a nondecreasing function, every solution $x(t)$ of Eq. (20) is bounded. If, in addition, $\det R \to \infty$ as $t \to \infty$, then at least one nontrivial solution tends to zero as $t \to \infty$.

This theorem is very close to the Sonin-Polya theorem and retains many of its desirable features. However, its applicability is limited because it requires the matrices $P(t)$ and $Q(t)$ to be commutative, a condition rarely encountered with in practice.

The following theorem by Shrivastava[22] encompasses a much wider class and is less severe in the requirements on the parameter matrices.

Theorem 16. Let $a_0(x)$, $a_2(x)$ be nonsingular and continuously differentiable on an interval I. If there exists a positive-definite symmetric matrix $S(x)$ such that $(Sa_0^{-1}a_2)$ is positive-definite symmetric (i.e., $a_0^{-1}a_2$ is symmetrizable),

$$S' \leqslant 0 \qquad \text{and} \qquad A = [(Sa_0^{-1}a_1) + (Sa_0^{-1}a_1)^T - (Sa_0^{-1}a_2)'] \geqslant 0$$

then the region of possible solutions of the coupled nonautonomous equation

$$a_2(x)z'' + a_1(x)z' + a_0(x)z = 0$$

is nonincreasing as x increases. If $A > 0$, the region will decrease asymptotically.

The theorem is simple for systems that do not possess circulatory terms. For circulatory systems it becomes necessary to determine the S matrix, for which an algorithm has not been formulated as yet.

An application of this theorem to torque-free rotating systems with variable inertia showed that stable rotations result if the body is symmetric, rotates about the axis of symmetry, and the inertia variations take place in such a way that the body retains symmetry.

Several theorems were proved by Shrivastava and Pradeep[16,23] on the stability of second-order linear multidimensional systems. We shall state some of the more important ones.

Theorem 17. Systems governed by the equation

$$M\boldsymbol{q}'' + (D + G)\boldsymbol{q}' + K\boldsymbol{q} = 0$$

are stable if

$$\left.\begin{array}{l} (i)\ K > 0 \\[2ex] (ii)\ M' \leqslant 0 \\[2ex] (iii)\ 2MK^{-1}(D+G) - (MK^{-1}M)' \geqslant 0 \end{array}\right\} \quad \text{for all} \quad t \in [0,\infty)$$

This theorem is applicable only to noncirculatory systems. However, as the following statement shows it can easily be modified to include circulatory terms.

Corollary 17.1 For systems governed by the equation

$$Mq'' + Bq' + Cq = 0 \tag{21}$$

if there exists a symmetric positive-definite matrix $S(t)$ such that $SM^{-1}C = (SM^{-1}C)^T$, then a set of sufficient conditions for the stability of Eq. (21) is

$$\left.\begin{aligned} &(i)\ SM^{-1}C > 0 \\ &(ii)\ S' \leqslant 0 \\ &(iii)\ 2SC^{-1}B - (SC^{-1}M)' \geqslant 0 \end{aligned}\right\} \text{ for all } t \in [0,\infty)$$

These theorems have close connections with results on constant coefficient systems. For instance, *Corollary* 17.1, when specialized to constant coefficient systems, agrees with *Theorem* 6.1 of Ref. 24, which in the notation used earlier states, "A circulatory system in the absence of velocity dependent forces is stable if and only if there exists a symmetric and positive definite matrix S_1 such that $S_1M^{-1}C$ is symmetric and positive definite."

The following theorem does not require matrix inversion and multiplication and therefore is quite attractive from the point of view of application:

Theorem 18. Systems governed by the equation

$$Mq'' + (D + G)q' + Kq = 0$$

are stable if

$$\left.\begin{aligned} &(i)\ K > 0 \\ &(ii)\ K' \leqslant 0 \\ &(iii)\ 2D - M' \geqslant 0 \end{aligned}\right\} \text{ for all } t \in [0,\infty)$$

A modification to accommodate circulatory terms can be carried out easily as in the preceding theorem.

When reduced to the constant coefficient system, this theorem is seen to be a particular case of the following well-known result.[25]

Consider the equation

$$Ax'' + 2Bx' + Cx = 0, \qquad x(0) = c_1, \qquad x'(0) = C_2$$

if A, B, C are positive-semidefinite and either A or C is positive-definite, then

$$\det(\lambda^2 A + 2\lambda B + C) = 0$$

has no roots with positive real parts. If A and C are positive-semi-definite and B is positive definite, then the only root with zero real part is $\lambda = 0$.

This theorem also agrees with the first part of the Kelvin-Tait-Chetayev (KTC) theorem, which states[26] "An equilibrium position which is stable under purely potential forces (nongyroscopic nondissipative system, or a system with $D = G = 0$) remains stable with the addition of gyroscopic and dissipative forces."

The stability theorems presented thus far are applicable to noncirculatory systems. For a nonsymmetric stiffness matrix there must exist a positive-definite symmetric matrix S that symmetrizes it. The S matrix is quite an elusive entity because of the absence of an algorithm to determine it; indeed, even more fundamental questions such as its existence and uniqueness are not yet answered. We state below criteria that circumvent determination of the S matrix. However, the new criteria are available at a cost: an investigation is required of a $(2n \times 2n)$ matrix for sign for sign definiteness, as opposed to the $(n \times n)$ matrices of the previous theorems. Two such theorems are given in the following:

Theorem 19. Systems governed by the equation

$$M\boldsymbol{q}'' + (D + G)\boldsymbol{q}' + (K + A)\boldsymbol{q} = 0$$

are stable if

(*i*) $K > 0$

(*ii*) $\left[\begin{array}{c|c} (-M)' & (MK^{-1}A)^T \\ \hline (MK^{-1}A) & 2MK^{-1}(D+G) - (MK^{-1}M)' \end{array}\right] \geqslant 0 \quad \text{for all } t \in [0,\infty)$

Theorem 20. Systems governed by the equation

$$M\boldsymbol{q}'' + (D + G)\boldsymbol{q}' + (K + A)\boldsymbol{q} = 0$$

are stable if

(*i*) $K > 0$

(*ii*) $\left[\begin{array}{c|c} (-K') & (-A) \\ \hline A & 2D - M' \end{array}\right] \geqslant 0 \quad \text{for all } t \in [0,\infty)$

Two instability theorems are also stated as follows[20,23]:

Theorem 21. Systems governed by the equation

$$M\boldsymbol{q}'' + D\boldsymbol{q}' + (K + A)\boldsymbol{q} = 0$$

are unstable if

$$D' - 2K - M'' > 0 \qquad \text{for all } t \in [0,\infty)$$

Theorem 22. Systems governed by the equation

$$M\boldsymbol{q}'' + (D + G)\boldsymbol{q}' + (K + A)\boldsymbol{q} = 0$$

where the elements of M are twice differentiable, are unstable if

$$\left[\begin{array}{c|c} D' - 2K - M'' & -G \\ \hline G & 2M \end{array}\right] > 0 \qquad \text{for all } t \in [0,\infty)$$

The extending simple pendulum was investigated using one of these theorems, and it has been confirmed that the motion is stable during extension. In another application the damped Mathieu equation has been investigated for stability, and it is seen that the results are superior to the existing results[27] for some range of parameter values.

Before conclusions are drawn, it is necessary for the following remarks to be made regarding the limitations of the theorems presented in this section:

1) In practice the application of the criteria to a space structure represented by a multiple-degree-of-freedom equation would turn out to be a difficult task.

2) The theorems represent sufficient conditions for stability. No study has been carried out to determine the extent of the parameter space that each theorem covers.

Conclusion

Time-variant systems are a recent introduction in the area of structural dynamics. Although they have been a subject or study for applied mathematicians, the understanding of time-variant systems is far from complete. The study of applied mathematicians has also not taken into account the special structure of the equations that arise in structural dynamic systems. The equations are either in second- or first-order form, which is slightly different from the equation usually used for analysis. The high dimension of the equations of these systems makes the task of conversion to the standard first-order form almost an impossibility. Some of the recent research treats the equations directly to circumvent these difficulties. This brief discussion has listed various criteria available for the analysis of time-varying structural dynamic systems, both conventional and the recent results. In retrospect, it is not unseemly to remark that the field of time-varying systems is wide open, with enormous potential for research.

References

[1]Liapunov, A. M., "Probleme Generalede la Stabilite du Mouvement, 1907," French translation of the Russian paper published in *Comm. Soc. Math. Kharkow*, 1882; also *Annals of Mathematical Studies*, Vol. 17, Princeton Univ. Press, Princeton, NJ, 1949.

[2]Cesari, L., *Asymptotic Behaviour and Stability Problems in Ordinary Differential Equations*, Academic, New York, 1963.

[3]Bellman, R., "On an Application of a Banach-Steinhaus Theorem to the Study of Boundedness of Solutions of Nonlinear Differential and Difference Equations," *Annals of Mathematics*, Vol. 2, No. 49, 1948, pp. 515–522.

[4]Mingori, D. L., "The Determination by Floquet Analysis of the Effects of Energy Dissipation on the Attitude Stability of a Dual-Spin Satellite," *AIAA Journal*, Vol. 7, Jan. 1969, pp. 20–27.

[5]Coleman, P., and Feingold, A., "Theory of Ground Vibrations of a Two-Blade Helicopter and Anisotropic Flexible Supports," NACA TN-1184, 1947.

[6]Bolotin, V. V., *The Dynamic Stability of Elastic Systems*, Holden-Day, San Francisco, CA, 1964.

[7]Von Kerczek, C., and Davis, S. H., "Linear Stability Theory of Oscillatory Stokes Layers," *Journal of Fluid Mechanics*, Vol. 62, No. 4, 1974, pp. 753–773.

[8]Coddington, E. A., and Levinson, N., *Theory of Ordinary Differential Equations*, McGraw-Hill, New York, 1955.

[9]Peters, D. A., and Hohenemser, K. H., "Application of the Floquet Transition Matrix to Problems of Lifting Rotor Stability," *Journal of the American Helicopter Society*, Vol. 16, No. 2, April 1971, pp. 25–33.

[10]Von Kerczek, C., and Davis, S. H., "Calculation of Transition Matrices," *AIAA Journal*, Vol. 13, Oct. 1975, pp. 1400–1403.

[11]Friedmann, P., Hammond, C. E., and Woo, T. H., "Efficient Numerical Treatment of Periodic Systems with Application to Stability Problems," *International Journal for Numerical Methods in Engineering*, Vol. 11, 1977, pp. 1117–1136.

[12]Gaonkar, G. H., Simha Prasad, D. S., and Sastry, D., "On Computing Floquet Transition Matrices of Rotorcraft," *Journal of the American Helicopter Society*, July 1981, pp. 56–61.

[13]Hsu, C. S., "On Approximating a General Periodic System," *Journal of Mathematical Analysis and Applications*, Vol. 45, 1974, pp. 234–251.

[14]Yakubovich, V. A., and Starzhinskii, V. M., *Linear Differential Equations with Periodic Coefficients*, Wiley, New York, 1975.

[15]Bellman, R., *Stability Theory of Differential Equations*, McGraw-Hill, New York, 1953.

[16]Pradeep, S., "On the Stability of Multidimensional Linear Systems with Time Varying Parameters," Ph.D. Dissertation, Indian Institute of Science, Bangalore, India, March 1987.

[17]Pradeep, S., and Shrivastava, S. K., "On the Asymptotic Behaviour and Boundedness of Systems with Time-Varying Coefficients," *Acta Astronautica*, Vol. 19, No. 10, 1989, pp. 787–795.

[18]Halmos, P. R., *Measure Theory*, Van Nostrand, Princeton, NJ, 1950.

[19]Kreider, D. L., Kuller, R. G., and Ostberg, D. R., *Elementary Differential Equations*, Addison-Wesley, Reading, MA, 1968.

[20]Sellappan, R., and Bainum, P. M., "Stability of Spacecraft During Asymmetrical Deployment of Appendages," *Journal of Guidance, Control, and Dynamics*, Vol. 1, Nov.–Dec. 1978, pp. 446–447.

[21]Coppel, W. A., *Stability and Asymptotic Behaviour of Differential Equations*, Heath, Boston, MA, 1965.

[22]Shrivastava, S. K., "Stability Theorems for Multidimensional Linear Systems with Variable Parameters," *ASME Journal of Applied Mechanics*, Vol. 48, March 1981, pp. 174–176.

[23]Shrivastava, S. K., and Pradeep, S., "On the Stability of Multidimensional Linear Systems," *Journal of Guidance, Control, and Dynamics*, Vol. 8, No. 5, Sept.–Oct. 1985, pp. 579–583.

[24]Huseyin, K., *Vibrations and Stability of Multiple Parameter Systems*, Sijthoff & Noordhoff, Alphen aan den Rijn, The Netherlands, 1978.

[25]Bellman, R., *Introduction to Matrix Analysis*, McGraw-Hill, New York, 1970, p. 255.

[26]Chetayev, N. G., *The Stability of Motion*, Pergamon, New York, 1961.

[27]Pradeep, S., and Shrivastava, S. K., "On the Stability of the Damped Mathieu Equation," *Mechanics Research Communications*, Vol. 15, No. 6, 1988, pp. 353–359.

Chapter 14

Robust Eigensystem Assignment for Second-Order Dynamic Systems

Jer-Nan Juang* and Peiman G. Maghami†
NASA Langley Research Center, Hampton, Virginia

Introduction

EIGENVALUE and eigenvector assignment has played a key role in the active control design of linear, time-invariant systems. To date, numerous methods and algorithms involving constant full-state or output feedback have been developed. Wonham[1] first related the controllability of linear multi-input and multi-output (MIMO) systems to their eigenvalue assignability. Moore[2] and Klien and Moore[3] identified the flexibility beyond eigenvalue assignment by characterizing the attainable eigenvector space for a set of desired closed-loop eigenvalues. Later, Davison,[4] Davison and Wang,[5] Kimura,[6] Srinathkumar,[7] and others extended the concept of eigensystem assignment to systems with partial-state feedback and established the limitations imposed by such feedbacks. Within the past decade several iterative and noniterative methods have been derived to exploit the freedom beyond eigenvalue assignment in order to improve the performance of the closed-loop system or to minimize the required control effort.

The problem of robust controller design with eigensystem assignment has received considerable attention recently. Among the many published works related to robust controller designs, several methods are identified, including the selection of closed-loop eigenvector sets to simultaneously satisfy modal insensitivity equations and eigenvalue constraint equations,[8] optimization of various measures of system robustness along with eigenvalue constraints using nonlinear programming techniques,[9,10] and iterative

*Principal Scientist, Spacecraft Dynamics Branch.
†Aerospace Engineer, Spacecraft Controls Branch.

methods that use an orthogonal projection technique to enhance various measures of closed-loop conditioning.[11] Of recent efforts an efficient sequential eigenvalue assignment technique by Maghami and Juang[12] using Schur decompositions and Givens rotations may be named, wherein a norm of the gain matrix and/or a measure of the closed-loop conditioning is minimized at every eigenvalue assignment. Moreover, Juang et al.[13] and Rew et al.[14] have developed a robust eigensystem assignment method in which the closed-loop eigenvectors are chosen to maximize their projection on the open-loop eigenvectors or the columns of its closest unitary matrix, in order to obtain a robust closed-loop design. The eigensystem assignment techniques developed to date generally are limited, in their developments, to the first-order linear systems. Furthermore, they usually require feedbacks in the form of velocity and/or displacement, rather than other forms of feedback combinations, such as acceleration and velocity or acceleration and displacement.

In this paper the basic approach presented in Refs. 13 and 14 is extended in three major directions. First, the method takes advantage of a second-order form of the system equations (instead of a first-order) that results in a considerable computational efficiency.[15] Second, the approach is extended to handle any forms of feedbacks (i.e., displacement, velocity, and acceleration). Third, a new sequential algorithm for the robust control design is developed using the concept of intersection of the attainable closed-loop eigenvector subspaces and the target space. Based on the second-order eigenvalue problem, the approach begins with the generation of the attainable closed-loop eigenvector space for each closed-loop eigenvalue using singular value or QR decompositions. Then a sequential algorithm is used to choose the closed-loop eigenvectors, such that they are as close as possible to the column space of a well-conditioned target matrix. This is achieved by computing the principal angles and vectors for the subspace pair of the attainable eigenvector space and the target matrix. A numerical example is given to demonstrate the proposed approach.

Robust Eigensystem Assignment Using Full-State Feedback

In the analysis and design of the dynamics and vibration control of flexible structures, a set of second-order linear, constant-coefficient, ordinary differential equations is frequently used:

$$M\ddot{x} + D\dot{x} + Kx = Bu \tag{1}$$

where x is the state vector of dimension n, and M, D, and K are the mass, damping, and stiffness matrices, respectively, which generally are symmetric and sparse. The $n \times p$ influence matrix B describes the actuator force distributions for the $p \times 1$ control force vector u.

Eigenvalue Assignment with Conventional Full-State Feedback

Assume that a full-state feedback is used to design the controller with two $p \times n$ gain matrices G_d and G_v such that

$$u = G_d x - G_v \dot{x} \tag{2}$$

The eigensolution of the closed-loop system for the second-order dynamic model [Eq. (1)] can then be written as

$$[\lambda_k^2 M + \lambda_k(D + BG_v) + (K + BG_d)]\boldsymbol{x}_k = 0; \qquad k = 1, \ldots, n \tag{3}$$

where λ_k is the kth eigenvalue, and $\boldsymbol{x}_k$ is the corresponding eigenvector. The subscript k refers to the mode number. Rearranging Eq. (3) in a compact form yields

$$[(\lambda_k^2 M + \lambda_k D + K)\, B]\begin{bmatrix} \boldsymbol{x}_k \\ (\lambda_k G_v + G_d)\boldsymbol{x}_k \end{bmatrix} \equiv \Gamma_k \boldsymbol{\phi}_k = 0 \tag{4}$$

If the closed-loop eigenvalues λ_k $(k = 1, \ldots, n)$ are assigned including their complex conjugates, Eq. (4) can be used to determine the gain matrices G_v and G_d. By observing that the vector $\boldsymbol{\phi}_k$ is in the null space of the matrix Γ_k, G_v and G_d may be determined when all of the null spaces of the matrices Γ_k $(k = 1, \ldots, n)$ are computed corresponding to the eigenvalues λ_k $(k = 1, \ldots, n)$. To obtain the nontrivial solution space of the homogeneous equation [Eq. (4)], singular value decomposition (SVD) is applied to the matrix[16,17] Γ_k, yielding

$$\Gamma_k = U_k S_k V_k^* = U_k \begin{bmatrix} \sigma_k & 0 \\ 0 & 0 \end{bmatrix} \begin{bmatrix} V_{\sigma k}^T \\ V_{ok}^T \end{bmatrix} \tag{5}$$

Since λ_k in Γ_k is a complex value, all of the quantities are complex except the diagonal matrix σ_k containing nonzero and positive singular values. Here the superscript * means transpose and complex conjugate. The matrix V_{ok} contains a set of orthogonal basis vectors spanning the null space of the matrix Γ_k such that

$$\Gamma_k \phi_k = \Gamma_k V_{ok} \boldsymbol{c}_k = 0 \tag{6}$$

where $\boldsymbol{c}_k$ is an arbitrary column vector of length p. If Γ_k is not close to a matrix of lesser (or higher) rank (which is easily found from the singular values, hence the advantage of using SVD), the previously given basis for null space, V_{ok}, can be computed more efficiently by taking the QR decomposition of Γ_k^T. If the matrix $[\lambda_k^2 M + \lambda_k D + K]$ is invertible, the vector $[\bar{\boldsymbol{\phi}}_k^T \; \hat{\boldsymbol{\phi}}_k^T]^T$, where $\hat{\boldsymbol{\phi}}_k$ is an arbitrary vector of length p and $\bar{\boldsymbol{\phi}}_k^T = [\lambda_k^2 M + \lambda_k D + K]^{-1} B\hat{\boldsymbol{\phi}}_k$, should be in the null space of the matrix Γ_k.

The SVD decomposition for the matrix Γ_k as shown in Eq. (5) is the most time-consuming computation for the algorithm developed in this paper since it has to be repeated n times if n eigenvalues are to be assigned. However, the matrix Γ_k as observed from Eq. (4) is sparse because M, D, and K are sparse, and so is their combination $[\lambda_k^2 M + \lambda_k D + K]$. The sparsity of the matrix Γ_k should ease the SVD computation. A special solution procedure for this type of equation [Eq. (5)] may be developed to significantly advance the state-of-the-art computational technology, particularly for the eigensystem assignment for mechanical systems, which are typically represented by second-order ordinary differential equations.

The freedom provided by multiple inputs is inherent in the choice of the coefficients c_k that determine the closed-loop eigenvectors. This freedom can be exploited to reduce the control effort, to enhance the robustness of the closed-loop system, etc. Here the coefficients c_k are chosen to achieve a robust closed-loop system by improving a conditioning index, defined as the condition number of the closed-loop modal matrix (matrix of the eigenvectors), which will be shown later.

To obtain an expression for the gain matrices G_v and G_d, choose a particular set of vectors, ϕ_k $(k = 1, ..., n)$ satisfying Eq. (6), corresponding to some choice c_k, and partition the vector ϕ_k into two components such that

$$\Gamma_k \phi_k \equiv \Gamma_k \begin{bmatrix} \bar{\phi}_k \\ \hat{\phi}_k \end{bmatrix} = \begin{bmatrix} \bar{V}_{ok} \\ \hat{V}_{ok} \end{bmatrix} c_k = 0; \qquad k = 1, ..., n \tag{7}$$

Comparison of Eqs. (4) and (7) yields

$$G_v \bar{\Phi} \Lambda + G_d \bar{\Phi} = \hat{\Phi} \tag{8}$$

where the $n \times n$ matrix $\bar{\Phi} = [\bar{\phi}_1, \bar{\phi}_2, ..., \bar{\phi}_n]$, the $p \times n$ matrix $\hat{\Phi} = [\hat{\phi}_1, \hat{\phi}_2, ..., \hat{\phi}_n]$, and the $n \times n$ matrix $\Lambda = \mathrm{diag}[\lambda_1, \lambda_2, ..., \lambda_n]$. To solve the gain matrices G_v and G_d (both matrices are real), decomposing Eq. (8) into real part and imaginary part yields

$$G_v[\bar{\Phi}_r \Lambda_r - \bar{\Phi}_i \Lambda_i] + G_d \bar{\Phi}_r = \hat{\Phi}_r \tag{9a}$$

$$G_v[\bar{\Phi}_r \Lambda_i + \bar{\Phi}_i \Lambda_r] + G_d \bar{\Phi}_i = \hat{\Phi}_i \tag{9b}$$

or, in matrix form,

$$G\bar{\Psi} \equiv [G_v \quad G_d] \begin{bmatrix} \bar{\Phi}_r \Lambda_i + \bar{\Phi}_i \Lambda_r & \bar{\Phi}_r \Lambda_r - \bar{\Phi}_i \Lambda_i \\ \bar{\Phi}_i & \bar{\Phi}_r \end{bmatrix} = [\hat{\Phi}_i \quad \hat{\Phi}_r] \equiv \hat{\Psi} \tag{10}$$

where the subscripts r and i, respectively, refer to the real and imaginary parts of the associated quantities. From Eq. (9) the gain matrices G_v and G_d can be solved by

$$G = [G_v \quad G_d] = \hat{\Psi}\bar{\Psi}^{-1} = [\hat{\Phi}_i \quad \hat{\Phi}_r] \begin{bmatrix} \bar{\Phi}_r \Lambda_i + \bar{\Phi}_i \Lambda_r & \bar{\Phi}_r \Lambda_r - \bar{\Phi}_i \Lambda_i \\ \bar{\Phi}_i & \bar{\Phi}_r \end{bmatrix}^{-1} \tag{11}$$

A matrix inversion is required in the computation of the gain matrices G_v and G_d. However, if the number of eigenvalues to be assigned is less than the number assignable, n, Eq. (11) becomes underdetermined, which leads naturally to a minimum gain solution. In order to ensure that the preceding matrix is well conditioned for inversion, the condition number of the matrix, $\bar{\Psi}$, should be the smallest possible. Interestingly, the previously given numerical requirement for well conditioning of the matrix inversion problem corresponds exactly to the eigenvalue conditioning problem since

$\bar{\Psi}$ consists of eigenvalues and eigenvectors. For a sufficiently small damping D and small real part of the closed eigenvalue Λ_r, $\bar{\Phi}_i$ in Eq. (10) approaches zero, since all of the null spaces corresponding to Γ_k ($k = 1, ..., n$) are nearly in real space. The matrix $\bar{\Psi}$ in Eq. (10) can then be approximated by $\bar{\Psi} = \text{diag}[\bar{\Phi}_r \Lambda_i, \bar{\Phi}_r]$. In terms of structural terminology, it means that the real part of the closed-loop eigenvector matrix $\bar{\Phi}_r$ dominates the conditioning of the matrix $\bar{\Psi}$. With this discussion in mind, we can determine the closed-loop eigenvectors as follows.

Eigenvector Assignment for Robustness

Define an $n \times n$ well-conditioned matrix Q_0, with vectors $\boldsymbol{q}_{01}, \boldsymbol{q}_{02}, ..., \boldsymbol{q}_{0n}$ representing its columns. Then the closed-loop eigenvectors are chosen to be as close as possible to the range space of the columns of matrix Q_0 in order to achieve a robust closed-loop design. If the open-loop conditioning is good to begin with, then the columns of the matrix Q_0 may be chosen to correspond to the open-loop eigenvectors, i.e.,

$$q_{0k} = \psi_{ok}; \qquad k = 1,2, ..., n \tag{12}$$

In general, the choice of the open-loop eigenvector ψ_{ok} is arbitrary, as long as the resulting closed-loop eigenvectors are linearly independent. If the control system is used only to provide active damping, and the closed-loop damped frequencies are quite close to their open-loop values, then it is best to use the open-loop eigenvector corresponding to an eigenvalue with the same (or similar) frequency for the vector ψ_{ok} in Eq. (12).

Alternatively, the matrix Q_0 may be chosen to be an arbitrary unitary matrix (with perfect conditioning) or the closest unitary matrix to the open-loop eigenvectors. In the latter case the matrix Q_0 is then the solution of the constrained least-squares problem minimizing

$$|\Psi_o - Q_0|$$

subject to $Q_0^* Q_0 = I$, which leads to

$$Q_0 = \boldsymbol{U}\boldsymbol{W}^* \tag{13}$$

where $\boldsymbol{U}$ and $\boldsymbol{W}$ are, respectively, the left and right singular vectors of the open-loop eigenvector matrix Ψ_o, I is an $n \times n$ identity matrix, and the superscript * denotes a complex conjugate transpose.

Having defined Q_0, we compute the closed-loop eigenvectors $\bar{\phi}_k$, $\bar{\phi}_k = \bar{V}_{ok} c_k$, $k = 1,2, ..., n$, and the corresponding coefficient vectors c_k are computed through the following three steps.

Step (1) Obtain the vector in the attainable closed-loop eigenvector space, $\bar{V}_{o1}$ [see Eq. (7)], corresponding to the first closed-loop eigenvalue, which is as close as possible to the range space of the columns of matrix Q_0. The vector can be obtained from the algorithm described by Golub and Van Loan[16] for the computation of principal angles and vectors of a subspace pair. Expand both matrices $\bar{V}_{o1}$ and Q_0 in terms of their QR decompositions, i.e.,

$$\bar{V}_{o1} = Q_{\bar{V}_{o1}} R_{\bar{V}_{o1}}; \qquad Q_0 = Q_{Q_0} R_{Q_0} \tag{14}$$

where $Q_{\bar{V}_{o1}}$ and Q_{Q_0} are orthonormal matrices of dimensions $n \times p$ and $n \times n$, respectively, and $R_{\bar{V}_{o1}}$ and R_{Q_0} are $p \times p$ and $n \times n$ upper triangular matrices, respectively. Project the vectors $Q_{\bar{V}_{o1}}$ unto Q_{Q_0} to obtain

$$H = Q^T_{\bar{V}_{o1}} Q_{Q_0} \tag{15}$$

The singular values of matrix H are cosines of the principal angles of the subspace pair $\{\mathscr{R}(Q_{V_{o1}}), \mathscr{R}(Q_{Q_0})\}$,[15] where $\mathscr{R}(\;\;)$ denotes the range space of $(\;\;)$. Taking the singular value decomposition of H gives

$$H = YSZ^T \tag{16}$$

in which Y and Z are unitary matrices, and the matrix S contains the singular values of H, i.e.,

$$S = \begin{bmatrix} \sigma_1 & \cdots & \cdots & 0 \\ \vdots & \sigma_2 & & \vdots \\ \vdots & & \ddots & \vdots \\ \vdots & & & \sigma_p \\ \vdots & & & \vdots \\ \vdots & & & \vdots \\ 0 & \cdots & \cdots & 0 \end{bmatrix} \equiv \begin{bmatrix} \cos\theta_1 & \cdots & \cdots & 0 \\ \vdots & \cos\theta_2 & & \vdots \\ \vdots & & \ddots & \vdots \\ \vdots & & & \cos\theta_p \\ \vdots & & & \vdots \\ \vdots & & & \vdots \\ 0 & \cdots & \cdots & 0 \end{bmatrix} \tag{17}$$

Note that the singular values $\sigma_1, ..., \sigma_p$ are all positive and less than or equal to 1 since matrix H is formed by orthonormal matrices as shown in Eq. (15). The angles $\theta_1, ..., \theta_p$ are the principal angles of the subspace pair. Bearing in mind that $1 \geqslant \sigma_1 \geqslant \sigma_2 \geqslant \cdots \sigma_p$, the vector $\bar{\phi}_1$ in the orthonormal columns of $Q_{V_{o1}}$, which is closest to the column space of Q_{Q_0}, is then

$$\bar{\phi}_1 = Q_{\bar{V}_{o1}} y = \bar{V}_{o1} R^{-1}_{\bar{V}_{o1}} y \tag{18}$$

in which y represents the first column of matrix Y. Choosing the first closed-loop eigenvector to be $\bar{\phi}_1$, the coefficients c_1 of Eq. (6) become

$$c_1 = R^{-1}_{\bar{V}_{o1}} y \tag{19}$$

Note that $R_{\bar{V}_{o1}}$ defined in Eq. (14) is a nonsingular matrix.

Step (2) Reduce the column space of matrix Q_0 by the vector $\bar{\phi}_1$, in order to ensure the linear independence of the closed-loop eigenvectors. If $\bar{\phi}_1$ is not in the column space of Q_0, then reduce the column space by the closest vector in the space to the vector $\bar{\phi}_1$. The new subspace is spanned by the columns of matrix Q_1 defined as

$$Q_1 = [Q_0 - Q_{Q_0} z z^* Q_{Q_0}^* Q_0] \tag{20}$$

where z is the first column of matrix Z. Note that the columns of matrix Q_1 are orthogonal to the vector $Q_{Q_0} z$, i.e., $z^* Q_{Q_0}^* Q_1 = 0$. Equation (20) may not be computationally efficient. It is shown here just for simplicity and clarity and other computational procedures to compute Q_1 may be used instead.

Step (3) Repeat steps (1) and (2) for the remaining $n - 1$ eigenvalues. For the ith eigenvalue the column space of matrix Q_{i-1} is reduced by the vector $Q_{Q_{i-1}} z$ resulting in a new matrix Q_i,

$$Q_i = [Q_{i-1} - Q_{Q_{i-1}} z z^* Q_{Q_{i-1}}^* Q_{i-1}] \tag{21}$$

In summary, following the developments outlined in steps (1–3), the closed-loop eigenvectors are chosen to be as close as possible to the column space of a well-conditioned matrix Q_0, thereby resulting in a robust closed-loop design in the sense that its gain matrix is less sensitive to the structural uncertainties. In addition to the well-conditioned matrix Q_0, the elements of gain matrix G can be minimized by replacing the matrix Q_0 with a new matrix R_0 defined as

$$R_0 = \begin{bmatrix} Q_0 \\ N_0 \end{bmatrix}$$

where N_0 is a $p \times n$ null matrix. It means that the closed-loop eigenvectors are chosen to be as close as possible to the column space of a well-conditioned matrix while the elements of gain matrix G are as small as possible.

Eigenvalue Assignment with Acceleration Feedback

In structural analysis accelerometers are often used as the measurement device for structural identification to determine the dynamics of the system. The question arises whether the acceleration measurements can also be used for active control feedback. The answer is affirmative. Assume that three $p \times n$ gain matrices G_a, G_v, and G_d are available for the controller design such that

$$\boldsymbol{u} = -G_d \boldsymbol{x} - G_v \dot{\boldsymbol{x}} - G_a \ddot{\boldsymbol{x}} \tag{22}$$

Obviously, G_a is the gain matrix for acceleration feedback. In fact, any pair $[G_v\ G_d]$, $[G_a\ G_v]$, or $[G_a\ G_d]$ is sufficient for the full-state feedback if the system is controllable, which is determined by the matrices M, D, and K and the input influence matrix B. For the purpose of generality, all of the three gain matrices are included in deriving formulations for the closed-loop eigenvalue assignment with acceleration feedback. The eigensolution of the closed-loop system for the second-order dynamic model [Eq. (1)], with the insertion of Eq. (22), can then be written as

$$[\lambda_k^2(M + BG_a) + \lambda_k(D + BG_v) + (K + BG_d)]\boldsymbol{x}_k = 0; \quad k = 1, \ldots, n \tag{23}$$

where λ_k is the kth eigenvalue and $\boldsymbol{x}_k$ the corresponding eigenvector. Again, the subscript k refers to the mode number. Equation (23) can be rewritten in a compact matrix form:

$$[(\lambda_k^2 M + \lambda_k D + K) \quad B]\begin{bmatrix} \boldsymbol{x}_k \\ (\lambda_k^2 G_a + \lambda_k G_v + G_d)\boldsymbol{x}_k \end{bmatrix} \equiv \Gamma_k \boldsymbol{\phi}_k = 0 \tag{24}$$

If the closed-loop eigenvalues λ_k $(k = 1, \ldots, n)$ are assigned including their complex conjugates, Eq. (24) can be used to determine the gain matrices G_a, G_v, and G_d. By observing that the vector $\boldsymbol{\phi}_k$ is in the null space of the matrix Γ_k, the matrices G_a, G_v, and G_d may be determined when all of the null spaces of the matrices Γ_k $(k = 1, \ldots, n)$ corresponding to the eigenvalues λ_k $(k = 1, \ldots, n)$ are computed. Observation of Eqs. (7) and (24) yields

$$G_a\bar{\Phi}\Lambda^2 + G_v\bar{\Phi}\Lambda + G_d\bar{\Phi} = \hat{\Phi} \tag{25}$$

where the $n \times n$ matrix $\bar{\Phi}$ and the $p \times n$ matrix $\hat{\Phi}$ are defined in Eq. (8). To solve the gain matrices G_a, G_v, and G_d in real domain, Eq. (25) is decomposed into real and imaginary parts, producing

$$G_a[\bar{\Phi}_r(\Lambda_r^2 - \Lambda_i^2) - 2\bar{\Phi}_i\Lambda_r\Lambda_i] + G_v[\bar{\Phi}_r\Lambda_r - \bar{\Phi}_i\Lambda_i] + G_d\bar{\Phi}_r = \hat{\Phi}_r \tag{26a}$$

$$G_a[\bar{\Phi}_i(\Lambda_r^2 - \Lambda_i^2) - 2\bar{\Phi}_r\Lambda_r\Lambda_i] + G[\bar{\Phi}_r\Lambda_i + \bar{\Phi}_i\Lambda_r] + G_d\bar{\Phi}_i = \hat{\Phi}_i \tag{26b}$$

or, in matrix form,

$$G\bar{\Psi} \equiv [G_a \quad G_v \quad G_d]\begin{bmatrix} \bar{\Phi}_i(\Lambda_r^2 - \Lambda_i^2) - 2\bar{\Phi}_r\Lambda_r\Lambda_i & \bar{\Phi}_r(\Lambda_r^2 - \Lambda_i^2) - 2\bar{\Phi}_i\Lambda_r\Lambda_i \\ \bar{\Phi}_r\Lambda_i + \bar{\Phi}_i\Lambda_r & \bar{\Phi}_r\Lambda_r - \bar{\Phi}_i\Lambda_i \\ \bar{\Phi}_i & \bar{\Phi}_r \end{bmatrix}$$
$$= [\hat{\Phi}_i \quad \hat{\Phi}_r] \equiv \hat{\Psi} \tag{27}$$

Observe that Eq. (27) is underdetermined in the sense that there are more unknowns than equations. In other words, there is a total of $3(p \times n)$ unknowns, i.e., elements of the matrices G_a, G_v, and G_d, but only $2(p \times n)$ equations. As a result, there is an infinite number of solutions that satisfy Eq. (27), i.e.,

$$G = [\hat{\Psi}\bar{\Psi}^\dagger + \alpha(I - \bar{\Psi}\Psi^\dagger)] \tag{28}$$

where α is a $p \times 3n$ arbitrary matrix, I is a $3n \times 3n$ identity matrix, and the superscript dagger denotes a pseudoinverse. Setting $\alpha = \boldsymbol{0}$ yields a minimum norm solution for the gain matrix G. With the absence of any matrix G_a, G_v, or G_d, $\bar{\Psi}$ becomes a $2n \times 2n$ square matrix. This simply means that any pair $[G_v \ G_d]$, $[G_a \ G_v]$, or $[G_a \ G_d]$ is sufficient for Eq. (27) to have a unique solution, if the $2n \times 2n$ matrix $\bar{\Psi}$ is invertible. Equation (11) is derived for the pair $[G_v \ G_d]$ without G_a, which is the conventional full-state feedback without the acceleration feedback.

With the absence of G_v, the equation for solving the pair $[G_a \ G_d]$ can be obtained by deleting the middle row of $\bar{\Psi}$ in Eq. (27), i.e.,

$$G\bar{\Psi} \equiv [G_a \quad G_d]\begin{bmatrix} \bar{\Phi}_i(\Lambda_r^2 - \Lambda_i^2) - 2\bar{\Phi}_r\Lambda_r\Lambda_i & \bar{\Phi}_r(\Lambda_r^2 - \Lambda_i^2) - 2\bar{\Phi}_i\Lambda_r\Lambda_i \\ \bar{\Phi}_i & \bar{\Phi}_r \end{bmatrix}$$
$$= [\hat{\Phi}_i \quad \hat{\Phi}_r] \equiv \hat{\Psi} \tag{29}$$

The gain matrices G_a and G_d can then be computed by

$$G \equiv [G_a \quad G_d] = [\hat{\Phi}_i \quad \hat{\Phi}_r]$$
$$\times \begin{bmatrix} \bar{\Phi}_i(\Lambda_r^2 - \Lambda_i^2) - 2\bar{\Phi}_r\Lambda_r\Lambda_i & \bar{\Phi}_r(\Lambda_r^2 - \Lambda_i^2) - 2\bar{\Phi}_i\Lambda_r\Lambda_i \\ \bar{\Phi}_i & \bar{\Phi}_r \end{bmatrix}^{-1}$$
$$= \hat{\Psi}\bar{\Psi}^{-1} \tag{30}$$

It is necessary to invert the matrix $\bar{\Psi}$ to compute the gain matrices G_a and G_d. Note that, if the number of eigenvalues to be assigned is less than the number assignable, n, Eq. (30) becomes underdetermined, which naturally leads to a minimum gain solution. For the matrix $\bar{\Psi}$ to be well conditioned for inversion, the condition number of the matrix should be the smallest possible. The procedure described in the above subsection may be applied here to obtain a well-conditioned matrix $\bar{\Psi}$.

Let us examine some special cases. Assume that there is no damping in the system equation [Eq. (1)], $D = \mathbf{0}$; thus, the open-loop model has pure imaginary eigenvalues. In this case the matrix Γ_k in Eq. (24) with either λ_k or $-\lambda_k$ for the closed-loop eigenvalue is identical and so is its corresponding null space. It simply means that the matrix $\hat{\Psi}$ in Eq. (29) is identical for either Λ_r and Λ_i or $-\Lambda_r$ and $-\Lambda_i$. Examination of Eq. (29) also reveals that the matrix Ψ with Λ_r and Λ_i is identical to the one with $-\Lambda_r$ and $-\Lambda_i$. As a result, λ_k and $-\lambda_k$ should be considered as a pair of closed-loop eigenvalues. If λ_k is an eigenvalue of the closed-loop system, then so is $-\lambda_k$. In other words, there is no way to make the closed-loop system stable. This leads to the conclusion that, for a structural system without passive damping, there does not exist any controller with deflection and acceleration feedbacks that make the closed-loop system asymptotically stable. Another case is worth mentioning wherein the matrix Λ_r in Eq. (29) is absent. Since Γ_k in this case is a real matrix, which makes its null space in real domain, the matrices $\bar{\Phi}_i$ and $\hat{\Phi}_i$ should be zero. Equation (29) would then yield that $G_a[-\bar{\Phi}_r\Lambda_i^2] + G_d\bar{\Phi}_r = \hat{\Phi}_i$. This is an underdetermined equation, which possesses infinite number of solutions for matrices G_a and G_d. Physically, it means that, for a structural system without passive damping and velocity feedback, only pure imaginary eigenvalues are assignable for the closed-loop system with deflection and acceleration feedbacks.

Now, with the absence of G_d, the equation for solving the pair $[G_a\ G_v]$ can be obtained by deleting the last row of Ψ and G_d in Eq. (27), which yields

$$G\bar{\Psi} \equiv [G_v \quad G_a]\begin{bmatrix} \bar{\Phi}_r\Lambda_i + \bar{\Phi}_i\Lambda_r & \bar{\Phi}_r\Lambda_r - \bar{\Phi}_i\Lambda_i \\ \bar{\Phi}_i(\Lambda_r^2 - \Lambda_i^2) - 2\bar{\Phi}_r\Lambda_r\Lambda_i & \bar{\Phi}_r(\Lambda_r^2 - \Lambda_i^2) - 2\bar{\Phi}_i\Lambda_r\Lambda_i \end{bmatrix}$$
$$= [\hat{\Phi}_i \quad \hat{\Phi}_r] \equiv \hat{\Psi} \tag{31}$$

Note that the positions for G_a and G_v in Eq. (31) have been switched for convenience in the following discussion. From Eq. (30) the gain matrices G_a and G_v can be obtained from

$$[G_v\ G_a] = [\hat{\Phi}_i\ \hat{\Phi}_r]\begin{bmatrix} \bar{\Phi}_r\Lambda_i + \bar{\Phi}_i\Lambda_r & \bar{\Phi}_r\Lambda_r - \bar{\Phi}_i\Lambda_i \\ \bar{\Phi}_i(\Lambda_r^2 - \Lambda_i^2) - 2\bar{\Phi}_r\Lambda_r\Lambda_i & \bar{\Phi}_r(\Lambda_r^2 - \Lambda_i^2) - 2\bar{\Phi}_i\Lambda_r\Lambda_i \end{bmatrix}^{-1}$$
$$= \hat{\Psi}\bar{\Psi}^{-1} \tag{32}$$

Again, a matrix inversion is involved in the computation of the gain matrices G_a and G_v. However, if the number of eigenvalues to be assigned is less than the number assignable, n, Eq. (31) becomes underdetermined, which naturally leads to a minimum gain solution. In order to ensure that the preceding matrix is well conditioned for inversion, the condition number of the matrix, $\bar{\Psi}$, should be the smallest possible. For sufficiently small damping D and small real part of the closed-loop eigenvalue Λ_r, $\bar{\Phi}_i$ in Eq. (31) approaches zero, since all of the null spaces corresponding to Γ_k ($k = 1, ..., n$) are nearly in real domain. The matrix in Eq. (29) can be approximated by $\bar{\Psi} = \text{diag}[\bar{\Phi}_r\Lambda_i, \bar{\Phi}_r\Lambda_r^2]$. In terms of structural terminology, this means that, for small damping and small real part of the eigenvalues assigned for the closed-loop model, the real part of the closed-loop eigenvector matrix $\bar{\Phi}_r$ dominates the conditioning of the matrix $\bar{\Psi}$. It is immediately seen that the characteristics of Eq. (11) is indeed similar to that of Eq. (32). Thus, theoretically, the acceleration feedback should perform as well as the deflection feedback.

Eigensystem Assignment Using Output Feedback

Although controllers implemented with full-state feedback have been successfully applied to small-order problems, the use of full-state feedback becomes impractical as the order of the system increases. As today's space structure grows in size and complexity, control strategies that only require direct output feedback play a more viable role in practical implementations. With output feedback the control input is defined, in general terms, as

$$\boldsymbol{u}(t) = -G\boldsymbol{y}(t) \tag{33}$$

where $\boldsymbol{y}(t)$ is an $m \times 1$ output measurement vector, and G is an $m \times n$ constant-gain matrix. The measurement vector $\boldsymbol{y}$ is given as

$$\boldsymbol{y} = C_d\boldsymbol{x} + C_v\dot{\boldsymbol{x}} + C_a\ddot{\boldsymbol{x}} \tag{34}$$

in which C_d, C_v, and C_a are output influence matrices corresponding to deflection, velocity, and acceleration, respectively. Although the most general form of output feedback includes displacement, velocity, and accelerations, for the sake of simplicity the eigensystem assignment technique would be developed for velocity and acceleration measurements only. These developments can easily be extended for the other forms of output feedbacks.

For an output velocity and acceleration feedback, the control input is written as

$$\boldsymbol{u} = -G[C_v\dot{\boldsymbol{x}} - C_a\ddot{\boldsymbol{x}}] \tag{35}$$

Assume that ℓ pairs of complex conjugate eigenvalues are to be assigned, and the number of measurements m is greater than or equal to the number of inputs p. The eigensolution of the closed-loop system for the second-order dynamic model [Eq. (1)] can then be written as

$$[\lambda_k^2(M + BGC_a) + \lambda_k(D + BGC_v) + K]\boldsymbol{x}_k = 0; \qquad k = 1, ..., \ell \tag{36}$$

where λ_k is the kth eigenvalue and $\boldsymbol{x}_k$ the corresponding eigenvector. The subscript k refers to the mode number. Rearranging Eq. (36) in a compact form yields

$$[(\lambda_k^2 M + \lambda_k D + K) \quad B]\begin{bmatrix} \boldsymbol{x}_k \\ (\lambda_k^2 GC_a + \lambda_k GC_v)\boldsymbol{x}_k \end{bmatrix} \equiv \Gamma_k \boldsymbol{\phi}_k = 0 \tag{37}$$

If the closed-loop eigenvalues λ_k ($k = 1, ..., \ell$) are assigned including their complex conjugates, Eq. (37) can be used to determine the gain matrix G. By observing that the vector $\boldsymbol{\phi}_k$ is in the null space of the matrix Γ_k, G may be determined when all of the null spaces of the matrices Γ_k ($k = 1, ..., \ell$) are computed corresponding to the eigenvalues λ_k ($k = 1, ..., \ell$). Comparison of Eqs. (7) and (37) yields

$$G[C_a\bar{\Phi}\Lambda^2 + C_v\bar{\Phi}\Lambda] = \hat{\Phi} \tag{38}$$

where the matrix $\bar{\Phi} = [\bar{\boldsymbol{\phi}}_1, ..., \bar{\boldsymbol{\phi}}_\ell]$, the matrix $\hat{\Phi} = [\hat{\boldsymbol{\phi}}_1, ..., \hat{\boldsymbol{\phi}}_\ell]$, and the matrix $\Lambda = \text{diag}[\lambda_1, ..., \lambda_\ell]$. Bearing in mind that the gain matrix G is real, decompose Eq. (38) into real and imaginary parts to yield

$$G\{C_a[\bar{\Phi}_r(\Lambda_r^2 - \Lambda_i^2) - 2\bar{\Phi}_i\Lambda_r\Lambda_i] + C_v[\bar{\Phi}_r\Lambda_r - \bar{\Phi}_i\Lambda_i]\} = \hat{\Phi}_r \tag{39a}$$

$$G\{C_a[\bar{\Phi}_i(\Lambda_r^2 - \Lambda_i^2) - 2\bar{\Phi}_r\Lambda_r\Lambda_i] + C_v[\bar{\Phi}_r\Lambda_i + \bar{\Phi}_i\Lambda_r]\} = \hat{\Phi}_i \tag{39b}$$

or, in matrix form,

$$G\bar{\Psi} \equiv G[C_a \quad C_v]\begin{bmatrix} [\bar{\Phi}_i(\Lambda_r^2 - \Lambda_i^2) - 2\bar{\Phi}_r\Lambda_r\Lambda_i] & [\bar{\Phi}_r(\Lambda_r^2 - \Lambda_i^2) - 2\bar{\Phi}_i\Lambda_r\Lambda_i] \\ [\bar{\Phi}_r\Lambda_i + \bar{\Phi}_i\Lambda_r] & [\bar{\Phi}_r\Lambda_r - \bar{\Phi}_i\Lambda_i] \end{bmatrix}$$

$$= [\hat{\Phi}_i \quad \hat{\Phi}_r] \equiv \hat{\Psi} \tag{40}$$

Observe that Eq. (40) is underdetermined, unique, or overdetermined, depending on the number of eigenvalues assigned, i.e.,

$$2\ell < m \qquad \text{underdetermined}$$

$$2\ell = m \qquad \text{unique}$$

$$2\ell > m \qquad \text{overdetermined}$$

Subsequently, the solution for the gain matrix can be expressed in terms of the psuedoinverse for all three conditions:

$$G = \hat{\Psi}\bar{\Psi}^{\dagger} \tag{41}$$

It is interesting to note that if G, C_v, and C_a are partitioned such that

$$G = [G_v \quad G_a]; \qquad C_v = \begin{bmatrix} I \\ \mathbf{0} \end{bmatrix}; \qquad C_a = \begin{bmatrix} \mathbf{0} \\ I \end{bmatrix}$$

the output feedback formulation [Eq. (40)] converges to the full-state feedback development of Eqs. (31) and (32). If the number of independent measurements is equal to $2n$, i.e., the matrix $[C_a \; C_v]$ is full rank, the same approach as shown in the last section for the full-state feedback may be applied to obtain a robust closed-loop system. Otherwise, the coefficients are simply chosen to form a set of linearly independent closed-loop eigenvectors for the assignable eigenvalues. It should be noted that the number of pairs of eigenvalues, ℓ, that can be assigned via output feedback is limited using this approach, i.e., $2\ell = \max(m,p)$.

Numerical Example

The eigensystem assignment technique is applied to the flexible truss structure shown in Fig. 1. The structure is a test article for NASA's controls-structures-interaction program. It is composed of a L-shaped bus, a reflector, a laser feed, and two suspension cables used to simulate on-orbit conditions. The original finite-element model is composed of 350 grid points with six degrees of freedom per grid point, resulting in a 2100-order system. However, for preliminary control designs a reduced-order model comprised of the first nine modes of the structure, covering a bandwidth of 0–5 Hz, is used. These modes include four pendulum modes (modes due to suspension effects) and five flexural as well as torsional modes. The actuation of the control forces is provided through six proportional gas thrusters located at various points on the body as illustrated in Fig. 1. Nine rate gyros and nine accelerometers provide eighteen measurements of angular velocity and linear accelerations at nine locations along the body, as indicated in Fig. 1.

The open-loop eigenvalues and the desired closed-loop eigenvalues are summarized in Table 1. The closed-loop eigenvalues are chosen such that each mode is provided with 5% damping, i.e., damping ratio $\zeta = 0.05$. The

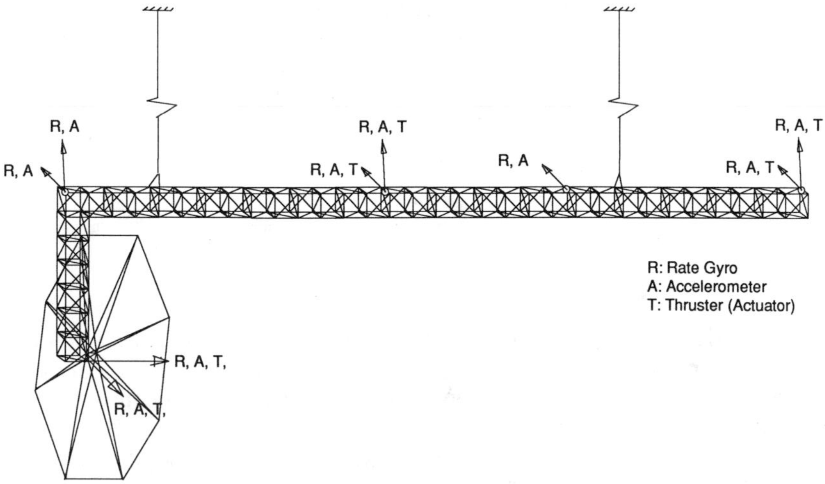

Fig. 1 NASA Controls-structures interaction test article.

closed-loop damped frequencies (imaginary part of the eigenvalues) are larger than the open-loop values for the first three modes but are the same for the remaining modes. The results of the eigenvalue assignment in terms of the Forbenius[16] norm of the gain matrix, $|G|_F$, and the condition number of the modal matrix of the closed-loop state matrix, $c(\Psi)$, is presented in Table 2 for three different desired closed-loop eigenvector matrices, represented by the matrix Q_0.

In these cases the matrix Q_0 is set equal to the open-loop eigenvector matrix Ψ_o, the unitary matrix closest to the open-loop eigenvector matrix Ψ_o, and an identity matrix, respectively. Although, the norm of the gain

Table 1 Open- and closed-loop eigenvalues

Open-loop eigenvalues	Closed-loop eigenvalues
$0.0000 + 0.7746i$	$-0.0501 + 1.0000i$
$0.0000 + 0.8019i$	$-0.0601 + 1.2000i$
$0.0000 + 0.8043i$	$-0.0751 + 1.5000i$
$0.0000 + 3.9237i$	$-0.1964 + 3.9237i$
$0.0000 + 13.4800i$	$-0.6748 + 13.4800i$
$0.0000 + 15.2194i$	$-0.7619 + 15.2194i$
$0.0000 + 18.9833i$	$-0.9504 + 18.9833i$
$0.0000 + 19.4861i$	$-0.9755 + 19.4861i$
$0.0000 + 23.4345i$	$-1.1732 + 23.4345i$

Table 2 Comparison of different desired eigenvector spaces

Case	$\|G\|_F$	$c(\Psi)$
Ψ_o	224.26	32.91
UW^H	100.96	24.65
I	189.21	29.28

matrix varies considerably for each of the cases, the condition number $c(\Psi)$ is moderately close for all three. Moreover, comparing the condition numbers with 23.43, the open-loop condition number indicates that the proposed algorithm for robust eigenvalue assignment is quite effective and relatively insensitive to the choice of target matrix Q_0. The choice of the closest unitary matrix to the open-loop eigenvalue results in both the smallest gain matrix norm and the smallest condition number. On the other hand, the choice of the open-loop eigenvector space gives the worst results in terms of the gain matrix norm and closed-loop conditioning. It appears that the proposed robust eigenstructure assignment technique produces better results when the desired closed-loop eigenvector space is chosen to be orthonormal. In particular, the closest unitary matrix to the open-loop eigenvectors seems to be the best choice.

Concluding Remarks

The conventional approach for dealing with miltivariable control problems of a flexible structure is to convert the second-order ordinary differential equation into a first-order state-space form. The basic formulations in this paper were derived while retaining the typical second-order form. This preserves the symmetry and sparsity of the mass, damping, and stiffness matrices of a flexible structure for possible exploitation. The proposed approach can assign closed-loop eigenvalues and admissible eigenvectors via velocity and displacement feedbacks or acceleration and velocity feedbacks. The ability to assign eigenvalues using acceleration feedback is particularly attractive for practical implementation because accelerometers are frequently used in structural dynamics testing. The singular value decomposition is the most time consuming computation of the algorithm developed in this paper since it has to be repeated n times if n eigenvalues are to be assigned. However, the sparsity of the matrices to be decomposed should ease the computational burden of singular value decomposition. A special solution procedure for this type of formulation should be developed to significantly advance the state-of-the-art computational technology particularly for the eigensystem assignment for mechanical systems that are typically represented by second-order ordinary differential equations.

References

[1]Wonham, W. M., "On Pole Assignment in Multi-Input, Controllable Linear Systems," *IEEE Transactions on Automatic Control*, Vol. AC-12, No. 6, Dec. 1967, pp. 660–665.

[2]Moore, B. C., "On the Flexibility Offered by State Feedback in Multivariable Systems Beyond Closed-Loop Eigenvalue Assignments," *IEEE Transactions on Automatic Control*, Vol. AC-21, No. 5, Oct. 1976, pp. 689–692.

[3]Klien, G., and Moore, B. C., "Eigenvalue-Generalized Eigenvector Assignment with State Feedback," *IEEE Transactions on Automatic Control*, Vol. AC-22, No. 1, Feb. 1977, pp. 140–141.

[4]Davison, E. J., "On Pole Assignment in Linear Systems with Incomplete State Feedback," *IEEE Transactions on Automatic Control*, Vol. AC-15, No. 3, June 1970, pp. 348–351.

[5]Davison, E. J., and Wang, H., "On Pole Assignment in Linear Systems Using Output Feedback," *IEEE Transactions on Automatic Control*, Vol. AC-20, No. 4, Aug. 1975, pp. 516–518.

[6]Kimura, H., "Pole Assignment by Gain Output Feedback," *IEEE Transactions on Automatic Control*, Vol. AC-20, No. 4, Aug. 1975, pp. 509–518.

[7]Srinathkumar, S., "Eigenvalue/Eigenvector Assignment Using Output Feedback," *IEEE Transactions on Automatic Control*, Vol. AC-23, No. 1, Feb. 1978, pp. 79–81.

[8]Howze, J. W., and Cavin, R. K. III, "Regulator Design with Modal Insensitivity," *IEEE Transactions on Automatic Control*, Vol. AC-24, No. 3, June 1979, pp. 466–469.

[9]Bhattacharyya, S. P., and DeSouza, E., "Pole Assignment via Sylvester's Equations," *Systems and Controls Letters*, Vol. 1, No. 4, 1982, pp. 261–263.

[10]Cavin, R. K. III, and Bhattacharyya, S. P., "Robust and Well-Conditioned Eigenstructure Assignment via Sylvester's Equations," *Optimal Control Applications and Methods*, Vol. 4, No. 3, Oct. 1983, pp. 205–212.

[11]Kautsky, J., Nichols, N. K., and Van Dooren, P., "Robust Pole Assignment in Linear State Feedback," *International Journal of Control*, Vol. 41, No. 5, July 1985, pp. 1129–1155.

[12]Maghami, P. G., and Juang, J.-N., "Efficient Eigenvalue Assignment for Large Space Structures," Proceedings of the 30th Structures, Structural Dynamics and Materials Conference, Mobile, AL, April 3–5, 1989; also *Journal of Guidance, Control, and Dynamics* (to be published).

[13]Juang, J.-N., Lim, K. B., and Junkins, J. L., "Robust Eigensystem Assignment for Flexible Structures," *Journal of Guidance, Control, and Dynamics*, Vol. 12, No. 3, May–June 1989, pp. 381–387.

[14]Rew, D. W., Junkins, J. L., and Juang, J.-N., "Robust Eigenstructure Assignment by a Projection Method: Applications Using Multiple Optimization Criteria," *Journal of Guidance, Control, and Dynamics*, Vol. 12, No. 3, May–June 1989, pp. 396–403.

[15]Belvin, K. B., "Simulation and Interdisciplinary Design Methodology for Control-Structure Interaction Systems," Ph.D. Dissertation, University of Colorado at Boulder, Boulder, CO., July 1989.

[16]Golub, G. H., and Van Loan, C. F., *Matrix Computations*, Johns Hopkins University Press, Baltimore, MD, 1983.

[17]Klema, V. C., and Laub, A. J., "The Singular Value Decomposition: Its Computation and Some Applications," *IEEE Transactions on Automatic Control*, Vol. AC-25, No. 2, April 1980, pp. 164–176.

Chapter 15

Minimum Sensitivity Design Method for Output Feedback Controllers

John L. Junkins* and Youdan Kim†
Texas A&M University, College Station, Texas

Introduction

IN recent years considerable attention has been focused on developing methods for control of large space structures (LSS). Three major problems arise when designing controllers for LSS.

First, many of the available controller design methods assume that the full-state vector (or at least the estimated full-state vector) is available. In most practical situations these control laws encounter difficulties when only a low-dimensioned output vector is available. The real-time implementation of high-dimensioned state estimators, although perhaps feasible, remain computationally unattractive.

Second, it is difficult and very expensive to design controllers for high-order systems such as a flexible structure. Therefore, it is usually desirable to design a controller based on a reduced-order model, which includes only the most important subset of the vibration modes, although we may retain the high-order model for validation purposes. The reduced-order-based controller may destabilize higher-order modes due to truncation and spillover effects.[1] Therefore, we must include consideration of truncation and spillover effects when designing controllers based on a reduced-order model for LSS.

Third, there exists a modeling error for any mathematical model of an LSS. These unknown, unmodeled effects can potentially destabilize the closed-loop system. In order to tolerate model errors, we need to optimize

*George J. Eppright Chair Professor, Department of Aerospace Engineering.
†Ph.D. Candidate, Department of Aerospace Engineering.

the robustness (tolerance of model errors and disturbances) by judicious design of the control law.

In this chapter we introduce an optimization method to design robust, globally stable linear output feedback controllers. Because of the specific parameterization used for the feedback gain matrices, the output feedback controller guarantees stability; thus, truncation and spillover effects cannot destabilize the system, even though 1) we may not have chosen a good reduced-order model, and 2) finite model errors are present. So that we do not sacrifice system performance on the altar of global stability, we develop a method for optimizing a closed-loop sensitivity measure with respect to model parameter variations. Numerical demonstrations are given to illustrate the salient features and practicality to this approach to design of robust output feedback controllers.

Symmetric Output Feedback Controller Design

Consider the class of dynamic systems modeled by the equation of motion

$$M\ddot{x} + C\dot{x} + Kx = Du \tag{1}$$

where $x \in R^n$ and $u \in R^m$ are configuration and control vectors, respectively; M is the $n \times n$ positive-definite symmetric mass matrix; C is the $n \times n$ positive-semidefinite symmetric structural damping matrix; K is the $n \times n$ positive-semidefinite stiffness matrix; D is the $n \times m$ control influence matrix; and the overdot represents differentiation with respect to time.

Motivated by the recent developments,[2–5] introduce the symmetric or "structural" output feedback form of the control law (collocated sensors and actuators):

$$u = -(G_1 D^T x + G_2 D^T \dot{x}) \tag{2}$$

where G_1 and G_2 are $m \times m$ positive-definite symmetric gain matrices. The closed-loop equations can now be written as

$$M\ddot{x} + (C + DG_2D^T)\dot{x} + (K + DG_1D^T)x = 0 \tag{3}$$

For all members of this special class of output feedback controllers, we can verify that the asymptotic stability of the closed-loop system is guaranteed. Consider the following Liapunov function:

$$V = \tfrac{1}{2}\dot{x}^T M\dot{x} + \tfrac{1}{2}x^T(K + DG_1D^T)x \tag{4}$$

By differentiating the preceeding equation with respect to time and using the closed-loop equation [Eq. (3)], $\dot{V}$ can be written as

$$\dot{V} = -\dot{x}^T(C + DG_2D^T)\dot{x} \tag{5}$$

It is clear from Eqs. (4) and (5) that the Liapunov function V is positive-definite and $\dot{V}$ is negative-semidefinite. We can introduce arbitrary finite variations in M, C, K, in Eq. (3) and verify that (as long as definiteness properties are maintained), we still have Liapunov stability for all positive-definite G_i.

Therefore, judicious selection of the gain matrices from a stable family (i.e., the positive-definiteness of the G_i matrices) guarantees the global stability of the closed-loop system in the Liapunov sense. The most familiar member of the family is the so-called direct feedback case of diagonal G_1 and G_2. More generally, fully populated G_i matrices allow every actuator to linearly operate on all sensor measurements and thereby provide more controller design freedom. This class of controllers is attractive because a stable family of controllers is guaranteed even in the presence of model errors, and it is not necessary to implement a state observer. However, it is necessary to have all sensors located at actuator stations, and this requirement presents a stumbling block in some applications. For this reason generalizations and modifications of this approach have recently been proposed[3] to circumvent these difficulties.

We admit the most general family of positive-semidefinite (PSD) gain matrices by introducing the Cholesky decomposition gain parameterizations

$$G_1 = L_1 L_1^T, \qquad G_2 = L_2 L_2^T \tag{6}$$

Note that G_1 and G_2 are guaranteed to be PSD for all real elements (q_{ij}, r_{ij}) assigned to the following lower triangular matrices:

$$L_1 = \begin{bmatrix} q_{11}^2 & 0 & 0 & \cdots & 0 \\ q_{21} & q_{22}^2 & 0 & \cdots & 0 \\ q_{31} & q_{32} & q_{33}^2 & \cdots & 0 \\ \vdots & \vdots & \vdots & \ddots & \vdots \\ q_{m1} & q_{m2} & q_{m3} & \cdots & q_{mm}^2 \end{bmatrix},$$

$$L_2 = \begin{bmatrix} r_{11}^2 & 0 & 0 & \cdots & 0 \\ r_{21} & r_{22}^2 & 0 & \cdots & 0 \\ r_{31} & r_{32} & r_{33}^2 & \cdots & 0 \\ \vdots & \vdots & \vdots & \ddots & \vdots \\ r_{m1} & r_{m2} & r_{m3} & \cdots & r_{mm}^2 \end{bmatrix} \tag{7}$$

Let the corresponding gain parameter vector be

$$\boldsymbol{p} = [q_{11} \quad q_{21} \quad \cdots \quad q_{m1} \quad q_{22} \quad \cdots \quad q_{mm} \quad r_{11} \quad r_{21} \quad \cdots \quad r_{mm}]^T \tag{8}$$

We consider the gain matrices G_1 and G_2 to be functions of $\boldsymbol{p}$, through Eqs.

(6) and (7), it is clear that

$$G_1 = G_1(\boldsymbol{p}) \quad \text{and} \quad G_2 = G_2(\boldsymbol{p}) \tag{9}$$

and that, for example, the partial derivatives of the gain matrices with respect to the elements of the parameter vector $\boldsymbol{p}$ can be constructed easily.

The closed-loop system can be rewritten as

$$M\ddot{x} + \hat{C}\dot{x} + \hat{K}x = 0 \tag{10}$$

where the closed-loop system damping and stiffness matrices are

$$\hat{C} = C + DG_2D^T \quad \text{and} \quad \hat{K} = K + DG_1D^T \tag{11}$$

It is apparent from Eqs. (9) and (11) that the closed-loop damping and stiffness matrices are simple algebraic functions of the parameter vector $\boldsymbol{p}$, i.e.,

$$\hat{C} = \hat{C}(\boldsymbol{p}) \quad \text{and} \quad \hat{K} = \hat{K}(\boldsymbol{p}) \tag{12}$$

Considering a first-order state-space form of the system differential equations that is equivalent to the second-order closed-loop system of Eq. (10), we write

$$B\dot{z} = Az \tag{13}$$

where

$$z = \begin{pmatrix} x \\ \dot{x} \end{pmatrix}, \quad A = \begin{bmatrix} 0 & M \\ -\hat{K} & -\hat{C} \end{bmatrix}, \quad B = \begin{bmatrix} M & 0 \\ 0 & M \end{bmatrix} \tag{14}$$

It is evident that

$$A = A(\boldsymbol{p}) \tag{15}$$

The right and left eigenvalue problems [associated with $z = \phi\, e^{\lambda t}$ solutions of Eq. (13)] are

Right:

$$\lambda_i B\phi_i = A\phi_i, \qquad i = 1, 2, ..., 2n \tag{16a}$$

Left:

$$\lambda_i B^T\psi_i = A^T\psi_i, \qquad i = 1, 2, ..., 2n \tag{16b}$$

where the conventional normalization[6] of the biorthogonality conditions for the eigenvectors are adopted as

$$\phi_i^T B \phi_i = 1, \qquad i = 1, 2, ..., 2n \tag{17a}$$

$$\psi_j^T B \phi_i = \delta_{ij}, \qquad i, j = 1, 2, ..., 2n \tag{17b}$$

so that

$$\psi_j^T A \phi_i = \lambda_i \delta_{ij}, \qquad i, j = 1, 2, ..., 2n \tag{18}$$

Since $A = A(\boldsymbol{p})$, it is natural that we consider the eigenvalues and eigenvectors to be functions of parameter vector $\boldsymbol{p}$.

Our goal is to find the parameter vector $\boldsymbol{p}$ that satisfies the design objectives. To focus the discussion, suppose that our design objectives are 1) to move some of the eigenvalues into an acceptable region of the complex plane and 2) to maximize a robustness measure for the closed-loop system. More generally, we can impose inequality constraints to enforce actuator saturation limits and other admissible control conditions.[3] The desired region for the eigenvalues can be defined by using both equality and inequality constraints on eigenvalues, and optimization can be approached by adopting a robustness measure as our performance index. It is well known that conditioning of the eigenvectors is a direct measure of the sensitivity of the corresponding eigenvalues to perturbations in the elements of the closed-loop system matrices.[7] The condition number of the modal matrix of closed-loop eigenvectors can be used to establish finite bounds on the system matrices within which stability can be guaranteed.[8] Therefore, we adopt the condition number of $\mathscr{K}(\Phi^c(\boldsymbol{p}))$ of the closed-loop modal matrix Φ^c as our performance measure.

The condition number of the general matrix Φ is related to the singular values as

$$\mathscr{K}(\Phi) = \frac{\sigma_1(\Phi)}{\sigma_{2n}(\Phi)} \tag{19}$$

where σ_1 and σ_{2n} denote the maximum and minimum singular values of the matrix Φ. The *singular value decomposition* is a matrix factorization that is very closely related to the eigenvalue problem. Consider a square $2n \times 2n$ complex-valued matrix Φ. *Any* matrix Φ can be decomposed as follows:

$$\Phi = U \Sigma V^H \tag{20}$$

where U and V are $2n \times 2n$ unitary matrices normalized so that

$$U^H U = I \tag{21}$$

$$V^H V = I \tag{22}$$

The matrix $\Sigma = \text{diag}(\sigma_1, \sigma_2, ..., \sigma_{2n})$ contains as its elements the real *singular values* of Φ ordered as $\sigma_1 \geqslant \sigma_2 \geqslant ... \geqslant \sigma_{2n} \geqslant 0$. The superscript H denotes the conjugate transpose.

Since U and V are unitary, we can use Eqs. (21) and (22) to rearrange Eq. (20) as

$$\Sigma = U^H \Phi V \tag{23}$$

Equations (20–23) can be expressed as conditions on the $2n \times 1$ complex column vectors of $\boldsymbol{U} = [\boldsymbol{U}_1 \boldsymbol{U}_2 \; ... \; \boldsymbol{U}_{2n}]$ and $\boldsymbol{V} = [\boldsymbol{V}_1 \boldsymbol{V}_2 \; ... \; \boldsymbol{V}_{2n}]$ as

$$\Phi \boldsymbol{V}_i = \boldsymbol{U}_i \sigma_i, \qquad i = 1, 2, ..., 2n \tag{24}$$

$$\Phi^H \boldsymbol{U}_i = \boldsymbol{V}_i \sigma_i, \qquad i = 1, 2, ..., 2n \tag{25}$$

$$\boldsymbol{U}_i^H \boldsymbol{U}_j = \delta_{ij}, \qquad i,j = 1, 2, ..., 2n \tag{26}$$

$$\boldsymbol{V}_i^H \boldsymbol{V}_j = \delta_{ij}, \qquad i,j = 1, 2, ..., 2n \tag{27}$$

$$\sigma_i \delta_{ij} = \boldsymbol{U}_i^H \Phi \boldsymbol{V}_j, \qquad i,j = 1, 2, ..., 2n \tag{28}$$

Using Eqs. (26–28) in Eqs. (24) and (25), it is easy to verify that $\sigma_i^2, \boldsymbol{V}_i$ are eigenvalue, eigenvector pairs of the symmetric matrix $\Phi^H\Phi$, whereas $\sigma_i^2, \boldsymbol{U}_i$ are eigenvalue, eigenvector pairs of the symmetric matrix $\Phi\Phi^H$. Although we are concerned with the case of a general Φ, certain details presented here apply only to the case of distinct singular values. The generalizations for the repeated singular value cases can likely be pursued in a fashion analogous to the developments in Ref. 9.

Now, let us return to our original objectives. Our problem can be stated as an optimization problem in which we seek to impose specified eigenspace equality and inequality constraints and minimize the condition number of the closed-loop modal matrix. By considering all constraints and the performance index as function of the parameter vector $\boldsymbol{p}$, we can formulate our problem as follows:

$$\underset{p \in \Omega}{\text{minimize}}\; J(\boldsymbol{p}) = \mathscr{K}(\Phi^c(\boldsymbol{p})) = \frac{\sigma_1(\Phi^c(\boldsymbol{p}))}{\sigma_{2n}(\Phi^c(\boldsymbol{p}))} \tag{29}$$

subject to

$$f_i(\boldsymbol{p}) = \omega_i^o - \omega_i(\boldsymbol{p}) = 0, \qquad i = 1, ..., m \tag{30}$$

$$g_i(\boldsymbol{p}) = \omega_i(\boldsymbol{p})\zeta_i(\boldsymbol{p}) \geqslant T_i, \qquad i = 1, ..., m \tag{31}$$

where

$\omega_i(\boldsymbol{p}) = \mathrm{Im}(\lambda_i)$

$\zeta_i(\boldsymbol{p}) = -\dfrac{\mathrm{Re}(\lambda_i)}{|\lambda_i|}$

Ω = set of admissible $\boldsymbol{p}$ vectors

σ_1, σ_{2n} = maximum and minimum singular values of the right modal matrix Φ^c

ω_i^o = objective values for the frequencies

T_i = constants related to the system time constants

In order to apply gradient-based optimization algorithms, it is useful to compute analytical partial derivatives of the eigenvalues, eigenvectors, and the singular values. Differentiating Eqs. (16) and using Eqs. (17), we can obtain the following results[6]:

$$\frac{\partial \lambda_i}{\partial p_k} = \psi_i^T \left(\frac{\partial A}{\partial p_k} - \lambda_i \frac{\partial B}{\partial p_k} \right) \phi_i, \qquad i = 1, 2, ..., 2n \tag{32a}$$

$$\frac{\partial \phi_i}{\partial p_k} = \sum_{j=1}^{2n} a_{ji}^k \phi_j \tag{32b}$$

$$\frac{\partial \psi_i}{\partial p_k} = \sum_{j=1}^{2n} b_{ji}^k \psi_j \tag{32c}$$

where

$$\begin{aligned} a_{ji}^k &= \frac{1}{\lambda_i - \lambda_j} \psi_j^T \left(\frac{\partial A}{\partial p_k} - \lambda_i \frac{\partial B}{\partial p_k} \right) \phi_i, \qquad j \neq i \\ &= -\frac{1}{2} \left[\phi_i^T \frac{\partial B}{\partial p_k} \phi_i + \sum_{\substack{l=1 \\ l \neq i}}^{2n} a_{li}^k \phi_l^T (B + B^T) \phi_i \right], \qquad j = i \end{aligned} \tag{33a}$$

$$\begin{aligned} b_{ji}^k &= \frac{1}{\lambda_i - \lambda_j} \psi_i^T \left(\frac{\partial A}{\partial p_k} - \lambda_i \frac{\partial B}{\partial p_k} \right) \phi_j, \qquad j \neq i \\ &= -\psi_i^T \frac{\partial B}{\partial p_k} \phi_i - a_{ii}^k, \qquad j = i \end{aligned} \tag{33b}$$

Except for isolated events of multiple eigenvalues and root bifurcations, we can consider the eigenvalues and eigenvectors to be smooth differentiable functions of the gain parameter vector $\boldsymbol{p}$.

Similarly, by differentiating Eqs. (24) and (25) and using Eqs. (24–28), we can obtain the gradient of the singular values[10]:

$$\frac{\partial \sigma_i}{\partial p_k} = \frac{1}{2} \left(U_i^H \frac{\partial \Phi}{\partial p_k} V_i + V_i^H \frac{\partial \Phi^H}{\partial p_k} U_i \right) = \mathscr{R} \left(U_i^H \frac{\partial \Phi}{\partial p_k} V_i \right) \tag{34}$$

where $\mathscr{R}(\ )$ denotes the real part of ( ).

The analytical formulation of the second gradient of the singular values is also derived in Ref. 10.

For a high-order system such as a flexible structure, it is usually desirable and often necessary to develop a reduced-order model to save on computational time when designing control laws. The high-order model is generally retained to verify the resulting design. The most common order reduction method is the modal truncation method that introduces a partition into the modal matrix to select a subset of low-frequency modes determined to be of most importance in the particular application. The general modal coordinate transformation is

$$x(t) = \Phi\eta(t) = [\Phi_1 \quad \Phi_2]\begin{pmatrix}\eta_1(t)\\ \eta_2(t)\end{pmatrix} \tag{35}$$

where Φ is the modal matrix of open-loop eigenvectors, η_1 is the retained subset of n_r modal coordinates that we are interested in including in the reduced-order model, and η_2 is the residual modal coordinates.

Then the reduced-order equation of motion becomes

$$\tilde{M}\ddot{\eta}_1 + \tilde{C}\dot{\eta}_1 + \tilde{K}\eta_i = \tilde{D}\boldsymbol{u} \tag{36}$$

where

$$\tilde{M} = \Phi_1^T M \Phi_1 = I$$

$$\tilde{C} = \Phi_1^T C \Phi_1 = \mathrm{diag}(2\zeta_1\omega_1, \ldots, 2\zeta_{n_r}\omega_{n_r})$$

$$\tilde{K} = \Phi_1^T K \Phi_1 = \mathrm{diag}(\omega_1^2, \ldots, \omega_{n_r}^2)$$

$$\tilde{D} = \Phi_1^T D$$

For "structural feedback" control, the control force vector takes the following form:

$$\boldsymbol{u} = -(\tilde{G}_1\tilde{D}^T\eta_1 + \tilde{G}_2\tilde{D}^T\dot{\eta}_1) \tag{37}$$

where $\tilde{G}_1$ and $\tilde{G}_2$ are positive gain matrices for the reduced-order model. After manipulating the preceding equations, we can easily verify that the gain matrices G_1 and G_2 for the original high-order system are identical to the gain matrices $\tilde{G}_1$ and $\tilde{G}_2$ for the reduced-order model; i.e., on introducing the order reduction approximation $\boldsymbol{x} \approx \Phi_1\eta_1$ into Eqs. (10) and (11), they become identical to Eqs. (36) and (37) with $G_i = \tilde{G}_i$. Even though we design the gain matrices by using the reduced-order model, we can use them for the original system without any further computation. Although some approximations are introduced in any order reduction, it is important to note that the stability of the full-order closed-loop system is guaranteed by the preceding approach; this is because the approximations merely select particular gain matrices from a stable family (G_1 and G_2 must be positive-definite). Although stability is guaranteed, of course, other performance measures may not be satisfactorily predicted by the reduced-order model.

In the following section we will discuss how to solve the nonlinear constrained optimization problem.

Homotopic Nonlinear Programming

The central feature of our nonlinear programming algorithm is the use of minimum norm differential correction to iteratively modify the design variables. This is a generalized Newton process for solving a system of underdetermined nonlinear equations; on each iteration we seek to minimize the norm (sum square) of the correction vector required to satisfy the specified constraints. The constraint functions are locally linearized to provide a system of linear algebraic equations to solve for the correction vector; the process is iterated with the derivatives (the Jacobian matrix) locally evaluated. Later we discuss the procedure used to deal with inequality as well as equality constraints.

Of course, the key requirements underlying successful convergence of any Newton-like algorithm are 1) the existence of a feasible solution and 2) the availability of a sufficiently close starting iterative. As we show later, the homotopic continuation method deforms the constraints by controllable, arbitrarily small amounts and thereby remains controllably close to a neighboring converged solution; this effectively guarantees that "arbitrarily good" starting iteratives are available for each local iteration. An embedding method is used to define a continuous family of neighboring problems that contain two important members: 1) a trivial problem (one for which the solution is available) and 2) the problem of interest.

The continuous one-parameter family of problems is constructed in such a way that an imbedding parameter ($0 \leqslant \gamma \leqslant 1$) may be set to define any member of the family with $\gamma = 0$ generating the trivial problem and $\gamma = 1$ generating the problem of interest. By sweeping γ and judiciously assigning the γ increments, we control how closely spaced the neighboring solutions lie (in the space of the constraint functions), and the γ increments can be assigned adaptively based on convergence progress. Thus, we can remain as near to neighboring converged solutions as is necessary to maintain sufficiently good starting iteratives. Therefore, if a local convergence cannot be achieved (for some γ value), we have eliminated the most common problem (of having a poor starting iterative), and we can focus on other, more interesting convergence issues, such as the nonexistence of feasible solutions, local linear dependence of the linearized constraint equations, turning points, etc. For example, comparing the active constraint sets and the corresponding Jacobian rows between the apparently unreachable set (the one for which convergence cannot be achieved) and the neighboring (previous) converged solution will reveal which constraints are in competition and will usually suggest avenues for revision of the problem and/or provide insights as to why the convergence failure occurred.

We now discuss the algorithm, beginning with the statement of the parameter optimization problem we are addressing: We seek a find $\boldsymbol{p}^*$, which minimizes a function

$$J = J(\boldsymbol{p}) \tag{38}$$

subject to the constraints

$$f_i(\boldsymbol{p}) = f_i^o, \qquad i = 1, ..., m \tag{39a}$$

$$g_j(\boldsymbol{p}) \geqslant T_j, \qquad j = 1, ..., m \tag{39b}$$

where f_i^o is the objective value of the ith equality constraint, and T_j represents boundaries on the feasible region.

Converging from an approximate starting solution $\boldsymbol{p}_{\text{start}}$ to a feasible solution satisfying the constraints of Eqs. (39) may be approached by locally considering all of the equality constraints and only the active (locally violated) subset of inequality constraints. The locally violated subset of inequality constraints are simply considered as additional equality constraints of the form

$$f_j(\boldsymbol{p}) = T_j, \qquad j = m + 1, ..., r \tag{40}$$

The subset of inequality constraints $g_j(\boldsymbol{p}) \geqslant T_j$ included in Eq. (40), as local equality constraints, obviously depends on which constraints are locally violated and thus will typically change during the differential correction process. As long as r (the number of locally violated constraints plus the m equality constraints) does not exceed N (the number of elements in $\boldsymbol{p}$), then we can proceed with our discussion; otherwise, only the $N - m$ "most important" inequality constraints will be locally imposed, assuming the original ordering represents the "importance hierarchy" of the inequality constraints. Adopting this approach permits the specification of greater than $N - m$ inequality constraints, although not more than $N - m$ can be active on any iteration, especially the last one. Note that it is often the case that final convergence to a feasible solution satisfying Eqs. (39) is achieved even when local iterations encounter more than $N - m$ locally active constraints (i.e., it is typical that only a small subset of the specified inequality constraints are active when convergence to a feasible solution is achieved, and only under special circumstances will this finally active subset of constraints be invariant during the local iterations). These ideas must also be viewed in the context of the generalized homotopy procedure, which we discuss later.

Thus, our restated optimization problem is to minimize J of Eq. (38) subject to

$$\boldsymbol{f}(\boldsymbol{p}) = \boldsymbol{f}_o \tag{41}$$

where $\boldsymbol{f}(\boldsymbol{p})$ is the $r \times 1$ vector of m equality constraints and the $r - m$ active inequality constraints.

We initially address the problem of achieving a feasible solution near our starting solution and delay considering minimization of J until the first feasible design has been achieved. To enhance convergence, we introduce a "portable objective" vector $\boldsymbol{f}_p(\gamma)$ as the one parameter family of objective constraint values

$$\boldsymbol{f}_p(\gamma) = \gamma \boldsymbol{f}_o + (1-\gamma)\boldsymbol{f}(\boldsymbol{p}_{\text{start}}), \qquad 0 \leqslant \gamma \leqslant 1 \tag{42}$$

where γ is the homotopy parameter. The linear homotopy map $\boldsymbol{H}(\boldsymbol{p}(\gamma),\gamma)$ is generated by replacing $\boldsymbol{f}_o$ in Eq. (41) with $\boldsymbol{f}_p$ from Eq. (42) to obtain

$$\boldsymbol{H}(\boldsymbol{p}(\gamma),\gamma) \equiv \gamma \boldsymbol{f}_o + (1-\gamma)\boldsymbol{f}(\boldsymbol{p}_{\text{start}}) - \boldsymbol{f}(\boldsymbol{p}(\gamma)) = 0 \tag{43}$$

Notice the γ boundary conditions satisfied by $\boldsymbol{H}(p(\gamma),\gamma)$ in Eq. (43):

$$\begin{aligned} &\text{at} \quad \gamma = 0\text{: trivial problem} \qquad (\boldsymbol{p}(0) \equiv \boldsymbol{p}_{\text{start}}) \\ &\text{at} \quad \gamma = 1\text{: problem of interest} \qquad (\boldsymbol{p}(1) = \boldsymbol{p}_{\text{feasible}}) \\ &\qquad\qquad (\equiv \text{a solution satisfying} \quad \boldsymbol{f}_o - \boldsymbol{f}(\boldsymbol{p}(1)) = 0) \end{aligned} \tag{44}$$

It is clear that an arbitrary guess $\boldsymbol{p}_{\text{start}}$ satisfies the homotopy map of Eq. (43) for $\gamma = 0$, and if we can obtain a solution for $\gamma = 1$, then we have a feasible solution $\boldsymbol{p}_{\text{feasible}}$. It is also apparent that sweeping γ at a suitably small increment generates a sequence of neighboring problems. We solve this sequence of problems by using the neighboring converged solutions to generate starting iteratives for each subsequent problem.

The local corrections (for each γ specification) are done by linearizing Eq. (43) about the local $\boldsymbol{p}$ iterative and computing the minimum norm differential correction that will satisfy the linearized constraint equation. Thus, we minimize $\Delta\boldsymbol{p}^T W \Delta\boldsymbol{p}$, subject to $\boldsymbol{H}(\boldsymbol{p}(\gamma),\gamma) + A\,\Delta\boldsymbol{p} = 0$, where

$$A \equiv \frac{\partial \boldsymbol{H}}{\partial \boldsymbol{p}} = -\frac{\partial \boldsymbol{f}(\boldsymbol{p}(\gamma))}{\partial \boldsymbol{p}}$$

and W is a suitable positive-definite weighting matrix. The solution provides the minimum norm differential correction result[11]

$$\Delta\boldsymbol{p} = -W^{-1}A^T(AW^{-1}A^T)^{-1}\boldsymbol{H}(\boldsymbol{p}(\gamma)) \tag{45}$$

and we use the differential corrections

$$\boldsymbol{p}_{\text{new}} = \boldsymbol{p}_{\text{old}} + \Delta\boldsymbol{p} \tag{46}$$

to iterate until Eq. (43) is satisfied for each local γ value. On achieving a local convergence, γ is incremented by a prescribed amount. If local convergence is not achieved, the γ increment is reduced and γ is assigned nearer the largest γ value for which convergence was previously achieved. The process is halted if 1) convergence to feasible solution is achieved (at $\gamma = 1$), or 2) local convergence cannot be achieved when we increment γ by some small tolerance $\Delta\gamma \leqslant \varepsilon$. In the latter case the user should reevaluate the problem statement in view of the active constraint set and other considerations from evaluation of the largest γ solution achieved.

Having achieved convergence to a feasible solution, we now consider minimization of the performance index J subject to the local equality constraint of Eq. (43). Assuming there are $r < N$ active constraints, this can readily be accomplished[11] by simply introducing an objective value for the performance index J_o and treating it as the $(r+1)$th equality constraint. The corresponding minimum norm differential correction, in lieu of Eq. (45), is

$$\Delta \boldsymbol{p} = -W^{-1}\tilde{A}^T(\tilde{A}W^{-1}\tilde{A}^T)^{-1}\tilde{\boldsymbol{H}}(\boldsymbol{p}(\gamma)) \tag{47}$$

where

$$\tilde{A} \equiv \begin{bmatrix} & A & \\ \cdots & \cdots & \cdots \\ \dfrac{\partial J}{\partial p_1} & \dfrac{\partial J}{\partial p_2} \quad \cdots & \dfrac{\partial J}{\partial p_N} \end{bmatrix}, \qquad \tilde{\boldsymbol{h}}(\boldsymbol{p}(\gamma)) \equiv \begin{pmatrix} \boldsymbol{H}(\boldsymbol{p}(\gamma)) \\ \cdots\cdots \\ J(\boldsymbol{p}(\gamma)) - J_o \end{pmatrix} \tag{48}$$

Instead of computing the explicit inverse indicated in Eq. (45) [or Eqs. (47)], we make use of the more numerically robust SVD pseudoinverse method[12] to solve

$$\tilde{\boldsymbol{H}} = -\tilde{A}\,\Delta \boldsymbol{p} \tag{49}$$

Generalizing the square matrix result of Eq. (20), any $(r+1) \times N$ matrix $\tilde{A}$ can be decomposed as follows:

$$\tilde{A} = [U_1 \quad U_2]\begin{bmatrix} \Sigma_1 & 0 \\ 0 & 0 \end{bmatrix}\begin{bmatrix} V_1^H \\ V_2^H \end{bmatrix} \tag{50}$$

where Σ_1 is an $(r+1) \times (r+1)$ diagonal matrix, (U_1, V_1) are unitary matrices spanning the (row, column) space of $\tilde{A}$, and (U_2, V_2) span the corresponding null spaces of $\tilde{A}$.

By using the properties of the unitary matrices [Eqs. (21) and (22)], the Moore-Penrose pseudoinverse of $\tilde{A}$ is given by

$$\tilde{A}^\dagger = V_1 \Sigma_1^{-1} U_1^H \tag{51}$$

It can be proven, for the present underdetermined case, that $\tilde{A}^\dagger$ is algebraically (but not computationally) equivalent to $\tilde{A}^T(\tilde{A}\tilde{A}^T)^{-1}$, if $\tilde{A}$ is of maximum rank. Therefore, the minimum norm solution of the problem (49) can be computed as

$$\Delta \boldsymbol{p} = -\tilde{A}^\dagger \tilde{\boldsymbol{H}} \tag{52}$$

If needed, of course, we can easily modify the pseudoinverse solution to accommodate a nonidentity weighting matrix. For moderate-dimensioned and well-conditioned problems, Eq. (47) is usually numerically well behaved.

On achieving convergence to the feasible region, we reset $\gamma = 0$ and subsequently use Eq. (47) to compute the differential corrections. For the iterations subsequent to the first feasible solution, a new "constrained optimization homotopy process" is established to drive the objective function to its minimum value, or alternatively, drive it to a "goal" value J_{goal}. The objective value $J_o(\gamma)$ is assigned the one-parameter family of values:

$$J_o(\gamma) = \gamma J_{goal} + (1 - \gamma)J(\boldsymbol{p}_{feasible}) \tag{53}$$

where $\boldsymbol{p}_{feasible}$ is the first feasible solution [it satisfies Eq. (43) with $\gamma = 1$].

The value for J_{goal} may be an actual goal, but is more typically interpreted as "the best one could possibly hope for"; it is often taken as zero for minimization problems. We increment γ from zero toward unity, with the size of the increments dictated by convergence progress of the differential corrections for each γ value. When convergence cannot be achieved for a small tolerance increase in γ, we adopt the solution corresponding to the largest γ for which convergence was achieved as the constrained minimum. We have rigorously proven[11] that this process is mathematically equivalent to a gradient projection with appropriate strategies for correction step size control. However, this algorithm is superior to the gradient projection method because 1) the correction formulas are more easily programmed, and 2) the one parameter homotopic continuation process has been found much more attractive than step size control, because it is inherently self-starting. Of course, if the problem under consideration has multiple constrained minima, one will find that the preceding problem formulation does not save one from the local vs global convergence difficulties. Also, turning points are occasionally encountered, but these can be accommodated by using the Chow–Yorke algorithm.[13]

In our analytical and numerical studies, we have found that reachability of the feasible region (for our gain parameterization) is not guaranteed. We have recently found that the degree of controllability[14–17] affects dramatically the extent to which gain variations can reshape the eigenstructure. The degree of controllability is a function not only of the physical system, but also depends on actuator placement. This is an important area that needs further research.

In the following section we will design robust output feedback controllers for flexible structures by using homotopic nonlinear programming, which was discussed in this section. We also examine the performance of the designed controllers in the presence of model uncertainty.

Numerical Examples

To illustrate the application of the robust output feedback controller approach developed in the previous sections, we consider two specific physical systems. The first example is a three coupled oscillator system of sixth order. The second example is a sixth-degrees-of-freedom finite-element model of a planar grid structure.

First we consider the low-order system (undamped case), which has three degrees of freedom and two actuators. The mass, stiffness, and control influence matrices are

$$M = \begin{bmatrix} 1 & 0 & 0 \\ 0 & 1 & 0 \\ 0 & 0 & 1 \end{bmatrix}, \quad K = \begin{bmatrix} 20 & -10 & 0 \\ -10 & 30 & -20 \\ 0 & -20 & 20 \end{bmatrix}, \quad D = \begin{bmatrix} 1 & 0 \\ 0 & 0 \\ 0 & 1 \end{bmatrix} \tag{54}$$

We adopt the condition number of the closed-loop modal matrix as our performance measure and consider two equality and two inequality constraints on eigenvalues placement [Eqs. (29–31)]. The objective values for the closed-loop frequencies, w_i^o, are taken as their corresponding zero-gain values. The inequality constraints are introduced to bound the slowest system time constant, and we adopted $T_i = 0.3$, $i = 1, 2$.

Utilizing the minimum norm differential correction algorithm developed in the previous section, we successfully solved the preceding problem. The optimization process was initiated by simply adopting identity matrices for the G_i starting iteratives. Ten homotopy steps were required with three or four minimum norm differential gain corrections on each step. The final converged position and velocity gain matrices are

$$G_1 = \begin{bmatrix} 1.1591 & 0.0253 \\ 0.0253 & 1.0557 \end{bmatrix}, \qquad \|G_1\|_f = 1.5682 \tag{55a}$$

$$G_2 = \begin{bmatrix} 2.2539 & 0.2834 \\ 0.2834 & 2.6918 \end{bmatrix}, \qquad \|G_2\|_f = 3.5336 \tag{55b}$$

where $\|G\|_f$ is the Frobenius norm of the gain matrix G.

The characteristics of the open-loop and closed-loop systems are summarized in Table 1. As is evident, we succesfully imposed the system time constants and also modestly improved the conditioning of the closed-loop eigenvectors as compared to the open-loop system. In order to evaluate the performance of this controller in comparison to competing robust gain

Table 1 The characteristics of the open-loop and closed-loop system

Open loop		Feasible solution	
Eigenvalues	Condition no.	Eigenvalues	Condition no.
$\pm 1.4447i$ $\pm 4.4721i$ $\pm 6.9219i$	9.4903	$-0.8833 \pm 1.4447i$ $-1.0574 \pm 4.4721i$ $-0.5322 \pm 6.7872i$	8.1291

design methods, we design a full-state feedback controller for the same system. It is well known that there are extra degrees of freedom when we design a full-state feedback controller. We use pole placement gain design, which exploits the extra design freedom to reduce the sensitivity of the poles to the perturbations in the closed-loop system.[18] We take the closed-loop eigenvalues obtained from the preceding output feedback controller (Table 1) as the target eigenvalues of the full-state feedback controller and use the open-loop eigenvector matrix as target closed-loop eigenvectors. The resulting full-state feedback gain matrix is given as follows:

$$G = \begin{bmatrix} 6.2450 & 3.2694 & 7.7150 & 2.2386 & 0.8388 & 0.2675 \\ 6.0412 & 1.7898 & -3.9661 & 0.5140 & -0.2674 & 2.7071 \end{bmatrix}$$

$$\|G\|_f = 13.3450 \tag{56}$$

The condition number of the closed-loop eigenvector matrix of this case is 8.4026, which is slightly larger than the preceding optimum output feedback case.

Now let us perturb all elements of the mass and the stiffness matrices by about 10% and use the controllers that were designed in order to examine the robustness of both controllers. The changed mass and stiffness matrices are as follows:

$$M = \begin{bmatrix} 1.1 & 0 & 0 \\ 0 & 0.9 & 0 \\ 0 & 0 & 1.05 \end{bmatrix}, \qquad K = \begin{bmatrix} 22 & -11 & 0 \\ -11 & 29 & -18 \\ 0 & -18 & 18 \end{bmatrix} \tag{57}$$

It is obvious that the changed mass and stiffness matrices remain positive-definite. As long as the mass and stiffness matrices are positive-definite, we know that the symmetric output feedback controller guarantees stability. Therefore, we can expect that the closed-loop system with the stable output feedback controller will at least remain stable, even though it was designed for the unperturbed system. On the other hand, the performance of the system will definitely be modified by model perturbations. The closed-loop eigenvalues of the perturbed system with the controllers that were designed for the unperturbed system are summarized in Table 2. As we can see in Table 2, both controllers remain stable for the perturbed system, and both are not very sensitive to the perturbation. By comparing Eqs. (55) with Eq.

Table 2 The closed-loop eigenvalues (perturbed system)

Output feedback	Full-state feedback
$-0.8351 \pm 1.5358i$	$-0.9216 \pm 1.4695i$
$-1.0432 \pm 4.7783i$	$-0.9734 \pm 4.3725i$
$-0.5946 \pm 6.5030i$	$-0.4116 \pm 6.8147i$

(56), however, we find that the output feedback controller uses less energy than the full-state feedback controller. Also, of course, the performance of the full-state feedback controller must also take into account the additional performance degradation due to state estimation errors not considered in this discussion. Therefore, output feedback controller appears to be a better selection for this particular case.

As a second example, consider a 60-degree-of-freedom model of a grid structure. A finite-element analysis of the grid was performed. Nodes were placed at each overlapping joint on the grid. Three degrees of freedom (one normal displacement and two, transverse and vertical rotations) appropriate for motion normal to the nominal plane of the grid were considered for each node. Figure 1 shows the flexible grid experimental configuration. The material properties for this model are listed in Table 3.

Angular velocity feedback torques are provided by three reaction wheel actuators. The actuator axes (about which torques applied) lie in the plane of the grid as indicated by the arrows in Fig. 1. The actuators (Clifton Precision Motors) have 60-Hz bandwidth and ±20 oz.-in. (±0.1412 N-m) saturation. The grid angular velocity is measured about the same axes with

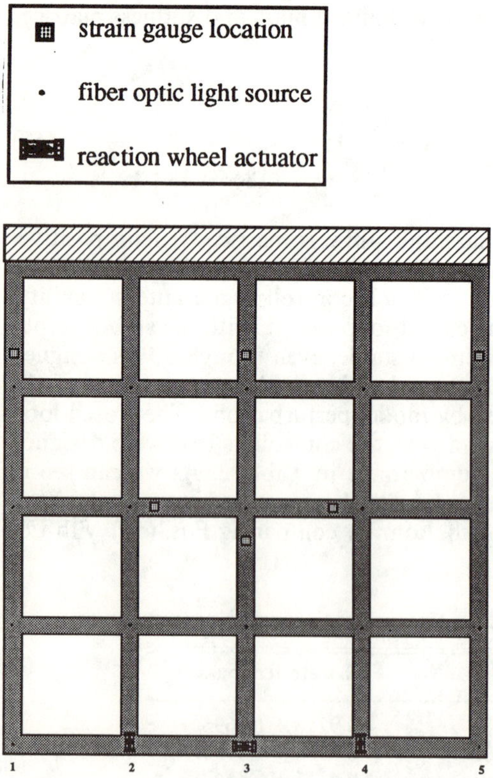

The motion of the twenty light sources are measured using a stereo pair of high speed (200 frames/sec) video cameras; stereo triangulation is used to determine the inertial deflection trajectory of each point. These measurements, along with the strain gauge measurements, are used for post experiment evaluation.

The reaction wheels allow torques to be imposed about in-plane axes (±) parallel to the vectors shown. Co-located angular rate measuring devices determine the angular velocity and, via integration, the angular deflection about axes parallel to the actuator torque axis. These signals are the output operated upon by the control gains to generate the feedback control.

Fig. 1 Texas A&M flexible grid identification and control experiment.

Table 3 Material properties of grid model

EI	$3255.2 \text{ lb} \cdot \text{in.}^2$	$9.34982 \text{ N} \cdot \text{m}^2$
GJ	$4943.0 \text{ lb} \cdot \text{in.}^2$	$14.18550 \text{ N} \cdot \text{m}^2$
ρA	$0.0000648 \text{ lb} \cdot \text{s}^2/\text{in.}^2$	0.44678 kg/m
ρJ	$0.0000217 \text{ lb} \cdot \text{s}^2$	$0.0000965 \text{ kg} \cdot \text{m}$

solid-state sensors. The particular position and orientation of the sensors and actuators shown were one of the six most controllable configurations that can be chosen by computing the controllability measures of Refs. 14 and 15. The admissible locations considered were the 20 grid locations and the direction of the actuators were permitted to be either vertical or horizontal. We confirmed that other actuator and sensor placements resulted in larger condition numbers and larger control gains.

A reduced-order model is used for the design process; in this case we adopted the first 10 lowest-frequency modes. Structural damping was also considered while the controller was designed. In this example the formulation of the problem is a bit different. We adopted the condition number of the closed-loop modal matrix as our performance measure and imposed constraints on the first three closed-loop frequencies and time constants. Thus, all the constraint equations are inequality type as follows:

$$\underset{p \in \Omega}{\text{minimize}}\ J(p) = \mathscr{K}(\Phi^c(p)) = \frac{\sigma_1(\Phi^c(p))}{\sigma_{2n}(\Phi^c(p))} \tag{58}$$

subject to

$$\begin{aligned} \omega_{oi} - \Delta\omega_i \leqslant \omega_i(p) \leqslant \omega_{oi} + \Delta\omega_i, \qquad i = 1, ..., m \\ \omega_i(p)\zeta_i(p) \geqslant T_i, \qquad i = 1, ..., m \end{aligned} \tag{59}$$

where $\Delta\omega_i$ is adopted arbitrarily as 1% of the open-loop natural frequency ω_{oi}.

We can solve the above problem by using the same kind of techniques discussed in the previous section. The preceding formulation [Eqs. (58) and (59)] is more desirable than the previous one [Eqs. (29–31)] in the engineering sense, since we do not have to retain the open-loop frequencies as the desired frequencies (this is an unnecessarily hard constraint). This idea, to move the selected eigenvalues into an acceptable region of the complex plane, is more attractive from design point of view, because it provides design freedom, which can be exploited to increase the robustness of the closed-loop system.

We adopt the following time constants for lower bounds on the first three closed-loop eigenvalues: $T_1 = 0.2$, $T_2 = 0.2$, and $T_3 = 0.25$. The optimization process was initiated by simply adopting the following vector for the parameter starting iteratives:

$$p_{\text{start}} = [1 \quad 0 \quad 0 \quad 1 \quad 0 \quad 1 \quad 1 \quad 0 \quad 0 \quad 2 \quad 0 \quad 1]^T$$

The final converged position and velocity gain matrices are

$$\tilde{G}_1 = \begin{bmatrix} 0.4468 & 0.0009 & -0.0665 \\ 0.0009 & 1.4699 & -0.0027 \\ -0.0665 & -0.0027 & 0.4364 \end{bmatrix}, \qquad \|\tilde{G}_1\|_f = 1.5999 \tag{60a}$$

$$\tilde{G}_2 = \begin{bmatrix} 1.1366 & 0.2102 & -0.0030 \\ 0.2102 & 7.3446 & -1.2589 \\ -0.0030 & -1.2589 & 1.1389 \end{bmatrix}, \qquad \|\tilde{G}_2\|_f = 7.7324 \tag{60b}$$

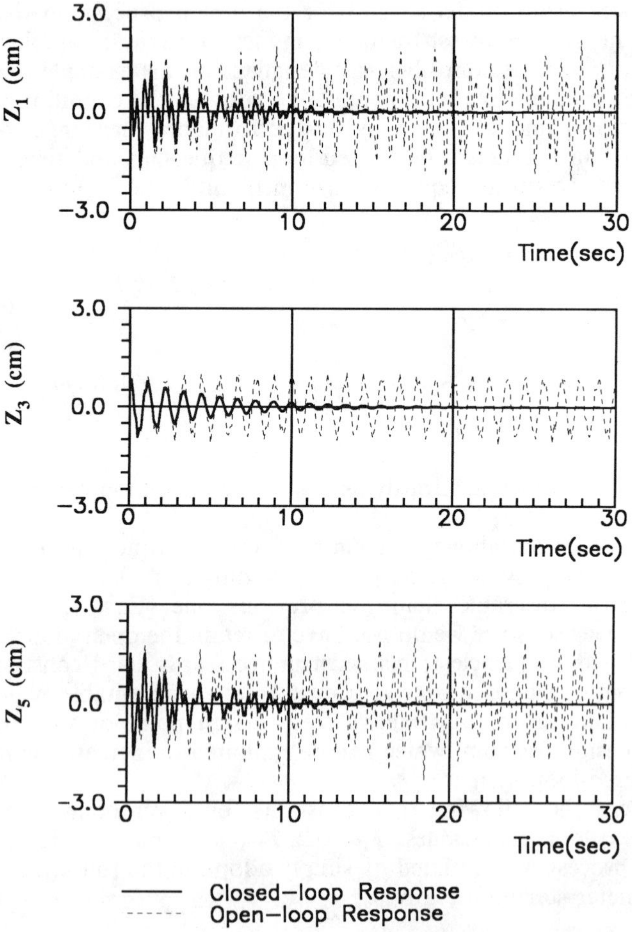

Fig. 2 Open- and closed-loop response of the flexible grid to an initial disturbance.

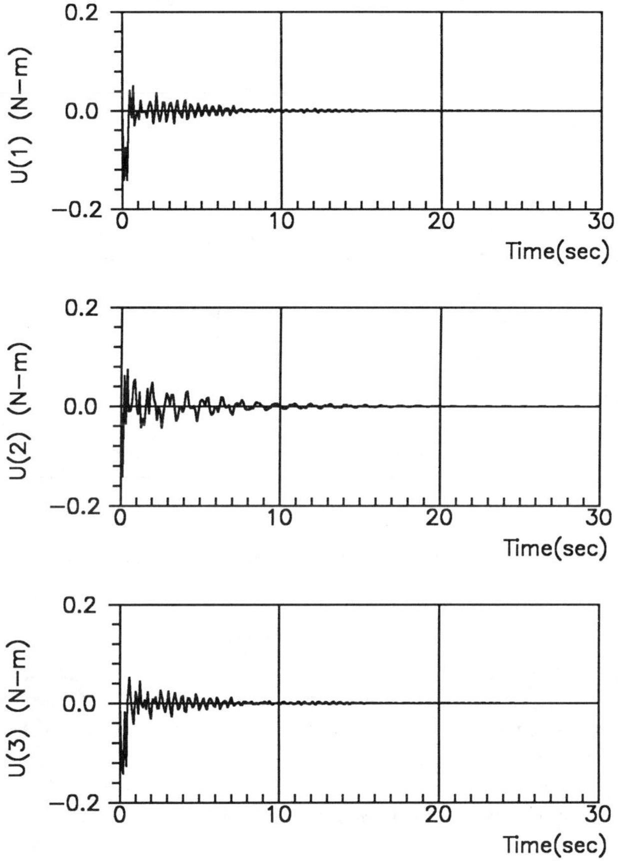

Fig. 3 Control torque input history.

The characteristics of the open-loop and feasible solution are summarized in Table 4. In this case it is obvious that we have successfully imposed the specified constraints on the closed-loop eigenvalues, with only a modest change in the eigenvector condition number. The open-loop and closed-loop response histories of the normal displacement to the grid plane (at three nodes of the bottom of the grid structure) due to a typical set of initial conditions are shown in Fig. 2, and the control input histories are displayed in Fig. 3. The simulations were done using both the reduced-order and full-order system with negligible differences in this case. We have imposed control saturation bounds (20 oz.-in.; 0.1412 N-m) to keep the reaction wheel speeds to modest levels. All the states of the full-order system are successfully controlled, and the amplitudes of the normal displacement of the closed loop (Fig. 2) are almost zero after 15 s.

Table 4 Characteristics of the open-loop and closed-loop system

Open loop		Feasible solution		Optimal solution	
ω_i	$\zeta_i\omega_i$	ω_i	$\zeta_i\omega_i$	ω_i	$\zeta_i\omega_i$
5.649	0.006	5.659	0.282	5.661	0.200
14.683	0.015	14.687	0.233	14.688	0.232
30.459	0.030	30.545	3.545	30.563	2.493
37.785	0.038	37.795	0.745	37.769	0.778
48.852	0.049	48.852	0.219	48.852	0.218
80.470	0.080	80.455	0.897	80.489	0.755
82.958	0.083	81.334	9.746	81.987	7.639
83.658	0.084	83.658	0.182	83.657	0.176
102.454	0.102	102.453	0.213	102.450	0.211
125.740	0.126	125.736	0.372	125.732	0.370
Condition no.	125.7396	Condition no.	129.5280	Condition no.	128.7071
		$\|\tilde{G}_1\|_f =$	0.6812	$\|\tilde{G}_1\|_f =$	1.5999
		$\|\tilde{G}_2\|_f =$	10.6028	$\|\tilde{G}_2\|_f =$	7.7324

Concluding Remarks

A robust output feedback controller design method for second-order dynamic systems has been developed and applied to two numerical examples. We demonstrate methods for parameterizing a stable family of controllers using collocated sensors and actuators and symmetric, fully populated, positive-definite gain matrices. The gains are designed by a nonlinear programming algorithm using a reduced-order model, yet the stability of the corresponding full-order system is guaranteed. We have found that the existence of the feasible solution is affected by the degree of controllability, which is in turn affected by the positions and orientation of the actuators. The latter truth offers exciting avenues for further research.

References

[1]Balas, M. J., "Trends in Large Space Structures Control Theory; Fondest Hopes, Wildest Dreams," *IEEE Transactions on Automatic Control*, Vol. AC-27, No. 3, June 1982, pp. 522–535.

[2]Rew, D. W., and Junkins, J. L., "Multi-Criterion Approaches to Optimization of Linear Regulators," AIAA Paper 86-2198-CP, Aug. 1986.

[3]Creamer, N. G., and Junkins, J. L., "A Pole Placement Technique for Vibration Suppression of Flexible Structures," AIAA Paper 88–4254–CP, Aug. 1988.

[4]Juang, J. N., and Pappa, R. S., "An Eigensystem Realization Algorithm (ERA) for Modal Parameter Identification and Model Reduction," *Journal of Guidance, Control, and Dynamics*, Vol. 8, No. 5, Sept.–Oct., 1985, pp. 620–627.

[5]Aubrun, J. N., "Theory of Control of Structures by Low-Authority Controllers," *Journal of Guidance, Control, and Dynamics*, Vol. 3, No. 5, Sept.-Oct. 1980, pp. 444–451.

[6]Lim, K. B., Junkins, J. L., and Wang, B. P., "Re-Examination of Eigenvector Derivatives," *Journal of Guidance, Control, and Dynamics*, Vol. 10, No. 6, 1987, pp. 581–587.

[7]Wilkinson, J. H., *The Algebraic Eigenvalue Problems*, Oxford Univ. Press, Oxford, UK, 1965.

[8]Patel, R. V., and Toda, M., "Quantitative Measures of Robustness for Multivariable Systems," *Proceedings of the Joint Automatic Control Conference*, IEEE, New York, TP8-A, Vol. I, 1980.

[9]Juang, J. N., and Lim, K. B., "On the Eigenvalue and Eigenvector Derivatives of a Non-Defective Matrix," AIAA Paper 88-2352, 1988.

[10]Junkins, J. L., and Kim, Y., "First and Second Order Sensitivity of the Singular Value Decomposition," Preprint, to appear in *Journal of the Astronautical Sciences*, 1989.

[11]Junkins, J. L., "Equivalence of the Minimum Norm and Gradient Projection Constrained Optimization Techniques," *AIAA Journal*, Vol. 10, July 1972, pp. 927–929.

[12]Dongarra, J. J., Moler, C. B., Bunch, J. R., and Stewart, G. W., *LINPACK User's Guide*, Society for Industrial & Applied Mechanics, Philadelphia, PA, 1979.

[13]Dunyak, J. P., Junkins, J. L., and Watson, L. T., "Robust Nonlinear Least Square Estimation Using the Chow–York Homotopy Method," *Journal of Guidance, Control, and Dynamics*, Vol. 7, No. 6, 1984, pp. 752–755.

[14]Moore, B. C., "Principal Component Analysis in Linear Systems: Controllability, Observability, and Model Reduction," *IEEE Transactions on Automatic Control*, Vol. AC-26, No. 1, 1981, pp. 17–32.

[15]Hamdan, A. M. A., and Nayfeh, A. H., "Measure of Modal Controllability and Observability for First- and Second-Order Linear Systems," *Journal of Guidance, Control, and Dynamics*, Vol. 12, No. 3, 1989, pp. 421–428.

[16]Viswanathan, C. N., Longman, R. W., and Likins, P. W., "A Degree of Controllability Definition: Fundamental Concepts and Application of Modal Systems," *Journal of Guidance, Control, and Dynamics*, Vol. 7, No. 2, 1984, pp. 222–230.

[17]Lindberg, R. E., and Longman, R. W., "Optimization of Actuator Placement via Degree of Controllability Criteria Including Spillover Considerations," AIAA Paper 82-1435, San Diego, CA, Aug. 1982.

[18]Laub, A. J., and Little, J. N., "*MATLAB User's Guide—Control System Toolbox*, The Mathworks, Sherborn, MA, 1986.

Chapter 16

Component Model Reduction by Component Cost Analysis

R. E. Skelton*
Purdue University, West Lafayette, Indiana
and
R. Singh,† and J. Ramakrishnan‡
Dynacs Engineering Company, Palm Harbor, Florida

Introduction

LARGE-SCALE mechanical systems are composed of N interconnected dynamic components. This chapter seeks to simplify the model of each dynamic component so that the overall system is a manageable size. Models of components are reduced in order to approximate the response of the overall system and the nonworking constraint forces acting on each component. The technique is an extension of component cost analysis and decomposes the norm of the response vector and the vector of nonworking constraint forces into contributions of each coordinate of each component. In this way the coordinates can be ranked according to their contribution and deleted accordingly.

Let

$$M_i(\boldsymbol{q}_i)\ddot{\boldsymbol{q}}_i + H_i(\boldsymbol{q}_i,\dot{\boldsymbol{q}}_i) = \boldsymbol{T}_i + Z_i^*(\boldsymbol{q})\boldsymbol{f} \tag{1a}$$

$$Z(\boldsymbol{q})\dot{\boldsymbol{q}} + b(\boldsymbol{q},t) = 0 \tag{1b}$$

$$\boldsymbol{y} = P\boldsymbol{q} + R\dot{\boldsymbol{q}} \tag{1c}$$

describe the dynamics of the ith body in an N-body system, where $\boldsymbol{q}^*$ denotes the transpose of $\boldsymbol{q}$, $\boldsymbol{q}^* = (\boldsymbol{q}_1^*,\boldsymbol{q}_2^*, \ldots, \boldsymbol{q}_N^*)$, $\boldsymbol{q}_i \in \boldsymbol{R}^{N_i}$, $\boldsymbol{y}$ is an n_y vector

*Professor, School of Aeronautics and Astronautics.
†President.
‡Senior Project Engineer.

of outputs, and $\boldsymbol{T}_i$ is a N_i vector of forces and torques acting on a body i. For a tree-topological structure $\boldsymbol{f}$ is an $N-1$ vector of nonworking constraint forces and torques, and Eq. (1b) represents the constraint equations. Now linearize Eqs. (1) about $\bar{\boldsymbol{q}}(t)$ [we assume that $\bar{\boldsymbol{q}}(t)=0$ for convenience in notation] to get

$$M_i\ddot{\boldsymbol{q}}_i + D_i\dot{\boldsymbol{q}}_i + K_i\boldsymbol{q}_i = B_i\boldsymbol{w} + Z_i^*\boldsymbol{f} \tag{2a}$$

and differentiate Eq. (1b) with respect to time to get

$$Z\ddot{\boldsymbol{q}} + \dot{\boldsymbol{b}} = 0 \tag{2b}$$

where

$$D_i \triangleq \frac{\partial H_i}{\partial \dot{\boldsymbol{q}}_i}(\boldsymbol{q}_i,\dot{\boldsymbol{q}}_i)\big|_{q(t)}$$

$$K_i \triangleq \frac{\partial H_i}{\partial \boldsymbol{q}_i}(\boldsymbol{q}_i,\dot{\boldsymbol{q}}_i)\big|_{q(t)}, \qquad \boldsymbol{T}_i = B_i\boldsymbol{w}$$

$$Z = [Z_1 \ldots Z_N], \qquad q^* = [q_1^* \ldots q_N^*]$$

where for the purposes of open-loop analysis $\boldsymbol{w}(t)$ is an n_u vector of colored noise disturbance in the actuators. Actuators and their noise sources are not infinite-bandwidth devices. It is important to include their finite-bandwidth properties in any model prediction problem. The colored noise in our problem description is described by the Gauss-Markov model

$$\boldsymbol{w} = P_w\boldsymbol{q}_w + R_w\dot{\boldsymbol{q}}_w \tag{3}$$

$$M_w\ddot{\boldsymbol{q}}_w + D_w\dot{\boldsymbol{q}}_w + K_w\boldsymbol{q}_w = B_w\boldsymbol{w}'$$

where $\boldsymbol{w}'(t)$ is a zero mean white noise.

There may be redundant variables appearing in the component model (2a). This makes the overall model unobservable or uncontrollable. Elimination of these variables is a straightforward procedure in any finite-element code or any N-body simulation code.[1,2] This subject will not be expanded here. We shall assume that redundant variables have already been eliminated in Eqs. (2). The physical meaning of the N_i vector $\boldsymbol{q}_i$ is chosen by the analyst. If the analyst chooses to model body i by modes, then $[\{M_i = I,\ D_i = 2\zeta_i\Omega_i,\ K_i = \Omega_i^2\}]$, and B_i is made up of mode shapes or mode slopes of the chosen type (fixed-fixed, fixed-free, free-free, etc.). If the analyst chooses to model body i as a continuum, then

$$M_i = \int_{E_i} \Psi_i^*(r)\rho_i\Psi_i(r)\,\mathrm{d}r \tag{4a}$$

$$K_i = \int_{E_i} \Psi_i^*(r)\underline{K}_i(r)\Psi_i(r)\,\mathrm{d}r \tag{4b}$$

where the displacement function satisfies the Ritz-type approximation

$$\mu_i(r,t) \approx \Psi_i(r)\boldsymbol{q}_i(t) \tag{4c}$$

where $\underline{K}_i(r)$ is the stiffness operator, $\boldsymbol{q}_i$ is a finite-dimensional N_i vector, and $\Psi_i(r)$ is a matrix of shape functions, selected a priori. Quite often N_i must be reduced so that the overall system of order $n \triangleq \Sigma_{i=1}^{N} N_i$ is a reasonable number for simulation or analytical purposes. Moreover, since the components of the system (bodies N_i, $i = 1,2, \ldots, N$) are often manufactured by different companies using different types of models for each component, it is desirable to have model reduction methods that can select which of the $(\boldsymbol{q}_i)_k$, $i = 1,2, \ldots, N_i$ to delete, considering the dynamic interactions among components. No coordinate transformations are allowed in this case.

One may also choose coordinate transformations $\boldsymbol{T}_i$ ($\boldsymbol{q}_i = \boldsymbol{T}_i\boldsymbol{\eta}_i$) and in these new coordinates delete from the N_i vector η_i the least significant variables.

Of course, the difference in these two approaches is the freedom to select different coordinates than $\boldsymbol{q}_i$ before making truncation decisions. Given this additional freedom, we would expect a better answer, but it comes with the price of additional numerical work, the coordinate transformations $\boldsymbol{T}_i$.

This chapter provides some improvements to model reduction by component cost analysis.[3–6] First we add actuator dynamics. To truncate in the original coordinates, we use the component cost analysis with augmented outputs to include the nonworking constraint forces $\boldsymbol{f}$. To truncate in transformed coordinates, we use modal cost analysis[3,4] with the extended output including $\boldsymbol{f}$. This requires transformation of the component to modal coordinates (unless component models are given in their modal form). Case studies are provided.

We shall restrict our attention to constant Z and b in Eq. (1c). From Eq. (2a) write the complete system in the matrix form

$$M\ddot{\boldsymbol{q}} + D\dot{\boldsymbol{q}} + K\boldsymbol{q} = B\boldsymbol{w} + Z^*\boldsymbol{f} \tag{5}$$

$$\boldsymbol{y} = P\boldsymbol{q} + R\dot{\boldsymbol{q}} \tag{6}$$

$$Z\dot{\boldsymbol{q}} = 0 \tag{7}$$

where $\boldsymbol{y}$ is an n_y vector of outputs of interest in the system:

$$M = \text{block diag}[\ldots\, M_i \ldots]$$

$$K = \text{block diag}[\ldots\, K_i \ldots]$$

$$B = \text{block}\begin{bmatrix} \vdots \\ B_i \\ \vdots \end{bmatrix}$$

$$Z = \text{block}[\ldots\, Z_i \ldots]$$

$$[P,\, R] = [\ldots\, P_i \ldots,\ \ldots\, R_i \ldots]$$

Solving Eqs. (5) and (7) for f yields

$$f = (ZM^{-1}Z^*)^{-1}[ZM^{-1}(D\dot{q} + Kq - Bw)] \tag{8}$$

the inverse $(ZM^{-1}Z)^{-1}$ exists if the constraints (7) are linearly independent. In this case Eqs. (5–8) lead to the system description

$$M\ddot{q} + \hat{D}\dot{q} + \hat{K}q = \hat{B}w \tag{9a}$$

$$y \triangleq \begin{pmatrix} y \\ f \end{pmatrix} = \hat{P}q + \hat{R}\dot{q} + Jw \tag{9b}$$

where

$$\hat{D} \triangleq [I - Z^*[ZM^{-1}Z^*]^{-1}ZM^{-1}]D$$

$$\hat{K} \triangleq [I - Z^*[ZM^{-1}Z^*]^{-1}ZM^{-1}]K$$

$$\hat{B} \triangleq [I - Z^*[ZM^{-1}Z^*]^{-1}ZM^{-1}]B$$

$$\hat{P} = \begin{bmatrix} P \\ [ZM^{-1}Z^*]^{-1}ZM^{-1}K \end{bmatrix}, \quad \hat{R} = \begin{bmatrix} R \\ [ZM^{-1}Z^*]^{-1}ZM^{-1}D \end{bmatrix}$$

$$J = \begin{bmatrix} 0 \\ -[ZM^{-1}Z^*]^{-1}ZM^{-1}B] \end{bmatrix}$$

The test for controllability and observability are given by the following known results.

Theorem 1. Let $M > 0$; then Eq. (9) is controllable if and only if

$$\text{rank}[\lambda^2 M + \lambda\hat{D} + \hat{K}, \hat{B}] = n \tag{10a}$$

for all eigenvalues. System (9) is observable if and only if

$$\text{rank}\begin{bmatrix} \lambda^2 M + \lambda\hat{D} + \hat{K} \\ \hat{P} + \lambda\hat{R} \end{bmatrix} = n \tag{10b}$$

for all eigenvalues λ. The ith mode of the system is controllable (observable) if and only if Eq. (10) holds for λ = eigenvalue of the ith mode.[7]

A state space realization of Eq. (9) is given by

$$\dot{x} = Ax + Bw \tag{11a}$$

$$y = Cx + Jw \tag{11b}$$

where

$$A=\begin{bmatrix} 0 & I \\ -M^{-1}\hat{K} & -M^{-1}\hat{D} \end{bmatrix}, \quad x=\begin{pmatrix} q \\ \dot{q} \end{pmatrix}, \quad B=\begin{bmatrix} 0 \\ \hat{B} \end{bmatrix}$$

$$C=[\hat{P}\hat{R}]$$

Component Cost Analysis

Most of the available model reduction theories are posed for the overall system and do not address the question of component model reduction (see, for example, Refs. 8–14). Some exceptions are the use of canonical correlation analysis[15] and the component cost analysis.[3–6] Also, the analysis of constrained dynamic systems has received little attention in the literature on controls and model reduction.[17]

The idea of the component cost analysis is to decompose a norm of the response into contributions from each coordinate of the given state space realization

$$\dot{x}=Ax+Dw \tag{12a}$$

$$y=Cx \tag{12b}$$

where $w(t)$ is a zero-mean white noise with intensity W, and the response norm is given by the expected value of the Euclidean norm of the output,

$$V=\lim_{t\to\infty} Ey^*Qy \tag{13}$$

The decomposition of V into contributions of x_i is defined by

$$V_i=\lim_{t\to\infty} E\frac{\partial y^*Qy}{\partial x_i}x_i \tag{14}$$

and V_i is finite if there are no unstable modes observable in either y or x_i. In this case V_i is given by

$$V_i=[XC^*QC]_{ii} \tag{15}$$

where X is the state covariance satisfying

$$0=XA^*+AX+DWD^* \tag{16}$$

The x_i in Eq. (14) can be the entire state of the ith component (say, the ith body in an N-body problem), or the ith state variable within a particular component.

This paper extends these results in two ways. First the nonworking constraint forces f are added as outputs so that V takes the form

$$V=\lim_{t\to\infty} E(y^*Q_yy+f^*Q_ff) \tag{17}$$

The choice of weight $Q_f = 0$ yields the traditional result, and the choice of large Q_f yields reduced models that focus on the preservation of the nonworking constraint forces $\boldsymbol{f}$. It is the attention to $\boldsymbol{f}$ that makes the new approach more appropriate to the N-body problem. By extending the output definition to include $\boldsymbol{y}$ and $\boldsymbol{f}$, Eqs. (12) become

$$\dot{x} = Ax + Bw \tag{18a}$$

$$y = \begin{pmatrix} y \\ f \end{pmatrix} = Cx \tag{18b}$$

The second extension of the component cost is to replace the C matrix by the matrix Q_q defined by

$$Q_q \triangleq \begin{bmatrix} CA^{-1} \\ C \\ CA \\ \vdots \\ CA^{q-1} \end{bmatrix} \tag{19}$$

In this way, both controllability (through X) and observability information (through Q_q) dictate V_i. The choice of $\boldsymbol{q}$ weights the observability information used in the model reduction decisions. The inclusion of CA^{-1} includes information about the steady-state response to a unit step input (which is $y_{ss} = -CA^{-1}B$).

The $2n \times 2n$ Liapunov equation [Eq. (16)] must be solved to compute component costs. Analytical solutions of this equation are available in terms of modal data of the system. This result is summarized as follows.

Theorem 2. Assuming

$$Ew(t) = 0, \qquad Ew(t)w^*(\tau) = W\delta(t-\tau)$$

$$Ew(t)\begin{pmatrix} \eta(0) \\ \dot{\eta}(0) \end{pmatrix}^* = 0 \tag{20}$$

$$0 < \zeta_i \leqslant 1, \qquad \omega_i > 0 \qquad \text{for all} \qquad i = 1,2,\ldots$$

the system

$$\ddot{\eta}_i + 2\zeta_i\omega_i\dot{\eta}_i + \omega_i^2\eta_i = b_i^* w, \qquad i = 1 \rightarrow N \tag{21}$$

has covariance

$$X \triangleq Exx^*, \qquad x^* \triangleq (\eta_1,\dot{\eta}_1,\eta_2,\dot{\eta}_2,\eta_3,\dot{\eta}_3 \ldots \eta_N,\dot{\eta}_N) \tag{22}$$

as follows. The 2×2 diagonal blocks are

$$X_{\alpha\alpha} = \frac{b_\alpha^* W b_\alpha}{4\zeta_\alpha \omega_\alpha^3} \begin{bmatrix} 1 & 0 \\ 0 & \omega_\alpha^2 \end{bmatrix}, \qquad \alpha = 1,2, \ldots, N \tag{23a}$$

The 2×2 off-diagonal blocks of X are

$$X_{\alpha\beta} = \frac{b_\alpha^* W b_\beta}{\Delta_{\alpha\beta}} \begin{bmatrix} 2(\zeta_\alpha \omega_\alpha + \zeta_\beta \omega_\beta) & \omega_\alpha^2 - \omega_\beta^2 \\ -(\omega_\alpha^2 - \omega_\beta^2) & 2\omega_\alpha \omega_\beta (\zeta_\alpha \omega_\beta + \zeta_\beta \omega_\alpha) \end{bmatrix}$$

$$\alpha,\beta = 1,2, \ldots, N \tag{23b}$$

where

$$\Delta_{\alpha\beta} \triangleq 4\omega_\alpha \omega_\beta (\zeta_\alpha \omega_\alpha + \zeta_\beta \omega_\beta)(\zeta_\alpha \omega_\beta + \zeta_\beta \omega_\alpha) + (\omega_\alpha^2 - \omega_\beta^2)^2 \tag{24}$$

(See Ref. 19.)

Corollary to Theorem 2. Assuming Eq. (20), the system [Eq. (21)] has covariance $X = Exx^*$,

$$X = Exx^*, \qquad x^* \triangleq (\eta_1, \eta_2, \ldots, \eta_N, \dot{\eta}_1, \dot{\eta}_2, \ldots, \dot{\eta}_N) = (\eta^*, \dot{\eta}^*) \tag{25}$$

as follows. The $N \times N$ blocks are

$$X = \begin{bmatrix} X_{\eta\eta} & X_{\eta\dot{\eta}} \\ X_{\eta\dot{\eta}}^* & X_{\dot{\eta}\dot{\eta}} \end{bmatrix} \tag{26}$$

where the $N \times N$ matrices are

$$X_{\eta\eta} = \begin{bmatrix} \dfrac{b_1^* W b_1}{4\zeta_1 \omega_1^3} & \dfrac{2b_1^* W b_2 (\zeta_1 \omega_1 + \zeta_2 \omega_2)}{\Delta_{12}} & \dfrac{2b_1^* W b_3 (\zeta_1 \omega_1 + \zeta_3 \omega_3)}{\Delta_{13}} & \cdots \\ & \dfrac{b_2^* W b_2}{4\zeta_2 \omega_2^3} & \dfrac{2b_2^* W b_3 (\zeta_2 \omega_2 + \zeta_3 \omega_3)}{\Delta_{23}} & \cdots \\ \text{symmetric} & & \dfrac{b_3^* W b_3}{4\zeta_3 \omega_3^3} & \ddots \end{bmatrix} \tag{27}$$

$$X_{\eta\dot{\eta}} = \begin{bmatrix} 0 & \dfrac{b_1^* W b_2 (\omega_1^2 - \omega_2^2)}{\Delta_{12}} & \dfrac{b_1^* W b_3 (\omega_1^2 - \omega_3^2)}{\Delta_{13}} & \cdots \\ & 0 & \dfrac{b_2^* W b_3 (\omega_2^2 - \omega_3^2)}{\Delta_{23}} & \cdots \\ \text{skew-symmetric} & & 0 & \ddots \end{bmatrix} \tag{28}$$

$$X_{\dot\eta\dot\eta} = \begin{bmatrix} \dfrac{b_1^* W b_1}{4\zeta_1\omega_1} & \dfrac{2b_1^* W b_2(\zeta_1\omega_2+\zeta_2\omega_1)\omega_1\omega_2}{\Delta_{12}} & \dfrac{2b_1^* W b_3(\zeta_1\omega_3+\zeta_3\omega_1)\omega_1\omega_3}{\Delta_{23}} & \cdots \\ & \dfrac{b_2^* W b_2}{4\zeta_2\omega_2} & \dfrac{2b_2^* W b_3(\zeta_2\omega_3+\zeta_3\omega_2)\omega_2\omega_3}{\Delta_{23}} & \cdots \\ \text{Symmetric} & & \dfrac{b_3^* W b_3}{4\zeta_3\omega_3} & \ddots \end{bmatrix} \tag{29}$$

where $\Delta_{\alpha\beta}$ is defined by Eq. (24).

The proof of *Theorem* 2 and its corollary is by construction. A straightforward substitution into $0 = XA^* + AX + BWB^*$ will verify these expressions for X, where $\dot{x} = Ax + Bw$. The solution for X in the desired (new) coordinates if $T^{-1}XT^{-*}$, where X is from Eq. (22) or (26) and $x_{\text{old}} = Tx_{\text{new}}$ is the transformation to modal coordinates of Eq. (21). The operative assumption in *Theorem* 2 is the existence of modal data [Eq. (21)].

$$V = \text{tr}[XO_q^* Q_q O_q] = \sum_{i=1}^{N} V_i \tag{30}$$

where

$$V_i = V_{q_i} + V_{\dot{q}_i}$$

$$O_q = \begin{bmatrix} CA^{-1} \\ C \\ CA \\ \vdots \\ CA^{q-1} \end{bmatrix}, \qquad Q_q = \begin{bmatrix} Q\rho_0 & & & & \\ & Q\rho_1 & & & \\ & & Q\rho_2 & & \\ & & & \ddots & \\ & & & & Q\rho_q \end{bmatrix} \tag{31}$$

where ρ_i, $i = 0,1,2, \ldots, q$ are scalar multipliers of Q to weight each term, and where the CA^{-1} term in O_q helps preserve steady-state step responses. The C term helps preserve the mean-squared value of the output vector y in Eqs. (9) and (11). The terms $CA, \ldots, CA^{q-1}$ help preserve the first q time rate of changes in $y(t)$. The overall effect, of course, is a combination of each of these terms; from Eqs. (30) and (31),

$$V = \sum_{i=0}^{q} \rho_i \, \text{tr} CA^{i-1} X (CA^{i-1})^* Q = \sum_{i=1}^{n} V_i \tag{32a}$$

and

$$V_i = \sum_{\alpha=1}^{q} \rho_\alpha [XCA^{\alpha-1*} QCA^{\alpha-1}]_{ii} \tag{32b}$$

is the cost decomposition of the performance measure [Eq. (17)]. To solve model reduction problem I, we shall rank the coordinates of the components by the magnitude of V_i given by Eq. (32). In the next section we shall find an analytical expression for V_i in modal coordinates. This will require a coordinate transformation and hence will be applicable only to model reduction problem II.

The cost decomposition [Eq. (30)] is a very general one that reduces to other published results for special choices of weight Q_q and $\boldsymbol{q}$.

Lemma. Let the $n \times n$ matrix e^{At} be written in its series from

$$e^{At} = \sum_{i=0}^{n-1} \alpha_i(t) A^i$$

where $\alpha_i(t)$ may be computed by the Fadeeva algorithm.[18] Let $q = n$ in Eq. (30) with $\rho_i = 1$, $i = 1 \rightarrow \boldsymbol{q}$, in Eq. (31), and let

$$Q_q \triangleq \begin{bmatrix} 0_{n_y x n_y} & 0 \\ 0 & \Psi \end{bmatrix}, \quad \Psi \triangleq \int_0^\infty \begin{bmatrix} \alpha_0(t)I \\ \alpha_1(t)I \\ \vdots \\ \alpha_{n-1}(t)I \end{bmatrix} Q[\alpha_0(t)I, \ldots, \alpha_{n-1}(t)I] \, dt \tag{33}$$

then the component cost V_i in Eq. (30) becomes

$$V_i = [XK]_{ii} \tag{34}$$

where X and K satisfy

$$0 = XA^* + AX + DWD^* \tag{35}$$

$$0 = KA + A^*K + C^*QC \tag{36}$$

Proof. The proof requires only that we establish the fact

$$K = [C^*, A^*C^*, \ldots, A^{n-1*}C^*]\Psi \begin{bmatrix} C \\ \vdots \\ CA^{n-1} \end{bmatrix} \tag{37}$$

Returning to the definition of the observability grammian K,

$$K = \int_0^\infty e^{A^*t} C^*QC \, e^{At} \, dt, \qquad e^{At} = \sum_{i=0}^{n-1} \alpha_i(t) A^i \tag{38}$$

Substitution of e^{At} into the integral yields

$$K = \int_0^\infty \left[\sum_{i=0}^{n-1} CA(\alpha_i t)\right] Q \left[\sum_{i=0}^{n-1} CA^i\alpha_i(t)\right] \mathrm{d}t \tag{39}$$

Multiplying the three matrices in Eq. (37) yields Eq. (39), completing the proof.

The reader of model reduction literature will recognize that the balanced coordinates of Moore[9] are defined by the property $X = K = \mathrm{diag}[\ldots \sigma_i^2 \ldots]$. Hence, if the system were first placed in balanced coordinates, the component costs, V_i in Eqs. (30–32) with weight [Eq. (33)] are simply the singular values of XK. Hence, under the conditions of the lemma the reduced-order model obtained by component cost analysis [Eq. (30)] is precisely that given by the balancing method of Moore.[9] Equation (30) does *not* require balanced coordinates, however, and can be applied to *component* reduction. In the original (component) coordinates the system is not balanced and XK is not diagonal. Nonetheless, Eq. (34) yields component costs that may be useful for model reduction.

Model Cost Approximation to Including Nonworking Constraint Forces

Now suppose the given coordinates in Eqs. (2) are component modal coordinates. That is, suppose

$$M_i = I, \qquad D_i = 2\zeta_i\Omega_i, \qquad K_i = \Omega_i^2 = \mathrm{diag}[\ldots \omega_{i_\alpha}^2 \ldots] \qquad \zeta_i = \mathrm{diag}[\ldots \zeta_{i_\alpha} \ldots]$$

Then Eq. (9) becomes

$$\ddot{q} + 2\hat{\zeta}\hat{\Omega}\dot{q} + \hat{\Omega}^2 q = \hat{B}(u + w) \tag{40a}$$

$$y \triangleq \begin{pmatrix} y \\ f \end{pmatrix} = \hat{P}q + \hat{R}\dot{q} + J(u + w) \tag{40b}$$

where

$$\hat{D} = 2\hat{\zeta}\hat{\Omega} \triangleq [I - Z^*[ZZ^*]^{-1}Z]2\zeta\Omega$$

$$\hat{K} = \hat{\Omega}^2 \triangleq [I - Z^*[ZZ^*]^{-1}Z]\Omega^2$$

$$\hat{B} = [I - Z^*[ZZ^*]^{-1}Z]B$$

$$\hat{P} = \begin{bmatrix} P \\ [ZZ^*]^{-1}Z\Omega^2 \end{bmatrix}, \qquad \hat{R} = \begin{bmatrix} R \\ -[ZZ^*]^{-1}ZB \end{bmatrix} \qquad J = \begin{bmatrix} 0 \\ -[ZZ^*]^{-1}ZB \end{bmatrix}$$

The modal cost approximation is motivated by the following fact.

Theorem 3. Let

$$\lambda_i = -\zeta_i\omega_i \pm j\omega_i\sqrt{1-\zeta_i^2}$$

be the eigenvalues of

$$\ddot{\boldsymbol{q}} + D\dot{\boldsymbol{q}} + K\boldsymbol{q} = 0$$

Then, to a first-order approximation, the eigenvalues of the system

$$\ddot{\boldsymbol{q}} + [D + \Delta D]\dot{\boldsymbol{q}} + [K + \Delta K]\boldsymbol{q} = 0$$

are $\lambda_i + \Delta\lambda_i$, where

$$\Delta\lambda_i = -\frac{1}{2}[\Delta D]_{ii} \mp j\frac{1}{2\sqrt{1-\zeta_i^2}}\left[\zeta_i[\Delta D]_{ii} + \frac{1}{\omega_i}[\Delta K]_{ii}\right]$$

(See Ref. 7.)

Theorem 4. Consider the system with noise input $Eu = 0$, $Eu(t)u^*(\tau) = Q\delta(t-\tau)$,

$$\ddot{\boldsymbol{\eta}}_i + 2\zeta_i\omega_i\dot{\boldsymbol{\eta}}_i + \omega_i^2\boldsymbol{\eta}_i = b_i^*u$$

$$y = \sum_{i=1}^{N} b_i\boldsymbol{\eta}_i \qquad \text{or} \qquad y = \sum_{i=1}^{N} b_i\dot{\boldsymbol{\eta}}_i$$

with small damping ζ_i, cost function V, and modal cost V_i,

$$V = \lim_{t\to\infty} Ey^*Qy = \sum_{i=1}^{N} V_i$$

Now suppose that, instead of white noise, the control is output feedback

$$u = Qy$$

Then the first-order approximations of the change in eigenvalues from the open-loop values are related to the open-loop modal costs V_i by

$$|\Delta\lambda_i|^2 = \zeta_i\omega_i V_i$$

(See Ref. 7, p. 453.)

These two results suggest that a first-order approximation of the eigenvalues (of a closed-loop system with collocated output feedback) will also yield a first-order approximation to the modal cost. This connection between open- and closed-loop calculations provides the motivation for the modal cost approximation for selecting modes of body i. When O_q is just the C matrix, then the approximate modal cost of mode α of component i is

$$V_{i_\alpha} = \frac{[\hat{B}_i W B_i^*]_{\alpha\dot{\alpha}}[(\hat{P}_i^* Q \hat{P}_i)_{\alpha\alpha} + (\hat{\Omega}_i^2)_{\alpha\alpha}(\hat{R}_i^* Q \hat{R}_i)_{\alpha\alpha}]}{4\hat{\zeta}_{i_\alpha}\hat{\omega}_{i_\alpha}^3} \tag{41}$$

where

$$\hat{\Omega}_i^2 \triangleq [I - Z_i^*[ZZ^*]^{-1}Z_i]\Omega_i^2, \qquad \hat{\omega}_{i_\alpha} \triangleq (\hat{\Omega}_i^2)_{\alpha\alpha} \tag{42a}$$

$$2\hat{\zeta}_i\hat{\Omega}_i \triangleq [I - Z_i^*(ZZ^*)^{-1}Z_i]2\zeta_i\Omega_i \tag{42b}$$

$$\hat{\zeta}_\alpha \triangleq [I - Z_i^*(ZZ^*)^{-1}Z_i]\zeta_i\Omega_i\Omega_i^+ \tag{43}$$

where Ω_i^+ indicates a pseudoinverse of $\hat{\Omega}_i$ in case the inverse does not exist.

Now let the output equation of Eqs. (40) be rewritten to include steady-state and higher rates of $\boldsymbol{y}$ and $\boldsymbol{f}$ in the spirit of Eq. (31). In the limit, as the damping becomes arbitrarily small for $x^* = (\boldsymbol{q}^*\dot{\boldsymbol{q}}^*)$, Eqs. (31) and (40) lead to (where terms are shown up to $\boldsymbol{q} = 7$)

$$O_q = \begin{bmatrix} CA^{-1} \\ C \\ CA \\ CA^2 \\ CA^3 \\ CA^4 \\ CA^5 \\ CA^6 \\ \vdots \end{bmatrix} = \begin{bmatrix} \hat{R} & -\hat{P}\hat{\Omega}^{-2} \\ \hat{P} & \hat{R} \\ -\hat{R}\hat{\Omega}^2 & \hat{P} \\ -\hat{P}\hat{\Omega}^2 & -\hat{R}\hat{\Omega}^2 \\ \hat{R}\hat{\Omega}^4 & -\hat{P}\hat{\Omega}^2 \\ P\hat{\Omega}^4 & \hat{R}\hat{\Omega}^4 \\ -\hat{R}\hat{\Omega}^6 & \hat{P}\hat{\Omega}^4 \\ -\hat{P}\hat{\Omega}^6 & -\hat{R}\hat{\Omega}^6 \\ \vdots & \vdots \end{bmatrix} \tag{44}$$

Now compute V_i from Eq. (32). The α column of $\hat{P}_i$ is written p_α, and the α column of $\hat{R}_i$ is written r_α. The $(\hat{P}_{i\alpha})$ in Eqs. (40) and (41) is replaced by the α column of the ith block of columns in $\hat{P}$. For lightly damped systems with diagonally dominant $\hat{\Omega}^2$, Eq. (41) becomes

$$V_{i_\alpha}(p) = \frac{b_\alpha^* W b_\alpha p_\alpha^* Q p_\alpha \rho(\alpha)}{4\zeta_\alpha\omega_\alpha^3}$$

when $r_\alpha = 0$. When $p_\alpha = 0$,

$$V_{i_\alpha}(r) = \frac{b_\alpha^* W b_\alpha r_\alpha^* Q r_\alpha \rho(\alpha)}{4\zeta_\alpha\omega_\alpha}$$

where

$$\rho(\alpha) \triangleq \sum_{k=0}^{q} \omega_\alpha^{2(k-1)} \rho_k$$

When neither r_α nor p_α is zero, then the total modal cost for mode α is $V_i(p) + V_i(r)$.

The p,q-Markov COVER

Write the component model (2) in the form

$$\begin{aligned} \ddot{\boldsymbol{q}}_i &+ [M_i^{-1} - M_i^{-1} Z_i^* (ZM^{-1}Z^*)^{-1} Z_i M_i^{-1}] D_i \dot{\boldsymbol{q}}_i \\ &+ [M_i^{-1} - M_i^{-1} Z_i^* (ZM^{-1}Z^*)^{-1} Z_i M_i^{-1}] K_i \boldsymbol{q}_i \\ &= [M_i^{-1} - M_i^{-1} Z_i^* (ZM^{-1}Z^*)^{-1} Z_i M_i] B_i u \\ &+ \sum_{k \neq i} [M_i^{-1} - M_i^{-1} Z_i^* (ZM^{-1}Z^*)^{-1} Z_k M_k^{-1}][-D_k \dot{\boldsymbol{q}}_k - K_k \boldsymbol{q}_k + B_k(u + \boldsymbol{w})] \end{aligned} \tag{45}$$

$$y = [\hat{P}_i, \hat{R}_i] \begin{pmatrix} \boldsymbol{q}_i \\ \dot{\boldsymbol{q}}_i \end{pmatrix} + \sum_{k \neq i} [\hat{P}_k \quad \hat{R}_k] \begin{pmatrix} \boldsymbol{q}_k \\ \dot{\boldsymbol{q}}_k \end{pmatrix} + J(u + w)$$

where we have used Eq. (9). The form of this equation is

$$\ddot{\boldsymbol{q}}_i + \Gamma_i D_i \dot{\boldsymbol{q}}_i + \Gamma_i K_i \boldsymbol{q}_i = \Gamma_i B_i u + e_{ix} \tag{46}$$

$$y = [\hat{P}_i, \hat{R}_i] \begin{pmatrix} \boldsymbol{q}_i \\ \dot{\boldsymbol{q}}_i \end{pmatrix} + e_{iy}$$

with obvious definitions for Γ_i and e_x, e_y. Ignoring e_x and e_y, the state space model is

$$\dot{x}_i = A_i x_i + B_i u, \qquad A_i = \begin{bmatrix} 0 & I \\ -\Gamma_i K_i & -\Gamma_i D_i \end{bmatrix}, \qquad x_i = \begin{pmatrix} \boldsymbol{q}_i \\ \dot{\boldsymbol{q}}_i \end{pmatrix} \tag{47}$$

The purpose of ignoring e_{ix} and e_{iy} is to determine how to reduce the model of only component i. Hence, model (47) will be reduced for each $i = 1,2, \ldots, N$. We shall use the following result from Ref. 16.

Theorem 5. Let w be a zero mean white noise with intensity W and uncorrelated with $x(0)$. Then

$$\dot{x}_R = A_R x_R + B_R w$$

$$y_R = C_R x_R$$

matches the $2(p+q)$ matrix parameters

$$CA^kB = C_RA_R^kB_R, \qquad k = -p, ..., 0,1, ..., q-1$$

$$CA^kXC^* = C_RA_R^kX_RC_R^*$$

if

$$A_R = LAR, \qquad B_R = LB, \qquad C_R = CR$$

$$L \triangleq U^*O_{qp}, \qquad R = XL^*[LXL^*]^{-1}$$

$$O_{pq} \triangleq \begin{bmatrix} CA^{-p} \\ \vdots \\ C \\ CA \\ \vdots \\ CA^{q-1} \end{bmatrix} = [U \quad U_b]\begin{bmatrix} \Sigma & 0 \\ 0 & 0 \end{bmatrix}\begin{bmatrix} V_a^* \\ V_b^* \end{bmatrix}$$

where Σ contains the nonzero singular values of O_{pq}, and X satisfies

$$0 = XA^* + AX + BWB^*$$

The use of *Theorem* 5 on each body equation (47) yields reduced-order models of each linearized body while preserving the matrices

$$CA^kB = C_RA_R^kB_R, \qquad k = -p, ..., 0,1, ..., q-1$$

$$CA^kXC^* = C_RA_R^kX_RC_R^*$$

(This is the definition of a p,q-Markov COVER.) This is equivalent to matching the first q derivatives of the impulse response at $t = 0$ and the first p derivatives at $t = \infty$, and the first q derivatives of the autocorrelation $Ey(t+\tau)y^*(\tau)$ at $t = 0$ and the first p derivatives at $t = \infty$. As a consequence, the covariance matrix of the output vector y is preserved for any choice of $q \geqslant 1$, if the e_{ix} and e_{iy} terms are negligible.

Numerical Example

As a simple example of the procedure, consider two Euler-Bernoulli beams representing a two-body system. The first beam is pinned to the inertial frame, and the second beam is connected to the first beam via a frictionless pin (Fig. 1). The minimum dimension multibody analysis program TREETOPS is used for the time-history simulation. From the

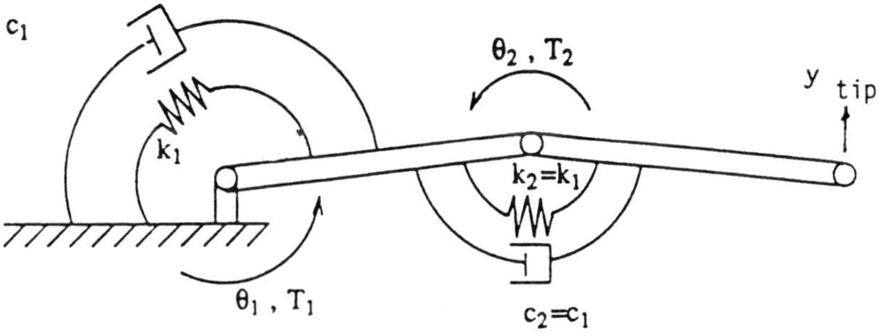

Fig. 1 Simple Multibody Model.

viewpoint of a TREETOPS model, the first beam is modeled using pinned-free modes and the second beam using free-free modes. The material constants and parameters of relevance are given in Table 1.

In the presence of an impulse excitation, the beam in Fig. 1 will undergo large (rigid-body) angle rotations. To avoid this, two soft springs and dashpots are introduced at the two pins. The spring constant is $k_{\theta 1} = 7.924$ N/rads and $k_{\theta 2} = 4.173$ N/rads, and the damping coefficients are $c_{\theta 1} = 0.162266$ N-s/rad and $c_{\theta 2} = 0.05893$ N-s/rad.

Each beam is modeled with 10 elastic modes. The linearized A,B,C,D quadruples are obtained from TREETOPS. Using the A,B,C,D quadruples, the design package ranks and selects the elastic coordinates to be retained. White noise process is a mathematical idealization. In order to make the design more realistic, the white noise is passed through a filter with a bandwidth equal to two-thirds the maximum system frequency to obtain a colored signal that is used to drive the actuator. At present, the constraint forces $\boldsymbol{f}$ do not form a part of the augmented output vector. However, the colored inputs are augmented with the system outputs to form the $\boldsymbol{y}$ vector. The inputs are moments applied at pins 1 and 2, and the outputs are the rotation at pin 1 (θ_1), the relative rotation at pin 2 (θ_2), and the tip deflection (y). The design is performed assuming white noise inputs. In the time-history simulation, however, the actuator inputs are impulses applied simultaneously. The design criteria together with the definitions are summarized in the following.

Table 1 Two-beam data

Parameter	Beam 1	Beam 2
EI, N-m^2	1000	1000
m, kg	1	1
ω_{low}, rad/s	487.5	707.5
ω_{high}, rad/s	32,787	34,409
L, m	1	1

Definitions:

$$\dot{x} = Ax + Bw$$

$$\tilde{y} = Cx \qquad C = \begin{bmatrix} C_y \\ C_u \\ C_f \end{bmatrix}$$

$$y = (\tilde{y}^T u^T f^T)^T, \qquad y \in R^{\bar{n}_y} = \bar{n}_y \triangleq n_y + n_u + n_f$$

$$V = \text{tr}[X O_{pq} Q_q^* O_{pq}]$$

$$O_{pq} = \begin{bmatrix} CA^{-p} \\ C \\ CA \\ \vdots \\ CA^{q-1} \end{bmatrix} \qquad Q_q = \begin{bmatrix} Q_{\rho_0} & & & \\ & Q_{\rho_1} & & \\ & & \ddots & \\ & & & Q_{\rho_q} \end{bmatrix}$$

$$Q \in R^{\bar{n}_y \times \bar{n}_y}, \qquad Q = \begin{bmatrix} Q_y & & \\ & Q_u & \\ & & Q_f \end{bmatrix}$$

To evaluate the reduced models, we compute the following:
1) steady-state error to step inputs:

$$\lim_{t \to \infty} [y(t) - y_R(t)]$$

2) covariance error:

$$|Y| = |Eyy^* - Ey_R y_R^*|$$

3) least-squares error:

$$\text{tr}\, YQ = \text{tr}[E(y - y_R)^T Q (y - y_R)]$$

4) impulse response of full and reduced-order models.

Table 2 is a list of cases considered in the analysis.

The ranking of the elastic modes for cases I and II are identical. It should be noted that the nonworking constraint forces f have not been penalized in the design. Cases I and II represent the standard component cost analysis with differing choices of weights. Cases III and IV represent component cost analysis with the modification suggested in Eq. (28). The elastic mode ranking is identical in cases 3 and 4. Tables 3 and 4 list the

Table 2 Design specifications

p	q	ρ_1	Qy	Qu	Qf	Method
0	1	1	I	I	0	CCA case I
0	1	1	$Q = \begin{bmatrix} (C_{yu}XC^*_{yu})^{-1} & 0 \\ 0 & 0 \end{bmatrix}$			CCA case II
1	2	1	I	0	0	CCA case III
1	2	1	$Q = \begin{bmatrix} (C_{yu}XC^*_{yu})^{-1} & 0 \\ 0 & 0 \end{bmatrix}$			CCA case IV

Table 3 Mode ranking (case I)

Body 1	Body 2
1	
3	
	1
5	
7	
	2
	3
9	
	4
	5
	6
	7
	8
	9
	10
2	
4	
6	
8	
10	

Table 4 Mode ranking (case III)

Body 1	Body 2
	3
	4
	2
	5
	1
5	
3	
	6
7	
	7
1	
	8
4	
6	
2	
	9
9	
8	
	10
10	

Table 5 Errors in the 14-mode model

	Case I	Case III
Impulse		Better
Steady-state error		
Input u_1	Better	
Input u_2		Better
Covariance error		Better
Least-squares error		Better

Table 6 Errors in the 10-mode model

	Case I	Case III
Impulse		Better
Steady-state error		Better
Input u_1		Better
Input u_2		Better
Covariance error		Better
Least-squares error		Better

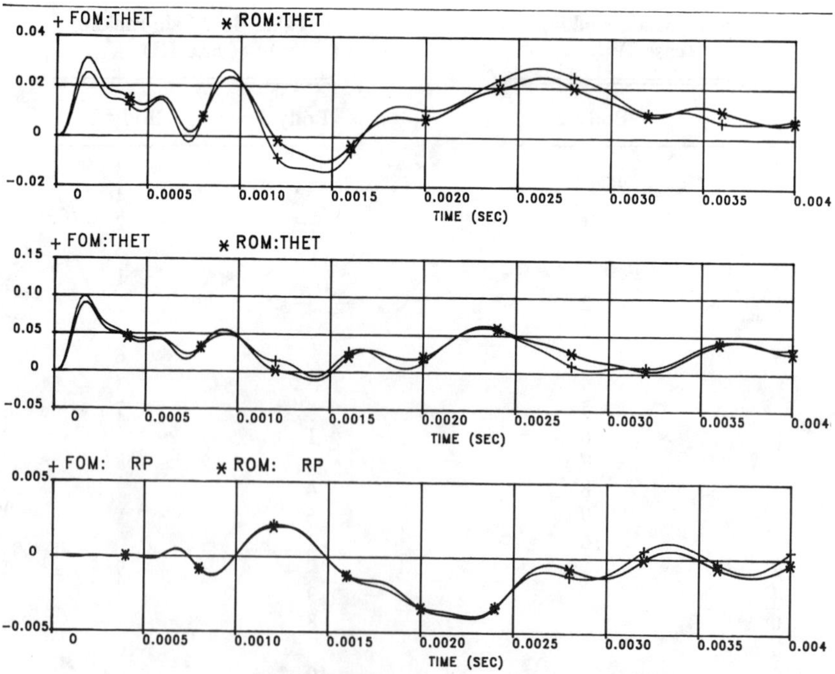

Fig. 2 Case I: Model reduction (20 vs 14 modes).

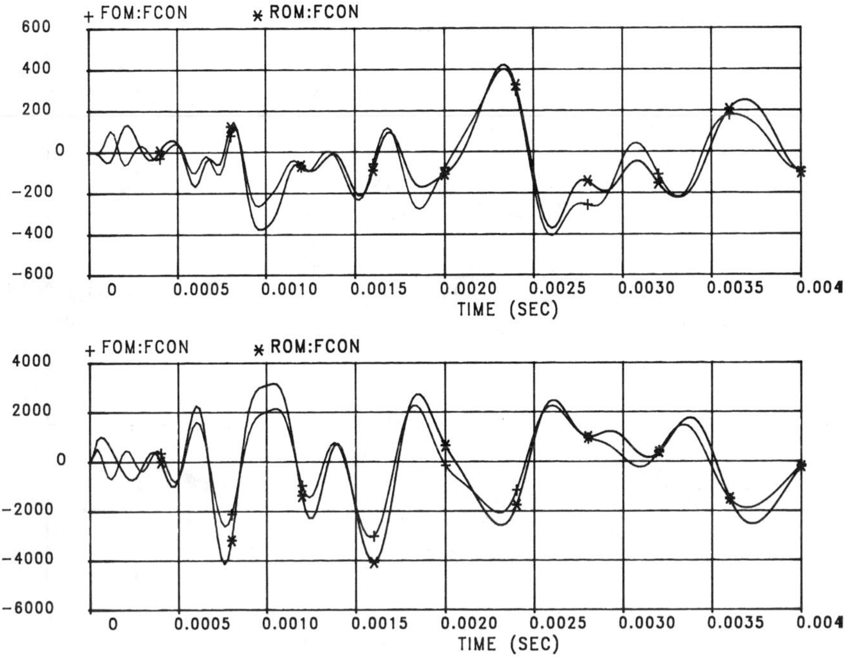

Fig. 3 Case I: model reduction (20 vs 14 modes).

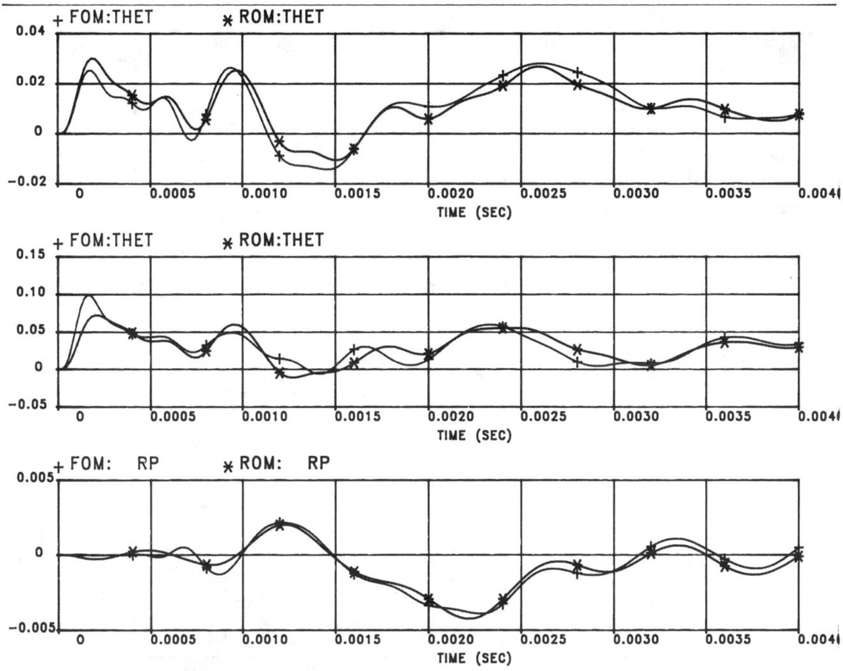

Fig. 4 Case I: model reduction (20 vs 10 modes).

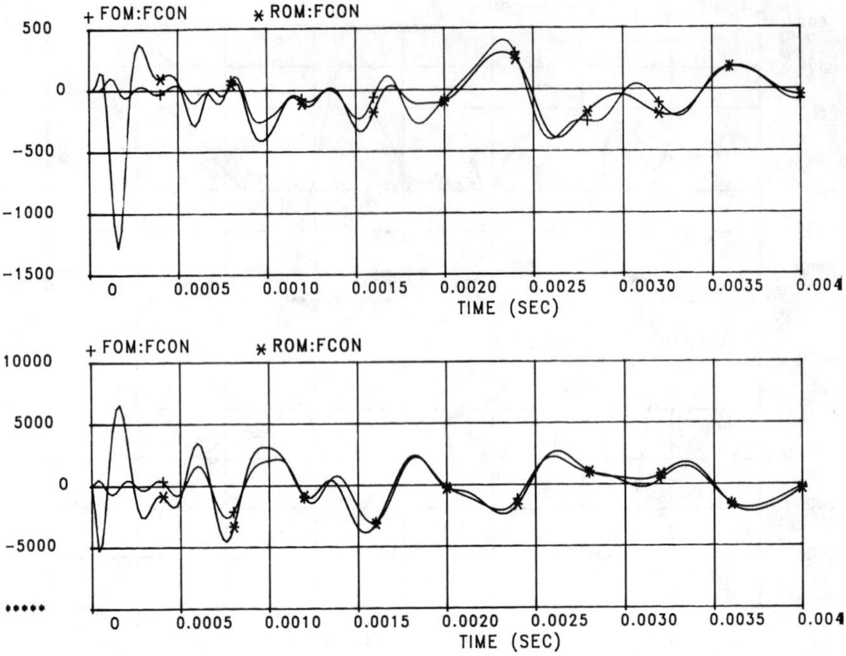

Fig. 5 Case I: model reduction (20 vs 10 modes).

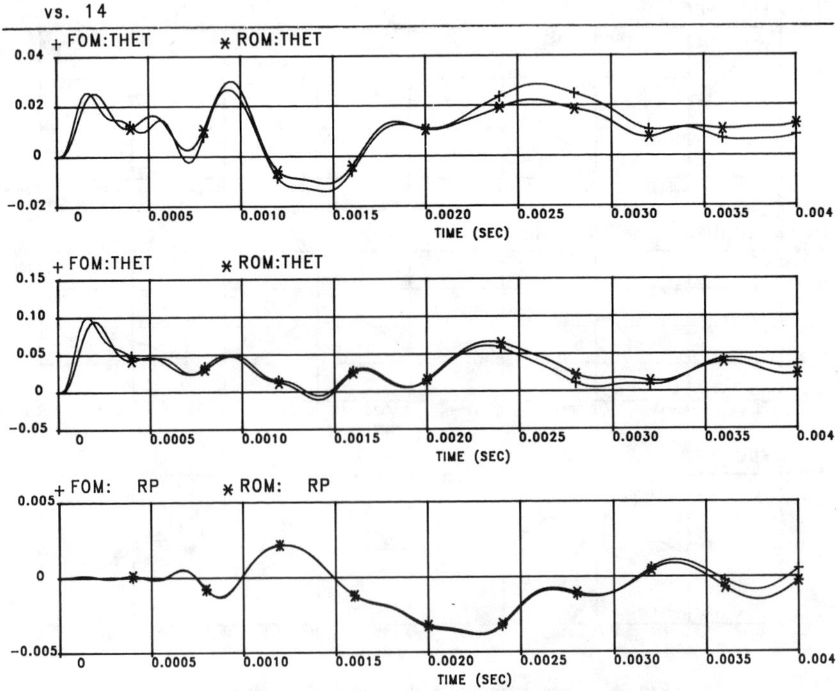

Fig. 6 Model truncation (with actuator dynamics): $q = 2$, 20 vs 14 modes.

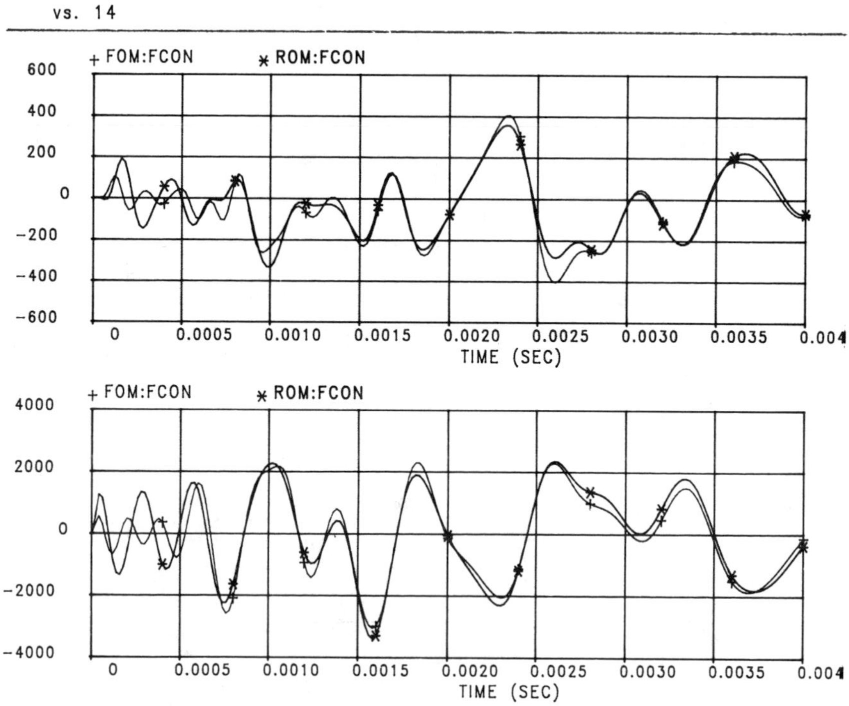

Fig. 7 Model truncation (with actuator dynamics): $q = 2$, 20 vs 14 modes.

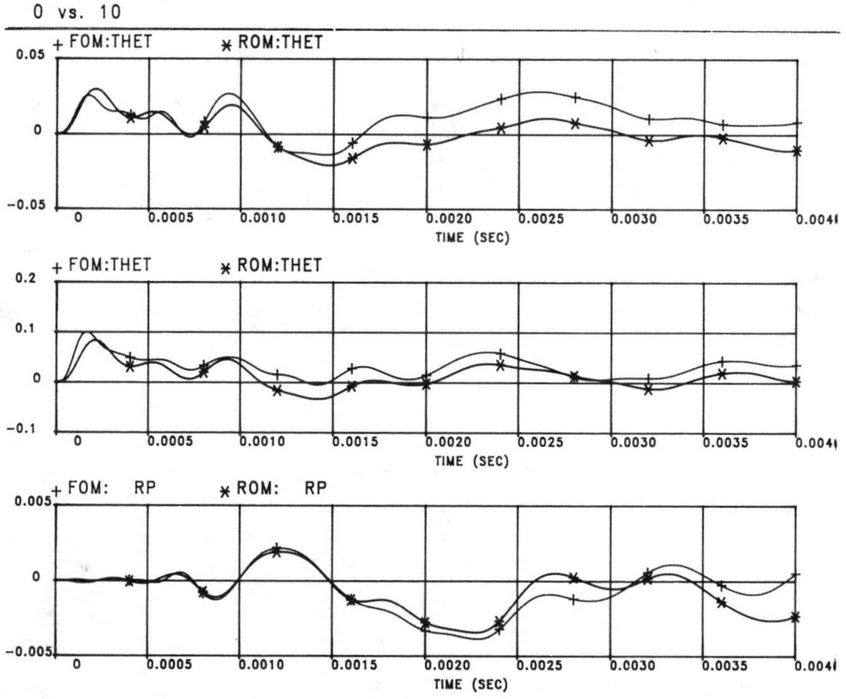

Fig. 8 Model truncation (with actuator dynamics): $q = 2$, 20 vs 14 modes.

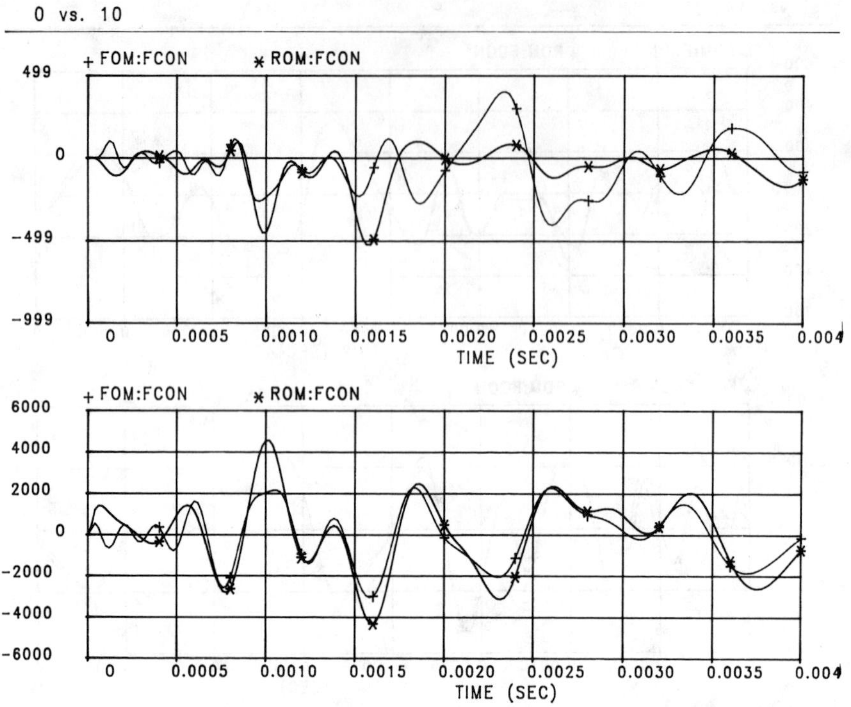

Fig. 9 Model truncation (with actuator dynamics): $q = 2$, 20 vs 10 modes.

coordinates according to their costs, and Tables 5 and 6 qualitatively indicate the errors between the full and reduced-order models. Figures 2 and 3 are the impulse response plots for case I with the reduced model having 14 elastic modes, and Figs. 4 and 5 are plots with the reduced-order model having 10 elastic modes. Figures 6–9 are the response plots for case III. Considering all of the outputs and the constraint forces, it is clear that models designed from cases III and IV are better. Penalizing the constraint forces will reduce the errors in these forces described in Figs. 3, 5, 7, and 9. These steps will be shown in future work.

Conclusions

In the context of multibody dynamics, reduced-order models derived by the standard component cost analysis and the generalized component cost analysis are analyzed. The two-beam numerical example indicates that the generalized component cost analysis gives better reduced-order models. The theory allows penalties to be placed on the constraint forces in addition to the system response.

References

[1]NASTRAN-NASA *Structural Analysis Program Users Manual*, NASA Publication NASA SP-222108, June 1986.

[2]Singh, R. P., Vandervoort, R. J., and Likins, P. W., "Dynamics of Flexible Bodies in Tree Topology—A Computer Oriented Approach," *Journal of Guidance, Control, and Dynamics*, Vol. 8, No. 5, Sept.–Oct. 1985, pp. 584–589.

[3]Skelton, R. E., and Yousuff, A., "Component Cost Analysis of Large Scale Systems," *International Journal of Control*, Vol. 37, No. 2, 1983, pp. 285–304.

[4]Yousuff, A., Wagie, D. A., and Skelton, R. E., "Linear Systems Approximations via Covariance Equivalent Realizations," *Journal of Mathematical Analysis and Applications*, Vol. 106, No. 1, 1985, pp. 91–114.

[5]Yousuff, A., and Skelton, R. E., "Controller Reduction by Component Cost Analysis," *IEEE Transactions on Automatic Control*, Vol. AC-29, No. 6, 1984, pp. 520–530.

[6]Wagie, D., and Skelton, R. E., "A Projection Approach to Covariance Equivalent Realizations of Discrete Systems," *IEEE Transactions on Automatic Control*, Vol. AC-31, No. 12, 1986, pp. 1114–1120.

[7]Skelton, R. E., *Dynamic Systems Control—Linear Systems Analysis and Synthesis*, Wiley, New York, 1988, Chap. V.

[8]Kokotovic, P. V., O'Malley, R. E., Jr., and Sannuti, P., "Singular Perturbations and Order Reduction in Control Theory—An Overview," *Automatica*, Vol. 12, 1976, pp. 123–132.

[9]Moore, B. C., "Principal Component Analysis in Linear Systems: Controllability, Observability and Model Reduction," *IEEE Transactions on Automatic Control*, Vol. AC-26, No. 1, 1981, pp. 17–32.

[10]Hutton, M. F., and Friedland, G. B., "Rough Approximations for Reducing Order of Linear Time-Invariant Systems," *IEEE Transactions on Automatic Control*, Vol. AC-20, No. 3, 1975, pp. 329–337.

[11]Anderson, B. D. O., "The Inverse Problem of Stationary Covariance Generation," *Journal of Statistical Physics*, Vol. 1, No. 1, 1969, pp. 133–141.

[12]Tse, E. C. Y., Medanic, J. V., and Perkins, W. R., "Generalized Hessenberg Transformations for Reduced-Order Modeling of Large Scale Systems," *International Journal of Control*, Vol. 27, 1978, pp. 493–512.

[13]Hyland, D. C., and Bernstein, D. S., "The Optimal Projection Approach to Model Reduction and the Relationship between the Methods of Wilson and Moore," *Proceedings of the 23rd IEEE CDC*, IEEE, New York, 1984.

[14]Glover, K., "All Optimal Hankel-Norm Approximations of Linear Multivariable Systems and Their L^∞ Error Bounds," *International Journal of Control*, Vol. 39, No. 6, 1984, pp. 1115–1193.

[15]Villemagne, C., and Skelton, R. E., "Controller Reduction Using Canonical Interactions," *IEEE Transactions on Automatic Control*, Vol. 33, Aug. 1988, pp. 740–750.

[16]Villemagne, C., and Skelton, R. E., "Model Reductions Using a Projection Formulation," *International Journal of Control*, Vol. 46, No. 6, 1987, pp. 2141–2169.

[17]Amirouche, F. M. L., and Huston, R. L., "Dynamics of Large Constrained Flexible Structures," *Transactions of ASME*, 78 Vol. 110, March 1988, pp. 78–83.

[18]Fadeeva, V. N., *Computational Methods in Linear Algebra*, Dover, New York, 1959.

[19]Skelton, R. E., "Cost Decomposition of Linear Systems with Application to Model Reduction," *International Journal of Control*, Vol. 32, No. 6, 1980, pp. 1031–1055.

References

Chapter 17

Controller Design by Eigenspace Assignment

G. L. Slater* and Q. Zhang†
University of Cincinnati, Cincinnati, Ohio

Introduction

EIGENSTRUCTURE assignment using constrained state or output feedback is a general approach to control design for linear multivariable systems. Initial work in the field of multivariable control concentrated primarily on the problem of pole placement and the necessary and sufficient conditions for the existence of the feedback matrix that yields a set of given eigenvalues.[1–8] However, for multivariable systems there is significant additional flexibility that can be used not only to place the eigenvalues but also to shape the entire eigenstructure of the system. Moore[9] categorized this design flexibility and proposed a design procedure for choosing not only the eigenvalues but also the components of the eigenvectors. This approach has been further developed by numerous authors in several other areas (see, for example, Refs. 10–20). One early application of eigenstructure assignment was in the area of decoupling control. For example, Andry et al.[13] considered a flight control example that decoupled the lateral modes of an aircraft. For a structural control problem the flexibility in choosing closed-loop eigenvectors was used to increase the robustness of the control system to model uncertainties.[11,12,17]

In this chapter we review the fundamental concepts of the eigenvalue placement problem and show the application toward flexible systems that 1) are of high (possible infinite) dimension and 2) may be modeled only through a limited set of experimental modal data. For this problem we will address the problem of eigenvector shaping to achieve desired modal characteristics while maintaining "reasonable" magnitude constraints on

*Professor, Department of Aerospace Engineering and Engineering Mechanics.
†Research Assistant Professor, Department of Mechanical and Industrial Engineering.

the control vector and also the problem of designing controls in order to reduce the response to harmonic or other type of excitations.

The particular concern in the present research is the application of eigenstructure assignment techniques to the active vibration control of flexible structures. The vibration motion of flexible structures can be represented by a modal model consisting of the eigenvalues and the right and left eigenvectors. Thus, vibration of flexible structures can be reduced or suppressed by assigning the closed-loop system eigenvalues and eigenvectors. Two controller design strategies are proposed in this chapter. The first is to combine the eigenstructure assignment technique with an optimization procedure in which the performance index is related to the amplitude of the control vector and the constraint functions are associated with desired eigenvalues and/or eigenvectors. In this procedure the desired eigenvalues may or may not be explicitly imposed. They can also be allowed to vary in a range fixed by the control system designer. For the desired right eigenvectors only a few elements are specified (they are generally specified to be zero), and the other elements will be determined subject to a minimum control effort criterion. The second problem to be considered is to suppress undesired inputs of linear system by shaping left eigenvectors.

The proposed methods are illustrated by several examples. Numerical simulation results and experimental results for the active vibration control of a laboratory beam are presented.

Review of Eigenvector Decomposition and Control Assignment

A linear time-invariant system with output feedback control can be described by the equations

$$\dot{x}(t) = Ax(t) + Ef(t) + Bu(t) \tag{1}$$

$$y(t) = Dx(t) \tag{2}$$

$$u(t) = Fy(t) \tag{3}$$

where $x \in R^N$, $f \in R^n$, $y \in R^r$, and $u \in R^m$ are state, disturbance input, output, and control vectors respectively, and A, E, B, D, and F are constant matrices of appropriate dimensions.

A general condition of assignability of eigenvalues and eigenvectors using output feedback was given by Srinathkumar.[16] For the controllable and observable system described earlier and with disturbance force $f \equiv 0$, then, if the matrices B and D are full rank, max(m,r) closed-loop eigenvalues can be assigned and max(m,r) eigenvectors (or reciprocal vectors by duality) can be partially assigned with min(m,r) entries in each vector arbitrarily chosen using output feedback. Juang et al.[11] showed how to generate an orthogonal basis that spans the admissible eigenvector space of the closed-loop system using the singular value decomposition. An optimal set of eigenvectors corresponding to assigned eigenvalues was then determined in this subspace in order to obtain both minimum conditioning and control gains for the eigenvalue placement problem. Fennell[12] also utilized the flexibility in eigenvector assignment to reduce control system sensitivity to changes in system parameters with application to symmetric flutter suppression.

Other authors chose the desired eigenvectors based on the requirement of a given problem rather than the optimal criteria mentioned earlier. Andry et al.[13] used (right) eigenvector assignment to alter the transient response of the system for decoupling flight control. They first determined the subspace that contains the admissible eigenvectors of the closed-loop system and then found in this subspace the closest eigenvectors to the desired ones. Other related work concerning eigenstructure assignment can be found in Fahmy and O'Reilly,[10] who gave a general mathematical treatment of the eigenspace assignment problem; Zhang et al.,[18] who developed a modal method and applied it to vibration control of flexible structures; and Becus and Sonmez,[19] who combined eigenstructure assignment with linear-quadratic optimal control design of a multivariable control system.

To examine this problem in modal form, consider the case where the control input $\boldsymbol{u} \equiv 0$. Then the response of the system described by Eq. (1) is

$$\boldsymbol{x}(t) = e^{At}\boldsymbol{x}(0) + \int_0^t e^{A(t-\tau)}E\boldsymbol{f}(r)\,\mathrm{d}\tau \tag{4}$$

Introduce the modal decomposition of A:

$$A = \Phi\Lambda L^T \tag{5}$$

where $\Lambda \in C^{N,N}$ is a diagonal matrix of eigenvalues, and $\Phi \in C^{N,N}$ is the matrix whose columns consists of the right eigenvectors of A determined by solving the standard eigenvalue problem

$$[A - \lambda_i I]\boldsymbol{\phi}_i = 0 \qquad \text{with} \qquad \Phi = [\ldots\ \boldsymbol{\phi}_i\ \ldots] \tag{6}$$

$L \in C^{N,N}$ is the matrix of the left eigenvectors of A. By definition, L^T is the inverse of Φ, or $L^T\Phi = I$.

The physical response $\boldsymbol{x}$ can be written as

$$\boldsymbol{x}(t) = \Phi\left[e^{\Lambda t}L^T\boldsymbol{x}(0) + \int_0^t e^{\Lambda(t-\tau)}L^T E\boldsymbol{f}(\tau)\,\mathrm{d}\tau\right] \tag{7}$$

Examination of Eq. (7) reveals the following:

1) The response of the system $\boldsymbol{x}(t)$ is a linear combination of the set of right eigenvectors $\boldsymbol{\phi}_i$, $i = 1, \ldots, N$. Hence, the system response can be controlled by shaping the $\boldsymbol{\phi}_i$.

2) The response of the system to external inputs is proportional to the projections of the left eigenvectors on the columns of the input matrix E.

In particular, if a left eigenvector l_i is orthogonal to the columns of the matrix E, the response in the ith mode is zero.

Eigenplacement and Gain Optimization

The problems of pole allocation and eigenstructure assignment using a gain optimization concept are presented in this section. The controller design is formulated based on a modal model. Because the authors' interest is to apply this to a high-order vibration problem, the assumption is made at the outset to consider only a limited number of modes in the control design. Assume that there are a total of r modes that we desire to control and that the desired closed-loop eigenvectors are a linear combination of this subset of open-loop eigenvectors (Note that this is also the number of outputs $\boldsymbol{y}$.) Arrange these open-loop eigenvectors as the matrix $\Phi_1 = [\boldsymbol{\phi}_1 \ldots \boldsymbol{\phi}_r]$. Similarly, let $L_1 = [l_1 \ldots l_r]$ be the matrix of (open-loop) left eigenvectors corresponding to Φ_1. A fundamental assumption we make at this point is that the desired closed-loop eigenvectors can be written as

$$\boldsymbol{\phi}_i^{(d)} = \Phi_1 \boldsymbol{q}_i \qquad \text{for} \qquad i = 1, \ldots, r \tag{8}$$

where $\boldsymbol{q}_i$, $i = 1, \ldots, r$ is a set of unknown vectors. This assumption allows us to use only the known or identified modal vectors in the analysis. The maximum number of specified elements of each desired mode must be less than r. Since we will generally assign less than r elements for each $\boldsymbol{\phi}_i$, Eq. (8) can generally be satisfied by some nonunique choice of $\boldsymbol{q}_i$. The closed-loop assignment problem, then, is to determine a gain matrix F such that

$$[A + BFD - \lambda_i^{(d)} I]\boldsymbol{\phi}_i^{(d)} = 0 \qquad \text{for} \qquad i = 1, \ldots, r \tag{9}$$

Substituting Eq. (8) into Eq. (9) and premultiplying by L_1^T, we obtain the reduced system

$$[\Lambda_1 + L_1^T BFD\Phi_1 - \lambda_i^{(d)} I]\boldsymbol{q}_i = 0 \qquad \text{for} \qquad i = 1, \ldots, r \tag{10}$$

where $\Lambda_1 = \text{diag}\{\lambda_i\}$, $i = 1, \ldots, r$ is the matrix of open-loop eigenvalues. Assuming that the open-loop eigenvalues are all changed, Eq. (10) can be rearranged as

$$\boldsymbol{q}_i = [\lambda_i^{(d)} I - \Lambda_1]^{-1} \mathrm{L}_1^T B\boldsymbol{m}_i \tag{11}$$

$$\text{where} \qquad \boldsymbol{m}_i = FD\boldsymbol{\phi}_i^{(d)} \tag{12}$$

Premultiplying Eq. (11) by Φ_1, the desired eigenvector $\boldsymbol{\phi}_i^{(d)}$ can be related to the vector $\boldsymbol{m}_i$ by

$$\boldsymbol{\phi}_i^{(d)} = T_i \boldsymbol{m}_i \qquad \text{where} \qquad T_i = \Phi_1 [\lambda_i^{(d)} I - \Lambda_1]^{-1} L_1^T B \tag{13}$$

The vector $\boldsymbol{m}_i$ has a particular role in the controller design and in fact is subsequently used as a design quantity in the analysis. The control vector $\boldsymbol{u}$ is related to the state vector $\boldsymbol{x}$ by

$$\boldsymbol{u} = F D \boldsymbol{x} \tag{14}$$

Assuming that $\boldsymbol{x}$ is a linear combination of the controlled closed-loop system eigenvectors, the control vector $\boldsymbol{u}$ can be expressed in the modal space as

$$\boldsymbol{u} = FD \sum_{i=1}^{r} \eta_i \boldsymbol{\phi}_i^{(d)} = FD\Phi^{(d)}\eta_1 \tag{15}$$

where η_i is the generalized coordinate of the response in the ith eigenvector direction, $\boldsymbol{\eta}_1$ is the controlled modal coordinate vector, and $\boldsymbol{\Phi}^{(d)}$ is the matrix of desired (closed-loop) eigenvectors. Using Eq. (12), however, Eq. (15) can be written as

$$u = \sum_{i=1}^{r} \eta_i \boldsymbol{m}_i = \hat{M}\boldsymbol{\eta}_1 \tag{16}$$

where $\hat{M}$ is the matrix of $\boldsymbol{m}_i$ vectors. This shows that the control vector $\boldsymbol{u}$ is a linear combination of the vectors $\boldsymbol{m}_i$. Since the vectors $\boldsymbol{m}_i$ link the control vector to the modal space of the closed-loop system, we call them "modal control vectors." For arbitrary $\boldsymbol{x}$ (hence η), the control effort can be minimized by minimizing the norm of $\hat{M}$. Since it is assumed that only r modes are important in controller design, the same number of modal control vectors will be taken into account. Rather than minimize the norm of $\hat{M}$ (or its maximum singular value), our approach in this paper is to formulate a procedure to determine each of the $\boldsymbol{m}_i$ subject to the eigenspace constraints. The algorithms for the computation of the modal control vectors will be described in the following sections. Once a set of modal vectors $\boldsymbol{m}_i$, $i = 1, \ldots, r$ are determined and arranged as the matrix $\hat{M}$, the feedback matrix F is obtained from

$$F = \hat{M}[D\Phi^{(d)}]^{-1} \tag{17}$$

Control Optimization for Pole Placement

Consider first the simple case of determining a control gain matrix to achieve a particular pole placement, while at the same time minimizing the control effort required to achieve this placement. We formulate this problem by writing the desired eigenvalues $\lambda_i^{(d)}$ as a function of the modal control vectors. Rewriting Eq. (10) as

$$\lambda_i^{(d)}\boldsymbol{q}_i = \Lambda_1 \boldsymbol{q}_i + L_1^T B \boldsymbol{m}_i \tag{18}$$

then premultiplying Eq. (18) by $\boldsymbol{q}_i^H$ and assuming without restriction that $\boldsymbol{q}_i$

is of unit length, then

$$\lambda_i^{(d)} = \boldsymbol{q}_i^H \Lambda_1 \boldsymbol{q}_i + \boldsymbol{q}_i^H L_1^T B \boldsymbol{m}_i \tag{19}$$

The superscript H denotes the conjugate transpose of the appropriate variable. The preceding equation is still redundant and in fact is a restatement of Eq. (11). To allow subsequent solution and to simplify the optimization problem, we choose to impose the additional constraint

$$\boldsymbol{q}_i^H \Lambda_1 \boldsymbol{q}_i = \lambda_i \tag{20}$$

This relationship simplifies Eq. (19) considerably and leads to the relation

$$\lambda_i^{(d)} = \lambda_i + \boldsymbol{q}_i^H L_1^T B \boldsymbol{m}_i \tag{21}$$

The assumption Eq. (20) can be justified by noting that, as the control energy goes to zero, the closed-loop system becomes consistent with the open-loop as $\boldsymbol{m}_i \to 0$, in which case $\boldsymbol{\phi}_i^{(d)} \to \boldsymbol{\phi}_i$, $\lambda_i^{(d)} \to \lambda_i$. Imposing this relation in general allows a significant simplification to the constraint relation and allows the optimization problem to be expressed independently of the $\boldsymbol{q}_i$. If we now substitute Eq. (11), then

$$\lambda_i^{(d)} = \lambda_i + \boldsymbol{m}_i^H [L_1^T B]^H [\Lambda_1 - \lambda_i^{(d)} I]^{-H} [L_1^T B] \boldsymbol{m}_i \tag{22}$$

which relates the desired eigenvalues to the modal model of the open-loop system and the modal control vectors to be determined. An optimization procedure to calculate the modal control vectors and to reduce control expenditures then can be formed as follows:

$$\min J = \sum_{i=1}^{r} \|\boldsymbol{m}_i\|^2 + w\left[\sum_{i=1}^{r} \sum_{j=i+1}^{r} \boldsymbol{m}_i^H T_i^H D^H D T_j \boldsymbol{m}_j + \sum_{i=1}^{r} (\boldsymbol{m}_i^H T_i^H D^H D T_i \boldsymbol{m}_i - 1)\right]$$

with the constraint functions

$$\lambda_i^{(d)} = \lambda_i + \boldsymbol{m}_i^H [L_1^T B]^H [\Lambda_1 - \lambda_i^{(d)} I]^{-H} [L_1^T B] \boldsymbol{m}_i, \qquad i = 1, \ldots, r$$

where w is a real positive weighting factor, and the optimization variables are the elements of $\boldsymbol{m}_i$. The objective function consists of three terms. The first term represents the minimization of the control effort, and the latter two terms attempt to enhance the robustness of the proposed algorithm against the model uncertainty. The accuracy of the modal control vectors $\boldsymbol{m}_i$ can be affected by uncertainties of the modal model of the open-loop system, and, if the matrix $D\Phi^{(d)}$ is ill-conditioned, the feedback gain matrix F derived from Eq. (17) may be inappropriate due to the inaccuracy of $\boldsymbol{m}_i$. The latter terms in the performance index attempt to weight the $\boldsymbol{m}_i$ vectors in order to minimize the condition number of $D\Phi^{(d)}$ by making it as a "near-orthogonal" matrix (remember that $\boldsymbol{\phi}_i^{(d)} = T_i \boldsymbol{m}_i$).

Control Optimization with Eigenstructure Assignment

For many applications eigenvalues and some elements of the closed-loop eigenvectors need to be assigned. In general we find that the choice of the closed-loop eigenvalues and eigenvectors may have a significant effect on the magnitude of the closed-loop gain matrix. For this paper, we treat only certain types of constraints that we have found to be effective for problems involving the vibration control of flexible structures. The specific constraints used here are as follows:

1) The modal damping factors (real part of the eigenvalues) of the closed-loop system are specified explicitly in order to guarantee the stability of the control system and to keep a reasonable stability margin for controller design.

2) Rather than specific frequencies (imaginary part of the eigenvalue), only a specified frequency bandwidth is given for the closed-loop system.

3) Only a few elements of each desired eigenvector are specified explicitly according to the given problem. An indirect constraint is assigned to the other unspecified elements by specifying that the norm of the desired eigenvectors is close to that of the open-loop system. Here $0.8\|\boldsymbol{\phi}_i\| \leqslant \|\boldsymbol{\phi}_i^{(d)}\| \leqslant 1.2\|\boldsymbol{\phi}_i\|$ is used in order to keep the value of the unspecified elements within a reasonable bound.

The optimization problem then is formulated as before but with additional constraint functions.

Objective function:

$$\min J = \sum_{i=1}^{r} \|\boldsymbol{m}_i\|^2 + w\left[\sum_{i=1}^{r}\sum_{j=i+1}^{r} \boldsymbol{m}_i^H T_i^H D^H D T_j \boldsymbol{m}_j + \sum_{i=1}^{r} (\boldsymbol{m}_i^H T_i^H D^H D T_i \boldsymbol{m}_i - 1)\right] \tag{23}$$

with equality constraint functions

$$\boldsymbol{\phi}_{iS}^{(d)} = T_{iS}\boldsymbol{m}_i, \qquad i = 1, \ldots, r \tag{24}$$

inequality constraint functions (group 1)

$$0.8\|\boldsymbol{\phi}_i\| \leqslant \|\boldsymbol{\phi}_i^{(d)}\| \leqslant 1.2\|\boldsymbol{\phi}_i\| \qquad (\boldsymbol{\phi}_i^{(d)} = T_i\boldsymbol{m}_i) \tag{25}$$

and (group 2)

$$\omega_{il} \leqslant \omega_i^{(d)} \leqslant \omega_{iu} \qquad \text{for} \qquad i = 1, \ldots, r \tag{26}$$

Here $\boldsymbol{\phi}_{iS}^{(d)}$ is the subvector of $\boldsymbol{\phi}_i^{(d)}$ containing the specified elements, and T_{iS} is the corresponding submatrix of T_i. The variables ω_{il} and ω_{iu} are, respectively, the lower and upper bounds of $\omega_i^{(d)}$.

In this process the modal frequencies of the closed-loop system $\omega_i^{(d)}$ are considered as optimization variables along with the modal control vectors. In fact, this optimization process selects the best pair of desired eigenvalues and eigenvectors in the sense of minimizing the control effort and increasing the robustness of the algorithm against the model uncertainty.

Discrete-Time Control Models

The previous analysis is based on a continuous-time model. Since most modern systems incorporate digital controls, the control algorithm is valid in practice, only if the sampling rate of the control cycle is much greater than the highest frequency of the controlled modes. Particularly for flexible structures, where many high frequencies exist, this is not possible. For these problems, a direct digital design may be desirable. To allow more general application, then, in this section the eigenstructure approach based on a discrete-time model is developed.

The linear system described by Eqs. (1) and (2) becomes for the discrete-time case

$$\boldsymbol{x}(k+1) = \bar{A}\boldsymbol{x}(k) + \bar{B}\boldsymbol{u}(k) + \boldsymbol{f}_k \tag{27}$$

$$\boldsymbol{y}(k) = D\boldsymbol{x}(k) \tag{28}$$

where $\bar{A} = e^{A\,\Delta t}$, $\bar{B} = A^{-1}[e^{A\,\Delta t} - I]B$ (assuming that A has no zero eigenvalues), $\boldsymbol{f}_k$ is the result of the disturbance force integrated over Δt, and Δt is the sampling period of the control cycle.

It is well known that the eigenvectors of $\bar{A}$ are the same as those of A and that the eigenvalues of $\bar{A}$ are related to λ_i, the eigenvalues of A, by

$$\bar{\lambda}_i = e^{\lambda_i\,\Delta t} \tag{29}$$

This implies the modal decomposition formulas

$$\bar{A} = \Phi\bar{\Lambda}L^T \tag{30}$$

and

$$\bar{B} = \Phi\Lambda^{-1}[e^{\Lambda\,\Delta t} - I]L^T B \tag{31}$$

Taking into account the time delay in the control cycle due to the time required by data acquisition, analog-to-digital (AD) and digital-to-analog (DA) conversion, and control computation, the control $\boldsymbol{u}(k)$ must realistically be expressed as a function of the delayed output as

$$\boldsymbol{u}(k) = \bar{F}\boldsymbol{y}(k-1) \tag{32}$$

Let the external force $\boldsymbol{f}_k$ be zero, and substitute Eq. (30) into Eq. (27):

$$\boldsymbol{x}(k+1) = \bar{A}\boldsymbol{x}(k) + \bar{B}\bar{F}D\boldsymbol{x}(k-1) \tag{33}$$

The eigenvalue problem associated with Eq. (33) leads to a linear homogeneous equation

$$\left[\bar{A} + \frac{1}{\bar{\lambda}_i^{(d)}} \bar{B}\bar{F}D - \bar{\lambda}^{(d)} I\right] \boldsymbol{\phi}_i^{(d)} = 0 \tag{34}$$

relating the desired eigenvalues and eigenvectors with the feedback gain matrix to be determined. Following a similar development as in the continuous case, we choose to characterize the eigenplacement problem in terms of a subset Λ_1, Φ_1, L_1 of the complete eigenvalues and eigenvectors.

Substituting Eqs. (8) and (31) into Eq. (34) and premultiplying by L_1^T yields

$$\left[\bar{\Lambda}_1 + \frac{1}{\bar{\lambda}_i^{(d)}} \Lambda_1^{-1}(e^{\Lambda_1 \Delta t} - I)(L_1^T B\bar{F}D\Phi_1) - \bar{\lambda}_i^{(d)}\right] \boldsymbol{q}_i = 0 \tag{35}$$

Similar to Eqs. (11) and (13), Eq. (35) can be rewritten as

$$\boldsymbol{q}_i = \frac{1}{\bar{\lambda}_i^{(d)}} [\bar{\Lambda}_1 - \bar{\lambda}_i^{(d)} I]^{-1} \Lambda^{-1} [\bar{\Lambda}_1 - I] L_1^T B \boldsymbol{m}_i \tag{36}$$

or

$$\phi_i^{(d)} = \bar{T}_i \boldsymbol{m}_i \tag{37}$$

with

$$\bar{T}_i = \frac{1}{\bar{\lambda}_i^{(d)}} \Phi_1 [\bar{\Lambda}_1 - \bar{\lambda}_i^{(d)} I]^{-1} \Lambda^{-1} [\bar{\Lambda}_1 - I][L_1^T B]$$

Based on these two fundamental equations, the modal control vector computation proceeds in the same way as that in the continuous-time case.

Illustrative Example

The algorithm described earlier is applied to the active vibration control of a laboratory beam. The control objective is to reduce the vibration motion of the free end of the beam which is subject to forced vibration. This simple test model is representative of practical applications such as the vibration control of a flexible robot arm or a space structure pointing experiment. The control law obtained is compared to a standard eigenvector placement approach and is tested by both numerical and experimental simulation.

The controller design is based on the discrete-time model. The effective control loop delay counted in the discrete time model is 4.4 ms, which is the minimum sampling period imposed by the digital computer (HP 1000) involved in the control loop (see Fig. 1).

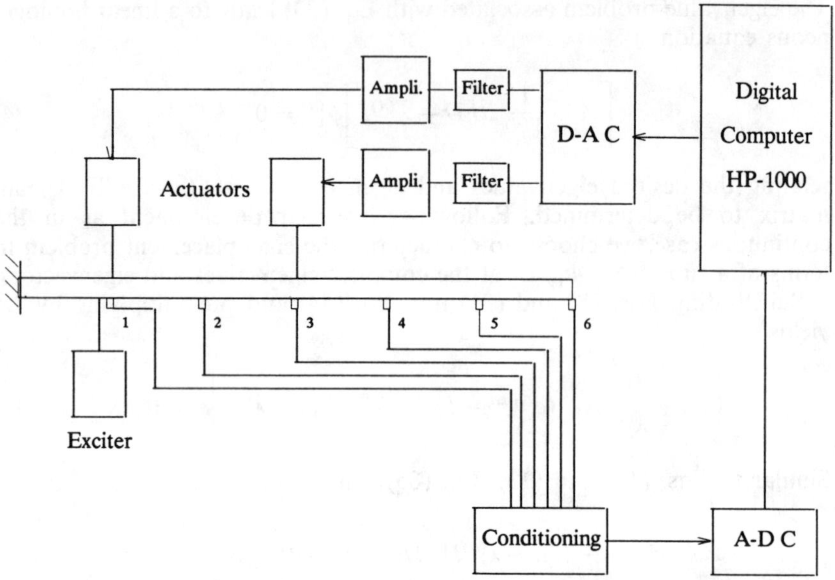

Fig. 1 Block diagram of control scheme.

The modeled beam is made of steel and has a length of 42.6 in. and a rectangular cross section of 2.75 × 0.125 in. A finite-element model with 12 beam elements is used to model the structure. This finite-element model is validated by an experimental modal test, and the experimentally determined modal damping factors are used to complete this analytical model. We assume that six sensors and two actuators are involved in the control scheme; i.e., the feedback gain matrix is a 2 × 6 matrix, and only the first three modes are to be controlled. Three different control configurations are discussed as follows:

Case 1

The only specified element of each desired eigenvector, which is constrained to be one-tenth that of the open-loop system, is that corresponding to the deflection at the free end of the beam. The other elements are subject to the constraint of the norm of the desired eigenvector mentioned earlier. In this example there is no explicit specification on modal damping factors or modal frequencies of the closed-loop system except the stability criterion. Considering the fact that the controller is designed based on the discrete-time model, the constraint imposed for the desired eigenvalues is $0 \leqslant |\bar{\lambda}_i^{(d)}| \leqslant 0.99$; i.e., the desired eigenvalues for the discrete time model will lie slightly within the unit circle. The feedback gain matrix and the first three closed-loop system equivalent continuous system eigenvalues [defined as $\lambda = (1/\Delta t)\, \ell n \bar{\lambda}$] determined through the optimization algorithm are listed in Table 1 and Table 2, respectively. Since there is no constraint on the modal frequencies of the closed-loop system, the poles of the closed-loop

Table 1 Equivalent continuous-time eigenvalues of open-loop and closed-loop systems

Open-loop		Closed-loop eigenvalues					
		Case 1		Case 2		Case 3	
Damping, %	Frequency, Hz	Damping, %	Frequency, Hz	Damping, %	Frequency, Hz	Damping, %	Frequency, Hz
3.74	2.13	2.12	11.06	9.5	2.5	2.12	11.06
0.60	13.3	2.13	11.09	1.8	13.5	2.13	11.09
0.40	37.17	0.46	39.9	0.63	37.5	0.46	39.9

Table 2 Feedback gain matrices

Case 1	$[\bar{F}] = \begin{bmatrix} 0.611 & 0.278 & -1.14 & 0.639 & -0.108 & -0.158 \\ 0.291 & -0.136 & 0.0099 & 0.239 & 0.281 & 0.146 \end{bmatrix}$
Case 2	$[\bar{F}] = \begin{bmatrix} 28.5 & -7.03 & -124 & 97.4 & -4.04 & 51.0 \\ -7.81 & 2.03 & 34.6 & -27.6 & 1.33 & -15.0 \end{bmatrix}$
Case 3	$[\bar{F}] = \begin{bmatrix} 0.808 & 0.368 & -1.49 & 0.847 & -0.121 & -0.290 \\ 1.50 & 0.294 & -2.23 & 1.55 & 0.0861 & -1.25 \end{bmatrix}$

system are free to move in the left half of the complex plane. From Table 1 we can see that $\omega_1^{(d)}$ has been greatly moved away from ω_1, whereas the changes in the second and third modal frequencies are relatively small. The corresponding mode shapes are shown in Fig. 2.

Case 2

As a comparison, a standard eigenstructure assignment problem[20] is applied to the same example. The method reported in Ref. 20 is an extension of the method proposed by Andry et al.[13] by introducing the modal transformation and by reformulating for the discrete-time model. Note that, for the standard eigenstructure assignment problem, if the number of controlled modes is equal to the number of sensors, and the number of arbitrarily specified elements of each desired eigenvector is equal to the number of actuators, the feedback gain matrix can be uniquely determined.

In this example the control loop includes two actuators. Thus, two components of each controlled mode are specified: the component corresponding to the deflection at point 12 (free end of the beam) is imposed to be one-tenth that of the open-loop system, and the component corresponding to the deflection at point 4 remains the same as that of the open-loop system. There is no constraint imposed on the norm of the desired eigenvectors. The desired eigenvalues are chosen as

$$\lambda_1^{(d)} = -1.5 + j15.7(2.5\ \text{Hz})$$

$$\lambda_2^{(d)} = -1.5 + j84.82(13.5\ \text{Hz})$$

$$\lambda_3^{(d)} = -1.5 + j235.6(37.5\ \text{Hz})$$

Note that the modal frequencies (imaginary part of the eigenvalues) are intentionally chosen close to those of the open-loop system.

The mode shapes of the closed-loop system are shown in Fig. 3, and the feedback gain matrix is listed in Table 2. Obviously, these feedback gains are much greater than those obtained previously using the optimization procedure, even though the closed-loop eigenvalues in this latter case are closer to those of the open-loop system.

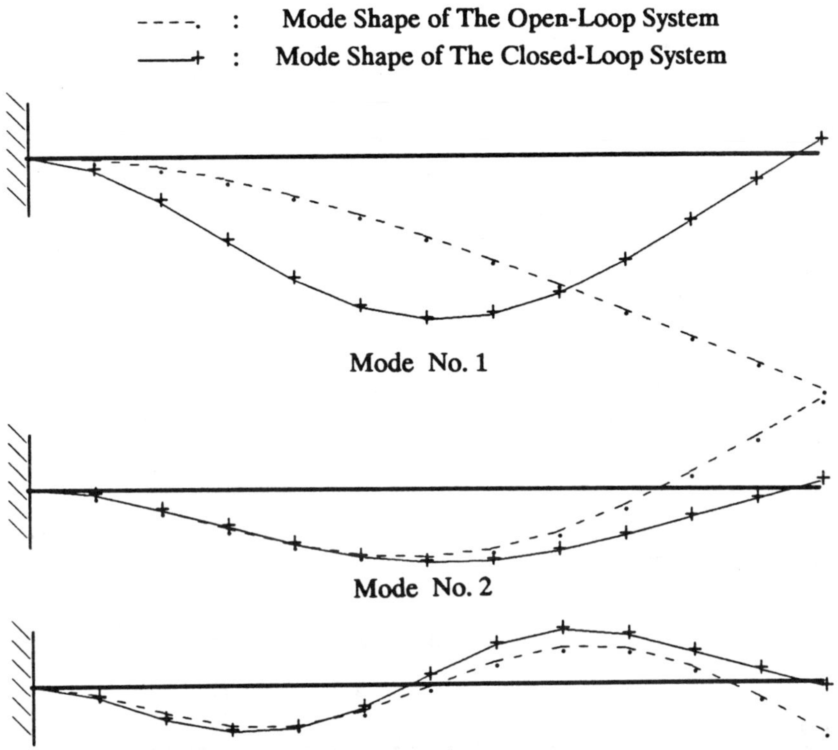

Fig. 2 Mode shapes (gains optimization).

Case 3

To better understand the importance of the consistancy of desired eigenvalues and eigenvectors, let us choose the "optimal" eigenvalues determined through the optimization process as the desired eigenvalues, and the desired eigenvectors are specified in the same manner as in case 2. As in case 2, the standard eigenstructure assignment technique is again used to determine the feedback gain matrix. The mode shapes corresponding to this control configuration are shown in Fig. 4. We note that the feedback gain matrix for this case approaches that found from optimization (case 1).

From these results some comments are in order:

1) When only the eigenvectors are required to change, the control effort is not necessarily minimized by forcing closed-loop eigenvalues to be close to the open-loop eigenvalues. In fact, as demonstrated by these cases, large eigenvector changes may require large movement of the eigenvalues in order to minimize the feedback gains.

2) Consistency between the desired eigenvalues and eigenvectors is very important for controller design. Mismatching the desired eigenvalues with

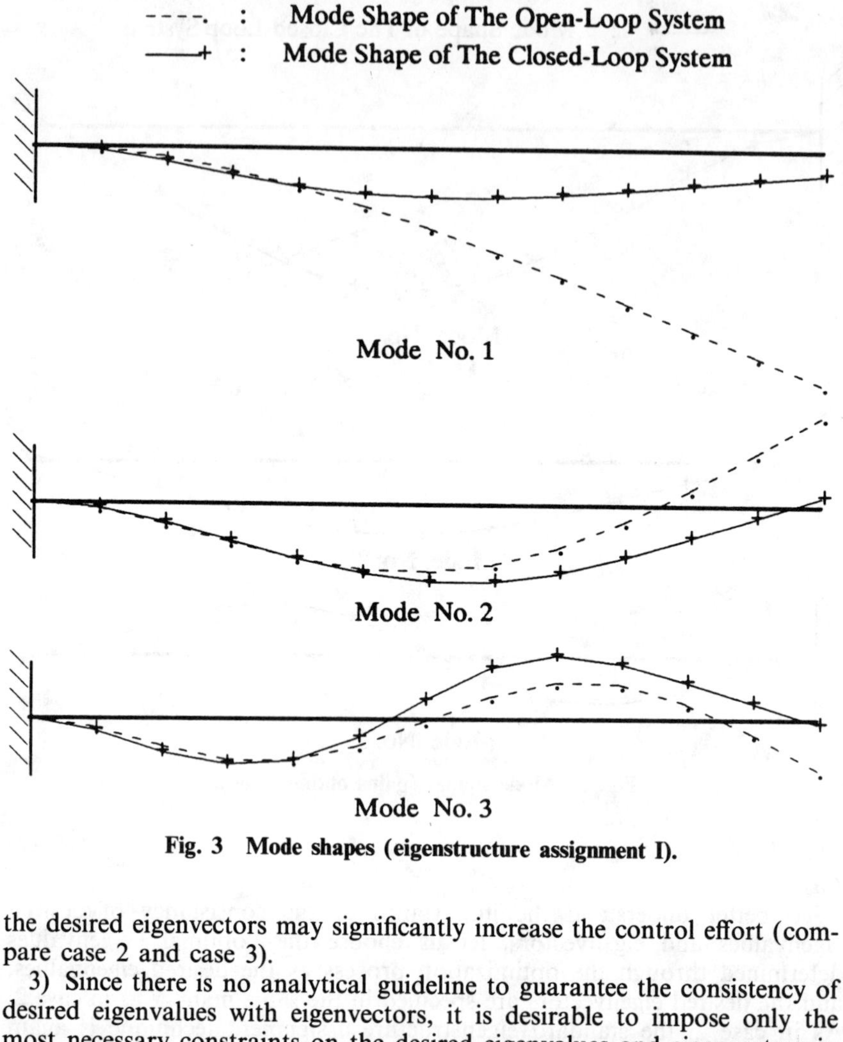

Fig. 3 Mode shapes (eigenstructure assignment I).

the desired eigenvectors may significantly increase the control effort (compare case 2 and case 3).

3) Since there is no analytical guideline to guarantee the consistency of desired eigenvalues with eigenvectors, it is desirable to impose only the most necessary constraints on the desired eigenvalues and eigenvectors in order to get a more "natural" control configuration. In this example the optimized control law only specifies one component of each desired eigenvector, the global constraints on the norm of the desired eigenvectors, and the stability of the controlled modes.

Experimental Implementation

The control scheme is shown in Fig. 1. Equipment used to control the beam consists of an HP 1000 digital computer, two actuators (electrodynamic shaker-type MB modal 50), and six sensors (PCB 307A accelerometers with sensitivity 0.1 v/g). The actuators are located at points 1 and 3, and sensors are uniformly implemented along the beam. An exciter is

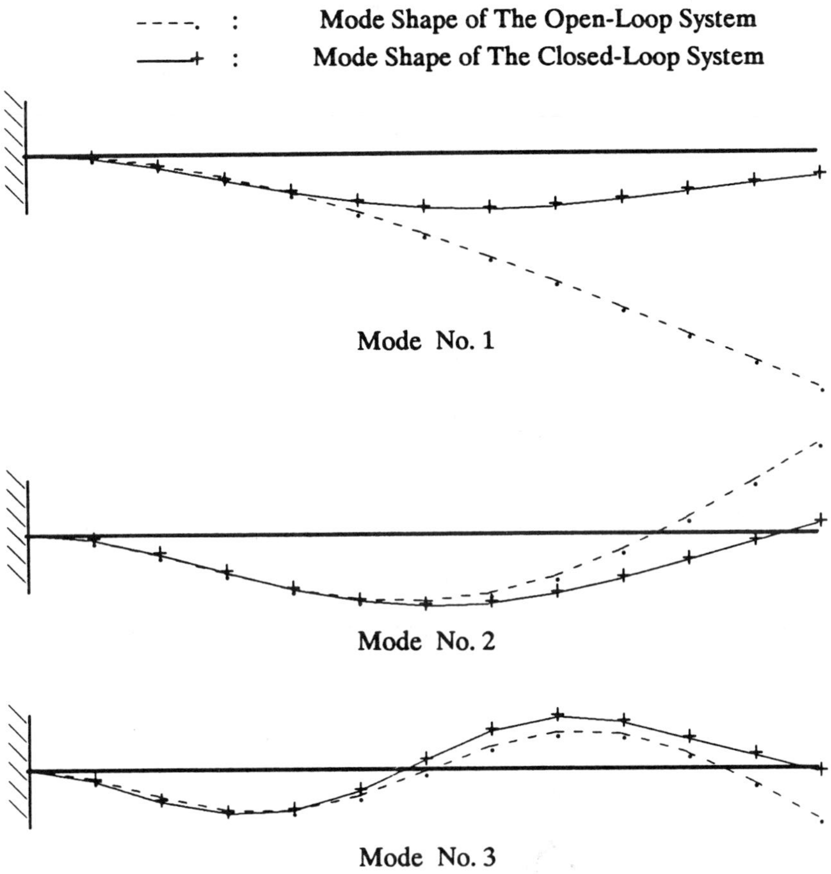

Fig. 4 Mode shapes (eigenstructure assignment II).

mounted near the fixed end of the beam in order to generate a sinusoidal excitation of 5 Hz. In order to avoid exciting the higher-frequency modes, two low-pass filters are used for filtering the high-frequency components in the control signals. A linear variable differential transformer (LVDT) is used to measure the displacements.

By forcing the closed-loop eigenvector to be close to zero at the beam tip, the controller should reduce the vibration motion at the free end of the beam subject to the excitation force. The system model used for the controller design includes a model of the actuator dynamics, and the optimization procedure is employed for computing the feedback gain matrix. Figure 5 shows the deflection of the free end of the beam with and without control. After the control loop is closed, the deflection is reduced to less than 30% of that without control. Figure 6 illustrates the stationary forced vibration shape with and without control.

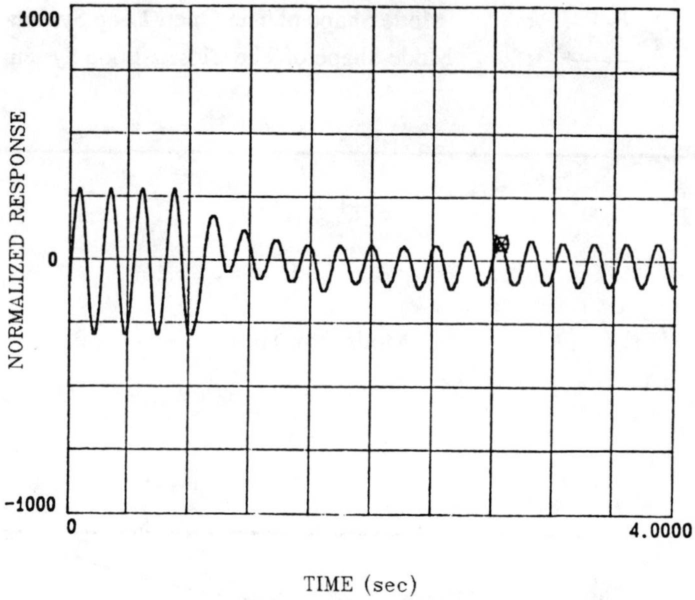

Fig. 5 Deflection of the free end of the beam without and with control (control comes on at $t = 1$ s).

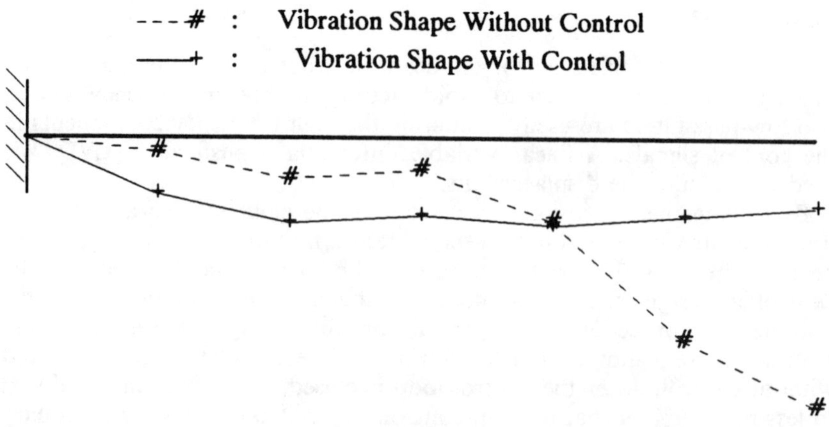

Fig. 6 Forced vibration shapes with and without control.

Eigenvector Shaping for Forced Vibration Control

The controller design in the previous section shaped the right eigenvectors in order to reduce the deflection in the controlled modes at certain points on the structure. This approach is most commonly applied to modification of the transient response of a structure to an unknown initial condition, but, as indicated in the experimental example, this approach will also help to reduce the response at fixed points to forced excitation. However, if the objective is specifically to reduce the overall response to forced excitation, then a more productive approach is to directly consider the forced response in the control design process. In this section we apply an eigenstructure assignment technique, specifically to the problem of minimization of forced response.

As mentioned in the introduction, the response to inputs is governed by the left eigenvectors of the system. From Eq. (7), if a left eigenvector of the system matrix is orthogonal to the columns of the input matrix E corresponding to undesired inputs, then the response in this mode is zero. In this section we demonstrate how to construct a closed-loop system whose dominant left eigenvectors possess this property. This method assumes that the undesired inputs are applied at certain fixed points on the structure and that the locations of these points are known.

From a mathematical point of view, the present problem is a special eigenspace assignment problem. Because this work requires placing of the eigenvalues and the left eigenvectors based on experimentally obtained data, the formulation and problem solution presented in this paper are quite different from those of the cited references.

Problem Statement

Our intent in this analysis is to choose the feedback gain matrix F in order to cause the dominant left eigenvectors to be (nearly) orthogonal to the columns of E. For the system of Eq. (1), we desire to find a feedback law that can reduce the effect of disturbances while stabilizing the system dynamics. Using the output feedback control law

$$\boldsymbol{u} = F\boldsymbol{y}$$

the new (closed-loop) system matrix $\hat{A}$ is defined as

$$\hat{A} = A + BFD \tag{38}$$

For the closed-loop matrix $\hat{A}$, denote the modal matrices analogous to Eqs. (5) and (6) as

$$\Lambda^{(d)} = \text{diag}\{\lambda_i^{(d)}\}, \qquad \Phi^{(d)} = [\ldots\ \boldsymbol{\phi}_i^{(d)}\ \ldots], \qquad L^{(d)} = [\ldots\ \boldsymbol{l}_i^{(d)}\ \ldots]$$

where the columns of $L^{(d)}$, $\boldsymbol{l}_i^{(d)}$ are determined from

$$[A^T + [BFD]^T - \lambda_i^{(d)} I]\boldsymbol{l}_i^{(d)} = 0 \tag{39a}$$

and

$$\boldsymbol{l}_i^{(d)T}\boldsymbol{\phi}_j^{(d)} = \delta_{ij} \tag{39b}$$

In this section we seek to determine the gain matrix F such that the dominant closed-loop left eigenvectors $l_i^{(d)}$ are approximately orthogonal to the known disturbance input matrix E. Because in this section we concentrate particularly on the left eigenvector problem, we will generally omit the word "left" when discussing these eigenvectors. Although this problem has many of the same features as those in the right eigenvector selection problem, the problem formulation and solution will differ due to the different approximations that are appropriate to this problem.

Feedback Gain Matrix Computation

For a large structural system it is neither possible nor desirable to control all the modes of a system. Assume that there is a total of r modes that we desire to control, where r is also the dimension of the output vector $\boldsymbol{y}$. Now, since the modal vectors span the N-dimensional space, it is reasonable to write the desired closed-loop (left) eigenvectors as a linear combination of the original open-loop (left) eigenvectors. We will in fact restrict the closed-loop eigenvectors further by assuming that the desired eigenvectors are a linear combination of a subset of $q \geqslant r$ open-loop eigenvectors, which we arrange as the matrix $L_1 = [l_1 \ldots l_q]$:

$$\boldsymbol{l}_i^{(d)} = L_1\boldsymbol{p}_i \qquad \text{for} \qquad i = 1, \ldots, r \tag{40}$$

This assumption allows us to use only the known or identified eigenvectors in the analysis and is justified by noting that the contribution of high-frequency (open-loop) modes to low-frequency (closed-loop) modes is generally negligible. This can be shown as follows. Consider the case where $q = N$, such that $L_1 = L$. Rewrite Eq. (39a) as

$$[A^T - \lambda_i^{(d)}I]\boldsymbol{l}_i^{(d)} = \boldsymbol{g}_i \qquad \text{where} \qquad \boldsymbol{g}_i = -[BFD]^T\boldsymbol{l}_i^{(d)} \tag{41}$$

Substituting Eqs. (40) into (41) and premultiplying by Φ^T yields

$$[\Lambda - \lambda_i^{(d)}I]\boldsymbol{p}_i = \Phi^T g_i = \hat{g}_i \tag{42}$$

The jth component of the vector $\boldsymbol{p}_i$, $\boldsymbol{p}_{ji}$ is found from Eq. (42) as

$$\boldsymbol{p}_{ji} = \frac{\hat{g}_{ji}}{\lambda_j - \lambda_i^{(d)}} \tag{43}$$

This implies that only when λ_j is close to $\lambda_i^{(d)}$ is l_j important in the closed-loop model. The truncation error due to omission of $\boldsymbol{p}_{ji}$ with $j > \boldsymbol{q}$ is allowable in practice, since it will be less than other possible errors such as model uncertainty and extraneous disturbances. In our experience the truncation error in general can be neglected, when $q > (1.5\text{–}2.0)r$. In some particular case, if the desired eigenvalues and eigenvectors are carefully chosen, we can even take $q = r$. It must be noted that Eq. (40) does not mean the L_1 spans the achievable eigenvector space of the closed-loop system. The achievability of the desired left eigenvectors will be discussed later.

To determine the feedback gain matrix F to achieve eigenvector placement, substitute Eq. (40) into Eq. (39) and premultiply by Φ_1^T. The result is

$$[\Lambda_1 + \Phi_1^T D^T F^T B^T L_1 - \lambda_i^{(d)} I]\,\boldsymbol{p}_i = 0 \tag{44}$$

where Λ_1 is an $(r \times r)$ diagonal matrix of open-loop eigenvalues of the controlled modes. Equation (44) can be written in a more convenient form:

$$[\lambda_i^{(d)} I - \Lambda_1]\,\boldsymbol{p}_i = [\Phi_1^T D^T F^T B^T L_1]\,\boldsymbol{p}_i \qquad \text{for} \qquad i = 1, \ldots, r \tag{45}$$

Equation (45) is the fundamental equation that relates the old and new eigenvectors, eigenvalues, and the gain matrix F. Note that only the identified eigenvalues and eigenvectors are required in this formulation. We are interested in what assignments are possible; plus, for a specified eigenspace we need an algorithm to solve for the gain matrix F.

In condensed form Eq. (45) can be rewritten as

$$\hat{P} = ZH \tag{46}$$

where

$$\hat{P} = [\ldots \hat{\boldsymbol{p}}_i \ldots], \qquad \hat{\boldsymbol{p}}_i = [\lambda_i^{(d)} I - \Lambda_1]\boldsymbol{p}_i, \qquad Z = [D\Phi_1]^T F^T, \qquad H = B^T L^{(d)}$$

The matrices Z and H are $(q \times m)$ and $(m \times r)$, respectively, and $\hat{P}$ is a matrix $(q \times r)$ with $\text{rank}[\hat{P}] \leqslant \min(m,r)$, since $q \geqslant r$.

It is well known that any (singular or regular) matrix can be decomposed as two arbitrary full rank matrices,[21] and if one of them is given, the other will be uniquely determined. For cases of practical interest we consider $m \leqslant r$. (In practice, the number of controls m is, in general, less than the number of outputs r.) If, for example, we pick $[\hat{P}]$ such that $\text{rank}[\hat{P}] = m$, then such H is a known full rank matrix. The matrix Z can then be uniquely determined from Eq. (15). Furthermore, $D\Phi_1$ is a known square matrix of order r and is assumed to be nonsingular. Hence, if Z is known, the feedback gain matrix F can be obtained as

$$F = Z^T[D\Phi_1]^{-1} \tag{47}$$

If the matrix Z were directly resolved from Eq. (46), the desired eigenvalues and left-hand eigenvectors must be chosen so that $\text{rank}[\hat{P}] = m$. Otherwise, there may be none or an infinite number of solutions of Z. To avoid this problem, we seek to formulate an algorithm that does not require this explicit calculation and leads to a unique solution of Z under some criteria. Reconsider Eq. (40) and rearrange as

$$\begin{aligned} \boldsymbol{l}_i^{(d)} &= L_1[\lambda_i^{(d)}I - \Lambda_1]^{-1}[\lambda_i^{(d)}I - \Lambda_1]\boldsymbol{p}_i \\ &= L_1[\lambda_i^{(d)}I - \Lambda_1]^{-1}\hat{\boldsymbol{p}}_i \\ &= \hat{L}_{1_i}Zh_i \qquad \text{for} \qquad i = 1, \ldots, r \end{aligned} \tag{48}$$

where

$$\hat{L}_i = L_1[\lambda_i^{(d)}I - \Lambda_1]^{-1}$$

and h_i is the ith column of H.

From Eq. (48) each element $l_{kr}^{(d)}$ of $\boldsymbol{l}_r^{(d)}$ can be expressed as a linear combination of components z_{lj} of the unknown matrix Z. These $N \times r$ linear algebraic equations can be arranged in the matrix form

$$\boldsymbol{l}^{(d)} = T\boldsymbol{z} \tag{49}$$

where the vector $\boldsymbol{z}$ contains the elements z_{lj} for $l = 1, \ldots, q$, $j = 1, \ldots, m$, of the unknown matrix Z. The variable $\boldsymbol{l}^{(d)}$ is an $N \times r$ vector containing all specified components of $\boldsymbol{l}_i^{(d)}$, $i = 1, \ldots, r$, and T is an $(N \times r, m \times q)$ known rectangular matrix.

In general this set will be overdetermined; however, a least squares solution of Eq. (49) can be obtained as

$$\boldsymbol{z} = T^{\dagger}\boldsymbol{l}^{(d)} \tag{50}$$

where $T^{\dagger}$ is the pseudoinverse of T, generally obtained by a singular value decomposition algorithm. From $\boldsymbol{z}$ we may form Z. The feedback gain matrix F is then derived from Eq. (47).

In Eqs. (47–50), it is assumed that r eigenvalues and $N \times r$ entries of left eigenvectors are assigned. This is in contradiction to the theorem shown by Srinathkumar.[16] In fact, Eq. (49) implies that, if the vector $\boldsymbol{l}^{(d)}$ formed from all entries of the desired eigenvectors lies precisely in the subspace spanned by the columns of T, they will be achieved exactly. Otherwise, only the projection of $\boldsymbol{l}^{(d)}$ onto the "achievability subspace" can be achieved. Denote $\boldsymbol{l}_A^{(d)} = TT^{\dagger}\boldsymbol{l}_i^{(d)}$ as the achieved eigenvector obtained by using the feedback gain matrix calculated from Eqs. (47–50). This, then, is the closest eigenvector to the desired one in the least-squares sense. That is,

$$\min J = \sum_{i=1}^{r} \| \boldsymbol{l}_i^{(d)} - \boldsymbol{l}_{iA}^{(d)} \|^2 \tag{51}$$

In most practical cases, only certain components of $\boldsymbol{l}_i^{(d)}$ are required to be specified. A typical situation is that each of the columns of E contains only m dominant components. In this case, where only m entries are to be specified for each eigenvector, only those components specified are used in Eq. (49). The vector $l^{(d)}$ consists of the $r \times m$ specified entries, and T becomes a square matrix of order $r \times m$. The feedback gain matrix derived from Eqs. (47–50) can guarantee these specified components to be totally achieved.

Implementation Summary

In this section we consider specifically the flexible structure problem and several practical issues in achieving control implementation are discussed. The data base required by the proposed method consists only of a subset of eigenvalues and left and right eigenvectors of the original (open-loop) structure. These modal parameters can be either from an analytical model of the structure, such as a finite-element model, or from modal identification of the desired modes.

The advantages of using a modal identification approach are that this method avoids the modeling errors due to discretization and uncertainty in structural characteristics and also that an on-line modal identification implementation can be used to build up an integrated identification/control procedure in a realistic flexible structure problem. This may be required if the system exhibits time variation or nonlinear characteristics, which may produce significant variation in modal characteristics from the nominal design. Considering recent advances in the modal identification area,[22–27] this approach may be practical in future space structure problems.

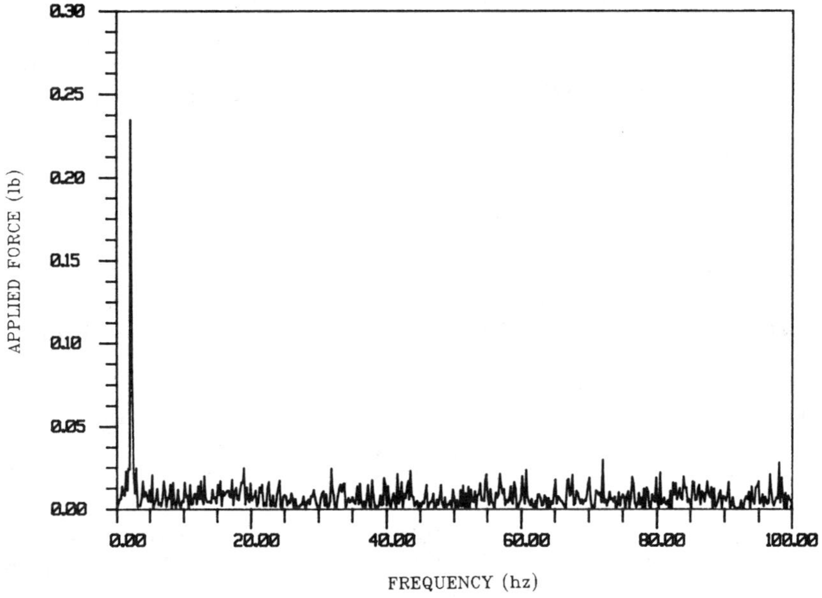

Fig. 7 Numerical generated spectrum of force configuration I.

Since only r modes can be controlled among total N modes, one has to first decide which modes to be controlled. Since the excitation of higher-order modes needs more energy than lower-order modes, a reasonable first approach is to consider only the lowest-frequency modes. For the present problem the selection of the controlled modes depends on the external forces. If the power spectrum of the external forces can be approximately estimated, the controlled modes must cover the frequency bandwidth where the main energy of the external forces is concentrated. The numerical example given in the next section will show how it is important to select controlled modes differently for different external force configurations.

Numerical Example

As an illustration, let us consider a cantilever beam similar to the beam considered in the previous section. The numerical model of a beam for this example consists of an eight-degree-of-freedom finite-element model with two actuators. A single external force is applied at the free end of the beam. The desired left eigenvector of the controlled modes $\boldsymbol{l}_i^{(d)}$ is chosen such that

$$\boldsymbol{l}_i^{(d)T} E = 0, \qquad i = 1, \ldots, r$$

Two different configurations of the external force are considered:

1) The external force is the sum of a sinusoidal function, whose frequency is 2 Hz, and a random (white noise) function (see Fig. 7).

2) The frequency of the sinusoidal function is at 25 Hz, and the random function is unchanged from 1 (see Fig. 8).

In the first configuration only the two lowest modes are controlled. Figure 9 shows the norm of the response to the external force described earlier without applying the active control. The forced response vector $\boldsymbol{x}(t)$ contains only the translations of the finite-element nodes, and the norm is defined as

$$\|\boldsymbol{x}\| = \sqrt{\boldsymbol{x}(t)^T \boldsymbol{x}(t)} \tag{52}$$

Figure 10 shows $\|\boldsymbol{x}\|$ with the active control. Comparing Fig. 10 with Fig. 9, we see that the global vibration of the structure is significantly reduced by using the proposed method. Note that, since the main energy of the external force is at 2 Hz, the contribution of higher modes on the response is very weak, and the effect of the external force can be greatly reduced by shaping the two lowest modes whose natural frequencies are 1.5 and 9.5 Hz, respectively.

In the second configuration we control successively the two lowest modes and then the four lowest modes. The results without and with control are shown in Figs. 11–13. From these figures we see that now controlling two modes is not enough to reduce the vibration of the structure, because these nodes do not cover the frequency bandwidth where the energy of the external force is concentrated.

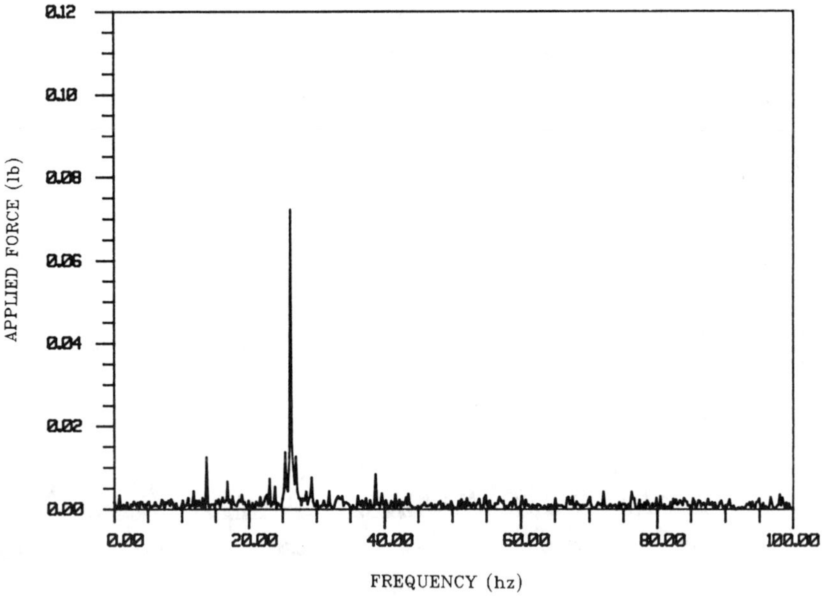

Fig. 8 Numerical generated spectrum of force configuration II.

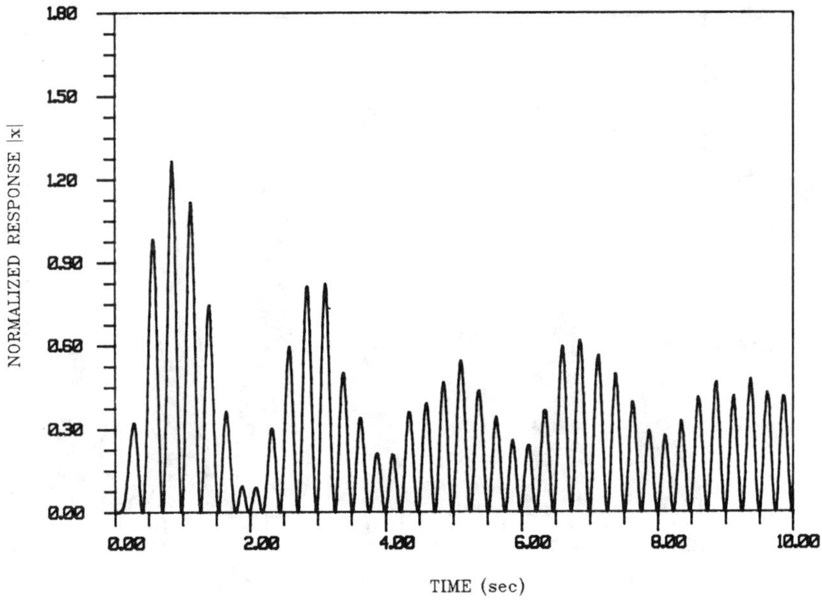

Fig. 9 Response without control (force configuration I).

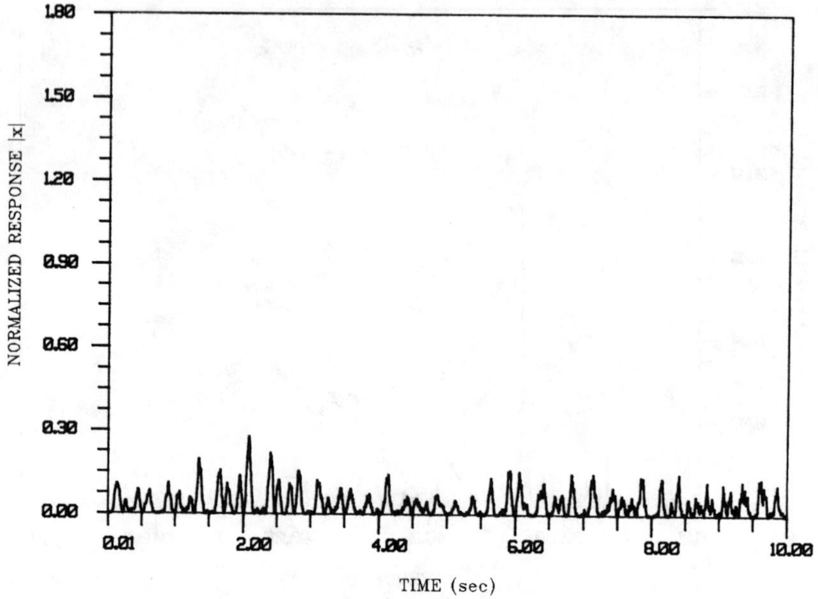

Fig. 10 Response with control (force configuration I).

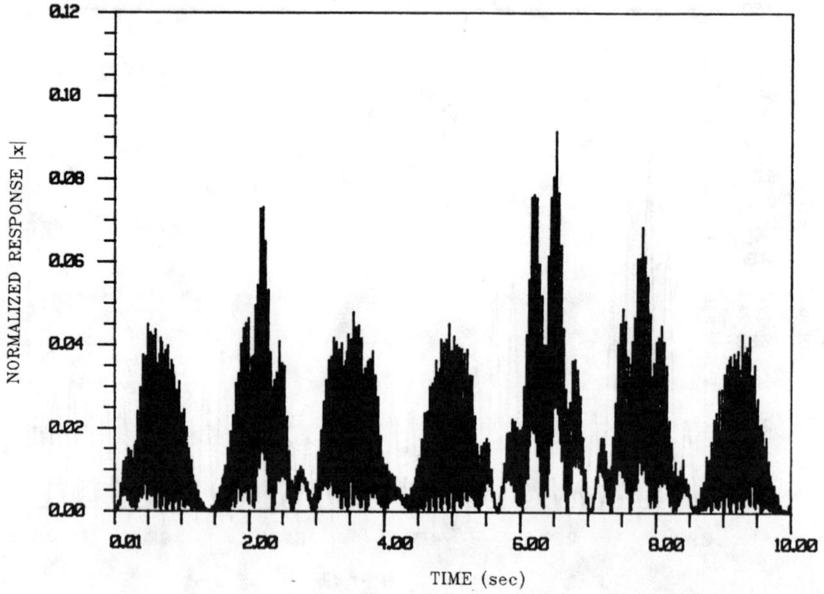

Fig. 11 Response without control (force configuration II).

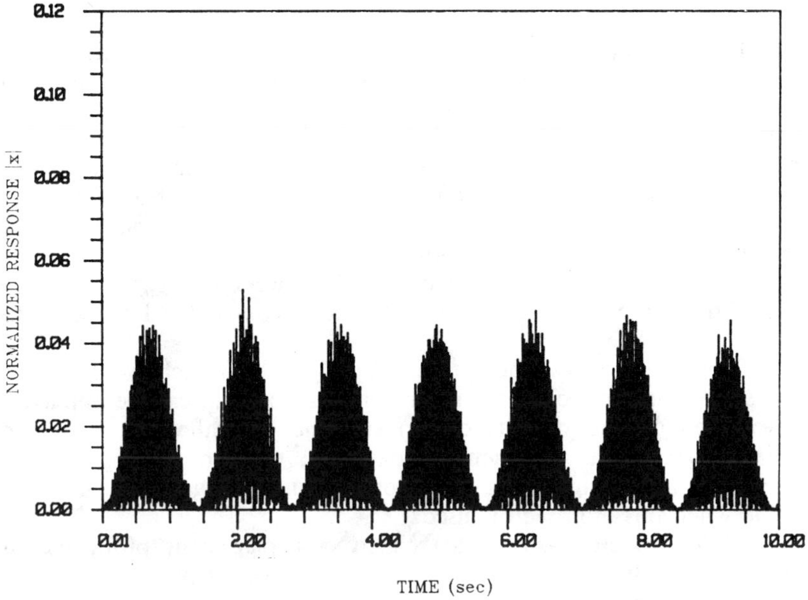

Fig. 12 Response with two controlled modes (force configuration II).

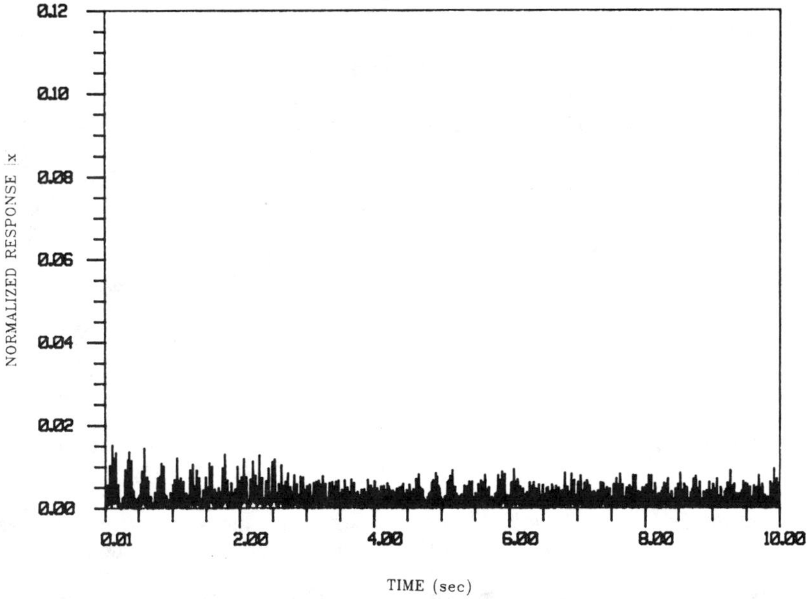

Fig. 13 Response with four controlled modes (force configuration II).

Concluding Remarks

The primary objective of this paper is to formulate an eigenstructure assignment algorithm that can be applied to large flexible systems whose dynamics are known only approximately and generally only within some specified frequency band. In addition to eigenplacement, it is equally important to keep control effort small; hence, an eigenplacement algorithm that also minimizes the control effort is an important contribution to the field. The synthesis technique formulated in this paper poses the eigenplacement problem as a parameter optimization problem for which standard numerical software is readily available. In addition, the algorithm as formulated requires only data that are generally available from a modal test.

This formulation allows the optimization algorithm to achieve the appropriate eigenstructure required for control minimization, can be applied to both continuous and discrete controllers, and also allows a variety of additional constraints (frequencies, damping, etc.) to be applied. The results show that a significant reduction in controller gain magnitudes is possible when this approach is used.

In addition to standard (right) eigenvector placement problems, the problem and application of left or reciprocal eigenvector placement are also examined and applied to an experiment in minimizing the forced response of a structure.

The tools for eigenplacement developed in this chapter indicate a useful and efficient way to achieve closed-loop control for large flexible systems.

References

[1]Wonham, W. M., "On Pole Assignment in Multi-Input Controllable Linear Systems," *IEEE Transactions on Automatic Control*, Vol. AC-12, No. 6, Nov. 1967, pp. 660–665.

[2]Jameson, A., "Design of a Single Input System for Specified Roots Using Output Feedback," *IEEE Transactions on Automatic Control*, Vol. AC-15, No. 3, May 1970, pp. 345–348.

[3]Davison, E. J., and Chatterjee, R., "A Note on Pole Assignment in Linear System with Incomplete State Feedback," *IEEE Transactions on Automatic Control*, Vol. AC-16, No. 1, 1971, pp. 98–99.

[4]Munro, N., and Vardulakis, A., "Pole Shifting Using Output Feedback," *International Journal of Control*, Vol. 18, No. 6, 1973, pp. 1267–1273.

[5]Kimura, H, "Pole Assignment by Gain Output Feedback," *IEEE Transactions on Automatic Control*, Vol. AC-20, No. 4, July 1975, pp. 509–516.

[6]Davidson, E. J., "On Pole Assignment in Linear Systems with Incomplete State Feedback," *IEEE Transactions on Automatic Control*, Vol. AC-15, No. 3, May 1970, pp. 348–351.

[7]Davidson, E. J., and Chow, S. G., "An Algorithm for the Assignment of Closed Loop Poles Using Output Feedback in Large Linear Multivariable Systems," *IEEE Transactions on Automatic Control*, Vol. AC-18, No. 1, 1973, pp. 74–75.

[8]Sridhar, B., and Lindorff, D. P., "Pole Placement with Constant Gain Output Feedback," *International Journal of Control*, Vol. 18, No. 5, 1973, pp. 993–1003.

[9]Moore, B. C., "On the Flexibility Offered by State Feedback in Multivariable Systems Beyond Closed Loop Eigenvalue Assignment," *IEEE Transactions on Automatic Control*, Vol. AC-21, No. 5, Sept. 1976, pp. 689–692.

[10]Fahmy, M. M., and O'Reilly, J., "On Eigenstructure Assignment in Linear Multivariable Systems," *IEEE Transactions on Automatic Control*, Vol. AC-27, No. 3, June 1982, pp. 690–693.

[11]Juang, J. N., Lim, K. B., and Junkins, J. L., "Robust Eigensystem Assignment for Flexible Structures," Proceedings of the 1987 Guidance, Navigation and Control Conference, Monterey, CA, Aug. 1987.

[12]Fennell, R. E., "An Application of Eigenspace Methods to Symmetric Flutter Suppression," Institute for Computer Applications in Science & Engineering, Rept. 88-9, 1988.

[13]Andry, A. N., Jr., Shapiro, E. Y., and Chung, J. C., "Eigenstructure Assignment for Linear Systems," *IEEE Transactions on Aerospace and Electronic Systems*, Vol. AES-19, No. 5, 1983, pp. 711–728.

[14]Zhang, Q., Allemang, R. J., and Slater, G. L., "Suppression of Undesired Inputs of Linear Systems by Eigenspace Assignment," *Journal of Guidence, Control, and Dynamics* (to be published).

[15]Zhang, Q., Shelley, S., Luo, X. N., Allemang, R. J., Brown, D. L., and Slater, G. L., "Experimental Study of Active Vibration Control of Flexible Structures Subject to Forced Vibration," Proceedings of the 7th International Modal Analysis Conference, Las Vagas, CA, Jan. 1989.

[16]Srinathkumar, S, "Eigenvalue/Eigenvector Assignment Using Output Feedback," *IEEE Transactions on Automatic Control*, Vol. AC-23, No. 1, 1978, pp. 79–81.

[17]Rew, D. W., Junkins, J. L., and Juang, J. N., "Robust Eigenstructure Assignment by a Projection Method: Applications Using Multiple Optimization Criteria," *Journal of Guidance, Control, and Dynamics*, Vol. 12, No. 3, 1989, pp. 396–403.

[18]Zhang, Q., Lui, J. Y. H., Allemang, R. J., and Tsuei, Y. G., "Active Vibration Control of Flexible Structure by Eigenstructure Assignment Technique," Proceedings of the 6th International Modal Analysis Conference, Orlando, FL, 1988.

[19]Bécus, G. A., and Sonmez, G., "Symbolic Manipulation and Numerical Optimization Techniques for Robust Eigenstructure Assignment in LQR Design of Multivariable Control Systems," *Proceedings of the IEEE CACSD Symposium*, 1989, pp. 102–109.

[20]Zhang, Q., Shelley, S., Luo, X. N., Allemang, R. J., and Brown, D. L., "Modal Control with Application to Vibration Suppression of Flexible Structures," Proceedings of the 13th International Seminar on Modal Analysis, Paper C-52, Louvain, Belgium, 1988.

[21]Lawson, C. L., and Handon, R. J., *Solving Least Squares Problems*, Prentice-Hall, Englewood Cliffs, NJ, 1974.

[22]Ibrahim, S. R., and Mikulcik, E. C., "A Method for the Direct Identification of Vibration Parameters from the Free Response," *Shock and Vibration Bulletin*, Vol. 47, Pt. 4, 1977.

[23]Brown, D. L., Zimmerman, R. D., Allemang, R. J., and Mergeay, R. J., "Parameter Estimation Techniques for Modal Analysis," *SAE Transactions*, Vol. 88, 1979, pp. 828–846 (Society of Automotive Engineers Paper 790221).

[24]Allemang, R. J., "Investigation of Some Multiple Input/Output Frequency Response Function Experimental Modal Analysis Techniques," Ph.D. Dissertation, Mechanical Engineering Department, University of Cincinnati, Cincinatti, OH, 1980.

[25]Vold, H., and Rocklin, T., "The Numerical Implementation of a Multi-Input Modal Estimation Algorithm for Mini-Computers," Proceedings of the 1st International Modal Analysis Conference, Orlando, FL, 1982, pp. 542–548.

[26]Fillod, R., Lallement, G., and Piranda, J., "Modal Identification of Non Perfect Linear Structures Using Harmonic Excitations," TR-82.S.1083, Applied Mechanics Lab., University of Besançon, Besançon, France, 1982.

[27]Juang, J. N., "Mathematical Correlation of Modal Parameter Identification Methods via System Realization Theory," *International Journal of Analytical and Experimental Modal Analysis*, Vol. 2, 1987, pp. 1–18.

Part 3. Controls/Structure Interactions: Analysis and Experiments

Chapter 18

Recent Literature on Experimental Structural Dynamics and Control Research

William L. Hallauer Jr.*
U.S. Air Force Academy, Colorado Springs, Colorado

Introduction

THIS survey covers literature published primarily during 1985–1989 that is relevant to experimental study of the dynamics and control of large spacecraft structures (LSS). Most of the references cited report research in the United States, but a few report research conducted abroad. All of the literature surveyed should be readily available in U.S. research libraries, or from standard sources such as the National Technical Information Service, the American Institute of Aeronautics and Astronautics, and the American Society of Mechanical Engineers. The references cited consist primarily of articles published in archival journals, papers presented at major conferences and printed in the proceedings, formal reports issued by research laboratories, and books.

The principal subjects surveyed are experimental studies, facilities, and methods relevant to LSS in the areas of structural dynamics, passive control, and active control, with an emphasis on the last area. Most of the references cited discuss specific experimental studies and present experimental data. The others either describe existing or planned experimental facilities and programs or discuss experimental methods that are already well established or that hold promise for future application to LSS.

Almost all of the experiments conducted to date have involved structures that were not truly LSS, but rather terrestrial laboratory structures, also called testbeds, intended to be representative dynamically in some respects of LSS. Some of the desirable characteristics of such laboratory

---

*Visiting Research Professor, F. J. Seiler Research Laboratory and Department of Engineering Mechanics.

structures include high flexibility, low inertia, light inherent damping, as many undamped rigid-body modes as possible, low natural frequencies, high modal density, some joint nonlinearity, space-qualified sensors, actuators and computers, onboard power, and a laser and pointing system. Probably no terrestrial structure will ever have all of these characteristics, but even a very simple laboratory structure can have at least a few of them. Indeed, relevant and valuable experimental research has been conducted on structures ranging from simple and inexpensive spring-inertia systems to highly complex and costly models of generic spacecraft. Consequently, no report of an experimental study has been excluded from being cited because the laboratory structure was considered too trivial.

However, many possibly relevant reports published during 1985–1989 are not cited. Among these are conference papers that eventually appeared substantially unchanged as journal articles, which are cited. Also not cited are less formal publications, such as slide/viewgraph presentations, that seem to be superseded by subsequent publications that are cited. Finally, researchers in this field publish their work in a great variety of journals, conference proceedings, and reports, and it has not been practical for the author to search all of these sources. The reader who checks almost any publication cited in this survey will find in its reference list relevant publications that are not cited here. However, an attempt has been made to cite at least one publication of every research group known to be doing experimental work, especially in active control, so that the reader will be informed of who is doing what type of work. The reader interested in a particular research group or type of work discussed here should then check the appropriate cited publications not only for details of the work but also for references to other related publications that are not cited in this survey.

Four sets of conference proceedings[1–4] are very helpful general references for LSS dynamics and control, particularly with regard to major U.S. Government experimental facilities and programs. These proceedings document recent government-sponsored control/structures interaction (CSI) conferences that are convened irregularly every one to two years. The facilities and programs are described in general terms by slide/viewgraph presentations reproduced in Refs. 1–4. The following is a short list of presentation topics, which may be of interest by itself and also will familiarize the reader with some of a long and confusing array of acronyms:

1) the Control of Flexible Structures (COFS) program[1–3];

2) the Large Space Structure Ground Test Facility at the NASA Marshall Space Flight Center (MSFC)[1,3];

3) the Spacecraft Control Laboratory Experiment (SCOLE) at NASA Langley Research Center (LaRC)[1];

4) the Solar Array Flight Dynamic Experiment (SAFDE or, more commonly, SAFE)[1];

5) the Flexible Structure Slew Testbed (CSDL testbed) at the Charles Stark Draper Laboratory (CSDL), sponsored by the U.S. Air Force (USAF) Astronautics Laboratory (AL)[2,3];

6) the Large Flexible Structure Test Facility at the Jet Propulsion Laboratory (JPL) of the California Institute of Technology (Caltech), sponsored jointly by USAF AL and NASA[2,3];

7) programs, facilities, and experiments at or sponsored by the USAF Flight Dynamics Laboratory (FDL), including Vibration Control of Space Structures (VCOSS II) and the Large Space Structures Technology Program (LSSTP)[2–4];

8) the Active Control Evaluation of Spacecraft (ACES) programs at NASA MSFC[2];

9) the Passive and Active Control of Space Structures (PACOSS) program at Martin Marietta Astronautics Group, sponsored by USAF FDL[2,4];

10) the USAF Joint Optics Structures Experiment (JOSE)[2];

11) the Advanced Space Structures Technology Research Experiments (ASTREX) Laboratory at USAF AL[3,4]; and

12) two flight experiments planned for the near future, the Controls, Astrophysics and Structures Experiment in Space (CASE), and the Low Power Atmospheric Compensation Experiment (LACE).[4]

Structural Dynamics

The literature cited in this section deals primarily with modal and other structural dynamics testing methods relevant to LSS, and with testing done to date on laboratory structures that are representative in some sense of LSS. Most of the testing methods discussed are applicable mainly to ground testing, but the subject of on-orbit testing has been explicitly addressed in a few studies.

The state of the art in modal testing of all kinds of structures has advanced steadily since the early 1970s, when the fast Fourier transform and small but powerful computers were first unleashed on the field. Several comprehensive references on modal testing have been published recently, of which the following are cited: the textbook by Ewins,[5] another textbook that emphasizes multiple-exciter methods by Zaveri,[6] an article by Stroud,[7] a six-volume contract report by Allemang and Brown,[8] and a handbook chapter also by Allemang and Brown.[9]

Ground testing of an LSS prior to launch into orbit will be required in order to validate mathematical models and to qualify the structure for flight. But an LSS may be so weak and flexible that it would collapse under its own weight if completely assembled on Earth and supported in a manner that permits the freedom of deformation possible in flight. Researchers are developing three general methods for solving this problem: multiple boundary condition testing (MBCT), experimental component mode synthesis (CMS), and testing of scale models.

References 10–12 are representative of a number of publications on MBCT by authors at JPL. The method is summarized as follows. First, support the fully assembled LSS at enough points so that its structural integrity is not jeopardized, and test the restrained structure. Next, change to another set of safe supports and test again. Repeat this procedure until enough data have been assembled that the unrestrained structure's stiffness and mass matrices can be inferred mathematically with confidence. To date, MBCT has been applied experimentally only to correct large errors intentionally introduced in the mathematical models of two beams, one uniform and the other nonuniform.[10]

The general approach in experimental CMS is first to isolate each major component, or substructure, of the structure and test it individually, and then to assemble mathematically the data from all the component tests into data characterizing the full structure.[8] One frequency-domain CMS method was proposed by Ewins[13] and applied by researchers at the Massachusetts Institute of Technology (MIT) to a structure consisting of two beam substructures.[14]

The third general method for ground testing of LSS is to test fully assembled, reduced-size scale models of an LSS, and then mathematically extrapolate the data to infer the characteristics of the actual structure. References 15 and 16 discuss scale model technology development centered at NASA LaRC and focusing on the proposed Freedom Space Station.

An important adjunct to an LSS scale model is a vertical suspension system that supports the model's weight without significantly changing the modes of vibration relative to those of the unrestrained structure. Two approaches were evaluated experimentally for the Freedom Space Station scale model: a zero-spring-rate mechanism (ZSRM) and a hybrid pneumatic-electromechanical device.[17]

The SAFE structural dynamics testing on Space Shuttle flight STS-41D in September 1984 was evidently the only relevant on-orbit experiment conducted to date. A 4 × 32 m solar array was extended from the Shuttle's cargo bay and was tested at two different deployment lengths. Some of the measurements are discussed in papers from the Lockheed Missiles and Space Company[18] and NASA LaRC.[19] Both papers focus on inherent damping characteristics.

Frequency-domain methods for on-orbit system identification of LSS are being developed at JPL.[20,21] The principal objectives are to identify only the information essential for the design of robust control and to do so using automated operations that fit within the constraints of flight. The methods have been tested on a very flexible, antennalike laboratory structure.

Several generic models of different types of space structures have been tested dynamically at NASA LaRC.[19,22–25] In most cases complementary mathematical analyses were conducted, and test results were used to refine the mathematical models. Planar, cable-stiffened frame structures are the subjects of Refs. 22 and 23. Laboratory hoop and radial-rib structures were dynamically similar to the frameworks of proposed space antennas and solar power stations, and the results of conventional linear finite-element analyses agreed well with modal test measurements.[22] A laboratory guyed boom was a pinned beam with two taut cables attached to its tip, a possible design for a spacecraft appendage.[23] Inertial loads on the guyed boom from, for example, a slew maneuver could cause cable slackening and lead to nonlinear transient response. This type of loading was simulated, and the nonlinear mathematical analysis was validated by good agreement with measured dynamic response.

References 24 and 25 describe analyses of two much more complex laboratory structures: a generic space station model consisting of five substructures, which ranged in stiffness from essentially rigid to extremely flexible[24]; and a cable-stiffened hoop-column antenna complete with a surface mesh.[25] Cable suspension systems were evaluated in both experiments. Each structure was tested within both atmospheric and near-vacuum air pressure levels. For both of these complex structures, extensive refinement of the mathematical model based on static measurements was required before reasonable agreement was achieved between predicted and measured modal parameters.

A modal test conducted at NASA LaRC on a beamlike space truss is discussed in Ref. 19. The truss members had relatively low bending stiffness; consequently, local member bending vibration modes coupled with and obscured the global truss modes. Sensors were mounted only on joints connecting the members; hence, the available response data were inadequate for accurate measurement of modal parameters. Exactly the same problem appeared in an independent experiment on a beamlike planar truss at Martin Marietta.[26]

Martin Marietta's testbed for the PACOSS program was designed to represent a flexible satellite.[27,28] It consists of several different types of substructures, and the entire assembly is supported by ZSRMs to simulate unrestrained flight. The results of extensive modal analysis and testing have been reported for several of the principal substructures[27] and for the full assembly.[28]

Researchers from Japan's Toshiba Corporation and Institute of Space and Astronautical Science built a space truss with the basic form of a 3.5-m-square by 0.7-m-thick flat plate.[29] The basic form can be adjusted into different shallow-shell forms with the use of variable-length truss members, so that this testbed can assume various different antenna shapes. Modal analysis and test results for five configurations are presented in Ref. 29.

Vertically suspended planar grids have been used often to represent LSS dynamics.[30–33] This type of structure is relatively simple and inexpensive to build, yet can have two-dimensional, very-low-frequency modes and high modal density. A planar grid at NASA LaRC was the testbed for several experiments, including a study of identification by least-squares lattice filters.[30] Reference 31, from Virginia Polytechnic Institute and State University (VPI&SU), presents a study of transient flexural waves in a planar grid. A planar grid at USAF AL was used in studies of modeling and identification techniques[32] and of the influence of gravity on accelerometers.[33] It was demonstrated that gravity can seriously contaminate low-frequency acceleration signals, and an estimation method to compensate for the contamination was tested on the planar grid.

Several publications describe structural dynamics identification methods designed to be applied eventually to LSS, but actually applied first to specific active control testbeds.[34–40] These reports are discussed in the context of the active control experiments in the section on active control.

The subject of nonlinear structural dynamics is obviously important, but the author has found only one recent experimental paper specifically oriented toward LSS applications: Ref. 14 assesses the effect of a sloppy joint on the application of experimental CMS to a joined-beam structure. It should be noted also that a book by Moon[41] describes in detail several experiments on chaotic vibration of beam structures. There is certainly much more literature on experimental nonlinear structural dynamics; thus, the interested reader should search elsewhere.

Passive Control

Some representative publications dealing with the passive control of global structural response are cited in this section. References 42 and 43, the proceedings of recent major conferences on damping, include many papers of this orientation. They also include many papers reporting the latest research on material damping properties, which is not considered here.

"Passive control," as used here, implies the dissipation of vibrational strain energy by mechanisms that do not require a source of power. Three general types of passive control are considered: the inherent damping that exists in any structure; damping that is designed into a structure with the use of special, highly dissipative materials; and energy absorption by mechanical or electromechanical components.

All structures have inherent damping due to mechanisms such as conversion of strain energy into heat and friction at joints and fasteners. Much effort has gone into the measurement of global inherent damping in structures and components related to LSS. The damping of the SAFE solar array was measured in space.[18,19] In a series of experiments conducted by Stanford University at NASA Ames Research Center, modal damping factors of aluminum and composite beams and plates and aluminum planar frames were measured in spacelike conditions: the test articles were catapulted into free flight in a vacuum, and their vibration responses were telemetered to recording units.[44] Researchers at Utah State University evaluated the inherent damping of a laboratory space truss due primarily to its joints; they found different levels of damping for different orientations and levels of static loading.[45]

An assumption shared by many researchers is that it will be necessary to design additional passive damping into most LSS. However, Ashley and Edberg[46] used data collected from spacecraft, aircraft, and laboratory experiments "to advance the claim that the modal damping naturally available in most large structures deployed in space will be more than adequate to permit the successful design of active control systems, without the use of special measures to augment passive energy dissipation."

Despite the arguments made in Ref. 46, research continues in the area of passive damping augmentation, particularly with the use of viscoelastic damping materials. In work by CSA Engineering for the PACOSS program, integral viscoelastic damping was used in an 18.3-m beamlike space truss; high levels of modal passive damping were measured for the truss, and the theoretical model accurately predicted the damping performance.[47] In a somewhat similar study JPL researchers tested several different viscoelastic member designs as the single integral damping member in a small beamlike space truss.[48] Viscoelastic materials designed into joints also can produce substantial passive damping, as was demonstrated by experiments conducted at the Georgia Institute of Technology.[49]

Martin Marietta's passive control experiments in the PACOSS program were probably the most ambitious attempted to date.[28,50] Several different types of viscoelastic passive damping components were used in the complex testbed, including rotational and extensional shear dampers, distributed constrained layer damping, and tuned-mass dampers. This research demonstrated that it is possible to design significant and predictable levels of passive damping into LSS.

A group at MIT has done extensive research on damping augmentation using passive vibration absorbers, including both tuned-mass dampers and piezoelectric truss members.[51–53] The types of absorbers used are capable also of providing actuation for active vibration control, and some were used both passively and actively in the MIT experiments. Accordingly, these studies are described in more detail in the next section.

Active Control

"Active control," as used here, implies closed-loop feedback control with the use of sensors to generate electrical signals representative of motion quantities; controllers (either analog or digital) to accept sensor signals, execute control algorithms, and generate command signals; and actuators to accept command signals and effect control forces and moments. Both control of vibration alone and simultaneous control of maneuvering and vibration are included in this definition.

If a translation or rotational sensor occupies physically the same motion degree of freedom as, respectively, a force or moment actuator, then the sensor and actuator are said to be collocated, or dual. "Active damping" is a subset of active control, the simplest form being direct feedback (without dynamic compensation) of velocity to a collocated actuator. Some authors refer to any form of velocity feedback as active damping.

There were a few active control experiments prior to 1980, but most were related to subjects other than LSS, such as acoustics. The first major U.S. Government program was the Active Control of Space Structures (ACOSS) Program started in 1978 and completed in 1984, and it produced many of the first experiments related to LSS. The September–October 1984 issue of the *Journal of Guidance, Control, and Dynamics* focused on the early experiments.[54] A general survey of LSS dynamics and control in that issue cited only 14 reports of active control experiments in a list of 194 references.[55] Another article in that issue[56] and an article published a year earlier[57] tabulated the experiments then known by their authors to be in existence; each identified fewer than a dozen experiments in the entire United States, and most of those had not yet been reported in the open literature.

However, the trickle of experimental studies that began in the late 1970s grew into a healthy stream by 1985, and in late 1989 they seem to be approaching flood level. Abundant evidence of this growth is presented in a recent survey of past, present, and future experiments and experimental facilities in the United States and Canada.[58]

A few publications give extensive descriptions of experimental facilities and programs for active control. Reference 59 describes a facility at CSDL, the primary component of which is the CSDL testbed. This structure consists of a rigid central hub mounted on a vertical-axis air bearing, with four horizontal flexible beams extending radially from the hub. The SCOLE facility[60] at NASA LaRC is based on a laboratory model of a rigid orbiter with rotational freedom about two axes, to which is attached an offset antenna reflector by means of a long, flexible beam. The Daisy facility[61] at the University of Toronto is based on a testbed consisting of a rigid hub with rotational freedom about three axes, to which 10 rigid spokes are flexibly attached. References 36, 62, and 63 describe the Flexible Structures Facility at the Ohio State University (OSU); experiments conducted there include vibration control of beams, slewing of one- and two-link flexible beams, and slewing of a satellite emulator. The Large Flexible Structure Test Facility[64] at JPL is based on an antennalike structure consisting of a central rigid hub with rotational freedom about two axes, a long, flexible beam hanging vertically from the hub, and 12 flexible ribs radiating horizontally from the hub. The ACES programs[65,66] consist of a series of activities at NASA MSFC involving several different complex flexible structures.

Any active control experiment has certain essential components: a laboratory structure, sensors, controllers and control techniques, actuators, and a system for data acquisition and analysis. It might seem that these components are all roughly equal in general importance and that any one except the last could be a logical basis for organization of a survey of the experimental active control literature. However, the practical experience of many researchers has been that actuators play the dominant role in determining the nature and success of an active control experiment. For this reason, the publications cited later are organized on the basis of the types of actuation used in the experiments reported.

An actuator capable of influencing motion of a spacecraft flying beyond the atmosphere obviously cannot be mechanically "grounded," i.e., reacted by the Earth or by atmospheric forces. Therefore, many researchers strive to use what are described as "structure-borne" or "space-realizable" actuators in their laboratory experiments. (The former term evidently was coined by Rockwell[67] many years ago, and the latter by Miller et al.[51] very recently.) However, for various reasons many experiments also have been conducted with the use of grounded actuators. Therefore, the two major subsections of this section on active control are, first, experiments based on grounded actuators and then experiments based on structure-borne actuators. This order of presentation follows from the experience in some cases that grounded-actuator experiments of a research group logically and chronologically preceded the same group's structure-borne-actuator experiments.

Experiments Based on Grounded Actuators

Active control experiments have been conducted with the following types of grounded actuators: force motors, torque motors, tendons, and airjets.

Active Vibration Control Using Grounded Force Motors

Researchers at the TRW Corporation have conducted experiments since 1984 on a laboratory structure consisting primarily of a heavy plate supported horizontally on top of a one-story, four-column frame structure. References 37 and 68–71 are representative publications, with Ref. 71 being a summary and retrospective of the work through mid-1988. The actuators used were magnet-coil force motors mounted in diagonal truss members attached to the grounded base and top of adjacent frame columns. The TRW experiments implemented advanced system identification and vibration control techniques and included testing of passive damping.

A two-story, three-column hanging frame structure is the testbed for experiments at Caltech.[39,72] As in the TRW experiments, grounded magnet-coil actuators are mounted to diagonal truss members of the first story.

A research group at VPI&SU has conducted several vibration control experiments using magnet-coil force motors. A principal objective of this work has been achievement of good agreement between theoretical predictions and experimental measurements. The laboratory structures used were a beam mounted on taut cables and a pendulous, skewed planar grid. Experimental results have been reported on different types of modal-space active damping,[73–75] the sensitivity of control system performance to structural modifications,[76,77] and comparisons of various active damping techniques with optimal control.[78,79]

Researchers at Italy's Politechnico di Milano used three independent control units (ICUs) on a 3.5-m beam suspended horizontally by cables.[80,81] Each ICU consisted of a velocity sensor, an analog controller, and a grounded electrodynamic shaker collocated with the sensor and serving as a force actuator. Direct feedback of velocity and position produced an active viscous dashpot and an active spring at each sensor-actuator location.

References 35 and 82–85 report experimental work at the German Aerospace Research Establishment (DFVLR), with Ref. 84 being an overview of studies through 1987. Electromagnetic force motors were used as actuators on two laboratory structures: a hanging cantilevered beam and a pendulous rectangular plate suspended from cables. (This was evidently the second experiment on a thin flat plate, the first being the Lockheed circular plate experiment.[56]) Vibration control was implemented using velocity feedback from sensors collocated with and dislocated from actuators. Also, wave-absorbing controllers were applied on the beam using control hardware located at its free end.[83]

Yet another experiment on a hanging plate was conducted at the Harris Corporation.[86] Grounded magnet-coil force motors were collocated with accelerometers to test a maximum entropy/optimal projection design for vibration control.

Some of the first experiments conducted on the JPL antennalike structure involved methods for active damping[64] and model reference adaptive vibration control.[64,87] Adaptive control is an important subject for LSS due to the likelihood of inaccurate mathematical models, slowly time-varying dynamics as LSS are constructed and modified in orbit, and in-service failures of control system components. In the JPL experiments, grounded magnet-coil force motors acted on the antenna hub through moment arms about a two-axis gimbal assembly, thus effecting rotational control moments. The control sensing used was essentially collocated with the actuation.

Studies conducted at the University of Cincinnati used electrodynamic shakers as force actuators for control of beam vibration. The researchers evaluated dislocation of control accelerometers from the single actuator supporting a beam at its center[88] and multivariable control of a cantilevered beam's tip response when the beam is excited by a disturbance having a single dominant frequency.[89] Reference 89 also describes a method to account for the time delay introduced by digital processing of the control algorithm. This important issue was addressed in the earlier articles on Lockheed experiments,[56,57] and one finds some discussion of it in most recent reports that describe applications of digital controllers.

Active Control Using Grounded Torque Motors

An article by Stanford researchers reports one of the earliest experimental studies of adaptive vibration control.[90] The laboratory structure was a two-degree-of-freedom spring-inertia system, and the control components were a rotational position sensor and a dislocated torque motor.

There have been several experiments in which flexible beams were rotated (slewed) about a vertical axis by torque motors. In general the control objectives were to slew a beam quickly through a prescribed large angle in a horizontal plane while minimizing excitation of vibration, and, simultaneously, to suppress vibration. In experiments at NASA LaRC comparison of theoretically predicted with measured motion indicated that aerodynamic drag significantly influenced the dynamic response.[91] In follow-on experiments, slewing response of a beam was measured both in atmospheric pressure and in an evacuated vacuum chamber, and mathematical modeling of the aerodynamic drag was validated.[92] In Ref. 93 authors from the Tokyo Metropolitan Institute of Technology demonstrated application of the "mission function control" algorithm for slewing a spacecraft with flexible appendages.

Slewing of beams is a subject of interest in robotics research, since it is often desirable to minimize the mass, and hence increase the flexibility of manipulator arms. Two articles by OSU researchers report experiments in which beam tip position accuracy (rather than global vibration control) was a principal control objective.[94,95] These articles are representative of many more relevant publications in the robotics literature that are not cited here.

Active Vibration Control Using Grounded Tendons

The use of tendons is considered promising for motion control of large civil structures, as discussed in Refs. 96 and 97 from the State University of New York at Buffalo (SUNY Buffalo). The authors of Ref. 97 document the sources of delay between sensors and actuators in the SUNY Buffalo control system due to all the hardware, not just the digital controller.

Researchers at Japan's University of Osaka Prefecture have proposed and tested a type of tendon control intended to suppress vibration and to control the static shape of a flexible beamlike appendage attached to a massive spacecraft.[40,98–100] The testbed was a hanging cantilevered beam, and the tendon actuator was a grounded magnet-coil force motor. Velocity feedback has been applied with both collocated and dislocated sensor-actuator pairs.

Active Control Using Grounded Airjets

A unique experiment at the University of Washington was based on a laboratory structure consisting of a rigid hub and six flexibly attached rigid spokes.[101] The hub rested on a flat disk that, in turn, floated on the air cushion of a pressurized air table, giving the structure freedom to translate and rotate in the horizontal plane with very little friction. Control force was provided by airjets flowing from nozzles fixed to the table and inpinging on fins attached to the structure.

Experiments Based on Structure-Borne Actuators

In contrast to most motion sensors and grounded actuators, structure-borne actuators appropriate for laboratory experiments generally are not available commercially; they must be designed, fabricated, and assembled in-house, usually with the use of purchased hardware components. Therefore, many researchers in this field have devoted major efforts to building and testing actuators. The types of structure-borne actuators used for active control range from conventional rigid-body control devices such as reaction wheels and gasjets to recently developed devices based on constitutive properties relating material deformation to electrical, magnetic, and thermal fields.

Active Vibration Control Using Active Members

An active member is defined as a member capable both of the standard structural function, i.e., passively reacting and transmitting loads, and of simultaneously generating active control forces or moments. Some authors define active members as being capable of sensing as well as actuation. The functioning of active members involves only internal, self-equilibrating actions; hence, active members cannot influence rigid-body motion.

Wada[102] presented a general description of research in the area of adaptive structures, which by definition consist at least partly of active members. Wada discussed the literature on all aspects of active members, including theoretical studies and experiments intended primarily to explore the characteristics of the active members themselves. We consider here only experiments in which active members have been applied to effect vibration control.

The most common form of active member used to date is a beam element with a composite cross section. Typically, the cross section consists of a metallic core and thin layers of active material bonded to both top and bottom surfaces of the core. When one active layer is caused to expand along the beam axis and the other is caused to contract, then a spatially distributed bending moment is induced in the beam. The expansion and contraction can be produced in conventional metallic active layers by heating and cooling. In fact, this simple thermal actuation was applied in an experiment at JPL to dampen the first mode of a cantilevered beam at about 2 Hz.[103]

Better and more diverse performance can be achieved with piezoelectric active layers, which are caused to expand or contract by a voltage difference across the layer thickness. Researchers from CSDL and MIT have used piezoelectric polymer active bending elements to apply, primarily, a type of nonlinear active damping similar to Coulomb friction. Their reports discuss spatially uniform[104,105] and nonuniform[106] actuating layers, active layers for both actuating and sensing,[107,108] and an extensive series of experiments that culminated in active damping of the complete CSDL structure.[109]

The spatially distributed character of piezoelectric bending elements is well suited for application of modal-space control techniques, as illustrated by three studies on vibration control of cantilevered beams. Researchers at JPL and Caltech reported the first experimental implementation of "positive position feedback" (PPF), a technique involving modal position feedback and dynamic compensation; as many as six bending modes were controlled simultaneously with the use of ceramic sensing and actuating layers.[110] A modified form of PPF was applied with ceramic actuators in subsequent experiments at the Catholic University of America.[111] Researchers at the IBM Corporation used polymer sensing and actuating layers to achieve critical active damping of the fundamental bending mode.[112,113]

Piezoelectric material can be embedded in truss members as well as beam members and, when activated by voltage differences, can generate control forces in a truss structure. The application of this mechanism to vibration control of laboratory space truss structures has been explored in four studies using several different passive and active truss member designs. In experiments at MIT the inner and outer surfaces of thin-walled tubular piezoelectric truss elements were connected electrically (shunted) through simple resistor-inductor networks; the electromechanical characteristics of the ceramic material combined with the shunting networks to increase passively the first-mode damping of a laboratory truss.[53] In two sets of experiments on different space truss testbeds, JPL researchers applied a patented active truss member design and single-input/single-output vibration control. In one study a digital controller was used to implement four different control techniques; the most successful at controlling three truss modes was PPF between a dislocated sensor-actuator pair.[38,114] In the other

JPL study three types of feedback compensation between a collocated sensor-actuator pair were implemented by analog circuitry, and effective control of the first truss mode was achieved.[115] In experiments at Sandia National Laboratories, signals from four piezoelectric strain sensors were fed through a digital controller back to four Sandia-designed active truss members, which were dislocated from the sensors. Several modes of a cantilevered space truss were controlled by implementation of linear-quadratic Gaussian[116,117] and optimal projection[117,118] techniques; the latter was more successful but required an extraordinarily accurate mathematical model of the structure.

Another type of active truss member is a jackscrew driven by a torque motor. Such an actuator can be designed to be a very stiff, extensible link in a "variable-geometry truss." Researchers at VPI&SU have used these active members in space truss[119] and planar truss[120] linkages to control the first-mode vibration of highly flexible beams attached to the trusses.

Active Vibration Control Using Structure-Borne Force Motors

Magnet-coil force motors were used as active truss members by researchers from Japan's Institute of Space and Astronautical Science and the Toshiba Corporation.[121] They collocated relative displacement sensors with four of these motors to suppress first-mode vibration of a beamlike laboratory space frame.

Magnet-coil motors applied directly (not through gears) are unsuitable for general purposes as primary load-carrying structural members because 1) without electric power, they have zero stiffness; and 2) even with power, they are almost certainly much more flexible than other geometrically similar structural members, including the other types of active truss members discussed earlier. However, such motors may be practical if they are mounted in parallel with primary load-carrying members so that the general structural integrity does not depend on the motors. For example, a force motor attached through posts to two points along the span of a beam can introduce equal and opposite control moments into the beam by virtue of its force axis being offset from the beam's neutral surface. This approach was demonstrated at the U.S. Air Force Academy (USAFA); the fundamental mode of a hanging cantilevered beam was suppressed by a magnet-coil force motor dislocated from a position sensor.[122]

In the JPL antennalike testbed, structurally parallel magnet-coil force motors are attached between the hub and an inboard point of each rib.[64] Four of these rib actuators were used in concert with two hub actuators and six collocated sensors to implement probably the most complex system to date for multivariable adaptive vibration control.[123]

Active Vibration Control Using Structure-Borne Torque Motors

The authors of Ref. 124, SUNY Buffalo and NASA LaRC researchers, report an experiment on a long, hanging, cantilevered beam with an interior pin joint. The joint served as an "active hinge," since it consisted of a hinge, a torque motor, and a collocated sensor connecting the ends of the two beam segments at the joint. Substantial active damping was achieved in the second and third modes of the beam.

The ACES-I testbed at NASA MSFC was designed to present a great number of challenges in vibration control representative of those expected for LSS.[65,66,125] The principal components of this complex structure included the following: an extremely light and flexible 14-m mast hanging from a grounded three-axis gimbal assembly; several flexible appendages attached to the mast's base and tip; and an optical system consisting primarily of a grounded laser source, a mirror, and a two-axis laser detector fixed to one of the tip appendages, and a mirror actuated by a two-axis gimbal assembly mounted on a base appendage. The laser ray reflected off both mirrors to reach the detector, and the control objective was to minimize deviation of the ray from its target on the detector caused by excitation at the mast's base. Torque motors in the two gimbal assemblies provided actuation for three general control methods that had been developed in the ACOSS program.

The other experiments in this category all followed earlier studies of single-beam slewing. They involved multibody structures and two or more torque motors, one motor grounded and actuating an "inboard" body, and the others connecting the "outboard" bodies to the inboard body through hinged joints. The general control objective was to maneuver the articulated flexible structure from an initial position to a specified final position while minimizing vibration.

In one robotics experiment at Stanford the inboard and outboard bodies were, respectively, a flexible beam and a short rigid link, and the specific control objectives were to achieve timely and accurate positioning of the link tip and to maintain proper contact force between the tip and a target object.[126] In another Stanford experiment the inboard and outboard bodies were, respectively, a long rigid link and a flexible beam.[127,128]

Experiments in progress at OSU include one in which both inboard and outboard bodies are flexible beams and another in which mirrors are mounted through motors onto a flexible beam, and the specific control objective is to maintain the position of a reflected laser ray during a slew maneuver.[63]

One laboratory testbed at NASA LaRC was a three-body structure consisting of a rigid inboard hub and two separate outboard flexible beams.[129] Complex large-angle slewing maneuvers with active vibration suppression were achieved. Another NASA LaRC structure consisted of a rigid trolley, actuated to translate horizontally in one direction by a motor-cable system, and an outboard flexible beam pinned to the trolley.[130] A maneuver combining trolley translation and beam rotation was executed for four different output feedback designs intended to suppress beam vibration.

Active Vibration Control Using Reaction-Mass Actuators

A reaction-mass actuator (RMA) exerts an inertial control force or moment on the structure bearing it. The reaction mass is attached to the structure by a spring (mechanical and/or electromechanical); thus, an RMA cannot effect control of rigid-body position. Other names used in the literature for actuators that use reaction masses include proof-mass actuator, linear momentum exchange device, and linear dc motor.

A research project at MIT focused on the application of moment-generating pivoted proof-mass actuators (PPMs) for vibration suppression of a 7.3-m beam suspended horizontally by wires.[51,52] Four PPMs, which included collocated motion sensors, were mounted on the beam. An RMA can function as a passive vibration absorber if its natural frequency and damping are tuned appropriately, and the MIT PPMs were used as both passive absorbers and active actuators. The principal control technique applied was direct output feedback. Significant levels of damping were produced in several beam modes.

The PPM design used at MIT was based on an earlier Lockheed design.[56] In subsequent studies at Lockheed a PPM was mounted on the tip of a cantilevered beam and used as the actuator in a control system intended to maintain pointing accuracy of a laser ray.[34,131] Reflecting mirrors attached to both the PPM and the beam midspan were disturbed by the beam's vibration. The control technique used was a combination of high-authority dynamic compensation and low-authority direct-rate feedback.

Three main types of linear, force-generating RMAs, called types 1–3 here, have evolved in the past few years. Types 1 and 2 are based on the force between a current-carrying coil attached to the structure and the field of a permanent magnet assembly, which is the reaction mass. Type 1 RMA (short coil, long field) was developed in a project involving USAF FDL, NASA MSFC, and their contractors,[132] and it has been applied in experiments at USAF FDL[133,134] and at Martin Marietta.[28,50] Type 2 RMA (long coil, short field) was developed jointly by researchers at NASA LaRC, the University of Virginia, and SUNY Buffalo,[135] and the design has been modified at MIT.[53] In the type 3 RMA, a development of the Harris Corporation, it appears that a magnetic reaction mass is accelerated by the field of an electromagnet attached to the structure.[86] Although all of the RMAs previously discussed weigh at least several pounds, a miniature RMA similar in principle to type 1 has been fabricated at OSU.[62,63]

Researchers from the Air Force Institute of Technology[133] and OSU[134] conducted experiments at USAF FDL on a hanging cantilevered beam with a platform attached to its tip. Four type 1 RMAs were mounted on the platform in an arrangement providing control authority over torsion and bending in two orthogonal directions. Several different control algorithms were implemented, including a modal decoupling technique[133] and optimal projection and decentralized linear-quadratic Gaussian techniques.[134] Martin Marietta used active control effected by six type 1 RMAs distributed over the PACOSS testbed to complement extensive viscoelastic passive damping; the combined passive-active control was predictable and very effective.[28,50]

SUNY Buffalo researchers used a single type 2 RMA in experiments to control vibration of a cantilevered beam; one of their main objectives was to demonstrate that the dynamics of an RMA are intrinsically coupled with the controlled structure's dynamics and that the system's mathematical model should account for this coupling.[135–137] MIT's type 2 RMAs were used either to excite or to passively dampen a 5-m space truss suspended horizontally by wires; the electromechanical design of the tuned-mass damper permits adjustment of both frequency and damping factor for optimal performance on different modes of vibration.[53]

Harris' type 3 RMA was mounted at the tip of a 5-m compound pendulum and served as the actuator for a maximum entropy/optimal projection control design. Harris has also developed a precision linear actuator that appears to be a type 1 or 2 RMA.[86]

Active Control Using Reaction-Wheel Actuators

A reaction-wheel actuator (RWA) exerts an inertial moment on the structure bearing it. Unlike a reaction mass, a reaction wheel is not attached to the structure by a spring; thus, an RWA has control authority over both vibration and rigid-body angular position.

RWAs were used by OSU researchers in an experiment implementing a form of decentralized model reference adaptive vibration control of the NASA LaRC planar grid.[138] Also at NASA LaRC, two different types of RWA were installed on the SCOLE testbed.[60] In one SCOLE experiment a single RWA was used to effect indirect rate-feedback control: the feedback signal was estimated angular velocity at the RWA location, as calculated by a Kalman filter from the signals of dislocated sensors.[139]

A laboratory structure at Texas A&M University is similar to the CSDL testbed, with a rigid central hub mounted on bearings and four horizontal flexible beams extending radially from the hub.[140] The hub is restricted to slewing about its vertical axis. The unique characteristic of this testbed is the single hub-mounted RWA, which is the actuator for both hub slew angle and suppression of beam vibration. The authors of Ref. 140 report the performances in large-angle slewing maneuvers of two control laws.

Active Control Using Control-Moment Gyros

A control-moment gyro (CMG) is, like an RWA, an inertial moment-generating actuator. However, CMGs are more complicated, expensive, and dangerous (to operate in a laboratory) than RWAs; thus, they have been used only rarely in active control experiments. One of the few applications reported was an early Lockheed experiment.[57] The SCOLE testbed was equipped with CMGs,[60] but they have not been used for active control.

Active Control Using Structure-Borne Gasjets

Gasjets, or thrusters, are relatively simple force actuators that can effect control of both rigid-body motion and vibration. However, they have not been used widely for active control. The conventional gasjet is a nonthrottleable on/off actuator, and its highly nonlinear functioning appears to be incompatible with the linear control methods that many researchers consider essential for precision control. Moreover, gasjets require nonrenewable compressed gas or chemicals; hence, they may not be as practical for flight operation as actuators that function on electrical power. Nevertheless, cold-gas gasjets have been applied in a few active control studies, in particular, on the CSDL and SCOLE testbeds.

In one of the first experiments on the CSDL testbed, nitrogen jets located at the tip of a flexible appendage provided actuation for large-angle slewing maneuvers.[59] The principal achievement of the study was establishing the feasibility of gasjet actuation for combined slewing and vibration control of LSS.

Airjets were installed on the SCOLE antenna reflector, which is attached to the rigid orbiter model by a flexible beam.[60] In one experiment a thruster was used to produce a large-angle slew without exciting the fundamental vibration mode.[141] The SCOLE thrusters were applied also in studies of methods for detecting failures in sensors and actuators.[142,143]

Active Control Using Two or More Types of Actuators

Several active control testbeds have been equipped with more than one type of actuator, but few experiments have been conducted in which two or more types were operating simultaneously. The following two studies are perhaps the only that have been reported.

In an extensive set of experiments on the CSDL testbed, various combinations of gasjets, RMAs, and a grounded hub torque motor were applied simultaneously.[144] The gasjets effected coarse actuation of large-angle slewing but were unsuitable for precision control. Integration of the linear actuators into the control system significantly improved the performance in fine pointing and tracking and in vibration suppression.

Finally, airjets and RMAs were used together to actively dampen the vibration of a 7.1-m cantilevered planar truss at the U.S. Air Force Academy.[145] The airjets were effective at low frequencies and the RMAs at higher frequencies; the objective of the study was to combine the two types in order to produce effective control over a greater frequency band than that of either type alone.

Acknowledgments

I am very grateful to the many researchers who sent me literature and to the U.S. Air Force Academy for providing the resources required for this endeavor. My residency at the Academy was sponsored by the Air Force Office of Scientific Research through the University Resident Research Program.

References

[1]Ryan, R. S., and Scofield, H. N. (eds.), "Structural Dynamics and Control Interaction of Flexible Structures," NASA CP-2467, Pts. 1 and 2, 1987.

[2]Wright, R. L. (ed.), "NASA/DOD Control/Structures Interaction Technology 1986," NASA CP-2447, Pt. 1, 1986 and Pt. 2, 1987.

[3]Swanson, A. D. (ed.), "NASA/DOD Control/Structures Interaction Technology 1987," AFWAL-TR-88-3052, 1988.

[4]Newsom, J. R. (ed.), "NASA/DOD Control/Structures Interaction Technology 1989," NASA CP-3041, 1989.

[5]Ewins, D. J., *Modal Testing: Theory and Practice*, Wiley, New York, 1984.

[6]Zaveri, K., *Modal Analysis of Large Structures—Multiple Exciter Systems*, Bruel & Kjaer, Naerum, Denmark, 1984.

[7]Stroud, R. C., "Excitation, Measurement, and Analysis Methods for Modal Testing," *Combined Experimental/Analytical Modeling of Dynamic Structural Systems*, American Society of Mechanical Engineers Publ. AMD-Vol. 67, 1985, pp. 49–78; also *Sound and Vibration*, Vol. 21, No. 8, Aug. 1987, pp. 12–27.

[8]Allemang, R. J., and Brown, D. L., "Experimental Modal Analysis and Dynamic Component Synthesis," Vols. I–VI, AFWAL-TR-87-3069, 1987.

[9]Allemang, R. J., and Brown, D. L., "Experimental Modal Analysis," *Handbook on Experimental Mechanics*, edited by A. S. Kobayashi, Prentice-Hall, Englewood Cliffs, NJ, 1987, pp. 659–738.

[10]Kuo, C. P., and Wada, B. K., "System Identification of a Truss Type Space Structure Using the Multiple Boundary Condition Test (MBCT) Method," AIAA Paper 87-0746, 1987.

[11]Kuo, C. P., and Wada, B. K., "Multiple Boundary Condition Test (MBCT): Identification with Mode Shapes," AIAA Paper 88-2353, 1988.

[12]Glaser, R. J., Kuo, C. P., and Wada, B. K., "Multiple Boundary Condition Testing Error Analysis," AIAA Paper 89-1162, 1989.

[13]Ewins, D. J., "Modal Test Requirements for Coupled Structure Analysis Using Experimentally Derived Component Models," *Combined Experimental/Analytical Modeling of Dynamic Structural Systems*, American Society of Mechanical Engineers Publ. AMD-Vol. 67, 1985, pp. 31–47.

[14]Blackwood, G. H., and von Flotow, A. H., "Experimental Component Mode Synthesis of Structures with Sloppy Joints," AIAA Paper 88-2411, 1988.

[15]Letchworth, R., McGowan, P. E., and Gronet, M. J., "Space Station: a Focus for the Development of Structural Dynamics Scale Model Technology for Large Flexible Space Structures," AIAA Paper 88-2483, 1988.

[16]Gronet, M. J., Crawley, E. F., and Allen, B. R., "Design, Analysis, and Testing of a Hybrid Scale Structural Dynamic Model of a Space Station," AIAA Paper 89-1340, 1989.

[17]Kienholz, D. A., Crawley, E. F., and Harvey, T. J., "Very Low Frequency Suspension Systems for Dynamic Testing," AIAA Paper 89-1194, 1989.

[18]Pinson, E. D., "Damping Characteristics of the Solar Array Flight Experiment," *Damping 1986 Proceedings*, AFWAL-TR-86-3059, 1986, pp. DC-1–DC-26.

[19]Pappa, R. S., and Juang, J.-N., "Some Experiences With the Eigensystem Realization Algorithm," *Proceedings of the 6th International Modal Analysis Conference*, Vol. II, 1988, pp. 1575–1581; also *Sound and Vibration*, Vol. 22, No. 1, Jan. 1988, pp. 30–34.

[20]Yam, Y., Bayard, D. S., Hadaegh, F. Y., Mettler, E., Milman, M. H., and Scheid, R. E., *Autonomous Frequency Domain Identification: Theory and Experiment*, Jet Propulsion Lab., Pasadena, CA, Publ. 89-8, April 1989.

[21]Bayard, D. S., Hadaegh, F. Y., Yam, Y., Scheid, R. E., Mettler, E., and Milman. M. H., "Frequency Domain Identification Experiment on a Large Flexible Structure," *Proceedings of the 1989 American Control Conference*, Vol. 3, 1989, pp. 2532–2542.

[22]Belvin, W. K., "Vibration Characteristics of Hexagonal Radial Rib and Hoop Platforms," *Journal of Spacecraft and Rockets*, Vol. 22, No. 4, July–August 1985, pp. 450–456.

[23]Housner, J. M., and Belvin, W. K., "Dynamic Response and Collapse of Slender Guyed Booms for Space Application," *Journal of Spacecraft and Rockets*, Vol. 23, No. 1, Jan.–Feb. 1986, pp. 88–95.

[24]Belvin, W. K., and Edighoffer, H. H., "Dynamic Analysis and Experiment Methods for a Generic Space Station Model," *Journal of Spacecraft and Rockets*, Vol. 24, No. 3, May–June 1987, pp. 270–276.

[25]Belvin, W. K., and Edighoffer, H. H., "15 Meter Hoop-Column Antenna Dynamics: Test and Analysis," *NASA/DOD Control/Structures Interaction Technology 1986*, NASA CP-2447, Pt. 1, 1986, pp. 167–185.

[26]White, C. W., "Dynamic Testing of a Two-Dimensional Box Truss Beam," NASA CR-4039, Jan. 1987.

[27]Gehling, R. N., "Large Space Structure Damping Treatment Performance: Analytic and Test Results," *The Role of Damping in Vibration and Noise Control*, American Society of Mechanical Engineers, New York, Publ. DE-Vol. 5, 1987, pp. 93–100.

[28]Gehling, R. N., "Modal Survey of the PACOSS DTA," *Damping 1989 Proceedings*, WRDC-TR-89-3116, 1989, pp. KCC-1–KCC-31.

[29]Kuwao, F., Yoshihara, M., Motohashi, S., Takahara, K., and Natori, M., "Vibration Characteristics and Shape Control of Adaptive Planar Truss Structures," AIAA Paper 89-1288, 1989.

[30]Montgomery, R. C., and Sundararajan, N., "Identification of the Dynamics of a Two-Dimensional Grid Structure Using Least Squares Lattice Filters," *Journal of the Astronautical Sciences*, Vol. 33, No. 1, Jan.–March 1985, pp. 35–47.

[31]Hallauer, W. L., Jr., and Trivedi, D. J., "Experimental Study of Transient Waves in a Plane Grid Structure," *AIAA Journal*, Vol. 26, Feb. 1988, pp. 228–234.

[32]Das, A., Strange, T. J., Schlaegel, W. T., and Ward, J. M., "Experiment in Modeling and Parameter Estimation of Flexible Structures," *NASA/DOD Control/Structures Interaction Technology 1987*, AFWAL-TR-88-3052, 1988, pp. 389–400.

[33]Norris, M. A., Thompson, R. C., and Das, A., "Low-Frequency Response for Observer Design in a Gravity Environment," *Journal of Guidance, Control, and Dynamics* (to be published).

[34]Sridhar, B., Aubrun, J.-N., and Lorell, K. R., "Identification Experiment for Control of Flexible Structures," *IEEE Control Systems Magazine*, Vol. 5, No. 2, May 1985, pp. 29–35.

[35]Schäfer, B., and Holzach, H., "Identification and Model Adjustment of a Hanging Plate Designed for Structural Control Experiments," *Structural Control*, edited by H. H. E. Leipholz, Nijhoff, The Hague, 1987, pp. 631–650.

[36]Yurkovich, S., "System Identification Experiments for Flexible Structural Control," *Dynamics and Control of Large Structures*, edited by L. Meirovitch, 1987, pp. 143–157.

[37]Roesler, M. D., Jabbari, F., Lukich, M. S., and Gibson, J. S., "Identification of a Flexible Truss Structure Using Lattice Filters," *Proceedings of the 1988 American Control Conference*, Vol. 2, 1988, pp. 1474–1482.

[38]Chen, J.-C., and Fanson, J. L., "System Identification Test Using Active Members," AIAA Paper 89-1290, 1989.

[39]Balas, G. J., and Doyle, J. C., "Identification for Robust Control of Flexible Structures," *Proceedings of the 1989 American Control Conference*, Vol. 3, 1989, pp. 2566–2571.

[40]Murotsu, Y., Okubo, H., and Senda, K., "Identification of the Tendon Control System for Flexible Space Structures," AIAA Paper 89-3568, 1989.

[41]Moon, F. C., *Chaotic Vibrations*, Wiley, New York, 1987.

[42]Rogers, L. (ed.), *Damping 1986 Proceedings*, AFWAL-TR-86-3059, 1986.

[43]Rogers, L. (ed.), *Damping 1989 Proceedings*, WRDC-TR-89-3116, 1989.

[44]Edberg, D. L., "Material Damping of Simple Structures in a Simulated Space Environment," *Journal of Spacecraft and Rockets*, Vol. 23, No. 3, May–June 1986, pp. 288–296.

[45]Folkman, S. L., and Redd, F. J., "Measurement and Modeling of Joint Damping in Space Structures," AIAA Paper 88-2449, 1988.

[46]Ashley, H., and Edberg, D. L., "On the Virtues and Prospects for Passive Damping in Large Space Structures," *Damping 1986 Proceedings*, AFWAL-TR-86-3059, 1986, pp. DA-1–DA-17.

[47]Johnson, C. D., and Kienholz, D. A., "Design and Testing of a Sixty-Foot Damped Generic Space Truss," *Damping 1986 Proceedings*, AFWAL-TR-86-3059, 1986, pp. HA-1–HA-17.

[48]Chen, G.-S., and Wada, B. K., "Passive Damping for Space Truss Structures," AIAA Paper 88-2469, 1988.

[49]Prucz, J., Reddy, A. D., Rehfield, L. W., and Trudell, R. W., "Experimental Characterization of Passively Damped Joints for Space Structures," *Journal of Spacecraft and Rockets*, Vol. 23, No. 6, Nov.–Dec. 1986, pp. 568–575.

[50]Richards, K. E. Jr., "PACOSS Program Status and Results," *NASA/DOD Control/Structures Interaction Technology 1989*, NASA CP-3041, 1989, pp. 31–65.

[51]Miller, D. W., Crawley, E. F., and Ward, B. A., "Inertial Actuator Design for Maximum Passive and Active Energy Dissipation in Flexible Space Structures," AIAA Paper 85-0777, 1985.

[52]Miller, D. W., and Crawley, E. F., "Theoretical and Experimental Investigation of Space-Realizable Inertial Actuation for Passive and Active Structural Control," *Journal of Guidance, Control, and Dynamics*, Vol. 11, No. 5, Sept.–Oct. 1988, pp. 449–458.

[53]Hagood, N. W., and Crawley, E. F., "Experimental Investigation into Passive Damping Enhancement for Space Structures," AIAA Paper 89-3436, 1989.

[54]Fraser, D. C., "Large Space Structure Control: Early Experiments," *Journal of Guidance, Control, and Dynamics*, Vol. 7, No. 5, Sept.–Oct. 1984, p. 513.

[55]Nurre, G. S., Ryan, R. S., Scofield, H. N., and Sims, J. L., "Dynamics and Control of Large Space Structures," *Journal of Guidance, Control, and Dynamics*, Vol. 7, No. 5, Sept.–Oct. 1984, pp. 514–526.

[56]Aubrun, J.-N., Ratner, M. J., and Lyons, M. G., "Structural Control for a Circular Plate," *Journal of Guidance, Control, and Dynamics*, Vol. 7, No. 5, Sept.–Oct. 1984, pp. 535–545.

[57]Bauldry, R. D., Breakwell, J. A., Chambers, G. J., Johansen, K. F., Nguyen, N. C., and Schaechter, D. B., "A Hardware Demonstration of Control for a Flexible Offset-Feed Antenna," *Journal of the Astronautical Sciences*, Vol. 31, No. 3, July–Sept. 1983, pp. 455–470.

[58]Sparks, D. W., Jr., Juang, J.-N., and Klose, G. J., "A Survey of Experiments and Experimental Facilities for Control of Flexible Structures," AIAA Paper 89-3567, 1989.

[59]Floyd, M. A., and Vander Velde, W. E., "Verification of RCS Controller Methods for Flexible Spacecraft," Vols. 1 and 2, AFRPL TR-84-092 and Charles Stark Draper Lab. Rept. CSDL-R-1735, 1984.

[60]Williams, J. P., and Rallo, R. A., "Description of the Spacecraft Control Laboratory Experiment (SCOLE) Facility," NASA TM-89057, 1987.

[61]Sincarsin, G. B., Sincarsin, W. G., Hughes, P. C., and Reynaud, A. H., "DAISY—A Laboratory Facility to Study the Control of Large Flexible Space-

craft," *Dynamics and Control of Large Structures*, edited by L. Meirovitch, 1987, pp. 185–200.

[62]Özgüner, Ü., Yurkovich, S., Martin, J., and Kotnik, P., "A Laboratory Facility for Flexible Structure Control Experiments," *Dynamics and Control of Large Structures*, edited by L. Meirovitch, 1987, pp. 235–250; also *IEEE Control Systems Magazine*, Vol. 8, No. 4, Aug. 1988, pp. 27–33.

[63]Yurkovich, S., and Özgüner, Ü., "Recent Developments in the OSU Flexible Structure Control Laboratory," *Dynamics and Control of Large Structures*, edited by L. Meirovitch, 1989.

[64]Vivian, H. C., Blaire, P. E., Eldred, D. B., Fleischer, G. E., Ih, C.-H. C., Nerheim, N. M., Scheid, R. E., and Wen, J. T., *Flexible Structure Control Laboratory Development and Technology Demonstration—Final Report*, Jet Propulsion Lab., Pasadena, CA, Publ. 88-29, 1987.

[65]Pearson, J., and Waites, H., "Advanced Control Evaluation for Structures (ACES) Programs," *Proceedings of the 1988 American Control Conference*, Vol. 2, 1988, pp. 1448–1452.

[66]Irwin, R. D., Jones, V. L., Rice, S. A., Tollison, D. K., and Seltzer, S. M., "Active Control Technique Evaluation for Spacecraft (ACES)," AFWAL-TR-88-3038, 1988.

[67]Rockwell, T. H., "Investigation of Structure-Borne Active Vibration Dampers," *Journal of the Acoustical Society of America*, Vol. 38, 1965, pp. 623–628.

[68]Major, C. S., and Shain, E. B., "Demonstration of Vibration Control of a Flexible Truss Structure," *Dynamics and Control of Large Structures*, edited by L. Meirovitch, 1985, pp. 183–190.

[69]Dailey, R. L., and Lukich, M. S., "Recent Results in Identification and Control of a Flexible Truss Structure," *Proceedings of the 1988 American Control Conference*, Vol. 2, 1988, pp. 1468–1473.

[70]Balas, G. J., Lukich, M., Dailey, R. L., and Doyle, J. C., "Robust Control of a Truss Experiment," AIAA Paper 88-4084, 1988.

[71]Lukich, M., "System Identification & Control of the Truss Experiment: a Retrospective," AIAA Paper 88-4152, 1988.

[72]Balas, G. J., and Doyle, J. C., "On the Caltech Experimental Large Space Structure," *Proceedings of the 1988 American Control Conference*, Vol. 2, 1988, pp. 1701–1702.

[73]Skidmore, G. R., and Hallauer, W. L., Jr., "Modal-Space Active Damping of a Beam-Cable Structure: Theory and Experiment," *Journal of Sound and Vibration*, Vol. 101, No. 2, July 1985, pp. 149–160.

[74]Hallauer, W. L., Jr., Skidmore, G. R., and Gehling, R. N., "Modal-Space Active Damping of a Plane Grid: Experiment and Theory," *Journal of Guidance, Control, and Dynamics*, Vol. 8, No. 3, May–June 1985, pp. 366–373.

[75]Skidmore, G. R., and Hallauer, W. L., Jr., "Experimental-Theoretical Study of Active Damping with Dual Sensors and Actuators," AIAA Paper 85-1921, 1985.

[76]Haftka, R. T., Martinovic, Z. N., and Hallauer, W. L., Jr., "Enhanced Vibration Controllability by Minor Structural Modifications," *AIAA Journal*, Vol. 23, Aug. 1985, pp. 1260–1266.

[77]Haftka, R. T., Martinovic, Z. N., Hallauer, W. L., Jr., and Schamel, G. C. II, "An Analytical and Experimental Study of a Control System's Sensitivity to Structural Modifications," *AIAA Journal*, Vol. 25, Feb. 1987, pp. 310–315.

[78]Martinovic, Z. N., Schamel, G. C. II, Haftka, R. T., and Hallauer, W. L., Jr., "Analytical and Experimental Investigation of Output Feedback vs Linear

Quadratic Regulator," *Journal of Guidance, Control, and Dynamics*, Vol. 13, Jan.–Feb. 1990, pp. 160–167.

[79]Schamel, G. C. II, and Haftka, R. T., "LQG and Direct Rate Feedback Control with Model Reduction on a Flexible Laboratory Grid Structure," *Dynamics and Control of Large Structures*, edited by L. Meirovitch, 1989, pp. 585–606.

[80]Finzi, A. E., Lanz, M., and Mantegazza, P., "Active Structural Control with Decentralized and Colocated Control Units," *Dynamics and Control of Large Structures*, edited by L. Meirovitch, 1985, pp. 487–501.

[81]Finzi, A. E., Lanz, M., and Mantagazza, P., "Active Structural Controllers Emulating Structural Elements by ICUs," *Mechanical Qualification of Large Flexible Spacecraft Structures*, AGARD CP-397, 1985, pp. 16.1–16.13.

[82]Schäfer, B. E., and Holzach, H., "Experimental Research on Flexible Beam Modal Control," *Journal of Guidance, Control, and Dynamics*, Vol. 8, No. 5, Sept.–Oct. 1985, pp. 597–604.

[83]von Flotow, A. H., and Schäfer, B., "Wave-Absorbing Controllers for a Flexible Beam," *Journal of Guidance, Control, and Dynamics*, Vol. 9, No. 6, Nov.–Dec. 1986, pp. 673–680.

[84]Schäfer, B. E., and Lange, T., "Dynamic and Control Experiments on Large Space Structures at DFVLR," *Control-Theory and Advanced Technology*, Mita Press, Tokyo, Vol. 4, No. 1, March 1988, pp. 91–119.

[85]Lange, T., "Generic Model Laboratory Tests for Large Flexible Structure Control," International Astronautical Federation Paper IAF-88-294, Oct. 1988.

[86]Greeley, S. W., Phillips, D. J., and Hyland, D. C., "Experimental Demonstration of the Maximum Entropy/Optimal Projection Design Theory for Active Vibration Control," *Proceedings of the 1988 American Control Conference*, Vol. 2, 1988, pp. 1462–1467.

[87]Ih, C.-H. C., Bayard, D. S., Wang, S. J., and Eldred, D. B., "Adaptive Control Experiment with a Large Flexible Structure," AIAA Paper 88-4153, 1988.

[88]Zhang, Q., Shelley, S., and Allemang, R. J., "Active Damping Design of Flexible Structures Based on SISO & SIMO Non-Collocated Sensor-Actuator Velocity Feedback," *Proceedings of the 59th Shock and Vibration Symposium*, 1988, pp. 117–138.

[89]Zhang, Q., Shelley, S., Luo, X. N., Allemang, R. J., Brown, D. L., and Slater, G. L., "Experimental Study of Active Vibration Control of Flexible Structures Subject to Forced Vibration," *Proceedings of the 7th International Modal Analysis Conference*, Vol. II, 1989, pp. 926–933.

[90]Chiang, W.-W., and Cannon, R. H., Jr., "The Experimental Results of a Self Tuning Adaptive Controller Using Online Frequency Identification," *Journal of the Astronautical Sciences*, Vol. 33, No. 1, Jan.–March 1985, pp. 71–83.

[91]Juang, J.-N., Horta, L. G., and Robertshaw, H. H., "A Slewing Control Experiment for Flexible Structures," *Journal of Guidance, Control, and Dynamics*, Vol. 9, No. 5, Sept.–Oct. 1986, pp. 599–607.

[92]Juang, J.-N., and Horta, L. G., "Effects of Atmosphere on Slewing Control of a Flexible Structure," *Journal of Guidance, Control, and Dynamics*, Vol. 10, No. 4, July–Aug. 1987, pp. 387–392.

[93]Fujii, H., "A Slew Maneuver Experiment of Mission Function Control," AIAA Paper 88-4226, 1988.

[94]Kotnik, P. T., Yurkovich, S., and Özgüner, Ü., "Acceleration Feedback for Control of a Flexible Manipulator Arm," *Journal of Robotic Systems*, Vol. 5, No. 3, June 1988, pp. 181–196.

[95]Yurkovich, S., and Pacheco, F. E., "On Controller Tuning for a Flexible-Link Manipulator with Varying Payload," *Journal of Robotic Systems*, Vol. 6, June 1989, pp. 233–254.

[96]Soong, T. T., Reinhorn, A. M., and Yang, J. N., "A Standardized Model for Structural Control Experiments and Some Experimental Results," *Structural Control*, edited by H. H. E. Leipholz, Nijhoff, The Hague, 1987, pp. 669–693.

[97]McGreevy, S., Soong, T. T., and Reinhorn, A. M., "An Experimental Study of Time Delay Compensation in Active Structural Control," *Proceedings of the 6th International Modal Analysis Conference*, Vol. I, 1988, pp. 733–739.

[98]Murotsu, Y., Okubo, H., and Terui, F., "Low-Authority Control of Large Space Structures by Using a Tendon Control System," *Journal of Guidance, Control, and Dynamics*, Vol. 12, No. 2, March–April 1989, pp. 264–272.

[99]Okubo, H., Murotsu, Y., Mitsuya, A., and Terui, F., "Research on Tendon Control System for Flexible Structures," *Distributed Parameter Systems: Modelling and Simulation*, edited by T. Futagami, S. G. Tzafestas, and Y. Sunahara, North-Holland, Amsterdam, 1989, pp. 417–424.

[100]Murotsu, Y., Okubo, H., Terui, F., Senda, K., and Shinoda, K., "Dynamics and Control of Experimental Tendon Control System for Flexible Space Structure," AIAA Paper 88-4154, 1988.

[101]Bossi, J. A., "A Laboratory Simulation of Flexible Spacecraft Control," AIAA Paper 87-2325, 1987.

[102]Wada, B. K., "Adaptive Structures," AIAA Paper 89-1160, 1989.

[103]Edberg, D. L., "Control of Flexible Structures by Applied Thermal Gradients," *AIAA Journal*, Vol. 25, June 1987, pp. 877–883.

[104]Bailey, T., and Hubbard, J. E., Jr., "Distributed Piezoelectric–Polymer Active Vibration Control of a Cantilever Beam," *Journal of Guidance, Control, and Dynamics*, Vol. 8, No. 5, Sept.–Oct. 1985, pp. 605–611.

[105]Plump, J. M., Hubbard, J. E., Jr., and Bailey, T., "Nonlinear Control of a Distributed System: Simulation and Experimental Results," *Journal of Dynamic Systems, Measurement, and Control*, Vol. 109, No. 2, June 1987, pp. 133–139.

[106]Burke, S. E., and Hubbard, J. E., Jr., "Active Vibration Control of a Simply Supported Beam Using a Spatially Distributed Actuator," *IEEE Control Systems Magazine*, Vol. 7, No. 4, 1987, pp. 25–30.

[107]Hubbard, J. E., Jr., "Distributed Transducers for Smart Structural Components," *Proceedings of the 6th International Modal Analysis Conference*, Vol. II, 1988, pp. 856–862.

[108]Miller, S. E., and Hubbard, J. E., Jr., "Smart Components for Structural Vibration Control," *Proceedings of the 1988 American Control Conference*, Vol. 3, 1988, pp. 1897–1902.

[109]Bailey, T., Gruzen, A., and Madden, P., "RCS/Piezoelectric Distributed Actuator Study," AFAL-TR-88-038 and Charles Stark Draper Lab. Rept. CSDL-R-2076, 1988.

[110]Fanson, J. L., and Caughey, T. K., "Positive Position Feedback Control for Large Space Structures," AIAA Paper 87-0902, 1987; also *AIAA Journal* (to be published).

[111]Baz, A., Poh, S., and Fedor, J., "Independent Modal Space Control with Positive Position Feedback," *Dynamics and Control of Large Structures*, edited by L. Meirovitch, 1989, pp. 553–568.

[112]Lee, C.-K., Chiang, W.-W., and O'Sullivan, T. C., "Piezoelectric Modal Sensors and Actuators Achieving Critical Active Damping on a Cantilever Plate," AIAA Paper 89-1390, 1989.

[113]Chiang, W.-W., and Lee, C.-K., "Critical Active Damping Control of a Flexible Slender Plate Using a Distributed Modal Actuator and Sensor," *Proceedings of the 1989 American Control Conference*, Vol. 1, 1989, pp. 700–705.

[114]Fanson, J. L., Blackwood, G. H., and Chu, C.-C., "Active-Member Control of Precision Structures," AIAA Paper 89-1329, 1989.

[115]Chen, G.-S., Lurie, B. J., and Wada, B. K., "Experimental Studies of Adaptive Structures for Precision Performance," AIAA Paper 89-1327, 1989.

[116]Lauffer, J. P., Allen, J. J., and Peterson, L. D., "Structural Dynamics Considerations for Structural Control," *Proceedings of the 7th International Modal Analysis Conference*, Vol. II, 1989, pp. 1079–1086.

[117]Peterson, L. D., Allen, J. J., Lauffer, J. P., and Miller, A. K., "An Experimental and Analytical Synthesis of Controlled Structure Design," AIAA Paper 89-1170, 1989.

[118]Peterson, L. D., "An Optimal Projection Controller for an Experimental Truss Structure," AIAA Paper 89-3434, 1989.

[119]Robertshaw, H. H., Wynn, R. H., Jr., Kung, H. F., Hendricks, S. L., and Clark, W. W., "Dynamics and Control of a Spatial Active Truss Actuator," AIAA Paper 89-1328, Pt. 3, 1989, pp. 1473–1479.

[120]Clark, W. W., Kimiavi, B., and Robertshaw, H. H., "Control of Flexible Beams Using a Free-Free Active Truss," Adaptive Structures American Society of Mechanical Engineers, Publ. AD-Vol. 15, 1989, pp. 61–68.

[121]Natori, M., Motohashi, S., Takahara, K., and Kuwao, F., "Vibration Control of Truss Beam Structures Using Axial Force Actuators," AIAA Paper 88-2273, 1988.

[122]Hallauer, W. L., Jr., and Wagie, D. A., "An Inexpensive Experiment for Undergraduates on Active Control of Structural Vibrations," *1989 ASEE Annual Conference Proceedings*, Vol. 2, 1989, pp. 433–435.

[123]Ih, C.-H. C., Bayard, D. S., Ahmed, A., and Wang, S. J., "Experiments in Multivariable Adaptive Control of a Large Flexible Structure," AIAA Paper 89-3571, 1989.

[124]Cudney, H. H., Jr., Horner, G. C., and Inman, D. J., "Vibration Control of Flexible Beams Using an Active Hinge," *Dynamics and Control of Large Structures*, edited by L. Meirovitch, 1985, pp. 455–470.

[125]Irwin, R. D., "Application of FAMESS to a Large Space Structure Ground Test Facility," *Proceedings of the 1988 American Control Conference*, Vol. 2, 1988, pp. 1456–1461.

[126]Tilley, S. W., and Cannon, R. H., Jr., "Experiments on End-Point Position and Force Control of a Flexible Arm with a Fast Wrist," AIAA Paper 86-1980, 1986.

[127]Oakley, C. M., and Cannon, R. H., Jr., "Initial Experiments on the Control of a Two-Link Manipulator with a Very Flexible Forearm," *Proceedings of the 1988 American Control Conference*, Vol. 2, 1988, pp. 996–1002.

[128]Oakley, C. M., and Cannon, R. H., Jr., "End-Point Control of a Two-Link Manipulator with a Very Flexible Forearm: Issues and Experiments," *Proceedings of the 1989 American Control Conference*, Vol. 2, 1989, pp. 1381–1388.

[129]Juang, J.-N., "Rapid Multi-Flexible-Body Maneuvering Experiments," *Proceedings of the 1988 American Control Conference*, Vol. 2, 1988, pp. 1009–1014.

[130]Juang, J.-N., Yang, L.-F., Huang, J.-K., and Macauley, R., "Rapid Rotational/Translational Maneuvering Experiments of a Flexible Steel Beam," *Proceedings of the 1989 American Control Conference*, Vol. 2, 1989, pp. 1403–1408.

[131]Parsons, E. K., "The Jitter Beam: an Experiment Demonstrating Pointing Control on a Flexible Structure," *Proceedings of the 1988 American Control Conference*, Vol. 1, 1988, pp. 61–68; also *IEEE Control Systems Magazine*, Vol. 9, No. 3, April 1989, pp. 79–86.

[132]Doane, G. B., III, Waites, H., and Edgemon, G. D., "Development and Use of a Linear Momentum Exchange Device," *NASA/DOD Control/Structures Interaction Technology 1986*, NASA CP-2447, Pt. 1, 1986, pp. 431–440.

[133]Cristler, T., and Calico, R. A., "Active Control Experiment Using Proof Mass Actuators," AIAA Paper 88-4307, 1988.

[134]Özgüner, Ü., Yurkovich, S., and Gordon, R. W., "A Case Study in Control Design and Implementation for Flexible Structure Systems," *Proceedings of the 1989 American Control Conference*, Vol. 3, 1989, pp. 2085–2090.

[135]Zimmerman, D. C., Horner, G. C., and Inman, D. J., "Microprocessor Controlled Force Actuator," *Journal of Guidance, Control, and Dynamics*, Vol. 11, No. 3, May–June 1988, pp. 230–236.

[136]Zimmerman, D. C., and Cudney, H. H., "Practical Implementation Issues for Active Control of Large Flexible Structures," *Dynamics and Control of Large Structures*, edited by L. Meirovitch, 1987, pp. 275–290.

[137]Zimmerman, D. C., and Inman, D. J., "On the Nature of the Interaction Between Structures and Proof-Mass Actuators," *Journal of Guidance, Control, and Dynamics*, Vol. 13, Jan.–Feb. 1990, pp. 82–88.

[138]Özgüner, Ü, Yurkovich, S., Martin, J. III, and Al-Abbass, F., "Decentralized Control Experiments on NASA's Flexible Grid," *Proceedings of the 1986 American Control Conference*, Vol. 2, 1986, pp. 1045–1051.

[139]Ghosh, D., Montgomery, R. C., Norris, M. A., and Meirovitch, L., "Linear Estimation and Control Studies of Vibration Suppression of the SCOLE Flexible Mast," *Dynamics and Control of Large Structures*, edited by L. Meirovitch, 1989, pp. 165–178.

[140]Rahman, Z., Junkins, J., Pollock, T., and Bang, H., "Large Angle Maneuvers with Vibration Suppression: Analytical and Experimental Results," *Dynamics and Control of Large Structures*, edited by L. Meirovitch, 1989, pp. 653–668.

[141]Williams, J. P., "Slew Maneuvers on the SCOLE Laboratory Facility," *NASA/DOD Control/Structures Interaction Technology 1986*, NASA CP-2447, Pt. 2, 1987, pp. 851–867.

[142]Mercadal, M., and Vander Velde, W. E., "Experimental Failure Detection Results Using the SCOLE Facility," *NASA/DOD Control/Structures Interaction Technology 1987*, AFWAL-TR-88-3052, 1988, pp. 457–483.

[143]Mercadal, M., "Sensor Failure Detection Using Generalized Parity Relations for Flexible Structures," *Journal of Guidance, Control, and Dynamics*, Vol. 12, No. 1, Jan.–Feb. 1989, pp. 125–127.

[144]Bailey, T. Gruzen, A., and Madden, P., *RCS/Linear Discrete Actuator Study*, AFAL-TR-88-039 and CSDL-R-2075, 1988.

[145]Hallauer, W. L., Jr., and Lamberson, S. E., "Experimental Active Vibration Damping of a Plane Truss Using Hybrid Actuation," AIAA Paper 89-1169, 1989.

Chapter 19

Large-Angle Maneuver Experiments in Ground-Based Laboratories

Alok Das*
Edwards Air Force Base, California

Introduction

MANY of the civilian and military space missions planned for the late 1990s and beyond will require orbiting large deployable or erectable space structures with very precise pointing and shape control requirements. A major source of disturbance for a number of these systems comes from the mission-derived need for rapidly reorienting the flexible structure. This includes retargeting and tracking maneuvers for space surveillance systems, often different subsystems on the spacecraft requiring independent pointing; rapid retargeting followed by very precise pointing/tracking for directed energy space defense platforms[1]; and precision pointing of large antennae.[2]

Large-angle maneuver with vibration suppression for flexible structures has been an area of active research over the past 10–15 years, and many different theoretical approaches have been suggested. The need for ground experimentation for technology validation was realized early in the 1980s, and subsequently a number of experiments have been performed. This chapter summarizes previous and current ground experiments in large-angle maneuver of flexible structures.

Survey of Ground Experiments

In this review only the facilities and experiments that focus on retargeting and vibration suppression of flexible structures have been included. Recent experiments in robotics with single and multiple links have many

*Aerospace Engineer, Aerospace Vehicle Systems Division, Astronautics Laboratory, Air Force Systems Command.

common features with flexible space structure control experiments, and both disciplines stand to gain from each other's advances. However, to maintain the focus on space structures, robotics experiments have not been included in this review.

Table 1 chronologically lists the large-angle maneuver experiments and facilities beginning in 1981. The author could not locate any retargeting experiments that concern flexible structure control prior to this date. There has been much interest in such experiments over the last few years; of the 13 experiments listed in Table 1, 6 became operational in 1989 or 1990. These experiments vary in complexity from single-axis slew of a flexible beam to facilities with realistic three-dimensional structures with three degrees of rotational freedom. Later in this section many of these experiments are discussed in detail.

Initial work in this area occurred at Lockheed with Toysat[3,4] and proof of concept (POC)[5,6] experiments. It is interesting to note that as early as 1982 the POC experiment used a realistic three-dimensional test structure similar in configuration to an offset feed antenna. The experiment was supported on a spherical air bearing, thus providing the structure with three degrees of rotational freedom. It was indeed unfortunate that the planned comparison of slewing techniques could not be performed and the testing was limited to vibration suppression. The control moment gyroscopes (CMGs) available for the experiment were inadequate to carry out the slew maneuvers.

In the rest of this section six major experiments in retargeting of flexible structures are described. First, the Air Force Astronautics Laboratory/Charles Stark Draper Laboratory (AFAL/CSDL) Flexible Satellite Testbed successfully completed its testing phase in 1987 and has since been dismantled. The following two experiments at NASA Langley Research Center are ongoing at the time of this writing and are expected to be operational for the next few years. Next, a series of simple experiments developed at Ohio State University are described. The fifth experiment (at Texas A&M University) has only recently come on-line and should be active for a number of years. Finally, the Advanced Space Structure Technology Research Experiments (ASTREX) facility at the Astronautics Laboratory is currently under development and achieved initial operational capability in early 1990.

AFAL/CSDL Flexible Satellite Testbed

In early 1980 a survey of future Department of Defense missions revealed that, for certain missions, the large moment of inertia of the spacecraft coupled with rapid retargeting requirements resulted in slew torques (10^5-10^6 Nm) that exceeded the capacity of momentum storage devices such as CMGs. These mission objectives in retargeting could be achieved by using reaction jets only; however, the available thrusters operated only in a bang-bang mode and could not be throttled to produce an arbitrary torque profile. At that time the problem of retargeting flexible structures using on/off thrusters had not been adequately addressed. Later

Vander Velde and He[20] developed a technique of using on/off thrusters for rapid retargeting and vibration control of a dynamic model with an arbitrary number of vibratory modes.

Based on these considerations, an experimental facility was developed jointly by the AFAL and the CSDL to demonstrate the use of on/off thrusters for rapid retargeting and vibration suppression of flexible structures.[7] The structure consisted of a rigid hub mounted on an air bearing, enabling it to rotate freely about the vertical axis. Four 1.2-m-long aluminum beams ("arms") were attached to the hub. Control torque was provided by four 0.23-kg force (0.5 lb) cold gas thrusters mounted at the tip of two opposite arms in opposing pairs. Sensing was provided by an angle resolver on the hub and accelerometers, one at the end of each flexible arm. Figure 1 shows the final configuration of the experiment, with additional low-authority actuators for vibration suppression and precision pointing.

Initially two control schemes were implemented on the structure: the method of Vander Velde and He[20] and the single-step scheme of Floyd and Vander Velde.[7] The control objective consisted of retargeting the structure through 15 deg in minimum time with minimum control effort resulting in a zero terminal state. Vander Velde and He[20] developed an approximate solution to the optimal control problem by first dividing the problem into subproblems, one for each mode, and then independently solving each of these subproblems. In Floyd's single-step optimal control scheme,[7] the controller minimizes the performance index at the next time step only, ignoring the overall control problem. Subsequent experiments by Bailey et al.[8] using a new, reduced friction air-bearing table demonstrated that both of these controllers had the tendency to limit cycle at the end of the maneuver. Bailey showed that by including terminal constraints on the thrusters, limit cycling could be contained. These constraints included the following: 1) a dead band around the target point to prevent thruster firing when the pointing accuracy was within the band; and 2) thrusters that were precluded from firing in the direction of local tip velocity or displacement.

In the next phase of experiments low-authority linear actuators were added to the structure. Earlier simulations by Floyd and Vander Velde indicated that use of linear actuators in conjunction with on/off thrusters can significantly improve the terminal pointing accuracy and vibration suppression. Proof mass actuators (PMAs) with collocated linear variable differential transformers were added at the tip of the four flexible arms. In addition, the air-bearing table torque motor was used as a hub torque actuator.

A large number of tests were conducted with different actuators and constraints active during the maneuver. Each test consisted of a 15-deg slew, and Table 2 compares the performance of the controllers at two intermediate times during the maneuver. In each case the bang-bang rigid-body slew is completed at $t = 4$ s. In general the PMAs were very effective in vibration suppression but did not directly improve pointing performance. However, the resulting increase in damping reduced the limit cycle behavior of the thrusters, allowing a reduction in the dead-band terminal constraint on the thrusters. This indirectly improved the hub

Table 1 Large-angle maneuver experiments and facilities

Facility/experiment name (operational dates)	Brief description	Sensors/actuators	Demonstration
Lockheed Toysat[3,4] (1981–1983)	Suspended rigid-body, 1.6-m cantilever beams, Al	Accelerometers, linear electrodynamic actuators	Digital high-authority structural control
Lockheed POC[5,6] (1982–1984)	Offset feed antenna, Al, 3-m-diam reflector, 4.5-m metering truss, spherical air bearing	Accelerometers, rate gyroscope, laser vibration sensor, CMGs, PMAs	Vibration suppression of flexible spacecraft in presence of disturbances
AFAL/CSDL Flexible Satellite Testbed[7–9] (1983–1987)	Rigid hub on air bearing, four 1.2-m-long flexible beams	Accelerometers, hub angle resolver, thrusters, hub motor, PMAs, piezo film	Slew and vibration suppression using thrusters, hub motor, PMAs, and piezo film
Univ. of Washington/ Boeing Lab. Exp. in CSI[10,11] (1985–1986)	Rigid core on flat air bearing, six flexible truss arms	Optical sensor, airjets	Rigid-body control in presence of structural flexibility
NASA/LaRC SCOLE[12,13] (1987–present)	Rigid Shuttle platform, 3-m-long mast, 1-m-diam offset reflector, suspended by cable	Accelerometers, rate and optical sensors, cold gas thrusters, CMGs, reaction wheels	Minimum-time slew with precision pointing

NASA/LaRC[14,15] 3-body rapid maneuvering exp. (1985–present)	Rigid hub, two flexible horizontal panels	Strain gages, angular rate sensor, motors	Single-axis slew with vibration suppression
OSU[16] rotating flex link (1986–present)	Rigid hub, 1-m flexible beam	Accelerometer, optical encoder, dc motor	Slewing with vibration suppression
OSU[16] two-link configuration (1989–present)	Rigid hub, 0.75-m flexible beam, 0.5-m hinged beam	Accelerometers, optical encoder, hub and hinge motors	To demonstrate slew and vibration suppression of two-link system
OSU[16] multiple-mirror slewing structure (1989–present)	0.75-m beam mounted on servomotor, two rotating mirrors mounted on beam	Linear array camera, optical encoders, hub and mirror motors	To demonstrate tracking and pointing of beam during slew maneuvers
OSU[16] satellite emulator (1989–present)	Hub with 2 flexible panels, 2-axis hub rotation, Independent panel rotation	Motor shaft position, motors	To demonstrate slew in multiple directions independently
NASA/LaRC[17] Rotational/Translational Maneuvering Exp. (1989–present)	Rigid trolly on horizonal track One 1.1-m flexible beam	Strain gages, tachometer angular potentiometer, trolly and beam motors	Vibration suppression during rotational and translational maneuvers
AFOSR/TAMU[18] Maneuver Exp. (1989–present)	Rigid hub on ceramic bearing Four 1-m flexible arms	Strain-gage bridges, hub angle encoder hub reaction wheel	Near-minimum time, large slew maneuvers
Astronautics Lab. ASTREX Facility[19] (Initial operational capability in early 1990)	Spherical air bearing SBL 3-mirror beam expander, 5.5 m diam, 5 m long metering truss	Rigid-body angles trans. and rotat. acc. throttleable thrusters reaction wheels, PMAs	To serve as a testbed for demonstrating controls/structures interaction technology

Fig. 1 AFAL/CSDL flexible satellite testbed.

pointing error by over a factor of four. On the other hand, the hub torquer assumes the precision pointing task in the terminal phase, thereby significantly improving the overall pointing performance. When used in combination with PMAs, this resulted in a hub angle error of under 0.05 deg, a 20-fold improvement over using thrusters alone. The major drawback during this experiment was the contamination of the tip accelerometers due to the unmodeled twisting of the flexible arms. The twisting was caused by the mass imbalance at the tip of the beams due to the motion of the proof mass in the PMAs.

In the final phase of experiments, Bailey et al.[9] investigated the use of piezoelectric distributed actuators for structural damping augmentation. Building on the work of Bailey and Hubbard[21] proof of concept demonstrations were done on a 0.15-m scaled steel beam and on one arm of the AFAL/CSDL experiment. The piezoelectric material used was a PVF_2 film bonded directly onto the structure. Control law was developed using Liapunov's direct method, resulting in a bang-bang feedback control. The control required the application of a constant voltage to the film, which was opposite in sign to the tip velocity of the beam. This Liapunov controller caused the vibration amplitude to decay linearly with time. Thus, the effective loss factor (and hence the damping) changed with the amplitude of vibration, increasing as the amplitude decreased. In a typical set of experiments the controller reduced the settling time from over 2 min to 8 s

Table 2 15-deg Slew maneuvers results

Test configuration	Hub angle error, deg		Tip acceleration, g		Remarks
	$t = 10$ s	$t = 20$ s	$t = 10$ s	$t = 20$ s	
Thrusters[a] only	1	1.2	0.12	0.18	Excessive limit cycling, unstable
Thrusters[b] only	1	1	0.09	0.06	Some limit cycling still occurs
Thrusters[b] and PMAs in terminal phase	0.6	0.25	0.06	<0.01	PMAs active after 6 s, thrusters cease firing after 13 s
Thrusters[b] and PMAs active throughout	0.25	<0.25	0.03	<0.01	No thruster firing after 10 s
Thrusters[b] and hub torquer in terminal phase	1	0.5	0.07	0.06	No thruster firing after 6 s
Thrusters,[b] PMAs active throughout, and hub torquer in terminal phase	—	<0.05	0.02	<0.01	Thrusters cease firing after 6 s, best pointing accuracy

[a]No terminal phase constraints.
[b]With terminal phase constraints.

for the scaled beam and from over 400 s to 126 s for the full-scale arm. The piezoelectric film actuator with the Liapunov controller was shown to be most effective in keeping resonant vibrations from building up because it provides a high level of damping at low vibration. The PVF_2 actuator was not used during a slew maneuver. It was determined that the vibrations resulting from the maneuver were too large for the low-authority actuator.

NASA/LaRC Spacecraft Control Laboratory Experiment

In 1983 the NASA Langley Research Center (NASA/LaRC) initiated the Spacecraft Control Laboratory Experiment (SCOLE) program to "serve as a focus of a design challenge for the purpose of comparing directly different approaches to control synthesis, modeling, order reduction, state estimation and system identification."[12] SCOLE was developed by NASA/LaRC with the advice and council of the IEEE Subcommittee on Large Space Structures. The design challenge consisted of two principal parts: 1) a mathematical model of a Shuttle-attached offset feed antenna used to design a controller for rapid retargeting followed by precise pointing of the antenna; and 2) the validation of the methodology on a subscale ground experiment at NASA/LaRC.

The laboratory experiment consists of a rigid, 228–kg Shuttle platform to which a 3-m-long mast, made of stainless steel tubing, is attached. To the other end of the mast, a 1-m-diam aluminum offset antenna is attached (see Fig. 2). Together the mast and the reflector weigh 4.2 kg. The experiment is well instrumented with spacecraft-type sensors and actuators. The Shuttle plate is equipped with three accelerometers, a three-axis rotational rate sensor, and an optical sensor, and a pair of two-axis CMGs for control. An additional three-axis rate sensor package is located at the junction of the reflector and the mast. Control torque on the reflector is provided by 0.1-kgf cold gas thrusters located at its center. In addition, two pairs of reaction wheels are mounted on the mast. The complete structure is suspended from a 3.35-m cable connected to the experiment center of mass via a universal joint with a two-degree-of-freedom gimbal, allowing limited rotations in pitch and roll. The suspension cable provides the yaw rotations. The control computer is a Motorola M68000 microprocessor-based system with eight 12-bit digital-to-analog converters, 64 12-bit analog-to-digital converters, and one 4-bit discrete output channel.[13]

The SCOLE laboratory experiment achieved the initial operation capability by early 1987 and became fully operational in mid 1990. The first part of the SCOLE design challenge has been very successful in focusing theoretical research in retargeting of flexible structures; much of this research is documented in the proceedings of the annual NASA SCOLE workshops. The second part of the challenge requiring implementation of control schemes on the laboratory hardware has not been as successful. To date, retargeting experiments have been limited to the implementation of rigid-body minimum time slew.[22,23] Investigators active in the field have failed to adequately utilize the SCOLE laboratory.

Fig. 2 NASA/LaRC spacecraft control laboratory experiment.

NASA/LaRC Three-Body Rapid Maneuvering Experiment

A single-axis, three-body experiment for flexible structure slewing and vibration suppression was developed at NASA/LaRC in 1985. This experiment was used by Juang et al.[14] to demonstrate control of flexible structures with high-authority linear actuators. The structure consists of a rigid hub to which two flexible panels are hinged. The hub can be rotated in the horizontal plane by an electric motor. Furthermore, each flexible panel can individually rotate in the horizontal plane via electric gear motors located at each hinge. Each panel is equipped with an angular potentiometer and three strain-gage bridges located at the root, 22%, and 50% of the panel length. The control loop is closed through a EAI 2000 analog computer. The experiment has been used in two configurations: 1) two 1-m-long steel beams are used as panels; and 2) the beams are replaced by two 3.9-m aluminum solar panel simulators. Figure 3 is a schematic of the experiment in the former configuration.

The experiment has served as a testbed for a number of open- and closed-loop control schemes. Table 3 summarizes the results of 30-deg, closed-loop slew maneuvers performed in a 3.5 s using the solar panel simulators as appendages. The control law selected was an optimal terminal control scheme developed by Juang et al.[24,25] and was implemented using three different sensor combinations. The structural vibration of the

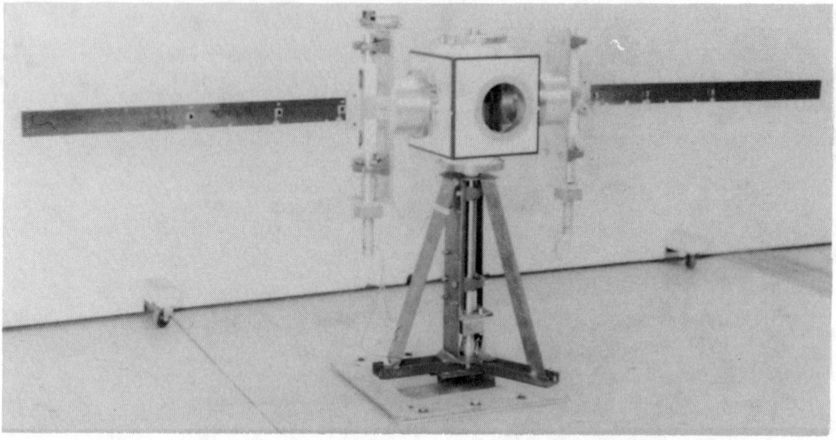

Fig. 3 NASA/LaRC three-body rapid maneuvering experiment.

panel after the maneuver, as indicated by the root strain, shows gradual reduction as additional sensors are included in the feedback law. The inclusion of a low-pass prefilter for shaping the command signal to the torque actuators produces a dramatic reduction in the structural vibrations. The torque-shaping filter effectively reduces any high-frequency component in the torque command signal that may be generated by the feedback law, consequently reducing structural excitation. Juang et al.[14] recommend it as a very effective practical technique for control implementation.

During the initial experiments it was noticed that the damping predictions were consistently lower than those observed in experiments, especially for the first mode. This discrepancy was attributed to air damping. In a subsequent set of tests, Juang and Horta[15] performed the experiments in air and then repeated them in a vacuum to quantify the effect of air damping. The air damping of the panels was modeled by adding two terms, one proportional to the instantaneous velocity and the other proportional to the square of the instantaneous velocity, to the dynamic model. Inclusion of these terms significantly improved the correspondence between theory and experimental results.

Besides the control experiments described earlier, the test setup was used by Garcia and Inman[26] to verify the matrix exponential method (MEM) of parameter estimation. The MEM uses free decay time histories to adjust the values of selected parameters in the system model in order to improve the model. The viscous damping coefficient was selected as the parameter to be identified. Two separate runs with MEM, each taking only a few iterations to converge, reduced the model error by an order of magnitude.

OSU Flexible Structure Control Laboratory

Since 1985 a number of simple experimental testbeds have been developed at the Ohio State University (OSU) Flexible Structures Control

Table 3 30-deg Slew maneuver with solar panel

	Normalized root strain			
	No torque shaping		With torque shaping	
Feedback	$t = 6$ s	$t = 12$ s	$t = 6$ s	$t = 12$ s
Angle	2.2	1.0	0.4	0.2
Angle and root strain	1.6	0.6	0.4	0.2
Angle and three strains	1.5	0.4	0.2	0.1

Laboratory to address generic problems in large flexible space structures and robotic manipulators, including rapid retargeting. Two of these experiments are described later. Much of the effort to date has focused on the rapid retargeting of a 1-m-long aluminum beam mounted horizontally on a rigid hub.[16] Retargeting and vibration control actuation is provided by a dc motor mounted at the hub. Sensing is provided via a tip accelerometer and an optical shaft encoder at the hub. In addition, a linear array camera provides the beam tip position. Modeling, identification, and retargeting experiments have been performed on the structure. More recently, a PID controller with real-time parameter estimation was implemented for the case of an unknown point mass loading on the structure.[27]

A second experiment models the pointing and tracking function associated with a number of proposed precision space structures.[16,28] As for the rotating flexible link experiment, the structure is a flexible aluminum beam mounted horizontally on to a large motor. In addition, two small mirrors are attached to the beam. The mirrors can be rotated independently about their vertical axis. A laser beam is directed by the two mirrors onto a target screen (Fig. 4). The control objectives are to track a predetermined trajectory via the mirrors and to maintain a point on the screen while slewing the beam. Initial experimentation on the setup involves implementation of decentralized control laws for vibration compensation using the mirror actuators.

AFOSR/TAMU Maneuver Experiment

A recent addition to the family of slew facilities is an experiment sponsored by the Air Force Office of Scientific Research (AFOSR) at Texas A&M University (TAMU).[18] This experiment is somewhat similar to the AFAL/CSDL four flexible arm slew experiment, except that it uses a reaction wheel for rapid retargeting, instead of on/off gasjets. The structure consists of four identical arms cantilevered to a rigid hub that is pivoted on a ceramic bearing, allowing it to rotate freely in the horizontal plane (see Fig. 5). Control torques are provided by a dc motor with a reaction wheel located at the hub. The wheel can be biased to serve as either a reaction wheel or a momentum wheel. Sensing is provided by an angle encoder to

Fig. 4 OSU multiple mirror slewing structure experiment.

Fig. 5 AFOSR/TAMU maneuver experiment.

measure the hub angle; two strain-gage bridges located near the root of an appendage to provide bending moment and shear forces on the appendage; and a tachometer, which gives the wheel speed. The control computer is a Packard Bell 286 system with appropriate input/output boards.

The initial experimentation in the facility began in mid-1989 and has focused on the implementation of a control scheme for near-minimum-time, large-angle maneuvers of flexible structures.[29] In principle, the approach uses a reference trajectory for rigid-body maneuvers and a tracking-type control law to reduce departures from that reference. One can also perform tradeoffs between maneuver time, robustness, and vibration suppression by appropriate adjustments to the reference trajectory.

Advanced Space Structure Technology Research Experiment (ASTREX)

ASTREX Laboratory is being developed at the Astronautics Laboratory as a versatile facility for demonstrating advanced structures and control technologies on realistic scaled models of high-performance spacecraft.[19] Central to the ASTREX facility will be a three-axis large-angle retargeting capability and a powerful real-time control computer with a user-friendly interface. The facility will be housed in a $12 \times 12 \times 12$ m environmentally controlled enclosure, allowing temperature control and still air conditions during experiments.

Experiments in the ASTREX facility will be supported by a 48-cm-diam spherical air bearing with a payload capability of 6600 kg. The air bearing sits on a 4.5-m-long pedestal, providing the test article with freedom to rotate about the three axes. Included in the air-bearing system is a double-gimbal, three-axis cable follower that is slaved to follow the test article angular rotations. This allows providing the test article with electrical and pneumatic utilities without introducing significant disturbance torques. The cable follower is equipped with sensors to provide rigid-body attitude measurements about the three axis with an accuracy of 5 μrads (1 arcsec). The ASTREX real-time control and data acquisition computer (CDAC) is an Intel 80386/Weitek 3167 based multiprocessor system with a minimum sustained speed of 11 million floating-point operations per second under various computational loads. The sampling speed is user selectable and is limited only by the overall computational speed of the computer. Initially the CDAC is configured for 32 analog input and 32 analog outputs. The CDAC can be programmed via a high-level control design and analysis software package.

The initial test article is a dynamically scaled model of a generic three-mirror space-based laser beam expander (Fig. 6). The beam expander was selected because it poses a combination of challenging problems in retargeting, precision pointing, and shape control to be experienced by the next generation of precision space systems. The structure consists of a 5.5-m-diam primary mirror support truss structure made of graphite epoxy tubes. The modular design allows one or more tubes to be removed and replaced with alternate tubes without disassembling the structure. The secondary mirror support metering truss is made up of three 5.2-m-long,

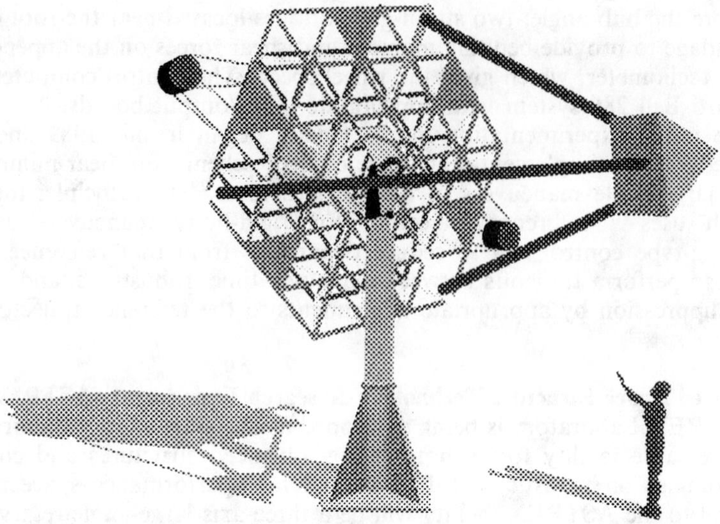

Fig. 6 Advanced space structure technology research experiments.

12.7-cm-diam graphite epoxy tubes. The complete structure is supported by the air bearing at its center of mass and can rotate ± 20 deg about the two horizontal axes and ± 180 deg about the vertical axis.

A variety of sensors and actuators will be available on the structure. Rapid retargeting torque will be provided using a complement of 12 cold-gas thrusters with a peak thrust of 900 N (200 lbf). The thrusters can be operated either in a bang-bang mode or they can be throttled for shaping the torque profile. A number of reaction wheels and proof mass actuators of sufficient authority will also be available for vibration control. A set of translational and rotational accelerometers will be available for use at various feasible locations throughout the structure. The facility is currently under development and achieved the initial operational capability in early 1990.

Concluding Remarks

This chapter has reviewed the recent ground experiments in large-angle retargeting of flexible structures. For a number of future spacecraft, rapid maneuvers will be one of the major causes of structural vibrations in the space environment. Currently a number of ground experiments are ongoing to test and verify the theoretical results.

Dynamic characteristics of sensors and specialty actuators have a significant impact on the control law implementation. Thus, in ground experimentation it is imperative to use devices with similar characteristics to those that will eventually be used on the spacecraft.

Finally, it will be essential to experimentally test and validate advanced control schemes prior to their use on spacecraft. Although the utility of simple conceptual experiments should not be underestimated, the final validation will have to be performed on realistic three-dimensional scaled or full-scale structural models of spacecraft.

References

[1]Johnson, J. R., "DOD Controls/Structures Interaction Technology: An Overview," NASA/DOD Controls-Structures Interaction Technology 1989 Conference, San Diego, CA, Jan.–Feb. 1989.

[2]Venneri, C. L., "Overview of NASA CSI Program," NASA/DOD Controls-Structures Interaction Technology 1989 Conference, San Diego, CA, Jan.–Feb. 1989.

[3]Strunce, R., Motyka, P., Schley, B., et al., "An Investigation of Enabling Technologies for Large Precision Space Systems: Vol. III—Survey of Enabling Technologies," Air Force Rocket Propulsion Lab., Edwards AFB, CA, AFRPL-TR-82-074, Sept. 1982.

[4]Stroud, R. C., Hamma, G. A., Smith, S., and Lyons, M. G., "Developments Towards Active Control of Space Structures," Society of Automotive Engineers TP-801234, 1980.

[5]ACOSS TWELVE (Active Control of Space Structures), Rome Air Development Center, Griffiss AFB, New York, Interim Rept. RADC-TR-82-320, Dec. 1982.

[6]ACOSS TWELVE (Active Control of Space Structures), Rome Air Development Center, Griffiss AFB, New York, Final Rept. RADC-TR-84-28, Feb. 1984.

[7]Floyd, M. A., and Vander Velde, W. E., "Verification of RCS Controller Methods for Flexible Spacecraft," Vols. I and II, Air Force Rocket Propulsion Lab., Edwards AFB, CA, AFRPL TR-84-092, Dec. 1984.

[8]Bailey, T., Gruzen, A., and Madden, P., "RCS/Linear Discrete Actuator Study," Air Force Astronautics Lab., Edwards AFB, CA, AFAL-TR-88-039, Aug. 1988.

[9]Bailey, T., Gruzen, A., and Madden, P., "RCS/Piezoelectric Distributed Actuator Study," Air Force Astronautics Lab., Edwards AFB, CA, AFAL-TR-88-038, Aug. 1988.

[10]Bossi, J. A., and Tsou, J. W., "A Laboratory Experiment in Control/Structure Interaction," *Proceedings of the 1986 American Control Conference*, June 1986, pp. 1034–1038.

[11]Bossi, J. A., "A Laboratory Simulation of Flexible Spacecraft Control," AIAA Paper 87-2325, 1987, pp. 389–394.

[12]Taylor, L. W., Jr., and Balakrishnan, A. V., "A Laboratory Experiment Used to Evaluate Control Laws for Flexible Spacecraft—NASA/IEEE Design Challenge," *Proceedings of the Fourth VPI&SU/AIAA Symposium on Dynamics and Control of Large Structures*, June 1983, pp. 311–318.

[13]Williams, J. P., "Slew Maneuvers on the SCOLE Laboratory Facility," NASA CP-2447, 1986, Pt. 2, pp. 851–867.

[14]Juang, J.-N., Horta, L. G., and Robertshaw, H. H., "A Slewing Control Experiment for Flexible Structures," *Journal of Guidance, Control, and Dynamics*, Vol. 9, No. 5, Sept.–Oct. 1986, pp. 599–607.

[15]Juang, J.-N., and Horta, L. G., "Effects of Atmosphere on Slewing Control of a Flexible Structure," *Journal of Guidance, Control, and Dynamics*, Vol. 10, No. 4, July–Aug. 1987, pp. 387–392.

[16]Yurkovich, S., and Ozguner, U., "Recent Developments in the OSU Flexible Structure Control Laboratory," Seventh VPI&SU Symposium on Dynamics and Control of Large Structures, Blacksburg, VA, May 1989.

[17]Juang, J.-N., Yang, L.-F., Huang, J.-K., and Macauley, R., "Rapid Rotational/Translational Maneuvering Experiments of a Flexible Steel Beam," 1989 American Control Conference, Pittsburgh, PA, June 1989.

[18]Rahman, Z., Junkins, J., Pollock, T., and Bang, H., "Large Angle Maneuver with Vibration Suppression: Analytical and Experimental Results," Seventh VPI&SU Symposium on Dynamics and Control of Large Structures, Blacksburg, VA, May 1989.

[19]Norris, G. A., "Initial Operational Capability of the ASTREX Space Structures Testbed," *Proceedings of the NASA/DOD Controls-Structures Interaction Technology Conference*, 1989, pp. 507–522.

[20]Vander Velde, W. E., and He, J., "Optimal Control of Large Space Structures Using On-Off Thrusters," *Journal of Guidance, Control, and Dynamics*, Vol. 6, No. 1, Jan.–Feb. 1983, pp. 53–60.

[21]Bailey, T., and Hubbard, J. E., "Piezoelectric-Polymer Active Vibration Control of a Cantilever Beam," *Journal of Guidance, Control, and Dynamics*, Vol. 8, No. 5, Sept.–Oct. 1985.

[22]Shenhar, J., Sparks, D., Williams, J. P., and Montgomery, R. C., "Attitude Control System Testing on SCOLE," Sixth VPI&SU/AIAA Symposium on Dynamics and Control of Large Structures, Blacksburg, VA, June–July 1987.

[23]Montgomery, R. C., Shenhar, J., and Williams, J. P., "On-Line Identification and Attitude Control for SCOLE," AIAA Paper 87-9138, August 1987.

[24]Juang, J.-N., Turner, J. D., and Chun, H. M., "Closed-Form Solutions of Control Gains for a Terminal Controller," *Journal of Guidance, Control, and Dynamics*, Vol. 8, Jan.–Feb. 1985, pp. 38–43.

[25]Juang, J.-N., Turner, J. D., and Chun, H. M., "Closed Form Solutions for a Class of Optimal Quadratic Regulator Problems with Terminal Constraints," *ASME Journal of Dynamics, Measurements and Control*, Vol. 108. March 1986, pp. 44–48.

[26]Garcia, E., and Inman, D. J., "Identification of System Parameters in a Slewing Control Experiment," *Proceedings of the Damping 89 Conference* (to be published).

[27]Yurkovich, S., and Pacheco, P. E., "On Controller Tuning for a Flexible-Link Manipulator with Varying Payload," *Journal of Robotic Systems*, Vol. 6, No. 3, June 1989.

[28]Barbieri, E., Ozguner, U., and Yurkovich, S., "Vibration Compensation in Optical Tracking Systems," *Journal of Guidance, Control, and Dynamics*, Vol. 12, No. 4, July–Aug. 1989, pp. 585–592.

[29]Junkins, J., Rahman, Z., Bang, H., and Hecht, N., "Near-Minimum-Time Feedback Control of Distributed Parameter Systems: A Liapunov Control Law Design Method," Seventh VPI&SU Symposium on Dynamics and Control of Large Structures, Blacksburg, VA, May 1989.

[30]Quartararo, R., and Harris, J., "ASTREX—A Facility for Integrated Structures and Control Research," Wright-Patterson AFB, OH, AFWAL-TR-88-3052, June 1988.

Chapter 20

Control/Structure Interaction: Effects of Actuator Dynamics

Daniel J. Inman*
State University of New York at Buffalo, Buffalo, New York

Introduction

TO date, most control laws developed for vibration suppression of flexible structures ignore the dynamics of the actuator hardware used to implement the control.[1] This chapter discusses the effects of including the actuator dynamics in the formulation of simple control laws for vibration suppression. The inclusion of actuator dynamics results in a theoretical prediction of a substantial interaction between the actuator dynamics, the structural dynamics, and the choice of control laws. This defines the nature of the control/structure interaction (CSI) problem considered here.

In this chapter the problem of closed-loop stability is considered in discussing these phenomena. It is assumed that the structure is modeled by an asymptotically stable system and that the purpose of the control is vibration suppression. It is further assumed that the structure is well modeled by some finite-dimensional linear set of equations with viscous damping (such as a finite-element model). In addition, it is assumed that the actuators used for control have second-order dynamics and that the sensors and controllers are perfect. It is well known that feedback control can greatly enhance disturbance rejection, but does so at the risk of causing instabilities. It has been shown previously[2] that these instabilities result from large values of feedback gain and are not predicted by models that exclude actuator dynamics. The work presented in Ref. 2 is expanded and generalized here to provide stability conditions for such combined systems.

*Professor and Chair, Mechanical Systems Laboratory, Department of Mechanical and Aerospace Engineering, School of Engineering and Applied Sciences.

The intent of this chapter is to present a simple stability theory for combined structure/actuator systems with simple state feedback. The theory is developed by using linear algebraic results to provide direct conditions for asymptotically stable operation. These conditions are stated in terms of the structure's parameters, the actuator's physical parameters, and the feedback control gains, and provide explicit formulas for their interaction. Stability conditions are derived and then numerically checked. In addition, experimental evidence is given to verify the conjectured instabilities.

Actuator Dynamics

The actuator used here to demonstrate the nature of CSI is a linear actuator composed of a solenoid-like arrangment of a mass moving in an electric field. This proof mass actuator (PMA) is a reaction-type force actuator, which creates a force by reacting against an inertial mass. Such actuators are also called reaction mass actuators (RMAs). The PMA used in this study was developed for the Spacecraft Dynamics Branch at NASA Langley Research Center by researchers at the University of Virginia[3,4] and the University at Buffalo.[5,6] This actuator has been extensively tested[6] and is capable of generating an arbitrary (but bounded) control force. The actuator consists of a movable "proof mass," a fixed coil, two collocated sensors, a digital microcontroller, and a power amplifier as described in detail in Ref. 6. All of the actuator components are mounted on a single compact fixture. Power lines are the only external connection required to operate the actuator unless uncollocated control is used. In the uncollocated case, the actuator accepts signals from sensors at other locations on the structure. In addition, an optional analog or digital input can be used as the control law for design purposes.

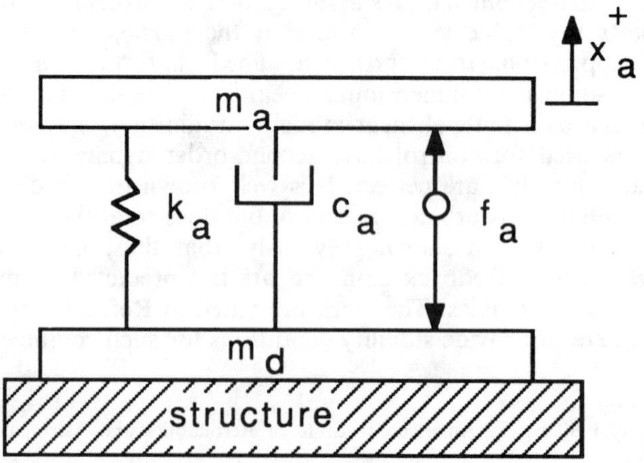

Fig. 1 Dynamics of a proof mass actuator.

As illustrated in Ref. 6, the PMA's transfer function clearly dictates a second-order model of the form of a single-degree-of-freedom oscillator. Laboratory bench tests and tests with the actuator mounted on a variety of structures indicate that the model of Fig. 1 provides an accurate description of the actuator dynamics. In the figure m_a represents the internal moving mass, or proof mass, of the actuator, and x_a is the time-dependent position of the actuator relative to the base. The quantity m_d is the dead, or nonmoving, mass of the actuator (i.e., the housing, transducers, magnets, control electronics, fixed to the structure). The damping coefficient c_a is determined experimentally and represents an equivalent viscous damping coefficient for the internal friction of the actuator. The quantity k_a, also determined experimentally, is the electronic stiffness force required to keep the proof mass centered in its housing. This force, as well as the internal damping force, is required for stability of the actuator. The control force is generated between the proof mass and the structure and is denoted as f_a. References 2 and 6 report the dynamic modeling in detail.

The particular physical parameters for the laboratory actuator are given in Table 1. The quantity ω_b in Table 1 is the natural frequency of the actuator, which is shown experimentally in Ref. 6 to be about 1.7 Hz. This agrees well with the natural frequency of the second-order system used to model the actuator. The stiffness k_a and hence ω_b can be adjusted electronically.

This actuator is, in concept, a space-realizable actuator. It does not require connections to ground and, except for a power source, is completely self-contained. The investigation presented here uses this particular actuator as an example and for comparison with experimental data. However, the results are generic for any actuator that can be successfully modeled by a second-order, ordinary differential equation. For instance, many electric motors and linear actuators have second-order transfer functions (see, for instance, Ref. 7 for other examples).

The nature of the dynamics of the actuator modeled in Fig. 1 results in an asymptotically stable structure (see the section in which stability theorems are discussed). However, once the feedback control force f_a is implemented, the resulting combined system (structure and actuator together) is potentially unstable. The following sections are concerned with illuminating this instability and deriving conditions that guarantee asymptotically stable operation.

Table 1 Experimentally determined parameter values for a PMA

m_d	0.73 kg
m_a	0.232 kg
c_a	0.76 N-s/m
k_a	108.27 N/m
ω_b	1.7 Hz

The Problem

This section introduces the nature of the interaction between the actuator and the structure by examining a simple single-degree-of-freedom structure. Once the basic phenomenon is presented, a general theory for any order system is presented in the following sections. To this end, consider the actuator attached to a single-degree-of-freedom system as illustrated in Fig. 2. In the figure k is the stiffness of the structure under consideration, c is the structure's natural viscous damping coefficient, and m represents the mass of the structure plus the dead mass of the actuator.

Note that the system of Fig. 2 is very similar to a vibration absorber. The notable difference is the presence of the control force f_a. The parameters k_a, m_a, and c_a could in principle be tuned to protect the motion of the structure mass m. This is the approach taken in Ref. 7. To provide active control for the sake of further improving the performance of the structure, the force f_a can be chosen in a variety of ways (see, for instance, Chapter 7 of Ref. 8). Vibration suppression is often accomplished by simple velocity feedback. Taking this approach, the velocity of the structure's mass, $\dot{x}$, is measured, and the force f_a is taken to be $-g\dot{x}$, where g is the constant gain constituting the control law. The desire is to choose g such that the response $x(t)$ decays as rapidly as possible.

Suppose, first, that the actuator dynamics are excluded, as is common in theoretical treatments of control. The feedback control problem then becomes one of finding the gain g such that

$$m\ddot{x} + c\dot{x} + kx = -g\dot{x} \tag{1}$$

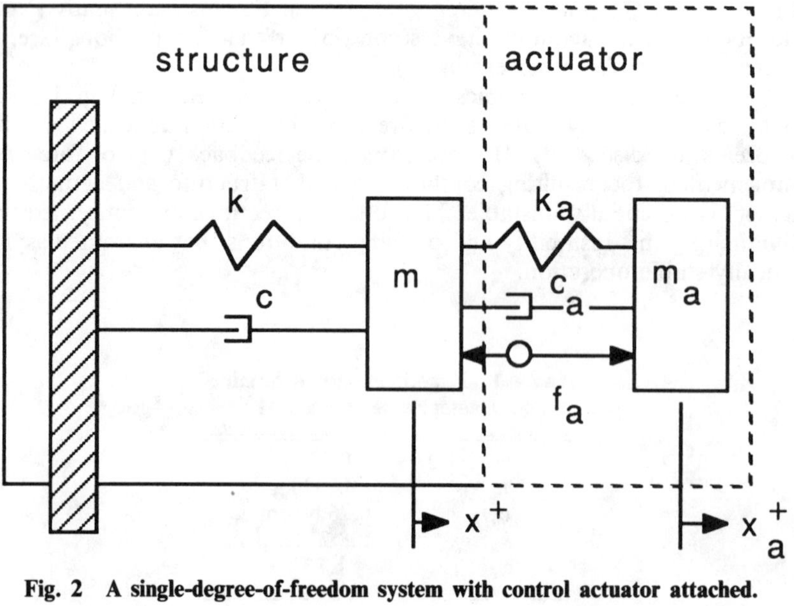

Fig. 2 A single-degree-of-freedom system with control actuator attached.

has a desired damping ratio. Rearranging Eq. (1) into the standard form yields

$$\ddot{x} + \frac{c+g}{m}\dot{x} + \frac{k}{m}x = 0 \tag{2}$$

Using the usual definitions of natural frequency $[\omega = \sqrt{(k/m)}]$ and damping ratio $[\zeta = c/(2\sqrt{(km)})]$ yields for the closed-loop system of Eq. (2)

$$\zeta = \frac{c+g}{2\omega m} \tag{3}$$

Two things are obvious from Eq. (3). Since m, c, and k are all positive, the solution of Eq. (2) will be asymptotically stable for any choice of positive gain (negative feedback). Second, the higher the gain, the higher the damping ratio, and the better the vibration suppression.

Next consider the same scenario with the actuator dynamics included. From Fig. 2 and a simple summation of forces, the equations of motion are

$$\begin{bmatrix} m & 0 \\ 0 & m_a \end{bmatrix}\begin{bmatrix} \ddot{x} \\ \ddot{x}_a \end{bmatrix} + \begin{bmatrix} c+c_a & -c_a \\ -c_a & c_a \end{bmatrix}\begin{bmatrix} \dot{x} \\ \dot{x}_a \end{bmatrix} + \begin{bmatrix} k+k_a & -k_a \\ -k_a & k_a \end{bmatrix}\begin{bmatrix} x \\ x_a \end{bmatrix} = \begin{bmatrix} -f_a \\ f_a \end{bmatrix} \tag{4}$$

Using the same simple velocity feedback control of the form $f_a = g\dot{x}$ yields the homogeneous coupled equations

$$\begin{bmatrix} m & 0 \\ 0 & m_a \end{bmatrix}\begin{bmatrix} \ddot{x} \\ \ddot{x}_a \end{bmatrix} + \begin{bmatrix} c+c_a+g & -c_a \\ -g-c_a & c_a \end{bmatrix}\begin{bmatrix} \dot{x} \\ \dot{x}_a \end{bmatrix} + \begin{bmatrix} k+k_a & -k_a \\ -k_a & k_a \end{bmatrix}\begin{bmatrix} x \\ x_a \end{bmatrix} = \begin{bmatrix} 0 \\ 0 \end{bmatrix} \tag{5}$$

The stability of this system can be analyzed by examining the characteristic equation associated with Eq. (5). If the characteristic equation has roots with positive real parts, the homogeneous solution will grow without bound and be unstable. If all of the roots have negative real parts, the homogeneous solution will exponentially decay to zero and hence be asymptotically stable. If there is a root with a zero real part, the solution will oscillate with constant amplitude, and this is called marginally stable (or just stable).

Consider a dimensionless configuration of an actuator and a structure such that $m = c = k = 1$ and $m_a = c_a = k_a = 0.1$. In this case the structure has an order of magnitude more mass than the actuator. Substituting these values into Eq. (5) and assuming that the solution is of the form $x = a\,e^{\lambda t}$, $x_a = b\,e^{\lambda t}$, where a and b are constant, yields the characteristic equation

$$\lambda^4 + (g+2.1)\lambda^3 + 3.1\lambda^2 + 2\lambda + 1 = 0 \tag{6}$$

Applying the Hurwitz[9] test to the coefficients in Eq. (6) yields a system that may be asymptotically stable if

$$g + 2.1 > 0 \tag{7}$$

To determine whether the corresponding system is asymptotically stable, further analysis is required. In this simple low-order case, the instability condition can be achieved by using the Routh array[9] and examining the Hurwitz determinant given in Table 2. The Routh-Hurwitz test simply states that the roots of the polynomial all have negative real parts if and only if the elements of column one are all positive. With $g > 0$ the elements are all positive if $(g + 1.3685)(3.3685 - g) > 0$, and there will be at least one root with positive real part if (and only if) this is violated, i.e., if

$$g > 3.3685 \tag{8}$$

This represents an instability condition for the closed-loop system with actuator dynamics included; i.e., if the gain is increased past the value of 3.3685, the closed-loop response of Eq. (5) will become unstable.

For this simple example of a single-degree-of-freedom structure, the effect of including the actuator dynamics is drastic. Without actuator dynamics the velocity feedback law characterized by Eq. (3) states clearly that the best performance results if the gain is made as large as possible. However, including the actuator dynamics shows clearly that an instability results if the gain becomes too large. Since the actuator dynamics are always present, it is therefore extrmely important to consider this source of instability and to compensate for it in the design of control laws.

Physically, the instability can be thought of in the context of a vibration absorber. As the actuator gain increases, it clearly adds damping to the structure. However, it does this at the expense of the stability of the proof mass. Hence, like an absorber mass, the proof mass oscillates to protect the mass of the main structure against vibration. As the gain is electronic, it can be adjusted past the point of stability. At the critical value of gain, the proof mass is marginally stable and vibrates rapidly with constant amplitude. This is illustrated experimentally in Ref. 2. It is the presence of the reaction force (from the structure to the actuator) and the form of the feedback law that allows this instability to occur.

Table 2 Routh array for Eq. (6)

	Column 1	Column 2	Column 3
λ^4	1	3.1	1
λ^3	$g + 2.1$	2	0
λ^2	$3.1g + 4.51$	$(g + 2.1)$	0
λ	$-100g^2 + 200g + 465$	0	0
λ^0	1	0	0

For this low-order example, the critical values of the gain g can be calculated with fixed physical parameters. However, for larger-order models of the structure, such as those more commonly used to represent large flexible spacecraft, and for multiple actuators, the Routh-Hurwitz approach quickly loses its usefulness. The problem posed here is to develop conditions stated in terms of the physical parameters of the structure, the physical constants of the actuator, and the control gains, which will guarantee stable operation of the closed-loop system in the presence of actuator dynamics. The relationship between these various parameters constitutes the interaction between control law, actuator, and structures (CSI) presented here.

System Model

The structures considered here are assumed to be finite dimensional with linear viscous damping. They typically result from a finite-element analysis with some experimentally determined damping added.[10] Let the number of degrees of freedom of the model be denoted by n and let $R_{\text{sym}}^{n \times n}$ denote the set of all $n \times n$ real symmetric matrices. Let M, C, $K \in R_{\text{sym}}^{n \times n}$ be the mass, damping, and stiffness matrices of the structure under study. The equation of motion of the structure is then

$$M\ddot{\boldsymbol{x}}(t) + C\dot{\boldsymbol{x}}(t) + K\boldsymbol{x}(t) = \boldsymbol{f}(t) \tag{9}$$

where $\boldsymbol{x}(t)$ is an $n \times 1$ vector of displacements [say, one for each dof of each node of a finite-element model (FEM)], $\dot{\boldsymbol{x}}(t)$ is the $n \times 1$ vector of velocities, and $\ddot{\boldsymbol{x}}(t)$ is the $n \times 1$ vector of accelerations. Here, $\boldsymbol{f}(t)$ denotes an $n \times 1$ vector of applied forces. The structure might also be subjected to an initial displacement, denoted by the constant vector $\boldsymbol{x}_0$ and initial velocity denoted by the constant vector $\dot{\boldsymbol{x}}_0$. It is further assumed that the system (9) under consideration is asymptotically stable so that the matrices M, C, and K are also positive-definite (formal definitions are reviewed in the next section).

The actuators used in this study are all assumed to be of the inertial type with second-order dynamics of the form suggested by Eq. (4). Hence, attaching an actuator to the structure given by Eq. (9) increases the order n of the system by an additional degree of freedom. Thus, the order of the system with actuator dynamics is $m = p + n$, where p is the number of actuators added to the structure. The combined system then has the dynamic equation

$$M_c\ddot{z}(t) + C_c\dot{z}(t) + K_c z(t) = \boldsymbol{F}(t) \tag{10}$$

where M_c, C_c, and $K_c \in R_{\text{sym}}^{m \times m}$ are augmented mass damping and stiffness matrices containing parameters of both the structure and the actuators. Here the vector $z(t)$ is $m \times 1$ and consists of the coordinates of each of the p proof masses (one for each actuator) imbedded in the $n \times 1$ position vector $\boldsymbol{x}$. The disturbance vector $\boldsymbol{F}(t)$ also becomes $m \times 1$ and consists of

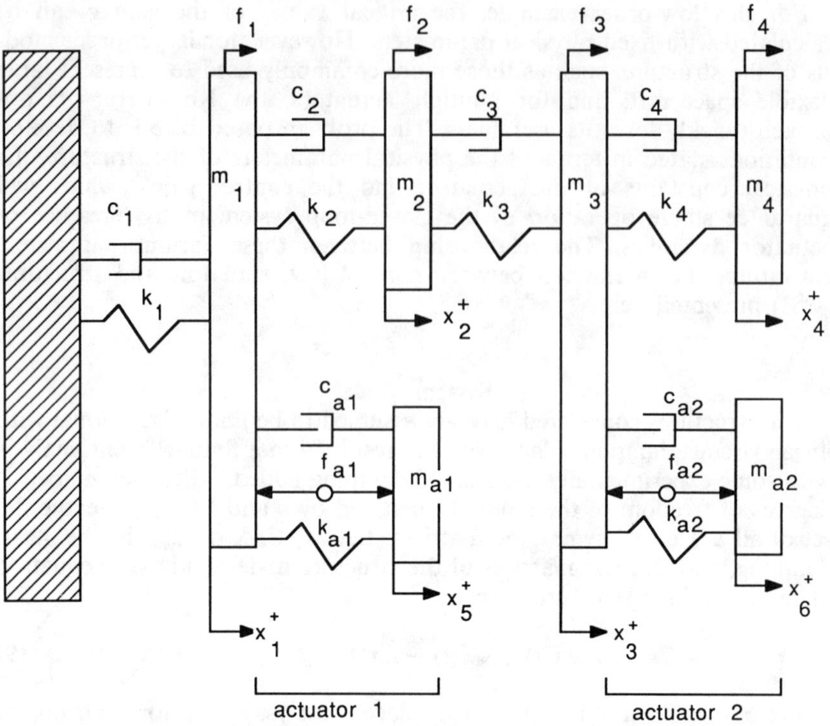

Fig. 3 A four-degree-of-freedom structural model with two actuators attached.

the $n \times 1$ vector $\boldsymbol{f}(t)$ augmented by the (action/reaction) control law forces $f_{aj}(t)$, $j = 1, 2, ..., p$. It is important to note that the term $f_{aj}(t)$ will appear in the vector $\boldsymbol{F}(t)$ in pairs with opposite signs.

In order to visualize these augmented matrices, consider a four-degree-of-freedom structure model ($n = 4$) with two actuators ($p = 2$) attached, as illustrated in Fig. 3. In the figure the structural parameters m_i, c_i, and k_i have the obvious definition of mass damping and stiffness associated with the position coordinates x_i, $i = 1, 2, 3, 4$. The similar actuator parameters are denoted m_{aj}, c_{aj}, and k_{aj}, $j = 1, 2$. The external forces are f_i, $i = 1, 2, 3, 4$, and the control forces are f_{aj}, $j = 1, 2$. The coordinates x_5 and x_6 represent the proof mass coordinates. The position vector of Eq. (9) is just $\boldsymbol{x} = [x_1 \ x_2 \ x_3 \ x_4]^T$, and the applied force vector is $\boldsymbol{f}(t) = [f_1 \ f_2 \ f_3 \ f_4]^T$. The matrix coefficients of Eq. (9) are the banded symmetric matrices $M = \text{diag}(m_1, m_2, m_3, m_4)$,

$$C = \begin{bmatrix} c_1 + c_2 & -c_2 & 0 & 0 \\ -c_2 & c_2 + c_3 & -c_3 & 0 \\ 0 & -c_3 & c_3 + c_4 & -c_4 \\ 0 & 0 & -c_4 & c_4 \end{bmatrix}$$

and

$$K = \begin{bmatrix} k_1+k_2 & -k_2 & 0 & 0 \\ -k_2 & k_2+k_3 & -k_3 & 0 \\ 0 & -k_3 & k_3+k_4 & -k_4 \\ 0 & 0 & -k_4 & k_4 \end{bmatrix}$$

With the actuator dynamics added to the structure, the combined system of Eq. (10) has coefficient matrices

$$M_c = \text{diag}[m_1, m_{a1}, m_2, m_3, m_{a2}, m_4]$$

$$C_c = \begin{bmatrix} c_1+c_2+c_{a1} & -c_{a1} & -c_2 & 0 & 0 & 0 \\ -c_{a1} & c_{a1} & 0 & 0 & 0 & 0 \\ -c_2 & 0 & c_2+c_3 & -c_3 & 0 & 0 \\ 0 & 0 & -c_3 & c_3+c_4+c_{a2} & -c_{a2} & -c_4 \\ 0 & 0 & 0 & -c_{a2} & c_{a2} & 0 \\ 0 & 0 & 0 & -c_4 & 0 & c_4 \end{bmatrix}$$

$$K_c = \begin{bmatrix} k_1+k_2+k_{a1} & -k_{a1} & -k_2 & 0 & 0 & 0 \\ -k_{a1} & k_{a1} & 0 & 0 & 0 & 0 \\ -k_2 & 0 & k_2+k_3 & -k_3 & 0 & 0 \\ 0 & 0 & -k_3 & k_3+k_4+k_{a2} & -k_{a2} & -k_4 \\ 0 & 0 & 0 & -k_{a2} & k_{a2} & 0 \\ 0 & 0 & 0 & -k_4 & 0 & k_4 \end{bmatrix}$$

where the elements m_1 and m_3 are the old values of m_1 and m_3 from the structure plus the dead mass of the actuators (m_d). In this case the vector z becomes $z = [x_1 \; x_5 \; x_2 \; x_3 \; x_6 \; x_4]^T$, which contains the elements of $\boldsymbol{x}$, as well as the actuator coordinates. The vector $\boldsymbol{F}(t)$ becomes the augmented 6×1 vector

$$\boldsymbol{F}(t) = \begin{bmatrix} f_1(t) - f_{a1}(t) \\ f_{a1}(t) \\ f_2(t) \\ f_3(t) - f_{a2}(t) \\ f_{a2}(t) \\ f_4(t) \end{bmatrix} \tag{11}$$

and contains the original external disturbance vector $\boldsymbol{f}(t)$.

Next consider limiting the control action to velocity feedback. Assuming perfect measurements, the control forces f_{aj}, $j = 1,2$ simplify to

$$f_{a1}(t) = g_1 \dot{x}_1(t) \qquad \text{and} \qquad f_{a2}(t) = g_2 \dot{x}_3(t) \tag{12}$$

for the case of "collocated" control without consideration of the actuator dynamics. However, this is actually not collocated when the actuator dynamics are included as they are here. Furthermore, assuming that the distrubance is impulsive and can be described by initial conditions (i.e., $f_1 = f_2 = f_3 = f_4 = 0$), substitution into Eq. (10) results in the closed-loop system becoming

$$M_c \ddot{z} + \tilde{C}_c \dot{z} + K_c z = \mathbf{0} \tag{13}$$

where $\tilde{C}_c$ now has the form

$$\tilde{C}_c = \begin{bmatrix} c_1+c_2+c_{a1}+g_1 & -c_{a1} & -c_2 & 0 & 0 & 0 \\ -c_{a1}-g_1 & c_{a1} & 0 & 0 & 0 & 0 \\ -c_2 & 0 & c_2+c_3 & -c_3 & 0 & 0 \\ 0 & 0 & -c_3 & c_3+c_4+c_{a2}+g_2 & -c_{a2} & -c_4 \\ 0 & 0 & 0 & -c_{a2}-g_2 & c_{a2} & 0 \\ 0 & 0 & 0 & -c_4 & 0 & c_4 \end{bmatrix} \tag{14}$$

Note that the velocity coefficient matrix $\tilde{C}_c$ is no longer symmetric. This matrix loses its symmetry because of the choice of control law (inertial velocity feedback) and because of the inclusion of actuator dynamics.

It should be noted here that, if the relative velocity between the reaction mass and the local structural mass (i.e., $\dot{x}_5 - \dot{x}_1$ and $\dot{x}_6 - \dot{x}_3$) is used as the feedback signal, then the augmented damping matrix $\tilde{C}_c$ will be symmetric and positive-definite and the stability will be maintained in the presence of actuator dynamics (see, for instance, Refs. 7 and 11). In this circumstance the problem becomes identical to that of a multiple-degree-of-freedom absorber problem. By feeding back the local inertial velocities, as well as the relative velocities, versatility is gained and performance is enhanced.[11] However, this increase in performance is accompanied by a risk of driving the system unstable. The following sections discuss conditions useful in avoiding this instability.

In the section on stability with velocity feedback, the stability of a system of the form of Eq. (13) is considered by exploiting the special form of the actuator dynamics, illustrated by the four-degree-of-freedom two-actuator example given here.

Some Stability Theorems

Most structures are asymptotically stable because of internal damping. The exceptions are those undergoing rigid-body motion, or those subjected to some sort of external force, such as the control forces given here, aerodynamic forces, or preloading. Those structural systems with rigid-body motions can often be separated into the rigid-body motion superimposed on an asymptotically stable vibration. In this chapter the structures are assumed to be of the form of Eq. (13) and asymptotically stable. This section reviews various stability definitions and conditions.

First it is important to precisely define stability of the solutions of Eq. (13). The simplest way to define stability for a linear system is to consider the eigenvalues of the system. The eigenvalues of the combined system of Eq. (13) are the complex scalars λ that satisfy

$$\det(\lambda^2 M_c + \lambda \tilde{C}_c + K_c) = 0 \tag{15}$$

where det denotes the determinant. Equation (15) is a polynomial of order $2m$ and hence has $2m$ roots λ_i satisfying this condition. An eigenvector $\boldsymbol{u}_i$ is a complex valued $m \times 1$, nonzero vector satisfying

$$(\lambda_i^2 M_c + \lambda_i \tilde{C}_c + K_c)\boldsymbol{u}_i = \boldsymbol{0}, \qquad i = 1, 2, ..., m \tag{16}$$

The solutions to Eq. (9) can all be written in the form

$$\boldsymbol{x}(t) = \sum_{i=1}^{2m} c_i \boldsymbol{u}_i \, e^{\lambda_i t} \tag{17}$$

where c_i are modal participation constants, as long as the matrix in Eq. (15) is simple (see, for instance, Ref. 8, page 60).

The combined system of Eq. (10) is defined to be stable if and only if it has no eigenvalues with positive real part. Furthermore, the system is defined to be asymptotically stable if and only if all of its eigenvalues have negative real parts (no zero real parts allowed). In the structures literature a stable system may have eigenvalues with zero real parts. This is called marginally stable in the controls literature and is sometimes considered as unstable. The controls literature also refers to those systems that have all eigenvalues with negative real parts as stable. Here the following definitions are used:

1) $\mathrm{Re}\lambda_i > 0$ for any i is *unstable.*
2) $\mathrm{Re}\lambda_i \leqslant 0$ for all $i = 1, ..., m$ is *stable* (or *marginally stable*).
3) $\mathrm{Re}\lambda_i < 0$ for all $i = 1, ..., m$ is *asymptotically stable.*

The definiteness of the coefficient matrices in a mechanical system can be used to state some stability conditions. A real matrix A is symmetric if $A^T = A$ and skew symmetric if $A = -A^T$. An $m \times m$ symmetric matrix A is

1) *positive-definite* if $\boldsymbol{x}^T A \boldsymbol{x} > 0$ for all nonzero real $m \times 1$ vectors $\boldsymbol{x}$;
2) *positive-semidefinite* if $\boldsymbol{x}^T A \boldsymbol{x} \geqslant 0$ for all real nonzero $m \times 1$ vectors $\boldsymbol{x}$; and
3) *indefinite* if $\boldsymbol{x}^T A \boldsymbol{x} \boldsymbol{y}^T A \boldsymbol{y} < 0$ for some $m \times 1$ real vectors $\boldsymbol{x}$ and $\boldsymbol{y}$.

If A is symmetric, its eigenvalues are all real, and it is positive-definite if and only if all of its eigenvalues are positive (semidefinite if they are all nonnegative). It is well known that, if M_c, $\tilde{C}_c$, and K_c are symmetric and positive-definite, the system of Eq. (10) is asymptotically stable (i.e., $\mathrm{Re}\lambda_i < 0$ for all i).

A matrix A is also positive-definite if and only if the determinant of each of its leading principal minors is positive. This yields a very useful condition providing inequalities in the matrix entries (physical parameters). Each of the symmetric matrices presented earlier is positive-definite.

It is well known that damped gyroscopic systems of the form

$$M\ddot{x} + (C + G)\dot{x} + Kx = \mathbf{0} \tag{18}$$

where $G = -G^T$, are also asymptotically stable.[8] Several other less known stability results are presented next.

Consider a general asymmetric system of the form

$$M\ddot{x} + (C + G)\dot{x} + (K + E)x = \mathbf{0} \tag{19}$$

where both G and E are skew symmetric; i.e., $G = -G^T$ and $E = -E^T$. This can also be rewritten as

$$\ddot{x} + A\dot{x} + Bx = \mathbf{0} \tag{20}$$

where $A = M^{-1}(C + G)$ and $B = M^{-1}(K + E)$ are asymmetric of the type introduced by using both velocity and position feedback with second-order actuator dynamics.

It has been shown in Ref. 12 that, if B is symmetrizable, i.e., if the matrix B is similar to a symmetric matrix, and has positive eigenvalues, then the response is asymptotically stable if the symmetric matrix

$$AT_1 + T_1A^T \tag{21}$$

is positive-definite. Here T_1 is a symmetric factor of the matrix B. In this case the assumption on B requires that $B = T_1T_2$, where T_1 and T_2 are both symmetric and positive-definite. This result is referred to as the symmetrizable case and forms a somewhat restrictive but interesting subclass of asymmetric systems.

A more general treatment of asymmetric systems, which is less restrictive than the symmetrizable case, is given in Ref. 13. Namely, the system is asymptotically stable if the symmetric matrices

$$B + B^T \tag{22}$$

$$A^TB + B^TA \tag{23}$$

and

$$A + A^T - (B - B^T)(A^TB + B^TA)^{-1}(B - B^T) \tag{24}$$

are positive-definite. If the matrix given by Eq. (24) is only positive-semidefinite, then the system is stable but not necessarily asymptotically stable.

It is also shown in Ref. 13 that the response of Eq. (20) is asymptotically stable if the symmetric matrix

$$A + A^T - (B + B^T) \tag{25}$$

is positive-definite and the matrix

$$A^T B - B - B^T B \tag{26}$$

is symmetric and positive-definite. If the matrix of Eq. (25) is nonnegative, then the system response is stable but not necessarily asymptotically stable.

Note that there are many systems for which the matrix of Eq. (26) is symmetric. However, if Eq. (26) is not symmetric, then the stability conditions given by matrices (22–24) must be used. Examples are given in Ref. 13.

The preceding conditions are in general easier to check than calculating the solution or using the Routh criteria. Furthermore, for lower-order systems the definiteness conditions can be manipulated to yield inequalities in the system parameters that indicate regions of stable operation. In the next section these results are applied to the CSI problem.

Stability with Velocity Feedback

This section examines the use of the stability conditions presented in the previous section for developing constraints on velocity feedback gains for stable control. In this section systems of the form given by Eq. (13) are considered. Criteria that establish relationships between the actuator parameters, the structure's parameters, and control gains are formulated.

An immediate result is established simply by examining the stability condition given for Eq. (18). The velocity coefficient matrix $\tilde{C}_c$ of Eq. (13) is asymmetric and as such can be written as the sum of a symmetric matrix, denoted $(\tilde{C}_c)_{\text{sym}}$, and a skew symmetric matrix denoted $(\tilde{C}_c)_{ss}$, given by

$$(\tilde{C}_c)_{\text{sym}} = \frac{\tilde{C}_c + \tilde{C}_c^T}{2} \tag{27}$$

$$(\tilde{C})_{ss} = \frac{\tilde{C}_c - \tilde{C}_c^T}{2} \tag{28}$$

These simple relationships yield the first CSI result presented here.

Theorem 1. If $(\tilde{C}_c)_{\text{sym}}$ is positive-definite, then the response of Eq. (13) is asymptotically stable.

Application of this result to low-order systems yields simple inequalities in the control gain, actuator parameters, and structural parameters. For example, consider the simple single-degree-of-freedom structure, single-actuator system of Eq. (5). Computing the matrix $(\tilde{C}_c)_{\rm sym}$ yields

$$(\tilde{C}_c)_{\rm sym} = \begin{bmatrix} c + c_a + g & -\dfrac{g}{2} - c_a \\ -\dfrac{g}{2} - c_a & c_a \end{bmatrix} \tag{29}$$

A stable response will be ensured by requiring this to be positive-definite. Assuming negative feedback (so that $g > 0$), this yields

$$4cc_a > g^2 \tag{30}$$

Hence, if the gain is increased to just less than the value $2(cc_a)^{1/2}$, the structure's damping will be increased by the feedback control and the response will remain asymptotically stable. Inequality (30) illustrates clearly an interaction between the structure's parameter c, the actuator dynamics c_a, and the control law gain g.

Note that the inequality (30) is substantially easier to calculate than calculating the polynomial required for the Routh test. However, the inequality obtained by the matrix approach is also much more conservative [recall inequality (8)]. In addition, as the order m of the combined system increases, the ability to calculate an analytical matrix determinant decreases and becomes impractical. However, because of the special form of the actuator dynamics, this simple single-degree-of-freedom result can be extended to multiple actuator systems for a large-degree-of-freedom structure by looking at some low-order inequalities.

To this end, consider the following partition of a general square matrix A. Let A_i denote the $i \times i$ matrix formed from the matrix A by deleting $n - i$ rows and columns of the matrix A. The determinant of the symmetric matrix A_i, denoted $|A_i|$, is called a principal minor of A. The principal minors of a symmetric matrix A essentially determine the definiteness of the matrix. In fact, a symmetric matrix is positive-definite if and only if all of its principal minors are positive.[14] Application of this concept to a multiple-actuator, multiple-degree-of-freedom system of Eq. (13) yields the following results.

Theorem 2. The response of the combined system of Eq. (13) is asymptotically stable if the principal minors of the symmetric matrix $(\tilde{C}_c)_{\rm sym}$ that contain the control gains are positive.

This follows directly from the following considerations. The matrices M_c and K_c are positive-definite; thus, by *Theorem* 1, if $(\tilde{C}_c)_{\rm sym}$ is positive-definite, the system is asymptotically stable. However, $(\tilde{C}_c)_{\rm sym}$ consists of the positive-definite matrix C_c plus a sparse matrix consisting of zeros with 2×2 symmetric blocks placed along the diagonal located at the index of the actuator location. This can be seen by examining the four-degree-of-freedom, two-actuator system of Fig. 3 and the corresponding matrix of

Eq. (14). The positive-definite matrix C_c has, by virtue of the principal minor theorem, all positive principal minors. When the sparse block diagonal matrix, consisting of the symmetric part of the feedback control matrix, is added to C_c to form $\tilde{C}_c$, the only principal minors of C_c affected are the 2×2 blocks added at the position of the actuator coordinates. The other 2×2 principal minors are all positive; hence, only those principal minors altered by the addition of nonzero elements need to be checked. The changed minors are positive-definite by hypothesis. Hence, $\tilde{C}_c$ is positive-definite, and the combined system is asymptotically stable.

The result allows the determination of the interaction between control gains, structural parameters, and actuator dynamics for combined systems of larger order by examining a series of decoupled inequalities that result from the determinant of a 2×2 symmetric matrix.

As an illustration of this result, consider the four-degree-of-freedom two-actuator system of Fig. 3. Referring to Eqs. (14) and (27), the symmetric part of the closed-loop velocity coefficient becomes

$$(\tilde{C}_c)_{\text{sym}} =$$

$$\begin{bmatrix} c_1+c_2+c_{a1}+g_1 & -c_{a1}-(g_1/2) & -c_2 & 0 & 0 & 0 \\ -c_{a1}-(g_1/2) & c_{a1} & 0 & 0 & 0 & 0 \\ -c_2 & 0 & c_2+c_3 & -c_3 & 0 & 0 \\ 0 & 0 & -c_3 & c_2+c_3+c_a+g_2 & -c_{a2}-(g_2/2) & -c_4 \\ 0 & 0 & 0 & -c_{a2}-(g_2/2) & c_{a2} & 0 \\ 0 & 0 & 0 & -c_4 & 0 & c_4 \end{bmatrix} \tag{31}$$

Examination of the 2×2 principal minors of $(\tilde{C}_c)_{\text{sym}}$ yields the following submatrices:

$$C_{11} = \begin{bmatrix} c_1+c_2+c_{a1}+g_1 & -c_{a1}-\dfrac{g_1}{2} \\ -c_{a1}-\dfrac{g_1}{2} & c_{a1} \end{bmatrix} \tag{32}$$

$$C_{22} = \begin{bmatrix} c_{a1} & 0 \\ 0 & c_2+c_3 \end{bmatrix}, \qquad C_{33} = \begin{bmatrix} c_1+c_3 & -c_3 \\ -c_3 & c_2+c_3+c_{a2}+g_2 \end{bmatrix} \tag{33}$$

$$C_{44} = \begin{bmatrix} c_2+c_3+c_{a2}+g_2 & -c_{a2}-\dfrac{g_2}{2} \\ -c_{a2}-\dfrac{g_2}{2} & c_{a2} \end{bmatrix} \tag{34}$$

$$C_{55} = \begin{bmatrix} c_{a2} & 0 \\ 0 & c_4 \end{bmatrix} \tag{35}$$

Obviously, the determinants of C_{22}, C_{33}, and C_{55} are all positive (recall that $g_1, g_2 > 0$ here). Requiring the determinant of C_{11} and C_{44} to be positive yields

$$4(c_1 + c_2)c_{a1} > g_1^2 \tag{36}$$

and

$$4(c_2 + c_3)c_{a2} > g_2^2 \tag{37}$$

This simple inequality expresses regions of stable operation of the combined system and can be used to design the structure, actuator, and control law to increase damping without the loss of stability.

The preceding presents a useful CSI result based on simple stability conditions for mechanical systems. It is important to note that these conditions are, in general, more restrictive than needed. For instance, if either of the inequalities (36) or (37) is violated, the system may or may not become unstable. This is discussed further in the next section on stability with position feedback, which gives examples. The main restriction in the velocity feedback gains of Eqs. (36) and (37) results from the fact that both the structure's internal damping (c_1, c_2, and c_3) and the PMA internal damping are typically small, forcing the gains to be small. The immediate conclusion is that, if the structure were first treated with a passive damping layer, the damping coefficients c_i would become larger, and hence more control gain could be used.

The remainder of the stability conditions presented in the previous section are more applicable to systems with position feedback or both position and velocity feedback (state variable feedback). These are applied in the next section.

Stability with Position Feedback

In this section the stability theory for general asymmetric systems of Eq. (20) is applied to the feedback control problem with second-order actuator dynamics. In this case the matrices A and B contain the structures' properties, the actuator dynamics, and the control law. For a given level of feedback gains, the resulting matrices A and B can be computed and substituted into the various stability conditions to check for asymptotic stability of the closed-loop system. However, a more exciting prospect is to attempt to use these stability criteria to develop some sort of inequalities in the system's parameters useful for design. Several scenarios are suggested here.

First consider an application of the symmetrizability conditions. The condition of Eq. (21) can be used to check the stability of a system under position feedback by first choosing a position feedback control such that the matrix $T_1^{-1}B$ is symmetric and positive-definite. The symmetric part of the resulting damping matrix can be checked for definiteness. Consider a position feedback control law of the form

$$F(t) = Uz(t) \tag{38}$$

where the matrix U represents the location of the PMAs and the associated position feedback gains h_i. That is, each PMA is programmed to produce a force of the form $-h_i z_j(t)$, $i = 1, 2, ..., p$. In this case no velocity feedback is considered. As an example of U, consider again the four-degree-of-freedom two-actuator system of Fig. 3. With $f_{a1} = -h_1 z_1(t)$ and $f_{a2} = -h_2 z_3(t)$ the matrix U becomes

$$U = \begin{bmatrix} h_1 & 0 & 0 & 0 & 0 & 0 \\ -h_1 & 0 & 0 & 0 & 0 & 0 \\ 0 & 0 & 0 & 0 & 0 & 0 \\ 0 & 0 & 0 & 0 & h_2 & 0 \\ 0 & 0 & 0 & 0 & -h_2 & 0 \\ 0 & 0 & 0 & 0 & 0 & 0 \end{bmatrix} \tag{39}$$

The homogeneous form of the closed-loop combined system becomes

$$M_c\ddot{z}(t) + C_c\dot{z}(t) + (K_c + U)z(t) = \mathbf{0}$$

where M_c, C_c, K_c, and $z(t)$ are as defined in Eq. (10). Multiplying by M_c^{-1} yields

$$\ddot{z}(t) + A_p\dot{z}(t) + B_p z(t) = \mathbf{0} \tag{40}$$

where $A_p = M_c^{-1}C_c$ and $B_p = M_c^{-1}(K_c + U)$ are both asymmetric. This formulation, along with the stability results summarized in the section in which some stability theorems are given, yields the following result.

Theorem 3. If B_p is symmetrizable, with positive eigenvalues (i.e., if $B_p = T_1 T_2$ and T_2 is positive-definite), then the response of Eq. (40) is asymptotically stable if the symmetric matrix $A_p T_1 + T_1 A_p^T$ is positive-definite.

This result can be used to generate inequalities in the feedback gains h_i that yield asymptotically stable operation when satisfied. The matrix T_1^{-1} is first calculated to make B_p symmetrizable. This places some constraints on the gains. Next, the definiteness of the symmetric matrix $A_p T_1 + T_1 A_p^T$ is calculated, yielding further restrictions on the gains. This is illustrated in the next section on examples.

Next consider the case of both position and velocity feedback. In this case the combined system of Eq. (10) is subjected to a feedback force of the form

$$F(t) = V\dot{z}(t) + Uz(t) \tag{41}$$

where V has the same form as U, with the h_i replaced by the velocity feedback gains g_i. The equivalent homogeneous system then becomes

$$\ddot{z}(t) + A_c\dot{z}(t) + B_c z(t) = \mathbf{0} \tag{42}$$

where $A_c = M_c^{-1}(C_c + V)$ and $B_c = M^{-1}(C_c + U)$ are both asymmetric. Relating the coefficient matrices A_c and B_c to Eqs. (22–26) yields the following theorem.

Theorem 4. The response of the combined system described by Eq. (42) is asymptotically stable if the matrices

$$B_c + B_c^T$$

$$A_c^T B_c + B_c^T A_c$$

and

$$(A_c + A_c^T) - (B_c - B_c^T)(A_c^T B_c + B_c^T A_c)^{-1}(B_c - B_c^T)$$

are each positive-definite.

Theorem 5. The response of the combined system described by Eq. (42) is asymptotically stable if the symmetric matrix

$$A_c + A_c^T - (B_c + B_c^T)$$

is positive-definite and if the matrix

$$A_c^T B_c - B_c - B_c^T B_c$$

is symmetric and positive-definite.

These conditions can be used to check the stability of various state feedback control laws. They can also be used to help design asymptotically stable closed-loop systems. These conditions illustrate the effects of actuator dynamics and the interaction between the control and the structure.

Examples

This section examines the use of velocity and position feedback as presented in the previous two sections by considering the system of Fig. 3 in some detail. In addition, an experimental result that illustrates the CSI concept presented here is reviewed.

First consider the effect of velocity feedback on stability by recalling the system of Fig. 3 with two PMAs attached. The values of the coefficients are given in Table 3, which corresponds to the entries of the matrices of Eq. (13). These values are taken from the prototoype PMA described in Ref. 5. In the table, m_1 and m_3 include a mass of 0.73 kg, representing the dead mass of the actuator. Applying the results of *Theorem* 2, i.e., inequalities (36) and (37), it can be seen that the response of the closed-loop system will

Table 3 Parameter values for a four-degree-of-freedom two-actuator system

$m_1 = 5$ kg	$k_1 = 5000$ N/m	$c_1 = 30$ Ns/m
$m_2 = 6$ kg	$k_2 = 6000$ N/m	$c_2 = 35$ Ns/m
$m_3 = 6$ kg	$k_3 = 5000$ N/m	$c_3 = 35$ Ns/m
$m_4 = 8$ kg	$k_4 = 8000$ N/m	$c_4 = 40$ Ns/m
$m_{a1} = m_{a2} = 0.232$ kg	$k_{a1} = k_{a2} = 108$ N/m	$c_{a1} = c_{a2} = 0.76$ Ns/m

be asymptotically stable if

$$g_1 < 14.057 \tag{43}$$

and

$$g_2 < 14.587 \tag{44}$$

These represent constraints that are conservative and do not imply instability when violated.

As an illustration of the effect of various choices of the control gains g_1 and g_2, several cases are presented. In order to compare various cases, natural frequencies and damping ratios of the various closed-loop systems are presented. The damping ratios are used as a measure of stability, as well as closed-loop performance. As the damping ratios of the response increase, better vibration suppression is achieved. As the damping ratios decrease, the system becomes "less" stable. A zero damping ratio denotes a stable response that is not asymptotically stable. A negative damping ratio denotes instability. Case 1 consists of the system with no actuator attached. Case 2 consists of the structure with two actuators added; the natural frequencies and damping ratios are given in Table 4. Note that the effect of the actuators is to lower the spectrum somewhat. No appreciable damping increase appears due to the presence of the actuators. No instability occurs either. Next consider setting the gains as large as possible. Case

Table 4 Case 1: no actuators attached; case 2: actuators are attached with zero gain

Case 1			Case 2		
n	ω_n	ζ_n	n	ω_n	ζ_n
1	9.808	0.0289	1	9.524	0.0282
2	30.568	0.0878	2	21.349	0.0747
3	51.322	0.1404	3	21.702	0.0762
4	59.494	0.1695	4	30.013	0.0873
			5	49.150	0.1341
			6	57.581	0.1644

Table 5 Case 1: maximum gain; case 2: large gain

Case 3			Case 4		
$g_1 = 14.057$, $g_2 = 14.87$			$g_1 = 24$, $g_2 = 30$		
n	ω_n	ζ_n	n	ω_n	ζ_n
1	9.521	0.0179	1	9.508	0.0076
2	21.312	0.0464	2	21.147	0.0272
3	21.724	0.0928	3	21.787	0.1098
4	30.090	0.1172	4	30.375	0.1368
5	49.094	0.1551	5	48.980	0.1735
6	57.581	0.1644	6	57.508	0.1882

3 is then the combined structure/actuator system with velocity feedback defined by $g_1 = 14.057$ and $g_2 = 14.87$. The natural frequencies and damping ratios are given in Table 5. For comparison the gains are increased past the accepted point suggested by *Theorem* 1. This forms case 4. The resulting closed-loop natural frequencies and damping ratios are given in Table 5.

Note that, as g_1 and g_2 are increased from zero to about 14, the first two damping ratios decrease, whereas the last four increase. This represents the actuators adding damping to the structural response at the expense of damping of the actuator masses. As the gains are increased further, say, to $g_1 = 24$ and $g_2 = 30$ of case 4 in Table 4, the trend continues with the first two damping ratios decreasing and the last four increasing.

Cases 5 and 6 consist of gains of $g_1 = g_2 = 40.019$, and $g_1 = 40$ and $g_2 = 50$, respectively. Again, the natural frequencies and damping ratios are tabulated and listed in Table 6. Note that, as the gains are increased to around 40, the real part of one of the eigenvalues becomes zero and the

Table 6 Marginally stable and unstable closed-loop systems

Case 5			Case 6		
$g_1 = g_2 = 40.019$			$g_1 = 40$, $g_2 = 50$		
n	ω_n	ζ_n	n	ω_n	ζ_n
1	9.508	0.	1	9.508	−0.0059
2	21.148	0.0015	2	21.148	0.0004
3	21.375	0.1216	3	21.788	0.1318
4	30.375	0.1648	4	30.375	0.1645
5	48.980	0.1937	5	48.980	0.2002
6	57.508	0.1993	6	57.508	0.2043

corresponding damping is zero, yielding a stable but not asymptotically stable response (marginal stability). If g_2 is further increased to 50, the damping ratio becomes negative as the system becomes unstable. This clearly indicates that instability is achieved if the gains are too high. If the gains are both increased slightly to $g_1 = g_2 = 42$, then both modes 1 and 2 have negative damping ratios.

The combined system of this example is clearly not decoupled, nor can it be decoupled (i.e., $\tilde{C}_c M_c^{-1} \tilde{K}_c \neq \tilde{K}_c M^{-1} \tilde{C}_c$; see Ref. 8, for instance). Hence, it is not possible to clearly identify which mode corresponds to the actuators and which modes correspond to the structures. However, computation of the system eigenvectors in each case shows that the lower two modes ($n = 1,2$) are dominated by the actuator position.

It appears, then, that as the gains are increased, the system eventually becomes marginally stable, then unstable by negative damping appearing in the actuators. At the same time, the structure's modes ($n = 3,4,5,6$) are becoming more highly damped, providing the desired vibration suppression. Note that, as g increases in the preceding cases, the lowest structure damping ratio (ζ_1 in case 1) increases from 0.0289 to 0.0762 (ζ_3 in case 2) with the addition of the actuator ($g = 0$) and to 0.1098 (ζ_4 in case 4) in the closed-loop case ($g_1 = 24$, $g_2 = 30$). As the gain becomes fairly large, the damping in the structure increases dramatically. However, it does so at the expense of the damping in the actuator; the actuators are eventually driven to be unstable. Hence, the inclusion of the actuator dynamics limits the gain and hence the performance (damping ratio) obtainable from the closed-loop system. This important observation is only possible if the actuator dynamics are considered in the closed-loop control system design.

This result has been experimentally verified in principle by Zimmerman and Inman.[2] In this experiment a PMA is used to control the free vibration of a cantilevered beam. The measured time response of this combined system clearly indicates that, as the PMA gain is increased, the response of the structure dies out very rapidly (high closed-loop damping ratio). However, the response of the actuator becomes marginally stable, illustrating a constant amplitude oscillation near the break frequency of the actuator.

In this experiment a single PMA was attached to a 1 m-long cantilevered beam made of a quasi-isotropic composite material. The PMA was positioned at the free end across from an accelerometer that was used to provide feedback, giving the appearance of collocated control. A rate feedback control also was implemented by digitally integrating the accelerometer signal and multiplying by the appropriate gain g. The electronic centering force gain k_a was selected such that the break frequency of the actuator was lower than the first mode frequency of the cantilevered beam. The details of the experimental arrangement, digital control, etc., can be found in Refs. 2 and 6.

The frequency response of the open-loop system with the control off (with the actuator locked) and the closed-loop system with the control on were both measured. Using two different methods, a circle fit technique[15]

Table 7 Measured natural frequencies and damping ratios

Open loop			Closed loop ($g = 1.2$ Ns/m)		
n	ω_n	ζ_n	n	ω_n	ζ_n
0	—	—	0	3.19 Hz	0.002
1	3.49 Hz	0.002	1	4.03 Hz	0.094
2	29 Hz	0.01	2	28 Hz	0.019

and the eigensystem realization algorithm,[16] both the same frequencies and damping ratios were obtained (as listed in Table 7).

Note that the first structural modal damping ratio ($n = 1$) is increased from 0.002 to 0.094, but that in the actuator mode ($n = 0$) damping is still very low at 0.002. This indicates that both the open-loop system and closed-loop system are close to being marginally stable. In fact, the time response of the open-loop system oscillates for well over 16 s, whereas the closed-loop time response of the structure dies out after 2.5 s (see Ref. 2). However, the actuator mass still oscillates with small amplitude, indicating marginal stability.

This marginally stable or potentially unstable behavior is predicted by the theory presented here. The measured values of the structure's damping, assuming a single-degree-of-freedom model, yields $c = 0.2$ Ns/m. The PMA's internal damping is measured to be 0.76 Ns/m (see Table 1). Substituting these values into Eq. (30) of *Theorem* 1 yields

$$4(0.76)(0.2) > g^2 \tag{45}$$

Thus, asymptotic stability of the response is guaranteed if $g < 0.77$ Ns/m. Hence, the gain used in the experiment of 1.2 Ns/m, which is larger than that suggested by *Theorem* 1; as a result, it is not unreasonable to expect some unstable behavior.

In fact, the experiment shows directly that damping in the actuator is decreased, whereas damping in the structure is increased as the result of larger values of velocity feedback gain. If the motion of the actuator is completely decoupled from that of the combined system, this situation might be acceptable. However, it has been shown both analytically and experimentally such combined systems cannot be decoupled.[17,18] This means that the essentially undamped oscillation of the actuator will eventually cause an undamped marginally stable oscillation of the structure's coordinates.

For the last example consider again a single-degree-of-freedom system with one actuator attached. In this case, the actuator is used to provide both position and velocity feedback. The closed-loop equation [Eq. (40)] becomes

$$\begin{bmatrix} m & 0 \\ 0 & m_a \end{bmatrix}\ddot{z} + \begin{bmatrix} c + c_a + g & -c_a \\ -g - c_a & c_a \end{bmatrix}\dot{z} + \begin{bmatrix} k + k_a + h & -k_a \\ -h - k_a & -k_a \end{bmatrix} z = \mathbf{0} \tag{46}$$

where g and h are the velocity and position feedback gains, respectively. Here the values of m, c, k, k_a, c_a, and m_a are chosen to match those of the experiment reported in the section on actuator dynamics. This yields

$$\begin{bmatrix} 1.1537 & 0 \\ 0 & 0.232 \end{bmatrix}\ddot{z} + \begin{bmatrix} 0.96+g & -0.76 \\ -g-0.76 & 0.76 \end{bmatrix}\dot{z} + \begin{bmatrix} 663.02+h & -108.27 \\ -h-108.27 & 108.27 \end{bmatrix}z = \mathbf{0} \tag{47}$$

Applying the conditions of *Theorem* 4 to Eq. (47) yields regions of asymptotically stable operation.

These regions are obtained by enforcing the matrix definiteness criteria of *Theorem* 4. This requires the use of a symbolic manipulation code (MACSYMA[19]). Some observations resulting for this particular case are that only the first and second conditions of *Theorem* 4 are active. This happens because the third condition yields the same regions as the second condition. The regions of guaranteed asymptotic stability are given in Fig. 4.

The values of g and h that fall inside the parallelogram of the plot in Fig. 4 yield asymptotically stable responses. Note that this implies a substantial interaction between the velocity gain and the position gain, as well as with the structural parameters.

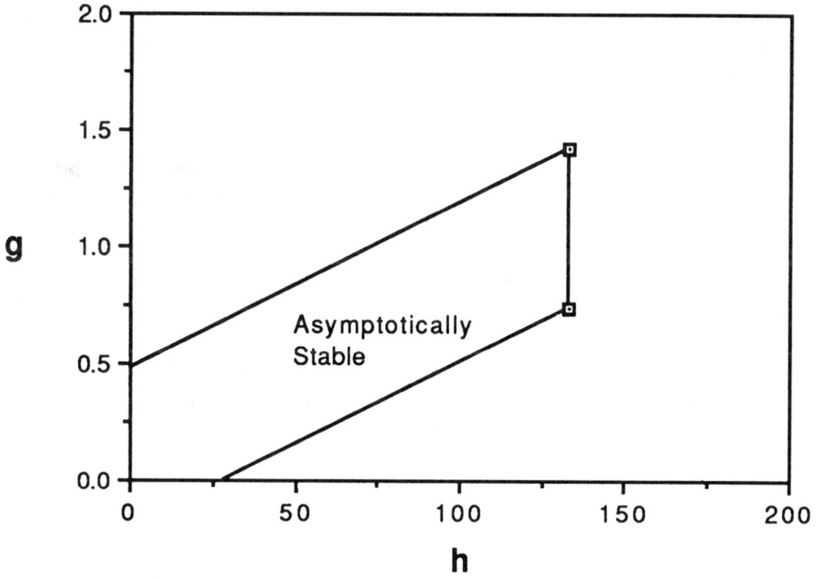

Fig. 4 Regions of acceptable positive values of position and velocity gains for asymptotically stable responses.

Previous Work

In the CSI results previously given, perfect sensors are assumed. However, as documented in several papers,[20,21] sensor dynamics can have a destabilizing effect on closed-loop control. First, if not properly filtered, sensors can result in observation spillover.[23] More importantly, Spanos[21] takes into consideration the dynamics of the sensors and their effects on the CSI problem. In particular, his work examines a proportional-derivative control of a flexible structure using sensor roll-off dynamics inside the feedback loop. The Routh criteria is used to obtain inequalities in the system parameters, providing stability conditions. Although this approach is limited to a single input/single output modal description, the study provides important insights into the nature of instabilities caused by including the sensor bandwidth as a design parameter. It is shown that, with sensor dynamics included in the model, it is possible to drive a closed-loop structure's mode unstable. In particular, if low-bandwidth inertial sensors are used, stability problems may arise if the structural modes occur near the sensor's bandwidth. In addition, phase lag caused by sensor dynamics (second order, low pass) may also destabilize the combined system through phase lag.

The result of the Spano's paper clearly illustrates that it is important to include the dynamics of sensors in closed-loop analysis. If they are not included, an unstable closed-loop design may result. The results presented here and in Ref. 2 clearly indicate that control designs considered without regard for actuator dynamics may also result in unstable closed-loop response. Although the results presented here do not include sensor dynamics and those presented by Spanos do not include actuator dynamics, the conclusion that closed-loop designs need to include both sensor and actuator dynamics is obvious.

In many cases, including the PMA discussed here, as well as the reaction mass actuators discussed in Refs. 7, 22, and 23, the velocity of the actuator mass, $\dot{x}_{ai}$, can be measured and used in the feedback law to provide a stable closed-loop system. Hallauer and Lamberson[22] pointed this out, and both Juang[23] and Miller and Crawley[7] used this approach. The feedback law used in these papers is the relative velocity of the form

$$f_{ai} = g_i(\dot{x}_{i-1} - \dot{x}_{ai}) \tag{48}$$

rather than control law $g\dot{x}_{i-1}$ used here. This form of feedback provides a symmetric feedback matrix. Furthermore, with g having the correct sign, the matrix V is positive-semidefinite so that

$$\tilde{C}_c = C_c + V \tag{49}$$

is symmetric and positive-definite, and the closed-loop system is asymptotically stable for arbitrarily large gains. This provides an extremely simple solution to the CSI instability problem presented here.

However, the use of relative velocity feedback causes the actuator to behave like a highly damped vibration absorber. Better closed-loop performance can be obtained by also using a separate feedback path for the inertial velocity, i.e., feedback laws of the form

$$f_{ai} = g_{1i}\dot{x}_{i-1} - g_{2i}\dot{x}_{ai} \tag{50}$$

The ability to assign different velocity feedback gains to both the inertial mass and the reaction mass allows better closed-loop performance. The use of a separate gain is equivalent to the inertial velocity feedback law discussed here and also results in a potentially unstable closed-loop response. The stability bounds provided here can be used to analyze control laws of the form given in Eq. (50) by simply replacing the coefficients c_{ai} with $c_{ai} + g_{2i}$. The increase in performance obtained by using this asymmetric feedback law proposed here is documented, using optimization methods, in Ref. 11.

Summary

The control structure interaction phenomenon has been approached from the point of view of examining the effects of adding actuator dynamics to a state feedback, vibration suppression problem. The results presented are restricted to actuators with second-order dynamics. The addition of the actuator's dynamics by itself results in a combined system that is still asymptotically stable. However, because of the reaction force generated against the actuator by a feedback law, the closed-loop system is potentially unstable for feedback of structural coordinates.

Stability conditions have been presented which guarantee that the closed-loop response remains asymptotically stable. These conditions are stated in terms of the definiteness of certain combinations of the mass, damping, stiffness, and feedback gain matrices. These definiteness conditions are derived from stability criteria for second-order dynamic systems and lead to inequalities in the system parameters and control gains. These inequalities quantify the control/structure interaction problem as defined here.

The conditions for stability presented in this chapter are only necessary conditions and are not sufficient. Therefore, if the conditions are satisfied, the combined system is asymptotically stable. If the conditions are not satisfied, then the response may be stable, unstable, or asymptotically stable. The conservative nature of the stability criteria has been explored by numerical examples.

Acknowledgment

This work was supported by NASA Grant NAG-1-993 monitored by Dr. Ernest Armstrong of the Langley Research Center. In addition for being grateful for this financial support, the author is indebted to Christine Jerome for preparing the manuscript under adverse conditions and to Sally Farmer for proofreading and editing, both of the University at Buffalo. In addition, the examples and numerical calculations were provided by Joseph Slater, also of the University at Buffalo, for which the author expresses his appreciation. The author also thanks William Hallauer of Virginia Tech for a very helpful discussion and for pointing out several flaws in the original manuscript.

References

[1]Balas, M. J., "Trends in Large Space Structure Control Theory: Fondest Hopes, Wildest Dreams," *IEEE Transactions on Automatic Control*, Vol AC-27, No. 3, June 1982, pp. 522–535.

[2]Zimmerman, D. C., and Inman, D. J., "On the Nature of the Interaction Between Structures and Proof-Mass Actuators," *Journal of Guidance, Control, and Dynamics*, Vol. 13, No. 1, Jan.–Feb. 1990, pp. 82–87.

[3]Pilkey, W. D., and Haviland, J. K., "Large Space Structure Damping Design, Final Report," Univ. of Virginia, Charlottesville, 1983.

[4]Haviland, J. K., Lim, T. M., Pilkey, W. D., and Politansky, H., "The Control of Linear Dampers for Large Space Structures," *Proceedings of the 1987 AIAA Guidance and Control Conference*, AIAA, New York, 1987, pp. 106–116.

[5]Zimmerman, D. C., "Dynamic Characterization and Microprocessor Control of the NASA/UVA Proof-Mass Actuator," MS Thesis, Dept. of Mechanical and Aerospace Engineering, State Univ. of New York at Buffalo, Buffalo, June 1984.

[6]Zimmerman, D. C., Horner, G. C., and Inman, D. J., "Microprocessor Controlled Force Actuator," *Journal of Guidance, Control, and Dynamics*, Vol. 11, No. 3, May–June 1985, pp. 230–236.

[7]Miller, D. W., and Crawley, E. F., "Theoretical and Experimental Investigation of Space-Realizable Inertial Actuation for Passive and Active Structural Control," *Journal of Guidance, Control, and Dynamics*, Vol. 11, No. 5, Sept.–Oct. 1988, pp. 449–458.

[8]Inman, D. J., *Vibration with Control Measurement and Stability*, Prentice-Hall, Englewood Cliffs, NJ, 1989.

[9]Kuo, B. C., *Automatic Control Systems*, 5th ed., Prentice-Hall, Englewood Cliffs, NJ, 1988.

[10]Minas, C., and Inman, D. J., "Matching Finite Element Models to Modal Data," *ASME Journal of Vibration and Acoustics*, Vol. 112, Jan. 1990, pp. 84–92.

[11]Umland, J., "Proof-Mass Actuators and Vibration Absorption," Mechanical Systems Laboratory Rept. 89-10, State Univ. of New York at Buffalo, Buffalo, 1989.

[12]Inman, D. J., "Dynamics of Asymmetric Nonconservative Systems," *ASME Journal of Applied Mechanics*, Vol. 50, No. 3, Sept. 1983, pp. 199–203.

[13]Ahmadian, M., and Inman, D. J., "Some Stability Results for General Lumped-Parameter Dynamic Systems," *ASME Journal of Applied Mechanics*, Vol. 53, March 1986, pp. 10–14.

[14]Lancaster, P., *Theory of Matrices*, Academic, New York, 1969.

[15]Ewins, D. J., *Modal Testing: Theory and Experiment*, Research Studies Press, Wiley, New York, 1986.

[16]Juang, J.-N., and Puppa, R. S., "An Eigensystem Realization Algorithm for Modal Parameter Identification and Model Reduction," *Journal of Guidance, Control, and Dynamics*, Vol. 8, No. 5, Sept.–Oct. 1984, pp. 620–627.

[17]Bellos, J., and Inman, D. J., "Frequency Response of Non Proportionally Damped, Lumped Parameter Linear Dynamic Systems," *ASME Journal of Vibration and Acoustics*, Vol. 112, No. 2, April 1990, pp. 194–201.

[18]Bellos, J., *Theoretical and Experimental Analysis of Non Proportional Damping*, PhD Dissertation, State Univ. of New York at Buffalo, Buffalo, 1989.

[19]Bogen, R., *MACSYMA Reference Manual*, Version 10, Mathlab Group, Massachusetts Inst. of Technology, Cambridge, 1983.

[20]Cudney, H. H., and Inman, D. J., "Control Formulation for Flexible Structures Using Physical Coordinates," *Mechanics of Structures and Machines*, Vol. 16, No. 4, March 1989, pp. 111–130.

[21]Spanos, J. T., "Control-Structure Interaction in Precision Pointing Servo Loops," *Journal of Guidance, Control, and Dynamics*, Vol. 12, No. 2, March–April 1989, pp. 256–263.

[22]Hallauer, W. L., and Lamberson, S. E., "Experimental Active Vibration Damping of a Plane Truss Using Hybrid Actuation," AIAA Paper 89-1169, April 1989, pp. 80–90.

[23]Juang, J., "Optimal Design of a Passive Vibration Absorber for a Truss Beam," *Journal of Guidance, Control, and Dynamics*, Vol. 7, No. 6, Nov.–Dec. 1984, pp. 733–739.

Chapter 21

Controlled Component Synthesis: A CSI Approach to Decentralized Control of Structures

K. David Young*
University of California, Livermore, California

Introduction

STRUCTURAL control has been an active area of research and development for the last decade in response to an increasing demand in large space structure pointing and control requirements. Although there has been many innovations in large space structure (LSS) control technology, one of the most difficult hurdles is the communication gap between structural engineers and control engineers. The typical structural control problem begins with the structural engineer's finite-element modeling and model reduction efforts, from which a dynamic model of the structure is generated. The modeling phase is followed by a control design phase in which the structure model is further reduced to a manageable dimension for control design. Although this level of interaction has proven to be sufficient in the past for structures with a few dominant flexible modes in the control system bandwidth, it becomes inadequate when the complexity of the structural control problem increases as the control system bandwidth increases, the number of flexible modes inside this bandwidth increases as the structural material becomes lighter, and the physical size of the structure, as well as the number of actuators and sensors, grow by an order of magnitude. In the last decade the need emerged for a distinctive hybrid engineering discipline that has its origins in structural engineering and control system engineering. This new interdisciplinary field was designated control structure interaction (CSI) technology. The first technical conference solely dedicated to CSI was sponsored by NASA and DOD and was held in November 1986.

---

*Project Engineer, Advanced Concept Engineering Group, Lawrence Livermore National Laboratory.

One of the key objectives in CSI technology is to develop methods and approaches in structural modeling and control while eliminating the technical inconsistencies that are hidden in structural and control system engineering practices. Such development perhaps is most needed in the design of decentralized control for large space structures. Decentralizing or distributing LSS structural controllers means that control action is localized. Since the structure models are developed without any consideration of the controller's action, the opportunity to model the local structural dynamic behavior in order to benefit the design of the decentralized control laws is lost.

In this chapter a connection between structural dynamics and large-scale systems control is established to benefit the design of decentralized structural controllers for LSS. The method developed herein detours from the conventional control system design path, which begins with a model of the open-loop plant. Instead, the controlled plant is assembled from *controlled components* in which the modeling phase and the control design phase are integrated at the component level. The developed method is labeled as a controlled component synthesis (CCS) method to reflect that it is motivated by the well-developed component mode synthesis (CMS) methods—a collection of structural analysis methods that has been demonstrated to be effective for solving large complex structural analysis problems for almost three decades. The design philosophy behind CCS is closely related to that of the subsystem decomposition approach in decentralized control. The ideas behind CCS are stimulated by the subsystem decomposition viewpoint in large-scale system theory on one hand and by the component mode synthesis methods in structural analysis. Connections between CCS and existing large-scale system decomposition techniques are established herein to build a control theoretic foundation for the developed method.

Subsystem Decomposition Issues

Decentralized control is intuitively appealing for structural control of large flexible structures: It offers simplified control system implementations that only require the feedback of local measurements to close the control loop for systems that may have from a few hundred to thousands of control loops. Many decentralized control methods have been developed in the last decade, typically using one of these two approaches. The first is an information constraint approach, in which the control design uses the model of the large structure and seeks a feedback control solution to the problem, subject to constraints on the flow of the measurement information. The control action decentralization is inherent in any feasible solution of the problem satisfying the imposed constraints. In this approach the design utilizes a model of the large structure in an optimization problem of the order of the model dimension. The theoretical foundation of this approach has been established by E. J. Davison and co-workers in the 1970's. Decentralized control designs for large space structures using this approach can be found in the paper by West-Vukovich et al.[1] Because of the need to exercise a model of the large structure in the design, the designer is forced to reduce the model order to cope with computational resource limitations, at the expense of causing spillovers in the controlled structure.

The second is a subsystem decomposition approach, in which the large structure model is first decomposed or partitioned into an interconnection of subsystem models. Local control designs are produced by designing controllers for the local decoupled subsystem models that are derived from the large structure model by discarding the interconnections. These local designs are solved independently, thereby reducing the computational resource requirements for performing control designs for very large structures. The behavior of the locally controlled structure is analyzed using aggregation analysis methods: An aggregate variable is defined to represent the subsystem's dynamic interactions with the other subsystems. The order of the aggregate analysis problem is equal to the number of subsystems, which is determined by the decomposition scheme. This approach is developed by Šiljak.[2] The works of Young,[3,4] Kida and Ohkami,[5] and Kida et al.[6] adopt this approach for decentralized control designs.

Given a model of a large structure, a control designer is faced with the problem of decomposing the given model into an interconnection of subsystem models if a subsystem decomposition approach is to be adopted for decentralized control design. There are two ways to obtain the subsystem models. The first converts the given structure model from its second-order structure (SOS) model form to a state space (SSP) model, i.e., from the model

$$M\ddot{\boldsymbol{x}}^n + K\boldsymbol{x}^n = \boldsymbol{f} \tag{1}$$

where M and K denote the mass and stiffness matrices of the structure, respectively; $\boldsymbol{x}^n$ is a vector of displacements at the nodes of a finite-element model; and $\boldsymbol{f}$ is a vector of generalized forces applied at the nodes, to a SSP model

$$\dot{\boldsymbol{x}} = A\boldsymbol{x} + B\boldsymbol{f} \tag{2}$$

where

$$\boldsymbol{x} = \begin{bmatrix} \boldsymbol{x}^n \\ \dot{\boldsymbol{x}}^n \end{bmatrix}, \quad A = \begin{bmatrix} 0 & I \\ -M^{-1}K & 0 \end{bmatrix}, \quad B = \begin{bmatrix} 0 \\ M^{-1} \end{bmatrix} \tag{3}$$

An interconnected subsystem (ISS) model is obtained by decomposing the A and B matrices and the state vector $\boldsymbol{x}$ of the SSP model:

$$\text{ISS}^i\!: \dot{\boldsymbol{x}}^i = A^{ii}\boldsymbol{x}^i + B^{ii}\boldsymbol{f}^i + \sum_{j \neq i} A^{ij}\boldsymbol{x}^j + \sum_{k \neq i} B^{ik}\boldsymbol{f}^k, \qquad i = 1, \ldots, s \tag{4}$$

which shows the interconnections of s subsystems. The local decoupled subsystem models are readily obtained from Eq. (4) by discarding the summation terms:

$$SSP_D^i: \dot{x}^i = A^{ii}x^i + B^{ii}f^i, \qquad i = 1, ..., s \tag{5}$$

One of the attractive features of the subsystem decomposition approach is the use of local subsystem models, rather than the model of the large structure, in designing decentralized controllers. However, from the foregoing discussions it is clear that the large structure model must also be known to the designer in this approach. Thus, the use of local models in the local control designs is an option exercised by the designer to reduce the dimensionality of the associated computational problems and to improve the robustness of the controlled structure with respect to parametric perturbations.

A more radical departure from the conventional large-scale system design path calls for the development of the decoupled local subsystem models without the use of the model of the large structure. Controlled component synthesis adopts this new viewpoint on subsystem decomposition. Such a departure from large-scale system theoretic perspectives effectively coincides with the component mode synthesis viewpoints held by structural analysts for quite some time.

Component Mode Synthesis

The concept behind component mode synthesis (CMS) was first introduced by Hurty.[7] Craig[8] provided a survey of CMS methods developed between 1960 and 1976. (CMS is also referred to as substructure coupling. The term *component* instead of substructure will be used throughout this paper.) The fundamental idea behind CMS methods is that a dynamic model of a large complex structure is to be built from models of its components. The components are first modeled as individual structures; these component model data are then processed to form a model of the large structure. The component models are typically developed using a finite-element method.

A critical design parameter in CMS methods is the choice of appropriate models or modes for the components. The selection of component modes directly affects the accuracy to which the model of the coupled structure, which is synthesized from component model data, approximates the finite-element model of the coupled structure. The development of CMS methods in the past two decades has been focused on the specification and the choice of component modes and methods to handle the constraint equations. The recent work of Flashner,[9] in which a computational procedure for CMS using singular value decomposition is derived, offered an updated view of CMS. More details on CMS can be found in Meirovitch[10] and Craig.[11]

Two of the many approaches for choosing the component model data that will be considered in CCS are particularly noted here. Benfield and Hruda[12] introduced component modes using boundary stiffness and inertia (mass) loading in which the reduced mass and stiffness matrices of a component are reduced and "loaded" on the boundary coordinates of the adjacent component using the boundary compatibility condition. Craig and Bampton[13] proposed the use of constraint modes in which the boundary coordinates are retained as component mode coordinates.

Controlled Component Synthesis

Controlled component synthesis (CCS)[14] is a framework for an integrated, component-oriented, finite-element modeling and structural control design. Similar to CMS methods, a CCS method is developed on the premise that a large complex controlled structure is to be built from *controlled components*. The finite-element modeling and control design are performed for the individual components. The model of the large complex structure is assembled from the controlled components only for the purpose of performance evaluation and sensitivity studies.

The CMS methods would be a logical first choice for generating component models. However, if the component models developed by CMS are used in the design of the controller for the component, a subsequent issue remains: What model of the coupled structure should be used to assess the controlled structure performance? Although the CMS methods produce a synthesized model of the coupled structure, the accuracy and suitability of such models for closed-loop evaluations require further examinations. On the other hand, the methodologies developed in the subsystem decomposition approach can be used to analyze the controlled coupled structure's behavior, provided that models of the coupled structure can be developed such that the approximations due to the assembly of the component are suitably characterized for these methods. In the development of CCS methods we elect to evaluate the system performance of the controlled coupled structure with systems theory tools developed in the subsystem decomposition approach. A high-fidelity model of the coupled structure, expressible in the form of an interconnected second order structure (ISOS) model, i.e.,

$$\text{ISOS}^i\colon\ M^{ii}\ddot{x}^{ni} + K^{ii}x^{ni} + \sum_{j\neq i} M^{ij}\ddot{x}^{nj} + \sum_{j\neq i} K^{ij}x^{nj} = f^i, \qquad i = 1, \ldots, s \tag{6}$$

will be adopted as an initial candidate for the coupled structure model, to be used in the performance evaluation of the controlled coupled structure.

The CCS method developed herein adopts the following modeling and control design considerations at the component level. Instead of using either the boundary loading or the constraint modes approach as outlined earlier, we introduce a new approach for the development of component models based on boundary loading. For the design of controllers for the component, an interlocking control concept is developed to minimize the motion of the nodes that are adjacent to the boundary, thereby suppressing the transmission of mechanical disturbance from component to component in the coupled structure. A new synthesis procedure for deriving the coupled structure models from component models is also developed.

Component Modeling for CCS

A two-component structure, as shown in Fig. 1, will be used to outline the modeling and design procedure of CCS. Each of the structural components is composed of three finite elements. The finite elements are identified in the figure by Roman numerals, the element node points are identified by solid circles.

In CMS methods the nodal coordinates of a component are partitioned into a set of internal (interior) coordinates x_{is} and a set of boundary coordinates x_{bs}. In the CCS method the internal coordinates are further subdivided into a group of internal boundary coordinates x_{ibs} and a group of internal coordinates x_{is}. Figure 1 indicates how the three groups of coordinates are defined. The boundary coordinates are coordinates of the boundary element, such as element III of component 1, which are on the boundary. The remaining coordinates of the boundary element are designated the internal boundary coordinates. Hereon, we use IIB coordinates to identify coordinates that are not on the boundary. The remaining coordinates of the component are the internal coordinates.

Partitioning with respect to the three groups of coordinates, the finite-element model for a component has the form

$$\begin{bmatrix} M^s_{ii} & M^s_{iib} & 0 \\ M^s_{bii} & M^s_{ib} & M^s_{ibb} \\ 0 & M^s_{bib} & M^s_{bb} \end{bmatrix} \begin{bmatrix} \ddot{x}_{is} \\ \ddot{x}_{ibs} \\ \ddot{x}_{bs} \end{bmatrix} + \begin{bmatrix} K^s_{ii} & K^s_{iib} & 0 \\ K^s_{bii} & K^s_{ib} & K^s_{ibb} \\ 0 & K^s_{bib} & K^s_{bb} \end{bmatrix} \begin{bmatrix} x_{is} \\ x_{ibs} \\ x_{bs} \end{bmatrix} = \begin{bmatrix} f^s_i \\ f^s_{ib} \\ f^s_b \end{bmatrix} \tag{7}$$

The block bidiagonality of the mass and stiffness matrices is a direct

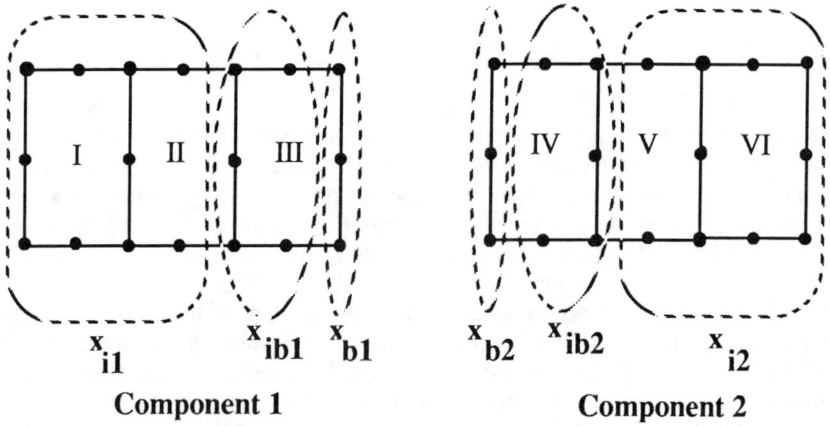

Fig. 1 A two-component structure modeled with finite elements.

consequence of the chosen finite-element mesh and the deliberate grouping of the coordinates for the component.

In developing a new approach to generate component models for CCS, we adopt the idea of boundary loading introduced by Benfield and Hruda,[12] yet we decide not to slave the IIB coordinates to the boundary coordinates. In boundary loading the stiffness matrix K^s of a component is reduced by eliminating the internal coordinates $\boldsymbol{x}_{is}$. Let the component coordinates and stiffness matrix be partitioned:

$$\boldsymbol{x}^s = \begin{bmatrix} \boldsymbol{x}_{is} \\ \boldsymbol{x}_{bs} \end{bmatrix}, \qquad K^s = \begin{bmatrix} K^s_{ii} & K^s_{ib} \\ K^s_{bi} & K^s_{bb} \end{bmatrix} \tag{8}$$

The reduced stiffness matrix with respect to the boundary coordinates is given by

$$K^s_b = T^{sT}_b K^s T^s_b \tag{9}$$

where

$$T^s_b = \begin{bmatrix} -K^{s-1}_{ii} K^s_{ib} \\ I \end{bmatrix} \tag{10}$$

The reduced mass matrix with respect to the boundary is derived following a similar process:

$$M^s_b = T^{sT}_b M^s T^s_b \tag{11}$$

In this new approach, instead of using the reduction matrix T^s_b, as given by Eq. (10), a truncation matrix that eliminates both the internal, and internal boundary degrees of freedom is used:

$$T^s_{bt} = \begin{bmatrix} 0 \\ 0 \\ I \end{bmatrix} \tag{12}$$

The modifications to the mass and stiffness matrices of the component finite-element model are computed using Eqs. (9–11) in the boundary-loading approach, but with T^s_{bt} replacing T^s_b in these equations. The resulting modified mass and stiffness matrices for loading from the jth component to the sth component are

$$M^s_{bb,iL} = M^s_{bb} + M^j_{bb} \tag{13}$$

$$K^s_{bb,iL} = K^s_{bb} + K^j_{bb} \tag{14}$$

Note that these modifications are restricted to the submatrices corresponding to the boundary coordinates; i.e., only the submatrices M^s_{bb} and K^s_{bb} are

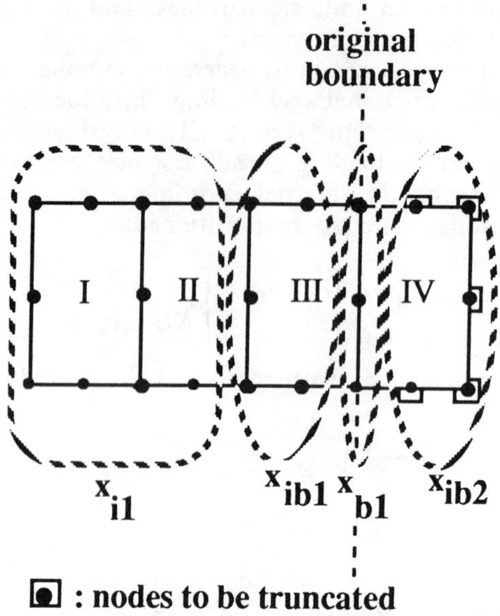

Fig. 2 Expanded component 1.

modified, and the other submatrices in Eq. (7) remain unchanged. We call this new approach to component modeling "isolated boundary loading." The modified mass and stiffness matrices that result from this approach can be obtained alternatively from the finite-element modeling of an expanded component; i.e., the original boundary of the component is extended one finite element into the adjacent component. The nodes of the expanded component consist of the original nodes of the component and the internal boundary coordinates of the adjacent component. This alternative derivation of the component model, which is physically explicit, results from the interpretation that transforming with T^s_{bt} is equivalent to the removal of the internal and internal boundary degree of freedom; i.e., $\boldsymbol{x}_{is} = 0$ and $\boldsymbol{x}_{ibs} = 0$. For isolated boundary loading from the jth to the sth component, the sth component is expanded as shown in Fig. 2. The modified mass and stiffness matrices are obtained from the mass and stiffness matrices of the expanded component by deleting the rows and columns corresponding to the nodes in the expanded portion.

Connections to Overlapping Decompositions

In addition to developing isolated boundary loading, we shall develop an alternative process, which is motivated by the following property of the modified matrices, for synthesizing a coupled structure model from the component models. The modified mass and stiffness submatrices are identical, irrespective of whether they are obtained by loading from the sth to the

jth component or from the jth to the sth component, i.e.,

$$M^s_{bb,iL} = M^j_{bb,iL} \tag{15}$$

$$K^s_{bb,iL} = K^j_{bb,iL}. \tag{16}$$

Referring to the two component structure, the coupled structure model is synthesized in two steps:

First, recombine the boundary coordinate equation in the model for the sth component,

$$M^s_{bb,iL}\ddot{x}_{bs} + M^s_{bib}\ddot{x}_{ibs} + K^s_{bb,iL}x_{bs} + K^s_{bib}x_{ibs} = f^s_b \tag{17}$$

and the model for the jth component,

$$M^j_{bb,iL}\ddot{x}_{bj} + M^j_{bib}\ddot{x}_{ibj} + K^j_{bb,iL}x_{bj} + K^j_{bib}x_{ibj} = f^j_b \tag{18}$$

into a single equation for the boundary coordinates, in which x_{bsj} denotes the coordinates on the boundary:

$$M^s_{bb,iL}\ddot{x}_{bsj} + M^s_{bib}\ddot{x}_{ibs} + M^j_{bib}\ddot{x}_{ibj} + K^j_{bib}x_{ibj} + K^s_{bb,iL}x_{bsj} + K^s_{bib}x_{ibs} = f^s_b + f^j_b \tag{19}$$

In executing this step we make use of superposition. Let

$$g^s_b = M^s_{bb,iL}\ddot{x}_{bs} + K^s_{bb,iL}x_{bs} \tag{20}$$

$$g^j_b = M^j_{bb,iL}\ddot{x}_{bj} + K^j_{bb,iL}x_{bj} \tag{21}$$

denote the forces due to the boundary motion of the sth component and the jth component, respectively; then

$$x_{bsj} = x_{bs} + x_{bj} \tag{22}$$

$$g^{sj}_b = g^s_b + g^j_b \tag{23}$$

i.e., the boundary motion x_{bsj}, as well as the forces due to this motion of the coupled structure, result from the superposition of the boundary forces and the boundary motion of the components, respectively. Solving for g^s_b and g^j_b in Eqs. (17) and (18), substituting them into Eq. (23), and making use of the boundary matrix identities [Eqs. (15) and (16)], yields the final boundary coordinate equation. Note that we have used superposition instead of the boundary compatibility condition to synthesize the equations for the boundary coordinates. The truncation matrix T^s_{bt} which performed the boundary loading was avoided in the synthesis process.

The second step in the process involves a straightforward assembly of the first and second rows of the matrix equation (7), which correspond to the internal and boundary coordinates, with Eq. (19). The assembled finite-element model, given by the following equation, is referred to as a high-fidelity model of the coupled structure,

$$M^{sj}\ddot{x}^{sj} + K^{sj}x^{sj} = f^{sj} \tag{24a}$$

where

$$M^{sj} = \begin{bmatrix} M^s_{ii} & M^s_{iib} & 0 & 0 & 0 \\ M^s_{bii} & M^s_{ib} & M^s_{ibb} & 0 & 0 \\ 0 & M^s_{bib} & M^s_{bb} + M^j_{bb} & M^j_{bib} & 0 \\ 0 & 0 & M^j_{ibb} & M^s_{ib} & M^j_{bii} \\ 0 & 0 & 0 & M^j_{iib} & M^j_{ii} \end{bmatrix}$$

$$K^{sj} = \begin{bmatrix} K^s_{ii} & K^s_{iib} & 0 & 0 & 0 \\ K^s_{bii} & K^s_{ib} & K^s_{ibb} & 0 & 0 \\ 0 & K^s_{bib} & K^s_{bb} + K^j_{bb} & K^j_{bib} & 0 \\ 0 & 0 & K^j_{ibb} & K^s_{ib} & K^j_{bii} \\ 0 & 0 & 0 & K^j_{iib} & K^j_{ii} \end{bmatrix}$$

$$x^{sj} = [x_{is}, x_{ibs}, x_{bsj}, x_{ibj}, x_{ij}]^T \tag{24b}$$

$$f^{sj} = [f^s_i, f^s_{ib}, f^{sj}_b, f^j_{ib}, f^j_i]^T, \qquad f^{sj}_b \doteq f^s_b + f^j_b \tag{24c}$$

This synthesized model is identical to a finite-element model of the coupled structure, derived with the same finite-element mesh and elements as those in the component model derivations, and finite-element methods.[15,16]

Thus, using the isolated boundary loading to generate the component models and the previously given superposition-based synthesis process, the exact model of the coupled structure is reproduced. Compared with the subsystem decomposition approach, in which interconnections are discarded in local decoupled subsystem models, such as in Eq. (5), the developed component model captures the dynamic interactions with its neighboring components. It is therefore reasonable to expect that the component controllers designed with these component models would produce better decentralized control actions.

The component models developed using this new approach have a direct connection with the subsystem decomposition approach. These models are identical to the decoupled subsystem models if an overlapping decomposition is applied to the high-fidelity model. This is a key connection, which allows the use of large-scale system analysis tools developed in the works of Ikeda and Šiljak[17] and Ikeda et al.[18] and in the report by Šiljak,[19] which contains a more elaborate exposition of the development of overlapping decomposition for decentralized control, to evaluate the performance of the controlled coupled structure after the controlled component designs have been completed. In the works of Young[3,4] decentralized control for truss structures has been investigated using decoupled subsystem models that are obtained by an overlapping decomposition of the structure. The performance analysis carried out in those earlier works can essentially be applied herein.

Interlocking Control Concept

The new insights gained from the newly developed component modeling approach in turn motivate a new component level control design concept that we call an interlocking control (ILC) concept. Again, we use a two-component structure to illustrate the idea. First, consider the model resulting from applying overlapping decomposition to the high-fidelity model. It is an ISOS model given by

$$\begin{bmatrix} M^p_{ii} & M^p_{iib} & 0 \\ M^p_{bii} & M^p_{ib} & M^p_{ibb} \\ 0 & M^p_{bib} & M^s_{bb,iL} \end{bmatrix} \begin{bmatrix} \ddot{x}_{ip} \\ \ddot{x}_{ibp} \\ \ddot{x}_{bp} \end{bmatrix} + \begin{bmatrix} K^p_{ii} & K^p_{iib} & 0 \\ K^p_{bii} & K^p_{ib} & K^p_{ibb} \\ 0 & K^p_{bib} & K^s_{bb,iL} \end{bmatrix} \begin{bmatrix} x_{ip} \\ x_{ibp} \\ x_{bp} \end{bmatrix}$$

$$= \begin{bmatrix} f^p_i \\ f^p_{ib} \\ f^p_b \end{bmatrix} + \begin{bmatrix} 0 \\ 0 \\ M^q_{bib}\ddot{x}_{ibq} + K^q_{bib}x_{ibq} + f^q_b \end{bmatrix} \tag{25}$$

for all p and q in the index set $(p,q) = \{(s,j),(j,s)\}$. Note that Eq. (25) with the second term on the right-hand side removed is identical to a component model for CCS. Interlocking control refers to a component level control design concept in which collocated actuators and sensors are placed at the internal boundary degree of freedom, and the control law is designed, using the developed component model for CCS, to minimize the internal boundary-coordinate motion. It is motivated by a closer examination of Eq. (25), which shows that such minimization would localize the dynamic interactions of the coupled structure in the components. The component control action is designed to lock up its own internal boundary to realize a boundary condition that better approximates the one assumed in the component modeling of its adjacent components.

A convenient control design technique for this concept is the linear-quadratic optimal regulator approach, in which the internal boundary coordinates are considered as regulated outputs of the component to be weighted together with the component control inputs in the quadratic performance index. The resulting component control law minimizes this index. Using this approach, the ILC concept translates into a two-step component control design process summarized in the following:

1) For the sth component use the component model

$$\begin{bmatrix} M^s_{ii} & M^s_{iib} & 0 \\ M^s_{bii} & M^s_{ib} & M^s_{ibb} \\ 0 & M^s_{bib} & M^s_{bb,iL} \end{bmatrix} \begin{bmatrix} \ddot{x}_{is} \\ \ddot{x}_{ibs} \\ \ddot{x}_{bs} \end{bmatrix} + \begin{bmatrix} K^s_{ii} & K^s_{iib} & 0 \\ K^s_{bii} & K^s_{ib} & K^s_{ibb} \\ 0 & K^s_{bib} & K^s_{bb,iL} \end{bmatrix} \begin{bmatrix} x_{is} \\ x_{ibs} \\ x_{bs} \end{bmatrix} = B^s u^s \tag{26a}$$

$$y^s = C^s \begin{bmatrix} x_{is} \\ x_{ibs} \\ x_{bs} \end{bmatrix} \tag{26b}$$

$$B^{sT} = C^s = [0 \quad I \quad 0] \tag{26c}$$

for control system design, where u^s and y^s denote the control force exerted by the actuators and the sensor outputs, respectively. For simplicity we assume collocated actuators and sensors at the internal boundary coordinates.

2) Derive the component control law by minimizing the performance index:

$$J^s_c = \frac{1}{2} \int_0^\infty (z^{sT} z^s + u^{sT} R^s u^s)\, dt, \qquad z^s = x_{ibs} \tag{27}$$

The resulting optimal component control is generally a state feedback control law:

$$u^s = [G^s_{di} \quad G^s_{dibs} \quad G^s_{db}] \begin{bmatrix} x_{is} \\ x_{ibs} \\ x_{bs} \end{bmatrix} + [G^s_{ri} \quad G^s_{ribs} \quad G^s_{rb}] \begin{bmatrix} \dot{x}_{is} \\ \dot{x}_{ibs} \\ \dot{x}_{bs} \end{bmatrix} \tag{28}$$

At first glance, the component control law is a decentralized control law, since it feeds back only measurements of the component. If we consider a state space whose state vector is composed of the components' coordinates and their time derivatives in the coupled structure, the component control law is indeed a decentralized one in this space. However, the state space of the coupled structure is a contraction[17] of this space; i.e., it is the state space with all of the duplicated boundary coordinates and their time derivatives removed. For the two-component structure whose model is given by Eq. (24), the component control laws given by Eq. (28) can be expressed in the contracted state space as

$$\begin{bmatrix} u^s \\ u^j \end{bmatrix} = \begin{bmatrix} G^s_{di} & G^s_{dibs} & G^s_{db} & 0 & 0 \\ 0 & 0 & G^j_{db} & G^j_{dibj} & G^j_{dij} \end{bmatrix} x^{sj} + \begin{bmatrix} G^s_{ri} & G^s_{ribs} & G^s_{rb} & 0 & 0 \\ 0 & 0 & G^j_{rb} & G^j_{ribj} & G^j_{rij} \end{bmatrix} \dot{x}^{sj} \tag{29}$$

showing that the boundary states are fed back to the adjacent component controllers.

The preceding design process, which utilizes collocated actuators and sensors at the internal boundary coordinates, is perhaps the most straightforward implementation of the ILC concept. However, actuators and sensors can be placed elsewhere in the component, provided that controllability and observability conditions are met. Although the linear-quadratic optimal regulator approach may require full component state feedback, other control design techniques which produce component controllers that use static and dynamic output feedback can also be used in the ILC concept, provided that the internal boundary-coordinate motion is minimized. Since internal boundary locking is the motivation behind the ILC concept, optimal control formulations may be more compatible than, for example, pole placement techniques for control design.

Applications to Truss Structure Control

A Six-Bay Planar Truss Example

The developed CCS method is applied to the design of structural control laws for a planar truss structure for a preliminary assessment of its feasibility toward solving more complex structural control design problems. This truss structure depicted in Fig. 3 has six bays, and the nodal coordinates are defined as the vertical and horizontal displacements at the joints. The internal boundary degrees of freedom at which collocated force actuators and displacement sensors are placed are marked by Δ.

External forces applied at the nodes are decomposed into orthogonal components. The assumptions made are that the truss members are subjected to axial forces alone, and not bending moments, and the members are uniform rods of identical lengths L, mass per unit length m, cross-sectional area per unit length A, and modulus of elasticity E.

The six-bay truss can be viewed as a structure consisting of three identical components: the left component, the center component, and the right component, which are composed of the leftmost, the middle, and the rightmost two bays, respectively. The six-bay/three-component truss structure is chosen to capture the essential characteristics of a truss consisting of an arbitrary number of identical components, i.e., a truss structure with an arbitrarily large number of two-bay components is composed of the three same types of components identified in the six-bay truss, with the center component duplicated as necessary. Thus, conclusions from the six-bay/three-component design apply equally well to the design of structural controls for a multiple-bay truss.

For CCS the component models are developed using the expanded component introduced in isolated boundary loading. The mass and stiffness matrices of the expanded component are derived using a finite-element method with the Ritz-Rayleigh approximation. The truss member mass and stiffness matrices expressed with respect to local coordinates and used in the assembly process are[20]

$$K_{\text{member}} = \frac{EA}{L}\begin{bmatrix} 1 & -1 \\ -1 & 1 \end{bmatrix}, \qquad M_{\text{member}} = \frac{mL}{6}\begin{bmatrix} 2 & 1 \\ 1 & 2 \end{bmatrix} \tag{30}$$

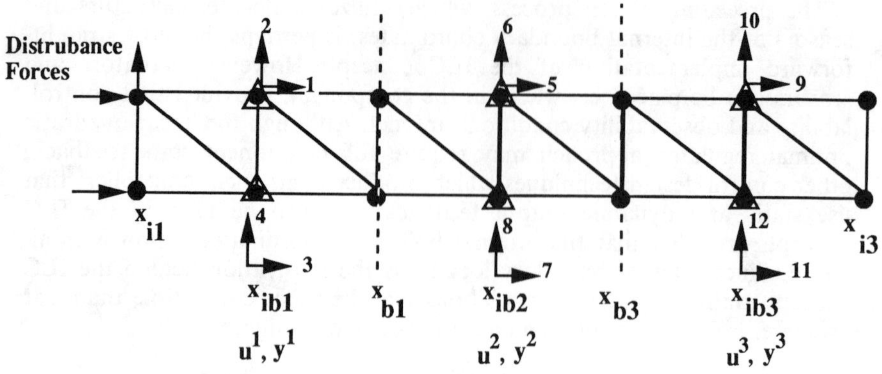

Fig. 3 Planar truss structure for CCS evaluation.

The component model is further scaled to remove the effects of the material properties. A new time variable

$$\tau = \left(\frac{m}{6EA}\right)^{1/2} Lt$$

is introduced, and the nodal forces are scaled by $L/(EA)$.

The three expanded components from which the component models are derived are shown in Fig. 4. The rows and columns of the expanded component's mass and stiffness matrices corresponding to the internal boundary nodes of the adjacent components are deleted to form the component's mass and stiffness matrices. These nodes are also identified in Fig. 4. Because of symmetry, the matrices for the leftmost and the rightmost components are the same after permutation.

Figure 5 illustrates how the coupled structure model can be synthesized from the component models. The labels *i*, *ib*, and *b* denote the submatrices of the internal, internal boundary, and boundary coordinates, respectively.

No damping is assumed for the truss structure; thus, the open-loop poles of the coupled structure and the three components all lie on the imaginary axis. In Fig. 6 the magnitudes of the imaginary parts of these poles are plotted against the mode numbers. Note that, since the coupled structure is a planar truss and is unconstrained, there are three rigid-body modes: the first three modes in this plot corresponding to the triple double poles at the origin of the complex plane. The absence of rigid-body modes in the component models, as indicated by the nonzero magnitudes of their poles, results from the boundary constraints imposed on the expanded components.

A linear-quadratic regulator approach to ILC, as outlined in the last section, is adopted in the design of the component controllers. An identical

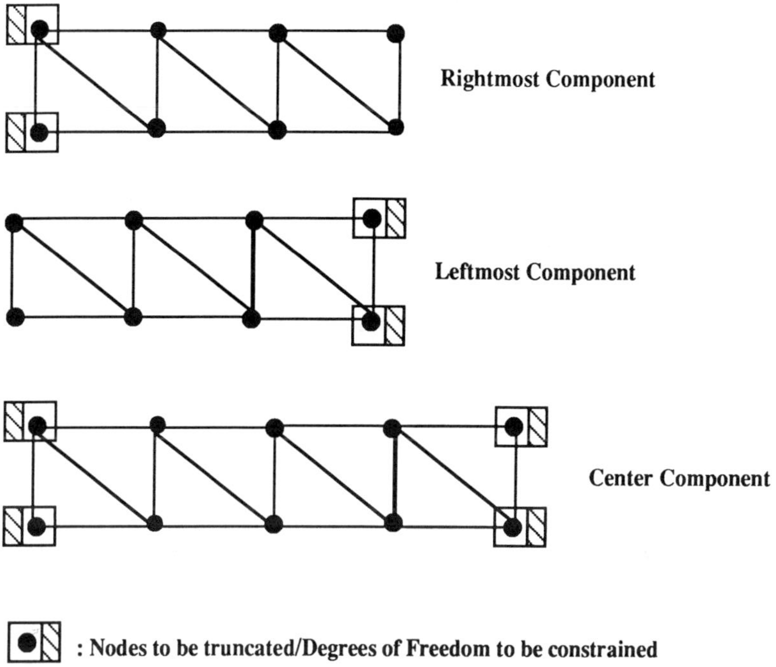

Fig. 4 The three expanded components.

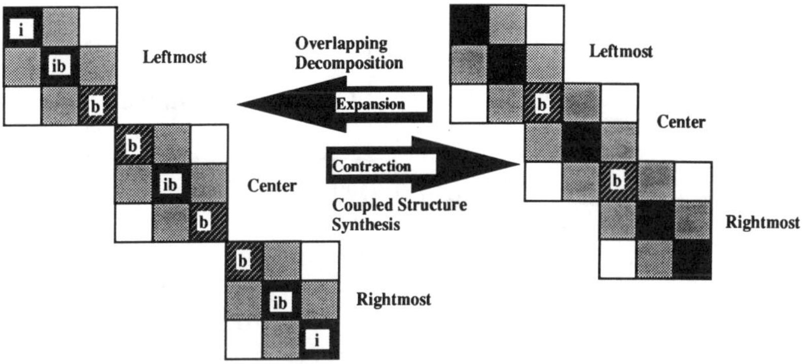

Fig. 5 Synthesis of six-bay truss model from component model.

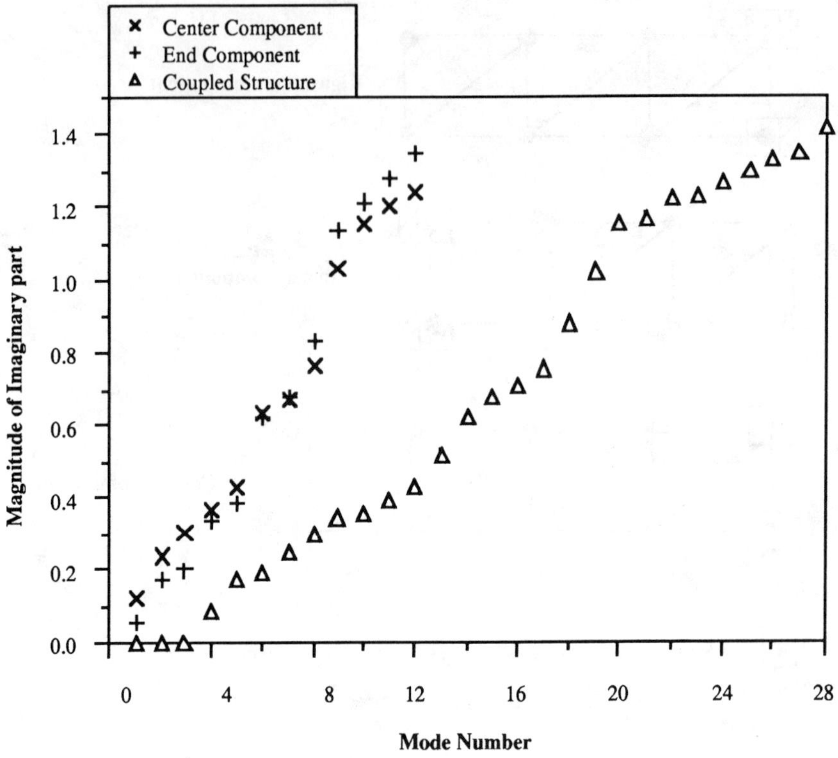

Fig. 6 Open-loop poles of the component and the coupled structure.

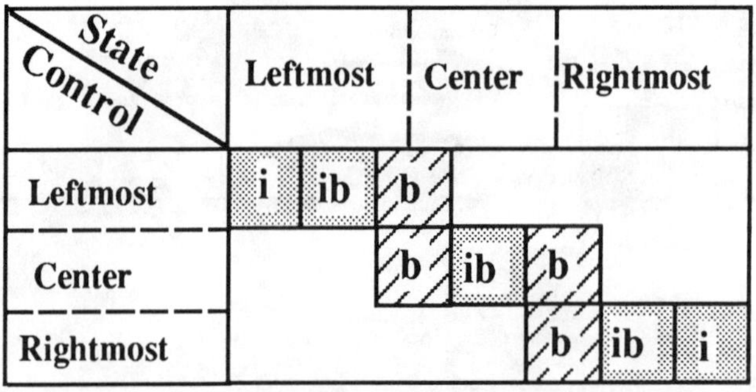

Fig. 7 Feedback matrix form for six-bay truss.

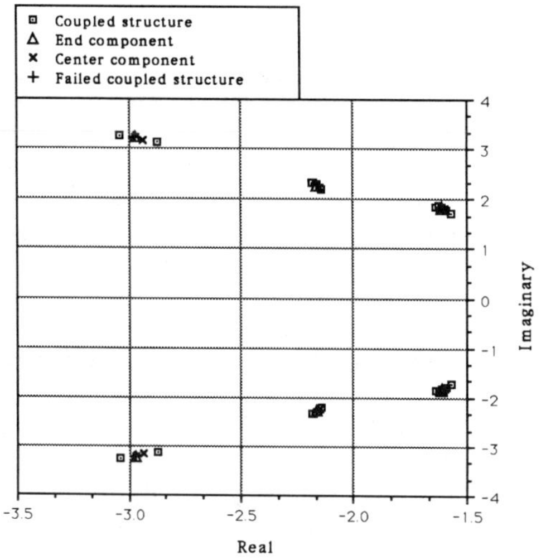

Fig. 8 Pole locations of the controlled components and the controlled truss: large real parts.

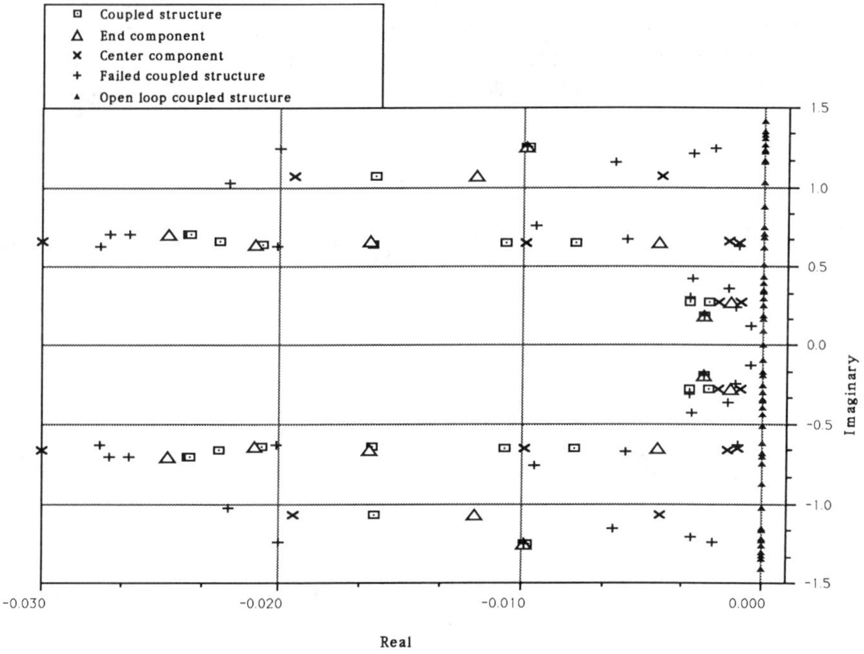

Fig. 9 Pole locations of the controlled components and the controlled truss: small real parts.

control weighting matrix $R = 0.001 I_{4\times 4}$ for all three components is chosen. The control designs for the leftmost and rightmost components are identical due to symmetry. Therefore, we only need to carry out a center component design and an end component control design. In each of these component control designs, a 24th-order Riccati equation is solved to compute the optimal feedback gains, in contrast to a 56th-order Riccati equation if an optimal centralized control approach is used. However, for an N bay truss, the computational burden of the component control design remains to be a 24th-order Ricatti equation, whereas in the centralized case the burden is an $8(N + 1)$-order Ricatti equation. Figure 7 illustrates how the feedback gain matrix for the six-bay truss is synthesized from the component controller gain matrices. The nonzero blocks of this matrix are highlighted.

The controlled components' poles, as well as the poles of the controlled truss structure, are plotted in Figs. 8 and 9. Since the leftmost and rightmost components are identical, we plot only the poles of one of them, which are denoted by end component poles in these figures. All of the poles of the controlled structure have negative real parts, indicating that the closed-loop system is asymptotically stable. The fact that the pole locations of the controlled components are close to those of the controlled coupled structure indicates that the component models developed for CCS are effective for this structural control design. The component controllers are also robust with respect to the unmodeled structural damping. The resulting closed-loop system eigenvalues shifted predictably to the left by the amounts of modal damping added to the uncontrolled structure.

For transient response studies we examine the response of the controlled structure to three unit amplitude disturbance force pulses of 0.5 duration,

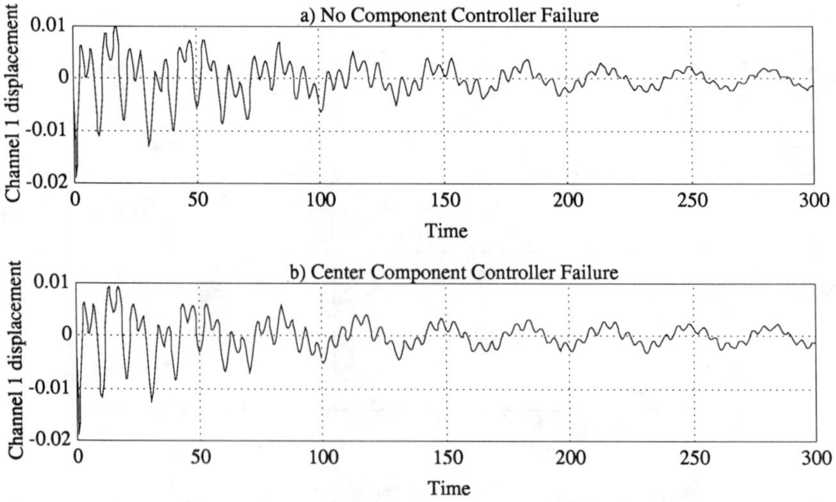

Fig. 10 Controlled channel 1 displacement time responses.

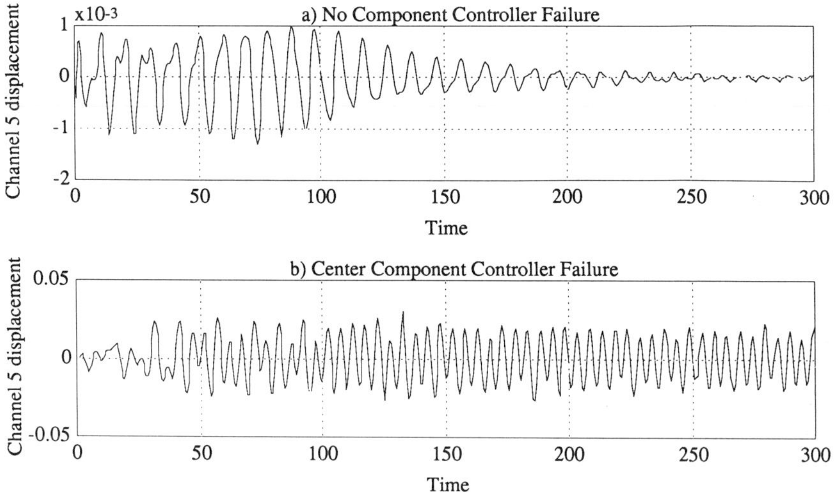

Fig. 11 Controlled channel 5 displacement time responses.

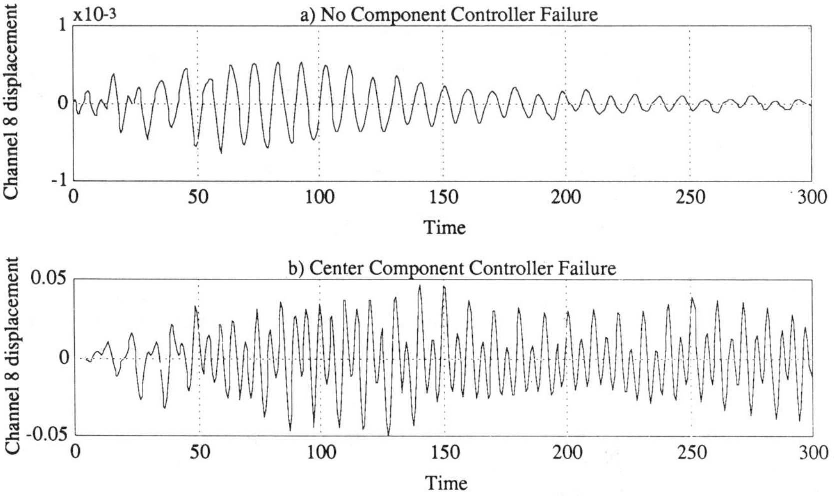

Fig. 12 Controlled channel 8 displacement time responses.

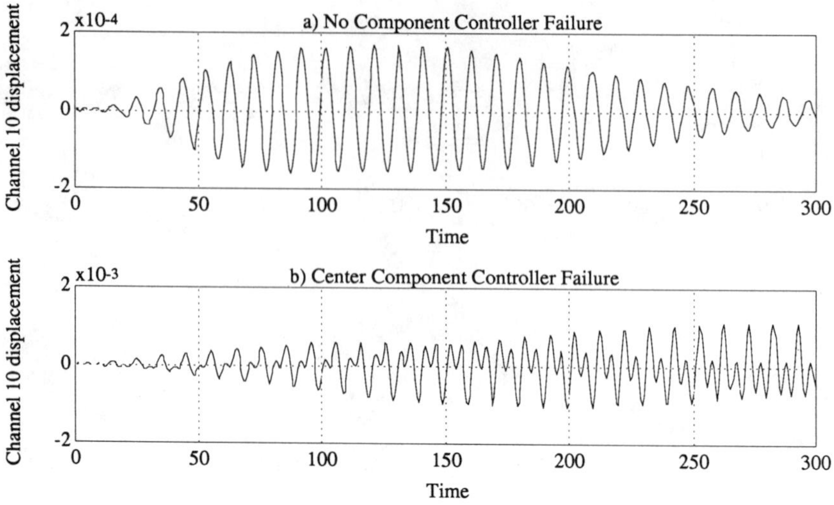

Fig. 13 Controlled channel 10 displacement time responses.

simultaneously applied to the leftmost nodes of the truss, as shown in Fig. 4. The coupled structure is assumed to be in static equilibrium initially in the simulation. A sample of the displacement sensor output time responses is shown in Figs. 10–13. Four of the 12 sensor channels, two horizontal (channels 1 and 5 in Fig. 4) and two vertical nodal displacements (channels 8 and 10) at the internal boundaries of the controlled components are selected. The magnitudes of the displacement response drop by an order of magnitude per component for nodes that are farther away from the disturbances. The delay effect of the force pulses on the displacements shown in Fig. 13a is typical for the rightmost component. The fact that the controlled components stabilize the rigid-body modes of the structure is illustrated with Fig. 14, which shows the displacement time responses of channels 1 and 10 for the uncontrolled structure.

The fact that the developed CCS method inherits the capability to withstand system failures from decentralized control developed using the subsystem decomposition approach is demonstrated with the following results for the controlled structure, in which the center component controller failed. Figures 8 and 9, in which the pole locations of the controlled structure with the component controller failure are also plotted, show a slight decrease in damping in a number of the modes near the imaginary axis. However, the controlled structure remains asymptotically stable. This failure scenario is simulated for the same disturbances and initial conditions as before. The same four-channel responses when the center component controller fails are also included in Figs. 10–13. Figures 11b and 12b show an order-of-magnitude performance degradation for the two displacements at the center component: from the $\mathcal{O}(10^{-3})$ peak displacement excursion for channels 5 and 8 in the no-failure state to $\mathcal{O}(10^{-2})$ in the

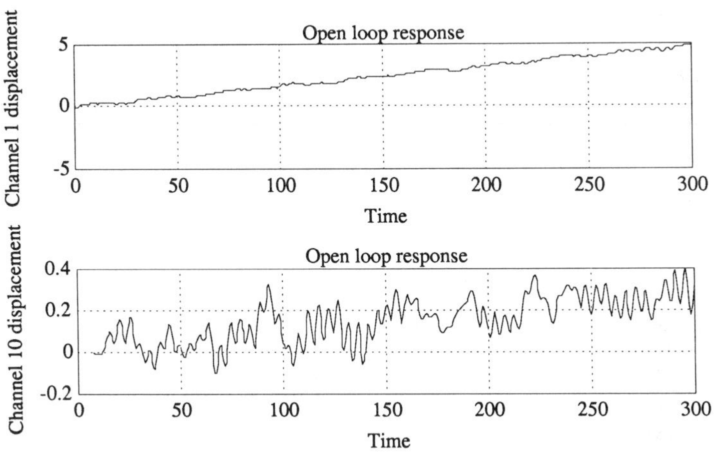

Fig. 14 Uncontrolled channels 1 and 10 displacement time responses.

failure state. However, despite the center component controller failure, the neighboring components stabilize the vibrations in the center component with interlocking controls.

A 12-m Truss Example

The second design example was carried out by Özgüner and Cagle[21] on a model of a 12-m truss, an experimental structure built to demonstrate large space structure technology at the Flight Dynamics Laboratory of the Wright-Patterson Research Center. (Data on this truss control design are furnished by Professor Ümit Özgüner of the Ohio State University.) This truss, which consists of a welded aluminum tube frame with bolt-in diagonal members of Lexan plastic, is cantilevered vertically as shown in Fig. 15. Eight collocated pairs of actuators and sensors are mounted on the structure. Proof mass actuators and accelerometers are employed. Two pairs of actuators are mounted at the tip of the truss to control both bending and torsion modes, and a pair of actuators are located at the three-quarter location up from the base and another pair at the midspan location to provide additional bending mode control. Figure 16 shows the layout of the actuator locations.

The initial finite-element model of the structure consists of 86 nodes. It is further decomposed into two component models: A model that represents the top half of the truss has 50 nodes, whereas the bottom half of the truss is modeled with 44 nodes. There are 8 interface nodes that are present in both of the substructure models. Although these component models are directly obtained from the finite-element model of the coupled structure, they are identical to the ones derived from the CCS modeling procedure. The connections between the CCS component models and the coupled

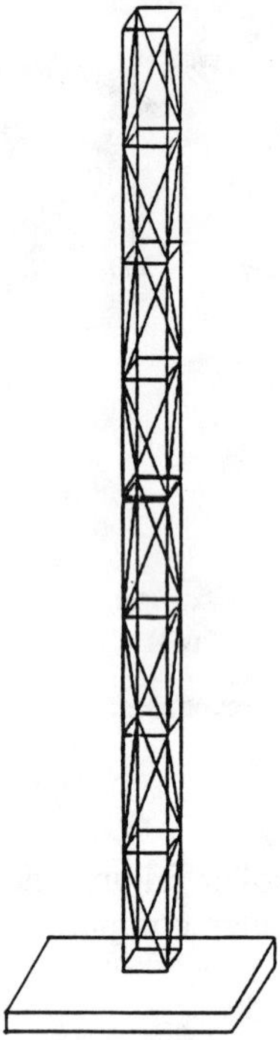

Fig. 15 Twelve-meter truss configuration.

structure model, which are expressed in terms of overlapping decomposition, are exercised to produce the component models.

For the component control design, the component model is further reduced to include only the lowest six modes, which are the first and second bending modes in the x and the y axes, and the first two torsional modes. The top component is equipped with three actuators and sensors on each axis, whereas the bottom component has one. A linear-quadratic performance index in which the modal velocities are weighted is optimized with respect to a collocated velocity feedback control law. For the top component the six-mode control design is further decomposed into an x-axis and a y-axis subsystem in which the velocities are fed back to drive the actuators in the respective directions. For the bottom component only the two bending modes are used in the design, since the torsional modes are weakly controllable. The resulting controller implements a decentralized velocity feedback law in which the feedback matrix is block diagonal; i.e., the three x-axis actuators feed back the three x-axis velocity measurements in the top component, and similarly in the y axis; the bottom component has single channel velocity feedback in the x and the y axes.

The modal damping achieved in this decentralized control design, compared with the inherent damping in the open-loop structure, is given in Fig. 17. The amount of damping achieved is comparable to that obtained using centralized control. The simulated time responses at selected locations at

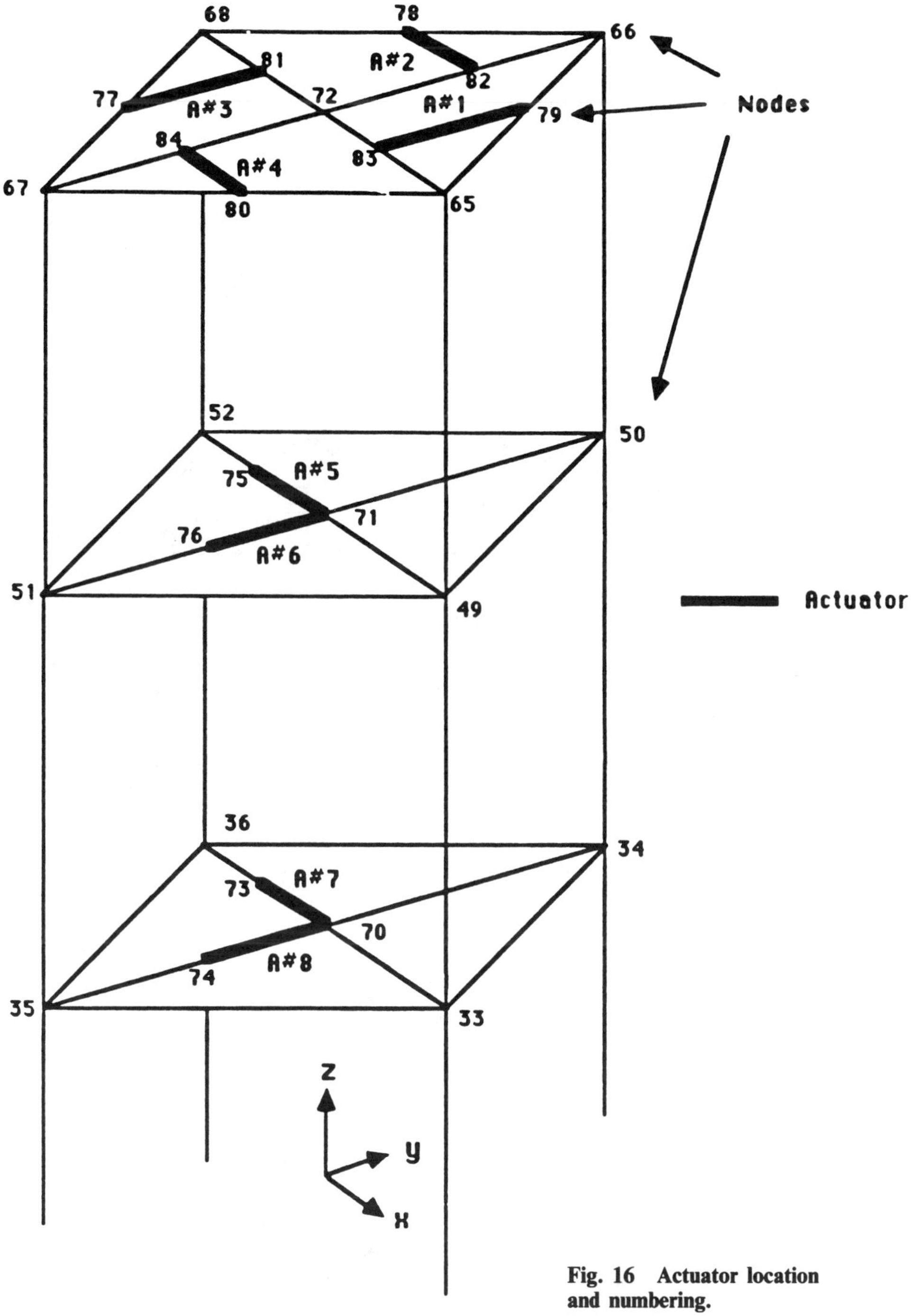

Fig. 16 Actuator location and numbering.

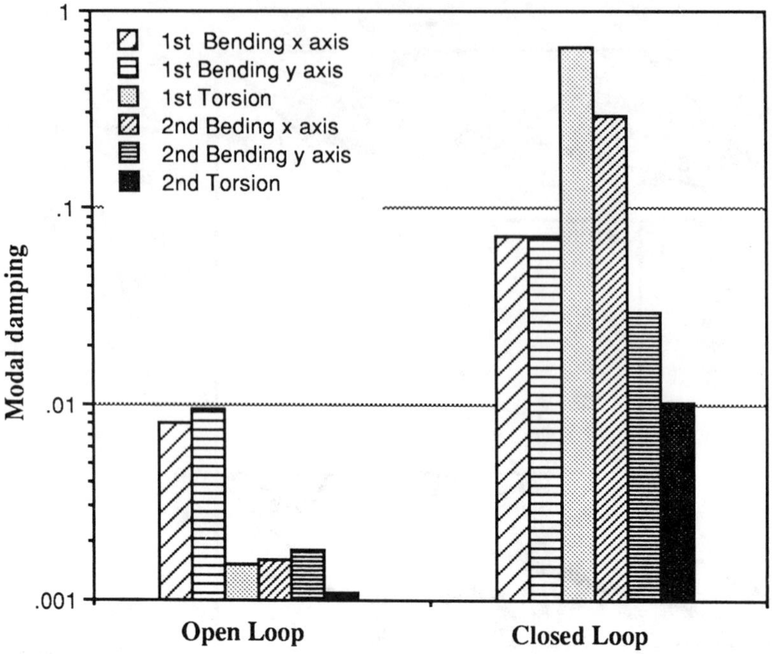

Fig. 17 Modal damping comparisons.

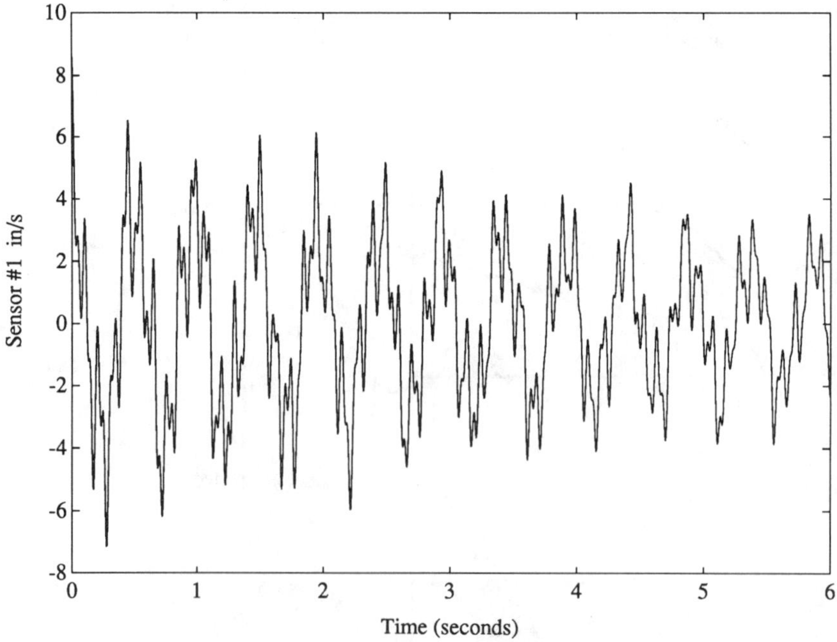

Fig. 18a Open-loop rate sensor time response, location A1 at the tip of the truss in the *y* direction.

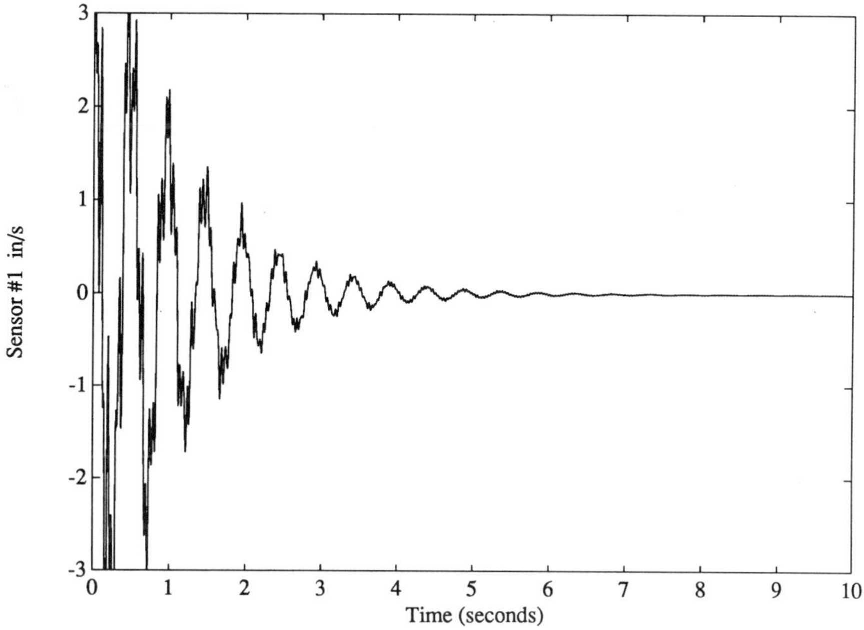

Fig. 18b Closed-loop rate sensor time response, location A1 at the tip of the truss in the *y* direction.

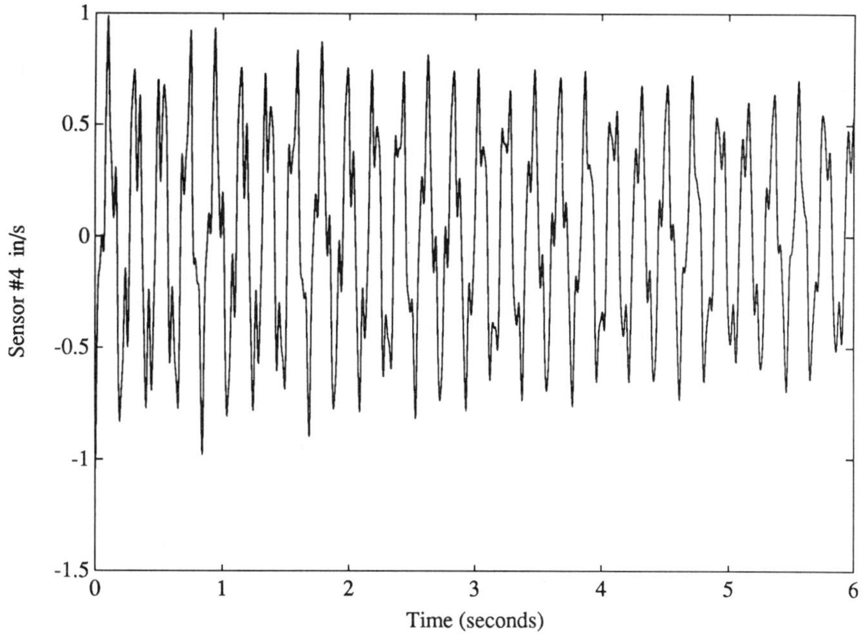

Fig. 19a Open-loop rate sensor time response, location A4 at the tip of the truss in the *x* direction.

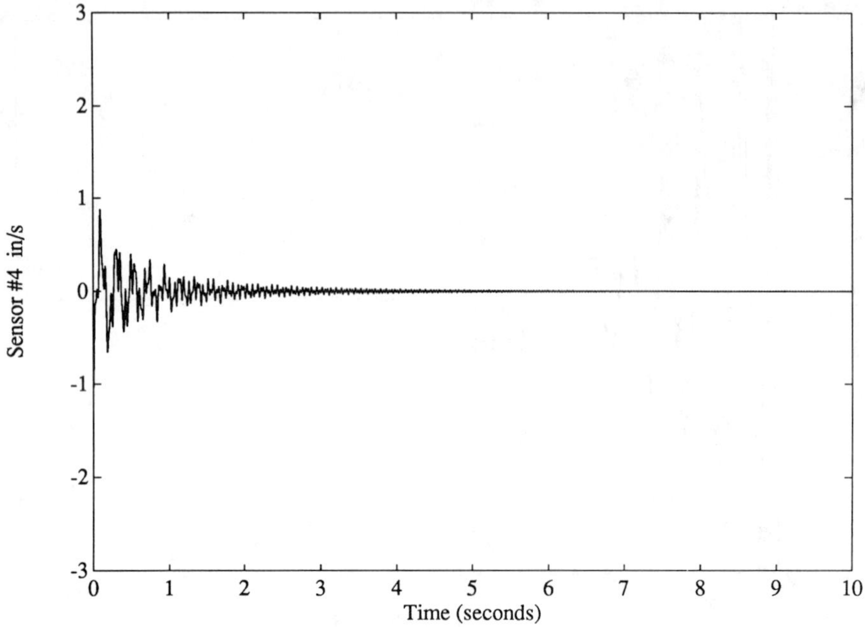

Fig. 19b Closed-loop rate sensor time response, location A4 at the tip of the truss in the *x* direction.

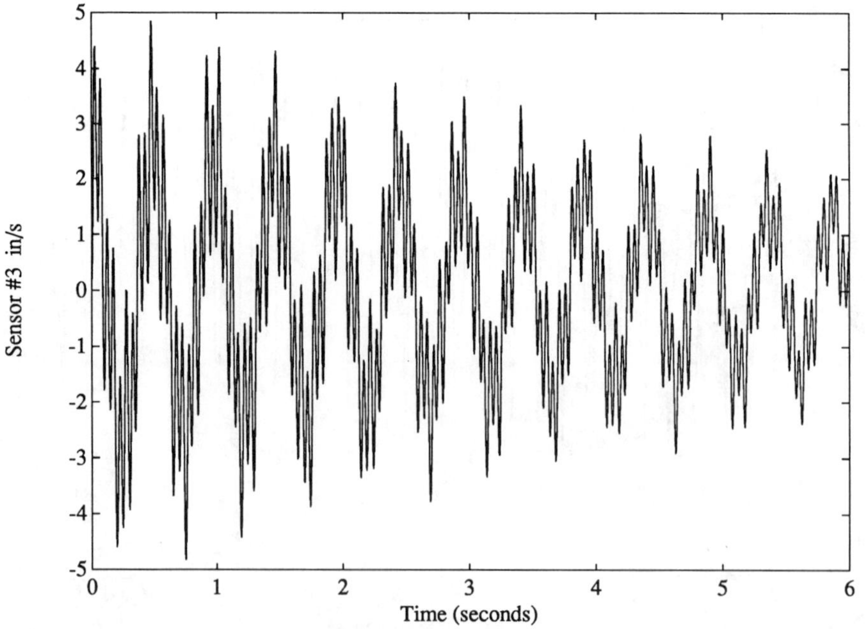

Fig. 20a Open-loop rate sensor time response, location A3 at the three quarters station of the truss in the *y* direction.

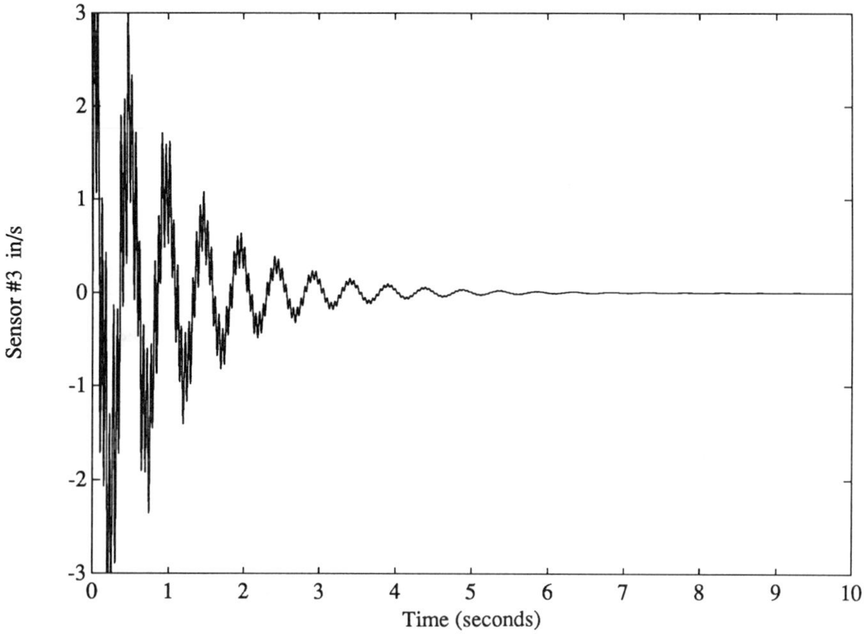

Fig. 20b Closed-loop rate sensor time response, location A3 at the three-quarters station of the truss in the *y* direction.

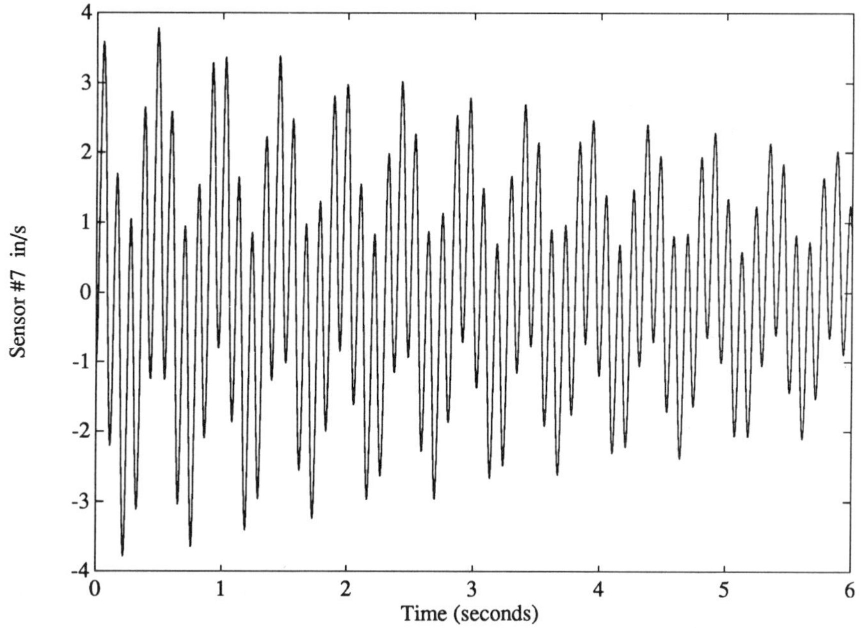

Fig. 21a Open-loop rate sensor time response, location A7 at the midspan station of the truss in the *y* direction.

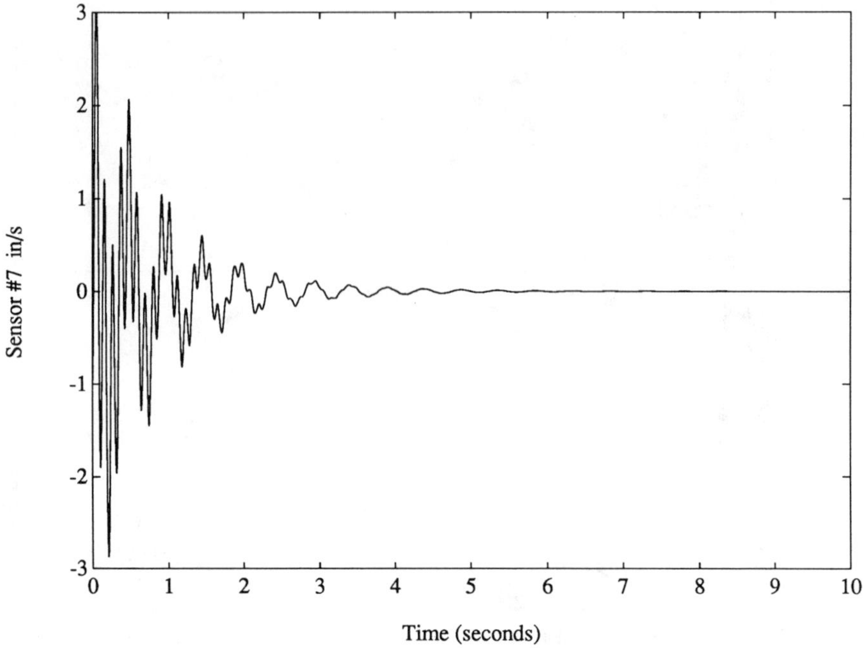

Fig. 21b Closed-loop rate sensor time response, location A7 at the midspan station of the truss in the *y* direction.

the tip, the three-quarter, and the midstations are shown in Figs. 18–21. In the simulation, an impulse is applied by actuator A1 in the y direction, which is located at the tip of the truss.

Conclusions

The controlled component synthesis method has been shown in this chapter to be an effective tool for the design of decentralized control for truss structures. A new component modeling approach, which is similar in many aspects to component mode synthesis but more appropriate for controlled component synthesis, is introduced for developing component models of complex structures. We also showed that the model of a large space structure is perhaps too large to be engaged in existing decentralized control methods, particularly those using the information constraint approach. The model dimension of the planar truss considered herein rapidly increases as the number of bays gets large: a 100-bay truss, similar to the one in Fig. 4, means 808 states. The proposed decentralized control method also deals with the spillover issue indirectly, yet effectively. For large complex structures with a large number of modes within the control system bandwidth, the order of the control system design model is constrained by the computability of the associated control design problem. Spillover often results from the use of reduced-order models in the control design while system dynamics well inside the design bandwidth are neglected. Through the development of component models and component control design, CCS provides the opportunity for the designer to model the components of the large structure with higher-fidelity models, while keeping the computational problem associated with the control design manageable.

The modeling and design of decentralized structural controllers for a planar truss demonstrates the feasibility of controlled component synthesis for structures with an open-chain topology. However, ring trusses, and other similar closed-chain structures, can also be handled by this approach using an assembly process that generates the controlled closed-chain structure from a hierarchy of open-chain supercomponents. Continued research in extending the proposed method has occurred in the directions of examining the robustness of the decentralized controllers to component modeling errors, pursuing alternative component control designs using the interlocking control concept, and developing a hierarchical controlled component approach for complex space structures.

Acknowledgment

This work was performed under auspices of the U.S. Department of Energy by the Lawrence Livermore National Laboratory under Contract W-7405-ENG-48.

References

[1]West-Vukovich, G. S., Davison, E. J., and Hughes, P. C., "The Decentralized Control of Large Flexible Space Structures," *IEEE Transactions on Automatic Control*, Vol. AC-29, 1984, pp. 886–879.

[2]Šiljak, D. D., *Large-Scale Dynamic Systems*, North-Holland, Amsterdam, 1978.

[3]Young, K. D., "An Application of Decomposition Techniques to Control of Large Structures," *Proceedings of the Fourth VPI&SU/AIAA Symposium on Dynamics and Control of Large Structures*, Virginia Polytechnic Institute and State University, Blacksburg, VA, 1983, pp. 201–214.

[4]Young, K. D., "Approximate Finite Element Models for Structural Control," *Proceedings of the 24th IEEE Conference on Decision and Control*, IEEE, New York, 1985, pp. 940–945.

[5]Kida, T., and Ohkami, Y., "A Preliminary Study on Decentralized Control of Large-Scale Flexible SPS," *Space Solar Power Review*, Vol. 5, 1985, pp. 171–176.

[6]Kida, T., Yamaguchi, I., and Ohkami Y., "An Approach to Local Decentralized Control for Large Flexible Space Structures," AIAA Paper 85-1924-CP, Aug. 1985.

[7]Hurty, W. C., "Vibration of Structural Systems by Component Mode Synthesis," *ASCE Journal of the Engineering Mechanics Division*, Vol. 85, 1960, pp. 51–69.

[8]Craig, R. R., Jr., and Bampton, M. C. C., "Coupling of Substructures for Dynamic Analyses," *AIAA Journal*, Vol. 6, No. 7, 1968, pp. 1313–1319.

[9]Flashner, H., "An Orthogonal Decomposition Approach to Modal Synthesis," *International Journal for Numerical Methods in Engineering*, Vol. 23, 1986, pp. 471–493.

[10]Meirovitch, L., *Computational Methods in Structural Dynamics*, Sijthoff & Noordoff, The Hague, 1981.

[11]Craig, R. R., Jr., *Structural Dynamics*, Wiley, New York, 1981.

[12]Benfield, W. A., and Hruda, R. F., "Vibration Analysis of Structures by Component Mode Substitution," *AIAA Journal*, Vol. 9, No. 7, 1971, pp. 1255–1261.

[13]Craig, R. R., Jr., "Methods of Component Mode Synthesis," *Vibration and Shock Digest*, Vol. 9, 1977, pp. 3–10.

[14]Young, K. D., "A Distributed Finite Element Modeling and Control Approach for Large Flexible Structures," AIAA Paper 88-4085-CP, Aug. 1988.

[15]Gallagher, R. H., *Finite Element Analysis Fundamental*, Prentice-Hall, Englewood Cliffs, NJ, 1975.

[16]Yang, T. Y., *Finite Element Structural Analysis*, Prentice-Hall, Englewood Cliffs, NJ, 1986.

[17]Ikeda, M., and Šiljak, D. D., "Overlapping Decompositions, Expansions, and Contractions of Dynamic Systems," *Large Scale Systems*, Vol. 1, 1980, pp. 29–38.

[18]Ikeda, M., Šiljak, D. D., and White, D. E., "Decentralized Control with Overlapping Information Set," *Journal of Optimization Theory and Applications*, Vol. 34, 1981, pp. 279–310.

[19]Šiljak, D. D., *The Inclusion Principle for Dynamic Systems*, Sept. 1977–April 30, 1982, University of Santa Clara, CA, Final Rept. DE-AC037ET29138-34.

[20]Meirovitch, L., *Elements of Vibration Analysis*, McGraw-Hill, New York, 1975.

[21]Özgüner, Ü, and Cagle, A., "Optimal Decentralized Static Output Feedback Control for a Truss Structure," AIAA Paper 89-3570-CP, Aug. 1989.

Chapter 22

Near-Minimum-Time Maneuvers of Flexible Vehicles: A Liapunov Control Law Design Method

John L. Junkins,* Z. Rahman,† and H. Bang‡
Texas A&M University, College Station, Texas

Introduction

CONSIDERABLE recent research has been directed toward the problem of maneuvering a flexible spacecraft. In Ref. 1 we present a fairly comprehensive treatment of this family of problems up through 1985. Other analytical and experimental investigations[2–12] have been carried out to evaluate 1) feedback control laws based on minimizing quadratic indices; 2) near-minimum-time and switching-type controllers; and 3) implementation issues. The main configuration of interest in the present discussion is similar to the structure of the Vander Velde and He study[12]; however, we address the use of variable torque actuation via a reaction wheel instead of on/off thrusters, utilize Liapunov control design approaches, and establish stable tracking-type control laws for distributed-parameter systems.

We present a method for generating globally stable feedback control laws for maneuvers of distributed-parameter structural systems. The method can accommodate nonlinearity, and our proof of stability does not rely on spatially discretizing distributed-parameter systems. The most fundamental developments are related to similar ideas presented in Refs. 3 and 6 and lead to controls that drive the system to a target fixed point in the state space. However, we extend these ideas to achieve near-minimum-time control. We present novel tracking-type control laws to null the departure motion of the system state from an arbitrary smooth target trajectory and develop the global conditions for stability of the closed-loop system. The

*George J. Eppright Professor, Department of Aerospace Engineering.
†Research Engineer, Department of Aerospace Engineering.
‡Research Assistant, Ph.D. Candidate, Department of Aerospace Engineering.

control law structure is such that the large motion controller (a tracking law) blends continuously into a constant-gain terminal controller. This "seamless" transition is qualitatively more attractive than, for example, gain scheduling approaches. The target trajectory (for the tracking law) may be any smooth path; i.e., the target trajectory can be selected as an exact or approximate solution of the system equations of motion. We show how this approach can be used to generate near-minimum-time feedback controllers that achieve global stability about a fixed terminal state. This approach accommodates general "torque shaping" to achieve a compromise between near-minimum-time and competing performance indices such as levels of sensitivity to model errors, disturbances, and implementation errors. Although we subsequently develop methods for controlling distributed-parameter systems governed by hybrid coupled sets of ordinary and partial differential equations, we first consider a system described by a sixth-order set of nonlinear, ordinary differential equations.

Motivation of the Approach Using Rigid-Body Maneuvers

The ideas are easily introduced by considering general three-dimensional nonlinear maneuvers of a single rigid body. The equations governing large motion can be written as[1]

$$I_1\dot{\omega}_1 = (I_2 - I_3)\omega_2\omega_3 + u_1 \tag{1a}$$

$$I_2\dot{\omega}_2 = (I_3 - I_1)\omega_3\omega_1 + u_2 \tag{1b}$$

$$I_3\dot{\omega}_3 = (I_1 - I_2)\omega_1\omega_2 + u_3 \tag{1c}$$

$$2\dot{q}_1 = \omega_1 - \omega_2 q_3 + \omega_3 q_2 + q_1(q_1\omega_1 + q_2\omega_2 + q_3\omega_3) \tag{1d}$$

$$2\dot{q}_2 = \omega_2 - \omega_3 q_1 + \omega_1 q_3 + q_2(q_1\omega_1 + q_2\omega_2 + q_3\omega_3) \tag{1e}$$

$$2\dot{q}_3 = \omega_3 - \omega_1 q_2 + \omega_2 q_1 + q_3(q_1\omega_1 + q_2\omega_2 + q_3\omega_3) \tag{1f}$$

where $(\omega_1,\omega_2,\omega_3)$ and (q_1,q_2,q_3) are the principal axis components of angular velocity and the Euler-Rodriguez parameters ("Gibbs vector"), respectively. Also, (I_1,I_2,I_3) and (u_1,u_2,u_3) are the principal moments of inertia and the principal axis components of the external control torque, respectively. For the case of zero control torque, it can be readily verified that total rotational kinetic energy is an exact integral of the motion, viz., $2T = (I_1\omega_1^2 + I_2\omega_2^2 + I_3\omega_3^2)$.

Motivated by this total system energy integral, we investigate the trial Liapunov function

$$U = \frac{1}{2}(I_1\omega_1^2 + I_2\omega_2^2 + I_3\omega_3^2) + A(q_1^2 + q_2^2 + q_3^2) \equiv \text{kinetic energy} + A\tan^2\frac{\phi}{2} \tag{2}$$

where ϕ is the instantaneous principal rotation angle (about the instantaneous Eulerian principal rotation axis, from the current angular position to the desired final angular position of the body[1]). It is apparent that the additive term $A(q_1^2+q_2^2+q_3^2)$ can be viewed as the potential energy stored in a conservative spring. We can therefore anticipate that the system dynamics will evolve such that U is constant if the only external torque is the associated conservative moment. Of course, we are not interested in preserving U as a constant, but rather we seek to drive it to zero. We therefore anticipate the necessity to determine an additional judicious control moment to guarantee that U is a decreasing function of time. It is obvious by inspection that U is positive-definite and vanishes only at the desired state $q_i = \omega_i = 0$. Differentiation of Eq. (2) and substitution of Eqs. (1) lead directly to the following expression for $\dot{U}$:

$$\dot{U} = \sum_{i=1}^{3} \omega_i[u_i + Aq_i(1+q_1^2+q_2^2+q_3^2)] \tag{3}$$

Of all of the infinity of possible control laws, we can see that any $u_i(q_1,q_2,q_3,\omega_1,\omega_2,\omega_3)$ that reduces the bracketed terms to a negative-definite function times ω_i will guarantee that $\dot{U}$ is globally negative-semidefinite. The simplest choice consists of the following: Select u_i so that the ith bracketed term becomes $-k_i\omega_i$. This gives $\dot{U} = -(k_1\omega_1^2+k_2\omega_2^2+k_3\omega_3^2)$, which is obviously a negative-definite function of angular velocity if we choose all $k_i > 0$. It is easy to verify from Eqs. (1) that $(\omega_1,\omega_2,\omega_3)$ are generally nonzero, except at an equilibrium state, and the only equilibrium point [where $(\omega_1,\omega_2,\omega_3) = (\dot{\omega}_1,\dot{\omega}_2,\dot{\omega}_3) = (0,0,0)$] of the closed-loop system is the origin of the state space: $(q_1,q_2,q_3,\omega_1,\omega_2,\omega_3) = (0,0,0,0,0,0)$. Thus, we obtain a globally stabilizing nonlinear feedback control law[2-4] for large-angle attitude control:

$$u_i = -[k_i\omega_i + Aq_i(1+q_1^2+q_2^2+q_3^2)], \qquad i = 1,2,3 \tag{4}$$

Since U is a positive-definite, decreasing function of time along all trajectories, and since it vanishes at the origin, then the necessary and sufficient conditions are satisfied for global Liapunov stability. We have implicitly excluded the geometric singularity $(q_i \to \infty)$ associated with this parameterization of rotational motion as $\phi \to n\pi$; we can use the quaternion or Euler parameter description of motion and avoid all geometric singularities as well. This more general path has been pursued in Ref. 13. The nonlinear feedback control law of Eq. (4) guarantees stability of the nonlinear closed-loop system under the assumption of zero model errors. In practice, of course, guaranteed stability in the presence of zero model error is not a sufficient condition to guarantee stability of the actual plant having arbitrary model errors and disturbances. On the other hand, rigorously defining a region in gain space guaranteeing global stability for our best model of the nonlinear system is an important step; it is reasonable to restrict the optimization of gains to this stable family of designs.

The determination of the particular gain values, selected from the space of globally stabilizing gains, is usually based on performance optimization criteria specified in consideration of the disturbance environment, sensitivity to model errors, desired system time constants, actuator saturation, and sensor/actuator bandwith limitations. For the purpose of gain optimization it is usually convenient to introduce spatial approximations and discretize partial differential equation systems to obtain a finite set of ordinary differential equations. It is also typically useful to linearize nonlinear differential equations about one or more points in the anticipated operation envelope. Using a generalization of the preceding example, we can rigorously determine a region of admissible gains that stabilizes globally a generally nonlinear, partial differential equation description of the system. Note that approximations subsequently introduced in the gain design process cannot destabilize the zero model error system if the gains are optimized over this region that is known to be stable. In the preceding rigid-body discussion, as long as we restrict $\{A,k_1,k_2,k_3\}$ to be positive, for example, approximations such as small-angle linearizations used to select particular values cannot destabilize the system. Of course, approximations introduced to facilitate the gain design process can indeed lead to undesired and unanticipated performance variations of the actual nonlinear distributed-parameter system, compared to predictions based on approximate models used to design the control gains.

Before generalizing the methodology to consider partial differential equation systems, it is important to reflect on the selection of the Liapunov function previously given. Notice that, if a system has no inherent stiffness with respect to rigid-body displacements, it is necessary to augment the open-loop energy integral by a pseudo-potential energy term [such as $A(q_1^2+q_2^2+q_3^2)$ in the preceding example]; generally speaking, the pseudo-energy term should be defined such that the candidate Liapunov function (U) is a positive-definite measure of motion that has its *global minimum at the desired target state*. Then the controls are usually selected as simply as possible (from an implementation point of view) to force pervasive dissipation ($\dot{U} \leqslant 0$) of the modified energy (Liapunov) function along all trajectories of the closed-loop system and thereby guarantee closed-loop stability.

Although the above insights are useful, the Liapunov function is generally not unique. We will see later that these ideas lead to an attractive strategy that defines the Liapunov function with relative weights on the portions of the total energy associated with structural subsystems. To accommodate discussion of path-type stability of tracking motion control, we will also introduce "error" energies measured from an a priori computed time-varying trajectory. The lack of uniqueness of the Liapunov function is not necessarily a disadvantage in practice because it is a source of user flexibility providing control design freedom qualitatively comparable to the freedom one has in selecting performance indices when applying optimal control theory.

Maneuvers and Vibration Control for Distributed Parameter Systems

In Figs. 1–3 we consider control of a rigid hub with four cantilevered flexible appendages. We consider the appendages to be identical uniform flexible beams and make the Euler-Bernoulli assumptions of negligible shear deformation and distributed rotary inertia. Each beam is cantilevered rigidly to the hub and has a finite tip mass. Motion is restricted to the horizontal plane, and a control torque $u(t)$ acting on the hub is the only external effect considered at this stage of the development. We subsequently address rejection of disturbances.

We are interested in a class of rest-to-rest maneuvers, and under the previously mentioned assumptions, we can show that the beams will deform in the antisymmetric fashion (Fig. 1) with the configuration's instantaneous mass center remaining at the hub's geometric center. Also, because of the antisymmetric deformation of the beams, we need to concern ourselves only with the deformation $y(x,t)$ of a single beam. We adopt the continuum viewpoint and avoid introducing spatial approximations in the application of Liapunov concepts; the resulting control law and stability arguments will therefore apply rigorously to the distributed

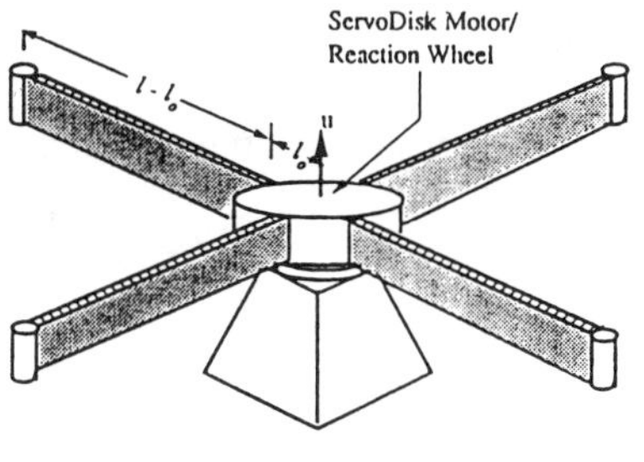

Figure 1a. Configuration Sketch

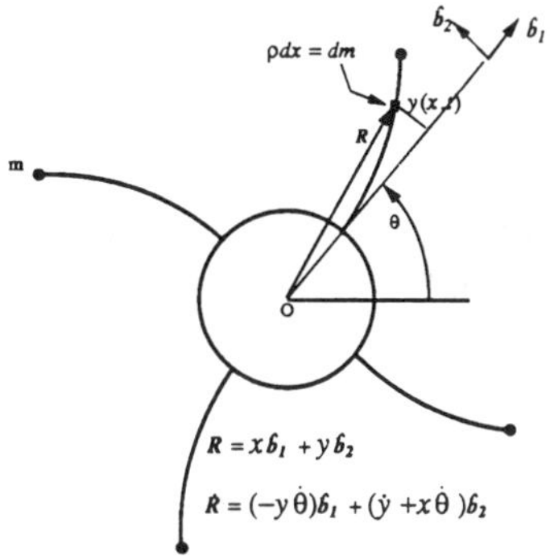

Figure 1b. Anti-Symmetric Deformation

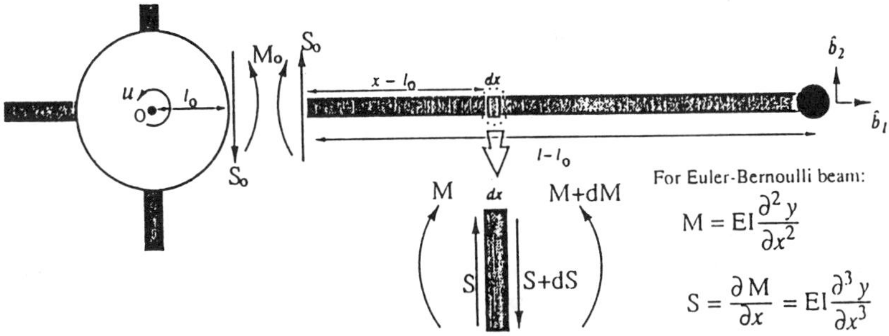

Figure 1c. Notation for Forces, Moments, and Euler-Bernoulli Constituitive Model

Fig. 1 Texas A&M maneuverable space structure experiment.

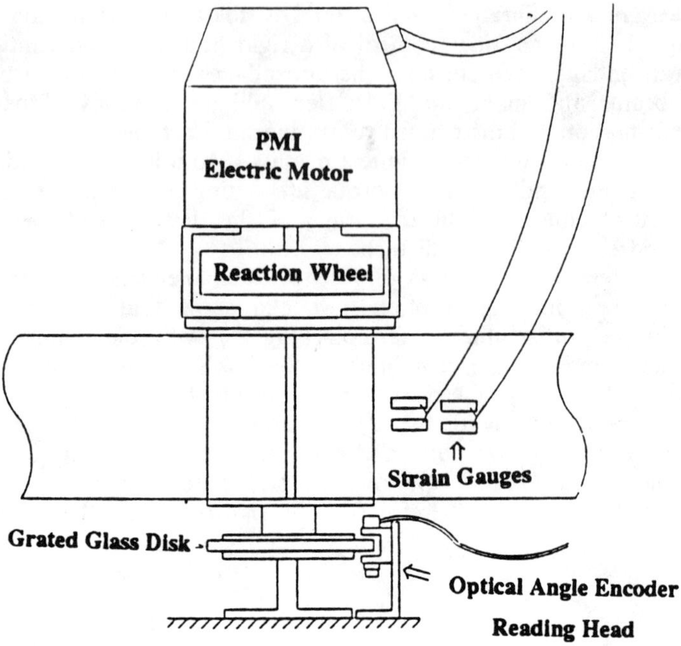

Fig. 2 Hub assembly.

parameter system. The hybrid system of ordinary and partial differential equations governing the dynamics of this system are

$$I_{\text{hub}}\frac{\mathrm{d}^2\theta}{\mathrm{d}t^2} = u + 4(M_0 - S_0 l_0) \tag{5a}$$

$$-(M_0 - S_0 l_0) = \int_{l_0}^{l} \rho x\left(\frac{\partial^2 y}{\partial t^2} + x\frac{\mathrm{d}^2\theta}{\mathrm{d}t^2}\right)\mathrm{d}x + ml\left(l\frac{\mathrm{d}^2\theta}{\mathrm{d}t^2} + \left.\frac{\partial^2 y}{\partial t^2}\right|_l\right) + \text{HOT} \tag{5b}$$

$$\rho\left(\frac{\partial^2 y}{\partial t^2} + x\frac{\mathrm{d}^2\theta}{\mathrm{d}t^2}\right) + EI\frac{\partial^4 y}{\partial x^4} = 0 + \text{HOT} \tag{5c}$$

where

ρ = assumed constant mass/unit length of the beams
EI = assumed constant bending stiffness of the beams
(M_0, S_0) = bending moment and shear force at the root of the beam
θ = hub inertial rotation
m = mass of the tip mass
(l, l_0) = distance from the hub center to the beam tip and the hub radius

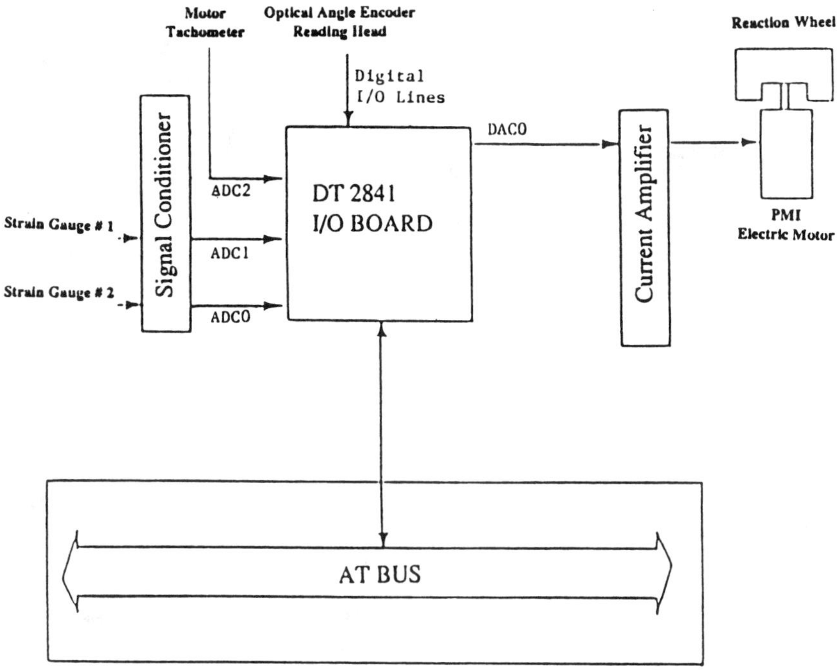

Fig. 3 System functional block diagram.

We denote higher-order by HOT to indicate other known linear and nonlinear effects (such as rotational stiffening, shear deformation, etc.). The most fundamental developments do not consider these higher-order effects; however, we selectively discuss these generalizations as well. The boundary conditions on Eqs. (5) are

$$\text{at} \qquad x = l_0\text{: } y(t,l_0) = \left.\frac{\partial y}{\partial x}\right|_{l_0} = 0$$

$$\text{at} \qquad x = l\text{: } \left.\frac{\partial^2 y}{\partial x^2}\right|_l = 0 \text{ (moment)}, \qquad \left.\frac{\partial^3 y}{\partial x^3}\right|_l = \frac{m}{EI}\left(l\frac{\mathrm{d}^2\theta}{\mathrm{d}t^2} + \left.\frac{\partial^2 y}{\partial t^2}\right|_l\right) \text{ (shear)} \tag{6}$$

The total energy of the system (constant in the absence of control or disturbances) is

$$2E = I_{\text{hub}}\left(\frac{\mathrm{d}\theta}{\mathrm{d}t}\right)^2 + 4\left[\int_{l_0}^{l} \rho\left(\frac{\partial y}{\partial t} + x\frac{\mathrm{d}\theta}{\mathrm{d}t}\right)^2 \mathrm{d}x + \int_{l_0}^{l} EI\left(\frac{\partial^2 y}{\partial x^2}\right)^2 \mathrm{d}x + m\left(l\frac{\mathrm{d}\theta}{\mathrm{d}t} + \left.\frac{\partial y}{\partial t}\right|_l\right)^2\right] \tag{7}$$

Motivated by the recent work of Fujii et al.[3] and Thompson et al.,[6] we investigate the Liapunov function

$$2U = a_1 I_{\text{hub}} \dot{\theta}^2 + a_2(\theta - \theta_f)^2$$

$$+ 4a_3 \left[\int_{l_0}^{l} \rho \left(\frac{\partial y}{\partial t} + x\dot{\theta} \right)^2 \mathrm{d}x + \int_{l_0}^{l} EI \left(\frac{\partial^2 y}{\partial x^2} \right)^2 \mathrm{d}x + m \left(l\dot{\theta} + \frac{\partial y}{\partial t}\bigg|_l \right)^2 \right] \tag{8}$$

where the weighting coefficients a_i are included to allow relative emphasis on the three contributors to the "error energy" of the system. We have added the "torsional spring energy" term $a_2(\theta - \theta_f)^2$ to make the final state

$$\left(\theta, \dot{\theta}, y(x,t), \frac{\partial y(x,t)}{\partial t} \right)_{\text{desired}} = (\theta_f, 0, 0, 0)$$

be the global minimum of U. It is obvious by inspection that imposing $a_i > 0$ guarantees that $U \geqslant 0$ and that indeed the global minimum of $U = 0$ occurs only at the desired state. Differentiation, substitution of the equations of motion [Eqs. (5) and (6)], and some calculus lead to

$$\dot{U} = \frac{\mathrm{d}U}{\mathrm{d}t} = \dot{\theta}[a_1 u + a_2(\theta - \theta_f) + 4(a_3 - a_1)(l_0 S_0 - M_0)] \tag{9}$$

Since we require that $\dot{U} \leqslant 0$ to guarantee stability, we set the term in brackets to $-a_4\dot{\theta}$, and this leads to the control law

$$u = -\frac{1}{a_1}[a_2(\theta - \theta_f) + a_4\dot{\theta} + 4(a_3 - a_1)(l_0 S_0 - M_0)] \tag{10}$$

or we see that the following *linear, spatially discrete* output feedback law globally stabilizes this distributed-parameter system:

$$u = -[g_1(\theta - \theta_f) + g_2\dot{\theta} + g_3(l_0 S_0 - M_0)];$$

$$g_1 \geqslant 0,\ g_2 \geqslant 0,\ g_3 \geqslant -4 \qquad \text{for stability} \tag{11}$$

This control law is elegant. Notice that controllers based on this law are easy to implement since no state estimation is required. The root shear and bending moment can be measured by (albeit somewhat noisy) conventional strain gages. The value and sign of the shear/moment feedback gain $g_3 = 4(a_3 - a_1)/a_1$ depends on whether we wish to emphasize dissipation of the beam energy (for $a_3 > a_1$), or the motion of the hub (for $a_3 < a_1$), as is evident from Eq. (8). Since $\dot{U} = -a_4\dot{\theta}^2$ is not an explicit, negative-definite function of the subset of state variables

$$\left[\theta, y(x,t), \frac{\partial y(x,t)}{\partial t} \right]$$

the stability arguments implicitly depend on the truth that all infinity of modes of motion of this structure, under the assumption that antisymmetric deflections have generally nonzero hub angular velocity ($\dot{\theta}$).

It is significant that this same linear feedback law of Eq. (11) maintains its globally stabilizing character even when the Euler/Bernoulli assumptions are relaxed to include additional linear and nonlinear effects. In particular, we have verified that closed-loop stability is maintained when we include the following: rotational stiffening, Coriolis kinematic coupling terms, aerodynamic drag, shear deformation, beam rotary inertia, and finite inertia of the tip mass. The verification of these truths requires appropriate modifications of the kinetic and potential energy functions, and of course, the differential equations of motion must be generalized consistently. *In short, global stability of the system using the simple linear control law of Eq.* (11) *has been found to be very forgiving of modeling assumptions and therefore modeling errors.*

On the other hand, the overall performance measures (time constants, required energy, rms output errors, etc.) of the system vs our simulations are generally not as forgiving as the most fundamental performance measure (stability). Although we are guaranteed stability for a large family of model errors, it should be apparent that *performance predictions are generally very sensitive to modeling errors.* However, with appropriate system modeling and selection of the feedback gains, it has been demonstrated experimentally[3,5] and in our results presented later that this feedback law works well over finite regions of the state and gain space and in the presence of typical model errors.

Although the constant-gain linear feedback works well for terminal pointing and vibration suppression, we have found it to be a poor law for carrying out both large-angle maneuvers and terminal pointing/vibration suppression. In fact, it is evident from both analytical and experimental studies that gain scheduling is needed to use the linear feedback law of Eq. (11) to efficiently control both the large maneuvering motions and the small terminal motions. This is because the large gains required for effective vibration suppression and disturbance rejection near the target state typically differ by several orders of magnitude from the smaller ones needed far from the target state (i.e., large gains appropriate for terminal vibration suppression, when used during a large-angle maneuver, typically result in actuator saturation and significant θ overshoots). To obtain a control law more appropriate for large-angle maneuvers with vibration suppression, we can modify the preceding developments. A stable tracking-type feedback control law is presented in the following discussion.

Near-Minimum-Time Maneuvers of Distributed-Paramter Systems

Consider first the near-minimum-time maneuver of a rigid body. We know that the strict minimum-time control is a bang-bang law,[1] which, for the rest-to-rest maneuver-to-the-origin case, saturates negatively during the first half of the maneuver and positively during the last half of the maneuver. From an implementation point of view, the instantaneous switches of bang-bang control are often troublesome for several reasons:

1) No torque-generating device exists that can in fact switch instantaneously.

2) When generalized and applied to flexible structures, the discontinuous class of controls will typically excite poorly modeled higher-frequency modes.

3) The predicted (model-derived) switch times and the response of the actual system are usually very sensitive to modeling errors.

As a consequence of these sources of difficulty, our analytical and experimental work indicates that bang-bang control of flexible body dynamics usually lacks robustness with respect to modeling errors.

Recently, an intuitively attractive family of controllably smooth approximations of the sign (sgn) function that arises in time-optimal control has been introduced. The smoothed sign function approximations presented in Ref. 6 have been studied in the context of optimization of near-minimum-time maneuvers considering multiple flexible-body modes. However, this approach requires expensive precalculation of the solution to a two-point boundary-value problem (tpbvp) or interpolation from an a priori computed map of tpbvp solutions. Our recent analytical and experimental work[4] indicates that much simpler piecewise continuous spline approximations of the sign function used in conjunction with analytical approximations of the optimum maneuvers and associated optimum controls are more attractive from an implementation point of view. Using this approach, a typical near-minimum-time control law (for single-axis, rest-to-rest maneuvers of a rigid body) has the form

$$I\ddot{\theta} = u = \pm u_{\max} f(\Delta t, t_f, t) \tag{12}$$

where $u_{\max}$ is the saturation torque, t_f is the maneuver time (we choose the plus sign if $\theta_f > \theta_0$), Δt is the rise time, and the smooth sgn function approximation adopted is

$$f(\Delta t, t_f, t)\begin{cases} = \left(\dfrac{t}{\Delta t}\right)^2\left[3 - 2\left(\dfrac{t}{\Delta t}\right)\right] & \text{for} \quad 0 \leqslant t \leqslant \Delta t \\ = 1 \quad \text{for} \quad \Delta t \leqslant t \leqslant \dfrac{t_f}{2} - \Delta t \equiv t_1 & \\ = 1 - 2\left(\left(\dfrac{t - t_1}{2\Delta t}\right)^2\left[3 - 2\left(\dfrac{t - t_1}{2\Delta t}\right)\right]\right) & \text{for} \quad t_1 \leqslant t \leqslant \dfrac{t_f}{2} + \Delta t \equiv t_2 \\ = -1 \quad \text{for} \quad t_2 \leqslant t \leqslant t_f - \Delta t \equiv t_3 & \\ = -1 + \left(\dfrac{t - t_3}{\Delta t}\right)^2\left[3 - 2\left(\dfrac{t - t_3}{\Delta t}\right)\right] & \text{for} \quad t_3 \leqslant t \leqslant t_f \end{cases}$$

Adopting the positive sign, Eq. (12) integrates to yield

$$\dot{\theta}(t) = \dot{\theta}_0 + \frac{u_{\max}}{I} \int_{t_0}^{t} f(\Delta t, t_f, \tau)\, d\tau \tag{13a}$$

$$\theta(t) = \theta_0 + \dot{\theta}_0(t - t_0) + \frac{u_{\max}}{I} \int_{t_0}^{t} \int_{t_0}^{\tau_1} f(\Delta t, t_f, \tau_2)\, d\tau_2\, d\tau_1 \tag{13b}$$

The integrations can be carried out in terms of elementary functions that are not presented here for brevity. Figure 4 shows the maneuver resulting from these integrations for a typical selection of parameters. For rest-to-rest maneuvers we can impose the boundary conditions

$$\begin{aligned} &\text{at} \quad t_0 = 0: \quad \theta(0) = \theta_0, \quad \dot{\theta}(0) = 0; \\ &\text{at} \quad t_f = T: \quad \theta(T) = \theta_f, \quad \dot{\theta}(T) = 0 \end{aligned} \tag{14}$$

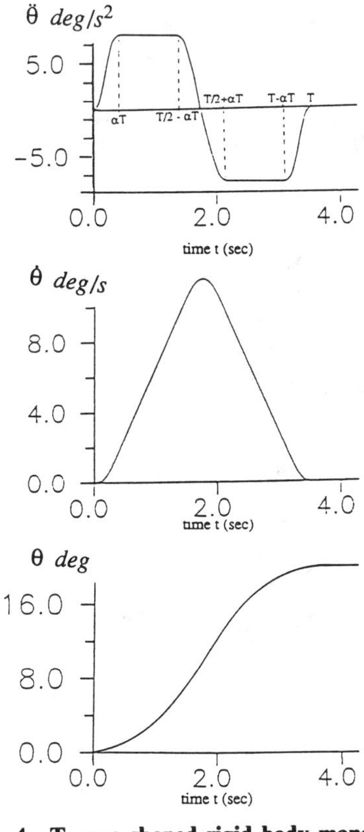

Fig. 4 Torque shaped rigid body maneuver.

and from carrying through the integrations of Eqs. (13), we obtain the useful relationship

$$\theta_f - \theta_0 = \frac{u_{\max}}{I}\left[\frac{1}{4} - \frac{1}{2}\alpha + \frac{1}{10}\alpha^2\right]T^2, \qquad \Delta t = \alpha T, \qquad 0 < \alpha < \frac{1}{4} \tag{15}$$

or

$$T = \left[\frac{I(\theta_f - \theta_0)}{u_{\max}\left(\frac{1}{4} - \frac{1}{2}\alpha + \frac{1}{10}\alpha^2\right)}\right]^{1/2} \tag{16}$$

In Eq. (16) we have the explicit rigid-body tradeoff between torque shaping (α), maneuver time (T), maneuver angle ($\theta_f - \theta_0$), and maximum angular acceleration ($u_{\max}/I$). Obviously, Eq. (15) can be inverted for any of these as a function of the remaining parameters. If we set $\alpha = \Delta t / T = 0$, of course, we obtain the special case result $\{T^2 = 4I(\theta_f - \theta_0)/u_{\max}\}$ expressing the well-known square root relationship between the minimum time, maneuver angle, inertia, and saturation torque for bang-bang control.

Figure 5 shows the rigid-body maneuver time T vs α and $u_{\max}$. Notice that the maneuver time is strongly dependent on $u_{\max}$ but varies much less as a function of α. However, we find that, when the torque $u_{\text{ref}}(t) = u_{\max} f(\alpha T, T, t)$ of Eq. (12) is applied to a flexible body, the reverse is true. To illustrate this point, we approximated the solution of Eqs. (5) by introducing a Galerkin expansion of the beam deflection in terms of the first 10 clamped-free cantilever mode shapes times time-varying generalized

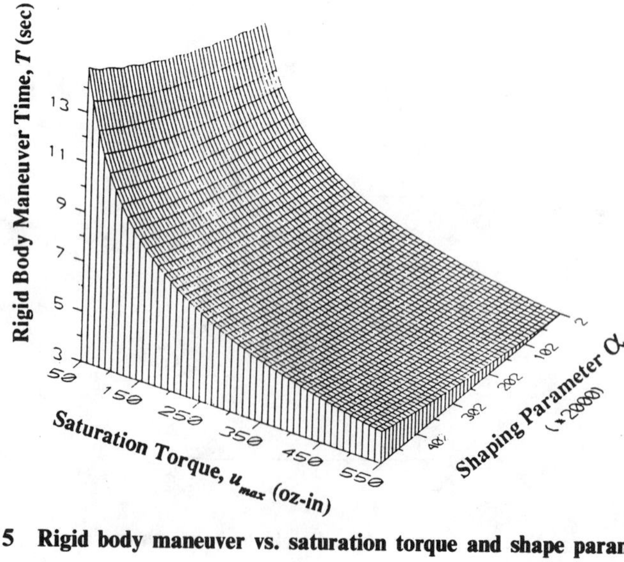

Fig. 5 Rigid body maneuver vs. saturation torque and shape parameter.

coordinates. The forced response of the resulting 20th-order system was judged adequately converged. Using this solution, we applied the same family of torques used for the rigid-body results of Fig. 5 and computed the flexible-body open-loop response for each choice of α and u_{max}. Figure 6 shows the total energy of the flexible body system at time T for this same family of maneuvers. As is evident, the vibration energy vanishes for sufficiently small u_{max} and is significantly reduced for any given u_{max} by increasing α. Notice the triangular region (large α, small u_{max}) for which the vibration energy at time T is negligible.

The preceding results and those of Ref. 4 support the intuitively obvious truth that applying judiciously "smoothed bang-bang" open-loop controls such as Eq. (12) to generate maneuvers of a flexible body results in moderate structural vibration for sufficiently slow and smooth maneuvers (small u_{max} and large α). Of course, unmodeled disturbances, control implementation errors, and model errors negate some of these apparent gains, especially without feedback. For relatively small penalties in maneuver time,[4,5] actual torque-shaped maneuvers enjoy several orders of magnitude reduction in residual vibration, even without feedback, although some offset in rigid-body position usually results. Thus, overall maneuver time, including vibration arrest, can be reduced dramatically by modest torque shaping, compared to simply using bang-bang control augmented by

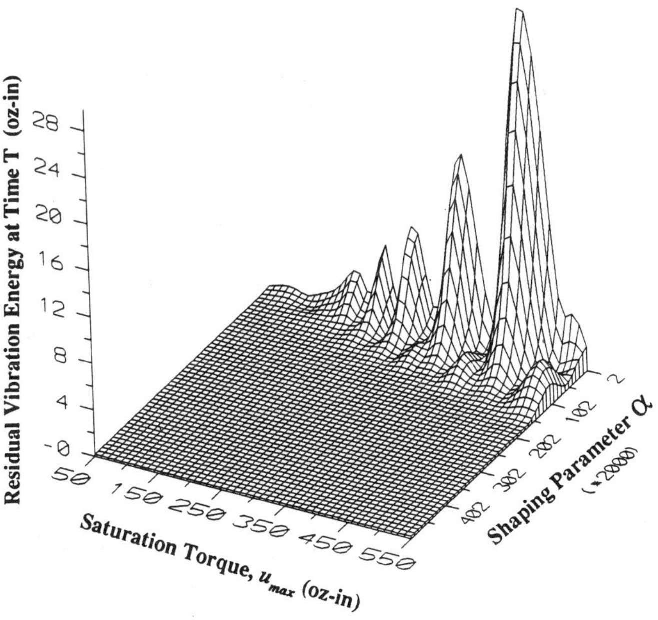

Fig. 6 Flexible body residual vibration energy vs. saturation torque and shape parameter.

vibration suppression. Also of significance, we find that it is usually desirable to select the torque profile (e.g., u_{max}, α, etc.) to consider the sensor and actuator dynamics and thereby make the commanded torque history more nearly physically achievable. More generally, however, we can use any reference maneuver (not necessarily a rigid-body torque-shaped maneuver). We will find merit in considering a certain class of flexible-body reference motions in our discussion given later.

Suppose we adopt a reference open-loop rigid-body maneuver denoted by the subscript ref as $\{\theta_{ref}(t), \dot{\theta}_{ref}(t), \ddot{\theta}_{ref}(t) = u_{ref}/I\}$ and satisfying Eqs. (12–16). Note that I is the undeformed moment of inertia, and we have implicitly selected the torque-shaping parameters α, u_{max} and thereby established the corresponding target maneuver time T from Eq. (16) for specified initial and final angle. Motivated by the issues discussed earlier and the quadratic regulator perturbation feedback controllers discussed in Refs. 1, 9, and 11, we hypothesize the following structure for the control law:

$$u = u_{ref}(t) - \{g_1(\theta - \theta_{ref}) + g_2(\dot{\theta} - \dot{\theta}_{ref}) + g_3[(l_0 S_0 - M_0) - (l_0 S_0 - M_0)_{ref}]\} \tag{17}$$

where it is easy to show that the root moment for the reference (rigid-body) motion is proportional to the angular acceleration:

$$(l_0 S_0 - M_0)_{ref} = [\rho(l^3 - l_0^3)/3 + ml^2]\ddot{\theta}_{ref}(t)$$

We wish to gain insight on the stability characteristics of the flexible body's closed-loop departure dynamics [from the target (ref) rigid-body motion (for $t < T$) and from the fixed target state (for $t \geqslant T$)], and we anticipate providing an analytical justification for the tracking control law of Eq. (17). Let us denote by the subscript r the state variables along the open-loop flexible-body solution of Eqs. (5) when driven by the torque $u_{ref}(t)$ of Eqs. (12–16). The instantaneous displacement of the open-loop flexible-body solution from the open-loop rigid-body motion is denoted $\Delta(\ ) = (\ )_r - (\ )_{ref}$ [e.g., $\Delta\theta(t) = \theta_r(t) - \theta_{ref}(t)$]. Variables without subscripts represent the "actual" instantaneous closed-loop controlled solution variables. Consider the candidate error energy Liapunov function

$$\begin{aligned} 2U &= a_1 I_{hub}(\dot{\theta} - \dot{\theta}_r)^2 + a_2(\theta - \theta_r)^2 \\ &\quad + 4a_3\left\{\int_{l_0}^{l} \rho\left[\left(\frac{\partial y}{\partial t} - \frac{\partial y_r}{\partial t}\right) + x(\dot{\theta} - \dot{\theta}_r)\right]^2 \right. \\ &\quad \left. + \int_{l_0}^{l} EI\left(\frac{\partial^2 y}{\partial x^2} - \frac{\partial^2 y_r}{\partial x^2}\right)^2 \mathrm{d}x + m\left[l(\dot{\theta} - \dot{\theta}_r) + \left(\left.\frac{\partial y}{\partial t}\right|_l - \left.\frac{\partial y_r}{\partial t}\right|_l\right)\right]^2\right\} \end{aligned} \tag{18}$$

Assuming that the actual control is some general, to-be-determined $u(t)$ and that the actual motion satisfies Eqs. (5), we investigate the time derivative of U of Eq. (18) and find that it is given by the following equation:

$$\dot{U} = -a_1(\dot{\theta} - \dot{\theta}_r) \times \left\{u - u_{\text{ref}} + \frac{a_2}{a_1}(\theta - \theta_r) + 4\frac{a_3 - a_1}{a_1}[(l_0 S_0 - M_0) - (l_0 S_0 - M_0)_r]\right\} \quad (19)$$

Analogous to the logic leading to Eq. (9), we see that we could determine a stabilizing $u(t)$ by setting the bracketed term to a positive quantity (g_2) times $(\dot{\theta} - \dot{\theta}_r)$ (i.e., this makes $\dot{U}$ a negative-definite function of the error in hub angular velocity). This gives the control law

$$u = u_{\text{ref}}(t) - \{g_1(\theta - \theta_r) + g_2(\dot{\theta} - \dot{\theta}_r) + g_3[(l_0 S_0 - M_0) - (l_0 S_0 - M_0)_r]\} \quad (20)$$

with $\dot{U} = -a_1 g_2(\dot{\theta} - \dot{\theta}_r)^2$, and we introduced the definitions $g_1 \equiv a_2/a_1$, $g_3 \equiv (a_3 - a_1)/a_1$. Notice that, to guarantee stability, g_1 and g_2 must be positive, but as shown earlier, g_3 must be greater than -4. The main difficulty with this control law is that it requires precomputing the flexible-body solution, and we would prefer a tracking law that requires a much simpler rigid-body maneuver to be precomputed. Let us pursue this objective. Since Eq. (19) holds for an arbitrary $u(t)$, we introduce the control law of Eq. (17) and investigate its stability characteristics. Making use of the notation $\Delta(\ ) = (\ )_r - (\ )_{\text{ref}}$, we are led to

$$\dot{U} = -a_1(\dot{\theta} - \dot{\theta}_r)\{g_2(\dot{\theta} - \dot{\theta}_r) - [g_1\Delta\theta + g_2\Delta\dot{\theta} + g_3\Delta(l_0 S_0 - M_0)]\} \quad (21)$$

Obviously, a sufficient condition characterizing the region where $\dot{U} \leqslant 0$ is the dominance of the first term in the braces of Eq. (21); this gives the inequality

$$|\dot{\theta} - \dot{\theta}_r| > \frac{1}{g_2}|g_1\Delta\theta + g_2\Delta\dot{\theta} + g_3\Delta(l_0 S_0 - M_0)| \equiv \mu \quad (22)$$

It is apparent that the Δ quantities on the right-hand side of Eq. (22) are finite and (pre-)computable differences between the open-loop flexible- and rigid-body motions; thus, an upper bound can be established directly by a priori simulation of the two open-loop motions and using a particular set of feedback gains. Note that the right-hand side of Eq. (22) is essentially a weighted measure of how nearly the target trajectory satisfies the flexible-body equations of motion.

A bounded-input/bounded-output viewpoint of stability can also be considered based on the departure motion differential equations obtained by differencing Eqs. (5) evaluated at each instant along the actual closed-loop trajectory and along the flexible-body open-loop $(\ )_r$ motion. Upon formulating these equations, we find that the departure motion is governed by an otherwise asymptotically stable system of differential equations forced by the known $\Delta(\ )$ terms that appear in Eqs. (22). Since these forcing terms can be bounded by direct calculation, the resulting departure motion can also be bounded. Since the actual numerical bounds on the Δ quantities can be made arbitrarily small (depending on how nearly the user defined reference trajectory is made to satisfy the open-loop equations of motion), we have a very attractive theoretical and practical situation vis-à-vis stability of the closed-loop tracking motion. *We see that the closed-loop motion is globally attracted to the controllably small region near the target trajectory that violates the inequality of Eq.* (22), *and considering the motions within this small region, we have bounded-input, bounded-output stability.*

The preceding discussion can probably be generalized for any smooth target trajectory, but we find that it is attractive to use a torque-shaped rigid-body reference trajectory, such as Eqs. (12–16). An attractive and significant feature of Eqs. (12–17) is that we have a C^1 continuous transition:

$$\{u_{\text{ref}}(t),\theta_{\text{ref}}(t),\dot{\theta}_{\text{ref}}(t),M_{0_{\text{ref}}}(t),S_{0_{\text{ref}}}(t)\} \Rightarrow \{0,\theta_f,0,0,0\} \qquad \text{as} \qquad t \Rightarrow T$$

so that for $t > T$ only the three feedback terms of Eq. (17) are contributing to the terminal fine-pointing/vibration arrest control. Thus, the controls blend continuously from the large-angle tracking law into a constant-gain controller identical to the globally stable fixed-point output feedback case.

Simulated Results for the AFOSR/Texas A&M University Large-Angle Maneuvers Experiment

Returning to the family of 40-deg open-loop maneuvers used to generate the energy surface of Fig. 6, we computed the velocity tracking bound μ for Liapunov stability [as given by Eq. (22)] and found the maximum value ($\mu_{\max}$) of $\mu(t)$ along each trajectory. Figure 7 displays this worst case tracking bound surface $\mu_{\max}$ ($\alpha,u_{\max}$) over the same ($\alpha,u_{\max}$) region as that in Figs. 5 and 6. It can be seen that the closed-loop tracking error bound has a behavior that is roughly analogous to the open-loop residual vibration energy surface of Fig. 6. Recall that, outside of the region bounded by the inequality of Eq. (22), we have guaranteed Liapunov stability when using the control law of Eq. (17) and the reference rigid-body torque given by Eqs. (12–16). From Fig. 6 it is clear that sufficiently small $u_{\max}$ and sufficiently large α result in arbitrarily small tracking errors, but the (small α) near-bang reference maneuvers cannot be tracked as accurately. By direct examination of the surfaces of Figs. 5–7, it is easy to see how a family of candidate ($\alpha,u_{\max}$) designs could be selected that satisfies specified inequalities on maneuver times, tracking errors, and residual vibration energy.

Our experiments with simulations (and in the actual hardware implementations presented in Ref. 5 and later in this chapter) support the conclusion that it is easy to use the previously mentioned surfaces, together with

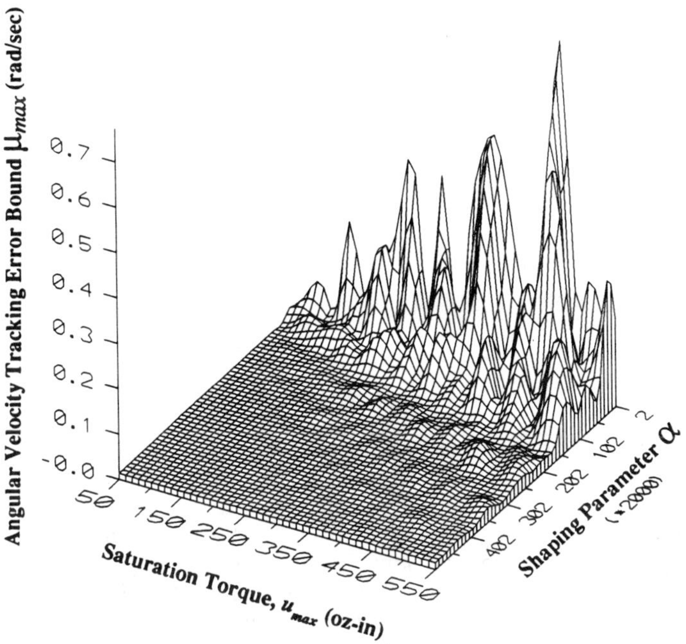

Fig. 7 Boundary of the Liapunov-stable tracking region vs. saturation torque and shape parameter.

disturbance rejection simulations to select the reference trajectory and the feedback gains to obtain well-behaved closed-loop control, to establish a large family of feasible designs. Before we discuss our experimental results we present some further simulations to show typical state and control variables histories along some of the trajectories of underlying the previously given surfaces. We include in these simulations the effects of disturbance torques to illustrate the effectiveness of the controls in the presence of unmodeled effects. The Appendix summarizes the physical parameters and hardware description of the maneuver experiment sketched in Fig. 1. For simplicity we consider only the case of a 40-deg rest-to-rest maneuver and set $u_{max} = 400$ oz-in. for all cases.

For our computational and experimental studies we consider two control laws: 1) the output feedback law (control law I)[3–5] of Eq. (11) and 2) the tracking-type feedback control law (control law II) of Eq. (17). Although control law II could be used with an arbitrary reference trajectory, we elect to investigate specifically the torque-shaped rigid-body trajectories of Eqs. (12–16). The torque-shaped open-loop control history $u_{ref}(t)$ is precomputed (in a fraction of a second!) from Eqs. (12–16) and stored. Note that the boundary conditions of Eq. (14) are enforced by using Eq. (16) to compute the target trajectory maneuver time as a function of the maneuver angle, saturation torque, and torque-shape parameter.

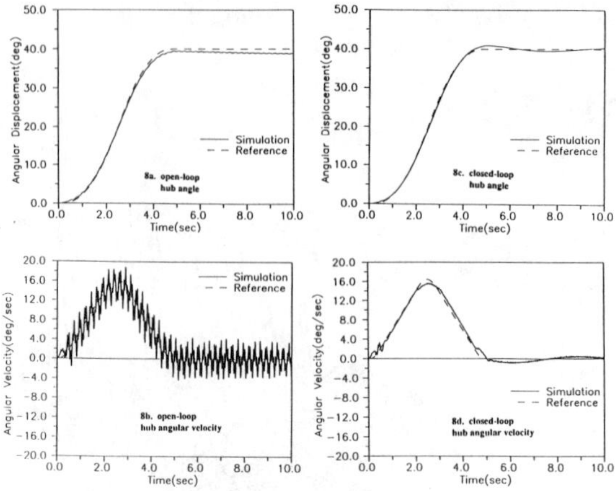

Fig. 8 Simulated open- and closed-loop 40 deg maneuver with disturbances.

We now discuss the simulation results using control law II, which obviously blends into control law I in the end game (for $t \geqslant T$). In the experimental results we report maneuvers carried out by both control laws. Both open-loop (all $g_i \equiv 0$) and closed-loop time histories of selected variables are shown in Figs. 8a–8d. On the left of Fig. 8 we show the hub angle and angular velocity for the case of an open-loop control and in the presence of substantial impulsive and quasirandom (5 oz-in., 1σ) disturbance torques. It is evident that the disturbance torque history is significant vis-à-vis exciting substantial rigid-body and flexural motions. On the right we show the closed-loop behavior of the system excited by the same disturbance torque history. The random component of this disturbance has been found to be typically pessimistic vis-à-vis our experimental hardware; however, certain nonrandom, nonlinear effects associated with the bearing friction cause disturbances that are highly correlated in time and are not well represented by the present white noise representation of the disturbance torques. Despite the substantial disturbance torques (Figs. 8a–8b), it is evident that we obtain a near-minimum-time rigid-body motion while effectively suppressing vibration (Figs. 8c–8d). This fact was also confirmed by evaluation of the energy distribution in the first six modes.

Description of Experimental Results

In all of the experiments discussed later, we set the target final angle to 40 deg and $u_{max} = 400$ oz-in. The values of the tracking gains (g_1, g_2, g_3) and of the torque-shape parameter (α) were assigned several feasible values to demonstrate their effects on the closed-loop behavior of the system. Moderate bearing friction and aerodynamic damping were present in our experiments.

Figure 9 shows the system response for a control law I [the constant-gain control law of Eq. (11)] maneuver with ($g_1 = 600$ oz-in./rad, $g_2 = 0$ oz-in./

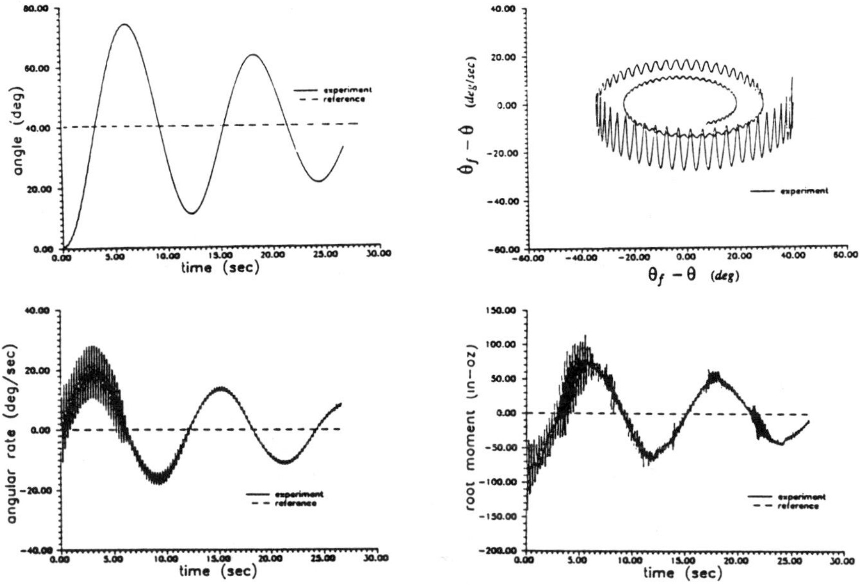

Fig. 9 Experimental results: control law I (control gains: $g_1 = 600$ oz-in./rad, $g_2 = 0$ oz-in./rad/s, $g_3 = 0$).

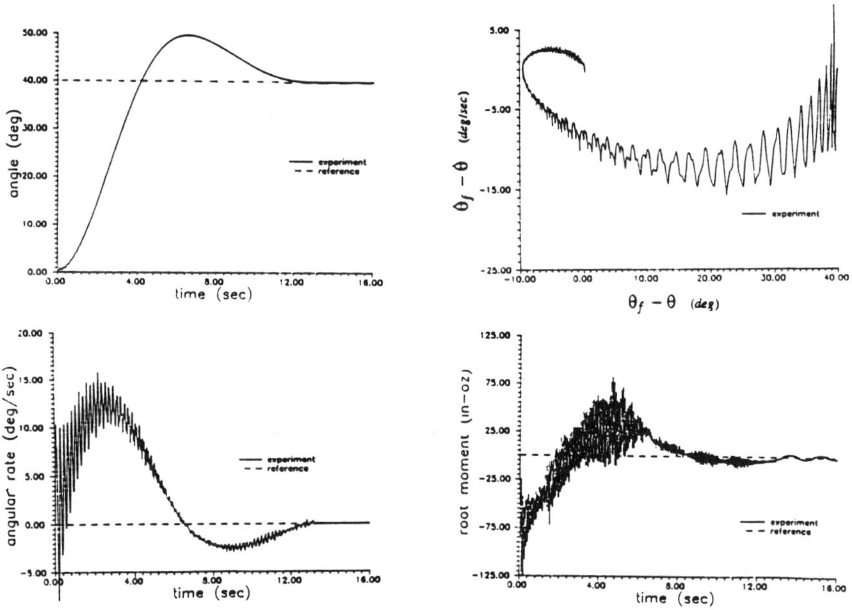

Fig. 10 Experimental results: control law I (control gains: $g_1 = 600$ oz-in./rad, $g_2 = 800$ oz-in./rad/s, $g_3 = 0$).

rad/s, and $g_3 = 0$). Since this initial position error is large, the maneuver starts from zero with an initial discontinuity to a large torque. For this gain selection we see a large overshoot (~ 34 deg) that had not settled when the control was terminated at 30 s. Figure 10 shows a similar experiment but with $g_2 = 800$ oz-in./rad/s, to illustrate the constructive effects of active damping during the maneuver. This is a representative set of "good" gains for control law I, but we will see that the maneuver time can be reduced by more than a factor of 2 by using control law II, whereas only about a 15% reduction can be achieved by optimizing gains for control law I (subject to a 400-oz-in. saturation constraint). Notice the large transient (induced by instantaneously applying the gain g_1 to the 40-deg initial angle error) decays after about 5 s, and the target rigid-body position is acquired with small residual vibrations in about 12 s.

A family of results of using control law II, Eq. (17) (with $\alpha = 0.1$ and 0.2; $g_1 = 3000$; $g_2 = 0$, 200, and 800; and $g_3 = 0$, 0.8, and -0.8), are shown in Figs. 11–16. These results demonstrate the effectiveness of control law II in comparison to control law I (e.g., compare Figs 10 and 14) and show the effects of varying the torque shape parameter (α) and the feedback gains (g_1, g_2, g_3). Because of the tracking type of control, the initial errors operated on by these gains are not as large for control law II as they are for control law I as a consequence of using a reference trajectory that departs smoothly from the initial state. Moreover, the smoothly increasing torque profile of control law II does not "ring" the structure with an initial jump discontinuity at time $t = 0$ to a large toque as in control law I. The

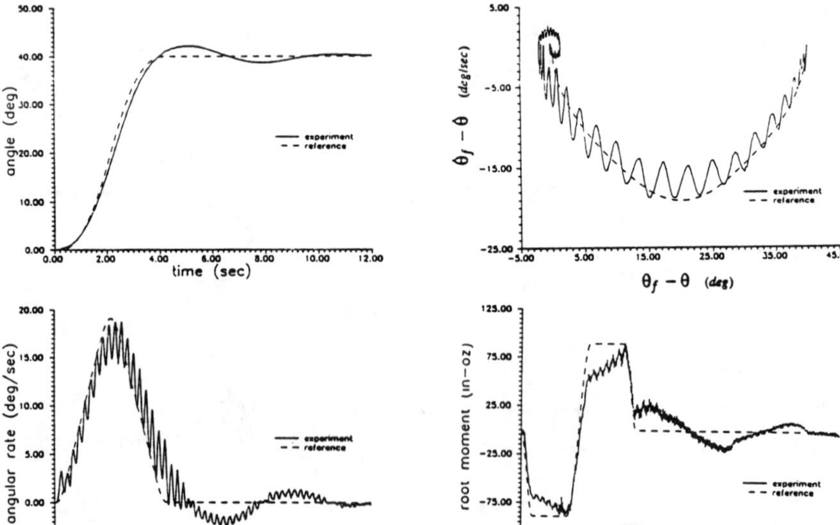

Fig. 11 Experimental results: control law II (torque shape parameter: $\alpha = 0.1$, control gains: $g_1 = 3000$ oz-in./rad, $g_2 = 0$ oz-in./rad/s, $g_3 = 0$).

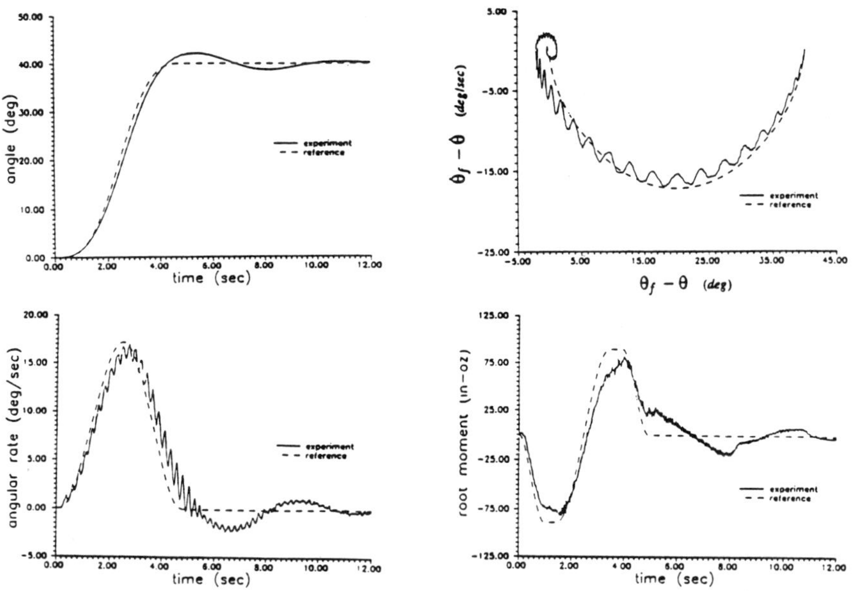

Fig. 12 Experimental results: control law II (torque shape parameter: $\alpha = 0.2$, control gains: $g_1 = 3000$ oz-in./rad, $g_2 = 0$ oz-in./rad/s, $g_3 = 0$).

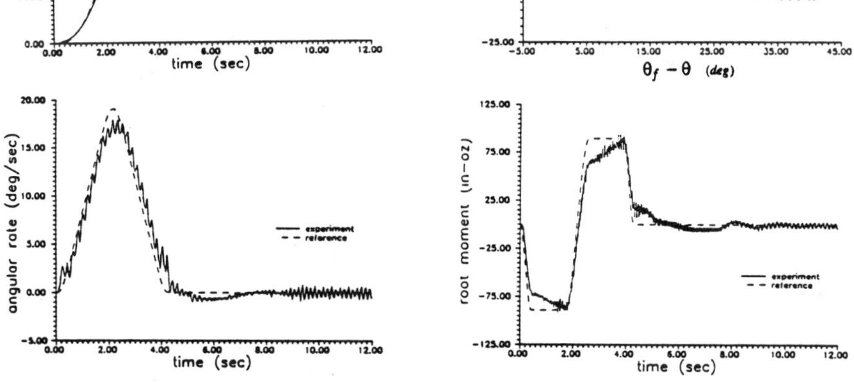

Fig. 13 Experimental results: control law II (torque shape parameter: $\alpha = 0.1$, control gains: $g_1 = 3000$ oz-in./rad, $g_2 = 800$ oz-in./rad/s, $g_3 = 0$).

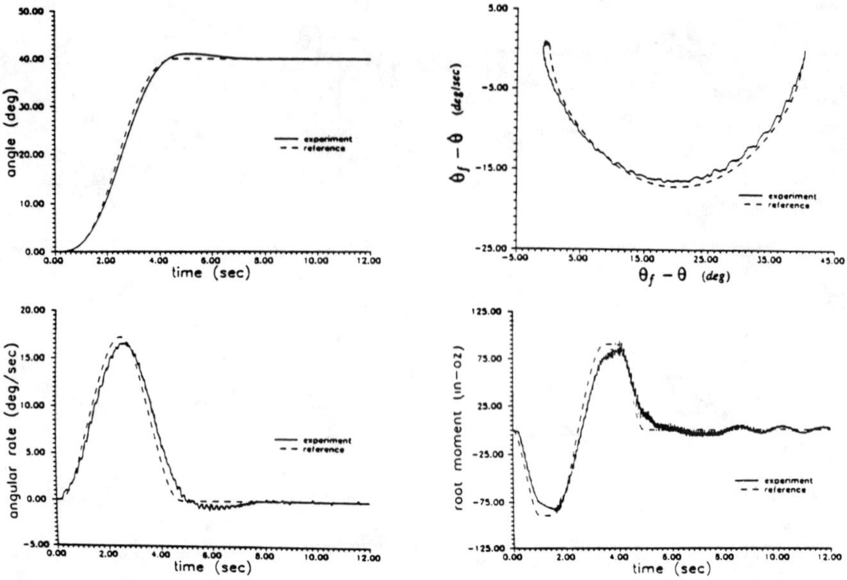

Fig. 14 Experimental results: control law II (torque shape parameter: $\alpha = 0.2$, control gains: $g_1 = 3000$ oz-in./rad, $g_2 = 800$ oz-in./rad/s, $g_3 = 0$).

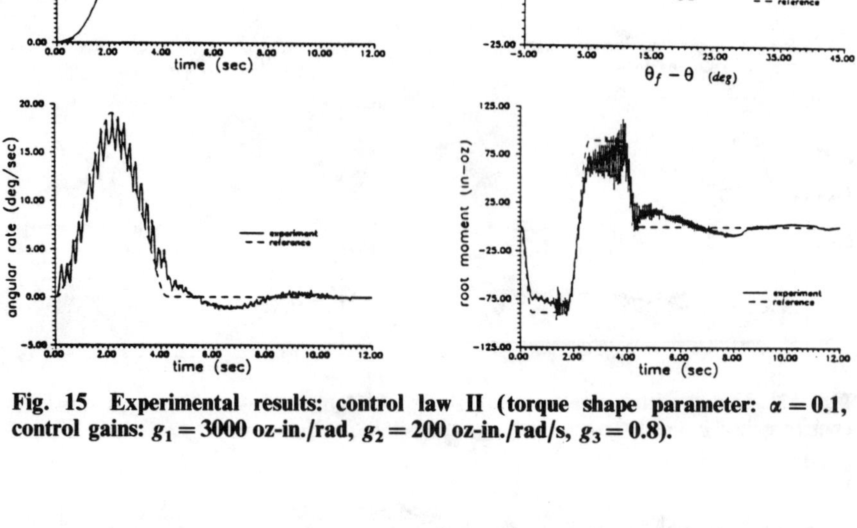

Fig. 15 Experimental results: control law II (torque shape parameter: $\alpha = 0.1$, control gains: $g_1 = 3000$ oz-in./rad, $g_2 = 200$ oz-in./rad/s, $g_3 = 0.8$).

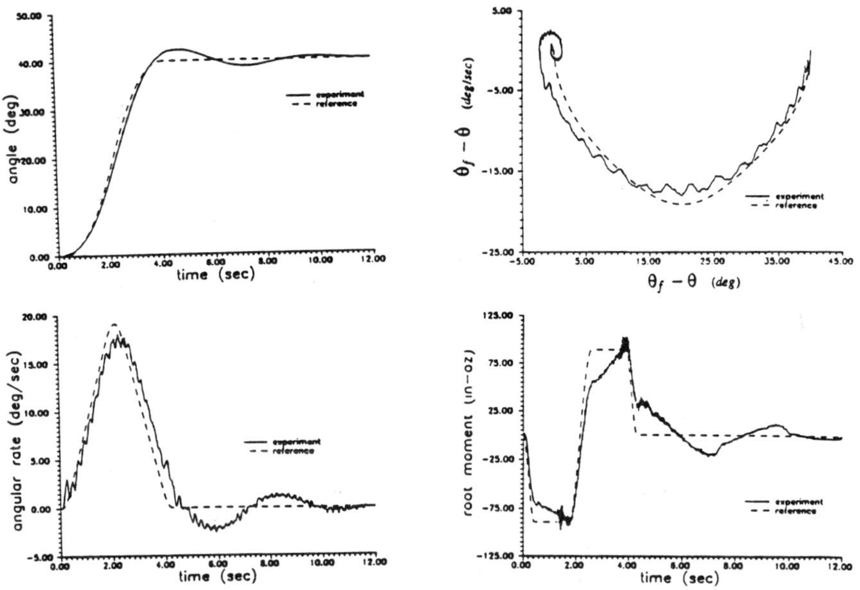

Fig. 16 Experimental results: control law II (torque shape parameter: $\alpha = 0.1$, control gains: $g_1 = 3000$ oz-in./rad, $g_2 = 200$ oz-in./rad/s, $g_3 = -0.8$).

effect of using a smooth, judiciously shaped reference torque history is evident if one compares the results in Fig. 14 with those of Fig. 10. Control law II produced much smaller overshoot (≈ 1.5 deg vs >10 deg) and shorter settling time (6 s vs over 12 s) and greatly reduced the severity of peak vibration. Notice in Figs. 9, 11, and 12 that significant passive energy dissipation exists due to the combined effects of bearing friction, aerodynamics, and structural mechanisms, because the velocity feedback gain is zero in these three experiments. On the other hand, the rate of passive energy dissipation is sufficiently low that active vibration control is required to satisfactorily arrest the structure. In comparing Figs. 11 and 12, notice that increasing α results in significantly reducing the amplitude of the structural vibration (which is primarily the first cantilever mode). As is evident in Figs. 13 and 14, a judicious selection of the velocity feedback gain (g_2) and the torque-shaping parameter (α) suppresses the structural vibration during the maneuver as well as in the end game.

Figures 15 and 16 provide some additional experimental results to contrast with the best design studied (Fig. 14). Notice that including feedback (g_3) on the root bending moment was not found to be effective in further refining our results. We did a number of experiments with other g_3 values without improving over the results of Fig. 14. The reason has been traced to the noisy derived angular rate from the angle encoder and the fact that the accelerometer acts as a noisy angular accelerometer. Our simulations indicated that noise-free root bending-moment feedback would be effective, over and above the hub angle and angular rate feedback, in suppressing the vibrations of all modes above 12 Hz. However, the effective bandwith of our derived angular velocity was only 0–10 Hz; therefore, we were unable to control the higher frequencies without injecting substantial noise-induced disturbance accelerations. We anticipate that enough attention to optimizing the design of analog and/or digital filters will lead to marginal improvements over the excellent results of Fig. 14; however, we feel that it would be more productive to invest in an accurate angular rate measuring device and eliminate the source of this paradox.

Based upon mechanical tests and analysis of our bearing hardware, it is evident that interaction of the structure with the bearing accounts for the overwhelming source of unmodeled disturbances. The bearing friction/stiction model developed from our analysis has the form

$$\tau_{\text{bearing}} = -c_1 \operatorname{sign}(\dot{\theta}) - c_2\dot{\theta} + \text{HOT} \tag{23}$$

where we find $c_1 \sim 20$ oz-in. and $c_2 \sim 0.001$ oz-in./rad/s. Thus the first (stiction) term of Eq. (23) dominates the bearing torque for moderate $\dot{\theta}$ and is about 5% of the peak commanded torque of 400 oz-in. Although we believe that Eq. (23) models the bearing friction well, it is difficult to use this model to compensate for bearing friction in real time, since our estimated value of $\dot{\theta}$ is uncertain due to angle encoder measurement noise. Thus, if we modify our control using Eq. (23), the commanded discontinuity (near the estimated time $\dot{\theta}$ changes sign) will not coincide exactly with the actual discontinuity; mistimed compensation torque discontinuities can actually worsen the disturbance! Although we experimented with bearing friction compensation schemes, we ultimately decided to simply consider this an anticipated and well-modeled disturbance. Our simulations (such as the results shown in Fig. 8) indicated that our control approach could easily tolerate disturbances of this magnitude, and our successful experiments (Figs. 9–16) certainly confirm that our implemented control laws are robust in the presence of the actual disturbances.

Concluding Remarks

We have presented a novel approach to design of feedback control laws for large maneuvers of distributed-parameter systems and have conducted successful experiments. The approach permits approximate imposition of actuator saturation constraints and a priori control shaping via user specification of a torque-shaped, optimized reference trajectory. The tracking-type control law is shown to result in Liapunov stability in the sense that all trajectories are globally attracted to a controllably small region near the reference trajectory. The tracking law automatically blends smoothly into a globally stable, constant-gain, terminal output feedback controller. We believe that this approach is much more attractive than, for example, gain scheduling, because the logical and implementation complications associated with discontinuous-gain change ("handoff") logic can be avoided altogether. We have considered in detail the case of single-axis maneuvers of a flexible-body system and a particular family of torque-shaped, near-minimum-time rigid-body reference trajectories. We demonstrated numerically the effects of torque shaping on maneuver time and established a bound on the usually small region near the target trajectory in which Liapunov-stable tracking cannot be guaranteed.

We described hardware experiments that successfully implemented near-minimum-time feedback control of single-axis maneuvers of a flexible structure, using a family of torque-shaped rigid-body reference trajectories. We presented new analytical developments, numerical simulations, and experimental results, all of which support the conclusion that robust, near-minimum-time control can be achieved by this approach. The simplicity of implementation makes this approach attractive; notice that no state estimator is required, and no truncation or spillover effects degrade our stability arguments. Our relatively low-budget experimental work afforded an excellent opportunity to demonstrate the validity and robustness of this feedback control approach: We were blessed (cursed?!) by difficulties associated with nonlinear friction/stiction phenomena in our hub bearing system and significant angle encoder sensor noise. In particular, most of our implementation difficulties arose from relatively crude ($\sim$0.5-deg/s) angular velocity estimates derived from the prediction/smoothing of the moderately precise ($\sim$0.01-deg) angle encoder measurements of hub rotation angle. The particular choice of the angle encoder filtering/smoothing/differentiation algorithms and the sample interval were found to impact the behavior of the control law, although we had no difficulty finding a large family of stable designs. Use of an accurate gyroscope to directly measure the hub angular velocity should eliminate this source of difficulty. It is also significant that more accurate angular rate measurements would permit correspondingly more accurate compensation for the bearing friction/stiction effects.

Of course, roughly analogous problems associated with the support mechanism (in our case, bearing friction) in any ground maneuver experiment but obviously disappear in actual 0-*g* on-orbit implementations. The nonideal sensor and actuator problems we encountered are judged to be typical of those associated with on-orbit implementations, although hindsight suggests that we should have invested in a sensor to directly measure hub angular velocity to eliminate the primary source of difficulty. Although the effect of departures from ideal sensor, actuator, and support system behavior are present in our experiments, especially near zero angular velocity, the control law nevertheless achieved excellent qualitative agreement with our simulated closed-loop response and routinely isolated the terminal state to within our sensor precision. Furthermore, our experiments were repeatable to within sensor precision. Our simulations and hardware experiments clearly establish that the unmodeled effects are well within the stability robustness margins of our control law.

We feel that we have conclusively demonstrated the feasibility of our analytical formulations and experimental approach. We are optimistic that these ideas can be extended to a significant family of multiple flexible-body systems and the maneuver control problems associated with multiple body reconfiguration, pointing/tracking, and deployment dynamics.

Appendix: TAMU/AFOSR Maneuver Experiment Hardware Description

Figures 1–2 show sketches of our experimental structure. The structure is comprised of four identical cantilevered flexible appendages, each having a finite tip mass. The actuator is a dc motor with a reaction wheel. The stator of the motor is rigidly connected to the top of the hub, keeping the axis of rotation aligned. The whole assembly is pivoted on a ceramic bearing so that it can rotate freely in the horizontal plane. Tipping motion is restrained by an adjustable slip bearing. An angle encoder, two full-bridge strain gages near the root of an appendage, and a motor tachometer are employed to measure, respectively, the hub angular position, the appendage bending moment and shear forces, and the reaction wheel speed as functions of time. Based on these measurements and an appropriate control law, a control signal is provided to a current amplifier that drives the dc motor to produce the necessary torque.

Figure 3 shows a block diagram of the overall system. The central unit for the system is a Packard Bell 286 (PB286) computer and Data Translation DT2841 input/output (I/O) board. The DT2841 performs digital I/O, analog-to-digital (A/D) conversion, digital-to-analog (D/A) conversion, and timing operations. Packard Bell 286 reads in data from the I/O board, does the necessary computations in its own CPU, and then sends out the resulting control signal to the amplifier through the I/O board.

The angle encoder (Teledyne Gurley model 8708) output is decoded and converted into a 16-bit digital number in hundredths of a degree by a logic converter (Teledyne Gurley model SCA CC) and counter (C-Tek model Lin-101-50-21-41-BE) combination. The angle encoder is accurate to 0.01 deg. This angle information is read into PB286 through the 16-bit digital port of DT2841. We utilize a sequential least-squares filter to extract smoothed angle and angular velocity estimates from the angle encoder measurements. Least-squares filters are also used to smooth the shear and moment measurements.

The motor used to apply the torque is a PMI servodisk dc motor equipped with a tachometer. The motor produces a torque proportional to the current in the armature as sent by the current amplifier. The commanded current is sent out to the amplifier by PB286 through D/A line (DAC0) of DT2841. The tachometer output is read by the computer through one A/D line (ADC2) and used in our control laws to compensate for the motor friction torque. A separate study has been carried out to characterize the motor bearing friction. Except near zero wheel speed, the friction is well modeled as a quadratic function of the wheel speed. A discontinuity characterizes the nonlinear stiction region at zero wheel speed.

The two full-bridge strain gages are employed near the root of one of the appendages at a finite distance apart from each other to determine the instantaneous bending moment and shear force. The strain-gage signals are read with the PB286 through two additional A/D channels (ADC0, ADC1). Least-squares filters are also used to smooth the shear and moment measurements. The bandwidth of the closed-loop system was found to be about 10 Hz; the limitation to low frequencies is a direct

Table 1 Hub/appendage experiment configuration parameters

Hub and appendages	Total system inertia, I	2128 oz-s^2-in.
	Hub center to gage center, l_0	5.5470 in.
	Hub center to tip mass, l	51.07 in.
	Tip mass, m	0.15627 oz-s^2/in.
	Appendage modules of elasticity, E	161.6 million oz/$in.^2$
	Inertia of bending section, I	0.000813 $in.^4$
	Mass density of appendage/length, ρ	0.003007 oz-s^2/$in.^2$
	Distance between two-gage set	1.365 in.
Motor (PMI, servodisk dc motor: JR16M 4 CH)	Torque constant, K_t	52.77 oz-in./amp
	Back electromotive force constant, K_e	39.77 V/1000 rpm
	Tachometer constant	3 V/1000 rpm
Power Supply (KEPCO BOP 36-12M)	Current gain	1.2 amp/commanded Volt

consequence of the errors introduced by deriving the angular rate approximation from the angle encoder measurements.

Four KEPCO bipolar-operational power supply (BOPs) in series are used to supply current to the motor. They can supply up to 12 amps of current for up to a compliance voltage of 150 V. The BOPs are remotely programmed by PB286 through one of the D/A channels (DAC0) of DT2841.

DT2841 from Data Translation performs all of the I/O (digital, A/D, and D/A) operations for PB286. The board has two programmable clocks (DACLK and ADCLK). DACLK and ADCLK are programmed to run at 75 and 4000 Hz, respectively; DACLK runs continuously during a whole control operation, and ADCLK runs in a burst mode (i.e., intermittently).

Every DACLK timing signal initiates the following:

1) converts a digital data into an analog signal to be used by the BOP to push the corresponding current through the motor; and

2) generates a hardware interrupt to initiate a new control cycle.

The new control cycle proceeds as follows:

1) Read angle encoder (through digital port).

2) Start ADCLK, complete reading the two strain gages and the tachometer values (through A/D channels 0, 1, and 2) during the next three ADCLK cycles, and then disable ADCLK.

3) Filter the data digitally.

4) Predict the current required at the beginning of the next cycle.

5) Store the data in random access memory (RAM) for time history and make the predicted data available to the D/A converter to be converted at the beginning of the next cycle.

6) Wait until the next DACLK signal, which converts the digital data into an analog signal and generates a hardware interrupt to initiate the next control cycle.

The control software is written in FORTRAN, except for the PB286's interrupt controller setting, the I/O board functions, and I/O routines, which are all written in assembly code. The control cycle is updated at the rate of 75 Hz (DACLK rate).

Acknowledgments

The authors are pleased to acknowledge the work of H. Fujii, S. R. Vadali, and A. Das; their earlier research provided inspiration for this work. We thank the following individuals for their contributions: T. C. Pollock for his contributions to all aspects of the design and development of the experiment hardware; Steve Morgan for identifying the structural parameters and for his assistance with the angle encoder; Johnny Hurtado for his assistance in carrying out the experiments; and Becky Masters for excellent support in preparing this manuscript. The support of the Air Force Office of Scientific Research (Contract F49620-87-C-0078) and A. K. Amos is appreciated. We also received support from the Texas Advanced Technology Program (Project 70110).

References

[1]Junkins, J. L., and Turner, J. D., *Optimal Spacecraft Rotational Maneuvers*, Elsevier, Amsterdam, 1986.

[2]Juang, J. N., Horta, L. G., and Robertshaw, H. H., "A Sewing Control Experiment for Flexible Structures," *Proceedings of the 5th VPI & SU Symposium on Dynamics and Control of Large Structures*, Virginia Polytechnic Institute and State University, Blacksburg, VA, 1985, pp. 547–551.

[3]Fujii, H., Ohtsuka, T., and Udou, S., "Mission Function Control for Slew Maneuver Experiment," Preprint (to be published in *Journal of Guidance, Control, and Dynamics*).

[4]Junkins, J., Rahman, Z., Bang, H., and Hecht, N., "Near-Minimum Time Feedback Control of Distributed Parameter Systems: A Liapunov Control Law Design Method," *Proceedings of the 7th VPI&SU Symposium on Dynamics and Control of Large Structures* (to be published).

[5]Rahman, Z., Junkins, J. L., Pollock, T. C., and Bang, H., "Large Angle Maneuvers with Vibration Suppression: Analytical and Experimental Results," *Proceedings of the 7th VPI&SU Symposium on Dynamics and Control of Large Structures* (to be published).

[6]Thompson, R. C., Junkins, J. L., and Vadali, S. R., "Near-Minimum-Time Open-Loop Slewing of Flexible Vehicles," *Journal of Guidance, Control, and Dynamics*, Vol. 12, No. 1, Jan.–Feb. 1989, pp. 82–88.

[7]Vadali, S. R., "Feedback Control of Flexible Spacecraft Large Angle Maneuvers Using Liapunov Theory," IEEE Paper 1674, 1984, pp. 1674–1678.

[8]Byers, R. M., Vadali, S. R., and Junkins, J. L., "Near-Minimum-Time Closed Loop Slewing of Flexible Spacecraft," *Journal of Guidance, Control, and Dynamics* (to be published).

[9]Meirovitch, L., and Quinn, R., "Maneuvering and Vibration Control of Flexible Spacecraft," *Journal of the Astronautical Sciences*, Vol. 35, No. 3, July–Sept. 1987, pp. 301–328.

[10]Singh, G., Kabamba, P., and McClamroch, N., "Planar Time Optimal Slewing Maneuvers of Flexible Spacecraft," *Journal of Guidance, Control, and Dynamics* (to be published).

[11]Breakwell, J. A., "Optimal Feedback Control for Flexible Spacecraft," *Journal of Guidance, Control, and Dynamics*, Vol. 4, No. 5, Sept.-Oct. 1981, pp. 472–479.

[12]VanderVelde, W., and He, J., "Design of Space Structure Control Systems Using On-Off Thrusters," *Journal of Guidance, Control, and Dynamics*, Vol. 6, Jan.–Feb. 1983, pp. 759–775.

[13]Wie. B., Weiss, H., and Araposthathis, A., "Quaternion Feedback for Spacecraft Eigenaxis Rotations," *Journal of Guidance, Control, and Dynamics*, Vol. 12, May–June 1989, pp. 375–380.

Chapter 23

Minimum-Time Maneuvers of Flexible Spacecraft

G. Singh,* P. T. Kabamba,† and N. H. McClamroch‡
The University of Michigan, Ann Arbor, Michigan

Introduction

ATTITUDE controllers for spacecraft have been based on the assumption that the bodies being controlled are rigid. However, future spacecraft may be quite flexible compared with their current counterparts. Many proposed applications require rotating these vehicles between two widely spaced quiescent attitudes. In some other cases it is required that, once the spacecraft has achieved orbit, it be given a finite spin about its major principal axis to spin-stabilize the vehicle, since a large spinning body is gyroscopically stiff and is stable in such a configuration. We will refer to these two problems as the rest-to-rest maneuver and spin-up maneuver problems, respectively. In some time-critical applications it is required that these maneuvers be performed as rapidly as possible. It is the design of a control system for such applications, using nonlinear control elements such as on/off actuators, that is considered here.

Flexible spacecraft rotational maneuver problems have received considerable attention in the past. Linear quadratic formulations have dominated most of the published work.[1-7] Slewing problems using on/off-type actuation (not necessarily time-optimal) have also received attention.[8-15] Near-optimal, feedback control (which minimizes a weighted combination of maneuver time and fuel consumption) to perform rest-to-rest rotational maneuvers was obtained in Ref. 8. Recently, the exact solution to this problem was obtained.[9] In another account[10] an open-loop control is obtained for a fixed maneuver time. Here the control was a priori assumed to be bang-zero-bang type with a fixed number of switchings, and the

*Member Technical Staff, Department of Aerospace Engineering.
†Associate Professor, Department of Aerospace Engineering.
‡Professor, Department of Aerospace Engineering.

intermediate switching times were selected such that the rest-to-rest maneuver boundary conditions are satisfied and the postmaneuver energy in a finite number of uncontrolled modes is minimized. Thompson et al.[11] have solved the planar, near-minimum-time, rest-to-rest slewing problem by shaping the control input profiles. The control shaping parameters are adjustable and allow for the inclusion of thruster rise and fall times. The resulting control is not bang-bang, and the maneuvers thus performed are not minimum time.

The problem of planar spinning up/despinning a rigid body with flexible appendages has also been studied. Both linear and nonlinear models describing the rotational dynamics of such systems have been considered. Turner and Junkins[4] have solved this problem by minimizing a cost functional involving quadratic penalties on the state and control variables over a fixed maneuver duration.

All of these accounts stop short of obtaining an exact optimal control that minimizes maneuver time. We have examined the minimum-time, planar, rotational maneuvers of a flexible spacecraft.[12–15] Two control problems are considered here: 1) the time-optimal rest-to-rest slewing problem, which brings the rigid-body mode and a finite number of vibrational modes to rest at the final time; and 2) the time-optimal spin-up problem, which brings a finite number of vibrational modes to rest and leaves the rigid-body mode with a finite velocity at the final time. The control histories are found to be bang-bang with multiple switches.

The original results of this work are the following:

1) recognition and proof that the time-optimal controls for both the aforementioned problems have important time-symmetry properties;

2) derivation of a system of nonlinear algebraic equations satisfied by the optimal switching times, the optimal final time, and the optimal co-states at midmaneuver (the solution is obtained by following a homotopy approach);

3) characterization of the *post-time-optimal slewing maneuver spillover* via residual energy, angular position, and angular velocity errors of the rigid central body;

4) derivation of upper bounds on the spillover measures, introduced earlier, for a finite-dimensional evaluation model (for a specific case, closed-form expressions are obtained for an *infinite-dimensional evaluation model*); and

5) observation that, for a scalar control input, the time-optimal control history is independent of the control input location.

Based on these results the numerical optimization problem can be reduced to a problem of solving a set of algebraic equations for only *one-half* of the switching times and the final time. In this work a homotopy method is proposed to obtain a solution for these algebraic equations.

In the following section we present the rotational equations of motion of a generic flexible spacecraft model. Various assumptions are made to simplify the mathematical model, which is then used to formulate a linear time-optimal control problem in the section on formulation of linear time-optimal control problems. In the section on properties of time-optimal control, we give the necessary and sufficient conditions for optimality in their classical form. The existence and uniqueness of solutions of these equations is well documented in the literature. These issues will not be addressed here. We then state various properties of the time-optimal controls for both problems. Based on these properties, the necessary and sufficient conditions for optimality can be cast in the form of a set of nonlinear algebraic equations. In the section on computational approach to the time-optimal control, we briefly discuss a homotopy approach to obtain the solution to the algebraic necessary and sufficient conditions. In the section on spillover analysis, we quantify the post-time-optimal maneuver spillover effects via postmaneuver residual energy, the postmaneuver pointing error of the rigid central body for the rest-to-rest case, and the postmaneuver error in angular velocity of the rigid central body for the spin-up case. Upper bounds on these effects are obtained next. When there is only one control input and it is applied at the rigid central body, these bounds can be stated in a closed form for an infinite-dimensional evaluation model. In the section on example slewing maneuvers, we discuss some example maneuvers to demonstrate the applicability of the results and conclude in the last section by emphasizing the original contributions of the present work, its shortcomings, and future research directions.

Equations of Motion

The generic physical configuration of the flexible structure to be considered here is shown in Fig. 1. We will consider an unconstrained rigid base to which $N (N \geqslant 2)$ identical flexible appendages are rigidly and symmetrically attached with uniform spacing between them. We will refer to the base as the rigid central body. We assume that the appendage deformations, deformation rates, and slopes are small and the appendages are inextensible. We further assume that the appendage sections have negligible rotatory inertia and that the shear deformations are small in comparison with the bending deformations. We assume that no structural damping is present and that the appendages deform antisymmetrically. The system is to be controlled by a single torque actuator located on the rigid central body and m torquers located on each of the N flexible appendages. These locations are *identical* on all appendages. Actuators located at the same location on all appendages act in unison. Last, we assume that the rigid central body rotation rate remains "small" at all times. In the formulation to follow the extension to a system controlled by force actuators or a combination of force actuators and torque actuators is straightforward and yields similar dynamic equations. Situations where only one appendage is attached to the rigid central body can also be considered, provided the rigid central body is constrained against translation.

Let $\theta(t)$ be the angular displacement of the central body and let $y(x,t)$ denote the relative transverse displacement of a differential appendage element at location x, the distance along the appendage measured from the appendage root, $0 \leqslant x \leqslant L$, where L is the overall length of an appendage.

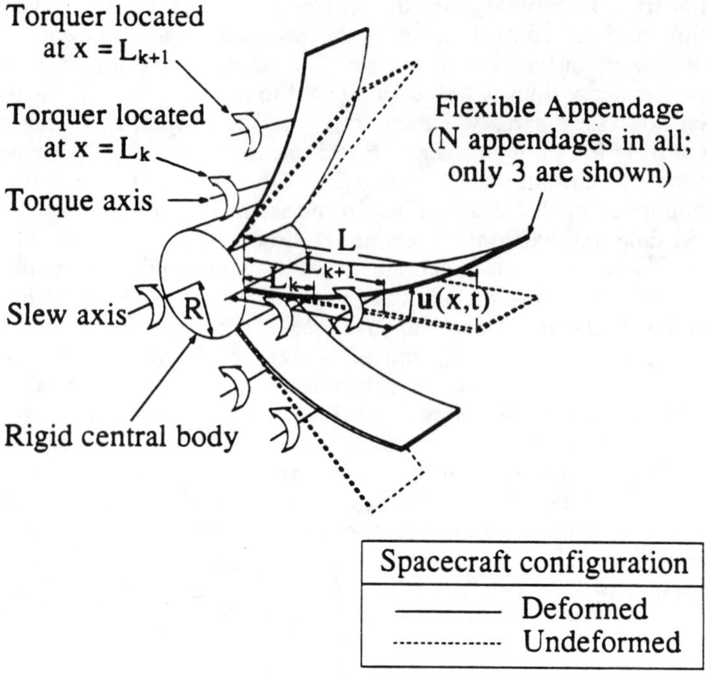

Fig. 1 Generic flexible spacecraft model.

Let J_o be the overall rotational inertia of the undeformed vehicle about the axis of rotation, R be the radius of the rigid central body (which is assumed to be cylindrical, with axis of symmetry coincident with the vehicle axis of rotation), $EI(x)$ be the appendage flexural rigidity, and $\rho(x)$ be the appendage material density at distance x from the appendage root. Upon application of Hamilton's extended principle,[18] the following coupled ordinary and partial differential equations of motion are obtained:

$$J_o\ddot{\theta} + N\int_0^L \rho(R+x)y_{,tt}\,\mathrm{d}x = T + N\sum_{j=1}^{m}\{T_j\} \tag{1}$$

$$\{EIy_{,xx}\}_{,xx} + \rho\{y_{,tt} + (R+x)\ddot{\theta}\} - \dot{\theta}^2\{(fy_{,x})_{,x} + \rho y\} = -\sum_{j=1}^{m}[\{\delta(x-L_j)\}_{,x}T_j] \tag{2}$$

where we have defined $f(x) \triangleq \int_x^L \rho(R+\xi)\,\mathrm{d}\xi$, the overdot denotes differentiation with respect to time, and $(\ )_{,x} \equiv \partial(\ )/\partial x$. Note that in writing Eq. (2) we have not yet used the assumption regarding small rigid-body rotation rates. We shall do so later in this section.

Assuming that the appendage relative displacement can be separated into products of space- and time-dependent variables, let the differential beam element lateral displacement be

$$y(x,t) = \sum_{i=1}^{M} \{\phi_i(x) q_i(t)\} \tag{3}$$

where $q_i(t)$ is the ith generalized coordinate; $\phi_i(x)$ is the ith assumed mode shape; $i = 1,2,3, \ldots, M$; and M is an arbitrary integer. The assumed mode shapes used in the expansion (3) satisfy the following geometric and natural boundary conditions:

$$\{EI\boldsymbol{\phi}_{i,xx}\}(L) = \{EI\boldsymbol{\phi}_{i,xx}\}_{,x}(L) = \{\boldsymbol{\phi}_{i,x}\}(0) = \boldsymbol{\phi}_i(0) = 0; \qquad i = 1,2, \ldots, M \tag{4}$$

We define the following vectors and matrices:

$$\boldsymbol{\phi} \triangleq \{\phi_1(x), \phi_2(x), \ldots, \phi_M(x)\}^T, \qquad \boldsymbol{q} \triangleq \{q_1(t), q_2(t), \ldots, q_M(t)\}^T \tag{5a}$$

$$\boldsymbol{m} \triangleq N \int_0^L \rho(x)(R + x)\boldsymbol{\phi}\, \mathrm{d}x, \qquad M \triangleq N \int_0^L \rho(x)(\boldsymbol{\phi}\ \boldsymbol{\phi}^T)\, \mathrm{d}x > 0 \tag{5b}$$

$$K \triangleq N \int_0^L \boldsymbol{\phi}([EI(x)\boldsymbol{\phi}_{,xx}]_{,xx})^T \mathrm{d}x, \qquad \overset{(4)}{=} N \int_0^L EI(x)(\boldsymbol{\phi}_{,xx}\ \boldsymbol{\phi}_{,xx}{}^T)\, \mathrm{d}x > 0 \tag{5c}$$

$$S \triangleq -N \int_0^L \boldsymbol{\phi}([f(x)\boldsymbol{\phi}_{,x}]_{,x})^T \mathrm{d}x, \qquad \overset{(4)}{=} N \int_0^L f(x)(\boldsymbol{\phi}_{,x}\boldsymbol{\phi}_{,x}{}^T)\, \mathrm{d}x \tag{5d}$$

Substituting Eq. (3) into Eqs. (1) and (2) and using the preceding definitions, we obtain the following coupled nonlinear ordinary differential equations:

$$J_o\ddot{\theta} + \boldsymbol{m}^T\ddot{\boldsymbol{q}} = T + N\boldsymbol{1}^T\boldsymbol{\tau} \tag{6}$$

$$M\ddot{\boldsymbol{q}} + [I + \dot{\theta}^2(S - M)K^{-1}]K\boldsymbol{q} + \boldsymbol{m}\ddot{\theta} = \Phi_x\boldsymbol{\tau} \tag{7}$$

where $\boldsymbol{\tau} \triangleq \{T_1, T_2, \ldots, T_m\}^T$ and $\boldsymbol{1} \triangleq \{1,1, \ldots, 1\}^T$ is a vector of appropriate dimension, and Φ_x is an $M \times m$ matrix defined as $\Phi_x \triangleq [N\boldsymbol{\phi}_{,x}(L_1), N\boldsymbol{\phi}_{,x}(L_2), \ldots, N\boldsymbol{\phi}_{,x}(L_m)]$.

Note that the quantity $-\boldsymbol{q}^T(S - M)\boldsymbol{q}$ can be shown to represent the change in the vehicle rotational inertia as it deforms. Since this change is always negative (since we have assumed inextensible appendages), the matrix $[S - M]$ must be positive-definite. The term $\dot{\theta}^2(S - M)\boldsymbol{q}$ in Eq. (7) therefore represents the nonlinear rotational "stiffening" effects. Using the matrix norm induced by the standard Euclidean norm on vectors, we expect that, if the maximum angular velocity attained by the rigid central body is such that

$$[\bar{\sigma}\{(S - M)K^{-1}\}]\max(\dot{\theta}^2) \ll 1 \tag{8}$$

it is possible to neglect the nonlinearity in Eq. (7). Here $\bar{\sigma}\{\ \}$ stands for the maximum singular value[23] of the matrix argument.

The quantity $\bar{\sigma}\{(S-M)K^{-1}\}$ is easily determined once a modal analysis has been carried out. Unfortunately, the maximum angular velocity attained by the rigid central body during a time-optimal rotational maneuver is not known a priori (except for a rigid spacecraft). If, in Eq. (8) we approximate the maximum rigid central body angular velocity for the flexible spacecraft case by the maximum angular velocity attained by an equivalent rigid body, we obtain the following for the rest-to-rest slewing case:

$$[\bar{\sigma}\{(S-M)K^{-1}\}]T\theta_f/J_o \ll 1 \tag{9}$$

where T is the maximum available torque about the slew axis, θ_f is the overall slew angle of the rigid central body, and J_o is the total rotational inertia about the slew axis. For the spin-up problem, we obtain

$$[\bar{\sigma}\{(S-M)K^{-1}\}]\Omega^2 \ll 1 \tag{10}$$

where Ω is the required change in the angular velocity of the rigid central body. In our analysis we will ignore this nonlinearity, for it simplifies the analysis considerably. However, we believe that, whenever "rapid" rotational maneuvers are considered, this nonlinearity does not *always* contribute significantly to vehicle dynamics, and in some cases it is possible to justify the omission of this nonlinearity. Let us now enforce the assumption regarding small rigid-body rotation rates. The following coupled, linear, ordinary differential equations are obtained as a result:

$$J_o\ddot{\theta} + \boldsymbol{m}^T\ddot{\boldsymbol{q}} = T + N\boldsymbol{1}^{\mathrm{T}}\boldsymbol{\tau} \tag{11}$$

$$M\ddot{\boldsymbol{q}} + K\boldsymbol{q} + \boldsymbol{m}\ddot{\theta} = \Phi_x\boldsymbol{\tau} \tag{12}$$

We note that, to verify the assumption of small angular rates of the rigid central body, we have applied time-optimal control histories based on a finite-dimensional linear control model to a finite-dimensional nonlinear evaluation model and found the degradation in performance to be acceptable in situations where condition (8) was satisfied.

Formulation of Linear Time-Optimal Control Problems

The planar rotational dynamics of flexible structures, assuming negligible rotational stiffening effects, are actually governed by a system of coupled ordinary and partial differential equations. The dynamic model is essentially infinite dimensional in nature, which has been truncated to yield Eqs. (11) and (12). The infinite-dimensional time-optimal control problems appear to be ill-posed.[12] Numerical results for the finite-dimensional problems formulated later in this work and numerical results for the time-optimal control of a linear hyperbolic system[16] suggest that there is, in general, no time-optimal solution within the class of amplitude-limited, piecewise-continuous control inputs. It has been observed, in solving the problem of time-optimal control of a linear hyperbolic system,[16] that, as the dimensionality of the problem is increased, the optimal final time converges to a limit from below, the number of switchings increases without limit, and therefore the time elapsed between switchings becomes smaller and smaller. The optimal control solutions for our finite-dimensional optimal control problems exhibit characteristics very similar to the problem treated by Goldwyn et al.[16]

Consider the coupled vector differential equations [Eqs. (11) and (12)]. Using Eq. (11) in Eq. (12), we obtain

$$[M - (1/J_o)\boldsymbol{m}\boldsymbol{m}^T]\ddot{\boldsymbol{q}} + K\boldsymbol{q} = -(T/J_o)\boldsymbol{m} + \{\Phi_x - (N/J_o)\boldsymbol{m}\mathbf{1}^T\}\boldsymbol{\tau} \tag{13}$$

Introduce a coordinate transformation $\boldsymbol{q} = U\boldsymbol{\eta}$, where $\boldsymbol{\eta}$ is a normal coordinate vector and U is a normalized matrix that satisfies

$$U^T[M - (1/J_o)\boldsymbol{m}\boldsymbol{m}^T]U = I,\ U^TKU = \Omega^2 = [\text{diag}(\omega_i^2);\ i = 1,2, ..., M].$$

Equation (13) can thus be written as

$$\ddot{\boldsymbol{\eta}} + \Omega^2\boldsymbol{\eta} = -(T/J_o)U^T\boldsymbol{m} + U^T\{\Phi_x - (N/J_o)\boldsymbol{m}\mathbf{1}^T\}\boldsymbol{\tau} \tag{14}$$

and condition (8) can be restated as follows:

$$[\bar{\sigma}\{\Omega^{-2}U^T(S - M)U\}]\ \max(\dot{\theta}^2) \ll 1 \tag{15}$$

Next, define the state variables as

$$x_1 \triangleq \theta + (1/J_o)\boldsymbol{m}^TU\boldsymbol{\eta} \tag{16}$$

$$x_2 \triangleq \dot{\theta} + (1/J_o)\boldsymbol{m}^TU\dot{\boldsymbol{\eta}} \tag{17}$$

$$x_3^i \triangleq \eta_i; \qquad i = 1,2, ..., M \tag{18}$$

$$x_4^i \triangleq \dot{\eta}_i/\omega_i; \qquad i = 1,2, ..., M \tag{19}$$

the control variables as

$$u_0 \triangleq T/J_o \tag{20}$$

$$u_j \triangleq (N/J_o)T_j; \qquad j = 1,2, ..., m \tag{21}$$

and the following parameters:

$$\beta_0^i \triangleq -\{\Omega^{-1}U^T\boldsymbol{m}\}_i; \qquad i = 1,2, ..., M \tag{22}$$

$$\beta_j^i \triangleq [\Omega^{-1}U^T\{(J_o/N)\Phi_x - \boldsymbol{m}\mathbf{1}^T\}]_{ij}; \qquad i = 1,2, ..., M, \qquad j = 1, ..., m \tag{23}$$

Here $[\ ]^{-1}$ denotes a matrix inverse.

Using Eqs. (16–23) in Eqs. (11) and (14), the following state space description of the plant dynamics is obtained:

$$\dot{x}_1 = x_2 \tag{24}$$

$$\dot{x}_2 = \sum_{j=0}^{m} \{u_j\} \tag{25}$$

$$\dot{x}_3^i = \omega_i x_4^i; \qquad i = 1,2,3, \ldots, M \tag{26}$$

$$\dot{x}_4^i = -\omega_i x_3^i + \sum_{j=0}^{m} \{\beta_j^i u_j\}; \qquad i = 1,2,3, \ldots, M \tag{27}$$

Time-Optimal Control Problems

The model (24–27) consists of one rigid-body mode and M undamped flexible modes. We require that only n of the M flexible modes are to be actively controlled. Defining the state vector as $\boldsymbol{x} \triangleq \{x_1, x_2, x_3^1, x_4^1, \ldots, x_3^n, x_4^n\}^T$ and the control vector as $\boldsymbol{u} \triangleq \{u_0, u_1, \ldots, u_m\}^T$, the equations of motion can be compactly written as

$$\dot{\boldsymbol{x}}(t) = A\boldsymbol{x}(t) + B\boldsymbol{u}(t); \qquad \boldsymbol{x}(t) \in \mathscr{R}^{2n+2}, \qquad \boldsymbol{u}(t) \in \mathscr{R}^{m+1} \tag{28}$$

where $A =$ block diagonal $[A_i;\ i = 0,1, \ldots, n]$ and $B =$ block column $[B_i;\ i = 0,1, \ldots, n]$ such that

$$A_0 = \begin{bmatrix} 0 & 1 \\ 0 & 0 \end{bmatrix}, \qquad A_i = \begin{bmatrix} 0 & \omega_i \\ -\omega_i & 0 \end{bmatrix}; \qquad i \geqslant 1,$$

$$B_i = \begin{bmatrix} 0 & 0 & \ldots & 0 \\ \beta_0^i & \beta_1^i & \ldots & \beta_m^i \end{bmatrix}; \qquad i \geqslant 0 \tag{29}$$

Note that $\beta_j^0 = 1$; $j = 0,1,2, \ldots, m$. The pair (A,B) is assumed to be normal.[19] We will refer to the model (28) and (29) as the "control model."

The following constraints are imposed on the $m + 1$ control inputs:

$$|u_j(t)| \leqslant U_j; \qquad j = 0,1,2, \ldots, m \tag{30}$$

We shall assume, without loss of generality, that the maneuvers are centered about the midmaneuver time, $t = 0$, and that the time interval over which the controls are active is $[-t_f/2, t_f/2]$. Outside of this interval all controls are turned off.

For the time-optimal rest-to-rest slewing problem, the boundary conditions are

$$\boldsymbol{x}(-t_f/2) = \{-\theta_f/2, 0, 0, 0, \ldots, 0, 0\}^T \tag{31a}$$

$$\boldsymbol{x}(t_f/2) = \{\theta_f/2, 0, 0, 0, \ldots, 0, 0\}^T \tag{31b}$$

where θ_f is the desired slew angle of the rigid central body, and we have supposed, without loss of generality, that the initial and terminal angular positions of the rigid central body are $-\theta_f/2$ and $\theta_f/2$, respectively. The finite-dimensional, time-optimal, rest-to-rest slewing problem (RTRSP) can be defined as follows: The finite-dimensional, time-optimal, rest-to-rest slewing problem is to find the optimal control vector $\boldsymbol{u}^*(t)$ that minimizes the maneuver time t_f while satisfying Eqs. (28–31).

We next formulate the time-optimal spin-up problem. Consider the motion of the system (28) and (29) between the following boundary conditions:

$$\boldsymbol{x}(-t_f/2) = \{0, -\Omega/2, 0,0, \ldots, 0,0\}^T, \ \boldsymbol{x}(t_f/2) = \{\theta_f, \Omega/2, 0,0, \ldots, 0,0\}^T \tag{32}$$

where θ_f, the final angular position of the rigid central body, is free, and Ω is the required change in the angular velocity of the rigid central body.

It can be shown that under suitable transformations[15] Eqs. (28), (29), and (32) describe motions of the system between two states such that in these states the appendage deformations and deformation rates are zero and between these two states the rigid body undergoes a *net change* in spin Ω, regardless of the initial and final spins of the rigid body (which must be such that the angular rates of the rigid body remain "small" during the maneuver). The finite-dimensional, time-optimal, spin-up problem (SUP) is thus stated as follows: The finite-dimensional, time-optimal, spin-up problem is to find the optimal control vector $\boldsymbol{u}^*(t)$ that minimizes the maneuver time t_f while satisfying Eqs. (28–30) and (32).

Properties of Time-Optimal Control

Necessary and Sufficient Conditions for Optimality

In the previous section we assumed the pair (A,B) to be normal. Normality of the pair is assured if the following conditions are met:

(i) $\omega_i \neq 0$,
(ii) $\omega_i \neq \omega_k$; $\quad i \neq k$
(iii) $\beta_j^i \neq 0$; $\quad (i,k) = 1,2, \ldots, n; \quad j = 0,1, \ldots, m$

Conditions (i) and (ii) are usually satisfied. Satisfaction of condition (iii) depends on actuator placements, which can be chosen a priori such that condition (iii) is also satisfied. Here assumption (iii) is not necessary, however. For both problems initial structural deformations and deformation rates are zero. As a result, the modes for which $\beta_j^i = 0; j = 0,1, \ldots, m$ do not contribute to the system dynamics since they have zero control input. Such modes may therefore be excluded from Eq. (28). Also, for the scalar control case, if modes that violate condition (ii) exist, then it is sufficient to include only one of them in the control model.

Pontryagin's maximum principle[20] is used to characterize the time-optimal control vector $\boldsymbol{u}(t)^*$. Define the Hamiltonian function as

$$\mathscr{H} \triangleq 1 + \boldsymbol{p}(t)^T\{A\boldsymbol{x}(t) + B\boldsymbol{u}(t)\}$$

where

$$\boldsymbol{p}(t) \triangleq \{p^0(t),q^0(t),p^1(t),q^1(t), \ldots, p^n(t),q^n(t)\}^T \in \mathbb{R}^{2n+2}$$

is a costate vector. For the RTRSP the necessary and sufficient conditions for optimality, along with Eqs. (28–31), are

$$\dot{\boldsymbol{p}}(t) = -\mathscr{H}^T_{,x} = -A^T\boldsymbol{p}(t); \qquad t \in [-t_f^*/2, t_f^*/2] \tag{33}$$

$$u_j(t)^* = -U_j \operatorname{sgn}\{\Gamma_j(t)\} \equiv -U_j \operatorname{sgn}\{\boldsymbol{p}(t)^T\boldsymbol{b}_j\}; \qquad t \in [-t_f^*/2, t_f^*/2] \tag{34}$$

$$\mathscr{H}(t) \equiv 0; \qquad t \in [-t_f^*/2, t_f^*/2] \tag{35}$$

where $\boldsymbol{b}_j$ is the jth column of the matrix B, and $\operatorname{sgn}(x)$ represents the signum function. For the SUP the final position of the rigid central body is also free. Therefore, for the SUP we have an additional transversality condition:

$$p^0(t_f/2) = 0 \tag{36}$$

For the SUP Eqs. (28–30) and (32–36) constitute the necessary and sufficient conditions for optimality.

The normality of (A,B) prohibits the existence of singular arcs for both problems. It can be shown[20] that for both problems a *unique* optimal control exists and is of the bang-bang variety with, possibly, multiple switches in each control variable.

The solution of the linear, time-invariant dynamic equation [Eq. (28)] is

$$\boldsymbol{x}(t) = \exp[A(t + t_f/2)]\boldsymbol{x}(-t_f/2) + \int_{-t_f/2}^{t} \exp[A(t - \tau)]B\boldsymbol{u}(\tau)\,\mathrm{d}\tau; \qquad t \geqslant -t_f/2 \tag{37}$$

Defining the costate vector at *midmaneuver* as

$$\boldsymbol{p}(0) \triangleq \{p_0^0, q_0^0, p_0^1, q_0^1, \ldots, p_0^n, q_0^n\}^T$$

Eqs. (33) can be solved in closed form:

$$p^0(t) = p_0^0 \tag{38}$$

$$q^0(t) = q_0^0 - p_0^0 t \tag{39}$$

$$p^i(t) = p_0^i \cos(\omega_i t) + q_0^i \sin(\omega_i t); \qquad i = 1,2, \ldots, n \tag{40}$$

$$q^i(t) = q_0^i \cos(\omega_i t) - p_0^i \sin(\omega_i t); \qquad i = 1,2, \ldots, n \tag{41}$$

This allows evaluation of all switching functions as explicit functions of time as

$$\Gamma_j(t) = q_0^0 - p_0^0 t + \sum_{i=1}^{n} [\beta_j^i \{q_0^i \cos(\omega_i t) - p_0^i \sin(\omega_i t)\}]; \qquad j = 0,1, ..., m \tag{42}$$

Properties of the time-optimal control are now stated. Proofs of these properties may be found in the appendices.

Characteristics of the RTRSP Time-Optimal Control

Theorem 1. For the RTRSP the optimal control $\boldsymbol{u}(t)^*$ is an *odd* function of time, i.e.,

$$\boldsymbol{u}(t)^* = -\boldsymbol{u}(-t)^*; \qquad t \in [-t_f^*/2, t_f^*/2] \tag{43}$$

where t_f^* is the optimal maneuver time.

Corollary 1. For the RTRSP the costate vector satisfying Eq. (33) is such that

$$\boldsymbol{p}(0) = \{p_0^0, 0, p_0^1, 0, ..., p_0^n, 0\}^T$$

Theorem 1 implies that each control variable must switch an odd number of times. They do not necessarily switch at the same times, but *all* must switch synchronously at time $t = 0$. Suppose that the jth optimal control variable, $u_j(t)^*$, switches $2k^j + 1$ times; $j = 0,1,2, ..., m$, where k^j is some nonnegative integer. These switches must occur at times $-t_{kj}^j, -t_{kj-1}^j, ..., -t_1^j, 0, t_1^j, ..., t_{kj-1}^j, t_{kj}^j$, where

$$0 < t_1^j < t_2^j < \cdots < t_{kj-1}^j < t_{kj}^j < t_f/2; \qquad j = 0,1,2, ..., m \tag{44}$$

The switching instants for each control input are distinct by definition. Furthermore, let

$$\alpha_j = \operatorname{sgn}(u_j(t)^*); \qquad -t_f/2 < t < -t_{kj}^j; \qquad j = 0,1,2, ..., m \tag{45}$$

represent the initial sign of the jth optimal control variable $u_j(t)^*$.

Let $\boldsymbol{x}(-t_f/2)$ be as given in Eqs. (31). Using the assumed control history in Eq. (37) and requiring that $\boldsymbol{x}(t_f/2) = \{\theta_f/2, 0, 0, 0, ..., 0, 0\}^T$, we obtain the following nonlinear algebraic equations in the switching times and the final time:

$$\sum_{j=0}^{m} [\alpha_j U_j \{(t_f/2)^2 - 2(t_{kj}^j)^2 + \cdots + 2(-1)^{kj}(t_1^j)^2\}] = \theta_f \tag{46}$$

$$\sum_{j=0}^{m} [\alpha_j \beta_j^i U_j \{\cos(\omega_i t_f/2) - 2\cos(\omega_i t_{kj}^j) + \cdots + 2(-1)^{kj}\cos(\omega_i t_1^j) + (-1)^{kj+1}]\}$$
$$= 0; \; i = 1,2, ..., n \tag{47}$$

Here it is noteworthy that, although $2n+2$ states are to be satisfied at the final time, the symmetry property of the optimal control automatically satisfies half of them (the velocity states). In order to satisfy the other half (the position states), the switching times and the final time must satisfy Eqs. (46) and (47).

Using *Corollary* 1 in Eq. (42), the switching function of the *j*th control variable can be expressed as

$$\Gamma_j(t) = -p_0^0 t - \sum_{i=1}^{n} \{\beta_j^i p_0^i \sin(\omega_i t)\}; \qquad j = 0,1, \ldots, m$$

The switching functions by definition must vanish at the respective switching times, i.e.,

$$p_0^0 t_r^j + \sum_{i=1}^{n} \{\beta_j^i p_0^i \sin(\omega_i t_r^j)\} = 0; \qquad 0 < t < t_f/2, \qquad 1 \leqslant r \leqslant k^j \tag{48}$$

To satisfy the transversality condition [Eq. (35)], the switching function must also satisfy

$$\sum_{j=0}^{m} \left[\alpha_j U_j \left\{ p_0^0 t_f/2 + \sum_{i=1}^{n} \{\beta_j^i p_0^i \sin(\omega_i t_f/2)\right\}\right] = -1 \tag{49}$$

Defining $\boldsymbol{p}_o \triangleq \{p_0^0, p_0^1, \ldots, p_0^n\}^T$, $\boldsymbol{c} \triangleq \{-2,0,0, \ldots, 0\}^T$, we can express the $\Sigma_{j=0}^m (k^j) + 1$ equations [Eqs. (48) and (49)] as follows:

$$H\boldsymbol{p}_o = \boldsymbol{c} \tag{50}$$

where $H =$

$$\begin{bmatrix}
2\sum_{j=0}^{m}\{\alpha_j U_j t_f/2\} & 2\sum_{j=0}^{m}\{\alpha_j \beta_j^1 U_j \sin(\omega_1 t_f/2)\} & \cdots & 2\sum_{j=0}^{m}\{\alpha_j \beta_j^n U_j \sin(\omega_n t_f/2)\} \\
-4\alpha_0 U_0 t_{k^0}^0 & -4\alpha_0 \beta_0^1 U_0 \sin(\omega_1 t_{k^0}^0) & \cdots & -4\alpha_0 \beta_0^n U_0 \sin(\omega_n t_{k^0}^0) \\
\cdots & \cdots & \cdots & \cdots \\
4(-1)^{k^0}\alpha_0 U_0 t_1^0 & 4(-1)^{k^0}\alpha_0 \beta_0^1 U_0 \sin(\omega_1 t_1^0) & \cdots & 4(-1)^{k^0}\alpha_0 \beta_0^n U_0 \sin(\omega_n t_1^0) \\
-4\alpha_1 U_1 t_{k^1}^1 & -4\alpha_1 \beta_1^1 U_1 \sin(\omega_1 t_{k^1}^1) & \cdots & -4\alpha_1 \beta_1^n U_1 \sin(\omega_n t_{k^1}^1) \\
\cdots & \cdots & \cdots & \cdots \\
4(-1)^{k^1}\alpha_1 U_1 t_1^1 & 4(-1)^{k^1}\alpha_1 \beta_1^1 U_1 \sin(\omega_1 t_1^1) & \cdots & 4(-1)^{k^1}\alpha_1 \beta_1^n U_1 \sin(\omega_n t_1^1) \\
\cdots & \cdots & \cdots & \cdots \\
\cdots & \cdots & \cdots & \cdots \\
-4\alpha_m U_m t_{k^m}^m & -4\alpha_m \beta_m^1 U_m \sin(\omega_1 t_{k^m}^m) & \cdots & -4\alpha_m \beta_m^n U_m \sin(\omega_n t_{k^m}^m) \\
\cdots & \cdots & \cdots & \cdots \\
4(-1)^{k^m}\alpha_m U_m t_1^m & 4(-1)^{k^m}\alpha_m \beta_m^1 U_m \sin(\omega_1 t_1^m) & \cdots & 4(-1)^{k^m}\alpha_m \beta_m^n U_m \sin(\omega_n t_1^m)
\end{bmatrix}$$

Define $\nu \triangleq \Sigma_{j=0}^{m} \{k^j\}$. Equations (46), (47), and (50) represent a system of $n+\nu+2$ nonlinear algebraic equations in $n+\nu+2$ unknowns, which consist of the following: ν switching times t_r^j, $1 \leqslant r \leqslant k^j$, $j=0,1, \ldots, m$; the final time $t_f/2$; and the $n+1$ costates at midmaneuver $p_0^0, p_0^1, \ldots, p_0^n$. Since these equations have been obtained, essentially, from Eqs. (28–31) and (33–35), after using the results stated earlier, they are necessary conditions for optimality. In addition, Eqs. (46), (47), and (50) are also sufficient conditions for optimality, provided that

$$p_0^0 + \sum_{i=1}^{n} \{\beta_j^i p_0^i \sin(\omega_i t)\} \neq 0; \qquad t \neq t_r^j; \qquad 0 < t < t_f/2, \qquad 1 \leqslant r \leqslant k^j, \qquad j=0,1, \ldots, m \tag{51}$$

which simply requires the switching function to vanish *only* at the candidate switching times.

Theorem 2. Suppose that for some integers k^j; $0 \leqslant j \leqslant m$, Eqs. (46), (47), and (50) admit a solution $(t_f/2, t_r^j, p_0^i;\ 1 \leqslant r \leqslant k^j;\ 0 \leqslant j \leqslant m;\ 0 \leqslant i \leqslant n)$ satisfying inequalities (44) and (51). Then

(i) the jth component of the optimal control solving the RTRSP switches exactly $2k^j+1$ times at times $-t_{kj}^j, -t_{kj-1}^j, \ldots, -t_1^j, 0, t_1^j, \ldots, t_{kj-1}^j, t_{kj}^j$; and
(ii) the optimal maneuver time of the RTRSP is exactly t_f.

When $\nu = n$, Eqs. (46) and (47) are a set of $n+1$ equations with $n+1$ unknowns. Under an additional assumption, the solution of Eqs. (46) and (47) completely characterizes the optimal control. We have the following:

Corollary 2. Suppose that integers k^j; $0 \leqslant j \leqslant m$ satisfy $\Sigma_{j=0}^{m} \{k^j\} = n$. Also, assume that Eqs. (46) and (47) admit a solution $(t_f/2, t_r^j; 1 \leqslant r \leqslant k^j; 0 \leqslant j \leqslant m)$, which is regular[21] and satisfies Eqs. (44), (50), and (51). Then claims (i) and (ii) of *Theorem* 2 are true.

Characteristics of the SUP Time-Optimal Control

Properties similar to the RTRSP time-optimal control properties exist for the SUP time-optimal control. Proofs of the SUP maneuver properties will be omitted here in the interest of brevity. These properties can be proven in a manner similar to the proofs constructed for the RTRSP.

Theorem 3. For the SUP the optimal control $\boldsymbol{u}(t)^*$ is an *even* function of time, i.e.,

$$\boldsymbol{u}(t)^* = \boldsymbol{u}(-t)^*; \qquad -t_f^*/2 \leqslant t \leqslant t_f^*/2 \tag{52}$$

Corollary 3. For the SUP the costate vector satisfying Eq. (33) is such that

$$\boldsymbol{p}(0) = \{0, q_0^0, 0, q_0^1, \ldots, 0, q_0^n\}^T$$

Theorem 3 implies that the jth optimal control variable, $u_j(t)^*$, must switch $2k^j$ times; $j=0,1,2,\ldots,m$, at times $-t^j_{kj}, -t^j_{kj-1},\ldots,-t^j_1, t^j_1,\ldots,t^j_{kj-1},t^j_{kj}$, where the switching instants satisfy Eq. (44). Like the RTRSP, we can restate the necessary and sufficient conditions for optimality in the form of the following algebraic equations:

$$\sum_{j=0}^{m}\{\alpha_j U_j[(t_f/2)-2(t^j_{kj})+\cdots+2(-1)^{kj}(t^j_1)]\}=\Omega/2 \tag{53}$$

$$\sum_{j=0}^{m}\{\alpha_j\beta^i_j U_j[\sin(\omega_i t_f/2)-2\sin(\omega_i t^j_{kj})+\cdots+2(-1)^{kj}\sin(\omega_i t^j_1)]\}=0$$

$$i=1,2,\ldots,n \tag{54}$$

$$G\boldsymbol{q}_o=\boldsymbol{d} \tag{55}$$

$$q^0_0+\sum_{i=1}^{n}\{\beta^i_j q^i_0\cos(\omega_i t)\}\neq 0;$$

$$t\neq t^j_r;\qquad 0<t<t_f/2;\qquad 1\leqslant r\leqslant k^j;\qquad j=0,1,\ldots,m \tag{56}$$

where $G=$

$$\begin{bmatrix}
\sum_{j=0}^{m}\{\alpha_j U_j\} & \sum_{j=0}^{m}\{\alpha_j\beta^1_j U_j\cos(\omega_1 t_f/2)\} & \cdots & \sum_{j=0}^{m}\{\alpha_j\beta^n_j U_j\cos(\omega_n t_f/2)\}\\
-2\alpha_0 U_0 & -2\alpha_0\beta^1_0 U_0\cos(\omega_1 t^0_{k0}) & \cdots & -2\alpha_0\beta^n_0 U_0\cos(\omega_n t^0_{k0})\\
\cdots & \cdots & \vdots & \cdots\\
2(-1)^{k0}\alpha_0 U_0 & 2(-1)^{k0}\alpha_0\beta^1_0 U_0\cos(\omega_1 t^0_1) & \cdots & 2(-1)^{k0}\alpha_0\beta^n_0 U_0\cos(\omega_n t^0_1)\\
-2\alpha_1 U_1 & -2\alpha_1\beta^1_1 U_1\cos(\omega_1 t^1_{k1}) & \vdots & -2\alpha_1\beta^n_1 U_1\cos(\omega_n t^1_{k1})\\
\cdots & \cdots & \vdots & \cdots\\
2(-1)^{k1}\alpha_1 U_1 & 2(-1)^{k1}\alpha_1\beta^1_1 U_1\cos(\omega_1 t^1_1) & \cdots & 2(-1)^{k1}\alpha_1\beta^n_1 U_1\cos(\omega_n t^1_1)\\
\cdots & \cdots & \vdots & \cdots\\
\cdots & \cdots & \vdots & \cdots\\
-2\alpha_m U_m & -2\alpha_m\beta^1_m U_m\cos(\omega_1 t^m_{km}) & \cdots & -2\alpha_m\beta^n_m U_m\cos(\omega_n t^m_{km})\\
\cdots & \cdots & \vdots & \cdots\\
2(-1)^{km}\alpha_m U_m & 2(-1)^{km}\alpha_m\beta^1_m U_m\cos(\omega_1 t^m_1) & \cdots & 2(-1)^{km}\alpha_m\beta^n_m U_m\cos(\omega_n t^m_1)
\end{bmatrix}$$

and we have used $\boldsymbol{q}_0\triangleq\{q^0_0,q^1_0,\ldots,q^n_0\}^T$ and $\boldsymbol{d}\triangleq\{-1,0,0,\ldots,0\}^T$. Note that Eq. (38) and *Corollary* 3 imply the satisfaction of Eq. (36). We have the following results for the SUP.

Theorem 4. Suppose that for some integers k^j, $0 \leqslant j \leqslant m$, Eqs. (53–55) admit a solution $(t_f/2, t_r^j, q_0^i;\ 1 \leqslant r \leqslant k^j;\ 0 \leqslant j \leqslant m;\ 0 \leqslant i \leqslant n)$ satisfying inequalities (44) and (56). Then

(i) the jth component of the optimal control solving the SUP switches exactly $2k^j$ times at times $-t_{kj}^j, -t_{kj-1}^j, \ldots, -t_1^j, t_1^j, \ldots, t_{kj-1}^j, t_{kj}^j$; and

(ii) the optimal maneuver time for the SUP is exactly t_f.

Corollary 4. Suppose that integers k^j, $0 \leqslant j \leqslant m$ satisfy $\Sigma_{j=0}^m \{k^j\} = n$. Also, assume that Eqs. (53) and (54) admit a solution $(t_f/2, t_r^j;\ 1 \leqslant r \leqslant k^j;\ 0 \leqslant j \leqslant m)$, which is regular[21] and satisfies Eqs. (44), (55), and (56). Then claims (i) and (ii) of *Theorem* 4 are true.

Scalar Control Case

Here we assume that only one *control input* is used to control the maneuver. This assumption holds if all torque actuators are activated by *one* control signal. We can also consider a spacecraft controlled by only one torque actuator located on the rigid central body as another special scalar control case. Equations (46), (47), (50), and (51) become

$$(t_f/2)^2 - 2(t_k)^2 + 2(t_{k-1})^2 - \cdots + 2(-1)^k (t_1)^2 = \theta_f/(\alpha_j U_j) \tag{57}$$

$$\cos(\omega_i t_f/2) - 2\cos(\omega_i t_k) + \cdots + 2(-1)^k \cos(\omega_i t_1) + (-1)^{k+1} = 0$$

$$i = 1,2, \ldots, n \tag{58}$$

$$\begin{bmatrix} 2t_f/2 & 2\sin(\omega_1 t_f/2) & \cdots & 2\sin(\omega_n t_f/2) \\ -4t_k & -4\sin(\omega_1 t_k) & \cdots & -4\sin(\omega_n t_k) \\ \cdots & \cdots & \cdots & \cdots \\ 4(-1)^k t_1 & 4(-1)^k \sin(\omega_1 t_1) & \cdots & 4(-1)^k \sin(\omega_n t_1) \end{bmatrix} \begin{Bmatrix} p_0^0 \\ \beta_j^1 p_0^1 \\ \cdots \\ \beta_j^n p_0^n \end{Bmatrix} = \begin{Bmatrix} -2/(\alpha_j U_j) \\ 0 \\ \cdots \\ 0 \end{Bmatrix} \tag{59}$$

$$p_0^0 t + \sum_{i=1}^n \{(\beta_j^i p_0^i)\sin(\omega_i t)\} \neq 0; \quad t \neq t_r; \quad 0 < t < t_f/2; \qquad r = 1,2, \ldots, k \tag{60}$$

Note that the index j is fixed here. Equations (57–60) are the necessary and sufficient conditions for optimality for the scalar control case. These equations admit a unique solution. Only β_j^i, $i = 1,2, ..., n$ are functions of actuator locations [see Eq. (23)], and in the above equations these parameters appear as $p_0^0, \beta_j^1 p_0^1, ..., \beta_j^n p_0^n$. It is obvious that, for the given optimal control problem with a scalar control input, if *only* the actuator locations are changed, then by appropriately scaling p_0^i; $i = 1,2, ..., n$ (this is possible so long as $\beta_j^i \neq 0$, which is prohibited by the normality assumption), the necessary and sufficient conditions for optimality [Eqs. (57–60)] can be satisfied for the *same* values of the switching times, the final time and the parameters $p_0^0, \beta_j^i p_0^i$; $i = 1,2, ..., n$. Uniqueness of the solution implies the following.

Proposition 1. The time-optimal control history for the RTRSP, where control is effected using only one control input, is independent of actuator location. The time-optimal control only depends on the modal frequencies, ω_i; $i = 1,2, ..., n$, and the parameter θ_f/U_0.

For the SUP we arrive at a similar conclusion.

Proposition 2. The time-optimal control history for the SUP, where control is effected using only one control input, is independent of actuator location. The time-optimal control only depends on the modal frequencies, ω_i; $i = 1,2, ..., n$, and the parameter Ω/U_0.

Computational Approach to the Time-Optimal Control

Based on the theorems stated in the previous section, the numerical optimization problem has been reduced to the problem of solving a set of nonlinear algebraic equations. The solution of these equations can be obtained using a homotopy approach.[21]

Homotopy Principle

Suppose we are given a nonlinear differentiable vector function $f: \mathscr{R}^n \to \mathscr{R}^n$, where $\mathscr{R}^n$ is n-dimensional real vector space. We seek a solution to the nonlinear equations:

$$\boldsymbol{f}(\boldsymbol{y}) = \boldsymbol{0} \tag{61}$$

Denote a solution by $\boldsymbol{y}^*$. To obtain $\boldsymbol{y}^*$ using a homotopy process, define a homotopy function: $\boldsymbol{h}(\boldsymbol{y},t): \mathscr{R}^{n+1} \to \mathscr{R}^n$, where an additional parameter, t, called the homotopy parameter, has been introduced such that

$$\boldsymbol{h}(\boldsymbol{y},0) = \boldsymbol{g}(\boldsymbol{y}) \quad \text{and} \quad \boldsymbol{h}(\boldsymbol{y},1) = \boldsymbol{f}(\boldsymbol{y}) \tag{62}$$

where the vector function $\boldsymbol{g}(\boldsymbol{y})$ is *chosen* such that a solution of $\boldsymbol{g}(\boldsymbol{y}) = \boldsymbol{0}$, denoted by $\boldsymbol{y}^o$, can be easily determined. Therefore, the solution of $\boldsymbol{h}(\boldsymbol{y},0) = \boldsymbol{g}(\boldsymbol{y}) = \boldsymbol{0}$ is $\boldsymbol{y}^o$, which is known a priori, and the solution of $\boldsymbol{h}(\boldsymbol{y},1) = \boldsymbol{f}(\boldsymbol{y}) = \boldsymbol{0}$ is $\boldsymbol{y}^*$, which we seek. Here the solution of $\boldsymbol{h}(\boldsymbol{y},t) = \boldsymbol{0}$, denoted by $\boldsymbol{y}(t)$, can be viewed as an implicit function of the homotopy parameter t such that $y(0) = y^o$ and $y(1) = y^*$. To obtain y^*, we follow a path in $y(t) - t$ space, starting at y^o and increasing t until $t = 1$.

The homotopy functions $h(y,t)$, which yield a path from $(y^o,0)$ to $(y^*,1)$, are not unique, and the path followed depends on the homotopy function selected. The set $\boldsymbol{h}^{-1}, \boldsymbol{h}^{-1} \triangleq \{(y,t): \boldsymbol{h}(y,t) = \mathbf{0}\}$, in general, could be arbitrary; it may consist of isolated points, bifurcations, infinite spirals, continuously differential paths, etc. The homotopy function $\boldsymbol{h}(y,t)$ is said to be *regular* if the $n \times (n+1)$ matrix, $H' = [\boldsymbol{h}_{,y}, \boldsymbol{h}_{,t}]$, the Jacobian of $\boldsymbol{h}(y,t)$, is of full rank for *all* $(y,t) \in \boldsymbol{h}^{-1}$. If $\boldsymbol{h}(y,t)$ is continuously differentiable and regular *everywhere*, then $\boldsymbol{h}^{-1}$ consists *only* of continuously differentiable paths. In practice it is extremely difficult to verify the regularity of $\boldsymbol{h}(y,t)$. However, the assumption that the function $\boldsymbol{h}(y,t)$ is regular everywhere is a weak one. Indeed, according to Sard's theorem[21] the regularity assumption holds almost all of the time.

Once the existence of a path in $y - t$ space has been established, it can be followed from $t = 0$ to $t = 1$ by numerically integrating the basic differential equation (BDE). Let p denote the distance moved along the path in the $n + 1$ dimensional $y - t$ space. In this space the path parameter p can be treated as the independent parameter. Further define

$$z(p) \triangleq \{z_1, z_2, \ldots, z_{n+1}\}^T$$

according to

$$z_i(p) \triangleq y_i(p); \qquad i = 1,2, \ldots, n, z_{n+1}(t) \triangleq t(p)$$

Therefore, as p varies, $z(p)$ described a path in $\boldsymbol{h}^{-1}$ such that $\boldsymbol{h}(z(p)) = \mathbf{0}$. Differentiating this equation with respect to p, we obtain

$$H'z' = \mathbf{0} \tag{63}$$

where $(\ )'$ denotes differentiation with respect to the path parameter p, and H' is the Jacobian of $\boldsymbol{h}(z)$. Let H'_{-i} represent H' with its ith column deleted. Equation (63) is a system of n equations in $n + 1$ unknowns. Equation (63) can be written as

$$z_i' = (-1)^i |H'_{-i}|; \qquad i = 1,2, \ldots, n+1 \tag{64}$$

These are the basic differential equations, which, when solved with initial conditions $(y^o,0)$, yield a path in $\boldsymbol{h}^{-1}$. If $\boldsymbol{h}(y,t)\colon \mathscr{R}^{n+1} \to \mathscr{R}$ is twice continuously differentiable and regular, then the solution to the BDE starting from $z^o = (y^o,0) \in \boldsymbol{h}^{-1}$ exists, is unique, and determines a smooth path in $\boldsymbol{h}^{-1}$.[21]

It is important for our purpose that the path which starts at $y^o \in \boldsymbol{h}^{-1}(0)$ passes through a point in $\boldsymbol{h}^{-1}(1)$, where $\boldsymbol{h}^{-1}(0) \triangleq \{y: \boldsymbol{h}(y,0) = \mathbf{0}\}$, $\boldsymbol{h}^{-1}(1) \triangleq \{y: \boldsymbol{h}(y,1) = \mathbf{0}\}$. The BDE theorem stated earlier does not guarantee it. Two scenarios might occur. The path starting at $y^o \in \boldsymbol{h}^{-1}(0)$ may not get to $y^* \in \boldsymbol{h}^{-1}(1)$. It may return to another point $y^1 \in \boldsymbol{h}^{-1}(0)$. This behavior cannot occur if $\boldsymbol{h}^{-1}(0)$ consists of a unique point y^o. Also, a solution to the BDE may have a finite escape time; i.e., a path starting at $y^o \in h^{-1}(0)$ may *never* get to a $\boldsymbol{y}^* \in \boldsymbol{h}^{-1}(1)$. This occurrence can be excluded if a *boundary-free* condition is satisfied. Let $\mathscr{I} = \{t: 0 \leqslant t \leqslant 1\}$ denote the unit interval, and let $\mathscr{R}^n \supset \mathscr{B}$ such that $f(y): \mathscr{B} \to \mathscr{R}^n$. Thus, $\boldsymbol{h}: (\mathscr{B} \times \mathscr{I}) \to \mathscr{R}^n$. Furthermore, let $\partial\mathscr{B}$ denote the boundary of the set $\mathscr{B}$. A homotopy function $\boldsymbol{h}$ is said to be *boundary free* if $\boldsymbol{y} \notin \partial\mathscr{B}\ \forall \boldsymbol{y} \in \boldsymbol{h}^{-1}(t)$, $t \in \mathscr{I}$. A homotopy function $\boldsymbol{h}$ is guaranteed to be boundary free if the set is $\mathscr{B}$ convex and compact.[21] However, in practice it is virtually impossible to verify whether an $\boldsymbol{h}$ is boundary free or not.

Determination of the Time-Optimal Control

A difficulty encountered in solving the algebraic necessary and sufficient conditions for optimality is that α_j and k^j; $1 \leqslant j \leqslant m$, are not known a priori. In our examples we have considered the scalar control cases and noticed that in all cases the initial control sign was positive. Furthermore, for the RTRSP we have also found that in most cases $v = n$, where v represents the number of control switchings on $(0, t_f/2)$. However, the case $v < n$ can occur if a mode that is initially not included in the control model but whose frequency satisfies Eq. (47) is included in the control model. The case $v > n$ can also occur, but seems to be restricted to some particular ranges of the parameter $\omega_1^2\theta_f/U_0$. Over such parameter ranges we have found that, in general, no real solutions exist for $v = n$. For the SUP the number of switchings also depend on the parameter $\omega_1\Omega/U_0$. We have found that, in general, for small values of the parameter $\omega_1\Omega/U_0$ ($\omega_1\Omega/U_0 < 2\pi$), $v = n$. As this parameter is increased, more and more switches are encountered. In particular, for the *one-mode case* we have found that

$$\begin{aligned} v &= k; \qquad (k-1)\pi < \omega_1\Omega/U_0 < k\pi; \qquad k = 1,2,\ldots \\ &= 0; \qquad \omega_1\Omega/U_0 = k\pi; \qquad k = 1,2,\ldots \end{aligned} \tag{65}$$

To obtain the optimal solution, we proceed as follows. We *assume* that the initial control sign is positive and that $v = n$ for RTRSP and SUP (if $\omega_1\Omega/U_0 < 2\pi$). *Corollaries* 2 and 4 are then used to determine if a solution of Eqs. (46), (47), and (50) or Eqs. (53–56) is optimal. If Eqs. (46), (47), and (50) or Eqs. (53–56) do not admit an optimal solution (for $v = n$), this implies that the supposition $v = n$ is inaccurate, and we explore the situation $v \neq n$ by increasing or decreasing v. We use *Theorems* 2 and 4 to determine if a candidate solution is optimal; if it is, then it is the only solution of the optimal control problem. For the SUP, for large values of parameter $\omega_1\Omega/U_0$ ($\omega_1\Omega/U_0 > 2\pi$), we *assume* the number of switchings according to Eq. (65) and use *Theorem* 4 to ensure the optimality of a solution. Details of construction of suitable homotopy functions can be found elsewhere.[13–15]

Spillover Analysis

The optimal control characterized in the section on properties of time-optimal control is based on a control model that has one rigid-body mode and n undamped flexible modes. This control law is now applied to an evaluation model (24–27), which has M flexible modes ($M \geqslant n$), in addition to the rigid-body mode. In general, these $M - n$ modes would have nonzero displacements and velocities when the controls are turned off at the end of the maneuver. Therefore, these effects result in free vibration of the system for the RTRSP and free vibrations superimposed on the constant angular velocity of the rigid central body for the SUP, after the maneuver has been completed. Here we consider these effects and their effect on system performance and quantify the performance degradation as a result of control spillover.

Consider the effect of control spillover on the $M - n$ residual modes whose dynamics are given by Eqs. (26) and (27). The solution of these equations for residual modes can be written as

$$x_3^i(t) = \left\{ \sum_{j=0}^{m} (\beta_j^i C_{ij}) \right\} \sin(\omega_i t) - \left\{ \sum_{j=0}^{m} (\beta_j^i S_{ij}) \right\} \cos(\omega_i t); \qquad t \geqslant t_f^*/2, \qquad i > n \tag{66}$$

$$x_4^i(t) = \left\{ \sum_{j=0}^{m} (\beta_j^i S_{ij}) \right\} \sin(\omega_i t) + \left\{ \sum_{j=0}^{m} (\beta_j^i C_{ij}) \right\} \cos(\omega_i t); \qquad t \geqslant t_f^*/2, \qquad i > n \tag{67}$$

where we have defined the following constants:

$$S_{ij} \triangleq \int_{-t_f^*/2}^{t_f^*/2} u_j(t)^* \sin(\omega_i t)\, \mathrm{d}t; \qquad i = n+1, n+2, \ldots, M; \qquad j = 0,1, \ldots, m \tag{68}$$

$$C_{ij} \triangleq \int_{-t_f^*/2}^{t_f^*/2} u_j(t)^* \cos(\omega_i t)\, \mathrm{d}t; \qquad i = n+1, n+2, \ldots, M; \qquad j = 0,1, \ldots, m \tag{69}$$

The total energy of the rotating system can be written as

$$\mathscr{E}(t) = \frac{1}{2}\left[J_o x_2(t)^2 + \sum_{i=1}^{M} [\omega_i^2 \{x_3^i(t)^2 + x_4^i(t)^2\}] \right]; \qquad t \geqslant -t_f^*/2 \tag{70}$$

Note that in both problems, for $t \geqslant t_f^*/2$, $x_2(t)$ is a specified constant and $x_3^i(t) = x_4^i(t) = 0$; $i \leqslant n$. The *constant* energy after the completion of the maneuver can therefore be written, after substitution of Eqs. (66) and (67) into Eq. (70), as

$$\mathscr{E} = \frac{1}{2}\left[J_o x_2^2 + \sum_{i=n+1}^{M} \omega_i^2 \left[\left\{ \sum_{j=0}^{m} (\beta_j^i C_{ij}) \right\}^2 + \left\{ \sum_{j=0}^{m} (\beta_j^i S_{ij}) \right\}^2 \right] \right] \tag{71}$$

Spillover Characterization for the RTRSP

For the RTRSP we note that, since $u_j(t)^*$; $j = 0,1, ..., m$ are odd functions of time, Eqs. (68) and (69) yield

$$S_{ij} = 2\int_0^{t_f^*/2} u_j(t)^* \sin(\omega_i t)\,dt; \qquad i = n+1, n+2, ..., M; \qquad j = 0,1, ..., m$$

$$= 2\alpha_j (U_j/\omega_i)\{\cos(\omega_i t_f^*/2) - 2\cos(\omega_i t^j_{kj*}) + \cdots + (-1)^{kj+1}\} \tag{72}$$

$$i = n+1, n+2, ..., M; \qquad j = 0,1, ..., m \tag{73}$$

$$C_{ij} = 0; \qquad i = n+1, n+2, ..., M; \qquad j = 0,1, ..., m \tag{74}$$

Since, for this problem, for $t \geqslant t_f^*/2$, $x_2(t) = 0$, the constant residual energy can be expressed, using Eq. (74) in Eq. (71), as

$$\mathscr{E}_r \triangleq \mathscr{E}(t \geqslant t_f^*/2) = \left(\frac{1}{2}\right) \sum_{i=n+1}^{m} \left[\omega_i^2 \left\{\sum_{j=0}^{m} (\beta_j^i S_{ij})\right\}^2\right] \tag{75}$$

Note that, according to Eq. (47), $\Sigma_{j=0}^{m} (\beta_j^i S_{ij}) = 0$; $i = 1,2, ..., n$. For the RTRSP we can also characterize spillover via a pointing error of the rigid central body. Using Eqs. (16), (18), (66), (74), and noticing that, for $t \geqslant t_f^*/2$, $x_1(t) = \theta_f/2$, $x_3^i = 0$; $i = 1,2, ..., n$, we obtain

$$|\theta_e(t)| \triangleq |\theta(t) - \theta_f/2| = \left|(1/J_o) \sum_{i=n+1}^{M} \left[\left\{\omega_i \beta_0^i \sum_{j=0}^{m} (\beta_j^i S_{ij})\right\} \cos(\omega_i t)\right]\right| \tag{76}$$

$$\leqslant (I/J_o) \sum_{i=n+1}^{M} \left\{\omega_i |\beta_0^i| \sum_{j=0}^{m} |\beta_j^i S_{ij}|\right\}; \qquad t \geqslant t_f^*/2 \tag{77}$$

Spillover Characterization for the SUP

For this problem the optimal control variables, $u_j(t)^*$; $j = 0,1, ..., m$, are even functions of time. Therefore, Eqs. (68) and (69) yield

$$S_{ij} = 0; \qquad i = n+1, n+2, ..., M; \qquad j = 0,1, ..., m \tag{78}$$

$$C_{ij} = 2\int_0^{t_f^*/2} u_j(t)^* \cos(\omega_i t)\,dt; \qquad i = n+1, n+2, ..., M; \qquad j = 0,1, ..., m \tag{79}$$

$$= 2\alpha_j (U_j/\omega_i)\{\sin(\omega_i t_f^*/2) - 2\sin(\omega_i t^j_{kj*}) + \cdots + 2(-1)^{kj} \sin(\omega_i t^j_{1*})\}$$

$$i = n+1, n+2, ..., M; \qquad j = 0,1, ..., m \tag{80}$$

The boundary conditions for this problem require that $x_2(t) = \Omega/2$; $t \geqslant t_f^*/2$. Therefore, the constant residual energy can be expressed, using Eq. (78) in Eq. (71), as

$$\mathscr{E}_r \triangleq \mathscr{E}(t \geqslant t_f^*/2) = \frac{1}{2} \sum_{i=n+1}^{M} \left[\omega_i^2 \left\{ \sum_{j=0}^{m} (\beta_j^i C_{ij}) \right\}^2 \right] \tag{81}$$

Note that, according to Eq. (54), $\Sigma_{j=0}^{m} (\beta_j^i C_{ij}) = 0$; $i = 1,2, \ldots, n$. Like the RTRSP, we can also characterize control spillover for the SUP via the error in the angular velocity of the rigid central body *after* the maneuver has been completed. Using Eqs. (17), (19), (67) and (78) and noticing that, for $t \geqslant t_f^*/2$, $x_2(t) = \Omega/2$, $x_4^i = 0$; $i = 1,2, \ldots, n$, we obtain

$$|\Omega_e(t)| \triangleq |\dot{\theta}(t) - \Omega/2| = \left| (1/J_o) \sum_{i=n+1}^{M} \left[\left\{ \beta_0^i \omega_i^2 \sum_{j=0}^{m} (\beta_j^i C_{ij}) \right\} \cos(\omega_i t) \right] \right| \tag{82}$$

$$\leqslant (1/J_o) \sum_{i=n+1}^{M} \left(|\beta_0^i| \omega_i^2 \sum_{j=0}^{m} |\beta_j^i C_{ij}| \right); \qquad t \geqslant t_f^*/2 \tag{83}$$

Upper Bounds on Control Spillover

Equations (75) and (76) for the RTRSP and Eqs. (81) and (82) for the SUP are exact expressions for control spillover for the respective problems. These can also be viewed as measures of performance degradation due to residual dynamics. Upper bounds on these quantities can be obtained. The RTRSP will be treated first.

Consider Eqs. (72) and (73). From these equations three distinct upper bounds can be obtained on S_{ij}; $i = n+1, n+2, \ldots, M$; $j = 0,1, \ldots, m$:

$$|S_{ij}| \leqslant U_j t_f^*; \qquad i = n+1, n+2, \ldots, M; \qquad j = 0,1, \ldots, m \tag{84}$$

$$|S_{ij}| \leqslant 4(U_j/\omega_i)(1 + k^j); \qquad i = n+1, n+2, \ldots, M; \qquad j = 0,1, \ldots, m \tag{85}$$

Yet another upper bound can be obtained on $|S_{ij}|$. Suppose that the optimal maneuver time is such that for the ith mode it satisfies $k_i\pi \leqslant \omega_i t_f^*/2 \leqslant (k_i + 1)\pi$, where k_i is some nonnegative integer. It follows that

$$|S_{ij}| \leqslant 2U_j \int_0^{t_f^*/2} |\sin(\omega_i t)| \, dt = 2U_j[2k_i + \{1 - \cos(\omega_i t_f^*/2 - k_i\pi)\}]/\omega_i$$

$$< (2U_j t_f^*/\pi)[1 + 2\pi/(\omega_i t_f^*)]j \quad i = n+1, n+2, \ldots, M; \; j = 0,1, \ldots, m \tag{86}$$

Substituting Eqs. (84–86) into Eqs. (75) and (77), we obtain the following upper bounds on the spillover energy and the maximum pointing error of the rigid central body:

$$\mathscr{E}_r \leqslant \frac{1}{2} \sum_{i=n+1}^{M} \left[(\omega_i t_f^*)^2 \left\{ \sum_{j=0}^{m} (\beta_j^i U_j) \right\}^2 \right] \tag{87}$$

$$\mathscr{E}_r \leqslant 2 \sum_{i=n+1}^{M} \left[\sum_{j=0}^{m} \{\beta_j^i U_j (2 + 2k^j)\} \right]^2 \tag{88}$$

$$\mathscr{E}_r \leqslant (2/\pi^2) \sum_{i=n+1}^{M} \left[(\omega_i t_f^* + 2\pi)^2 \left\{ \sum_{j=0}^{m} (\beta_j^i U_j) \right\}^2 \right] \tag{89}$$

$$|\theta_e(t)| \leqslant (t_f^*/J_o) \sum_{i=n+1}^{M} \left\{ |\omega_i \beta_0^i| \sum_{j=0}^{m} |\beta_j^i U_j| \right\}; \qquad t \geqslant t_f^*/2 \tag{90}$$

$$|\theta_e(t)| \leqslant (2/J_o) \sum_{i=n+1}^{M} \left\{ |\beta_0^i| \sum_{j=0}^{m} |\beta_j^i U_j (2 + 2k^j)| \right\}; \qquad t \geqslant t_f^*/2 \tag{91}$$

$$|\theta_e(t)| \leqslant (2/\pi J_o) \sum_{i=n+1}^{M} \left\{ |\beta_0^i| (\omega_i t_f^* + 2\pi) \sum_{j=0}^{m} |\beta_j^i U_j| \right\}; \qquad t \geqslant t_f^*/2 \tag{92}$$

As in the RTRSP case, the following distinct upper bounds are obtained on the SUP performance degradation measures:

$$\mathscr{E}_r \leqslant \frac{1}{2} \sum_{i=n+1}^{M} \left[(\omega_i t_f^*)^2 \left\{ \sum_{j=0}^{m} (\beta_j^i U_j) \right\}^2 \right] \tag{93}$$

$$\mathscr{E}_r \leqslant 2 \sum_{i=n+1}^{M} \left[\sum_{j=0}^{m} \{\beta_j^i U_j (1 + 2k^j)\} \right]^2 \tag{94}$$

$$\mathscr{E}_r \leqslant (2/\pi^2) \sum_{i=n+1}^{M} \left[(\omega_i t_f^* + 2\pi)^2 \left\{ \sum_{j=0}^{m} (\beta_j^i U_j) \right\}^2 \right] \tag{95}$$

$$|\Omega_e(t)| \leqslant (t_f^*/J_o) \sum_{i=n+1}^{M} \left\{ |\omega_i^2 \beta_0^i| \sum_{j=0}^{m} |\beta_j^i U_j| \right\}; \qquad t \geqslant t_f^*/2 \tag{96}$$

$$|\Omega_e(t)| \leqslant (2/J_o) \sum_{i=n+1}^{M} \left\{ \omega_i |\beta_0^i| \sum_{j=0}^{m} |\beta_j^i U_j (1 + 2k^j)| \right\}; \qquad t \geqslant t_f^*/2 \tag{97}$$

$$|\Omega_e(t)| \leqslant (2/\pi J_o) \sum_{i=n+1}^{M} \left\{ |\beta_0^i \omega_i| (\omega_i t_f^* + 2\pi) \sum_{j=0}^{m} |\beta_j^i U_j| \right\}; \qquad t \geqslant t_f^*/2 \tag{98}$$

These expressions are further simplified when we consider the following special case.

Special Case of a Single Actuator Located at the Rigid Central Body

For this particular case (note that $j = 0$ here), the upper bounds can be expressed in closed form for an *infinite-dimensional evaluation model*

($M = \infty$). Let

$$\mathscr{S}_1 \triangleq \sum_{i=1}^{\infty} (\beta_0^i)^2, \qquad \mathscr{S}_2 \triangleq \sum_{i=1}^{\infty} (\beta_0^i \omega_i)^2, \qquad \text{and} \qquad \mathscr{S}_3 \triangleq \sum_{i=1}^{\infty} \{(\beta_0^i)^2 \omega_i\}$$

Modal relations, which are proven in Appendix B, enable us to express the upper bounds in the following closed forms for the RTRSP:

$$\mathscr{E}_r \leqslant \frac{1}{2}(U_0 t_f^*)^2 \left\{ \mathscr{S}_2 - \sum_{i=1}^{n} (\beta_0^i \omega_i)^2 \right\} \tag{99}$$

$$\mathscr{E}_r \leqslant 2U_0^2(2 + 2k^0)^2 \left\{ \mathscr{S}_1 - \sum_{i=1}^{n} (\beta_0^i)^2 \right\} \tag{100}$$

$$\mathscr{E}_r \leqslant (2/\pi^2)(U_0 t_f^*)^2 \left\{ (2\pi/t_f^*)^2 \mathscr{S}_1 + \mathscr{S}_2 + (4\pi/t_f^*)\mathscr{S}_3 - \sum_{i=1}^{n} [(\beta_0^i)^2(\omega_i + 2\pi/t_f^*)^2] \right\} \tag{101}$$

$$|\theta_e(t)| \leqslant (t_f^* U_0/J_o) \left\{ \mathscr{S}_3 - \sum_{i=1}^{n} \{(\beta_0^i)^2 \omega_i\} \right\}; \qquad t \geqslant t_f^*/2 \tag{102}$$

$$|\theta_e(t)| \leqslant 2\{U_0(2 + 2k^0)/J_o\} \left\{ \mathscr{S}_1 - \sum_{i=1}^{n} (\beta_0^i)^2 \right\}; \qquad t \geqslant t_f^*/2 \tag{103}$$

$$|\theta_e(t)| \leqslant (2/\pi)(t_f^* U_0/J_o) \left\{ (2\pi/t_f^*)\mathscr{S}_1 + \mathscr{S}_3 - \sum_{i=1}^{n} [(\beta_0^i)^2(\omega_i + 2\pi/t_f^*)] \right\} \quad t \geqslant t_f^*/2 \tag{104}$$

Similarly, we obtain the following for the SUP:

$$\mathscr{E}_r \leqslant \frac{1}{2}(U_0 t_f^*)^2 \left\{ \mathscr{S}_2 - \sum_{i=1}^{n} (\beta_0^i \omega_i)^2 \right\} \tag{105}$$

$$\mathscr{E}_r \leqslant 2U_0^2(1 + 2k^0)^2 \left\{ \mathscr{S}_1 - \sum_{i=1}^{n} (\beta_0^i)^2 \right\} \tag{106}$$

$$\mathscr{E}_r \leqslant (2/\pi^2)(U_0 t_f^*)^2 \left\{ (2\pi/t_f^*)^2 \mathscr{S}_1 + \mathscr{S}_2 + (4\pi/t_f^*)\mathscr{S}_3 - \sum_{i=1}^{n} [(\beta_0^i)^2(\omega_i + 2\pi/t_f^*)^2] \right\} \tag{107}$$

$$|\Omega_e(t)| \leqslant (t_f^* U_0/J^*) \left\{ \mathscr{S}_2 - \sum_{i=1}^{n} (\beta_0^i \omega_i)^2 \right\}; \qquad t \geqslant t_f^*/2 \tag{108}$$

$$|\Omega_e(t)| \leqslant \{2U_0(1+2k^0)/J_o\}\left\{\mathscr{S}_3 - \sum_{i=1}^{n}\{\omega_i(\beta_0^i)^2\}\right\}; \qquad t \geqslant t_f^*/2 \tag{109}$$

$$|\Omega_e(t)| \leqslant (2/\pi)(t_f^* U_0/J_o)\left\{\mathscr{S}_2 + (2\pi/t_f^*)\mathscr{S}_3 - \sum_{i=1}^{n}[(\beta_0^i\omega_i)^2(1+2\pi/\omega_i t_f^*)]\right\}$$
$$t \geqslant t_f^*/2 \tag{110}$$

Example Slewing Maneuvers

In the following we have modeled the flexible spacecraft using the rigid-body mode and the first 10 elastic modes (not all of which are actively controlled). There is only one control, and it is located at the rigid central body (i.e., $j = m = 0$). For simplicity we have assumed uniform, homogeneous appendages. We have chosen two flexible spacecraft configurations; one has purposely been chosen to be "stiffer" than the other. It is found that the nonlinear rotational stiffening effects are less significant for the stiffer structure. In fact, we find that, in all slewing examples for which these effects are less pronounced, the inequality (15) is satisfied to a greater extent. The two configurations to be examined here are detailed in Table 1 and will be addressed as spacecraft A and spacecraft B, respectively. To discretize the hybrid set of equations [Eqs. (1) and (2)], we have chosen the normalized cantilever mode shapes as the assumed mode shapes in the expansion (3). Results of the modal analysis appear in Table 2.

Table 1 Spacecraft dimensions and appendage specifications

	Spacecraft A	Spacecraft B
Radius of the rigid central body R, m	1.000	1.000
Mass of the rigid central body, kg	50.000	400.000
Number of appendages N	2	2
Appendage length L, m	5.000	40.000
Appendage flexural rigidity EI, N-m^2	328.300	1500.000
Appendage material density ρ, kg/m	3.080	0.04096

Table 2 Modal quantities

Spacecraft A: ($\bar{\sigma}[\Omega^{-2}U^T\{S-M\}U] = 0.331$)										
ω_i	3.861	12.013	27.745	51.839	84.241	124.869	173.692	230.692	295.863	369.206
β_0^i	13.152	4.214	1.221	0.457	0.212	0.114	0.068	0.044	0.030	0.021
Spacecraft B: ($\bar{\sigma}[\Omega^{-2}U^T\{S-M\}U] = 3.055$)										
ω_i	1.151	3.009	7.529	14.546	23.962	35.752	49.908	66.429	85.312	106.557
β_0^i	98.283	24.403	3.946	1.102	0.428	0.203	0.110	0.065	0.041	0.028

Table 3 Spacecraft B: RTRSP maneuver specifications

Total slewing angle (θ_f)	90.00 deg
Maximum available torque (T^o)	150.00 N-m

Table 4 Spacecraft B: optimal solutions for RTRSP

n	α_0^*	$t_f^*/2$	t_n^*	t_{n-1}^*	t_{n-2}^*	t_{n-3}^*
1	+1	4.725066	0.511785			
2	+1	4.728901	0.780724	0.523999		
3	+1	4.729219	0.851772	0.705614	0.233237	
4	+1	4.729224	0.853039	0.707974	0.240651	0.052748
n		p_0^{n*}	p_0^{n-1*}	p_0^{n-2*}	p_0^{n-3*}	p_0^{n-4*}
1		0.024030	−2.563338			
2		−0.009014	0.027104	−2.468772		
3		0.002907	−0.009237	0.027201	−2.463996	
4		−0.001353	0.003059	−0.009303	0.027205	−2.463581

We now consider some rest-to-rest slewing maneuver examples for spacecraft B and spin-up slewing examples for spacecraft A. The RTRSP maneuver specifications for configuration B are stated in Table 3. Four situations are now considered in which the first, the first two, the first three, and the first four vibrational modes are used, in addition to the rigid-body mode, to define the control model and hence are actively controlled. The optimal solutions (the switching times, the final time, and the costates at midmaneuver) appear in Table 4. The maneuver times ($t_1^*, t_2^*, \ldots, t_f^*/2$) apply to a rest-to-rest slewing maneuver *centered* about the time $t = 0$. The optimal control histories and switching functions for spacecraft B are shown in Fig. 2.

We apply the control histories listed in Table 4 to higher-order evaluation models, which include the rigid-body mode and first 10 elastic modes. We have considered two evaluation models: evaluation model 1, which is linear [Eqs. (11) and (12)], and evaluation model 2, which is nonlinear [Eqs. (6) and (7)]. The rigid central body attitude and attitude rate time histories are shown in Fig. 3. Figure 4 depicts the total energy time histories for spacecraft B. As a result of the unmodeled dynamics, the rigid central body continues to oscillate after the maneuver has been completed. This results in a pointing error of the rigid central body. The specific actuator location allows us to evaluate the upper bounds on this error and the energy spillover for the infinite-dimensional evaluation model using Eqs. (99–104). The actual maximum pointing error and the actual spillover energy values are compared with the lowest upper bounds, obtained using these equations, in Table 5. The actual values for maximum pointing error

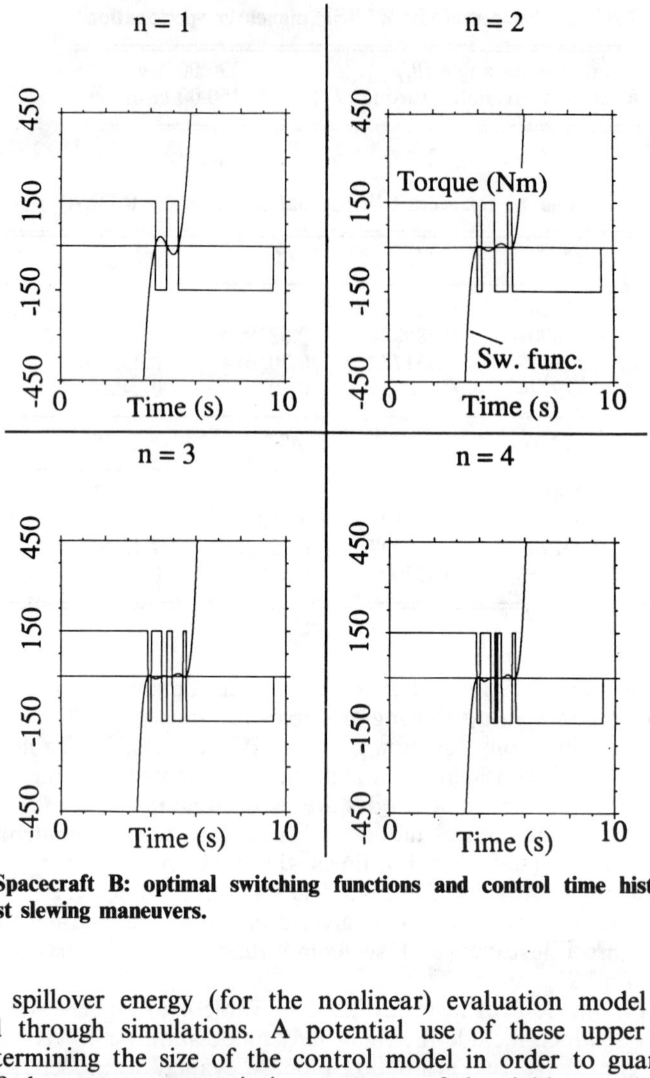

Fig. 2 Spacecraft B: optimal switching functions and control time histories for rest-to-rest slewing maneuvers.

and the spillover energy (for the nonlinear) evaluation model 2 were obtained through simulations. A potential use of these upper bounds is in determining the size of the control model in order to guarantee a prespecified postmaneuver pointing accuracy of the rigid central body or allowable postmaneuver energy.[14]

Next we consider time-optimal spin-up maneuver examples for spacecraft A. The maneuver specifications are stated in Table 6. As discussed earlier, four situations will be considered. The optimal solutions and related spillover calculations appear in Tables 7 and 8, respectively. In all of these cases the maneuvers are again found to have begun with $u(t) = +U_0$. As in RTRSP, the optimal maneuver times appear to approach a limit as we increase the size of the control model. Furthermore, in all of these SUP the number of times the control switches on $(0,t_f^*/2)$ is equal to n, the number of vibrational modes being actively suppressed at the final time. In

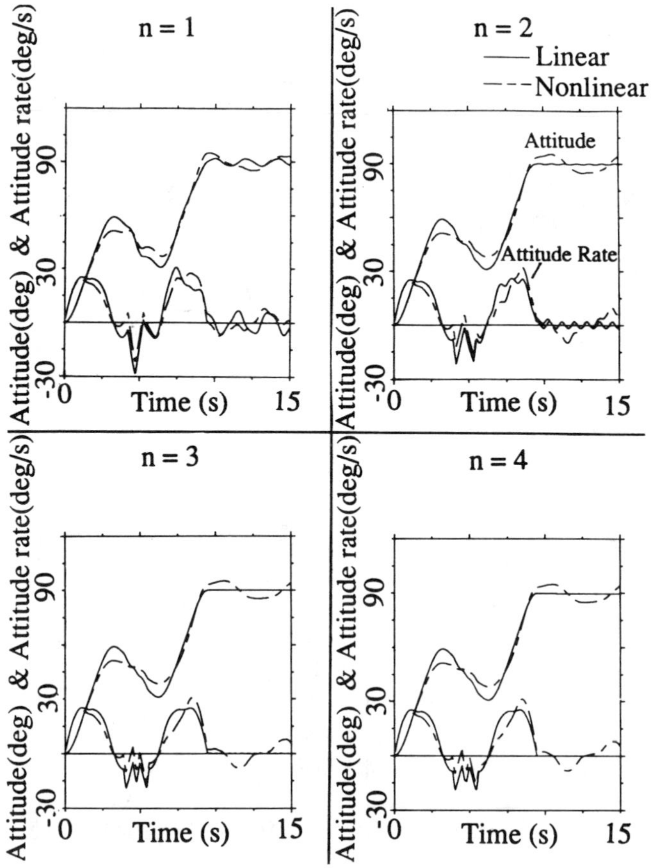

Fig. 3 Spacecraft B: rigid central body attitude and attitude rate time histories during rest-to-rest slewing maneuvers.

general, however, this is not always so. The optimal control histories and the corresponding switching functions appear in Fig. 5. The rigid central body attitude and attitude rate time histories for the two spacecrafts are shown in Fig. 6, and the total energy time histories for the SUP are shown in Fig. 7.

In the section on equations of motion, we offered a heuristic argument to justify the omission of the rotational stiffening nonlinearity from the vehicle equations of motion. It is noteworthy that, although the maximum maneuver angular velocities are not significantly different, approximately 23 deg/s for spacecraft A and 30 deg/s for spacecraft B, the contribution of the nonlinearity to the dynamics is markedly different for the two configurations. We note from Figs. 3 and 4 for the RTRSP and Figs. 6 and 7 for the SUP that, although the omission of the nonlinearity cannot be justified for spacecraft B, it can be for spacecraft A. In Table 9 we list the left-hand

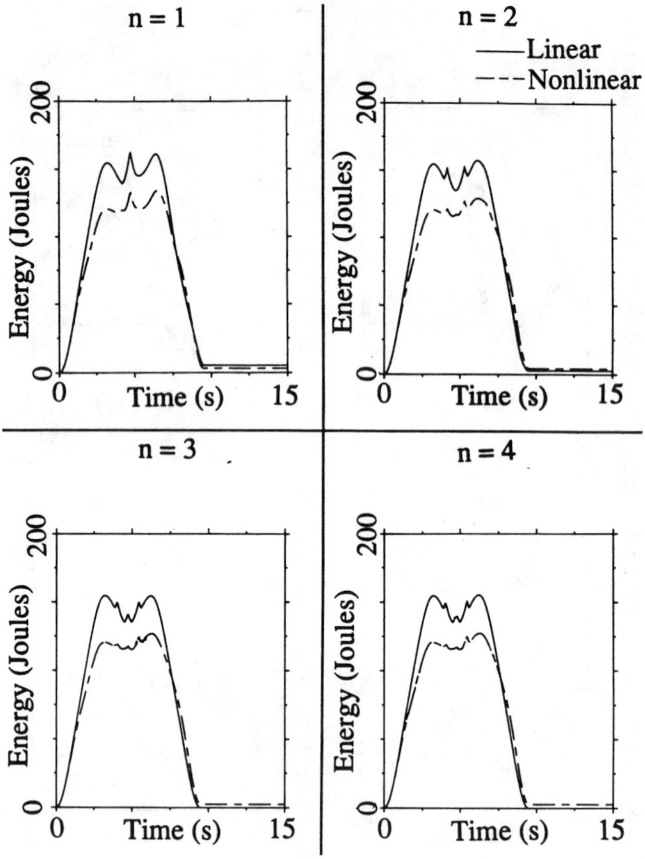

Fig. 4 Spacecraft B: total energy time histories during rest-to-rest slewing maneuvers.

side of the inequality (8) for various cases. The nonlinear evalution model values (the "actual" values) were used for max($\dot{\theta}^2$) and were obtained through simulations. Note that, although this inequality is satisfied for spacecraft A, it is clearly not the case for spacecraft B (for which the nonlinearity is seen to contribute significantly to the vehicle dynamics). We also note that, for spacecraft B, the nonlinear postmaneuver response is dominated by the *first* elastic mode.

Conclusions and Recommendations

In this chapter the problem of time-optimal slewing maneuvers of flexible spacecraft has been considered. Two distinct problems have been addressed: 1) the time-optimal rest-to-rest slewing problem and 2) the time-optimal spin-up problem. Control histories, which control the rigid-body mode and a finite number of undamped elastic modes, have been

Table 5 Spacecraft B: RTRSP spillover calculations

Maximum pointing error ($\max\lvert\theta_e(t)\rvert$; $t \geqslant t_f^*/2$), deg			
n	Exact: Evaluation model 1	Evaluation model 2	Upper bound
1	2.152763	5.012	9.716273
2	0.262894	5.678	0.405149
3	0.003752	5.123	0.046256
4	0.001899	5.025	0.009633
Spillover energy $\mathscr{E}_r$, J			
1	5.199126	2.903	101.748571
2	2.762780	4.205	6.364062
3	0.010969	2.512	0.968777
4	0.013577	2.635	0.252199

Table 6 Spacecraft A: SUP maneuver specifications

Initial angular velocity	Ω_1	0.00 deg/s
Final angular velocity	Ω_2	10.00 deg/s
Maximum available torque	T^o	80.00 N-m

Table 7 Spacecraft A: optimal solutions for the SUP

n	α_0^*	$t_f^*/2$	t_n^*	t_{n-1}^*	t_{n-2}^*	t_{n-3}^*
1	+1	0.657606	0.074385			
2	+1	0.660783	0.118730	0.042756		
3	+1	0.661042	0.136116	0.095215	0.035202	
4	+1	0.661081	0.143023	0.114233	0.066214	0.018882
n		p_0^{n*}	p_0^{n-1*}	p_0^{n-2*}	p_0^{n-3*}	p_0^{n-4*}
1		0.248650	−3.136201			
2		−0.100409	0.260871	−3.015703		
3		0.025009	−0.116302	0.264135	−3.011920	
4		−0.007146	0.027617	−0.118478	0.264681	−3.012618

Table 8 Spacecraft A: SUP spillover calculations

n	Exact: Evaluation model 1	Evaluation model 2	Upper bound
	Maximum pointing rate error ($\max\|\Omega_e(t)\|$; $t \geqslant t_f^*/2$), deg/s		
1	10.123090	8.645	49.899782
2	4.221229	3.735	27.427336
3	3.049957	2.901	17.268792
4	0.996945	1.345	12.286334
	Spillover energy $\mathscr{E}_r$, J		
1	0.844622	0.879	10.337596
2	0.305707	0.309	2.596384
3	0.335653	0.335	0.793348
4	0.044964	0.039	0.314761

Table 9 $\bar{\sigma}[\Omega^{-2}U^T\{S-M\}U] \times \max(\dot{\theta}^2)$ values

n	1	2	3	4
Spacecraft A	0.036	0.035	0.036	0.034
Spacecraft B	0.717	1.013	0.906	0.879

characterized through the control switching times and the final time. The optimal control time histories have been shown to possess important time symmetry properties. As a result, in order to solve the complete problem, it is sufficient to consider and solve only one-half of the maneuver problem. Typically, the optimal control histories have been found to exhibit the following pattern:

1) The optimal *maneuver time approaches a limit from below* as we increase the size of the control model.

2) The *number of control switchings increases as we increase the size* of the control model.

3) For the RTRSP, in between the initial time and the final time, all control switchings are concentrated around the midmaneuver time; actually, the numerical evidence suggests that, as the size of the control model is increased, the *control switchings are dense in the neighborhood of the midmaneuver time*.

4) For the SUP, in between the initial time and the final time, all control switchings are concentrated around and are dense in the neighborhood of the midmaneuver time as we increase the size of the control model *only* for small values of the SUP maneuver parameter. For a given control model the number of switchings increases as the maneuver parameter is increased, and, in general, additional control switchings are found *away* from the midmaneuver time.

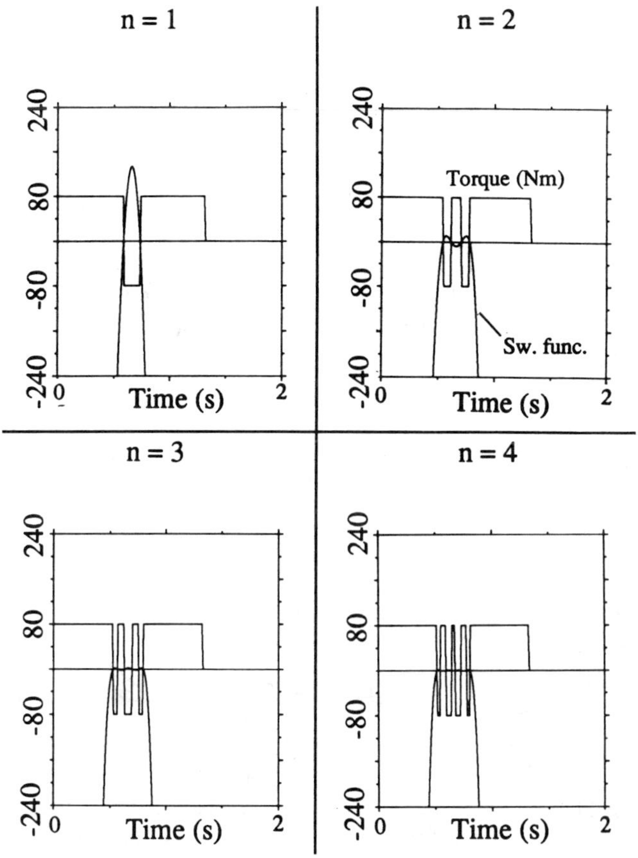

Fig. 5 Spacecraft A: optimal switching functions and control time histories for spin-up slewing maneuvers.

These observations lead us to believe that the infinite-dimensional time-optimal control problems are ill-posed. We have stated the necessary and sufficient conditions for optimality in the form of a system of nonlinear algebraic equations in costates at the midmaneuver time, the switching times, and the final time before using the homotopy principle on these equations to obtain the solution.

Control spillover appears naturally when the control, which is optimal for the finite-dimensional control model, is applied to the actual structure. We have defined postmaneuver spillover measures and obtained upper bounds on these measures. Where possible, we have also obtained closed-form expressions for the upper bounds on the *infinite-dimensional control spillover measures*. A potential application of these measures is in the determination of the *size* of the control model in order to meet the stated performance objective, e.g., the desired postmaneuver accuracy and allowable spillover energy.[14]

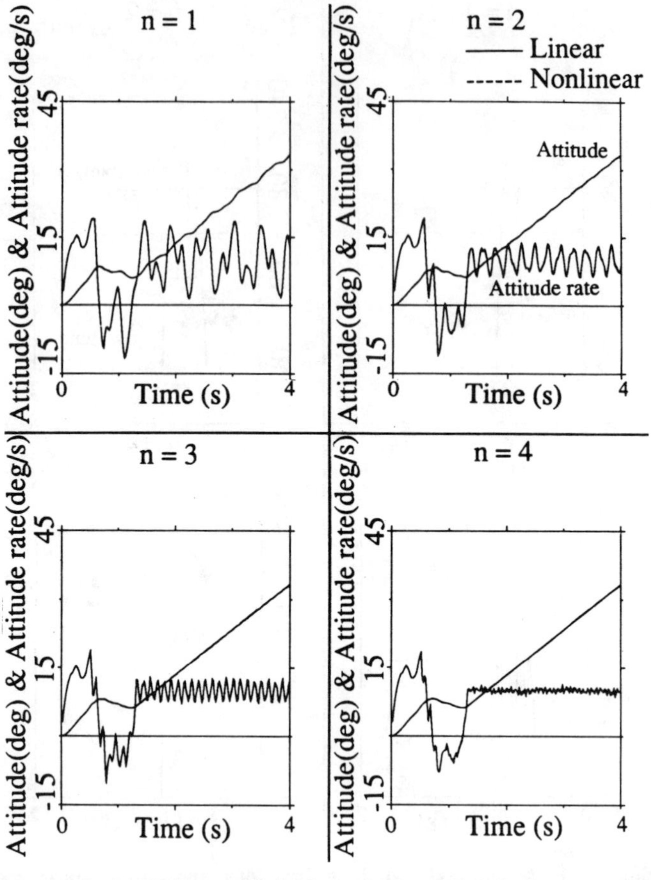

Fig. 6 Spacecraft A: rigid central body attitude and attitude rate time histories during spin-up slewing maneuvers.

A decoupling of the time-optimal control problem from the structural design problem is found to occur for the scalar control case. It is found that the control history is independent of (torque) actuator locations for this particular case.

It is a straightforward exercise to derive the sensitivity of the final state to structural parameter variations.[15] It is noteworthy that the higher the mode number the smaller are these sensitivities *if the actuator is located at the rigid central body*. This is fortunate because of the imprecise knowledge of structural parameters for higher modes.

In our numerical studies we have found for large values of the parameter $\omega_1^2\theta_f/U_0$ for the RTRSP and the parameter $\omega_1\Omega/U_0$ that, for the SUP, the time to slew the flexible spacecraft is not significantly different from the time to slew an equivalent rigid body. This is apparently true regardless of the number of modes actively suppressed at the final time.

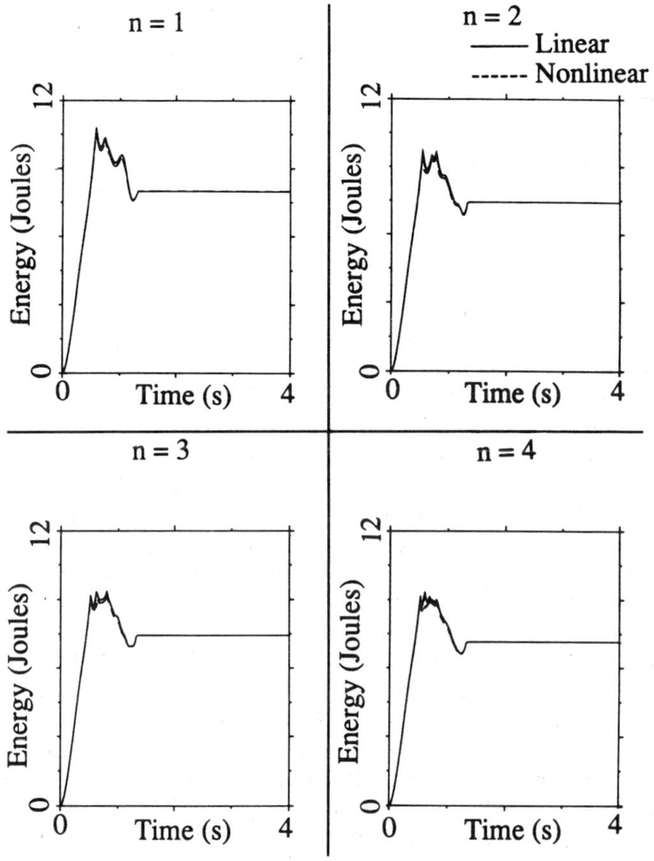

Fig. 7 Spacecraft A: total energy time histories during spin-up slewing maneuvers.

However, there are some limitations of the present work. The analysis here has been restricted to linear, elastic, undamped, nongyroscopic systems, which are fairly good models of flexible structures rotating *slowly* and without any rotating substructures. The small angular velocities of the rigid central body constitutes the main assumption of this work. This might fail to hold in practice where large angular velocities are attained during a maneuver. The question, therefore, arises concerning how large the angular velocities must be in order to contribute *significantly* to the overall motions through the nonlinear rotational stiffening effects. We have attempted to answer the question of "how large is *large*" by comparing the maximum angular velocity attained during a maneuver with the maximum singular value of a matrix whose entries depend on the appendage geometry, stiffness, and mass distributions. Roughly speaking, if the natural stiffness of a flexible structure is *large* compared to the additional stiffening of the flexible structure that takes place due to the angular motions, the dynamic stiffening will contribute little to the vehicle motions and therefore can be neglected. There are many physical situations where the results of the linearized analysis presented in this chapter are found to be valid and entirely applicable. By the same token, there are also situations where this is not the case. In these cases, for an accurate analysis the nonlinear dynamic stiffness terms must be retained in the vehicle equations of motion. In solving the resulting, more difficult, nonlinear time-optimal control problem, one would normally need a good initial guess for the eventual optimal control history. It is hoped that the results of the linearized analysis will provide precisely such a guess to initiate a numerical algorithm.

The other assumption regarding no natural damping was made, in principle, because the damping is small and poorly known at best. Moreover, the presence of damping destroys the symmetry property of the time-optimal control, and it is no longer sufficient to consider one-half of the maneuver problem in order to solve the complete problem. However, if the damping terms are to be included in the formulation, following the solution procedure outlined here, one can state the necessary and sufficient conditions for optimality in the form of nonlinear algebraic equations and subsequently apply the homotopy principle to obtain solutions. However, for a control model of the same order, there are approximately twice as many unknowns to be dealt with.

A potential difficulty encountered in solving the linear optimal control problems posed in this work has to do with the two quantities that need to be guessed to start the numerical algorithm: the initial control sign and the number of control switchings. These are not known a priori. In our scalar control examples we found that the initial control sign is positive and for the RTRSP the number of control switchings on one-half of the maneuver interval is equal to the number of vibrational modes suppressed at the final time. For the SUP this is the case only for small values of the SUP maneuver parameter. It is these two observations on which we have based the algorithm startup procedure. However, for multiple input-multiple mode cases we foresee complications because one would then need to *guess* the initial sign and the number of control switchings for *each control input*. However, the difficulty is allayed somewhat when one realizes that the solution to the necessary and sufficient conditions exists and is *unique*.

The aforementioned difficulties and limitations notwithstanding, some of the results obtained here have obvious implications in regard to future work. The following two research directions appear to be most promising. As a result of the separation of the time-optimal control problem from the structural optimization problem for the scalar control case, it makes sense to optimize the actuator location in order to optimize some structural excitation criterion (without affecting the optimal control history). If the actuators are placed at the "optimum" location, we will have minimized not only the maneuver time but also a measure of control spillover. The other implication has to do with the design of the flexible structure itself, under some feasible design constraints, to minimize the optimal maneuver time. This problem is obviously the more difficult of the two and relies on evaluation of design sensitivities (some of which can be borrowed from the existing literature in structural mechanics). A flexible structure optimized in this fashion will be superior to a nonoptimized structure in that it will require less time to slew.

In conclusion, we note that the intent of this work was to solve the problem of pure time-optimal control of planar slewing maneuvers of flexible spacecraft. Although the more general three-dimensional maneuver problem is much more difficult, the results of the present work, even though restricted somewhat for reasons we have underlined, are original and can be applied to many physical situations. From the application standpoint these results are applicable to all linear, undamped, nongyroscopic systems possessing a single rigid-body mode and a finite number of vibrational modes. It is our hope that the insight gained into the flexible spacecraft maneuver control problems, especially the *nature* of optimal solutions, will prove to be useful in future undertakings in this field.

Appendix A: Properties of Time-Optimal Control

Theorem 1. For the RTRSP the optimal control $\boldsymbol{u}(t)^*$ is an odd function of time, i.e.,

$$\boldsymbol{u}(t)^* = -\boldsymbol{u}(-t)^*; \qquad -t_f^*/2 \leqslant t \leqslant t_f^*/2 \tag{A.1}$$

where t_f^* is the optimal maneuver time.

Proof. Normality of the pair (A,B) guarantees that an optimal control exists[19,20] and is unique. The rest of the proof demonstrates that the unique optimal control $\boldsymbol{u}(t)^*$ satisfies the symmetry property (A.1).

Let $t_f^*, \boldsymbol{x}(t)^*, \boldsymbol{u}(t)^*, t \in [-t_f^*/2, t_f^*/2]$ denote the optimal time, optimal state, and the optimal control, respectively. There exists a vector function $\boldsymbol{p}(t) = \{p^0(t), q^0(t), p^1(t), q^1(t), \ldots, p^n(t), q^n(t)\}^T, t \in [-t_f^*/2, t_f^*/2]$ such that

$$\dot{\boldsymbol{x}}(t)^* = A\boldsymbol{x}(t)^* + B\boldsymbol{u}(t)^*; \qquad t \in [-t_f^*/2, t_f^*/2] \tag{A.2}$$

$$\dot{\boldsymbol{p}}(t) = -A^T\boldsymbol{p}(t); \qquad t \in [-t_f^*/2, t_f^*/2] \tag{A.3}$$

$$u_j(t)^* = -U_j \operatorname{sgn}\left[\sum_{i=0}^{n} \{\beta_j^i q^i(t)\}\right]; \quad t \in [-t_f^*/2, t_f^*/2]; \quad 0 \leqslant j \leqslant m \tag{A.4}$$

$$1 + \boldsymbol{p}(t)^T\{A\boldsymbol{x}(t)^* + B\boldsymbol{u}(t)^*\} \equiv 0; \qquad t \in [-t_f^*/2, t_f^*/2] \tag{A.5}$$

$$\boldsymbol{x}(-t_f^*/2)^* = \{-\theta_f/2, 0, 0, 0, \ldots, 0, 0\}^T \tag{A.6}$$

$$x(t_f^*/2)^* = \{\theta_f/2, 0, 0, 0, \ldots, 0, 0\}^T \tag{A.7}$$

Existence and uniqueness theorems[19,20] guarantee that Eqs. (A.2–A.8) have the unique solution $t_f^*, \boldsymbol{x}(t)^*, \boldsymbol{u}(t)^*, \boldsymbol{p}(t)^*;\ t \in [-t_f^*/2, t_f^*/2]$.

Now consider the related time-optimal control problem:
Minimize

$$\mathscr{I} = \int_{-t_f/2}^{0} \mathrm{d}t \tag{A.8}$$

subject to

$$\dot{\boldsymbol{x}}(t) = A\boldsymbol{x}(t) + B\boldsymbol{u}(t); \qquad t \in [-t_f/2, 0] \tag{A.9}$$

$$|u_j(t)| \leqslant U_j; \qquad j = 0,1,2,\ldots,m; \qquad t \in [-t_f/2, 0] \tag{A.10}$$

$$\boldsymbol{x}(-t_f/2) = \{-\theta_f/2, 0, 0, 0, \ldots, 0, 0\}^T \tag{A.11}$$

$$\boldsymbol{x}(0) \in \mathscr{X} \triangleq \{\{0, x_2^L, 0, x_4^{1L}, 0, x_4^{2L}, \ldots, 0, x_4^{nL}\}^T |\, x_2^L, x_4^{iL} \in \mathscr{R};\ i = 1,2,\ldots,n\} \tag{A.12}$$

where $x_2^L, x_4^{iL};\ i = 1,2,\ldots,n$, are *free*. As in the original problem, it can be shown that this problem also has a *unique* optimal solution. Denote this solution by $t_L^*, x_L(t)^*, \boldsymbol{u}_L(t)^*;\ t \in [-t_L^*/2, 0]$. Therefore, there exists a vector function $\boldsymbol{p}_L(t) \triangleq \{p_0^L(t), q_0^L(t), p_1^L(t), q_1^L(t), \ldots, p_n^L(t), q_n^L(t)\}^T$; $t \in [-t_L^*/2, 0]$ such that

$$\dot{\boldsymbol{x}}_L(t)^* = A\boldsymbol{x}_L(t)^* + B\boldsymbol{u}_L(t)^*; \qquad t \in [-t_L^*/2, 0] \tag{A.13}$$

$$\dot{\boldsymbol{p}}_L(t) = -A^T\boldsymbol{p}_L(t); \qquad t \in [-t_L^*/2, 0] \tag{A.14}$$

$$u_{jL}(t)^* = -U_j \operatorname{sgn}\left[\sum_{i=0}^{n} \{\beta_j^i q_L^i(t)\}\right]; \qquad t \in [-t_L^*/2, 0]; \qquad 0 \leqslant j \leqslant m \tag{A.15}$$

$$1 + \boldsymbol{p}_L(t)^T\{A\boldsymbol{x}_L(t)^* + B\boldsymbol{u}_L(t)^*\} \equiv 0; \qquad t \in [-t_L^*/2, 0] \tag{A.16}$$

$$\boldsymbol{p}_L(0) \in \{\{p_0^L, 0, p_1^L, 0, \ldots, p_n^L, 0\}^T |\, p_i^L \in \mathscr{R}; \qquad i = 0,1,\ldots,n\} \tag{A.17}$$

$$\boldsymbol{x}_L(-t_L^*/2)^* = \{-\theta_f/2, 0, 0, 0, \ldots, 0, 0\}^T \tag{A.18}$$

$$\boldsymbol{x}_L(0)^* \in \mathscr{X} \tag{A.19}$$

Define

$$\tilde{\boldsymbol{u}}(t) = \boldsymbol{u}_L(t)^*; \qquad t \in [-t_L^*/2, 0] \tag{A.20}$$

$$= -\boldsymbol{u}_L(-t)^*; \qquad t \in [0, t_L^*/2] \tag{A.21}$$

and

$$\tilde{x}(t),\tilde{p}(t) \triangleq \{\tilde{p}^0(t),\tilde{q}^0(t), \tilde{p}^1(t),\tilde{q}^1(t), \ldots, \tilde{p}^n(t),\tilde{q}^n(t)\}^T; \qquad t \in [-t_L^*/2, t_L^*/2]$$

by

$$\dot{\tilde{x}}(t) = A\tilde{x}(t) + B\tilde{u}(t); \qquad \tilde{x}(-t_L^*/2) = \{-\theta_f/2, 0, 0, 0, \ldots, 0, 0\}^T \quad \text{(A.22)}$$

$$\dot{\tilde{p}}(t) = -A^T p(t); \qquad \sim p(0) = p_L(0) \quad \text{(A.23)}$$

Upon integration the solutions $\tilde{x}(t),\tilde{p}(t);\ t \in [0, t_L^*/2]$ of Eqs. (A.22) and (A.23), satisfy the following:

$$\tilde{x}(t_L^*/2) = \{\theta_f/2, 0, 0, 0, \ldots, 0, 0\}^T$$

$$\tilde{u}_j(t) = -U_j \operatorname{sgn}\left[\sum_{i=0}^{n} \{\beta_j^i q^i(t)\}\right]; \qquad t \in [-t_L^*/2, t_L^*/2]; \qquad 0 \leqslant j \leqslant m$$

$$1 + \tilde{p}(t)^T\{A\tilde{x}(t) + B\tilde{u}(t)\} \equiv 0; \qquad t \in [-t_L^*/2, t_L^*/2]$$

By construction, therefore, $t_L^*, \tilde{x}(t), \tilde{u}(t), \tilde{p}(t);\ t \in [-t_L^*/2, t_L^*/2]$ satisfy Eqs. (A.2–A.7). The uniqueness of the solution of Eqs. (A.2–A.7) implies that

$$t_f^* = t_L^*$$

$$x(t)^* = \tilde{x}(t); \qquad t \in [-t_f^*/2, t_f^*/2]$$

$$p(t)^* = \tilde{p}(t); \qquad t \in [-t_f^*/2, t_f^*/2]$$

$$u(t)^* = \tilde{u}(t); \qquad t \in [-t_f^*/2, t_f^*/2]$$

Equations (A.20) and (A.21) then imply that $u(t)^* = -u(-t)^*$; $t \in [-t_f^*/2, t_f^*/2]$.

Corollary 1. For the RTRSP the optimal costate vector satisfying Eq. (33) is such that

$$p(0) = \{p_0^0, 0, p_0^1, 0, \ldots, p_0^n, 0\}^T \quad \text{(A.24)}$$

Proof. Equation (A.24) follows directly from *Theorem* 1, since $p(t)^* = \tilde{p}(t)$; therefore, $p(0)^* = \tilde{p}(0) = \{p_0^L, 0, p_1^L, 0, \ldots, p_n^L, 0\}^T$.

Theorem 2. Suppose that for some integers k^j; $0 \leqslant j \leqslant m$, Eqs. (46), (47), and (50) admit a solution ($t_r^j, t_f/2, p_0^i$; $1 \leqslant r \leqslant k^j$; $0 \leqslant j \leqslant m$; $0 \leqslant i \leqslant n$) satisfying inequalities (44) and (51). Then

(i) The jth component of the optimal control solving the RTRSP switches exactly $2k^j+1$ times at times $-t^j_{kj}, -t^j_{kj-1}, \ldots, -t^j_1, 0, t^j_1, \ldots, t^j_{kj-1}, t^j_{kj}$.

(ii) The optimal maneuver time of the RTRSP is exactly t_f.

Proof. Suppose that for some integers k^j; $0 \leqslant j \leqslant m$, Eqs. (46), (47), and (50) admit a solution $(t^j_r, t_f/2, p^i_0;\ 1 \leqslant r \leqslant k^j;\ 0 \leqslant j \leqslant m;\ 0 \leqslant i \leqslant n)$ satisfying inequalities (44) and (51). Using the solution for costates at midmaneuver time, i.e., p^i_0; $0 \leqslant i \leqslant n$, we obtain a costate vector $\boldsymbol{p}(t) \in \mathscr{R}^{2n+2}$ satisfying Eqs. (33) and (A.24). Using this $\boldsymbol{p}(t)$ we define $\boldsymbol{u}(t) \in \mathscr{R}^{m+1}$ by $u_j(t) = -U_j \operatorname{sgn}\{\boldsymbol{p}(t)^T \boldsymbol{b}_j\}$; $0 \leqslant j \leqslant m$; $t \in [-t^*_f/2, t^*_f/2]$. It is easily verified that the functions $\boldsymbol{p}(t)^T \boldsymbol{b}_j$ vanish exactly at instants $-t^j_{kj}, -t^j_{kj-1}, \ldots, -t^j_1, 0, t^j_1, \ldots, t^j_{kj-1}, t^j_{kj}$; $0 \leqslant j \leqslant m$. Moreover, between times $-t_f/2$ and $+t_f/2$, these functions have no other zeros because of Eq. (51). Therefore, $\boldsymbol{u}(t)$ constructed above is piecewise constant such that $u_j(t)$ switches only at instants $(0, \pm t^j_r;\ 1 \leqslant r \leqslant k^j)$. Integrating Eq. (37) with initial conditions [Eq. (31a)] yields

$$x_1(t_f/2) = \sum_{j=0}^{m} [\alpha_j U_j[(t_f/2)^2 - 2(t^j_{kj})^2 + \cdots + 2(-1)^{kj}(t^j_1)^2]] - \theta_f/2$$

$$x_2(t_f/2) = 0$$

$$x^i_3(t_f/2) = -(2/\omega_i) \sum_{j=0}^{m} [\alpha_j \beta^i_j U_j \{\cos(\omega_i t_f/2) - 2\cos(\omega_i t^j_{kj}) + \cdots + 2(-1)^{kj}\cos(\omega_i t^j_1) + (-1)^{kj+1}\}] \cos(\omega_i t_f/2); \qquad i = 1,2, \ldots, n$$

$$x^i_4(t_f/2) = (2/\omega_i) \sum_{j=0}^{m} [\alpha_j \beta^i_j U_j \{\cos(\omega_i t_f/2) - 2\cos(\omega_i t^j_{kj}) + \cdots + 2(-1)^{kj}\cos(\omega_i t^j_1) + (-1)^{kj+1}\}] \sin(\omega_i t_f/2); \qquad i = 1,2, \ldots, n$$

But since t^j_r; $1 \leqslant r \leqslant k^j$; $0 \leqslant j \leqslant m$, $t_f/2$, satisfy Eqs. (46) and (47), the preceding equations yield $\boldsymbol{x}(t_f/2) = \{\theta_f/2, 0, 0, 0, \ldots, 0, 0\}^T$, the desired final state. As a consequence, the $\boldsymbol{p}(t)$ and $\boldsymbol{u}(t)$ constructed earlier satisfy all necessary and sufficient conditions for optimality and are therefore the optimal costate and control, respectively.

Corollary 2. Suppose integers k^j; $0 \leqslant j \leqslant m$ satisfy $\Sigma^m_{j=0}(k^j) = n$. Also assume that Eqs. (46) and (47) admit a solution $(t^j_r, t_f/2;\ 1 \leqslant r \leqslant k^j;\ 0 \leqslant j \leqslant m)$, which is regular[21] and satisfies Eq. (50) and inequalities (44) and (51). Then claims (i) and (ii) of *Theorem* 2 are true.

Proof. Suppose Eqs. (46) and (47) admit a regular solution satisfying Eq. (44), as assumed. Consider the vector equation, Eq. (50).

Rewriting Eq. (47) as

$$(-2/\omega_i)\sum_{j=0}^{m}[\alpha_j\beta_j^i U_j\{\cos(\omega_i t_f/2)-2\cos(\omega_i t_{kj}^j)+\cdots+(-1)^{k^j+1}\}]=0;$$
$$i=1,2,\ldots,n \qquad \text{(A.25)}$$

it is seen that the coefficient matrix in Eq. (50) is exactly the transpose of the Jacobian of Eqs. (46) and (A.25). Regularity of the solution $t_f/2, t_r^j$; $1 \leqslant r \leqslant k^j$; $0 \leqslant j \leqslant m$, of Eqs. (46) and (47) implies existence and uniqueness of the solution of Eq. (50). Using this solution a costate vector $\boldsymbol{p}(t)$ satisfying Eq. (33) can be constructed. Satisfaction of Eq. (51) by the costate vector thus obtained implies that all of the necessary and sufficient conditions are met. The costate thus constructed is therefore the optimal costate, and the optimal control is characterized by the regular solution of Eqs. (46) and (47), which satisfies Eqs. (44), (50), and (51).

Appendix B: Modal Relations

Modal identities, used in deriving the infinite-dimensional upper bounds in the section on spillover analysis, are presented here. The equations of motion that govern the linearized rotational dynamics of a flexible spacecraft, controlled by a single torque actuator located at the rigid central body, consist of the following:

$$J_o\ddot{\theta}+\sum_{i=1}^{M}(m_i\ddot{q}_i)=T \qquad \text{(B.1)}$$

$$N(\ddot{q}_i+\tilde{\omega}_i^2 q_i)+m_i\ddot{\theta}=0; \qquad i=1,2,\ldots,M \qquad \text{(B.2)}$$

In writing Eq. (B.2) we have assumed that the assumed mode shapes, $\phi_i(x)$; $i=1,2,\ldots,M$, satisfy the following orthogonality relationships:

$$\int_0^L \rho(x)\phi_i(x)\phi_j(x)\,\mathrm{d}x=\delta_{ij}, \quad \int_0^L EI(x)\phi_{i,xx}(x)\phi_{j,xx}(x)\,\mathrm{d}x=\tilde{\omega}_i^2\delta_{ij};$$
$$i,j=1,2,\ldots$$

where δ_{ij} is a Kronecker delta. We further assume that the set of eigenfunctions $\{\phi_i\}$ is *complete*.[22] For a *complete orthonormal* system $\{\phi_i\}$, Hughes[17] has shown that the following modal identities exist for coefficients $m_i, \tilde{\omega}_i$, appearing in the models (B.1) and (B.2):

$$\sum_{i=1}^{\infty}(m_i/\tilde{\omega}_i)^2=N^2\int_0^L\left[\rho(y)(R+y)\left\{\int_0^L\rho(x)(R+x)\mathscr{F}(x,y)\,\mathrm{d}x\right\}\right]\mathrm{d}y \qquad \text{(B.3)}$$

$$\sum_{i=1}^{\infty}(m_i)^2=NJ_E \qquad \text{(B.4)}$$

where J_E is the total rotational inertia of the undeformed flexible appendages about the rotation axis, and $\mathscr{F}(x,y)$, a symmetric positive-definite function, is the flexibility kernel. Another proof of Eq. (B.4) is forwarded by Likins.[24]

Proceeding as in the section on formulation of linear time-optimal control problems, we can obtain the state space model of the plant [Eqs. (B.1) and (B.2)]. We let $\boldsymbol{q} = U\boldsymbol{\eta}$, where $\boldsymbol{\eta} \triangleq \{\eta_1, \eta_2, ..., \eta_M\}^T$ is a normal coordinate vector and the $M \times M$ matrix U satisfies the following:

$$U^T[NI - (1/J_o)\boldsymbol{m}\boldsymbol{m}^T]U = I \tag{B.5}$$

$$U^T K U = \Omega^2 = [\mathrm{diag}\{\omega_i^2; \qquad i = 1,2, ..., M\}] \tag{B.6}$$

Furthermore, let $\beta_0 \triangleq \{\beta_0^1, \beta_0^2, ..., \beta_0^M\}^T$. Definition (22) therefore implies that

$$\boldsymbol{\beta}_0 \triangleq -\Omega^{-1}U^T\boldsymbol{m} \tag{B.7}$$

Lemma 1. There exists the following relations,

(i)
$$\mathscr{S}_1 \triangleq \sum_{i=1}^{\infty} (\beta_0^i)^2$$
$$= N\int_0^L \left[\rho(y)(R+y)\left\{\int_0^L \rho(x)(R+x)\mathscr{F}(x,y)\,\mathrm{d}x\right\}\right]\mathrm{d}y \tag{B.8}$$

(ii)
$$\mathscr{S}_2 \triangleq \sum_{i=1}^{\infty} (\beta_0^i\omega_i)^2 = (1 + J_E/J_R)J_E \tag{B.9}$$

(iii)
$$\mathscr{S}_3 \triangleq \sum_{i=1}^{\infty} \{(\beta_0^i)^2\omega_i\} \leqslant \left\{\sum_{i=1}^{\infty} \{\beta_0^i\omega_i\}^2\right]^{1/2}\left[\sum_{i=1}^{\infty} \{\beta_0^i\}^2\right]^{1/2} \tag{B.10}$$

where (ii) holds only if $J_R \neq 0$.

Proof. In order to prove Eqs. (B.8) and (B.9), we need to show that the following hold:

$$\left(I + \frac{\boldsymbol{m}\boldsymbol{m}^T}{NJ_o - \boldsymbol{m}^T\boldsymbol{m}}\right)^{-1} = \left(I - \frac{\boldsymbol{m}\boldsymbol{m}^T}{NJ_o}\right); \qquad NJ_o \neq \boldsymbol{m}^T\boldsymbol{m} \tag{B.11}$$

$$\boldsymbol{\beta}_0^T\boldsymbol{\beta}_0 = \boldsymbol{m}^T K^{-1}\boldsymbol{m} \tag{B.12}$$

$$\boldsymbol{\beta}_0^T\Omega^2\boldsymbol{\beta}_0 = J_o\left(\frac{\boldsymbol{m}^T\boldsymbol{m}}{NJ_o - \boldsymbol{m}^T\boldsymbol{m}}\right); \qquad NJ_o \neq \boldsymbol{m}^T\boldsymbol{m} \tag{B.13}$$

Equation (B.11) follows immediately from the *matrix inversion lemma.*[23] In Eq. (B.11) the matrix on the left-hand side is invertible if $NJ_o \neq \boldsymbol{m}^T\boldsymbol{m}$, since $|I - \boldsymbol{m}\boldsymbol{m}^T/(NJ_o)| = 1 - \boldsymbol{m}^T\boldsymbol{m}/(NJ_o)$. Equality (B.12) follows from Eqs. (B.7) and (B.6):

$$\boldsymbol{\beta}_0^T\boldsymbol{\beta}_0 \overset{(B.7)}{=} \boldsymbol{m}^T U\Omega^{-2}U^T\boldsymbol{m} \overset{(B.6)}{=} \boldsymbol{m}^T K^{-1}\boldsymbol{m}$$

To prove Eq. (B.13), we proceed as follows:

$$UU^T \overset{(B.5)}{=} (1/N)[I - \boldsymbol{mm}^T/(NJ_o)]^{-1} \tag{B.14a}$$

$$\overset{(B.11)}{=} (1/N)[I + \boldsymbol{mm}^T/(NJ_o - \boldsymbol{m}^T\boldsymbol{m})] \tag{B.14b}$$

Hence,

$$\boldsymbol{\beta}_0^T\Omega^2\boldsymbol{\beta}_0 \overset{(B.6)}{=} \boldsymbol{m}^T UU^T\boldsymbol{m} \overset{(B.14)}{=} (1/N)\boldsymbol{m}^T[I + \boldsymbol{mm}^T/(NJ_o - \boldsymbol{m}^T\boldsymbol{m})]\boldsymbol{m}$$

$$= \left(\frac{J_o\boldsymbol{m}^T\boldsymbol{m}}{NJ_o - \boldsymbol{m}^T\boldsymbol{m}}\right)$$

We have thus far established the following for the finite-dimensional model

$$\sum_{i=1}^{M} (\beta_0^i)^2 \overset{(B.12)}{=} (1/N) \sum_{i=1}^{M} (m_i/\tilde{\omega}_i)^2 \tag{B.15}$$

$$\sum_{i=1}^{M} (\beta_0^i\omega_i)^2 \overset{(B.13)}{=} J_o\left[\sum_{i=1}^{M} (m_i^2)\right]\Bigg/\left[NJ_o - \sum_{i=1}^{M} (m_i^2)\right] \tag{B.16}$$

where M is an arbitrary, finite, positive integer. Now consider the partial sums $\mathscr{P}_M \triangleq \Sigma_{i=1}^M (m_i^2)$ and $\mathscr{R}_M \triangleq \Sigma_{i=1}^M (m_i/\tilde{\omega}_i)^2$. The infinite sums (B.3) and (B.4) imply that

$$\lim_{M\to\infty} \{\mathscr{R}_M\} = N^2 \int_0^L \rho(y)(R+y)\left\{\int_0^L \rho(x)(R+x)\mathscr{F}(x,y)\,\mathrm{d}x\right\}\mathrm{d}y \tag{B.17}$$

$$\lim_{M\to\infty} \{\mathscr{P}_M\} = NJ_E \tag{B.18}$$

Also,

$$\lim_{M\to\infty} \left(\frac{\Sigma_{i=1}^M (m_i^2)}{NJ_o - \Sigma_{i=1}^M (m_i^2)}\right) = \left(\frac{J_E}{J_o - J_E}\right) = \frac{J_E}{J_R} \tag{B.19}$$

which follows from Eq. (B.18) since the partial sums in the numerator and the denominator converge to NJ_E and NJ_R, respectively. To prove the validity of Eqs. (B.15) and (B.16) as $M \to \infty$, define the partial sums $\mathscr{Q}_M \triangleq \Sigma_{i=1}^M (\beta_0^i)^2$ and $\mathscr{S}_M \triangleq \Sigma_{i=1}^M (\beta_0^i\omega_i)^2$. It is obvious that $\mathscr{Q}_M = (1/N)\mathscr{R}_M$ and $\mathscr{S}_M = J_o\mathscr{P}_M/(NJ_o - \mathscr{P}_M)$ for *finite* M, but, since the partial sums $\mathscr{R}_M$ and $J_o\mathscr{P}_M/(NJ_o - \mathscr{P}_M)$ converge according to

Eqs. (B.17) and (B.19) as $M \to \infty$, so must $\mathcal{Q}_M$ and $\mathcal{S}_M$, which completes the proof of Eqs. (B.8) and (B.9). Since both series of $(\beta_0^i)^2$ and $(\beta_0^i \omega_i)^2$ converge, Eq. (B.10) is a direct consequence of *Cauchy's inequality*.[22]

References

[1]Markley, F. L., "Large Angle Maneuver Strategies for Flexible Spacecraft," Advances in Astronautical Sciences. Paper AAS 79-156, 1979.

[2]Farrenkopf, R. L., "Optimal Open Loop Maneuver Profiles for Flexible Spacecraft," *Journal of Guidance and Control*, Vol. 2, Nov.–Dec. 1979, pp. 491–498.

[3]Swigert, C. J., "Shaped Torque Techniques," *Journal of Guidance and Control*, Vol. 3, Sept.–Oct. 1980, pp. 460–467.

[4]Turner, J. D., and Junkins, J. L., "Optimal Large Angle Single Axis Rotational Maneuvers of Flexible Spacecraft," *Journal of Guidance and Control*, Vol. 3, Nov.–Dec. 1980, pp. 578–585.

[5]Turner, J. D., Chun, H. M., and Junkins, J. L., "Optimal Large Angle Maneuvers with Active Vibration Suppression," NASA CP-2258, Feb. 1983, pp. 177–215.

[6]Turner, J. D., and Chun, H. M., "Optimal Distributed Control of a Flexible Spacecraft during a Large Angle Maneuver," *Journal of Guidance, Control, and Dynamics*, Vol. 7, May–June 1984, pp. 257–264.

[7]Skaar, S. B., and Tucker, D., "Point Control of a One-Link Flexible Manipulator," *Journal of Applied Mechanics*, Vol. 53, March 1986, pp. 23–27.

[8]Vander Velde, W. E., and He, J., "Design of Space Structure Control System Using On-Off Thrusters," *Journal of Guidance and Control*, Vol. 6, Jan.–Feb. 1983, pp. 759–775.

[9]de Olivera e Souza, M. L., "Exactly Solving the Weighted Time/Fuel Optimal Control of an Undamped Harmonic Oscillator," *Journal of Guidance, Control, and Dynamics*, Vol. 11, Nov.–Dec. 1988, pp. 488–494.

[10]Skaar, S. B., Tang, L., and Yalda-Mooshabad, Y., "On-Off Attitude Control of Flexible Satellites," *Journal of Guidance, Control, and Dynamics*, Vol. 9, July–Aug. 1986, pp. 507–510.

[11]Thompson, R. C., Junkins, J. L., and Vadali, S. R., "Near Minimum Time Open Loop Control of Dynamic Systems," AIAA Paper 87-0958-CP, 1987.

[12]Singh, G., Kabamba, P. T., and McClamroch, N. H., "Time Optimal Slewing of a Rigid Body with Flexible Appendages," *Proceedings of the 26th Conference on Decision and Control, Los Anglees, California*, 1987.

[13]Singh, G., Kabamba, P. T., and McClamroch, N. H., "Planar Time Optimal Slewing Maneuvers of Flexible Spacecraft," *Journal of Guidance, Control, and Dynamics*, Vol. 12, Jan.–Feb. 1989, pp. 71–81.

[14]Singh, G., Kabamba, P. T., and McClamroch, N. H., "Bang-Bang Control of Flexible Spacecraft: Guaranteed Terminal Pointing Accuracy," *Journal of Guidance, Control, and Dynamics*, Vol. 13.

[15]Singh, G., "Time-Optimal Slewing Maneuvers of Flexible Spacecraft," Ph.D. Dissertation, Dept. of Aerospace Engineering, Univ. of Michigan, Ann Arbor, MI 1988.

[16]Goldwyn, R. M., Sriram, K. P., and Graham, M., "Time Optimal Control of a Linear Hyperbolic System," *International Journal of Control*, Vol. 12, No. 4, 1970, pp. 645–656.

[17]Hughes, P. C., "Modal Identities for Elastic Bodies, With Application to Vehicle Dynamics and Control," *Journal of Applied Mechanics*, Vol. 47, March 1980, pp. 177–184.

[18]Greenwood, D. T., *Classical Dynamics*, Prentice-Hall, Englewood Cliffs, NJ, 1977.

[19]Lee, E. B., and Marcus, L., *Foundations of Optimal Control Theory*, Wiley, New York, 1967.

[20]Hermes, H., and LaSalle, J., *Functional Analysis and Time Optimal Control*, Academic, New York, 1969.

[21]Garcia, C. B., and Zangwill, W. I., *Pathways to Solutions, Fixed Points and Equilibria*, Prentice-Hall, Englewood Cliffs, NJ 1981.

[22]Nikolsky, S. M., *A Course in Mathematical Analysis*, Vol. 2, Mir, Moscow, 1977.

[23]Stengel, R. F., *Stochastic Optimal Control—Theory and Applications*, Wiley, New York, 1986.

[24]Likins, P. W., "Dynamics and Control of Flexible Space Vehicles," Jet Propulsion Lab., California Institute of Technology, Pasadena, CA, TR-32-1329, Feb. 1969.

Chapter 24

Feedback Control of Space Structures: A Liapunov Approach

Srinivas R. Vadali*
Texas A&M University, College Station, Texas

Introduction

AS we envisage control of complicated space structures in demanding situations, one important observation is the fact that accurate mathematical models are usually nonlinear. It is no longer sufficient to neglect significant nonlinear effects arising from various mechanisms such as geometric and material nonlinearities, kinematics, and joint friction. Thanks to the current efforts underway to understand the behavior of nonlinear systems, we are in a better position to model and control them than ever before.

The customary approach for designing control laws for nonlinear systems has been to linearize about reference solutions and use neighboring perturbation controllers[1,2] for deviations. This approach typically involves determining an open-loop or feedforward solution of a nonlinear optimal control problem and then finding a feedback solution of a nonautonomous linear optimal control problem. If the departure from the nominal solution is beyond the linear range, the procedure is repeated about discrete nominal operating solutions to preserve optimality and/or good performance. This is gain scheduling. A generalization of this method, called extended linearization or pseudo linearization,[3] is based on linearization about a continuous family of nominal solutions. Much work still needs to be done in this area for multi-input systems.

Recently, other forms of local and global linearizations have been put forth.[4,5] These techniques are based on Lie algebras and coordinate and control transformations. The necessary and sufficient conditions under

*Assistant Professor, Aerospace Engineering.

which one can find suitable transformations do exist but are restrictive and not constructive for higher-order systems. The use of symbolic manipulation helps in this regrd to a certain extent. Other drawbacks are that the transformed coordinates often lose all physical meaning, and the control laws are extremely complicated to implement.

Another approach useful for weakly nonlinear systems is optimal control based on a Taylor series solution of either the Hamilton-Jacobi-Bellman equation[6-9] or the state-costate equations.[10] In this procedure the cost (or costates) is(are) represented by series expansion(s) in terms of the states, and higher-order gains are determined successively starting with linear gains. As with any series expansion, convergence is not guaranteed globally, specifically for problems that are highly nonlinear or nonconvex.

Last but not least in this discussion is the family of methods based on Liapunov's famous second method. We can, in many instances, consider control as stabilization by suitably reformulating the problem. Rather than asking whether the system is stable with the given control law, we ask what control can globally, asymptotically stabilize the system. Obtaining feedback control laws becomes easier by using this inverse process. Lest the reader get the feeling that the method works in all situations, we mention that finding the right candidate Liapunov function still remains the biggest challenge. On the other hand, based on the principles of analytical dynamics,[11] we can often find suitable positive-definite functions either in the energy or the Hamiltonian that suffice. The beauty of the method is that it is not based on linearization. This chapter will be devoted to vrious aspects of controlling rigid and flexible spacecraft using the Liapunov approach.

Liapunov Approach

Perhaps the earliest work utilizing the Liapunov approach to control design for nonlinear systems is due to Kalman and Bertram.[12] Later, Mortensen[13] utilized this approach for controlling rigid bodies. He used Rodriguez as well as Euler parameters to represent spacecraft attitude. Another important piece of work is that of Meyer,[14] who developed the idea of attitude control using the Euler's theorem on rotations. This school of thought is quite prevalent in Shuttle mission operations. More recently, Vadali and Junkins[15] used this technique for large-angle reorientation and tracking maneuvers of spacecraft with multiple reaction wheels. Wie and Barba[16] proposed the application of pulse width and pulse frequency modulation in conjunction with this approach. Carrington[17] proposed a Hamilton's canonical equation based generalized angular momentum feedback scheme. A Gibbs vector (Rodriguez parameters) closed-loop attitude estimation and control study has been performed by Tallant.[18]

Vadali[19] used this approach for large-angle maneuvers of flexible spacecraft of the so-called Rocket Propulsion Laboratory (RPL, now known as Astronautics Laboratory AFAL) type. The feedback scheme involves displacement and velocity feedback but is not based on discretization and modal feedback. Fujii and Ishijima[20] and Fujii[21] also used this approach and demonstrated its performance experimentally. Their scheme, "known mission function control," uses displacement, velocity, bending moment, and shear force feedback. Junkins et al.[22] extended Fujii's work by designing a tracking controller that follows a predetermined torque-shaped near-minimum-time command. Experimental verification of this idea shows good agreement with theory and excellent behavior of the controlled system.

The mission function approach has also been used for tether deployment and retrieval by Fujii and Ishijima.[23] This closed-loop deployment and retrieval scheme is based on tension control.

The Liapunov approach is briefly outlined next. Let the dynamic system be described by the system of nonlinear ordinary differential equations

$$\dot{\boldsymbol{x}} = \boldsymbol{f}(\boldsymbol{x},\boldsymbol{u},t) \tag{1}$$

where $\boldsymbol{x}$ is the state vector and $\boldsymbol{u}$ the control vector. Let V be a positive-definite function of $\boldsymbol{x}$. The time derivative of V can be written as

$$\dot{V} = \left(\frac{\partial V}{\partial \boldsymbol{x}}\right)^T \boldsymbol{f} + \frac{\partial V}{\partial t} \tag{2}$$

If the preceding identity can be solved for $\boldsymbol{u}$ as a function of $\boldsymbol{x}$ and t such that $\dot{V}$ is globally negative-definite, we obtain the feedback control law that will asymptotically stabilize the closed-loop system. Of course, the nature of the control law depends on the type of V chosen. If $\dot{V}$ can only be rendered negative semidefinite, it must be verified that $\dot{V}$ remains zero at the desired final state.

Rigid-Body Applications

For many applications a rigid-body assumption is invoked for a preliminary control design. Hence, this will be treated first. We will represent the attitude by a set of four Euler parameters $\beta(\beta_0,\beta_1,\beta_2,\beta_3)$. The differential equations for the time rate of change of these parameters is given by

$$\dot{\beta} = \tfrac{1}{2}[G(\omega)]\beta \tag{3}$$

where $\omega(\omega_1,\omega_2,\omega_3)$ is the spacecraft angular velocity vector with the components in a body-fixed axis system, and

$$[G(\omega)] = \begin{bmatrix} 0 & -\omega_1 & -\omega_2 & -\omega_3 \\ \omega_1 & 0 & \omega_3 & -\omega_2 \\ \omega_2 & -\omega_3 & 0 & \omega_1 \\ \omega_3 & \omega_2 & -\omega_1 & 0 \end{bmatrix}$$

The equations of rotational motion for a rigid body are

$$[I]\dot{\omega} = -[\tilde{\omega}]\,[I]\omega + \boldsymbol{u} \tag{4}$$

where $[I]$ is the inertia matrix, and $[\tilde{\omega}]$ is the angular velocity cross-product operator. The final orientation is assumed to be $\beta_f^T = [1,0,0,0]$. Any arbitrary final orientation can be transformed to the previously mentioned equation by suitably redefining the reference frame. For motion-to-rest maneuvers, the final angular velocity vector is $\boldsymbol{\omega}_f = \mathbf{0}$. Tracking maneuvers will be considered later. Now consider the following changes in variables:

$$\boldsymbol{e}_\beta = \beta - \beta_f \tag{5}$$

$$\boldsymbol{e}_\omega = \boldsymbol{\omega} - \boldsymbol{\omega}_f \tag{6}$$

where $\boldsymbol{e}$ represents the error coordinates. We define the following Liapunov function:

$$V = K\boldsymbol{e}_\beta^T\boldsymbol{e}_\beta + \tfrac{1}{2}K_0\boldsymbol{e}_\omega^T[I]\boldsymbol{e}_\omega \tag{7}$$

where K and K_0 are positive constants. The motivation behind such a choice for V is the desire to represent a suitable measure of the generalized energy with reference to the equilibrium point. It is easy to identify the kinetic energy as the second term in V. The first term is the "pseudo-potential energy of rotation." It is added to make V a positive-definite function of the rotational error coordinates also. It is not unique, but a judicious selection of this term is crucial for obtaining elegant feedback control laws.

The variable $\dot{V}$ can be evaluated using the skew symmetric property of G, as the following equation shows:

$$\dot{V} = K\boldsymbol{e}_\beta^T[G(\boldsymbol{e}_\omega)]B_f + K_0\boldsymbol{e}_\omega^T\boldsymbol{u} \tag{8}$$

It is now observed that the following linear control law renders $\dot{V}$ negative-semidefinite in the $\boldsymbol{e}_\omega$ coordinates:

$$\boldsymbol{u} = -K_1\boldsymbol{e}_\beta - K_2\boldsymbol{e}_\omega \tag{9}$$

where $K_1(=K/K_0)$ is a positive constant, and K_2 is a positive-definite matrix, preferably diagonal to avoid coupling between the axes of the spacecraft. This leads to the following equation:

$$\dot{V} = -K_0K_2\boldsymbol{e}_\omega^T\boldsymbol{e}_\omega \tag{10}$$

Since the only possible way that $\dot{V}$ can remain zero is $\boldsymbol{e}_\beta = \mathbf{0}$ and $\boldsymbol{e}_\omega = \mathbf{0}$, global asymptotic stability can be inferred. The gains K_1 and K_2 have to be selected for desired response and actuator limitations. A simple method for doing so is to linearize the closed-loop system about the origin near which the equations decouple into three linear oscillators, and classical concepts of natural frequency and damping ratio can be used. For critical damping there exists the following relationship between the gains:

$$K_{2_i}^2 = 2I_i K_1, \qquad i = 1,2,3 \tag{11}$$

where K_{2_i} and I_i are the entries of the diagonal matrix K_2 and the $[I]$ matrix, respectively.

A similar approach can also be utilized for spacecraft with reaction wheels. The control law can be written as follows:

$$[B]^T \boldsymbol{u} = \begin{Bmatrix} K_1\beta_1 + K_{2_1}\omega_1 \\ K_1\beta_2 + K_{2_2}\omega_2 \\ K_1\beta_3 + K_{2_3}\omega_3 \end{Bmatrix} \tag{12}$$

where $[B]^T \boldsymbol{u}$ represents the three body-axis components of the motor torque. If the number of reaction wheels is more than three, $\boldsymbol{u}$ can be obtained by the pseudo-inverse. The control design for spacecraft with control moment gyroscopes (CMGs) have been studied by Oh and Vadali.[24]

The control laws derived in Eq. (12) do not depend on spacecraft moments of inertia. Hence, stability is guaranteed regardless of parameter uncertainties. In a recent work Wie et al.[25] compared several methods for determining control gains and showed that, if the initial angular velocities are small, proper gain selection can lead to near-eigenaxis rotation, which takes the shortest angular path between two orientations.

Let us now consider tracking maneuvers. We assume that the target motion is given by the following model:

$$\dot{\beta}_f = \tfrac{1}{2}[G(\omega_f)]\beta_f \tag{13}$$

where β_f prescribes the attitude of the target frame with respect to the inertial frame. The angular velocity of the target frame ω_f can be time-varying, unlike that for slewing maneuvers. The same Liapunov function still applies. The time derivative of V is given by

$$\dot{V} = K\boldsymbol{e}_\beta^T[G(\boldsymbol{e}_\omega)]\beta + K_0\boldsymbol{e}_\omega^T[\tilde{\omega}[I]\omega + \boldsymbol{u} - [I]\dot{\omega}_f] \tag{14}$$

or

$$\dot{V} = -K\beta_f^T[G(\boldsymbol{e}_\omega)]\beta + K_0\boldsymbol{e}_\omega^T[\tilde{\omega}[I]\omega + \boldsymbol{u} - [I]\dot{\omega}_f] \tag{15}$$

Expanding the first term in Eq. (15), we can rewrite it as

$$\dot{V} = K_0\boldsymbol{e}_\omega^T[-K/K_0\alpha + \tilde{\omega}[I]\omega + \boldsymbol{u} - [I]\dot{\omega}_f] \tag{16}$$

where

$$\alpha = \begin{bmatrix} \beta_1 & -\beta_0 & -\beta_3 & \beta_2 \\ \beta_2 & \beta_3 & -\beta_0 & -\beta_1 \\ \beta_3 & -\beta_2 & \beta_1 & \beta_0 \end{bmatrix} \begin{Bmatrix} \beta_{0_f} \\ \beta_{1_f} \\ \beta_{2_f} \\ \beta_{3_f} \end{Bmatrix}$$

It is clear from Eq. (16) that the following feedback control law can be used:

$$\boldsymbol{u} = K_1\boldsymbol{\alpha} - \tilde{\boldsymbol{\omega}}[I]\boldsymbol{\omega} + [I]\dot{\boldsymbol{\omega}}_f - K_2\boldsymbol{e}_\omega \tag{17}$$

where K_1 and K_2 are as defined earlier. Feedback control laws for tracking maneuvers with reaction wheels or CMGs can also be obtained in a similar manner.

Torque Shaping

Since present-day spacecraft are becoming increasingly fragile and flexible, smooth torque profiles are desirable and in some cases mandatory. For open-loop maneuvers, torque-shaping techniques have been discussed by Vadali and Junkins[26] and Turner and Chun.[27] For the closed-loop maneuvers discussed earlier, it is relatively simple to shape the torque profiles.[15] A multiplier function is designed such that, when multiplied with the feedback gains, the initial control profile is smooth, with zero initial magnitude and slope. Such a multiplier is given by

$$m(t) = \tau^2(3 - 2\tau), \qquad \tau = t/T_1, \qquad t \le T_1 \tag{18a}$$

$$m(t) = 1, \qquad t > T_1 \tag{18b}$$

The rise time T_1 is selected to control sharpness (frequency content) of the multiplier function. The multiplier function is shown in Fig. 1.

As an example, we consider a slewing maneuver of a spacecraft with four reaction wheels in a NASA standard configuration shown in Fig. 2. The moments of inertia (kg-m^2) in a principal-axis system are given as

$$[I] = \begin{bmatrix} 87.19 & 0 & 0 \\ 0 & 85.95 & 0 \\ 0 & 0 & 114.45 \end{bmatrix}$$

The mass of each wheel is 5 kg, and the axial moment of inertia is 0.05 kg-m^2. The initial and final boundary conditions are given in Table 1. The gains are selected as follows:

$$K_1 = 10 \text{ N-m} \qquad \text{and} \qquad K_{2_1} = K_{2_2} = K_{2_3} = 50 \text{ N-m-s}$$

The rise time T_1 is 10 s. Figures 3 and 4 show the torque and wheel speed histories, respectively. The maneuver time is about 60 s.

The next example is a tracking maneuver using external torques. The spacecraft model is the same as previously considered except that the wheels are locked. It is desired to track a rotating coordinate frame with $\boldsymbol{\omega}^T = [0,0,0.5]$ rad/s. The target state is a pure spin about the third axis of the body as well as the inertial frame. The spacecraft initial conditions are the same as those given earlier. The initial conditions on the target attitude

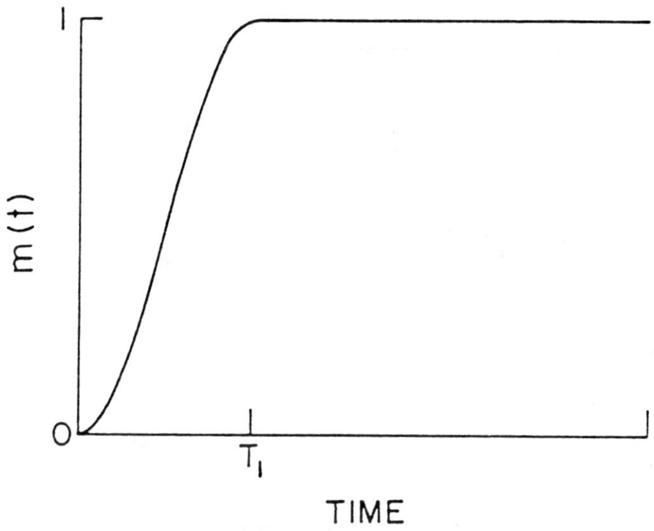

Fig. 1 Multiplier function.

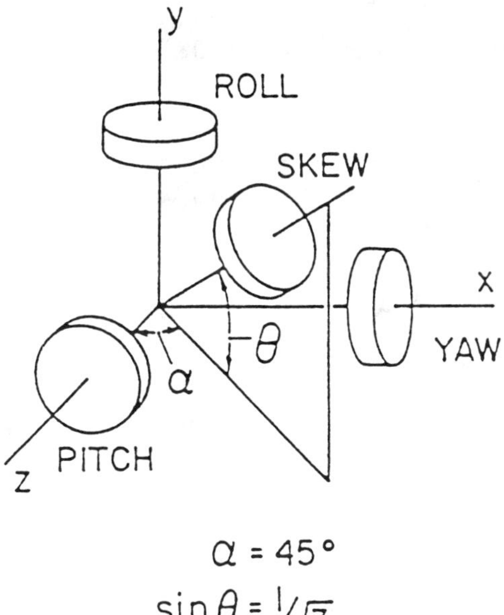

Fig. 2 NASA standard reaction wheel configuration.

Table 1 Boundary conditions

State	Initial conditions	Final conditions
β_0	0.64288	1
β_1	0.44226	0
β_2	0.44226	0
β_3	0.44226	0
ω_1	0.01	0
ω_2	0.005	0
ω_3	0.0001	0
Ω_1	0	Free
Ω_2	0	Free
Ω_3	0	Free
Ω_4	0	Free

are $\beta_f^T = [1,0,0,0]$. The analytical solutions to the target motion are given by the following equations:

$$\beta_f = [\cos(t/16),0,0,\sin(t/16)]^T$$

$$\omega_f = [0,0,0.5]^T$$

The gains selected are $K_1 = 30$ N-m and $K_{2_1} = K_{2_2} = K_{2_3} = 100$ N-m-s, and the rise time T_1 is 5 s. Figures 5 and 6 show the Euler parameter and angular velocity histories. It can be seen that, after the initial error subsides, tracking commences at about 50 s.

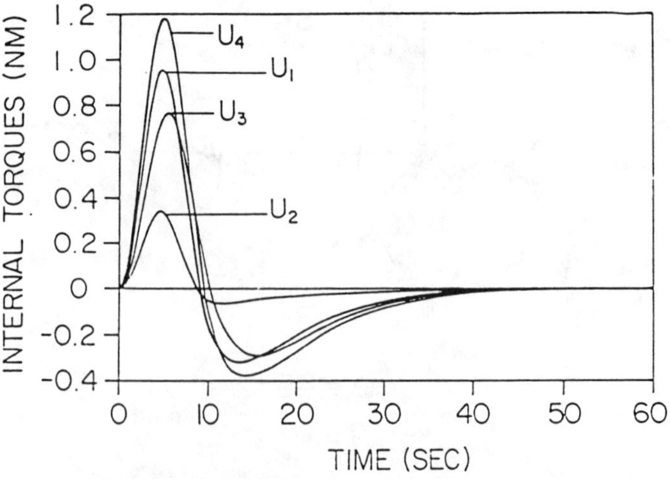

Fig. 3 Torques for the slew maneuver.

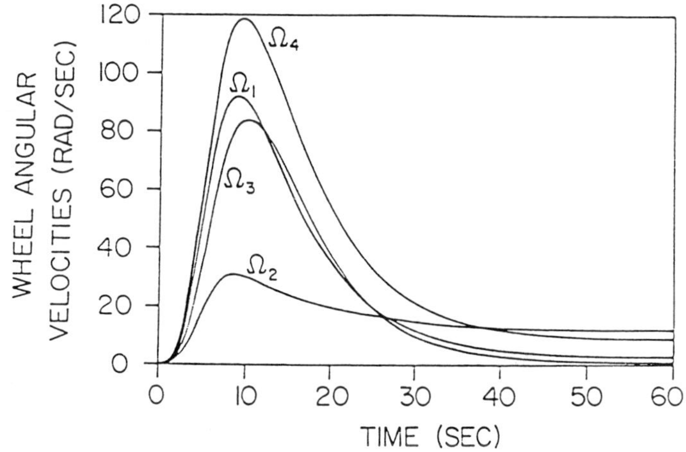

Fig. 4 Wheel angular velocities for the slew maneuver.

Flexible Spacecraft Maneuvers

Many large space structures may be regarded as central rigid bodies with long flexible appendages. The motion of these spacecraft is described by a set of hybrid coordinates consisting of rigid-body attitude and angular velocities as well as appendage displacement and velocity fields. Thus, the motion is described by a set of coupled partial and ordinary differential equations. The literature on this subject is quite voluminous, and solution procedures based on spatial discretization of the partial differential equations seem to be the more popular choice. In this section the Liapunov approach will be applied to determine feedback control laws for maneuvering flexible spacecraft.

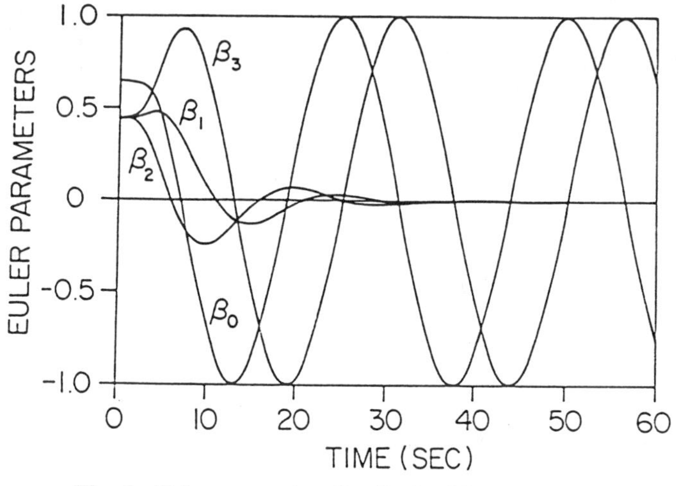

Fig. 5 Euler parameters for the tracking maneuver.

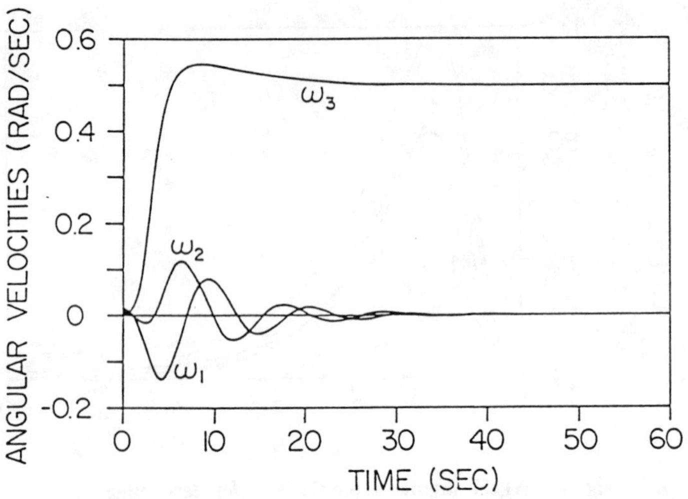

Fig. 6 Spacecraft angular velocities for the tracking maneuver.

Figure 7 shows the RPL (AFAL) spacecraft model. Assuming small deformations, the differential equations and boundary conditions for single-axis motion[28] are

$$EIy'''' + \rho(\ddot{y} + x\ddot{\theta}) = 0 \tag{19a}$$

$$I_T\ddot{\theta} + 4\rho \int_{R_i}^{R_0} x\ddot{y}\,\mathrm{d}x = u(t) \tag{19b}$$

$$y(R_i) = y'(R_i) = y''(R_0) = y'''(R_0) = 0 \tag{19c}$$

where I_T is the total (underformed) mass moment of inertia about the axis of rotation, and ρ is the mass per unit length of the appendages. The prime indicates the spatial (x) derivative.

By utilizing Eqs. (19) and boundary conditions, we can also obtain the following equation for the hub motion:

$$I_h\ddot{\theta} + 4EI[R_i y'''(R_i) - y''(R_i)] = u(t) \tag{20}$$

where I_h is the moment of inertia of the hub alone. We can clearly see the effects of the root bending moment and shear force on the hub motion.

The initial and desired final conditions are

$$y(0,x) = y_0(x), \qquad y(t_f,x) = 0 \tag{21a}$$

$$\dot{y}(0,x) = \dot{y}_0(x), \qquad \dot{y}(t_f,x) = 0 \tag{21b}$$

$$\theta(0) = \theta_0, \qquad \theta(t_f) = 0 \tag{21c}$$

$$\dot{\theta}(0) = \dot{\theta}_0, \qquad \dot{\theta}(t_f) = 0 \tag{21d}$$

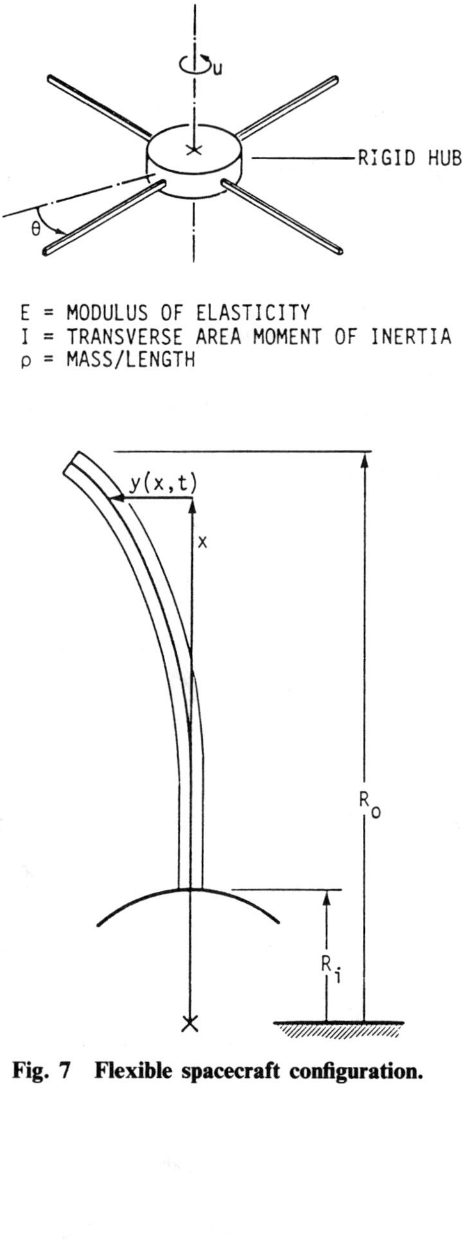

Fig. 7 Flexible spacecraft configuration.

where t_f is the final time. The energy of the system is

$$E = \frac{1}{2}\left[I_h\dot{\theta}^2 + 4\rho\int_{R_i}^{R_0}(\dot{y} + x\dot{\theta})^2\,\mathrm{d}x + 4EI\int_{R_i}^{R_0}(y'')^2\,\mathrm{d}x\right] \tag{22a}$$

or

$$E = \frac{1}{2}\left[I_t\dot{\theta}^2 + 4\rho\int_{R_i}^{R_0}(\dot{y}^2 + 2\dot{\theta}x\dot{y})\,\mathrm{d}x + 4EI\int_{R_i}^{R_0}(y'')^2\,\mathrm{d}x\right] \tag{22b}$$

where the first two terms combine to form the kinetic energy, and the last term is the strain energy. Since a rigid-body mode is present, a candidate Liapunov function is defined as

$$V = E + \tfrac{1}{2}K\theta^2 \tag{23}$$

where K is a positive constant. The additional term, the pseudo-rotational potential energy, is included to make V a positive-definite function of θ also. The time derivative of V is given by

$$\dot{V} = I_t\dot{\theta}\ddot{\theta} + 4\rho\int_{R_i}^{R_0}(\dot{y}\ddot{y} + \ddot{\theta}x\dot{y} + \dot{\theta}x\ddot{y})\,\mathrm{d}x + 4EI\int_{R_i}^{R_0}y''\dot{y}''\,\mathrm{d}x + K\theta\dot{\theta} \tag{24}$$

Using the equations of motion, it can be shown that

$$\dot{V} = \dot{\theta}u + K\theta\dot{\theta} \tag{25}$$

The control law can then be selected as

$$u = -K_1\theta - K_2\dot{\theta} \tag{26}$$

where K_1 and K_2 are positive constants.

Another form for the control law can be obtained by modifying the Liapunov function as follows:

$$V = \frac{1}{2}\left[K_1\theta^2 + K_2\dot{\theta}^2 + K_3\int_{R_i}^{R_0}\rho(\dot{y} + x\dot{\theta})^2\,\mathrm{d}x + K_4\int_{R_i}^{R_0}EI(y'')^2\,\mathrm{d}x\right] \tag{27}$$

where the Ks are positive constants to be selected to emphasize various components of the energy relative to each other. As shown by Fujii et al.,[20] this Liapunov function allows appendage root bending moment and shear force feedback also.

If we assume that a tip thruster is also present on each appendage, the following changes must be made to Eqs. (19):

$$EIy'''' + \rho(\ddot{y} + x\ddot{\theta}) = 0 \tag{28a}$$

$$I_T\ddot{\theta} + 4\rho\int_{R_i}^{R_0}x\ddot{y}\,\mathrm{d}x = u(t) + 4(R_0 - R_i)f \tag{28b}$$

$$y(R_i) = y'(R_i) = y''(R_0) = 0, \qquad EIy'''(R_0) = -f \tag{28c}$$

Using the Liapunov function used earlier in Eq. (23), we obtain the following equation:

$$\dot{V} = K_1\theta\dot{\theta} + \dot{\theta}u + 4f[\dot{y}(R_0) + (R_0 - R_i)\dot{\theta}] \tag{29}$$

It is clear that we can choose the following control laws:

$$u = -K_1\theta - K_2\dot{\theta} \tag{30}$$

$$f = -K_3[\dot{y}(R_0) + (R_0 - R_i)\dot{\theta}] \tag{31a}$$

or

$$f - K_3 \operatorname{sign}[\dot{y}(R_0) + (R_0 - R_i)\dot{\theta}] \tag{31b}$$

depending on whether thrust is used continuously or in bang bang mode. It is apparent that this choice leads to collocation of actuators and sensors, the robustness implications of which are well known. Here, we qualify this further by noting that the control laws do not depend on the spacecraft parameters.

Once the equations are discretized using modal transformations, either classical or modern control methods can be used for determining the gains. Since the structure of the control law is fixed, either parameter optimization or optimal output feedback seem appropriate. In order to simulate the closed-loop system and optimize the gains, the equations are discretized using the following assumed modes[27]:

$$\begin{aligned}\phi_j(x - R_i) = 1 - \cos[j\pi(x - R_i)/(R_0 - R_i)]\\ + \tfrac{1}{2}(-1)^{j+1}[j\pi(x - R_i)/(R_0 - R_i)]^2\end{aligned} \tag{32}$$

As an example, let us consider a 90-deg rest-to-rest maneuver. The spacecraft parameters are given in Table 2. Four assumed modes are used for the simulation.The gains selected using an optimal output feedback program are $K_1 = 16.8$ N-m and $K_2 = 627$ N-m/s. The rise time for the multiplier is 20 s. Figure 8 shows the maneuver angle, tip displacement, and torque. Figure 9 shows the same maneuver also using the tip thrusters. The gain K_3 is 0.001 N/m-s, and the other gains are the same as those given earlier. It is clear that the tip thrusters help in actively stiffening the appendages.

The design procedure outlined earlier can also be used if a dissipative constitutive relationship is assumed. As an example, let us consider the Kelvin solid (not a realistic material model but used for the sake of simplicity) which is modeled by the following equation:

$$\sigma = E\epsilon + E_1\dot{\epsilon} \tag{33}$$

where σ is the stress, ϵ is the strain, and E_1 is a material constant. This

Table 2 Flexible spacecraft data

I_t	7,000 kg-m^2
ρ	0.0004 kg/m
R_i	1 m
R_0	151 m
EI	1,500 kg-m^3/s^2

material model leads to the following change in Eq. (19a):

$$I[Ey'''' + E_1\dot{y}''''] + \rho(\ddot{y} + x\ddot{\theta}) = 0 \tag{34}$$

Starting with the same candidate Liapunov function as in Eq. (23), we obtain the following equation:

$$\dot{V} = \dot{\theta}u + K\theta\dot{\theta} - 4E_1 I \int_{R_i}^{R_0} (\dot{y}'')^2\, \mathrm{d}x \tag{35}$$

Thus, we observe that the same linear output feedback control law is also

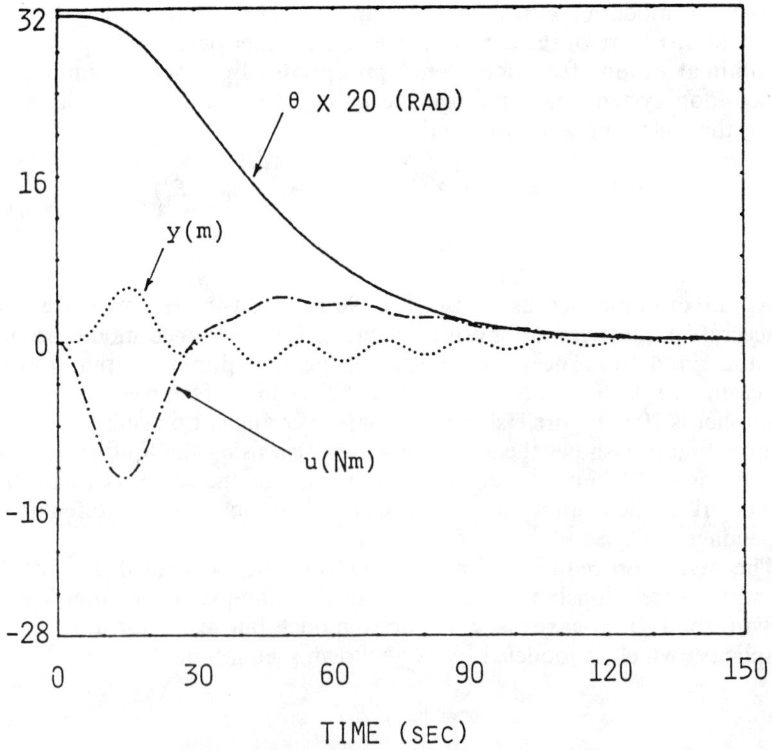

Fig. 8 Flexible spacecraft slew maneuver.

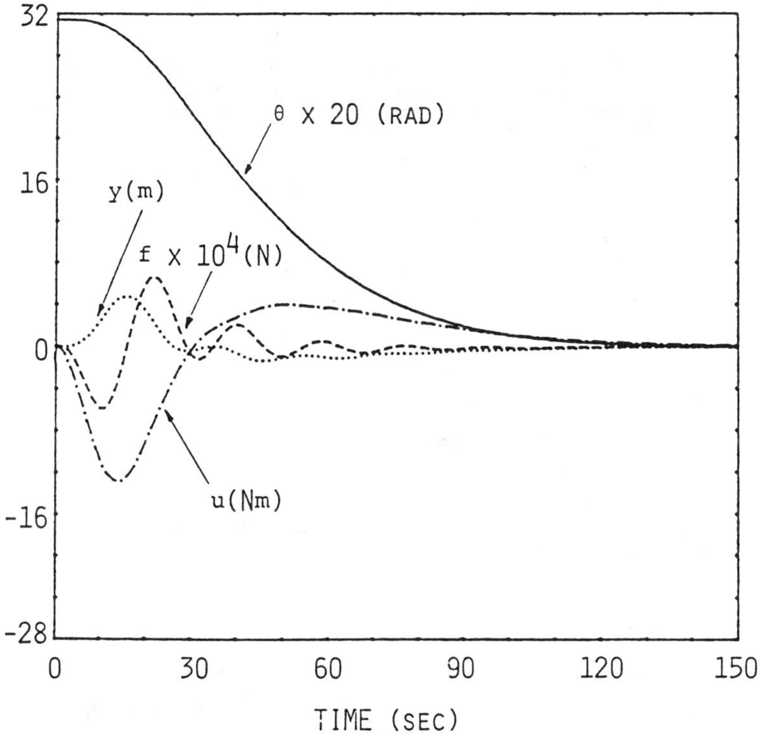

Fig. 9 Flexible spacecraft maneuver using additional tip thrusters.

applicable in this situation. In fact, this control law is also globally stabilizing in the presence of nonlinearities as long as they are either conservative or dissipative in nature. If the nonlinearities are nonconservative and nondissipative, they have to be compensated for. One of the ways to do this is to cancel these terms using a proper choice of the control law. This is called prefiltering. Another interesting nonlinearity is due to gyroscopics.[29,30] These terms do not contribute to the derivative of a properly chosen Liapunov function and in many cases need not be canceled.

Tether Deployment and Retrieval

The Liapunov design approach is also applicable for deployment and retrieval of tethered satellites. Much work in this area has been done by Rupp,[31] Misra and Modi,[32] Modi et al.,[33] Bainum and Kumar,[34] and Liangdong and Bainum.[35] These studies have modeled the effects of massive tethers, atmospheric drag, and tether flexibility. In this chapter we neglect these effects and assume that the tether mass is negligible and the tether itself remains straight. It has been shown in previous studies that including the mass of the tether in the dynamic model does not change the behavior qualitatively during deployment and retrieval. The Liapunov approach to tether operations in the orbital plane was first considered by Fujii et al.[23] We will present a more elegant treatment of the general problem that includes both in-plane as well as out-of-plane librations.[36,37]

The equations of motion of the system[33] are

$$\ddot{l} - l[\dot{\phi}^2 + \cos^2\phi(\dot{\theta} + \Omega)^2 - \Omega^2 + 3\Omega^2\cos^2\phi\cos^2\theta] = -T/m \tag{36a}$$

$$\ddot{\theta}\cos\phi + 2\left(\frac{\dot{l}}{l}\cos\phi - \dot{\phi}\sin\phi\right)(\dot{\theta} + \Omega) + 3\Omega^2\cos\phi\cos\theta\sin\theta = 0 \tag{36b}$$

$$\ddot{\phi} + 2\frac{\dot{l}}{l}\dot{\phi} + \cos\phi\sin\phi[(\dot{\theta} + \Omega)^2 + 3\Omega^2\cos^2\theta] = F/(ml) \tag{36c}$$

where l indicates the instantaneous tether length, θ is the pitch angle (in-plane), ϕ is the yaw angle (out of plane), Ω is the orbital rate, T is the tension, F is the out-of-plane thrust, and m is the mass of the subsatellite. These equations can be nondimensionalized by defining the following nondimensional variables:

$$\tau = \Omega t, \qquad \lambda = \frac{l}{L}, \qquad \hat{T} = \frac{T}{[m\Omega^2 L]}, \qquad \text{and} \qquad \hat{F} = \frac{F}{[m\Omega^2 L]}$$

where L is the reference tether length. The nondimensional equations are

$$\lambda'' - \lambda[\phi'^2 + \cos^2\phi(1 + \theta')^2 - 1 + 3\cos^2\phi\cos^2\theta] = -\hat{T} \tag{37a}$$

$$\theta''\cos\phi + 2\left(\frac{\lambda'}{\lambda}\cos\phi - \phi'\sin\phi\right)(1 + \theta') + 3\cos\phi\cos\theta\sin\theta = 0 \tag{37b}$$

$$\phi'' + 2\frac{\lambda'}{\lambda}\phi' + \cos\phi\sin\phi[(1 + \theta')^2 + 3\cos^2\theta] = \hat{F}/\lambda \tag{37c}$$

where the prime indicates derivative with respect to nondimensional time.

In-Plane Deployment and Retrieval

Assuming that out-of-plane motion is actively controlled, the orbit plane equations of motion reduce to the following:

$$\lambda'' - \lambda(1 + \theta')^2 + \lambda - 3\lambda\cos^2\theta = -\hat{T} \tag{38a}$$

$$\lambda\theta'' + 2\lambda'(1 + \theta') + 3\lambda\cos\theta\sin\theta = 0 \tag{38b}$$

We now investigate the possibility of obtaining a suitable Liapunov function and a feedback control law. In the absence of a suitable total energy integral, the following candidate Liapunov function is selected:

$$V = \tfrac{1}{2}[\lambda'^2 + K_1(\lambda - \lambda_f)^2 + (K_2 + 3\lambda^2)(\tfrac{1}{3}\theta'^2 + \sin^2\theta)] \tag{39}$$

where $\lambda_f > 0$ is the desired final value of λ. The variable K_1 is a positive constant, and K_2 can either be positive or zero. This Liapunov function has some similarity to that used by Fujii et al.[23] It is primarily a quadratic function; the quartic terms have been included to obtain mathematical simplifications. The above Liapunov function is related to the Hamiltonian of the system. The time derivative of V is given by

$$V' = \lambda'[3\lambda - \hat{T} + K_1(\lambda - \lambda_f) - \tfrac{2}{3}K_2\theta'(1 + \theta')/\lambda] \tag{40}$$

We select

$$\hat{T} = 3\lambda + K_1(\lambda - \lambda_f) - \tfrac{2}{3}K_2\theta'(1 + \theta')/\lambda + K_3\lambda' \tag{41}$$

as the nondimensional tension control law so that

$$V' = -K_3\lambda'^2 \tag{42}$$

To investigate the stability of the closed-loop system, we must ensure that, when V' is zero, V is also zero. We begin by assuming that $\lambda' = 0$. Then, from Eqs. (38) and (41), we obtain

$$\lambda[(1 + \theta')^2 + 3\cos^2\theta] = 4\lambda + K_1(\lambda - \lambda_f) - \tfrac{2}{3}K_2\theta'(1 + \theta')/\lambda \tag{43}$$

and

$$\theta'' + 3\cos\theta \sin\theta = 0 \tag{44}$$

By differentiating Eq. (43) and substituting Eq. (44), we can see that, since λ is a constant, θ' must also be a constant in Eq. (43). Then θ'' is zero in Eq. (44), which leads to θ being equal to zero, π, or $\pm\pi/2$. If θ reaches either of the first two values, it can be seen from Eq. (43) that λ reaches λ_f. The remaining values of θ are not equilibrium values and must also be avoided. If we consider downward deployment and upward retrieval, the desired local equilibrium point is $\lambda = \lambda_f$ and $\lambda' = \theta = \theta' = 0$.

In the following we consider deployment and retrieval of a tethered satellite from the Space Shuttle assumed to be in a circular orbit at an altitude of 220 km, with an orbital rate of 0.07068 rad/min. The tether is assumed to be 100 km long.

The initial conditions for the motion of the tether before deployment are assumed to be $\lambda = 0.01$, $\lambda' = 0.5$ (58.8 m/s), and $\theta = \theta' = 0$, and the desired final conditions are $\lambda = 1.0$ and $\lambda' = \theta = \theta' = 0$. The gains selected are $K_1 = 2.0$, $K_2 = 0$, and $K_3 = 4$. It is important to keep the tension positive. This requires a few trials to select the gains. Figure 10 shows the nondimensional tether length, pitch angle, and tension. It can be seen that deployment is essentially complete in about *one orbit*. The nondimensional tension settles down to its equilibrium value of 3.

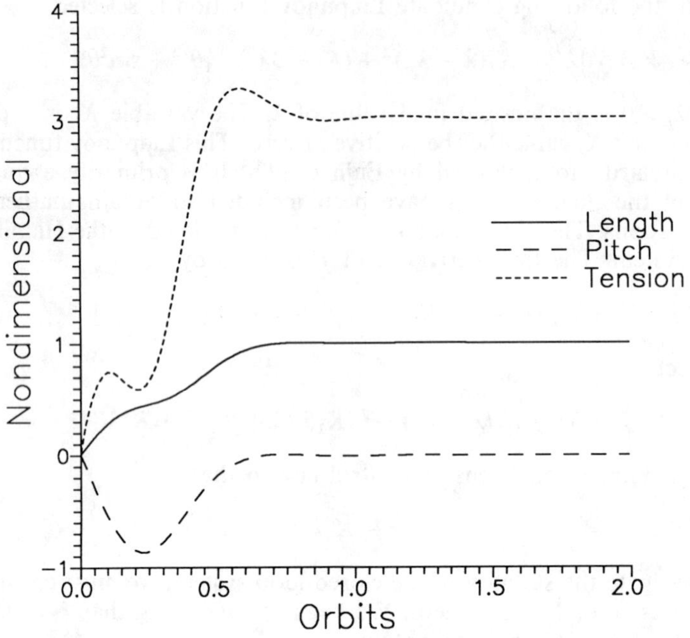

Fig. 10 In-plane deployment, gains $K_1 = 2.0$, $K_2 = 0$, and $K_3 = 4$.

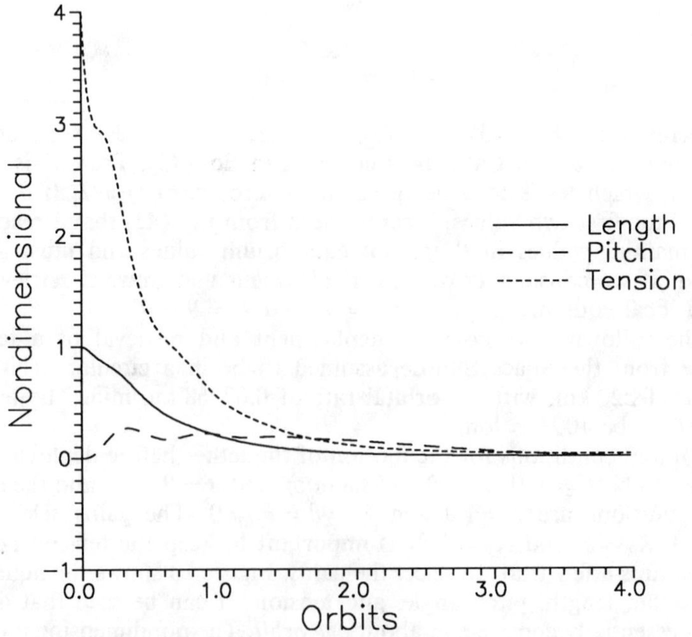

Fig. 11 In-plane retrieval, gains $K_1 = 1.0$, $K_2 = 0$, and $K_3 = 4$.

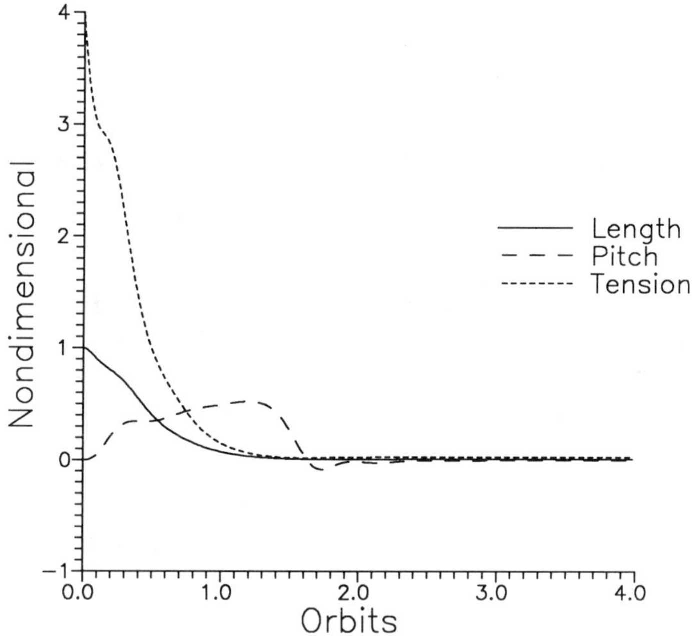

Fig. 12 In-plane retrieval, gains $K_1 = 1.0$, $K_2 = 0$, and $K_3 = 3$.

The preceding control law is also applicable for retrieval. The initial conditions are assumed to be $\lambda = 1.0$ and $\lambda' = \theta = \theta' = 0$, and the desired final conditions are $\lambda = 0.01$ and $\lambda' = \theta = \theta' = 0$. The gains selected are $K_1 = 1.0$, $K_2 = 0$, and $K_3 = 4$. Figure 11 shows the nondimensional tether length, pitch angle, and tension. It can be seen that retrieval to the desired length is completed in nearly *four orbits*. The gains and the initial velocity selected for this example result in an initial nondimensional tension of 4.0, which is 33% higher than the equilibrium value at the initial time. The initial velocity and the gains can be adjusted for a faster retrieval. As another example, Fig. 12 shows the retrieval process for the same initial conditions as before but with $K_1 = 1.0$, $K_2 = 0$, and $K_3 = 3$. It is clear from a comparison of Figs. 11 and 12 that, by allowing higher values of the pitch angle, a faster retrieval can be accomplished. The retrieval depicted in Fig. 12 is essentially complete in less than *two orbits*. On the other hand, the pitch angle should not be allowed to be too high because undesirable as well as unsafe equilibrium points may be reached. If a nonzero value of K_2 is used, the pitch angle remains small but decays very slowly.

It is also possible to hold the terminal pitch angle constant for a rapid retrieval. If the pitch angle is held constant, we notice[35] from Eq. (38b) that

$$2\lambda' + 3\lambda \cos\theta \sin\theta = 0 \tag{45}$$

The analytical solution to the preceding equation is

$$\lambda = \lambda_0 \exp[-0.75 \sin(2\theta)\tau] \tag{46}$$

where λ_0 is the initial value of λ. Therefore, the fastest retrieval rate occurs when $\theta = 45$ deg. Furthermore, it can be shown with the help of Eqs. (38) and (46) that, if the pitch angle is constant, the nondimensional tension is $3\lambda[\cos^2\theta - \frac{3}{16} \sin(2\theta)]$, which is positive for angles between 0 and 45 deg. With this knowledge a terminal control law can be designed to hold the pitch angle at a small positive value. The retrieval rate is obtained first by treating it as a control variable in Eq. (38b). The tension is then determined by using the retrieval rate in Eq. (38a). If a target pitch angle is set at θ_f, a suitable Liapunov function is

$$V = \tfrac{1}{2}K_1(\theta - \theta_f)^2 + \tfrac{1}{2}\theta'^2 \tag{47}$$

Carrying out the required mathematical manipulations, we obtain

$$\lambda' = \lambda[K_1(\theta - \theta_f) - 3 \sin\theta \cos\theta + K_2\theta']/[(2(1 + \theta')] \tag{48}$$

General Case

We now consider the presense of both in-plane (pitch) and out-of-plane (roll) motion. Previous studies have shown that, when both pitch and roll are present, the tether retrieval rate is not unidirectional but oscillatory. If we use out-of-plane thrusting to keep roll motion bounded, the procedure demonstrated previously can be applied to produce an almost unidirectional retrieval. We will demonstrate two examples[37] for designing feedback control laws. The initial conditions for the motion of the tether in both the examples are $\lambda = 1.0$, $\lambda' = 0$, $\theta = \phi = 5$ deg, and $\theta' = \phi' = 0$, and the final conditions are $\lambda = 0.01$ and $\lambda' = \theta = \theta' = \phi = \phi' = 0$.

The first example is based on a coordinate transformation that nearly uncouples the in-plane and out-of-plane motions. If we define $z = \lambda \cos\phi$ and $y = \lambda \sin\phi$, the differential equations [Eqs. (37)] transform to

$$z'' - z[(1 + \theta')^2 - 1 + 3\cos^2\theta] = -\frac{\hat{T}}{\lambda} z - \frac{\hat{F}}{\lambda} y \tag{49a}$$

$$\theta'' + 2\left(\frac{z'}{z}\right)(1 + \theta') + 3 \cos\theta \sin\theta = 0 \tag{49b}$$

$$y'' + y = -\frac{\hat{T}}{\lambda} y + \frac{\hat{F}}{\lambda} z \tag{49c}$$

It is interesting to note that Eqs.(49a) and (49b) are similar to Eqs. (38). Thus, the in-plane motion can be controlled as was shown earlier, and it is a simple matter to control the out-of-plane motion using derivative feedback.

Hence, the system can be controlled if

$$\begin{bmatrix} z & y \\ y & -z \end{bmatrix} \begin{Bmatrix} \dfrac{\hat{T}}{\lambda} \\ \dfrac{\hat{F}}{\lambda} \end{Bmatrix} = \begin{Bmatrix} 3z + K_1(z - z_f) - \tfrac{2}{3}K_2\theta'(1+\theta')/z + K_3 z' \\ K_3 y' \end{Bmatrix} \tag{50}$$

In terms of the original variables the control laws can be written as

$$\hat{T} = (3 + K_1)\lambda \cos^2\phi - K_1\lambda_f \cos\phi - \tfrac{2}{3}K_2\theta'(1+\theta')/\lambda + K_3\lambda' \tag{51}$$

and

$$\hat{F} = (3 + K_1)\lambda \cos\phi \sin\phi - K_1\lambda_f \sin\phi - \tfrac{2}{3}K_2\theta'(1+\theta')/\lambda \tan\phi + K_3\lambda\phi' \tag{52}$$

Figure 13 shows the variations of tether length, pitch angle, and tension, and Fig. 14 shows the roll angle and out-of-plane thrust variations during retrieval for gain settings $K_1 = 1.0$, $K_2 = 0$, and $K_3 = 3$. The results indicate that the tether length and pitch angle reach their respective final values in nearly *two orbits*, but the roll angle response is slower. The reason for this

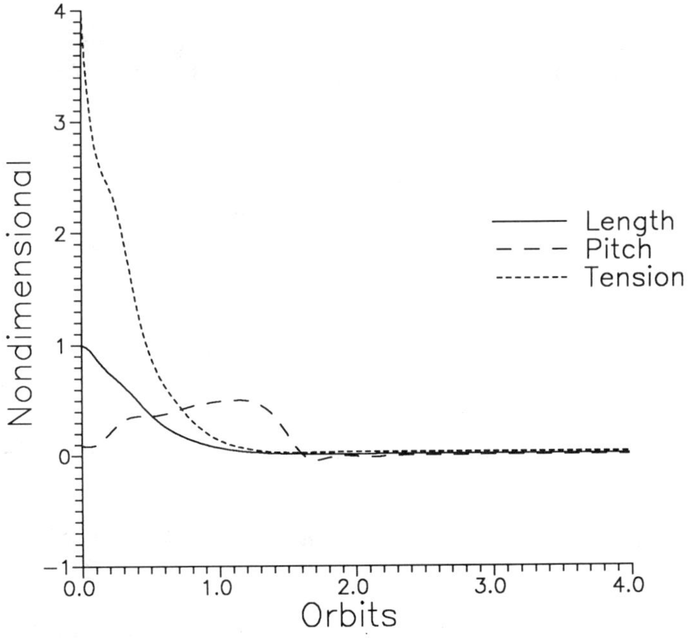

Fig. 13 General retrieval, example 1.

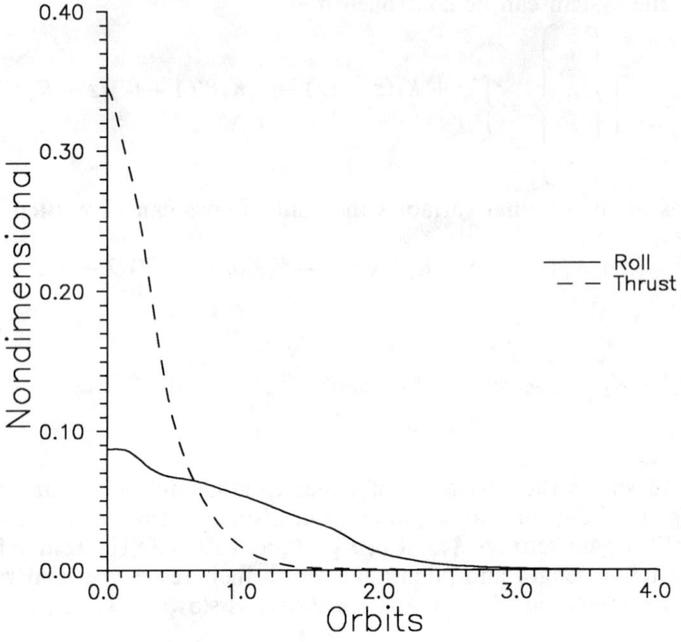

Fig. 14 General retrieval, example 1.

behavior can be explained as a result of the decoupled control design.

The second example is based on the following Liapunov function:

$$V = \tfrac{1}{2}[\lambda'^2 + K_1(\lambda - \lambda_f)^2$$
$$+ (K_2 + \lambda^2)(\theta'^2 \cos^2\phi + 3\sin^2\theta \cos^2\phi + \phi'^2 + 4\sin^2\phi)] \tag{53}$$

This Liapunov function is related to the system Hamiltonian. The time derivative of V is given by

$$V' = \lambda'\{3\lambda - \hat{T} + K_1(\lambda - \lambda_f) - 2K_2[\theta'(1+\theta')\cos^2\phi - \phi'^2]/\lambda\}$$
$$+ \phi'(k_2 + \lambda^2)\hat{F}/\lambda \tag{54}$$

If we assume that no out-of-plane thrusting is present, we can select the tension control law

$$\hat{T} = 3\lambda + K_1(\lambda - \lambda_f) - 2K_2[\theta'(1+\theta')\cos^2\phi + \phi'^2]/\lambda + K_3\lambda' \tag{55}$$

so that

$$V' = -K_3\lambda'^2 \tag{56}$$

Simulations using the preceding control law reveal that there is a significant interplay between the tether length and the out-of-plane motion. The tether retrieval rate is oscillatory, and convergence to the desired equilibrium point is extremely slow.

If we include out-of-plane thrusting, the tension control law need not be changed. Simple rate feedback is sufficient to enhance the stability of the closed-loop system significantly. The out-of-plane control law is selected as

$$\hat{F} = -K_4 \lambda \phi' \tag{57}$$

The use of the tension as well as the out-of-plane control laws lead to

$$V' = -K_3 \lambda'^2 - (K_2 + \lambda^2) K_4 \phi'^2 \tag{58}$$

It is easy to verify the stability of the closed-loop system. If λ' and ϕ' are both zero, the closed-loop system is given by the following equations:

$$-\lambda[\cos^2\phi(1+\theta')^2 - 1 + 3\cos^2\phi\cos^2\theta]$$
$$= -3\lambda - K_1(\lambda - \lambda_f) + 2K_2/\lambda[\theta'(1+\theta')\cos^2\phi] \tag{59a}$$

$$\theta''\cos\phi + 3\cos\phi\cos\theta\sin\theta = 0 \tag{59b}$$

$$\cos\phi\sin\phi[(1+\theta')^2 + 3\cos^2\theta)] = 0 \tag{59c}$$

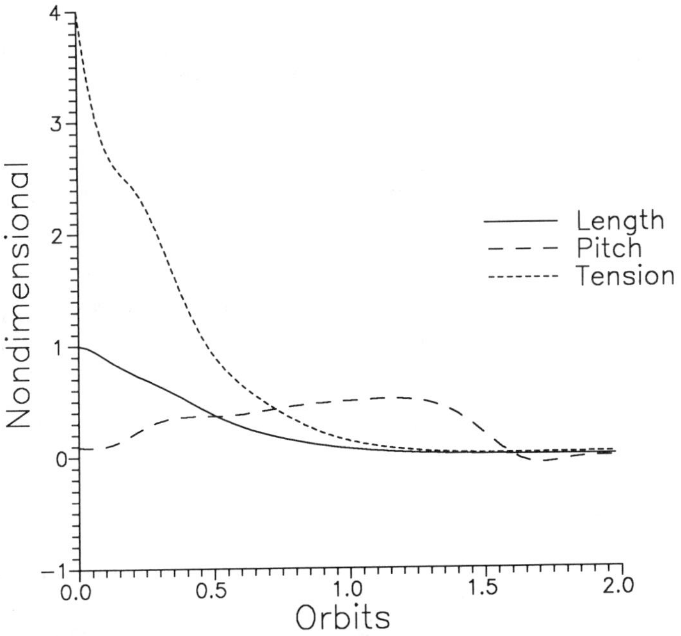

Fig. 15 General retrieval, example 2.

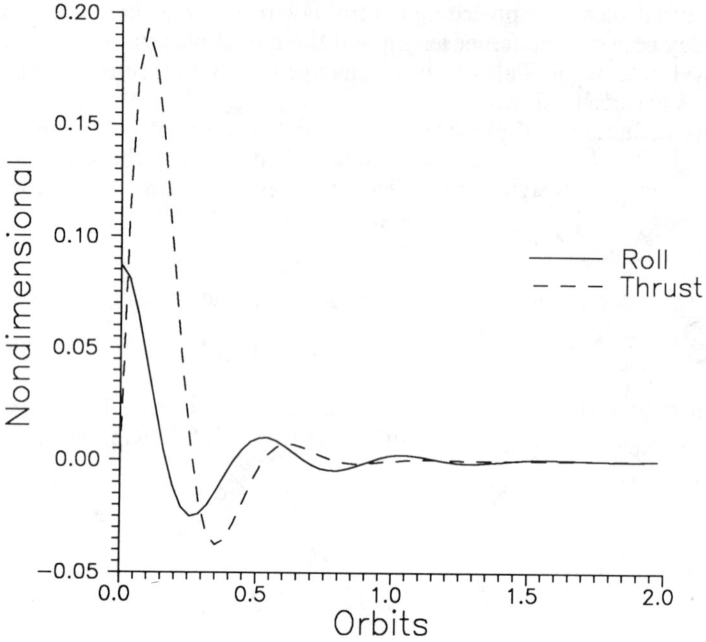

Fig. 16 General retrieval, example 2.

As seen earlier, it can be easily shown that a local equilibrium point is $\lambda = \lambda_f$ and $\lambda' = \theta = \theta' = \phi = \phi' = 0$. Simulation results using this control law with gains $K_1 = 1.0$, $K_2 = 0$, and $K_3 = 3$ are shown in Figs. 15 and 16. It is observed that the retrieval rate is almost unidirectional, and the required out-of-plane thrust is minimal.

Further comparison of the examples previously considered reveals another important result. The out-of-plane thrust requirement for the second example is much less than that for the first. This conclusion was strengthened by evaluating the index $\int |\hat{F}|\, d\tau$ over two orbits for each example. The index was approximately 0.9 for the first and 0.24 for the second. Finally, it is apparent that controlling the roll librations is essential for rapid retrieval.

Conclusions

We have discussed a variety of applications of the Liapunov approach to feedback control design for space structures. With the same Liapunov function it is possible to derive many different control laws. If the system is conservative, it is trivial to find a Liapunov function. Even though it is not always desirable, sometimes, by proper feedback, we can cancel the nonconservative, nondissipative effects by prefiltering. Prefiltering is always possible if the number of controls equals that of the states. It is important to note that, if the control law is independent of the system parameters, it naturally becomes robust with respect to their variations. Much work still needs to be done to utilize this approach in the presence of disturbances, parameter variations, noncollocated actuators and sensors, and actuator/ sensor dynamics in the mathematical models.

Acknowledgments

The author thanks Dr. J. L. Junkins, E.-S. Kim, and H.-S. Oh for many interesting discussions. This chapter is dedicated to the memory of Douglas W. Kobberdahl. This work has benefited from the support of the Texas Advanced Research and Technology Program as well as continuous AFOSR support for the last few years, with Dr. Anthony K. Amos as the monitor.

References

[1]Bryson, A. E., and Ho, Y.-C., *Applied Optimal Control*, Hemisphere, Washington, DC, 1975.

[2]Chun, H. M., and Turner, J. D., "Frequency-Shaped Large Angle Maneuvers," *Journal of Astronautical Sciences*, Vol. 36, No. 3, July–Sept. 1988, pp. 219–244.

[3]Rugh, W. J., "The Extended Linearization Approach for Nonlinear Systems Problems," *Algebraic and Geometric Methods in Nonlinear Control Theory*, edited by M. Fliess and M. Hazewinkel, Reidel, Boston, MA, 1986.

[4]Fliess, M., and Hazewinkel, M. (eds.), *Algebraic and Geometric Methods in Nonlinear Control Theory*, Reidel, Boston, MA, 1986.

[5]Dwyer, T. A. W. III, "Exact Nonlinear Control of Large Angle Rotational Maneuvers," *IEEE Transactions on Automatic Control*, Vol. AC-29, No. 5, Sept. 1984, pp. 769–774.

[6]Al'Brekht, E. G., "On Optimal Stabilization of Nonlinear Systems," *Journal of Applied Mathematics and Mechanics*, Vol. 25, 1962, pp. 1254–1266.

[7]Lukes, D. L., "Optimal Regulation of Nonlinear Dynamical Systems," *SIAM Journal of Control*, Vol. 7, 1969, pp. 75–100.

[8]Dabbous, T. E., and Ahmed, N. U., "Nonlinear Optimal Feedback Regulation of Satellite Angular Momenta," *IEEE Transactions on Aerospace Electronic Systems*, Vol. AES-18, No. 1, Jan. 1982, pp. 2–10.

[9]Dwyer, T. A. W. III, "The Control of Angular Momentum of Rigid Bodies," *IEEE Transactions on Automatic Control*, Vol. AC-27, No. 3, June 1982, pp. 686–688.

[10]Carrington, C. K., and Junkins, J. L., "Optimal Nonlinear Feedback Control for Spacecraft Attitude Maneuvers," *Journal of Guidance, Control, and Dynamics*, Vol. 9, No. 1, Jan.–Feb. 1986, pp. 99–107.

[11]Meirovitch, L., *Analytical Dynamics*, McGraw-Hill, New York, 1970.

[12]Kalman, R. E., and Bertram, J. E., "Control System Analysis and Design via the Second Method of Liapunov," *Journal of Basic Engineering*, Vol. 82, June 1960, pp. 371–400.

[13]Mortensen, R. E., "A Globally Stable Linear Attitude Regulator," *International Journal of Control*, Vol. 8, No. 3, 1968, pp. 297–302.

[14]Meyer, G., "On the Use of Euler's Theorem on Rotations for Synthesis of Attitude Control Systems," NASA TN D-3643, Sept. 1966.

[15]Vadali, S. R., and Junkins, J. L., "Optimal Open Loop and Stable Feedback Control of Rigid Spacecraft Attitude Maneuvers," *Journal of the Astronautical Sciences*, Vol. 32, No. 2, April–June 1984, pp. 105–122.

[16]Wie, B., and Barba, P. M., "Quaternion Feedback for Spacecraft Large Angle Maneuvers," *Journal of Guidance, Control, and Dynamics*, Vol. 8, No. 3, May–June 1985, pp. 360–365.

[17]Carrington, C. K., "Optimal Nonlinear Feedback Control of Spacecraft Attitude Maneuvers," Ph.D. Dissertation, Dept. of Engineering Science and Mechanics, Virginia Polytechnic Inst. and State Univ., Blacksburg, VA, Dec. 1983.

[18]Tallant, G. S., "A Kalman Filtering Technique for Spacecraft Attitude Determination and Control Using Gibbs Vector," M.S. Thesis, Dept. of Aerospace Engineering, Texas A&M Univ., College Station, TX, Aug. 1988.

[19]Vadali, S. R., "Feedback Control of Flexible Spacecraft Large-Angle Maneuvers Using Liapunov Theory," *Proceedings of the American Control Conference*, Vol. 3, 1984, pp. 1674–1678.

[20]Fujii, H., and Ishijima, S., "Mission Function Control Applied to Slew Maneuvers," International Aeronautical Federation Paper 87-354, 1987.

[21]Fujii, H., "A Slew Maneuver Experiment of Mission Function Control," AIAA Paper 88-4226-CP, 1988.

[22]Junkins, J. L., Rahman, Z., and Bang, H., "Near-Minimum-Time Maneuvers of Flexible Vehicles: A Liapunov Control Law Design Method," *Mechanics and Control of Large Space Structures*, edited by J. L. Junkins, AIAA, New York, 1990, Chap. 22.

[23]Fujii, H., and Ishijima, S., "Mission Function Control for Deployment and Retrieval of a Subsatellite," *Journal of Guidance, Control, and Dynamics*, Vol. 12, No. 2, March–April 1989, pp. 243–247.

[24]Oh, H.-S., and Vadali, S. R., "Feedback Control and Steering Laws for Spacecraft Using Single Gimbal Control Moment Gyros," AIAA Paper 89-3475, 1989; also *Journal of the Astronautical Sciences* (to be published).

[25]Wie, B., Weiss, H., and Arapostathis, A., "Quaternion Feedback Regulator for Spacecraft Eigenaxis Rotations," *Journal of Guidance, Control, and Dynamics*, Vol. 12, No. 3, May–June 1989, pp. 375–380.

[26]Vadali, S. R., and Junkins, J. L., "Spacecraft Large Angle Rotational Maneuvers with Optimal Momentum Transfer," *Journal of Astronautical Sciences*, Vol. 31, No. 2, April–June 1983, pp. 217–235.

[27]Turner, J. D., and Chun, H. M., "Optimal Distributed Control of a Flexible Spacecraft Using Control-Rate Penalties in the Controller Design," AIAA Paper 82-1438, Aug. 1982.

[28]Skaar, S. B., "Closed Form Optimal Control Solutions for Continuous Linear Elastic Systems," *Journal of the Astronautical Sciences*, Vol. 32, No. 4, Oct.–Dec. 1984, pp. 447–462.

[29]Cannon, R. H., Jr., "Gyroscopic Coupling in Space Vehicle Attitude Control Systems," *Journal of Basic Engineering*, Vol. 84, March 1962, pp. 41–53.

[30]Junkins, J. L., Rajaram, S, Baracat, W., and Carrington, C. K., "Precision Autonomous Satellite Attitude Control Using Momentum Transfer and Magnetic Torquing," *Journal of the Astronautical Sciences*, Vol. 30, No. 1, Jan.–March 1982, pp. 31–48.

[31]Rupp, C. C., and Laue, J. H., "Shuttle/Tethered Satellite System," *Journal of the Astronautical Sciences*, Vol. 26, No. 1, Jan.–March 1983, pp. 1–18.

[32]Misra, A. K., and Modi, V. J., "A Survey on the Dynamics and Control of Tethered Satellite Systems," *Tethers in Space, Advances in the Astronautical Sciences*, Vol. 62, American Astronautical Society, San Diego, CA, 1987, pp. 667–720.

[33]Modi, V. J., Chang-Fu, G., Misra, A. K., and Xu, D. M., "On the Control of the Space Shuttle Based Tethered Systems," *Acta Astronautica*, Vol. 9, June–July 1982, pp. 437–443.

[34]Bainum, P. M., and Kumar, V. K., "Optimal Control of the Shuttle-Tethered-Subsatellite System," *Acta Astronautica*, Vol. 7, May 1980, pp. 1333–1348.

[35]Liangdong, L., and Bainum, P. M., "Effect of Tether Flexibility on the Tethered Shuttle Subsatellite Stability and Control," *Journal of Guidance, Control, and Dynamics*, Vol. 12, Nov.–Dec. 1989, pp. 866–873.

[36]Vadali, S. R., "Feedback Tether Deployment and Retrieval," *Journal of Guidance, Control, and Dynamics* (to be published).

[37]Vadali, S. R., and Kim, E.-S., "Feedback Control of Tethered Satellites Using Liapunov Stability Theory," AIAA Paper 90-1197, April 1990.

[38]Arnold, D. A., "The Behavior of Long Tethers in Space," *Journal of the Astronautical Sciences*, Vol. 35, No. 1, Jan.–March 1987, pp. 3–18.

Chapter 25

Insights and Approximations in Dynamic Analysis of Spacecraft Tethers

Andreas H. von Flotow* and Norman M. Wereley†
Massachusetts Institute of Technology, Cambridge, Massachusetts

Introduction

THIS chapter[1,2] presents a critical discussion of selected physical effects in the dynamics of spacecraft tethers. Nonlinear extensional stress-strain behavior of the tether is identified as an important effect. Spectral separation is invoked to decouple the motion into a slow libration akin to that of a rigid body upon which fast oscillations involving tether deformation and end body attitude motions are superimposed. The effect of a slight quasiequilibrium tether curvature on this fast motion is illuminated via a linearized analysis. Eigenmotion, frequency-response, and wave-propagation behavior are investigated in the course of this linearized analysis. The chapter closes with a discussion of the effects of slow tether deployment and retrieval and with an investigation into the possibility for a thruster-driven dynamic instability.

Past analyses of the dynamics of spacecraft tethers have generally been of two types: either quite simple models were used, typically with the tether assumed to be straight and inextensible, or comprehensive simulations[3–5] were prepared for computer implementation. General agreement on the importance of various physical effects does not seem to have been reached. Each analyst has individually made the choice of which effects to model, sometimes quite arbitrarily. Reference 4 presents a broad summary of work published to date and includes 97 citations.

The work reported in this chapter was motivated by the philosophy that the appropriate approach to developing an engineering model of a physical

*Associate Professor, Department of Aeronautics and Astronautics.

†Research Assistant, Department of Aeronautics and Astronautics.

system is to include no more than the bare minimum number of effects required to reproduce the behavior of interest. Since this is necessarily an iterative procedure, this chapter builds on insights gleaned from the results of previous analyses. Spectral separation is invoked to reduce the dynamics to a relatively fast vibration, superimposed on the slow roll/yaw librations of the system. The slow libration establishes a slowly varying quasiequilibrium reference shape about which the fast motions occur. It is suggested (with experimental evidence) that, because of the low tensions experienced by spacecraft tethers, nonlinear extensional stress-strain behavior will be important.

Infinitesimal perturbations of the tether from its slowly varying quasiequilibrium are described in this chapter by a system of linear partial differential equations, in which longitudinal and lateral motion are coupled by slight curvature of the equilibrium shape. These equations are nondimensionalized and investigated with respect to wave propagation, revealing that, for wavelengths much smaller than equilibrium radius of curvature, lateral and longitudinal motions effectively decouple. Assumption of point-mass dynamics for the end bodies leads to an eigenanalysis that yields eigenfrequencies and shapes, the first few of which can be quite different from those of the classic cable approximation. Effects of tether stiffness, curvature, tension, length, and retrieval rate are discussed in terms of nondimensional parameters.

Several proposed implementations of tethered spacecraft employ tether-aligned thrusting on one end body to enhance tether tension. This chapter points out that such thrusting is a "follower force," potentially driving a dynamic instability. A linearized stability analysis indeed predicts instability for a representative system configuration.

The intent of this chapter is to provide simple conceptual models of the motion of tethered satellites, both to guide development of future simulations and to provide a basis against which simulation results can be compared.

Appropriate Models of Tether Elasticity

To date, models of tether elasticity have ranged from that of an inextensible chain to that of a beam with nonsymmetrical cross section and with torsional strain energy. It is easy to show that the bending length l_B beyond which the effects of bending stiffness EI are negligible compared to those of tension T is given by $l_B \approx \sqrt{(EI/T)}$. With tether diameter on the order of 1 mm and tether tension on the order of a few Newtons, this bending characteristic length is on the order of a few centimeters. Thus, for such tethers with lengths of many meters or kilometers, bending stiffness does not contribute significantly to global restoring forces, and a beam model is inappropriate.

Bending and torsional stiffness do, however, contribute nonlinearly to the extensional stress-strain behavior of such tethers. The unstressed form of a real wire is not straight. Residual stresses of unspecified origin and a spiral shape due to deployment from a reel will both be present. The

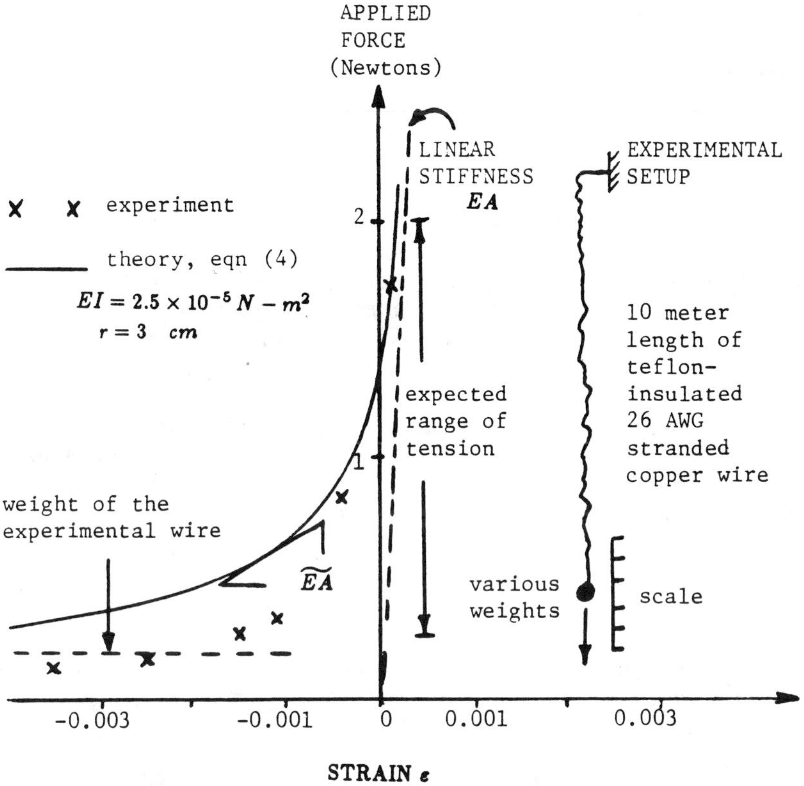

Fig. 1 Nonlinear stress-strain relationship of a wrinkled wire. A wire exhibits a nonlinear stress-strain relation due to straightening of initial wrinkles. This figure compares the results of one experiment with an analysis based on the assumption that the wire is unstressed when wound on a spool of radius r.

characteristic length of these imperfections may very well be comparable to the bending length. Figure 1, adapted from Ref. 6, presents relevant experimental results.

Nonlinear Extensional Stiffness of a Wire

Spacecraft tethers will often operate under vey low tension, even as low as a few Newtons for end masses of many hundreds of kilograms. We are thus interested in the extensional stress-strain behavior as tension approaches zero. The most common assumption has been bilinear: linear elastic for extensional strain and zero tension for compressive strain. The experimental results of Fig. 1, first reported in Ref. 6, show the behavior to be more complex. Even more complex behavior, including the effect of tension on torsional stiffness and on the zero-torque twist angle of a braided tether, has been reported by others.[7]

Two approximate analyses of the extensional force/deflection behavior of a wire with initial wrinkles can be offered. Both predict a similar nonlinear stress-strain behavior:

$$\epsilon = -\left(\frac{\kappa}{T}\right)^2 + \frac{T}{EA} \tag{1}$$

(valid for small strain, approximately $-0.05 < \epsilon < 0.01$). Thus, the strain ϵ is the sum of two contributions: a linear elastic extension proportional to the tension T and inversely proportional to the axial stiffness EA and a nonlinear shortening due to the wire's tendency to return to its unstressed shape as $T \rightarrow 0$. We ignore here the twisting motion mentioned in Ref. 7. The value of the constant κ depends on the assumptions made in the derivations.

If an inextensible wire's unstressed shape is assumed to be a helix of radius r, pitch angle $\alpha \approx 0$, and arc length L (the wire is tightly wound on a reel of radius r), then the tension T required to extend the wire to a length h (while preventing untwisting) is given by[8]

$$T = \frac{GJ}{r^2}\frac{h}{L} - \frac{EI}{r^2}\frac{h}{L}\left(1 - \frac{1}{\sqrt{1-(h/L)^2}}\right); \qquad h < L \tag{2}$$

where EI is the bending stiffness and GJ the torsional stiffness. If we define the strain with respect to the wire's arc length, $\epsilon = (h - L)/L$, and limit $|\epsilon| \ll 1$, then expression (2) reduces to

$$T = \frac{GJ}{r^2} - \frac{EI}{r^2}\frac{1}{\sqrt{-2\epsilon}}; \qquad \epsilon < 0 \tag{3}$$

which, since $GJ \approx EI$ for isotropic materials and solid convex sections, becomes

$$\epsilon = -\left(\frac{\kappa}{T}\right)^2; \qquad \kappa = EI/\sqrt{2}r^2 \tag{4}$$

An alternative derivation attempts to model a slightly wrinkled unstressed shape by introduction of a sinusoidal planar wrinkle with amplitude Δ and wavelength Λ. It is not difficult to show that, under this assumption, the tension/strain behavior of the axially inextensible wire becomes, as $h \rightarrow L$,

$$\epsilon = -\left(\frac{\kappa}{T}\right)^2; \qquad \kappa = 32\sqrt{6}EI\Delta/\Lambda^3 \tag{5}$$

Equation (1) is then the sum of the nonlinear shortening of Eq. (4) or (5) and the linear elastic extension T/EA.

The experimental points of Fig. 1 were measured for a length of Teflon-insulated stranded copper wire sold on a small plastic reel. This wire

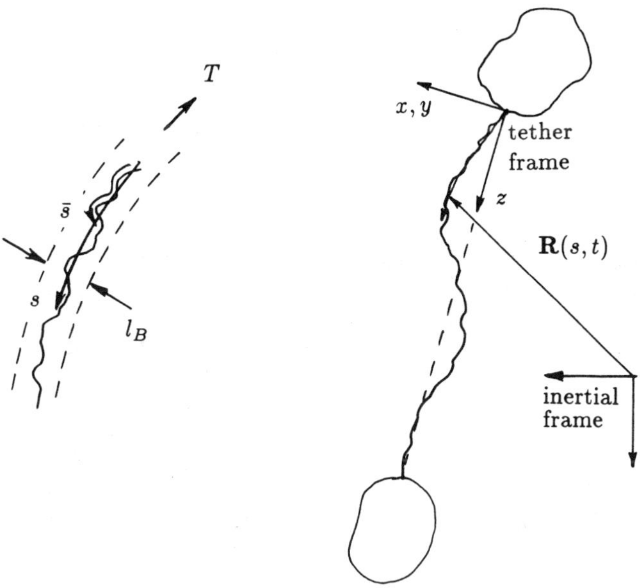

Fig. 2 Tether reference frame and kinematics. It is proposed to avoid detailed modeling of the tether "wrinkles" in favor of modeling only the approximate position and using a nonlinear stress-strain relation. The tether position would then be modeled with a resolution of approximately $l_B \approx \sqrt{(EI/T)}$.

was under consideration for use as a spacecraft tether.[6] Plausible values have been inserted into expressions (1) and (4) to calculate the theoretical curve in Fig. 1.

It would be interesting to speculate about (or to measure) the low-tension stress-strain behavior of other candidate tethers. Effects of braiding, stranding, twist, and other tether details would potentially add additional nonlinearity to that discussed here. The results summarized by Fig. 1 suggest that such nonlinear behavior will be much more important than many of the other nonlinear effects included in past analyses. For the mission studied in Ref. 6, the nonlinear stiffness $\widetilde{EA}(\epsilon)$ can be less than the linear stiffness by a factor of up to one thousand!

In principle it is possible to model this subtle effect of tether bending stiffness by using a beam model for the tether and resolving the details of the tether shape. The resolution demanded would be on the order of a few centimeters (the bending length) and thus would lead to a hopelessly large computer simulation for tethers of several kilometers length. The approach introduced in the preceding paragraphs would be to use an elastic cable model, with a nonlinear stress-strain behavior as in Eq. (1). Figure 2 indicates that the tether position would then be modeled to a precision on the order of the bending length, adequate for most purposes.

Reference Frames and Equations of Motion

The equations of motion of an elastic cable (no bending or torsional stiffness and no rotary inertia per unit length) have long been known[8] and are concisely expressed in vector notation,

$$\mu \frac{\partial^2 \boldsymbol{R}}{\partial t^2} = \frac{\partial}{\partial s}\left(T \frac{\partial \boldsymbol{R}}{\partial s}\right) + \boldsymbol{F} \tag{6}$$

where μ is the mass per unit strained length of the tether, and $\boldsymbol{R}(s,t)$ is the vector position of the tether as a function of time t and strained arc length s. The arc length is conveniently measured from one end of the tether, say, within end-body A. External forces per unit length are indicated by the vector $\boldsymbol{F}$. The tension T is given as a function of ϵ, where $\epsilon = \|\partial \boldsymbol{R}/\partial s\| - 1$, by Eq. (1). If the tether is assumed to be of constant length, Eq. (1) is replaced by the constraint $\|\partial \boldsymbol{R}/\partial s\| = 1$. (The issue of when an inelastic model of the tether is appropriate is discussed in a subsequent section.)

The boundary conditions required to complete Eq. (6) will depend on the dynamics of the end bodies of the tethered satellite system. The force exerted by the tether on the end-body A (at $s = s_A$) is just

$$\boldsymbol{F}_A = \left\{\left[T - \mu\left(\frac{\partial s_A}{\partial t}\right)^2\right]\frac{\partial \boldsymbol{R}}{\partial s}\right\}_{s=s_A} \tag{7}$$

Note the appearance of the (normally negligible) thrusting term due to tether deployment.

Reference 5 reports the development of a simulation based on a finite-difference discretization of Eq. (6) and boundary conditions built upon Eq. (7) and the assumption of suitable models for the end bodies. The reference frame chosen is Earth-centered inertial; thus, relative motion must be resolved as the small difference of two large numbers. In this reference frame Eqs. (6) and (7) are highly nonlinear in the three components of $\boldsymbol{R}$.

A more common approach has been to introduce a tether reference frame[3,4] and to express tether motion with respect to this frame. This tether frame has historically been chosen with one axis defined by the tether attachment points on the two bodies and the other axes defined via the orientation of the bodies with respect to the orbital plane (see Fig. 2). If tether deflections are expressed with respect to this frame, then the rotational motions of the reference frame add Coriolis, centripetal, and angular acceleration terms to the time derivatives of Eqs. (6) and (7), and the equations describing tether deflection are coupled to those describing system attitude. Since the angular velocity of the tether reference frame is defined by end-body motion, which depends (in part) on tether forces, the equations of motion governing system attitude are also coupled to the partial differential equations governing tether deflections.

The approach of Ref. 3 (and many works cited in Ref. 4) has been to numerically simulate this system of equations. The tether shape is discretized by the introduction of shape functions or by a lumped-mass model. The resulting ordinary differential equations can then be integrated forward in time. Experience with these simulations has revealed the system to be "stiff"; some motions occur much faster than others, and the integration time step must be chosen to be a very small fraction of the orbital period.

Approximation of Spectral Separation

An alternative approach to extracting information from a stiff system of ordinary differential equations is to introduce the approximation of spectral separation. This approximation is based on the realization that the effect of coupling between fast and slow dynamics of a system is well approximated by the introduction of two assumptions:

1) The slow dynamics define a quasiequilibrium from the point of view of the fast dynamics.

2) The fast dynamics modify the slow dynamics quasistatically; that is, they participate in the slow dynamics at their instantaneous equilibrium levels.

The best reference for the theory and practice of spectral separation, with application to many examples, is perhaps an unpublished book draft.[9]

Table 1 is a summary of estimates of the eigenfrequencies of a tethered system, both in symbolic and numerical form. The numbers used are

Table 1 Estimates of periods of vibration of a tethered satellite system

Description	Symbolic	Numerical[a]
System pitch	$\dfrac{2\pi}{\sqrt{3}\omega_o}$	3100 s
System roll	$\dfrac{\pi}{\omega_o}$	2700 s
Elastic bounce	$2\pi\sqrt{\dfrac{Lm_A m_B}{EA(m_A+m_B)}}$	61 s[b]
Tether modes		
First lateral	$2L\sqrt{\dfrac{\mu}{T}}$	440 s
First longitudinal	$2L\sqrt{\dfrac{\mu}{EA}}$	9 s[b]
End-body attitude	$2\pi\sqrt{\dfrac{I_o}{Td}}$	Orbiter: 1000 s
		Subsatellite: 13 s

[a]Numerical values used are orbital rate $\omega_o = 1.16 \times 10^{-3}$/s, (low Earth orbit), axial stiffness $EA = 10^5$ N, Orbiter mass $m_A = 10^5$ kg, subsatellite mass $m_B = 500$ kg, tether length $L = 20$ km, tension $T = 40$ N, tether mass density $\mu = 5 \times 10^{-3}$ kg/m, Orbiter inertia $I_o = 10^7$ kg-m^2, Orbiter attach point offset $d = 10$ m, subsatellite inertia $I_o = 125$ kg-m^2, subsatellite attach point offset $d = 0.75$ m.

[b]This period is based on a linear elastic model for tether axial stiffness. The nonlinear stress-strain behavior of the tether to be used for the Italian/American mission is not available.

representative of the Italian/American Shuttle-tethered satellite system. Note that the numerical estimates for this specific system indicate that most internal motions will be much faster than pitch and roll of the entire system. The one exception to this spectral separation is the pendular attitude oscillation of the Orbiter due to tether tension acting at the end of the tether deployment boom.

Quasiequilibrium Tether Shape

The attitude motion of a general rigid body in orbit, responding to environmental torques, is well understood.[10] The motion of a general flexible body in orbit and its interaction with the gravitational field is less well understood. Initial studies have shown that the rigid-body motion is essentially unperturbed by flexible deformations as long as these deformations remain small relative to system dimensions[11] and their natural frequencies remain perhaps a factor of 10 larger[12] than the orbital rate ω_o. A review of Table 1 reveals that, with the exception of orbiter attitude motion, the natural frequencies of the internal motion of the tethered system are indeed much faster than orbital rate. If slow end-body attitude motion can be ignored or is actively controlled, the attitude librations of the tethered system will occur much like those of a rigid body, as long as internal deflections remain small compared to tether length.

References 13 and 14 have considered the static equilibrium longitudinal strain and tension distributions in a tether, subject to the assumptions of linear elasticity and a constant gravitational gradient acting along the axis of a straight tether. These studies have mathematically confirmed the validity of a simple approximation that is exact for infinitesimal tether strain; tether tension is effectively constant if both end masses are much greater than tether mass and spatially parabolic if tether mass is significant and if the tether linear mass density is constant.

The quasiequilibrium shape of a tether is not necessarily straight. Lateral forces due to aerodynamic drag, gravity gradient, electrodynamic interaction with the Earth's magnetic field, and Coriolis effects due to retrieval or deployment of the tether from a massive end body will all deflect the tether laterally against tensile restoring forces. Each of these forces will vary with time and space, both in magnitude and direction, with much of the variation occurring at frequencies comparable to orbital rate. Order-of-magnitude estimates of these lateral forces are given in Table 2. Numerical values given in Table 1 have been used where needed.

The approximation of spectral separation permits calculation of an instantaneous tether shape by insisting on static equilibrium with these slow lateral forces. If one assumes that the tether experiences a linear force density $\boldsymbol{F}$ that is uniform in direction and magnitude along the wire, then the equilibrium curve is planar and is given by the solution of

$$\frac{\partial}{\partial s}\left(T\frac{\partial \boldsymbol{R}}{\partial s}\right)+\boldsymbol{F}=0 \tag{8}$$

together with an appropriate stress-strain relation. Since this curve will be

Table 2 Order-of-magnitude estimates of lateral forces

Aerodynamic drag (atmospheric density 10^{-14} to 10^{-13} g/cm^3)	10^{-6} to 10^{-5} N/m
Electordynamic (Earth's magnetic field in low Earth orbit)	2×10^{-5} N/m/amp
Gravity gradient (v is lateral deflection, in m)	$v \times 10^{-8}$ N/m
Coriolis ($\dot{L}$ is retrieval or deployment rate, in m/s)	$\dot{L} \times 10^{-5}$ N/m

very shallow, we will approximate it to be a portion of a circular arc of radius R, $R \gg L$, and since the tether mass is typically much less than end-body mass, we will introduce the approximation that the equilibrium tension T is independent of s. Basic equilibrium analysis then yields $R = T/F_\perp$ ($F_\perp$ is the component of $\boldsymbol{F}$ perpendicular to the line connecting the tether attach points) and midpoint sag of $\delta = L^2/8R$. Using a tension of $T = 40$ N and a lateral force density of 2×10^{-5} N/m yields a radius of curvature of $R = 2000$ km and a midpoint sag of $\delta = 25$ m for a tether length of 20 km. These values become $R = 200$ km and $\delta = 2.5$ m for a tether length of 2 km and a tension of 4 N. The nondimensional equilibrium radius of curvature, $\bar{R} = R/L$, is thus seen to be very large (approximately 100) and independent of length.

References 11 and 12 define the libration of a flexible system in terms of the attitude of a reference frame with origin at the deformed system mass center and with attitude defined by orthogonality of the system flexible and rigid-body motion. The tether frame satisfies this definition only approximately, even when the end bodies are modeled as point masses. Since the tether deflection from the straight condition is very small, the curved tether will couple the two end bodies much as a straight tether would, and system libration is essentially the same as the libration of the tether frame, particularly on slow time scales. Reference 12 comes to a similar conclusion.

The slowly varying equilibrium shape we will consider can thus be summarized as follows: A global pitch and roll libration of the system creates a spatially near-uniform tension in the tether, which varies slowly in time. The equilibrium tether shape is nearly straight, with a shallow planar sag that slowly changes amplitude and swings about the line connecting the two tether attachment points. This equilibrium is static when observed on the time scales of internal tether motions.

Fast Tether Motions

Fast tether motions are of interest for several reasons:

1) Simulations have shown[4] that such tether motions may grow during retrieval.

2) For maneuvers involving intermittent thrusting of the end bodies,[4,15] such fast tether motions will be excited directly.

3) The performance of kinetic isolation and subsatellite attitude control systems based on remote tethering of a sensitive payload to a noisy space station[16] depends strongly on these tether dynamics.

Prior studies[17] of the dynamics of slightly curved elastic cables have concluded that even very slight equilibrium curvature can have a great influence on perturbational motion. The fundamental symmetric mode of a pinned-pinned elastic catenary was shown to change character when a nondimensional parameter, $\lambda^2 = (L/R)^2 EA/T$, exceeded a critical value. The value of this parameter defines whether the tether behaves extensibly ($\lambda < 2\pi$) or inextensibly ($\lambda > 2\pi$) as it is perturbed from the equilibrium shape. The values of Table 1, together with the estimate of $R/L = 100$, yield a value of $\lambda = 2$, very near the critical value. Since these parameter values are only estimates, and since the boundary conditions on spacecraft tethers are very different from those treated in Ref. 17, an analysis is appropriate.

This section derives the equations of motion governing infinitesimal perturbations of a slightly curved tether. The equilibrium curve is taken to be planar, and the perturbations are defined within this tether equilibrium plane, and normal to it. The effects of small coupling due to the slow angular velocities (approximately equal to orbital rate) of the reference frame are ignored, which is consistent with the approximation of spectral separation.

Figure 3 defines the small perturbations $u(s,t)$, $v(s,t)$, $w(s,t)$, $\xi(s,t)$, and $\eta(s,t)$ of the tether with respect to its equilibrium shape, defined by $z(s)$ and $y(s)$, where s is the equilibrium arc length. This equilibrium arc length differs slightly from the natural arc length, $\bar{s}$ (see Fig. 2), depending on the equilibrium tension. The tension is taken to be the sum of the equilibrium value, $T(s)$, and a perturbation $\tau(s,t)$. Under the assumptions justified in the preceding paragraphs, and using the equilibrium defined by Eq. (8), Eq. (6) can be expanded to the following.

In the plane of tether equilibrium,

Axial:

$$\frac{\partial}{\partial s}\left(T\frac{\partial w}{\partial s} + \tau\frac{\mathrm{d}z}{\mathrm{d}s}\right) = \mu\ddot{w} \tag{9}$$

Lateral:

$$\frac{\partial}{\partial s}\left(T\frac{\partial v}{\partial s} + \tau\frac{\mathrm{d}y}{\mathrm{d}s}\right) = \mu\ddot{v} \tag{10}$$

Normal to the plane of tether equilibrium,

$$\frac{\partial}{\partial s}\left(T\frac{\partial u}{\partial s}\right) = \mu\ddot{u} \tag{11}$$

where terms of second and higher order in perturbational quantities have been dropped. The perturbational tension is given by

$$\tau = \widetilde{EA}\left(\frac{\mathrm{d}z}{\mathrm{d}s}\frac{\partial w}{\partial s} + \frac{\mathrm{d}y}{\mathrm{d}s}\frac{\partial v}{\partial s}\right) \tag{12}$$

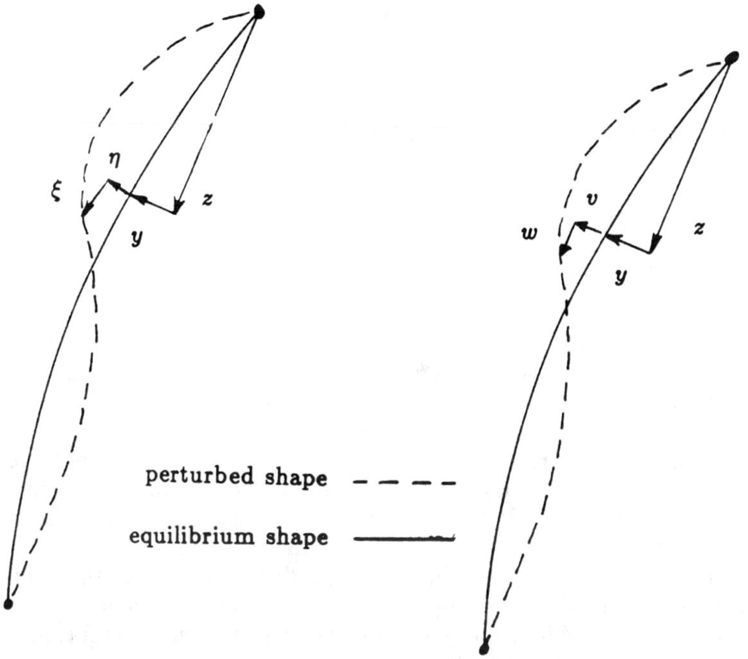

Fig. 3 Equilibrium tether shape and perturbations. The equations of motion governing infinitesimal tether deflections from the equilibrium curve can be expressed in several different coordinate systems. Curvilinear coordinates are useful for exposing the high-frequency wave-propagation limiting behavior and can be used for an eigenanalysis.

where $\widetilde{EA}$ is the effective extensional stiffness of the tether at the equilibrium strain (see Fig. 1).

Reference 15 makes further approximations, valid for very shallow pinned-pinned catenaries, to manipulate Eqs. (9), (10), and (12) into a single second-order partial differential equation for $v(s,t)$. We prefer the approach of Ref. 18, in which cable deflections are expressed in terms of components tangential and perpendicular to the equilibrium curve (see Fig. 3). Equations (9), (10), and (12) can be transformed, using the transformation

$$w = \xi \frac{\mathrm{d}z}{\mathrm{d}s} - \eta \frac{\mathrm{d}y}{\mathrm{d}s} \tag{13}$$

$$v = \xi \frac{\mathrm{d}y}{\mathrm{d}s} + \eta \frac{\mathrm{d}z}{\mathrm{d}s} \tag{14}$$

into an equivalent system.

In the plane of tether equilibrium,

Parallel to equilibrium tether:

$$\frac{\partial}{\partial s}\left(\widetilde{EA}\,\frac{\partial \xi}{\partial s}\right)-\mu\frac{\partial^2 \xi}{\partial t^2}=-\frac{T}{R}\left(\frac{\partial \eta}{\partial s}-\xi\frac{1}{R}\right)-\frac{\partial}{\partial s}\left(\frac{\widetilde{EA}}{R}\eta\right) \tag{15}$$

Perpendicular to equilibrium tether:

$$\frac{\partial}{\partial s}\left(T\frac{\partial \eta}{\partial s}\right)-\mu\frac{\partial^2 \eta}{\partial t^2}=\frac{\widetilde{EA}}{R}\left(\frac{\partial \xi}{\partial s}+\eta\frac{1}{R}\right)+\frac{\partial}{\partial s}\left(\frac{T}{R}\xi\right) \tag{16}$$

where $R=(\mathrm{d}y/\mathrm{d}s\ \mathrm{d}^2x/\mathrm{d}s^2-\mathrm{d}x/\mathrm{d}s\ \mathrm{d}^2y/\mathrm{d}s^2)^{-1}$ is the equilibrium radius of curvature of the tether. Equation (11) remains unchanged.

Equations (11), (15), and (16) now describe the perturbational motion of an elastic cable from its planar equilibrium curve. It is not required that the curve be shallow, nor is it required that the radius of curvature, R, be constant with s. It is clear that curvature couples the two components of the planar motion; Eqs. (15) and (16) are coupled by terms involving $1/R$. As $R\to\infty$, the coupling disappears, and the equations reduce to the familiar decoupled wave equations for axial and lateral motion of an elastic cable. Motion normal to the equilibrium plane remains unaffected by curvature.

Wave Propagation Along the Tether

Equations (15) and (16) are suitable for investigation of the intuition that high-frequency tether motion is effectively described by propagation of decoupled lateral and tangential waves. We assume that the equilibrium condition is independent of s and propose a solution of the form

$$\begin{pmatrix}\xi\\ \eta\end{pmatrix}=\begin{pmatrix}\xi_0\\ \eta_0\end{pmatrix}\exp[\gamma(\omega)s+i\omega t] \tag{17}$$

where $\gamma(\omega)=\alpha(\omega)+ik(\omega)$ is known as the propagation coefficient. Substitution of Eq. (17) into Eqs. (15) and (16) yields a pair of coupled second-order polynomials in γ and ω. These polynomials are satisfied along the lines sketched in Fig. 4. We see that wave propagation becomes nondispersive for $kR\gg 1$, that is, when the wavelength becomes small relative to the radius of curvature. Since the frequency has been nondimensionalized with respect to lateral wave speed, $\sqrt{(T/\mu)}$, three values of $T/\widetilde{EA}$ give three different values of nondimensional tangential wave speed. Investigation of the corresponding wave-mode eigenvector, $(\xi_0,\eta_0)^T$, reveals that the response consists of pure extensional $(\xi_0,\eta_0)=(1,0)$, or pure lateral $(\xi_0,\eta_0)=(0,1)$ motion for $kR\gg 1$. An order-of-magnitude analysis of the terms of Eqs. (15) and (16), as $|\gamma R|\to\infty$, will yield a similar insight.

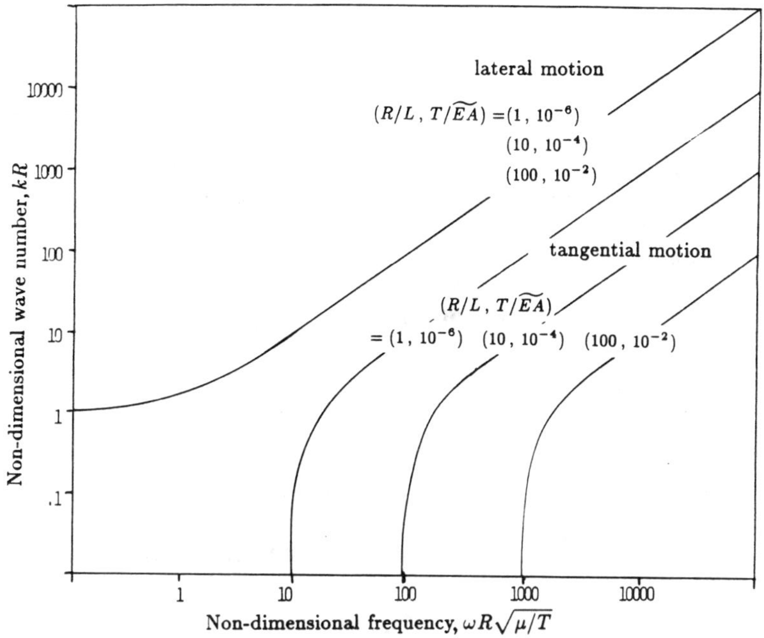

Fig. 4 Wave propagation along tether. Dispersion curves of lateral and tangential waves propagating along a tether with uniform equilibrium tension and radius of curvature.

Natural and Forced Motions of the Tethered System

Irvine and Caughey[17] report an approximate eigenanalysis of Eqs. (9), (10), and (12) for pinned-pinned boundary conditions and shallow parabolic catenaries. They conclude that the only mode strongly affected by slight curvature of an elastic cable is the first symmetric lateral mode. For analysis of a tethered spacecraft system, one is interested in different boundary conditions. A general end body of such a tethered system can provide a complex boundary condition, including effects of end-body flexibility and attitude motion. This section formulates the eigenanalysis for arbitrary boundary conditions and shows representative solutions for the simplest: one end effectively pinned by attachment to a very massive body, the other end connected to a point mass only several times more massive than the tether.

Introduction of the nondimensional cross-sectional vector

$$\mathbf{\Gamma} = \left(\xi/L, \widetilde{EA}/T\frac{\partial \xi}{\partial s}, \eta/L, \frac{\partial \eta}{\partial s} \right)^T \tag{18}$$

permits writing (Fourier-transformed) Eqs. (15) and (16) as

$$\Gamma' = \begin{bmatrix} 0 & \epsilon & 0 & 0 \\ -\bar{\omega}^2 + 1/\bar{R}^2 & 0 & 0 & -1/\bar{R}\epsilon \\ 0 & 0 & 0 & 1 \\ 0 & 1/\bar{R} & -\bar{\omega}^2 + 1/\bar{R}^2\epsilon & 0 \end{bmatrix} \Gamma \tag{19}$$

where $\epsilon = T/\widetilde{EA}$, $\bar{R} = R/L$, $\bar{\omega} = \omega L\sqrt{\mu/T}$, and the spatial derivative is taken with respect to a nondimensional distance $\Gamma' = L\,\partial\Gamma/\partial s$.

Boundary conditions on either end of the tether are conveniently expressed as a matrix equation,

$$B(\bar{\omega})\Gamma = f \tag{20}$$

where f is the vector of applied forces or enforced displacements, as appropriate. To define a pinned end at $s = 0$, Eq. (20) would employ

$$B = \begin{pmatrix} 1 & 0 & 0 & 0 \\ 0 & 0 & 1 & 0 \end{pmatrix}; \qquad f_0 = \begin{pmatrix} \xi/L \\ \eta/L \end{pmatrix}_{s=0} \tag{21}$$

and f_0 is an enforced nondimensional displacement. A point mass m at $s = L$ is described by

$$B = \begin{bmatrix} -\dfrac{m}{\mu L}\bar{\omega}^2 & 1 & 0 & 0 \\ 0 & 0 & -\dfrac{m}{\mu L}\bar{\omega}^2 & 1 \end{bmatrix}; \qquad f_1 = \begin{pmatrix} f_\xi/T \\ f_\eta/T \end{pmatrix}_{s=L} \tag{22}$$

where f_ξ and f_η are external forces applied to the point mass. The general form of the solution of Eq. (19) is

$$\Gamma(s,\bar{\omega}) = (\Gamma_1\, e^{\gamma_1 s/L} \quad \Gamma_2\, e^{\gamma_2 s/L} \quad \Gamma_3\, e^{\gamma_3 s/L} \quad \Gamma_4\, e^{\gamma_4 s/L}) \begin{bmatrix} C_1 \\ C_2 \\ C_3 \\ C_4 \end{bmatrix} \tag{23}$$

where Γ_i is the ith eigenvector of the matrix in Eq. (19), and $\gamma_i(\bar{\omega})$ is the corresponding eigenvalue. Application of boundary conditions at the two ends, each in the form of Eq. (20), permits evaluation of the constants C_i, $i = 1,2,3,4$ in terms of the force vector $(f_0,f_1)^T$. The response anywhere along the tether can then be expressed in terms of a matrix of transfer functions,

$$\Gamma(s,\bar{\omega}) = H(s,\bar{\omega})\begin{pmatrix} f_0 \\ f_1 \end{pmatrix} \tag{24}$$

These 16 transfer functions can be evaluated for any value of s, $0 < s < L$, and any value of $\bar{\omega}$, and relate the frequency response along the tether to the excitation at the two ends.

Transfer Functions. The transfer functions of Eq. (24) will vary slowly with time, since they depend on the current equilibrium condition. This equilibrium establishes values of tension T, axial strain ϵ, modulus $\widetilde{EA}$, and radius of curvature R, and defines the directions of ξ, η, and u. [Note that deflections normal to the plane of tether equilibrium, u, decouple from ξ and η. The relevant transfer functions can be derived from Eq. (11) and suitable boundary conditions.]

Figure 5 gives a plot of $H_{1,2}(L,\bar{\omega})$, parameterized with three nondimensional groups: $\bar{R} = R/L$, $\epsilon = T/\widetilde{EA}$, and $m/\mu L$. This transfer function shows the tangential motion of the far end of the tether ($s = L$) being excited an enforced lateral motion by a unit lateral acceleration at the near end ($s = 0$). If the equilibrium tether shape were straight, this response would be zero. Figure 5a shows, for a fixed equilibrium $T/\widetilde{EA}$ and a fixed subsatellite mass ratio, $m/\mu L$, the ratio of $\xi(s = L)$ and $\eta(s = 0)$ as a function of frequency. As expected, the response tends toward zero as the tether becomes straight, $R/L \to \infty$. It is clear that even slight tether curvature will couple these two motions, particularly at resonance. Figure 5b shows the same curves for an increased value of $T/\widetilde{EA}$. The natural frequencies of the first few resonances are strongly affected.

Such transfer functions are of prime importance to designers of tether-based vibration isolation systems and even more to those contemplating attitude control of a tethered subsatellite. One such design[16] has been based on the assumption that tether dynamics are that of a massless, straight, viscoelastic spring. The axis cross coupling of Fig. 5 may have a major effect on the performance and stability of such an attitude control scheme, particularly since cross coupling between orthogonal motions at the same end is also predicted.

Tether Modes. Figures 5a and 5b yield little insight into the character of the underlying vibration. Figure 6 shows the evolution of the first few natural frequencies and mode shapes as a function of two nondimensional parameters, R/L and $T/\widetilde{EA}$. R/L is held fixed at two values, $R/L =$ 10,100, whereas $T/\widetilde{EA}$ is varied. Strong coupling is predicted between pendular motion and lateral tether deflection whenever $\lambda^2 > 10$. At $\lambda^2 \approx 30$ (very close to the value predicted by Ref. 17 for fixed boundary conditions), the two lowest mode shapes become identical, and their contribution disappears from all transfer functions. A root locus of these poles of the model in the vicinity of $\lambda^2 = 30$ shows the two frequencies coalescing at $\lambda^2 = 32$ and becoming complex for greater values of λ^2. The implied instability, oscillatory with frequency $\omega \approx 2.5/L\sqrt{(T\mu)}$, is not understood and is suspected to be of mathematical rather than physical origin. The boundary-value problem defined by partial differential Eqs. (15) and (16) and boundary conditions (20) can be shown to be self-adjoint. Explanation of this unexpected instability remains a mystery. Lateral and longitudinal motion can also couple strongly. Figure 6b shows such coupling between the first few lateral modes and the longitudinal bounce mode. Tether curvature, even as low as $R/L = 100$, can greatly modify the behavior.

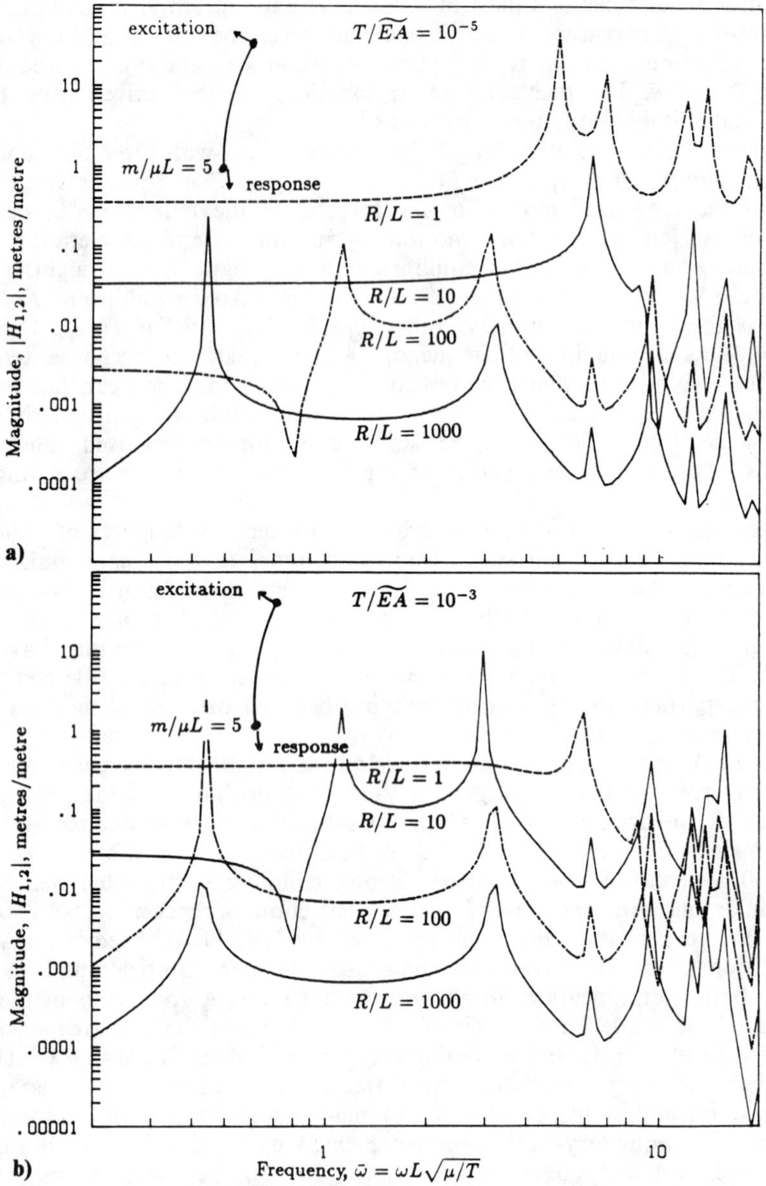

Fig. 5 System transfer functions. Transfer functions of the tangential response at one end of the tether to a lateral excitation at the other end, for two values of nondimensional tension and several values of nondimensional curvature.

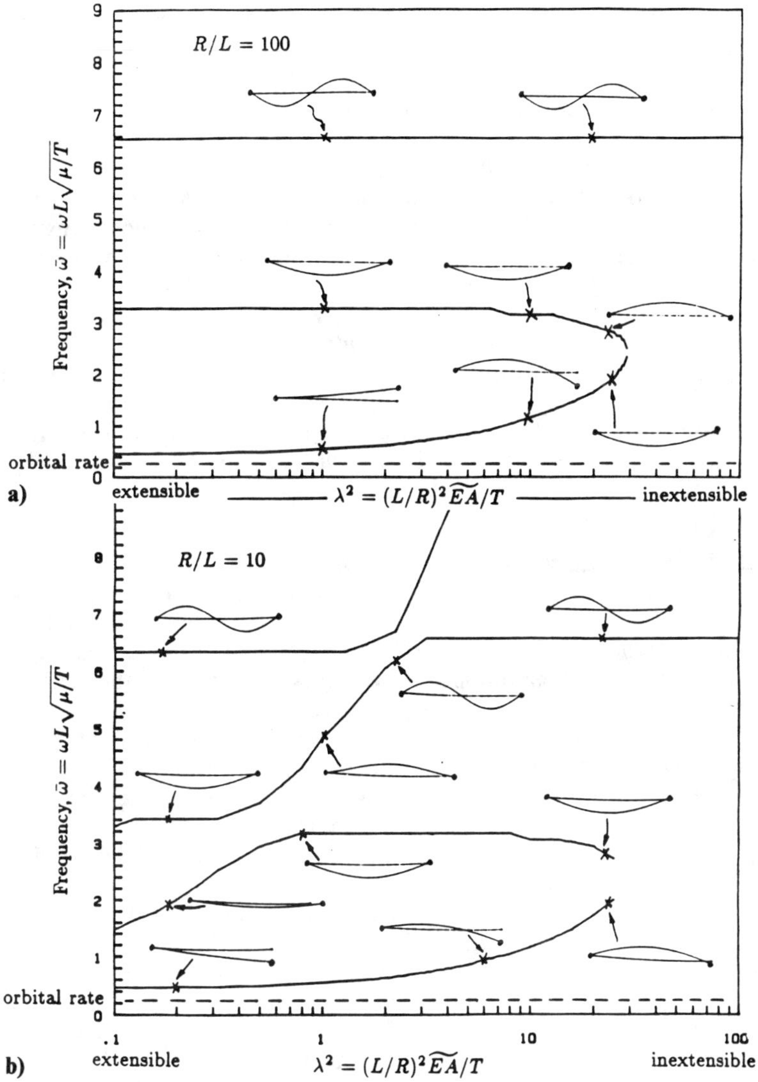

Fig. 6 System natural modes. Trends of the first few natural frequencies and mode shapes of a tethered system with slight equilibrium curvature, for two values of nondimensional curvature. One end mass is effectively infinite, the other has five times the tether mass, $m/\mu L = 5$.

Effect of Deployment or Retrieval

The preceding analysis has ignored effects of deployment and retrieval on the fast tether dynamics, both the Coriolis coupling due to a tangential velocity of the tether and the gradual length change.

Effect of Tether Translational Velocity

Retrieval of the tether to a massive end body will result in a translational velocity of the tether along its equilibrium curve. This translational velocity with respect to a Newtonian reference frame will add Coriolis coupling terms to Eqs. (15) and (16). Reference 19 has shown that such translational velocity will have a negligible effect on the perturbational motion of an elastic catenary as long as this velocity is much less than the classic lateral wave speed $\sqrt{(T/\mu)}$. Retrieval and deployment speed of tethered spacecraft systems is limited by attitude stability considerations to $\dot{L}/L \ll \omega_o$ (ω_o is orbital rate), which, on the basis of the spectral separation assumed, implies $\dot{L} \ll \sqrt{(T/\mu)}$. Ignoring the effect of this Coriolis coupling is thus consistent with the assumption of spectral separation on which the preceding analysis is based, and one would expect no change to the conclusions drawn if it were included.

Effect of Gradual Length Changes

The preceding paragraph argues that relative tether length changes will be small over the period of even the fundamental tether mode. When this is satisfied, tether mode shapes will be effectively those of the constant length system, but their amplitude will vary slowly with the length. An approximate expression for this dependence can be derived by assuming a given mode shape, of amplitude a and geometrically independent of length, and insisting on conservation of strain energy. This leads to the conclusion that $\widetilde{EA}\, a^2/L$ remain constant if the mode is primarily tangential and that Ta^2/L remain constant if the mode is primarily lateral. Since T is approximately proportional to length, lateral deflection amplitudes should be approximately independent of length. This agrees with the simulation results reported in Ref. 4. The dependence of $\widetilde{EA}$ on length is more difficult to approximate. If the nonlinear effects of Fig. 1 can be neglected, then $\widetilde{EA} = EA$ is independent of length, and tangential tether vibrations should decrease during retrieval according to $a \sim \sqrt{(L)}$. If the tether mode is mixed lateral and axial, then its dependence on length should be some weighted average of the individual dependencies.

The estimate that the amplitude of transverse tether vibrations remain constant during retrieval has generated some disagreement.[20,21] Published simulation results variously show slow growth or slow reduction of amplitude with length reduction. Heuristic arguments[22] based on conservation of angular momentum about the line connecting the tether attachment points of an hypothesized "skipping-rope" motion yield the estimate that the amplitude of lateral vibrations goes as $L^{-1/4}$. An approximate solution of the linear "piano string" equation in Ref. 23 yields the same $L^{-1/4}$ prediction. In this analysis the convective temporal derivatives in the governing partial differential equation due to axial retrieval velocity are retained, whereas the terms involving rotation of the tether frame are ignored. This is inconsistent, since, for the slow retrievals mandated by ridid-body attitude stability, both terms are of comparable magnitude.

A conservative future implementation of a tethered spacecraft will ensure that lateral tether motion does not grow during retrieval. Passive damping and/or active control mechanisms must be implemented to prevent such growth. The argument about whether the amplitude of such motion, in the absence of damping, grows slowly or remains constant during retrieval may thus never be experimentally settled.

Follower Forces and Dynamic Instability

Several studies[4] have proposed the use of a thruster on one end body and nominally aligned with the tether. This thruster is a means of enhancing tether tension, particularly during retrieval. This section points out that such a thruster is a potential source of dynamic instability and proposes a simple model to investigate system stability.

A force that in some sense follows the motion of a dynamic system is termed a follower force[24] in the literature of dynamic stability. Such a force is not conservative; the work done by it on the system during a deflection from one configuration to another depends on the path taken. Figure 7, adapted from Ref. 24, shows such a force compressing a slender column, but always directed tangentially to the tip of the column. This appears to be the first example of dynamic instability reported in the literature.[25]

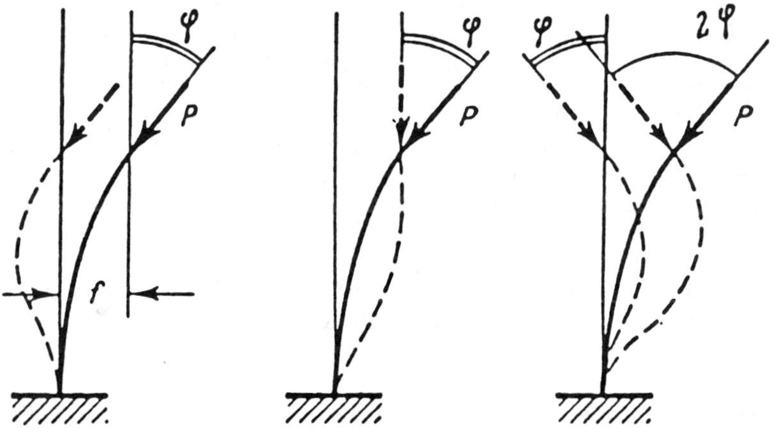

Fig. 7 Beck's dynamic stability problem. Prototypical system demonstrating dynamic instability due to a nonconservative follower force is the column compressed by a force applied at the tip and directed tangentially. This figure, taken from Ref. 24, sketches three deformation sequences with identical starting and finishing configurations, but with negative (a), zero (b), and positive (c) work done by the force in the course of the deflection.

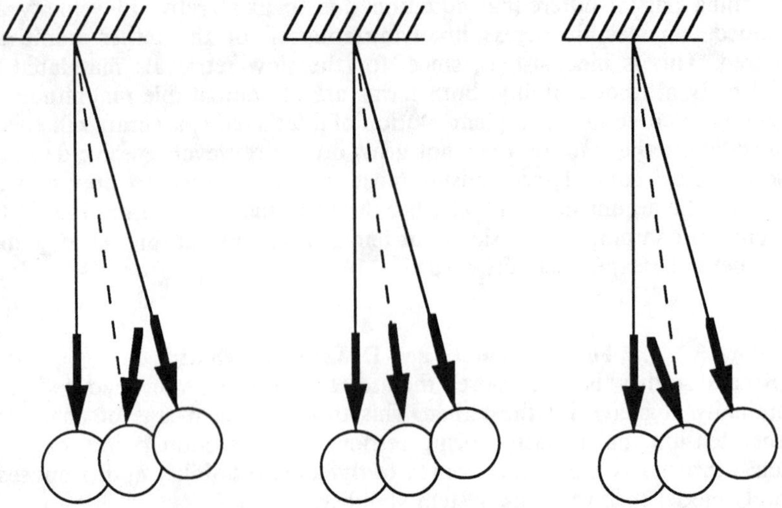

Fig. 8 A spacecraft thruster as a follower force. The tether-aligned thrust is a nonconservative follower force. This figure sketches three deformation sequences with identical starting and finishing configurations, but with negative (a), zero (b), and positive (c) work done by the force in the course of the deflection.

Figures 7a–7c show three sequences in which the column deflects from the straight reference configuration to a final configuration (the solid line). An intermediate configuration is shown as a dashed line. In Fig. 7a the work done by the force is negative, in Fig. 7b it is zero, and in Fig. 7c the work done by the force on the column is positive. The potential thus exists for the follower force to do positive work on the column during cyclic motion. Should the dynamics of the column permit such motion, it will be unstable. Analysis[24,25] has shown the column to be statically stable, whereas a load of $P = 20.05EI/L^2$ leads to an oscillatory instability. (The column is uniform, of length L, and bending stiffness EI.)

A thruster attached to the subsatellite and nominally aligned with the tether creates such a follower force. The direction of this thrust varies with subsatellite attitude. Figure 8 sketches three deflection sequences by which the subsatellite might move from its equilibrium configuration to a laterally deflected position. As in Fig. 7, these sequences differ in the work done by the thruster force on the system in the course of the deflection: negative in Fig. 8a, zero in Fig. 8b, and positive in Fig. 8c. This force is thus nonconservative and may cause an unstable dynamic response. Whether such an instability occurs will depend on whether the dynamics permit a cyclic motion that, over one cycle, extracts net energy from the thrust.

Models and Stability Analyses

This section employs two linearized models of system dynamics and investigates each model for dynamic stability. The models are developed with an eye toward the form of the expected response; this guides the choice of degrees of freedom. The interaction of several natural motions of similar natural frequency is typical in dynamic instability. Inspection of Table 1 suggests that system bounce, subsatellite attitude oscillation, and several tether piano-string modes will have similar natural frequencies. These are the degrees of freedom selected in this section.

The model developed is linear and is restricted to planar motion. One end of the tether is modeled as connected to a very massive end body, here modeled as fixed in a Newtonian reference frame. A further simplification is made by neglecting the Coriolis coupling arising from rotation of the equilibrium configuration at orbital rate ω_o. This approximation is based on the assumption of spectral separation; the response of interest in this analysis has a frequency of approximately 0.1 rad/s, the frequency of attitude oscillations of the subsatellite. For such fast motions an orbital rate of $\omega_o = 0.001$ rad/s can be ignored. A simple pendulum is used to model slosh of fuel in a spherical tank.

Figure 9 describes the deflected system and assigns generalized coordinates. For the sake of discussion we take the plane of the motion to be the

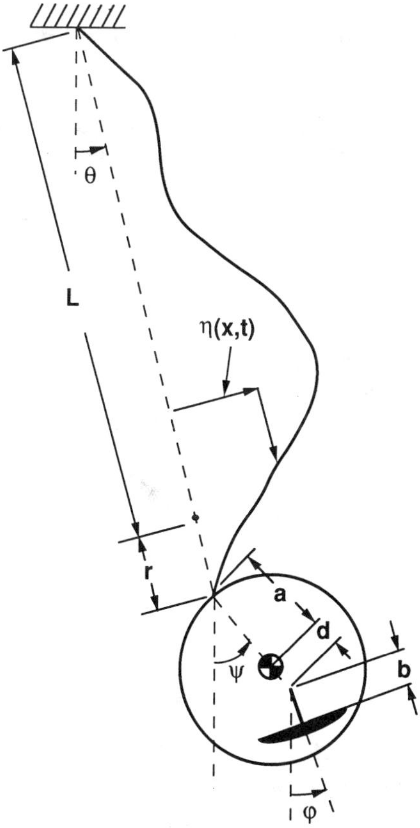

Fig. 9 System deformations and generalized coordinates. The system deflection from the reference configuration is described by several generalized coordinates. The tether may stretch a distance r and pitch through an angle θ. The subsatellite may pitch through an angle ψ and the fuel slug through an angle ϕ. Tether lateral deflection $\eta(x,t)$ is discretized by assumption of a finite number of deflection mode shapes.

orbital plane. With this choice, θ might be termed system pitch, ψ subsatellite pitch, and ϕ pitch angle of a fuel slosh mode. Tether deformation is described by uniform axial stretching and by a set of assumed modes for lateral deflection:

$$\eta(x,t) = \sum_{i=1}^{N} \phi_i(x) q_i(t) \tag{25}$$

The equilibrium tether shape is taken to be straight.

Four-Degree-of-Freedom Model with Damping

Damping can destabilize a nonconservatively forced system[24] and should thus be included in a stability analysis of a nonconservative system. A physical damping mechanism for system pitch and subsatellite pitch is difficult to justify. Before tether modes are added to the model and an attempt is made to approximate their damping, it may be useful to model the dynamics and the dissipation of liquid fuel sloshing in the spherical tank at the center of the subsatellite. This is accomplished by adding a pendulum model of this fuel, as depicted in Fig. 9. This pendulum is defined by a length b, a pivot offset from the subsatellite mass center d, a viscous damping coefficient c_f, a fuel mass m_f, and a fuel inertia $I_f = b^2 m_f \alpha_f$ about its mass center. The pendulum deflection is defined by an angle ϕ in Fig. 9.

The linearized equations of motion in the perturbational quantities $\boldsymbol{x} = (r,\theta,\psi,\phi)^T$ are then

$$M\ddot{\boldsymbol{x}} + C\dot{\boldsymbol{x}} + K\boldsymbol{x} = \boldsymbol{f} \tag{26}$$

where

$$M = \begin{bmatrix} m & 0 & 0 & 0 \\ 0 & mL^2 & (ma + m_f d)L & m_f bL \\ 0 & (ma + m_f d)L & ma^2(1+\alpha) + d(d+2a)m_f & (a+d)bm_f \\ 0 & m_f bL & (a+d)bm_f & b^2 m_f(1+\alpha_f) \end{bmatrix} \tag{27a}$$

$$C = \begin{bmatrix} 0 & 0 & 0 & 0 \\ 0 & 0 & 0 & 0 \\ 0 & 0 & +c_f & -c_f \\ 0 & 0 & -c_f & c_f \end{bmatrix} \tag{27b}$$

$$K = \begin{bmatrix} EA/L & 0 & 0 & 0 \\ 0 & L(mg+F) & -LF & 0 \\ 0 & 0 & mga + m_f gd & 0 \\ 0 & 0 & 0 & m_f gb \end{bmatrix} \tag{27c}$$

$$f = \begin{Bmatrix} F + mg \\ 0 \\ 0 \\ 0 \end{Bmatrix} \tag{27d}$$

where the total subsatellite mass, $m = m_s + m_f$, is expressed in terms of the portion that is rigid, $m_s = 420$ kg, and the portion due to the fuel, $m_f = 30$ kg. Other values used in the subsequent numerical stability analysis are fuel pendular length, $b = 0.2$ m; fuel moment of inertia, $I_f = m_f b^2 \alpha_f$, $\alpha_f = 1$; and fuel pivot point offset, $d = 0$ m. The fuel slosh damping parameter c_f was chosen to give the slosh mode a damping ratio of 0.05. These parameters completely specify the system matrices of Eq. (26) in terms of a single free parameter, the deployed tether length L:

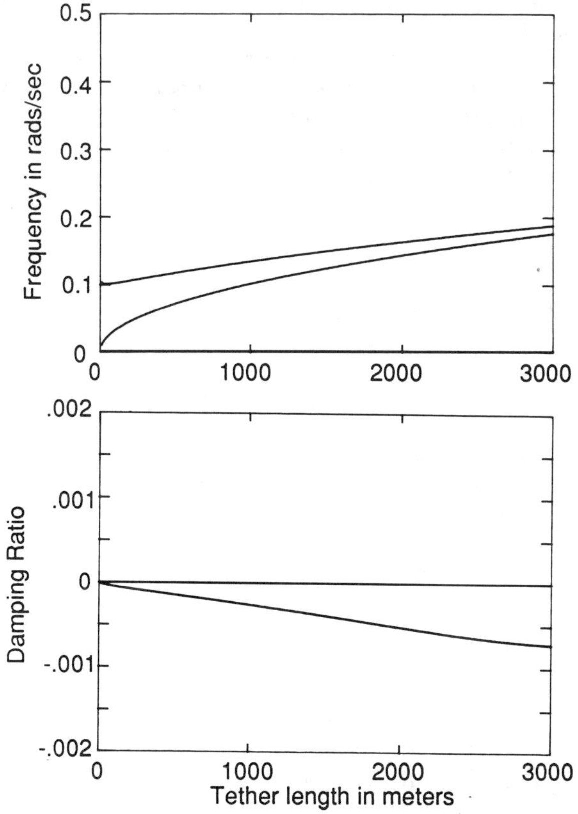

Fig. 10 Simple dynamic instability. Thrusting couples fuel slosh and subsatellite attitude motion. The interaction is unstable with the strength of the instability growing with tether length.

Figure 10 gives the dependence of system natural frequencies and damping ratios on tether length. Inclusion of the fuel-slosh mode has introduced instability, but this instability depends on the presence of slosh damping. Deletion of damping [setting $c_f = 0$ in Eq. (26)] leads to pure imaginary system eigenvalues. Thus, in this model both damping and nonconservative forcing must be present to cause instability. The instability is very weak, with a minimum damping ratio of -0.001, occurring at the maximum length. It is not known whether stronger fuel slosh damping or thrusting will strengthen the instability or over what range of values of F and c_f the instability will persist.

A Model Including Lateral Tether Motion

The previous model was based on the assumption that the tether remains straight, even as it stretches axially. This section investigates the possibility of interaction of the subsatellite attitude and fuel slosh with lateral tether motion. The shape functions $\phi_i(x) = \sin(i\pi L/x)$ satisfy the geometric boundary conditions and lead to simple integrations along the tether. With these shape functions describing the lateral tether deflection according to Eq. (25), the linearized equations of motion have the same form as Eq. (26), but the perturbational degrees of freedom now include the generalized tether deflection amplitudes $q_i(t)$; $\boldsymbol{x} = (r, \theta, \psi, \phi, q_1, \ldots, q_N)^T$.

The system matrices now become

$$M = \begin{bmatrix} m + \mu L/3 & 0 & 0 & 0 & 0 & \cdots & 0 \\ 0 & mL^2 + \mu L^3/3 & (ma + m_f d)L & m_f bL & \cdots & -\mu\cos(i\pi)L^2/i\pi & \cdots \\ 0 & (ma + m_f d)L & m_3 & (a+d)bm_f & 0 & \cdots & 0 \\ 0 & m_f bL & (a+d)bm_f & b^2 m_f(1+\alpha_f) & 0 & \cdots & 0 \\ 0 & \vdots & 0 & 0 & \ddots & \cdots & 0 \\ \vdots & -\mu\cos(i\pi)L^2/i\pi & \vdots & \vdots & \vdots & \mu L/2 & \vdots \\ 0 & \vdots & 0 & 0 & 0 & \cdots & \ddots \end{bmatrix} \tag{28a}$$

$$C = \begin{bmatrix} 0 & 0 & 0 & 0 & 0 & \cdots & 0 \\ 0 & 0 & 0 & 0 & 0 & \cdots & 0 \\ 0 & 0 & c_f & -c_f & 0 & \cdots & 0 \\ 0 & 0 & -c_f & c_f & 0 & \cdots & 0 \\ 0 & 0 & 0 & 0 & \ddots & \cdots & 0 \\ \vdots & \vdots & \vdots & \vdots & \vdots & c_i & \vdots \\ 0 & 0 & 0 & 0 & 0 & \cdots & \ddots \end{bmatrix} \tag{28b}$$

$$K = \begin{bmatrix} EA/L & 0 & 0 & 0 & 0 & \cdots & 0 \\ 0 & TL & -LF & 0 & \cdots & g\mu \frac{L}{(i\pi)}(1-\cos i\pi) & \cdots \\ 0 & 0 & mga + m_f gd & 0 & 0 & \cdots & 0 \\ 0 & 0 & 0 & m_f gb & 0 & \cdots & 0 \\ 0 & \vdots & 0 & 0 & \ddots & \cdots & 0 \\ \vdots & g\mu \frac{L}{(i\pi)}(1-\cos i\pi) & \vdots & \vdots & \vdots & T(i\pi/L)^2 L/2 & \vdots \\ 0 & \vdots & 0 & 0 & 0 & \cdots & \ddots \end{bmatrix}$$

$$f = \begin{bmatrix} F + mg \\ 0 \\ 0 \\ 0 \\ 0 \\ \vdots \\ 0 \end{bmatrix} \tag{28d}$$

where $m_3 = ma^2(1 + \alpha) + d(d + 2a)m_f$, the tether mass per unit length μ has been introduced, and the tether tension is approximated as uniform with a value of $T = mg + F + \mu gL/2$, chosen to be the average value. The tether damping constants c_i are chosen to yield a tether damping ratio (of unspecified origin) of 0.01. A value of μ representative of the Italian/American tethered satellite mission is $\mu = 6$ kg/km. Use of this value then completely specifies the system matrices, the only free parameter being the length of the unretrieved tether, L.

Figure 11 plots the system natural frequencies calculated with the tether motion discretized by the first six assumed modes and with damping neglected. Since the thrusting has been turned off, there is little interaction between the individual degrees of freedom, and the interaction is not unstable. When thrusting is present, the system responds in an unstable manner, even in the absence of damping. Figure 12a shows that in this situation the instability is limited to very specific lengths, those at which tether modes and the fuel slosh mode have nearly the same frequency. When tether damping of 1% is added, but the fuel slosh is undamped (Fig. 12b), the unstable interaction of Fig. 12a persists, and a stable interaction between tether modes and subsatellite attitude motion begins. When fuel-slosh damping of 5% is present, but tether modes are undamped (Fig. 12c), a weak instability is predicted for all tether lengths, and the stronger instability at specific tether lengths has disappeared. The presence of both tether and fuel-slosh damping (Fig. 12d) still permits a weakly unstable response over a wide range of tether lengths.

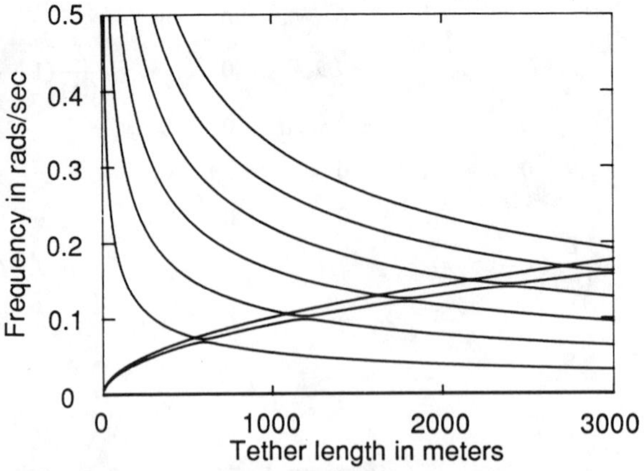

Fig. 11 System eigenloci with no thrusting. Tether modes to be included in the stability analysis are selected based on an estimate of their probability of participating in the response. Shown are estimates of the system natural frequencies with the tether motion discretized by the first six pinned-pinned sinusoids. This figure was obtained with the same software as Fig. 12, but with the thrust level set to zero and with zero damping.

The results of Fig. 12 suggest that fuel-slosh damping is destabilizing, whereas tether damping is stabilizing. It is interesting to speculate whether this conclusion is independent of details of the fuel-slosh model. The model developed in this section permits investigation of the importance of fuel-slosh pivot offset. Other investigations must await the development of further models.

Modeling Limitations

The insights gained in these analyses were achieved only at the cost of rather severe assumptions. Interest was restricted to tethered systems involving one tether and two end bodies, with a spectrum permitting the approximation of spectral separation. The results of Figs. 5 and 6 show that, for some choices of parameters, the assumption of spectral separation will be violated. The conclusions reached will not be applicable to very different systems such as tethers with negligible end masses, massive, tapered tethers, multiple tethers, rapidly spinning tethers, and many others.

The eigenanalyses presented are based on the assumption of small deflections. References 26 and 27 have considered larger-amplitude motion of cables. Reference 26 shows that the quadratic terms in the strain energy expression of a shallow catenary remain dominant only as long as deflections remain small compared to the equilibrium sag. Equations (15) and (16) are thus only valid for very small deflections (less than $\delta = L/8\bar{R} \approx 0.001L$). Reference 27 derives the equations of motion of a sequence of assumed deflection modes of a spacecraft tether, where the

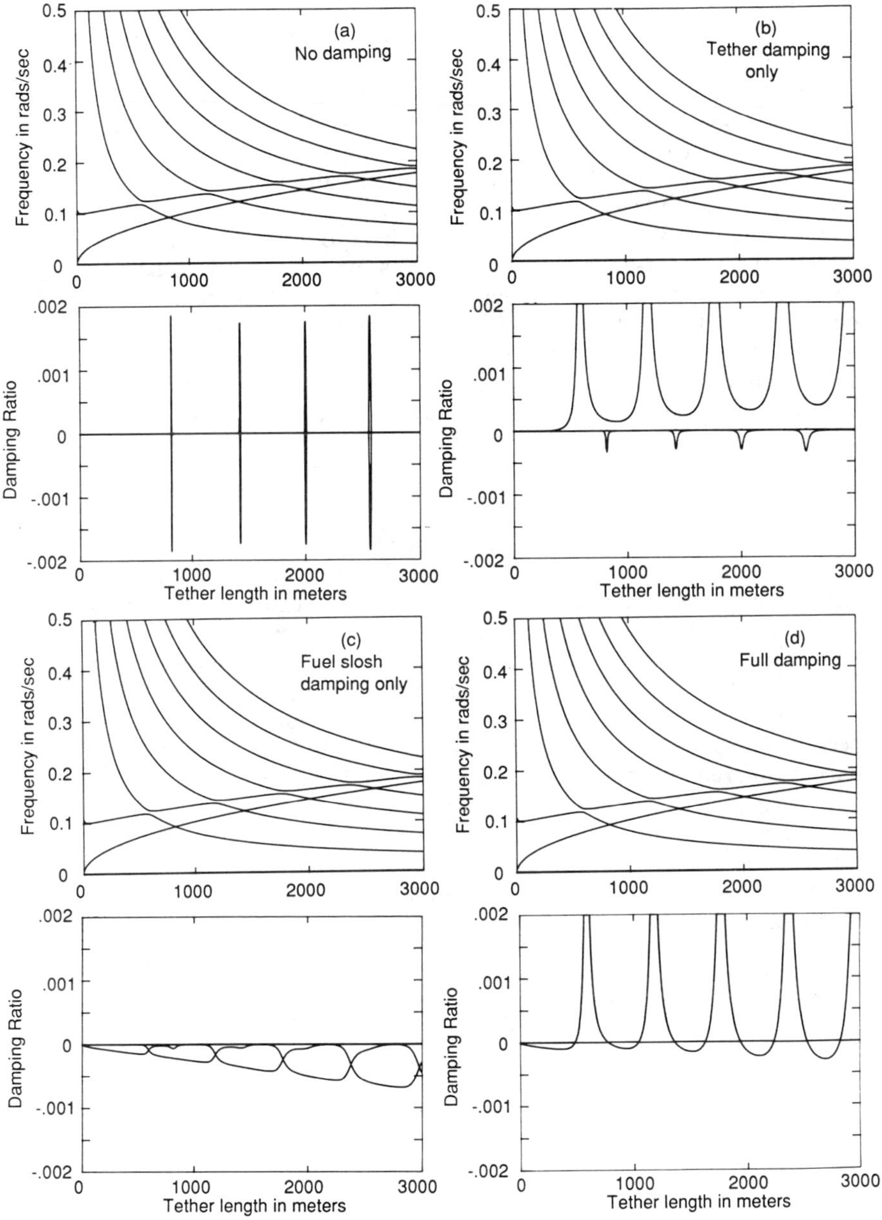

Fig. 12 System eigenloci with thrusting and damping. Shown are the dependence of system natural frequencies and damping ratios on tether length with tether lateral deflection discretized via six assumed modes. The type of damping included in the model is varied. Fuel-slosh damping appears to destabilize the motion, whereas tether damping stabilizes.

reference state is taken to be straight and axially unstrained. This work shows that nonlinear effects will couple axial and both directions of lateral deflection and by integration of the equations of motion shows this coupling to be significant when deflections become as large as $0.1L$.

A more questionable approximation, introduced by the assumption of infinitesimal perturbations, seems to be the implicit assumption that tension is nearly constant, with infinitesimal perturbations. Other studies[4] have shown retrieval to lead to large temporal variations of tether tension T. The tension fluctuations are severe enough to lead to loss of tension and are the reason that tether-aligned thrusting was originally proposed. A large component of the tension fluctuation is attributable to the system bounce mode and is thus dynamically coupled to the fast internal dynamics of the sections on reference frames and equations of motion and follower forces and dynamic instability. This effectively imposes an additional amplitude limit on the linearized analyses; perturbational tension τ must remain small compared to the equilibrium tension value T.

Summary

This chapter examined the dynamics of typical tethered spacecraft systems. The motion is shown to occur at two time scales: one comparable to orbital rate and the other much faster. Spectral separation is invoked to approximately decouple this motion. Fast tether vibrations occur with respect to a slowly varying quasiequilibrium. The equilibrium shape of the tether is estimated to be slightly sagged from a straight line, and the small perturbations from this equilibrium are described by a system of linear partial differential equations. Nondimensional parameter groups are identified that govern the character of the fast tether vibrations.

Linearized stability analyses in the presence of the nonconservative forcing of a subsatellite thruster has shown weak dynamic instability to be a pervasive characteristic of this mode of operation. The instabilities identified in these analyses have been very weak, with negative damping ratios bounded by -0.002. This would tempt one to the conclusion that dissipation will stabilize the interactions. This conclusion is not warranted. These analyses have shown that passive damping has an unpredictable effect on system stability. Two forms of damping were studied: one stabilized the interaction, whereas the other destabilized it.

Acknowledgments

The work summarized by this chapter is drawn directly from two prior studies given in Refs. 1 and 2 and was begun during the summer of 1986 during a research visit at the German Aerospace Research Establishment, DFVLR Oberpfaffenhofen. The visit was made possible by the hospitality of the DFVLR, in particular, that of Dr. Konrad Reinel, and by the generosity of the Alexander von Humboldt Foundation. The work was continued at the Massachusetts Institute of Technology under the sponsorship of the U.S. Air Force Office of Scientific Research with Dr. Anthony Amos as monitor. Some computations were performed by Jeff Morse under the auspices of the Undergraduate Research Opportunities Program at MIT.

References

[1]von Flotow, A. H., "Some Approximations for the Dynamics of Tethered Spacecraft," *Journal of Guidance, Control, and Dynamics*, Vol. 11, No. 4, 1988, pp. 357–364.

[2]von Flotow, A. H., and Wereley, N. M., "Coupling of Tether Modes with Sub-Satellite Attitude Motion," *Advances in the Astronautical Sciences: Proceedings of the AAS/AIAA Astrodynamics Conference Held August 10–13, 1987, Kalispell, Montana*, pp. 411–424. Vol. 65, American Astronautical Society, Paper AAS 87-433, 1988.

[3]Lang, D., "TOSS (Tethered Object System Simulation) Reference Manual," version C, May 1985, Private Communication, Jan. 1986.

[4]Misra, A. K., and Modi, V. J., "A Survey on the Dynamics and Control of Tethered Satellite Systems," *NASA/AIAA/PSN International Conference on Tethers in Space*, Washington, DC, 1986.

[5]Kohler, P., Maag, W., Wehrli, R., Weber, R., and Brauchli, H., "Dynamics of a System of Two Satellites Connected by a Deployable and Extensible Tether of Finite Mass," Vols. 1 and 2, European Space Agency Technical Rept., Contract 2992/76/NL/AK(SC), Oct. 1978.

[6]von Flotow, A. H., and Williamson, P. R., "Deployment of a Tethered Plasma Diagnostics Satellite into Low Earth Orbit," *Journal of the Astronautical Sciences*, Vol. 34, Jan.–March 1986, p. 65.

[7]Carroll, J. A., and Oldson, J. C., "ESL Report to TRW on Electrodynamic Tether Stability," TR, June 1988.

[8]Love, A. E. H., *A Treatise on the Mathematical Theory of Elasticity*, Dover, New York, 1944.

[9]Bryson, A. E., *Control of Spacecraft and Aircraft*, Stanford University Lecture Notes, Sept. 1985.

[10]Hughes, P. C., *Spacecraft Attitude Dynamics*, Wiley, New York, 1986.

[11]Diarra, C. M., and Bainum, P. M., "On the Accuracy of Modeling the Dynamics of Large Space Structures," *Proceedings of IAF 1985, 36th Congress of the International Astronautical Federation*, 1985.

[12]Ashley, H., "Observations on the Dynamic Behavior of Large Flexible Bodies in Orbit," *AIAA Journal*, Vol. 5, No. 3, pp. 460–469, 1967.

[13]Storch, J., and Gates, S., "The Dynamics of a Uniform Tether with a Tip Mass Subject to Gravitational Body Forces," Charles Stark Draper Lab., TR-CSDL-R-1791, June 1985.

[14]Misra, A. K., and Modi, V. J., "Frequencies of Longitudinal Oscillations of Tethered Satellite Systems," AIAA Paper 86-2274, Aug. 1986.

[15]Xu, D. M., Misra, A. K., and Modi, V. J., "Thruster Augmented Active Control of a Tethered Subsatellite System During its Retrieval," *Journal of Guidance, Control, and Dynamics*, Vol. 9, Nov.–Dec. 1986, pp. 663–672.

[16]Lemke, L. G., Powell, J. D., and He, X., "Attitude Control of Tethered Spacecraft," *Journal of the Astronautical Sciences*, Vol. 35, Jan.–March 1987, pp. 41–56.

[17]Irvine, H. M., and Caughey, T. K., "The Linear Theory of the Free Vibration of a Suspended Cable," *Proceedings of the Royal Society of London, Series A*, Vol. A341, 1974, pp. 299–315.

[18]Saxon, D. S., and Cahn, A. S., "Modes of Vibration of a Suspended Chain," *Quarterly Journal of Mechanics and Applied Mathematics*, Vol. 6, Pt. 3, 1953, pp. 273–285.

[19]Simpson, A., "On the Oscillating Motions of Translating Elastic Cables," *Journal of Sound and Vibration*, Vol. 20, No. 2, 1972, pp. 177–189.

[20]Misra, A. K., "Comment on Some Approximations for the Dynamics of Spacecraft Tethers," *Journal of Guidance, Control, and Dynamics,* Vol. 13, March–April 1990, p. 380.

[21]von Flotow, A. H., "Reply to A. K. Misra's Comment on 'Some Approximations for the Dynamics of Spacecraft Tethers,'" *Journal of Guidance, Control, and Dynamics*, Vol. 13, March–April 1990, p. 38.

[22]Arnold, D. A., "Tether Tutorial," International Conference on Tethers in Space, Venice, Sept. 1988.

[23]Kalaycioglu, S., and Misra, A. K., "Analytical Expressions for Vibratory Displacements of Deploying Appendages," AIAA Paper 88-3281, Aug. 1988.

[24]Bolotin, V. V., *Nonconservative Problems of the Theory of Elastic Stability*, Macmillan, New York, 1963.

[25]Beck, M., "Die Knicklast des Einseitig Eingespannten, Tangential Gedruckten Stables," *Zeitschrift fur Angewandte Mathematik und Physik*, Vol. 3, 1952.

[26]Hagedorn, P., and Schafer, B., "On Non-Linear Vibrations of an Elastic Cable," *International Journal of Nonlinear Mechanics*, Vol. 15, 1980, pp. 333–340.

[27]Misra, A. K., Xu, D. M., and Modi, V. J., "On Vibrations of Orbiting Tethers," *Acta Astronautica*, Vol. 13, No. 10, 1986, pp. 587–597.

Author Index

PROGRESS IN ASTRONAUTICS AND AERONAUTICS SERIES VOLUMES

*1. **Solid Propellant Rocket Research** (1960)
Martin Summerfield
Princeton University

*2. **Liquid Rockets and Propellants** (1960)
Loren E. Bollinger
Ohio State University
Martin Goldsmith
The Rand Corp.
Alexis W. Lemmon Jr.
Battelle Memorial Institute

*3. **Energy Conversion for Space Power** (1961)
Nathan W. Snyder
Institute for Defense Analyses

*4. **Space Power Systems** (1961)
Nathan W. Snyder
Institute for Defense Analyses

*5. **Electrostatic Propulsion** (1961)
David B. Langmuir
Space Technology Laboratories, Inc.
Ernst Stuhlinger
NASA George C. Marshall Space Flight Center
J.M. Sellen Jr.
Space Technology Laboratories, Inc.

*6. **Detonation and Two-Phase Flow** (1962)
S.S. Penner
California Institute of Technology
F.A. Williams
Harvard University

*7. **Hypersonic Flow Research** (1962)
Frederick R. Riddell
AVCO Corp.

*8. **Guidance and Control** (1962)
Robert E. Roberson, Consultant
James S. Farrior
Lockheed Missiles and Space Co.

*9. **Electric Propulsion Development** (1963)
Ernst Stuhlinger
NASA George C. Marshall Space Flight Center

*10. **Technology of Lunar Exploration** (1963)
Clifford I. Cummings
Harold R. Lawrence
Jet Propulsion Laboratory

*11. **Power Systems for Space Flight** (1963)
Morris A. Zipkin
Russell N. Edwards
General Electric Co.

*12. **Ionization in High-Temperature Gases** (1963)
Kurt E. Shuler, Editor
National Bureau of Standards
John B. Fenn, Associate Editor
Princeton University

*13. **Guidance and Control – II** (1964)
Robert C. Langford
General Precision Inc.
Charles J. Mundo
Institute of Naval Studies

*14. **Celestial Mechanics and Astrodynamics** (1964)
Victor G. Szebehely
Yale University Observatory

*15. **Heterogeneous Combustion** (1964)
Hans G. Wolfhard
Institute for Defense Analyses
Irvin Glassman
Princeton University
Leon Green Jr.
Air Force Systems Command

16. **Space Power Systems Engineering** (1966)
George C. Szego
Institute for Defense Analyses
J. Edward Taylor
TRW Inc.

17. **Methods in Astrodynamics and Celestial Mechanics** (1966)
Raynor L. Duncombe
U.S. Naval Observatory
Victor G. Szebehely
Yale University Observatory

18. **Thermophysics and Temperature Control of Spacecraft and Entry Vehicles** (1966)
Gerhard B. Heller
NASA George C. Marshall Space Flight Center

*Out of print.

*19. **Communication Satellite Systems Technology** (1966)
Richard B. Marsten
Radio Corporation of America

*20. **Thermophysics of Spacecraft and Planetary Bodies: Radiation Properties of Solids and the Electromagnetic Radiation Environment in Space** (1967)
Gerhard B. Heller
NASA George C. Marshall Space Flight Center

21. **Thermal Design Principles of Spacecraft and Entry Bodies** (1969)
Jerry T. Bevans
TRW Systems

22. **Stratospheric Circulation** (1969)
Willis L. Webb
Atmospheric Sciences Laboratory, White Sands, and University of Texas at El Paso

23. **Thermophysics: Applications to Thermal Design of Spacecraft** (1970)
Jerry T. Bevans
TRW Systems

24. **Heat Transfer and Spacecraft Thermal Control** (1971)
John W. Lucas
Jet Propulsion Laboratory

25. **Communication Satellites for the 70's: Technology** (1971)
Nathaniel E. Feldman
The Rand Corp.
Charles M. Kelly
The Aerospace Corp.

26. **Communication Satellites for the 70's: Systems** (1971)
Nathaniel E. Feldman
The Rand Corp.
Charles M. Kelly
The Aerospace Corp.

27. **Thermospheric Circulation** (1972)
Willis L. Webb
Atmospheric Sciences Laboratory, White Sands, and University of Texas at El Paso

28. **Thermal Characteristics of the Moon** (1972)
John W. Lucas
Jet Propulsion Laboratory

29. **Fundamentals of Spacecraft Thermal Design** (1972)
John W. Lucas
Jet Propulsion Laboratory

30. **Solar Activity Observations and Predictions** (1972)
Patrick S. McIntosh
Murray Dryer
Environmental Research Laboratories, National Oceanic and Atmospheric Administration

31. **Thermal Control and Radiation** (1973)
Chang-Lin Tien
University of California at Berkeley

32. **Communications Satellite Systems** (1974)
P.L. Bargellini
COMSAT Laboratories

33. **Communications Satellite Technology** (1974)
P.L. Bargellini
COMSAT Laboratories

34. **Instrumentation for Airbreathing Propulsion** (1974)
Allen E. Fuhs
Naval Postgraduate School
Marshall Kingery
Arnold Engineering Development Center

35. **Thermophysics and Spacecraft Thermal Control** (1974)
Robert G. Hering
University of Iowa

36. **Thermal Pollution Analysis** (1975)
Joseph A. Schetz
Virginia Polytechnic Institute
ISBN 0-915928-00-0

37. **Aeroacoustics: Jet and Combustion Noise; Duct Acoustics** (1975)
Henry T. Nagamatsu, Editor
General Electric Research and Development Center
Jack V. O'Keefe, Associate Editor
The Boeing Co.
Ira R. Schwartz, Associate Editor
NASA Ames Research Center
ISBN 0-915928-01-9

38. **Aeroacoustics: Fan, STOL, and Boundary Layer Noise; Sonic Boom; Aeroacoustics Instrumentation** (1975)
Henry T. Nagamatsu, Editor
General Electric Research and Development Center
Jack V. O'Keefe, Associate Editor
The Boeing Co.
Ira R. Schwartz, Associate Editor
NASA Ames Research Center
ISBN 0-915928-02-7

39. Heat Transfer with Thermal Control Applications (1975)
M. Michael Yovanovich
University of Waterloo
ISBN 0-915928-03-5

40. Aerodynamics of Base Combustion (1976)
S.N.B. Murthy, Editor
J.R. Osborn, Associate Editor
Purdue University
A.W. Barrows
J.R. Ward, Associate Editors
Ballistics Research Laboratories
ISBN 0-915928-04-3

41. Communications Satellite Developments: Systems (1976)
Gilbert E. LaVean
Defense Communications Agency
William G. Schmidt
CML Satellite Corp.
ISBN 0-915928-05-1

42. Communications Satellite Developments: Technology (1976)
William G. Schmidt
CML Satellite Corp.
Gilbert E. LaVean
Defense Communications Agency
ISBN 0-915928-06-X

43. Aeroacoustics: Jet Noise, Combustion and Core Engine Noise (1976)
Ira R. Schwartz, Editor
NASA Ames Research Center
Henry T. Nagamatsu, Associate Editor
General Electric Research and Development Center
Warren C. Strahle, Associate Editor
Georgia Institute of Technology
ISBN 0-915928-07-8

44. Aeroacoustics: Fan Noise and Control; Duct Acoustics; Rotor Noise (1976)
Ira R. Schwartz, Editor
NASA Ames Research Center
Henry T. Nagamatsu, Associate Editor
General Electric Research and Development Center
Warren C. Strahle, Associate Editor
Georgia Institute of Technology
ISBN 0-915928-08-6

45. Aeroacoustics: STOL Noise; Airframe and Airfoil Noise (1976)
Ira R. Schwartz, Editor
NASA Ames Research Center
Henry T. Nagamatsu, Associate Editor
General Electric Research and Development Center
Warren C. Strahle, Associate Editor
Georgia Institute of Technology
ISBN 0-915928-09-4

46. Aeroacoustics: Acoustic Wave Propagation; Aircraft Noise Prediction; Aeroacoustic Instrumentation (1976)
Ira R. Schwartz, Editor
NASA Ames Research Center
Henry T. Nagamatsu, Associate Editor
General Electric Research and Development Center
Warren C. Strahle, Associate Editor
Georgia Institute of Technology
ISBN 0-915928-10-8

47. Spacecraft Charging by Magnetospheric Plasmas (1976)
Alan Rosen
TRW Inc.
ISBN 0-915928-11-6

48. Scientific Investigations on the Skylab Satellite (1976)
Marion I. Kent
Ernst Stuhlinger
NASA George C. Marshall Space Flight Center
Shi-Tsan Wu
University of Alabama
ISBN 0-915928-12-4

49. Radiative Transfer and Thermal Control (1976)
Allie M. Smith
ARO Inc.
ISBN 0-915928-13-2

50. Exploration of the Outer Solar System (1976)
Eugene W. Greenstadt
TRW Inc.
Murray Dryer
National Oceanic and Atmospheric Administration
Devrie S. Intriligator
University of Southern California
ISBN 0-915928-14-0

51. Rarefied Gas Dynamics, Parts I and II (two volumes) (1977)
J. Leith Potter
ARO Inc.
ISBN 0-915928-15-9

52. Materials Sciences in Space with Application to Space Processing (1977)
Leo Steg
General Electric Co.
ISBN 0-915928-16-7

53. Experimental Diagnostics in Gas Phase Combustion Systems (1977)
Ben T. Zinn, Editor
Georgia Institute of Technology
Craig T. Bowman, Associate Editor
Stanford University
Daniel L. Hartley, Associate Editor
Sandia Laboratories
Edward W. Price, Associate Editor
Georgia Institute of Technology
James G. Skifstad, Associate Editor
Purdue University
ISBN 0-015928-18-3

54. Satellite Communications: Future Systems (1977)
David Jarett
TRW Inc.
ISBN 0-915928-18-3

55. Satellite Communications: Advanced Technologies (1977)
David Jarett
TRW Inc.
ISBN 0-915928-19-1

56. Thermophysics of Spacecraft and Outer Planet Entry Probes (1977)
Allie M. Smith
ARO Inc.
ISBN 0-915928-20-5

57. Space-Based Manufacturing from Nonterrestrial Materials (1977)
Gerard K. O'Neill, Editor
Brian O'Leary, Assistant Editor
Princeton University
ISBN 0-915928-21-3

58. Turbulent Combustion (1978)
Lawrence A. Kennedy
State University of New York at Buffalo
ISBN 0-915928-22-1

59. Aerodynamic Heating and Thermal Protection Systems (1978)
Leroy S. Fletcher
University of Virginia
ISBN 0-915928-23-X

60. Heat Transfer and Thermal Control Systems (1978)
Leroy S. Fletcher
University of Virginia
ISBN 0-915928-24-8

61. Radiation Energy Conversion in Space (1978)
Kenneth W. Billman
NASA Ames Research Center
ISBN 0-915928-26-4

62. Alternative Hydrocarbon Fuels: Combustion and Chemical Kinetics (1978)
Craig T. Bowman
Stanford University
Jorgen Birkeland
Department of Energy
ISBN 0-915928-25-6

63. Experimental Diagnostics in Combustion of Solids (1978)
Thomas L. Boggs
Naval Weapons Center
Ben T. Zinn
Georgia Institute of Technology
ISBN 0-915928-28-0

64. Outer Planet Entry Heating and Thermal Protection (1979)
Raymond Viskanta
Purdue University
ISBN 0-915928-29-9

65. Thermophysics and Thermal Control (1979)
Raymond Viskanta
Purdue University
ISBN 0-915928-30-2

66. Interior Ballistics of Guns (1979)
Herman Krier
University of Illinois at Urbana-Champaign
Martin Summerfield
New York University
ISBN 0-915928-32-9

*__67. Remote Sensing of Earth from Space: Role of "Smart Sensors"__(1979)
Roger A. Breckenridge
NASA Langley Research Center
ISBN 0-915928-33-7

68. Injection and Mixing in Turbulent Flow (1980)
Joseph A. Schetz
Virginia Polytechnic Institute and State University
ISBN 0-915928-35-3

69. Entry Heating and Thermal Protection (1980)
Walter B. Olstad
NASA Headquarters
ISBN 0-915928-38-8

70. Heat Transfer, Thermal Control, and Heat Pipes (1980)
Walter B. Olstad
NASA Headquarters
ISBN 0-915928-39-6

71. Space Systems and Their Interactions with Earth's Space Environment (1980)
Henry B. Garrett
Charles P. Pike
Hanscom Air Force Base
ISBN 0-915928-41-8

72. Viscous Flow Drag Reduction (1980)
Gary R. Hough
Vought Advanced Technology Center
ISBN 0-915928-44-2

73. Combustion Experiments in a Zero-Gravity Laboratory (1981)
Thomas H. Cochran
NASA Lewis Research Center
ISBN 0-915928-48-5

74. Rarefied Gas Dynamics, Parts I and II (two volumes) (1981)
Sam S. Fisher
University of Virginia
ISBN 0-915928-51-5

75. Gasdynamics of Detonations and Explosions (1981)
J.R. Bowen
University of Wisconsin at Madison
N. Manson
Université de Poitiers
A.K. Oppenheim
University of California at Berkeley
R.I. Soloukhin
Institute of Heat and Mass Transfer, BSSR Academy of Sciences
ISBN 0-915928-46-9

76. Combustion in Reactive Systems (1981)
J.R. Bowen
University of Wisconsin at Madison
N. Manson
Université de Poitiers
A.K. Oppenheim
University of California at Berkeley
R.I. Soloukhin
Institute of Heat and Mass Transfer, BSSR Academy of Sciences
ISBN 0-915928-47-7

77. Aerothermodynamics and Planetary Entry (1981)
A.L. Crosbie
University of Missouri-Rolla
ISBN 0-915928-52-3

78. Heat Transfer and Thermal Control (1981)
A.L. Crosbie
University of Missouri-Rolla
ISBN 0-915928-53-1

79. Electric Propulsion and Its Applications to Space Missions (1981)
Robert C. Finke
NASA Lewis Research Center
ISBN 0-915928-55-8

80. Aero-Optical Phenomena (1982)
Keith G. Gilbert
Leonard J. Otten
Air Force Weapons Laboratory
ISBN 0-915928-60-4

81. Transonic Aerodynamics (1982)
David Nixon
Nielsen Engineering & Research, Inc.
ISBN 0-915928-65-5

82. Thermophysics of Atmospheric Entry (1982)
T.E. Horton
University of Mississippi
ISBN 0-915928-66-3

83. Spacecraft Radiative Transfer and Temperature Control (1982)
T.E. Horton
University of Mississippi
ISBN 0-915928-67-1

84. Liquid-Metal Flows and Magnetohydrodynamics (1983)
H. Branover
Ben-Gurion University of the Negev
P.S. Lykoudis
Purdue University
A. Yakhot
Ben-Gurion University of the Negev
ISBN 0-915928-70-1

85. Entry Vehicle Heating and Thermal Protection Systems: Space Shuttle, Solar Starprobe, Jupiter Galileo Probe (1983)
Paul E. Bauer
McDonnell Douglas Astronautics Co.
Howard E. Collicott
The Boeing Co.
ISBN 0-915928-74-4

86. Spacecraft Thermal Control, Design, and Operation (1983)
Howard E. Collicott
The Boeing Co.
Paul E. Bauer
McDonnell Douglas Astronautics Co.
ISBN 0-915928-75-2

87. Shock Waves, Explosions, and Detonations (1983)
J.R. Bowen
University of Washington
N. Manson
Université de Poitiers
A.K. Oppenheim
University of California at Berkeley
R.I. Soloukhin
Institute of Heat and Mass Transfer, BSSR Academy of Sciences
ISBN 0-915928-76-0

88. Flames, Lasers, and Reactive Systems (1983)
J.R. Bowen
University of Washington
N. Manson
Université de Poitiers
A.K. Oppenheim
University of California at Berkeley
R.I. Soloukhin
Institute of Heat and Mass Transfer, BSSR Academy of Sciences
ISBN 0-915928-77-9

89. Orbit-Raising and Maneuvering Propulsion: Research Status and Needs (1984)
Leonard H. Caveny
Air Force Office of Scientific Research
ISBN 0-915928-82-5

90. Fundamentals of Solid-Propellant Combustion (1984)
Kenneth K. Kuo
Pennsylvania State University
Martin Summerfield
Princeton Combustion Research Laboratories, Inc.
ISBN 0-915928-84-1

91. Spacecraft Contamination: Sources and Prevention (1984)
J.A. Roux
University of Mississippi
T.D. McCay
NASA Marshall Space Flight Center
ISBN 0-915928-85-X

92. Combustion Diagnostics by Nonintrusive Methods (1984)
T.D. McCay
NASA Marshall Space Flight Center
J.A. Roux
University of Mississippi
ISBN 0-915928-86-8

93. The INTELSAT Global Satellite System (1984)
Joel Alper
COMSAT Corp.
Joseph Pelton
INTELSAT
ISBN 0-915928-90-6

94. Dynamics of Shock Waves, Explosions, and Detonations (1984)
J.R. Bowen
University of Washington
N. Manson
Université de Poitiers
A.K. Oppenheim
University of California at Berkely
R.I. Soloukhin
Institute of Heat and Mass Transfer, BSSR Academy of Sciences
ISBN 0-915928-91-4

95. Dynamics of Flames and Reactive Systems (1984)
J.R. Bowen
University of Washington
N. Manson
Université de Poitiers
A.K. Oppenheim
University of California at Bereley
R.I. Soloukhin
Institute of Heat and Mass Transfer, BSSR Academy of Sciences
ISBN 0-915928-92-2

96. Thermal Design of Aeroassisted Orbital Transfer Vehicles (1985)
H.F. Nelson
University of Missouri-Rolla
ISBN 0-915928-94-9

97. Monitoring Earth's Ocean, Land, and Atmosphere from Space – Sensors, Systems, and Applications (1985)
Abraham Schnapf
Aerospace Systems Engineering
ISBN 0-915928-98-1

98. Thrust and Drag: Its Prediction and Verification (1985)
Eugene E. Covert
Massachusetts Institute of Technology
C.R. James
Vought Corp.
William F. Kimzey
Sverdrup Technology AEDC Group
George K. Richey
U.S. Air Force
Eugene C. Rooney
U.S. Navy Department of Defense
ISBN 0-930403-00-2

99. Space Stations and Space Platforms – Concepts, Design, Infrastructure, and Uses (1985)
Ivan Bekey
Daniel Herman
NASA Headquarters
ISBN 0-930403-01-0

100. Single- and Multi-Phase Flows in an Electromagnetic Field: Energy, Metallurgical, and Solar Applications (1985)
Herman Branover
Ben-Gurion University of the Negev
Paul S. Lykoudis
Purdue University
Michael Mond
Ben-Gurion University of the Negev
ISBN 0-930403-04-5

101. MHD Energy Conversion: Physiotechnical Problems (1986)
V.A. Kirillin
A.E. Sheyndlin
Soviet Academy of Sciences
ISBN 0-930403-05-3

102. Numerical Methods for Engine-Airframe Integration (1986)
S.N.B. Murthy
Purdue University
Gerald C. Paynter
Boeing Airplane Co.
ISBN 0-930403-09-6

103. Thermophysical Aspects of Re-Entry Flows (1986)
James N. Moss
NASA Langley Research Center
Carl D. Scott
NASA Johnson Space Center
ISBN 0-930403-10-X

104. Tactical Missile Aerodynamics (1986)
M.J. Hemsch
PRC Kentron, Inc.
J.N. Nielsen
NASA Ames Research Center
ISBN 0-930403-13-4

105. Dynamics of Reactive Systems Part I: Flames and Configurations; Part II: Modeling and Heterogeneous Combustion (1986)
J.R. Bowen
University of Washington
J.-C. Leyer
Université de Poitiers
R.I. Soloukhin
Institute of Heat and Mass Transfer, BSSR Academy of Sciences
ISBN 0-930403-14-2

106. Dynamics of Explosions (1986)
J.R. Bowen
University of Washington
J.-C. Leyer
Université de Poitiers
R.I. Soloukhin
Institute of Heat and Mass Transfer, BSSR Academy of Sciences
ISBN 0-930403-15-0

107. Spacecraft Dielectric Material Properties and Spacecraft Charging (1986)
A.R. Frederickson
U.S. Air Force Rome Air Development Center
D.B. Cotts
SRI International
J.A. Wall
U.S. Air Force Rome Air Development Center
F.L. Bouquet
Jet Propulsion Laboratory, California Institute of Technology
ISBN 0-930403-17-7

108. Opportunities for Academic Research in a Low-Gravity Environment (1986)
George A. Hazelrigg
National Science Foundation
Joseph M. Reynolds
Louisiana State University
ISBN 0-930403-18-5

109. Gun Propulsion Technology (1988)
Ludwig Stiefel
U.S. Army Armament Research, Development and Engineering Center
ISBN 0-930403-20-7

110. Commercial Opportunities in Space (1988)
F. Shahrokhi
K.E. Harwell
University of Tennessee Space Institute
C.C. Chao
National Cheng Kung University
ISBN 0-930403-39-8

111. Liquid-Metal Flows: Magnetohydrodynamics and Applications (1988)
Herman Branover, Michael Mond, and Yeshajahu Unger
Ben-Gurion University of the Negev
ISBN 0-930403-43-6

112. Current Trends in Turbulence Research (1988)
Herman Branover, Michael Mond, and Yeshajahu Unger
Ben-Gurion University of the Negev
ISBN 0-930403-44-4

113. Dynamics of Reactive Systems Part I: Flames; Part II: Heterogeneous Combustion and Applications (1988)
A.L. Kuhl
R & D Associates
J.R. Bowen
University of Washington
J.-C. Leyer
Université de Poitiers
A. Borisov
USSR Academy of Sciences
ISBN 0-930403-46-0

114. Dynamics of Explosions (1988)
A.L. Kuhl
R & D Associates
J.R. Bowen
University of Washington
J.-C. Leyer
Université de Poitiers
A. Borisov
USSR Academy of Sciences
ISBN 0-930403-47-9

115. Machine Intelligence and Autonomy for Aerospace (1988)
E. Heer
Heer Associates, Inc.
H. Lum
NASA Ames Research Center
ISBN 0-930403-48-7

116. Rarefied Gas Dynamics: Space-Related Studies (1989)
E.P. Muntz
University of Southern California
D.P. Weaver
U.S. Air Force Astronautics Laboratory (AFSC)
D.H. Campbell
University of Dayton Research Institute
ISBN 0-930403-53-3

117. Rarefied Gas Dynamics: Physical Phenomena (1989)
E.P. Muntz
University of Southern California
D.P. Weaver
U.S. Air Force Astronautics Laboratory (AFSC)
D. Campbell
University of Dayton Research Institute
ISBN 0-930403-54-1

118. Rarefied Gas Dynamics: Theoretical and Computational Techniques (1989)
E.P. Muntz
University of Southern California
D.P. Weaver
U.S. Air Force Astronautics Laboratory (AFSC)
D.H. Campbell
University of Dayton Research Institute
ISBN 0-930403-55-X

119. Test and Evaluation of the Tactical Missile (1989)
Emil J. Eichblatt Jr.
Pacific Missile Test Center
ISBN 0-930403-56-8

120. Unsteady Transonic Aerodynamics (1989)
David Nixon
Nielsen Engineering & Research, Inc.
ISBN 0-930403-52-5

121. Orbital Debris from Upper-Stage Breakup (1989)
Joseph P. Loftus Jr.
NASA Johnson Space Center
ISBN 0-930403-58-4

122. Thermal-Hydraulics for Space Power, Propulsion and Thermal Management System Design (1989)
William J. Krotiuk
General Electric Co.
ISBN 0-930403-64-9

123. Viscous Drag Reduction in Boundary Layers (1990)
Dennis M. Bushnell
Jerry N. Hefner
NASA Langley Research Center
ISBN 0-930403-66-5

124. Tactical and Strategic Missile Guidance (1990)
Paul Zarchan
Charles Stark Draper Laboratory, Inc.
ISBN 0-930403-68-1

125. Applied Computational Aerodynamics (1990)
P.A. Henne
Douglas Aircraft Company
ISBN 0-930403-69-X

126. Space Commercialization: Launch Vehicles and Programs (1990)
F. Shahrokhi
University of Tennessee Space Institute
J.S. Greenberg
Princeton Synergetics Inc.
T. Al-Saud
Ministry of Defense and Aviation Kingdom of Saudi Arabia
ISBN 0-930403-75-4

127. Space Commercialization: Platforms and Processing (1990)
F. Shahrokhi
University of Tennessee Space Institute
G. Hazelrigg
National Science Foundation
R. Bayuzick
Vanderbilt University
ISBN 0-930403-76-2

128. Space Commercialization: Satellite Technology (1990)
F. Shahrokhi
University of Tennessee Space Institute
N. Jasentuliyana
United Nations
N. Tarabzouni
King Abulaziz City for Science and Technology
ISBN 0-930403-77-0

129. Mechanics and Control of Large Flexible Structures (1990)
John L. Junkins
Texas A&M University
ISBN 0-930403-73-8

(Other Volumes are planned.)